Statistical Modeling: Regression, Survival Analysis, and Time Series Analysis

Lawrence M. Leemis

Department of Mathematics

William & Mary

Williamsburg, Virginia

Library of Congress Cataloging-in-Publication Data

Leemis, Lawrence M.
      Statistical Modeling: Regression, Survival Analysis, and Time Series Analysis /
         Lawrence M. Leemis
      Includes bibliographic references and index.
      ISBN 978-0-9829174-3-5
      1. Statistics
      QA 273.L44 2023

The author and publisher of this book have used their best efforts in preparing this book. These efforts include the development, research, and testing of the mathematics and computer programs to determine their effectiveness. The author and publisher make no warranty of any kind, expressed or implied, with regard to the mathematics or programs or the documentation contained in this book. The author and publisher shall not be liable in any event for incidental or consequential damages in connection with, or arising out of, the furnishing, performance, or use of the mathematics or programs.

Printed in the United States of America

10 9 8 7 6 5 4 3 2 1

ISBN 978-0-9829174-3-5

Statistical Modeling:
Regression, Survival Analysis, and Time Series Analysis

Short Contents

Detailed Contents

Preface

This book provides a brief introduction to three statistical modeling techniques: regression, survival analysis, and time series analysis. My motivation for writing this book came from a recent article in *Nature* that indicated that the paper introducing the product–limit estimator by American statisticians Edward Kaplan and Paul Meier in 1958 and the paper introducing the proportional hazards model written by British statistician David Cox in 1972 were the two most cited papers in the statistical literature. Yet most undergraduates majoring in applied mathematics, statistics, data science, systems engineering, and management science do not encounter the statistical models developed in either of these two pivotal papers. This book provides an elementary introduction to these two statistical procedures, and many others.

This book is designed as a one-semester introduction to regression, survival analysis, and time series analysis for advanced undergraduates or first-year graduate students. The pre-requisites for this book are (a) a course in linear algebra, (b) a calculus-based introduction to probability, and (c) a course in mathematical statistics that covers point estimation, interval estimation, and hypothesis testing. The book is not comprehensive and is not a replacement for a full-semester class on each of the topics. It contains only brief introductions to the three topics.

Three chapters are devoted to each of the three topics. The initial two chapters move at about the pace one would expect in a full-semester course. The third chapter on each of the topics is like a "further reading" section which briefly introduces some topics that would be covered in depth in a full-semester course. An instructor might choose to skip or expand on these topics.

The material in the book can be covered at the ambitious pace of one chapter per week. An instructor could also choose to move more slowly if some of this material is part of a course covering another topic.

Most of the data sets that are used for examples in the book are given as clear text on the website `www.math.wm.edu/~leemis/data/topics`.

The text is organized into chapters, sections, and subsections. When there are several topics within a subsection, they are set off by **boldface** headings. Definitions and theorems are boxed; examples are indented; proofs are terminated with a box, like this: $\square$. Proofs are included when they are instructive to the material being presented. Exercises are numbered sequentially at the end of each chapter. Computer code is set in monospace font, and is not punctuated. Indentation is used to indicate nesting in code and pseudocode. An index is included. Italicized page numbers in the index correspond to the primary source of information on a topic.

The term *estimator* is used to describe a point estimator in the abstract or as a random variable; the term *estimate* is used to describe a point estimator that assumes a specific value estimated from a realization of data values. In some instances the case is altered to highlight this distinction. The sample mean $\bar{X}$, for example, is a point estimator for the population mean μ. A numerical value of the sample mean calculated from data values is sometimes denoted by the point estimate $\bar{x}$.

The R language is used throughout the text for graphics, computation, and Monte Carlo simulation. In many of the examples involving computations, the results are computed arithmetically, then confirmed in R, and then computed a third time using an R built-in function (such as lm for computing the coefficients in a regression model, coxph from the survival package for computing the regression coefficients in a Cox proportional hazards model, survfit from the survival package to calculate the step heights in the Kaplan–Meier product–limit estimator, or arima to fit a univariate time series). This three-step process is used to avoid treating R functions as black boxes without considering what goes on underneath the hood. R can be downloaded for free at r-project.org.

There are no references cited in the text for readability. The sources of materials in the various chapters are cited in the paragraphs below.

Chapter 1 notes: The quote by George Box is from page 202 of the book chapter: Box, G.E.P. (1979), "Robustness in the Strategy of Scientific Model Building," from Robustness in Statistics, edited by R.L. Launer and G.N. Wilkinson, New York: Academic Press, pages 201–236. The data pairs associated with the boiling points and barometric pressures in Example 1.11 are from Forbes, J. (1857), "Further Experiments and Remarks on the Measurement of Heights and Boiling Point of Water," *Transactions of the Royal Society of Edinburgh*, Volume 21, Issue 2, pages 235–243.

Chapter 2 notes: The four sets of data pairs known as *Anscombe's quartet* are from Anscombe, F.J. (1973), "Graphs in Statistical Analysis," *The American Statistician*, Volume 27, Number 1, pages 17–21. The housing data set in Example 2.9 is from De Cock, D. (2011), "Ames, Iowa: Alternative to the Boston Housing Data as an End of Semester Regression Project," *Journal of Statistics Education*, Volume 19, Number 3, pages 1–15. The Shapiro–Wilk test for normality (and related tests) are overviewed in Razali, N., and Wah, Y.B. (2011), "Power Comparisons of Shapiro–Wilk, Kolmogorov–Smirnov, Lilliefors and Anderson–Darling Tests," *Journal of Statistical Modeling and Analytics*, Volume 2, Number 1, pages 21–33.

Chapter 3 notes: The chemical data from Example 3.1 is from Bennett, N.A., and Franklin, N.L. (1954), *Statistical Analysis in Chemistry and the Chemical Industry*, New York: Wiley. Cook's distances are derived in Cook, R.D. (1977), "Detection of Influential Observations in Linear Regression," *Technometrics*, Volume 19, Number 1, pages 15–18. The U.S. National debt over time is from https://www.thebalance.com/national-debt-by-year-compared-to-gdp-and-major-events-3306287. The original paper introducing ridge regression is Hoerl, A.E., and Kennard, R.W. (1970), "Ridge Regression: Biased Estimation for Nonorthogonal Problems," *Technometrics*, Volume 12, Number 1, pages 55–67.

Chapter 4 notes: Early references on the Weibull distribution include Fisher, R.A., and Tippett, L.H.C. (1928), "Limiting Forms of the Frequency Distribution of the Largest or Smallest Member of a Sample," *Proceedings of the Cambridge Philosophical Society*, Volume 24, Issue 2, pages 180–190, Weibull, W. (1939), "A Statistical Theory of the Strength of Materials," *Ingeniors Vetenskaps Akademien Handlingar*, Number 153, and Weibull, W. (1951), "A Statistical Distribution Function of Wide Applicability," *Journal of Applied Mechanics*, Volume 18, pages 293–297. The moment ratio diagrams given in Section 4.5 are adapted from those given in Vargo, E., Pasupathy, R., and Leemis, L. (2010), "Moment-Ratio Diagrams for Univariate Distributions," *Journal of Quality Technology*, Volume 42, Number 3, pages 1–11. The Cox proportional hazards model was formulated in Cox, D.R. (1972), "Regression Models and Life-Tables" (with discussion), *Journal of the Royal Statistical Society B*, Volume 34, Number 2, pages 187–220.

Chapter 5 notes: The ball bearing failure times from Example 5.5 are from Lieblein, J., and Zelen, M. (1956), "Statistical Investigation of the Fatigue Life of Deep-Groove Ball Bearings," *Journal of Research of the National Bureau of Standards*, Volume 57, Number 5, pages 273–316. The 48.48 data value in the ball bearing data set is given as 48.40 on page 99 of Lawless, J.F. (2003), *Statistical Models and Methods for Lifetime Data*, Second Edition, Hoboken, NJ: John

Wiley & Sons, Inc., and page 4 of Meeker, W.Q., and Escobar, L.A. (2022), *Statistical Methods for Reliability Data*, Second Edition, New York: John Wiley & Sons, Inc., but is listed as 48.48 in Caroni, C. (2002), "The Correct 'Ball Bearings' Data," *Lifetime Data Analysis*, Volume 8, Number 4, pages 395–399. The 6–MP data set is from Gehan, E.A. (1965), "A Generalized Wilcoxon Test for Comparing Arbitrarily Singly-Censored Samples," *Biometrika*, Volume 52, Parts 1 and 2, pages 203–223. The automotive a/c switch failure times are from pages 253–254 of Kapur, K.C., and Lamberson, L.R. (1977), *Reliability in Engineering Design*, New York: John Wiley & Sons, Inc. The initial estimator for the Weibull shape parameter κ from a complete data set is given by Menon, M.V. (1963), "Estimation of the Shape and Scale Parameters of the Weibull Distribution," *Technometrics*, Volume 5, Number 2, pages 175–182.

Chapter 6 notes: The Clopper–Pearson confidence interval was introduced by Clopper, C.J., and Pearson, E.S. (1934), "The Use of Confidence or Fiducial Limits Illustrated in the Case of the Binomial," *Biometrika*, Volume 26, Number 4, pages 404–413. The Wilson–Score confidence interval was introduced by Wilson, E.B. (1927), "Probable Inference, the Law of Succession, and Statistical Inference," *Journal of the American Statistical Association*, Volume 22, Number 158, pages 209–212. The Jeffreys confidence interval is described by Brown, L.D., Cai, T.T., and Das-Gupta, A. (2001), "Interval Estimation for a Binomial Proportion," *Statistical Science*, Volume 16, Number 2, pages 101–133. The Agresti–Coull confidence interval was introduced by Agresti, A., and Coull, B.A. (1998), "Approximate is Better than 'Exact' for Interval Estimation of Binomial Proportions," *The American Statistician*, Volume 52, Number 2, pages 119–126. The product–limit estimator was devised by Kaplan, E.L., and Meier, P. (1958), "Nonparametric Estimation from Incomplete Observations," *Journal of the American Statistical Association*, Volume 53, Number 282, pages 457–481. The earliest reference to Greenwood's formula comes from Greenwood, M. (1926), "The Natural Duration of Cancer," *Reports on Public Health and Medical Subjects*, Her Majesty's Stationery Office, London, Volume 33, pages 1–26. The proof of Theorem 6.1 is given in Appendix C of Leemis, L.M. (2009), *Reliability: Probability Models and Statistical Methods*, Second Edition, Lightning Source. A lucid presentation of Poisson processes and nonhomogeneous Poisson processes is given by Ross, S.M. (2019), *Introduction to Probability Models*, Twelfth Edition, London: Academic Press.

Chapter 7 notes: The monthly international airline traveler counts from Example 7.2 are Series G from Box, G.E.P., and Jenkins, G.M. (1976), Time Series Analysis: Forecasting and Control, Oakland, CA: Holden–Day. Introductions to the basics of time series analysis are given in Chatfield, C. (2004), *The Analysis of Time Series: An Introduction*, Sixth Edition, Boca Raton, FL: Chapman & Hall/CRC, and Brockwell, P.J., and Davis, R.A. (2016), *Introduction to Time Series and Forecasting*, Third Edition, Springer International Publishing Switzerland.

Chapter 8 notes: The details associated with the Box–Pierce test are given in Box, G.E.P., and Pierce, D.A. (1970), "Distribution of Residual Auto-Correlations in Autoregressive-Integrated Moving Average Time-Series Models," *Journal of the American Statistical Association*, Volume 65, Number 332, pages 1509–1526. The details associated with the Ljung–Box test are given in Ljung, G.M., and Box, G.E.P. (1978), "On a Measure of Lack of Fit in Time Series Models," *Biometrika*, Volume 65, Number 2, pages 297–303. The turning point test was first devised by Bienaymé, I.-J. (1874), "Sur Une Question de Probabilités," Bulletin de la Société Mathématique de France, Volume 2, pages 153–154. The Akaike Information Criterion was formulated in Akaike, H. (1974), "A New Look at the Statistical Model Identification," *IEEE Transactions on Automatic Control*, Volume 19, Number 6, pages 716–723. The corrected Akaike Information Criterion was formulated in Hurvich, C.M., and Tsai, C–L. (1989), "Regression and Time Series Model Selection in Small Samples," *Biometrika*, Volume 76, Number 2, pages 297–307.

Chapter 9 notes: The graph displaying the stationarity region in terms of ϕ_1 and ϕ_2 shown in

Figure 9.13 is adapted from a figure in Stralkowski, C.M. (1968), "Lower Order Autoregressive-Moving Average Stochastic Models and Their Use for the Characterization of Abrasive Cutting Tools," PhD Thesis, The University of Wisconsin. Rom Lipscus from the Virginia Institute of Marine Science helped me find the source of the Lake Huron lake level data given in Example 9.14. Lauren M. Fry from the NOAA Great Lakes Environmental Research Laboratory provided the source of the Lake Huron levels on pages 151–154 of

`https://tidesandcurrents.noaa.gov/publications/Monthly_Annual_Averages_IGLD55_1860thru1985.PDF`

The time series of 210 consecutive chemical production yields from Example 9.35 are from pages 120–121 of Box, G.E.P., Hunter, J.S., and Hunter, W.G. (2005), *Statistics for Experimenters: Design, Innovation, Discovery*, Second Edition, Hoboken, NJ: John Wiley & Sons. The data values are IGLD55, which means that they are the sea level (in feet) above the level of the Atlantic Ocean. The shortened version of the lynx pelt sales from the Hudson's Bay Company that was fit to the transformed AR(3) model in Example 9.29 was suggested by Wei, W.W.S. (2006), *Time Series Analysis: Univariate and Multivariate Methods*, Second Edition, Boston: Pearson/Addison–Wesley. An early comprehensive treatment of ARIMA modeling is given in Box, G.E.P., and Jenkins, G.M. (1976), Time Series Analysis: Forecasting and Control, Oakland, CA: Holden–Day.

There are dozens of people to thank for making this book possible. Carrie Cooper, Lisa Nickel, Tami Back, Rosie Liljenquist, and all of the librarians and generous donors at the Swem Library at William & Mary made this book possible through their Library Scholar position. Thanks also goes to the William & Mary statisticians Ed Chadraa, Flip deCamp, Greg Hunt, Ross Iaci, Rui Pereira, Heather Sasinowska, and Guannan Wang, who have helped brainstorm about the topics and their sequencing in the book. I am grateful for Olivia Ding, Kexin Feng, Robert Jackson, Yuxin Qin, Chris Weld, and Hailey Young taking the time to read all or portions of an early draft of the text and providing helpful feedback. Barry Lawson from Bates College helped with the inset and lines in Figure 7.25. Five special people have made extraordinary contributions to this book: Heather Sasinowska edited the regression and time series chapters, Robert Lewis edited most of the entire textbook, Raghu Pasupathy provided keen insight concerning the presentation of the time series material and the moment ratio diagrams, Rosie Liljenquist edited the first two chapters just before having her sixth baby, and my wife Jill helped me push the book over the finish line. Finally, thanks goes to Drea George for the handsome book cover.

Since this is an open educational resource, this book is a work in progress. Please e-mail any typographical errors or suggested alterations that you spot to me. Thank you.

Williamsburg, VA   *Larry Leemis*
                                                                                    January 2023

Part I

REGRESSION

Chapter 1

Simple Linear Regression

Regression is a statistical technique that involves describing the relationship between one or more *independent variables* and a single *dependent variable*. For simplicity, assume for now that there is just a single independent variable. To establish some notation, let

- X be an independent variable, also called an explanatory variable, predictor variable, or regressor, which is typically assumed to take on fixed values (that is, X is *not* a random variable) which can be observed without error, and

- Y be a dependent variable, also called a response variable, which is typically a continuous random variable.

The relationship between the independent variable X and the dependent variable Y is often established by collecting n data pairs denoted by (X_1, Y_1), (X_2, Y_2), ... , (X_n, Y_n), plotting these pairs on a pair of axes, and looking for a pattern that can be translated to a mathematical form. This process establishes an empirical mathematical model for the underlying relationship between the independent variable X and the dependent variable Y.

1.1 Deterministic Models

Regression analysis establishes a functional relationship between X and Y. The simplest type of relationship between X and Y is a *deterministic* relationship $Y = f(X)$. In this rare case, the value of Y can be determined without error once the value of X is known, so Y is not a random variable when the relationship between X and Y is deterministic. The *deterministic model* is described by $Y = f(X)$. Deterministic relationships are uncommon in real-world applications because there is typically uncertainty in the dependent variable. If data pairs (X_1, Y_1), (X_2, Y_2), ... , (X_n, Y_n) are collected and the deterministic relationship $Y = f(X)$ establishes the correct functional relationship between X and Y, then all of the data pairs will fall on the graph of the function $Y = f(X)$.

Example 1.1 Bob is a salesman. The independent variable X is the *number* of sales that he makes per week. Bob receives a \$50 commission for each sale, regardless of the amount of each sale. The dependent random variable Y is the total weekly commission that Bob receives. Find the deterministic relationship between X and Y.

In this setting, the independent variable X is a fixed constant which is measured without error, and the deterministic relationship between X and Y is

$$Y = f(X) = 50X.$$

This deterministic relationship expresses Y as a linear function of X. If the next three weeks of Bob's sales activity result in the three data pairs

$$(X_1, Y_1) = (6, 300), \qquad (X_2, Y_2) = (8, 400), \qquad \text{and} \qquad (X_3, Y_3) = (2, 100),$$

then all three of these data pairs will fall on the graph of the deterministic relationship $Y = f(X) = 50X$. The X_i values are distinct for these data pairs, but this need not necessarily be the case. Bob could have had weeks in which he made the same number of sales multiple times. Figure 1.1 shows the deterministic relationship and the three data values that fall on the line. Notice that the graph of $Y = f(X) = 50X$ passes through the origin, $(0, 0)$, because zero weekly sales results in no weekly commissions. In this particular example, a line is plotted even though X can only take on integer values.

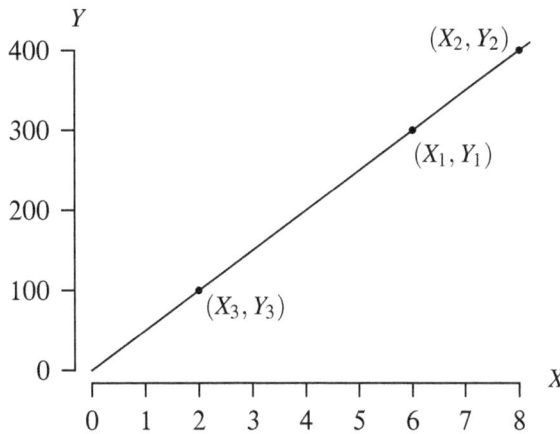

Figure 1.1: A deterministic linear relationship between X and Y.

Determining the relationship between the number of sales per week X and the commissions paid per week Y did not require the collection of any data to determine the function $Y = f(X)$. That linear relationship was implicit in the problem statement. Other cases can arise, such as (a) the relationship is deterministic but requires data to determine its functional form, or (b) the relationship is deterministic, but unlike the relationship in the previous example, it is not linear. The following example illustrates a nonlinear deterministic relationship between the independent variable X and the dependent variable Y.

Example 1.2 Alice purchases a five-year certificate of deposit paying 8% annually with an initial deposit of $1000. Let the independent variable X be the number of months that the certificate of deposit has been held at a bank. Let the dependent variable Y be the associated balance. Find the deterministic relationship between X and Y assuming that the interest on the certificate of deposit is compounded monthly.

Under these assumptions, the balance on Alice's certificate of deposit at month X is

$$Y = f(X) = 1000 \left(1 + \frac{0.08}{12} \right)^X.$$

(This relationship between X and Y makes three somewhat minor simplifying assumptions: (1) $Y = f(X)$ gives the instantaneous value of the CD after X months have passed even though interest is paid monthly, making this a continuous function rather than a step function, (2) all 12 months are assumed to have the same number of days, and (3) all years have the same number of days, which is not the case because of leap years. The violation of these assumptions are minor, and the relationship given here is very close to the balance Y after X months have passed.)

The curve in Figure 1.2 associated with the deterministic relationship is concave upward because of compounding. The three points plotted on the curve are

$$(X_1, Y_1) = (0, 1000), \quad (X_2, Y_2) = (12, 1083.00), \quad \text{and} \quad (X_3, Y_3) = (60, 1489.85).$$

The first data pair corresponds to the initial $1000 deposit into the certificate of deposit at $X = 0$. The second data pair corresponds to the account balance after one year, or $X = 12$ months. The balance after 12 months is slightly more than the annual simple interest balance $1000 \cdot (1 + 0.08) = \$1080$ because of the monthly compounding. The third data pair corresponds to the final balance of $1489.85 after 60 months. As was the case with the sales commissions in the previous example, the three data pairs were not necessary to establish the deterministic relationship between the independent variable X and the dependent variable Y. Their relationship is implicit in the problem statement. In both examples, the three data pairs fall on the graph of the deterministic relationship.

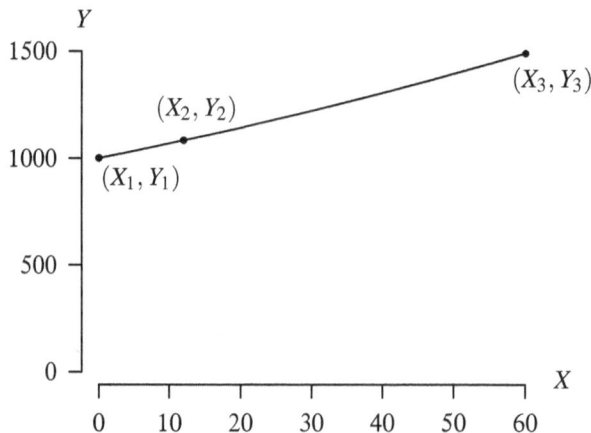

Figure 1.2: A deterministic nonlinear relationship between X and Y.

In most applications, the relationship between the independent variable X and the dependent variable Y is not deterministic because Y is typically a random variable. The next section introduces some of the thinking behind the development of a *statistical model* that describes the relationship between X and Y.

1.2 Statistical Models

The goal in constructing a statistical model is to write a formula that adequately captures the governing probabilistic relationship between an independent variable X and a dependent variable Y. This formula might be used subsequently for prediction or some other form of statistical inference. In this section, we assume that the dependent variable Y is a continuous random variable that can assume a range of values associated with a particular setting of the independent variable X. The relationship

$$Y = f(X)$$

that was used in the previous section is no longer adequate because X is assumed to be observed without error, and this formula results in a value of Y which is deterministic rather than random. One way of overcoming this problem is to replace the left-hand side of this equation by the expected value of Y, which is a constant, resulting in

$$E[Y] = f(X).$$

To be a little more careful about what is meant by this statistical relationship, the left-hand side is actually a conditional expectation, namely

$$E[Y \,|\, X = x] = f(x).$$

In words, given that the independent variable X assumes the value x, the transformation $f(x)$ gives the conditional expected value of the dependent variable Y. Notice that this statistical model does not specify the *distribution* of the random variable Y for a particular value of X; it only tells us the expected value of Y for a particular value of X. This statistical regression model defines a *hypothesized* relationship between the observed value of X on the right-hand side of the model and the conditional expected value of Y on the left-hand side of the model. The hypothesized relationship might be adequate for modeling or it might need some refining. There is typically no model that perfectly captures the relationship between X and Y. This was recognized by George Box, who wrote:

> All models are wrong; some models are useful.

In a statistical model that involves parameters, the estimation of the model parameters will be followed by assessments to determine whether the model holds in an empirical sense. If the model needs refining, the new set of parameters are estimated and new assessments are made to see if the refined model is an improvement over the previous model. Regression modeling is an iterative process.

There is a second way to write a statistical model that is equivalent to the statistical model described in the previous paragraph. The model can be written as

$$Y = f(X) + \varepsilon,$$

where the error term ε (also known as the "noise" or "disturbance" term) is a random variable that accounts for the fact that the independent variable cannot predict the dependent variable with certainty. This term makes the relationship between X and Y a random (or *statistical* or *stochastic*) relationship rather than a deterministic relationship. If the probability distribution of the error term is specified, then not only is the expected value of Y conditioned on the value of X determined, but also the entire probability distribution of Y conditioned on the value of X is specified. It is common practice to assume that the expected value of ε is zero. The probability distribution of ε establishes the nature and magnitude of the scatter of the data values about the regression function. When the population variance of ε is small, the values of Y are tightly clustered about the regression function $f(X)$; when the population variance of ε is large, the values of Y stray further from the regression function $f(X)$.

Regression modeling involves determining the functional form of $f(X)$ from a data set of n data pairs $(X_1, Y_1), (X_2, Y_2), \ldots, (X_n, Y_n)$. The statistical model for X and Y in a general sense also applies to each of the data points, so

$$Y_i = f(X_i) + \varepsilon_i$$

for $i = 1, 2, \ldots, n$. The sign of ε_i indicates whether the observed data pair (X_i, Y_i) falls above ($\varepsilon_i > 0$) or below ($\varepsilon_i < 0$) the conditional expected value of Y_i, for $i = 1, 2, \ldots, n$.

The function $f(X)$ is called the *regression function*, and was first referred to in print as such by Sir Francis Galton (1822–1911), a British anthropologist and meteorologist, in his 1885 paper titled "Regression Toward Mediocrity in Hereditary Stature" published in the *Journal of the Anthropological Institute*. He established a regression function relating the adult height of an offspring, Y, as a function of an average of the parent's heights, X, which had been adjusted for gender.

The regression function $Y = f(X)$ can be either linear or nonlinear. The next section focuses on the easier case, a linear regression function. In this case, the model is typically referred to as a *simple linear regression* model, which is often abbreviated as an SLR model. The model is *simple* because there is only one independent variable X that is used to predict the dependent variable Y. The model is *linear* because the regression function $f(X) = \beta_0 + \beta_1 X$ is assumed to be linear in the parameters β_0 and β_1. The more complicated cases of multiple linear regression, which involve more than one independent variable, and nonlinear regression, in which $f(X)$ is not a linear function, will be introduced later.

1.3 Simple Linear Regression Model

A simple linear regression model assumes a linear relationship between an independent variable X and a dependent variable Y. In this section, the more general regression model

$$Y = f(X) + \varepsilon$$

is reduced to the simple linear regression model given in the definition below.

Definition 1.1 A *simple linear regression model* is given by

$$Y = \beta_0 + \beta_1 X + \varepsilon,$$

where

- X is the independent variable, assumed to be a fixed value observed without error,

- Y is the dependent variable, which is a continuous random variable,

- β_0 is the population intercept of the regression line, which is an unknown constant,

- β_1 is the population slope of the regression line, which is an unknown constant, and

- ε is the error term, a continuous random variable with population mean zero and positive, finite population variance σ^2 that accounts for the randomness in the relationship between X and Y.

Stating the simple linear regression model in this fashion will not seem natural from probability theory. As a non-regression illustration from probability theory, $W \sim N\left(\mu, \sigma^2\right)$ indicates that W has a normal distribution with population mean μ and population variance σ^2. Although much less compact, the probability distribution of W can also be written as $W = \mu + \varepsilon$, where $\varepsilon \sim N\left(0, \sigma^2\right)$. This illustration reflects the essence behind writing the simple linear regression model in the form $Y = \beta_0 + \beta_1 X + \varepsilon$ in Definition 1.1.

The formulation of the simple linear model from Definition 1.1 involves a random variable ε on the right-hand side of the model. In some settings, this model might be viewed as a transformation of a random variable, but this is not the correct interpretation of the model in this setting. The simple linear regression model defines a hypothesized relationship between the random variable on the left-hand side of the model and terms on the right-hand side of the model. This probability model is hypothesized to govern the relationship between X and Y. The goal in constructing a simple linear regression model is to determine if it adequately captures the probabilistic relationship between X and Y. Estimation of the model parameters will be followed by assessment to see if the model holds in an empirical sense.

The assumption that the random variable ε has population mean zero and population variance σ^2 in the most basic simple linear regression model in Definition 1.1 allows for mathematically tractable statistical inference. In models that allow for confidence intervals and hypothesis testing concerning the estimated slope and intercept, the error term is assumed to have a specific distribution, which is typically the normal distribution. The error term models all sources of variation, both known and unknown, other than the variation in Y associated with the particular level of X. Notice that σ^2 is constant over all values of X.

The assumption that the independent variable X is not subject to random variability is not always satisfied in practice. The fitting procedure becomes more complicated when X is considered to be a random variable. For this reason, we assume that the observed value of X is either exact or that the variation of X is small enough so that its observed value can be assumed to be exact.

The assumption of a *linear relationship* between X and Y might also be flawed. In some cases it might not be a perfectly linear relationship, but a linear relationship provides a close enough approximation between X and Y to be useful for associated statistical inference. In other cases, a linear relationship might be appropriate for some range of values of X, known as the *scope* of the model, but not others. One important step in establishing a simple linear regression model is to specify the values of X for which the simple linear regression model is valid.

The procedure for establishing a simple linear regression model that relates the dependent variable Y to the independent variable X is given below.

1. **Collect the data pairs**. The data pairs are denoted by (X_1, Y_1), (X_2, Y_2), $\ldots$, (X_n, Y_n). In some settings, it is possible to exert some control over the X_i values. As will be seen later, there are advantages to having the X_i values spread out as much as possible in terms of the precision of the fitted regression model.

2. **Make a scatterplot of data pairs**. A *scatterplot* is just a plot of the points (X_1, Y_1), (X_2, Y_2), $\ldots$, (X_n, Y_n) on a set of axes. The purpose of the scatterplot is to see if the linear relationship between X and Y is appropriate and to visually assess the spread of the data values about the regression function. With modern statistical software, scatterplots are easy to generate.

3. **Inspect the scatterplot**. Although this step is subjective, it is important to visually assess (*a*) whether the relationship between X and Y appears to be linear or nonlinear, (*b*) whether the spread of the data pairs about the regression function is small or large, and (*c*) whether the

variability of the data pairs about the regression function remains constant over the range of X values that have been collected.

4. **State the regression model**. In this chapter, the regression model is assumed to be the simple linear regression model $Y = \beta_0 + \beta_1 X + \varepsilon$. Nonlinear regression models, such as the quadratic model $Y = \beta_0 + \beta_1 X + \beta_2 X^2 + \varepsilon$, and multiple regression models with more than one independent variable, such as $Y = \beta_0 + \beta_1 X_1 + \beta_2 X_2 + \varepsilon$, will be considered later.

5. **Fit the regression model to the data pairs**. The method of *least squares*, which will be described in the next section, is commonly used to estimate the parameters in the regression model. The least squares criterion is to choose the regression model that minimizes the sum of the squares of the vertical differences between data points and the fitted regression model.

6. **Assess the adequacy of the fitted regression model**. Visual assessment techniques for assessing the fitted regression model include superimposing the fitted regression model onto the scatterplot of the data pairs and examining a plot of the residuals. A residual is the signed vertical distance between a data pair and its associated value on the regression function. In addition, there are statistical methods that can be applied to the fitted regression model to see if it adequately describes the relationship between X and Y.

7. **Perform statistical inference**. Once the fitted regression model is deemed an acceptable approximation to the relationship between X and Y, it can be used for statistical inference. One simple example of statistical inference that occurs often in practice is the prediction of a future value of Y for a particular level of X.

The seven steps for establishing a regression model are not necessarily performed in the order given here. Many times the fitted regression model is rejected in Step 6, and it is necessary to return to Step 4 in order to formulate an alternative model. Steps 4 through 6 might need to be repeated several times before arriving at an acceptable model for statistical inference.

The simple linear regression model given in Definition 1.1 implies that all of the (X_i, Y_i) pairs also follow the simple linear regression model:

$$Y_i = \beta_0 + \beta_1 X_i + \varepsilon_i$$

for $i = 1, 2, \ldots, n$, where

- (X_i, Y_i) are the data pairs, for $i = 1, 2, \ldots, n$,

- X_i is the value of the independent variable for observation i, which is observed without error, for $i = 1, 2, \ldots, n$,

- Y_i is the value of the dependent variable for observation i, which is a continuous random variable, for $i = 1, 2, \ldots, n$,

- β_0 is the population intercept of the regression line,

- β_1 is the population slope of the regression line, and

- ε_i is the random error term for observation i which satisfies

   - $E[\varepsilon_i] = 0$ for $i = 1, 2, \ldots, n$,
   - $V[\varepsilon_i] = \sigma^2$ for $i = 1, 2, \ldots, n$,

– the random ε_i values are mutually independent random variables, which implies that their variance–covariance matrix is diagonal.

When the simple linear regression model is stated in this fashion, four properties become apparent. First, Y_i is a random variable that can be broken into two components: a deterministic component $\beta_0 + \beta_1 X_i$, and a random component ε_i, for $i = 1, 2, \ldots, n$. Second, Y_i has population mean

$$E[Y_i] = E[\beta_0 + \beta_1 X_i + \varepsilon_i] = \beta_0 + \beta_1 X_i$$

for $i = 1, 2, \ldots, n$ and population variance

$$V[Y_i] = V[\beta_0 + \beta_1 X_i + \varepsilon_i] = V[\varepsilon_i] = \sigma^2$$

for $i = 1, 2, \ldots, n$. Using slightly different notation, it would be reasonable to write the population mean and variance as the conditional expectations

$$E[Y_i \,|\, X_i] = \beta_0 + \beta_1 X_i \qquad \text{and} \qquad V[Y_i \,|\, X_i] = \sigma^2$$

for $i = 1, 2, \ldots, n$. The property that the variance does not change with X_i is known as *homoscedasticity*. Temporarily dropping the subscripts, the line

$$E[Y] = \beta_0 + \beta_1 X,$$

with β_0 and β_1 replaced by the associated estimated values $\hat{\beta}_0$ and $\hat{\beta}_1$, is oftentimes superimposed onto the scatterplot to visualize the fitted regression model. Third, each data pair (X_i, Y_i) has a Y_i value that misses the regression function by the error term ε_i, for $i = 1, 2, \ldots, n$. Fourth, the values of the observed dependent variables $Y_1, Y_2, \ldots, Y_n$ must be mutually independent random variables because the error terms $\varepsilon_1, \varepsilon_2, \ldots, \varepsilon_n$ are mutually independent random variables.

1.4 Least Squares Estimators

We now turn to the question of estimating the intercept β_0 and the slope β_1 by the method of least squares. German mathematician Carl Friedrich Gauss (1777–1855) invented the least squares method and French mathematician Adrien–Marie Legendre (1752–1833) first published the method in 1805. The least squares method determines the values of β_0 and β_1 that minimize the sum of the squares of the errors, where the error is the vertical distance between the Y_i value and the fitted regression line. The term *estimator* will be used here to refer to a generic formula for $\hat{\beta}_0$ or $\hat{\beta}_1$; the term *estimate* will be used to refer to a specific numeric value for $\hat{\beta}_0$ or $\hat{\beta}_1$.

One bit of notation that will make the expressions of the point estimators more compact is

$$
\begin{aligned}
S_{XY} &= \sum_{i=1}^{n} \left(X_i - \bar{X} \right)\left(Y_i - \bar{Y} \right) \\
&= \sum_{i=1}^{n} \left(X_i Y_i - X_i \bar{Y} - \bar{X} Y_i + \bar{X}\bar{Y} \right) \\
&= \sum_{i=1}^{n} X_i Y_i - n\bar{X}\bar{Y} - n\bar{X}\bar{Y} + n\bar{X}\bar{Y} \\
&= \sum_{i=1}^{n} X_i Y_i - n\bar{X}\bar{Y}.
\end{aligned}
$$

Similarly,

$$S_{XX} = \sum_{i=1}^{n} (X_i - \bar{X})^2 = \sum_{i=1}^{n} X_i^2 - n\bar{X}^2$$

and

$$S_{YY} = \sum_{i=1}^{n} (Y_i - \bar{Y})^2 = \sum_{i=1}^{n} Y_i^2 - n\bar{Y}^2.$$

This new notation allow us to express nS_{XY}, nS_{XX}, and nS_{YY} as

$$nS_{XY} = n\sum_{i=1}^{n} X_i Y_i - \sum_{i=1}^{n} X_i \sum_{i=1}^{n} Y_i,$$

$$nS_{XX} = n\sum_{i=1}^{n} X_i^2 - \left(\sum_{i=1}^{n} X_i\right)^2,$$

and

$$nS_{YY} = n\sum_{i=1}^{n} Y_i^2 - \left(\sum_{i=1}^{n} Y_i\right)^2.$$

Using this notation, the least squares estimators for the slope and intercept of the model, denoted by $\hat{\beta}_1$ and $\hat{\beta}_0$, are given in the following theorem. Notice that the term *normal equations* in the theorem is not related to the normal distribution.

Theorem 1.1 Let $(X_1, Y_1), (X_2, Y_2), \ldots, (X_n, Y_n)$ be n data pairs with at least two distinct X_i values. The *least squares estimators* of β_0 and β_1 minimize the sum of the squared deviations between Y_i and the associated fitted value $\hat{\beta}_0 + \hat{\beta}_1 X_i$ in the simple linear regression model. The least squares estimators are the solution to the simultaneous *normal equations*

$$n\hat{\beta}_0 + \hat{\beta}_1 \sum_{i=1}^{n} X_i = \sum_{i=1}^{n} Y_i$$

$$\hat{\beta}_0 \sum_{i=1}^{n} X_i + \hat{\beta}_1 \sum_{i=1}^{n} X_i^2 = \sum_{i=1}^{n} X_i Y_i$$

and are given by

$$\hat{\beta}_1 = \frac{S_{XY}}{S_{XX}}$$

and

$$\hat{\beta}_0 = \bar{Y} - \hat{\beta}_1 \bar{X},$$

where $\bar{X}$ and $\bar{Y}$ are the sample means

$$\bar{X} = \frac{X_1 + X_2 + \cdots + X_n}{n} \qquad \text{and} \qquad \bar{Y} = \frac{Y_1 + Y_2 + \cdots + Y_n}{n}.$$

Proof The deviation of Y_i from the associated value on the population regression line is

$$Y_i - (\beta_0 + \beta_1 X_i),$$

for $i = 1, 2, \ldots, n$. The sum of the squared deviations is

$$S = \sum_{i=1}^{n} (Y_i - \beta_0 - \beta_1 X_i)^2.$$

The least squares estimators are those that minimize S respect to β_0 and β_1; that is,

$$(\hat{\beta}_0, \hat{\beta}_1) = \operatorname*{argmin}_{\beta_0, \beta_1} \sum_{i=1}^{n} (Y_i - \beta_0 - \beta_1 X_i)^2.$$

Using calculus to minimize S with respect to β_0 and β_1 requires taking the partial derivatives of S with respect to β_0 and β_1:

$$\frac{\partial S}{\partial \beta_0} = -2 \sum_{i=1}^{n} (Y_i - \beta_0 - \beta_1 X_i) = 0$$

$$\frac{\partial S}{\partial \beta_1} = -2 \sum_{i=1}^{n} X_i (Y_i - \beta_0 - \beta_1 X_i) = 0.$$

Simplifying and using the hat notation to denote the estimators results in the simultaneous normal equations

$$n\hat{\beta}_0 + \hat{\beta}_1 \sum_{i=1}^{n} X_i = \sum_{i=1}^{n} Y_i$$

$$\hat{\beta}_0 \sum_{i=1}^{n} X_i + \hat{\beta}_1 \sum_{i=1}^{n} X_i^2 = \sum_{i=1}^{n} X_i Y_i.$$

The normal equations are a system of two linear equations in the two unknowns $\hat{\beta}_0$ and $\hat{\beta}_1$. Solving these equations simultaneously yields the point estimator for the slope

$$\hat{\beta}_1 = \frac{\sum_{i=1}^{n} (X_i - \bar{X})(Y_i - \bar{Y})}{\sum_{i=1}^{n} (X_i - \bar{X})^2} = \frac{S_{XY}}{S_{XX}}.$$

Dividing the first normal equation by the sample size n yields the point estimator for the intercept

$$\hat{\beta}_0 = \bar{Y} - \hat{\beta}_1 \bar{X}.$$

The next step is to show that the *least* squares estimators $\hat{\beta}_1$ and $\hat{\beta}_0$ *minimize S*. This will be done by showing that the Hessian matrix is positive definite. The Hessian matrix **H** is the matrix of second partial derivatives of S with respect to β_0 and β_1:

$$\mathbf{H} = \begin{bmatrix} \dfrac{\partial^2 S}{\partial \beta_0^2} & \dfrac{\partial^2 S}{\partial \beta_0 \partial \beta_1} \\ \dfrac{\partial^2 S}{\partial \beta_1 \partial \beta_0} & \dfrac{\partial^2 S}{\partial \beta_1^2} \end{bmatrix} = \begin{bmatrix} 2n & 2 \sum_{i=1}^{n} X_i \\ 2 \sum_{i=1}^{n} X_i & 2 \sum_{i=1}^{n} X_i^2 \end{bmatrix}.$$

The **H** matrix is unchanged when evaluated at the least squares estimators $\hat{\beta}_0$ and $\hat{\beta}_1$. To show that this matrix is positive definite at the least squares estimators, it is sufficient to show that the upper-left-hand element and the determinant of **H** are both positive. The upper-left-hand element is positive for all values of the sample size n. The determinant of $\mathbf{H}$ is

$$|\mathbf{H}| = \begin{vmatrix} 2n & 2\sum_{i=1}^{n} X_i \\ 2\sum_{i=1}^{n} X_i & 2\sum_{i=1}^{n} X_i^2 \end{vmatrix} = 4n\sum_{i=1}^{n} X_i^2 - 4\left(\sum_{i=1}^{n} X_i\right)^2.$$

This expression is positive when there are at least two distinct X_i values by the Cauchy–Schwartz inequality. The Cauchy–Schwartz inequality (a special case of the triangle inequality) states that for real numbers $a_1, a_2, \ldots, a_n$ and $b_1, b_2, \ldots, b_n$,

$$\left(a_1^2 + a_2^2 + \cdots + a_n^2\right) \cdot \left(b_1^2 + b_2^2 + \cdots + b_n^2\right) \geq (a_1 b_1 + a_2 b_2 + \cdots + a_n b_n)^2,$$

where equality is satisfied if and only if $a_1 = a_2 = \cdots = a_n$ and $b_1 = b_2 = \cdots = b_n$. Letting $a_i = 1$ and $b_i = x_i$ in the Cauchy–Schwartz inequality indicates that the determinant of $\mathbf{H}$ is positive when there are at least two distinct X_i values. Hence, the Hessian matrix $\mathbf{H}$ is positive definite and the least squares estimators $\hat{\beta}_0$ and $\hat{\beta}_1$ minimize S. $\square$

The requirement that there are at least two distinct X_i values in Theorem 1.1 is consistent with intuition. Figure 1.3 shows $n = 5$ data pairs in which the independent variable assumes the same value for each pair: $X_1 = X_2 = X_3 = X_4 = X_5 = 3$. It is not possible to estimate the slope of the regression line in this particular setting. This is the geometric reason for the requirement that there are at least two distinct X_i values. In addition, the denominator in $\hat{\beta}_1 = S_{XY}/S_{XX}$ is zero when all X_i values are equal, which gives the associated algebraic reason for the requirement. From this point forward, whenever the simple linear regression model is used, it is assumed that the associated data pairs $(X_1, Y_1), (X_2, Y_2), \ldots, (X_n, Y_n)$ have at least two distinct X_i values.

Figure 1.4 shows the geometric interpretation associated with the estimated intercept $\hat{\beta}_0$ and estimated slope $\hat{\beta}_1$. The $n = 9$ data pairs $(X_1, Y_1), (X_2, Y_2), \ldots, (X_9, Y_9)$ are plotted as points, along with the associated estimated regression line $Y = \hat{\beta}_0 + \hat{\beta}_1 X$. The y-intercept of the graph $\hat{\beta}_0$ is the height of the estimated regression line at $X = 0$. The "rise over run" interpretation of the slope is illustrated by the right triangle with legs consisting of dotted lines.

Figure 1.3: Identical independent variable values for all $n = 5$ data pairs.

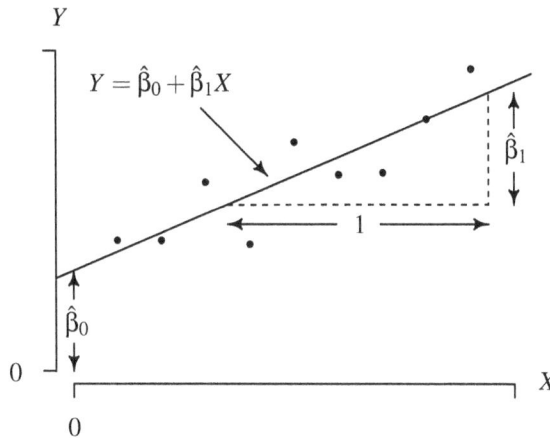

Figure 1.4: Geometry associated with $\hat{\beta}_0$ and $\hat{\beta}_1$.

The next example illustrates the mechanics associated with calculating the least squares estimates $\hat{\beta}_0$ and $\hat{\beta}_1$. In order to focus on the calculations performed by hand, a small sample size of $n = 3$ data pairs is used. The numbers have been handpicked in order to make the resulting parameter estimates come out to whole numbers. A sample size of $n = 2$ is too simplistic in that two points determine a line, and the estimated regression line will always pass through those two points.

Example 1.3 Cheryl sells farm equipment and supplies. Let X be the number of sales she completes in a week, which will serve as the independent variable in this example. Each sale that she completes results in an associated random amount of revenue to the company that can be attributed to Cheryl's sales prowess. The dependent random variable Y is the associated total revenue to the company from Cheryl's sales for that week, in thousands of dollars. The data pairs for the past $n = 3$ weeks are

$$(X_1, Y_1) = (6, 2), \qquad (X_2, Y_2) = (8, 9), \qquad \text{and} \qquad (X_3, Y_3) = (2, 2).$$

Find the least squares estimates of the population intercept β_0 and population slope β_1 for the simple linear regression model from these data pairs and plot the fitted regression line and the data pairs on a single plot.

A scatterplot for this data set is generated using the `plot` function in the R commands

```
x = c(6, 8, 2)
y = c(2, 9, 2)
plot(x, y, xlim = c(0, 8), ylim = c(0, 9))
```

and is displayed in Figure 1.5. Your immediate reaction to the scatterplot might be to conclude that this is certainly not a linear relationship between X and Y. But this conclusion might not be warranted because of the small number of data pairs collected. One thing that is unusual about this data set is that Cheryl generated six sales in the first week, resulting in just $2000 in revenue, and then two sales in the third week, also resulting in $2000 in revenue. Clearly the sales transacted during the first week were

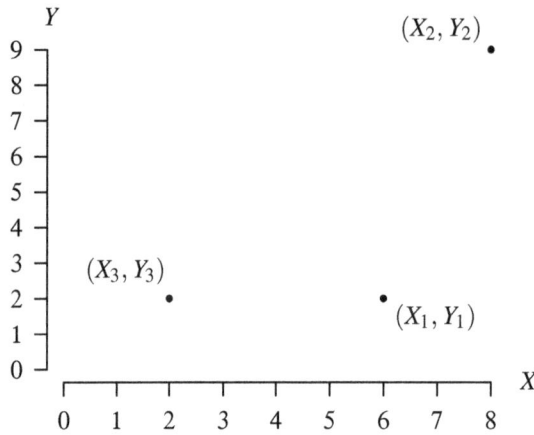

Figure 1.5: A scatterplot of the sales data pairs.

much smaller in size, on average, than those in the third week. Since the purpose of this example is to illustrate the calculations for computing $\hat{\beta}_0$ and $\hat{\beta}_1$, we will proceed as if the linear model were appropriate. Assessing a simple linear regression model with only $n = 3$ data pairs is nearly impossible.

The least squares estimates for β_0 and β_1 will be calculated in three different fashions. First, they will be calculated by hand, with all of the calculations displayed here. Second, they will be calculated in R using an approach that mirrors the hand calculations.

Third, they will be calculated in R using the lm (for linear model) function, which automates the process of estimating β_0 and β_1.

Table 1.1 contains the data pairs and calculations necessary to compute the estimated slope and intercept of the regression line. The sample means of the independent and dependent variables are

$$\bar{X} = \frac{16}{3} \qquad \text{and} \qquad \bar{Y} = \frac{13}{3}.$$

Although $\bar{X}$ and $\bar{Y}$ are set in upper case, it is important to remember that the X_i values are observed without error and the Y_i values are the associated random responses. The

Observation number i	Number of sales X_i	Total revenue Y_i	$(X_i - \bar{X})^2$	$(X_i - \bar{X})(Y_i - \bar{Y})$
1	6	2	$(6-16/3)^2$	$(6-16/3)(2-13/3)$
2	8	9	$(8-16/3)^2$	$(8-16/3)(9-13/3)$
3	2	2	$(2-16/3)^2$	$(2-16/3)(2-13/3)$
Sum	16	13	$168/9$	$168/9$

Table 1.1: Data pairs and calculated values for estimating β_0 and β_1.

sums in the bottom row of Table 1.1 give the sums of squares

$$S_{XX} = \sum_{i=1}^{3}(X_i - \bar{X})^2 = \frac{168}{9} \qquad \text{and} \qquad S_{XY} = \sum_{i=1}^{3}(X_i - \bar{X})(Y_i - \bar{Y}) = \frac{168}{9}.$$

The fact that $S_{XX} = S_{XY}$ is coincidental, and is typically not the case in practice. Using Theorem 1.1, the least squares estimates of β_1 and β_0 are

$$\hat{\beta}_1 = \frac{S_{XY}}{S_{XX}} = \frac{168/9}{168/9} = 1 \qquad \text{and} \qquad \hat{\beta}_0 = \bar{Y} - \hat{\beta}_1 \bar{X} = \frac{13}{3} - 1 \cdot \frac{16}{3} = -1.$$

A second way to calculate the least squares estimates $\hat{\beta}_1$ and $\hat{\beta}_0$ uses the R code below to implement the formulas given in Theorem 1.1. The code is generic in the sense that once the two vectors x and y are defined using the first two commands, the last four commands will calculate the point estimates $\hat{\beta}_1$ and $\hat{\beta}_0$ for any number of (X_i, Y_i) pairs.

```
x         = c(6, 8, 2)
y         = c(2, 9, 2)
sxx       = sum((x - mean(x)) ^ 2)
sxy       = sum((x - mean(x)) * (y - mean(y)))
beta1hat  = sxy / sxx
beta0hat  = mean(y) - beta1hat * mean(x)
```

This code also returns the point estimates

$$\hat{\beta}_1 = 1 \qquad \text{and} \qquad \hat{\beta}_0 = -1.$$

As you might imagine, these calculations are performed so often by statisticians that R has a built-in function to estimate β_1 and β_0.

A third way to calculate the least squares estimates of β_1 and β_0 via use of the R lm function.

```
x = c(6, 8, 2)
y = c(2, 9, 2)
lm(y ~ x)$coefficients
```

The lm function takes a formula for an argument, in this case $y \sim x$, and returns a list. One component of the list returned by lm is named coefficients, and it contains the estimated regression coefficients $\hat{\beta}_1 = 1$ and $\hat{\beta}_0 = -1$.

The fitted regression line is added to the scatterplot in Figure 1.6 using the R code below. The plot function plots the data pairs, the lm function estimates the intercept and slope of the regression line via least squares, and the abline function plots the fitted regression line. The labels on the data pairs can be added with the text function. The regression line plotted in Figure 1.6 is the line which minimizes the sum of the squares of the vertical distances between the points associated with the data pairs and the fitted regression line.

```
x = c(6, 8, 2)
y = c(2, 9, 2)
plot(x, y, xlim = c(0, 8), ylim = c(-1, 9))
fit = lm(y ~ x)
abline(fit$coefficients)
```

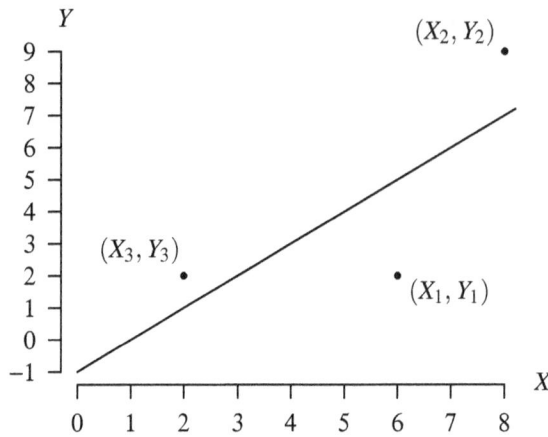

Figure 1.6: A scatterplot of the sales data pairs with the fitted regression line.

The fitted regression line has intercept $\hat{\beta}_0 = -1$ and slope $\hat{\beta}_1 = 1$. The fact that the intercept is $\hat{\beta}_0 = -1$ rather than $\hat{\beta}_0 = 0$ (because $X = 0$ sales in a week should result in $Y = 0$ revenue in that week) is due to random sampling variability. Section 3.1 investigates how to force a regression line through the origin, which would be appropriate in this setting. The interpretation of the estimated slope $\hat{\beta}_1 = 1$ is that the average amount of revenue generated from each sale that Cheryl completes is \$1000.

Figure 1.7 makes two embellishments to Figure 1.6. First, the axes have been adjusted so that the length of one unit on the vertical axis is the same as the length of one unit on the horizontal axis. Second, three shaded squares have been added to the plot. Each square has one vertex at a data pair, and a second vertex at the associated point on the fitted regression line. The numbers in each square give the area of the square. For these data pairs, the total area is the sum of squares

$$S = (Y_1 - \hat{\beta}_0 - \hat{\beta}_1 X_1)^2 + (Y_2 - \hat{\beta}_0 - \hat{\beta}_1 X_2)^2 + (Y_3 - \hat{\beta}_0 - \hat{\beta}_1 X_3)^2$$
$$= (2 + 1 - 6)^2 + (9 + 1 - 8)^2 + (2 + 1 - 2)^2$$
$$= 9 + 4 + 1$$
$$= 14.$$

The fitted least squares line is unique in the following sense. The squares illustrated in Figure 1.7 for any line having an intercept and/or slope that differ from $\hat{\beta}_0 = -1$ and $\hat{\beta}_1 = 1$ will have a total area that exceeds $S = 14$. The fitted least squares line is that line which minimizes S. If a different line were selected and plotted, some of the squares might become smaller, but at least one of the squares would become larger, and the total area of the squares would exceed 14.

Another way to view the minimization of S is to consider contours, or level surfaces, of the sum of squares as a function of the intercept β_0 and the slope β_1. The point

$$\left(\hat{\beta}_0, \hat{\beta}_1\right) = (-1, 1)$$

in Figure 1.8 corresponds to the fitted least squares line with a sum of squares $S = 14$ for the three data pairs. The concentric contours corresponding to $S = 15$, $S = 18$, $S = 23$,

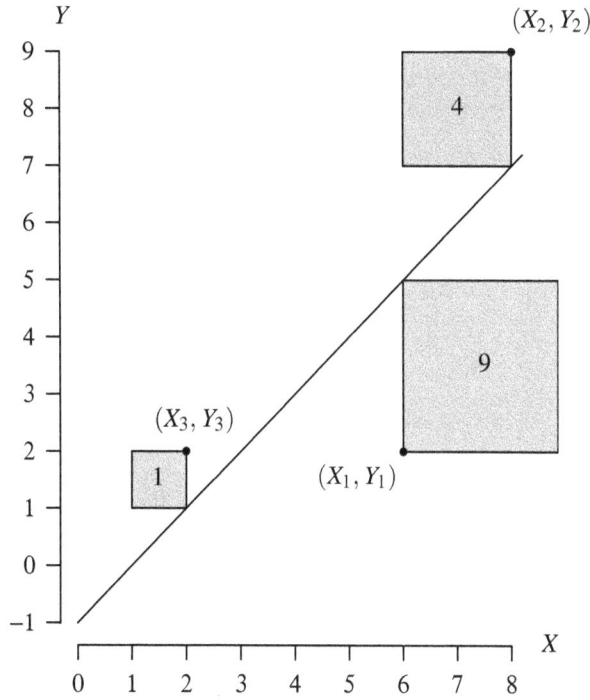

Figure 1.7: A scatterplot of the sales data pairs with the fitted regression line.

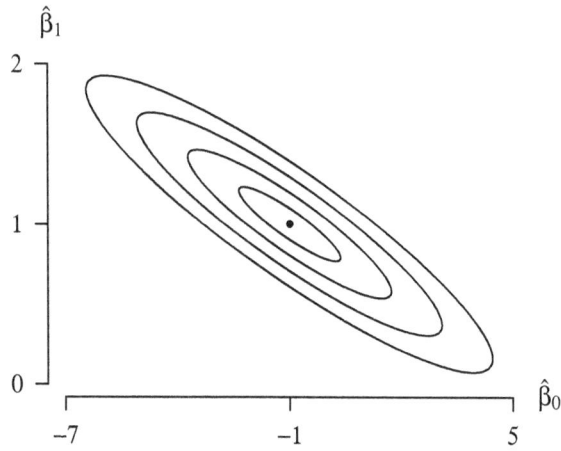

Figure 1.8: Level surfaces of the sum of squares.

and $S = 30$ show how the sum of squares increases as the intercept and slope stray from the least squares estimates.

1.5   Properties of Least Squares Estimators

The least squares estimators of β_0 and β_1 possess several properties which are important for statistical inference. The four properties established in this section are:

- the least squares estimators $\hat{\beta}_0$ and $\hat{\beta}_1$ are unbiased estimators of β_0 and β_1,

- the least squares estimators $\hat{\beta}_0$ and $\hat{\beta}_1$ can be written as linear combinations of the dependent variables $Y_1, Y_2, \ldots, Y_n$,

- the variance–covariance matrix of $\hat{\beta}_0$ and $\hat{\beta}_1$ can be written in closed form, and

- the least squares estimators $\hat{\beta}_0$ and $\hat{\beta}_1$ have the smallest population variance among all unbiased estimators that can be expressed as linear combinations of the dependent variables.

Proofs of the associated results are included in each of the following subsections.

1.5.1 $\hat{\beta}_0$ and $\hat{\beta}_1$ are Unbiased Estimators of β_0 and β_1

A key property associated with the least squares estimators $\hat{\beta}_0$ and $\hat{\beta}_1$ is that their expected values equal the associated population values β_0 and β_1. The next result establishes the unbiasedness of the two point estimators.

Theorem 1.2 The least squares estimators $\hat{\beta}_0$ and $\hat{\beta}_1$ in the simple linear regression model are unbiased estimators of β_0 and β_1, respectively.

Proof To show that $\hat{\beta}_1$ and $\hat{\beta}_0$ are unbiased estimators of β_1 and β_0, it is sufficient to show that

$$E\left[\hat{\beta}_1\right] = \beta_1 \qquad \text{and} \qquad E\left[\hat{\beta}_0\right] = \beta_0.$$

The denominator of the expression for $\hat{\beta}_1$, which is S_{XX}, is a constant because the values of the independent variables $X_1, X_2, \ldots, X_n$ are assumed to be observed without error in the simple linear regression model. Thus, the expected value of $\hat{\beta}_1$ is

$$
\begin{aligned}
E\left[\hat{\beta}_1\right] &= E\left[\frac{S_{XY}}{S_{XX}}\right] \\
&= E\left[\frac{\sum_{i=1}^{n} X_i Y_i - n\bar{X}\bar{Y}}{\sum_{i=1}^{n} X_i^2 - n\bar{X}^2}\right] \\
&= \frac{\sum_{i=1}^{n} X_i E\left[Y_i\right] - n\bar{X} E\left[\bar{Y}\right]}{\sum_{i=1}^{n} X_i^2 - n\bar{X}^2} \\
&= \frac{\sum_{i=1}^{n} X_i \left(\beta_0 + \beta_1 X_i\right) - n\bar{X}\left(\beta_0 + \beta_1 \bar{X}\right)}{\sum_{i=1}^{n} X_i^2 - n\bar{X}^2} \\
&= \frac{\beta_0 \sum_{i=1}^{n} X_i + \beta_1 \sum_{i=1}^{n} X_i^2 - \beta_0 \sum_{i=1}^{n} X_i - n\beta_1 \bar{X}^2}{\sum_{i=1}^{n} X_i^2 - n\bar{X}^2} \\
&= \beta_1.
\end{aligned}
$$

The expected value of $\hat{\beta}_0$ is

$$E\left[\hat{\beta}_0\right] = E\left[\bar{Y} - \hat{\beta}_1 \bar{X}\right] = \beta_0 + \beta_1 \bar{X} - \beta_1 \bar{X} = \beta_0.$$

Therefore, $\hat{\beta}_1$ and $\hat{\beta}_0$ are unbiased estimators of β_1 and β_0.   □

The fact that the least squares estimators of the slope and intercept of the regression line are unbiased will be supported by a Monte Carlo simulation experiment in the next example. Unlike the typical simple linear regression setting in which data pairs (X_1, Y_1), (X_2, Y_2), ..., (X_n, Y_n) are used to estimate the *unknown* parameters β_0 and β_1, the simulation will generate data pairs and associated regression lines for *known* parameters β_0 and β_1.

Example 1.4 Consider the simple linear regression model

$$Y = \beta_0 + \beta_1 X + \varepsilon,$$

where

- the population intercept is $\beta_0 = 1$,
- the population slope is $\beta_1 = 1/2$, and
- the error term ε has a $U(-1, 1)$ distribution.

The population parameters have been chosen arbitrarily. The error term distribution has population mean zero and finite population variance, so it satisfies the conditions of a simple linear regression model from Definition 1.1. The uniform error term distribution is not likely to occur in practice, however, because it cuts off at -1 and 1. Probability distributions with tails, such as the normal distribution, are used more often in practice. Conduct a Monte Carlo simulation with 5000 replications that analyzes the probability distribution of the estimated intercept $\hat{\beta}_0$ and estimated slope $\hat{\beta}_1$ for $n = 10$ data pairs. Assume that the X_i values are equally likely to be one of the integers 0, 1, 2, ..., 9. The independent variable X happens to assume discrete values in this example, but it would pose no difficulty if it took on continuous values.

One problem that might arise in the Monte Carlo experiment is that the X_i values might all be equal. This would violate the assumption in Theorem 1.1 that at least two X_i values must be distinct. Even though this event occurs with the low probability

$$10 \cdot \left(\frac{1}{10}\right)^{10} = 10^{-9},$$

an `if` statement will be included in the Monte Carlo simulation code to catch this problem if it occurs.

The R code below conducts 5000 replications of the Monte Carlo experiment. The commands prior to the `for` loop set the number of replications to 5000, set the sample size to $n = 10$, set the population intercept to $\beta_0 = 1$, set the population slope to $\beta_1 = 1/2$, define the vectors `beta0hat` and `beta1hat` to hold the simulated estimated intercepts and slopes, and establish the random number stream with the `set.seed` function with arbitrary argument. Within the `for` loop, `x` contains the values of the independent variables, `y` contains the values of the associated dependent variables, and `fit` is the list that stores the results of the regression analysis generated by the call to the `lm` function.

```
nrep     = 5000
n        = 10
```

```
beta0    = 1
beta1    = 1 / 2
beta0hat = numeric(nrep)
beta1hat = numeric(nrep)
set.seed(100)
for (i in 1:nrep) {
  x = sample(0:9, n, replace = TRUE)
  if (min(x) == max(x)) stop("All x values are equal")
  y = beta0 + beta1 * x + runif(n, -1, 1)
  fit = lm(y ~ x)
  beta0hat[i] = fit$coefficients[1]
  beta1hat[i] = fit$coefficients[2]
}
```

Figure 1.9 shows the scatterplot and the fitted regression line for the first replication of the simulation. Notice that having tied values for the independent variables poses no difficulty for calculating the estimates of the intercept and slope of the fitted regression line. This first fitted regression line has intercept $\hat{\beta}_0 = 1.398$ which exceeds the population intercept $\beta_0 = 1$; this first fitted regression line has slope $\hat{\beta}_1 = 0.399$ which is less than the population slope $\beta_1 = 0.5$. Each of the 5000 replications will yield unique values of $\hat{\beta}_0$ and $\hat{\beta}_1$. Since $\hat{\beta}_0$ and $\hat{\beta}_1$ are unbiased estimators of β_0 and β_1 by Theorem 1.2, the 5000 simulated point estimates will hover around their population counterparts.

Figure 1.10 contains four lines. The thick, solid line is the population regression line with intercept $\beta_0 = 1$ and slope $\beta_1 = 1/2$. The other three dashed lines correspond to the fitted regression lines for the first three replications of the simulation. As expected, the estimated intercepts and slopes differ from the associated population values from one replication to the next.

When the simulation is run for all 5000 replications, there are 5000 $(\hat{\beta}_0, \hat{\beta}_1)$ pairs generated. The *additional* R commands below plot a histogram of the 5000 $\hat{\beta}_0$ values on

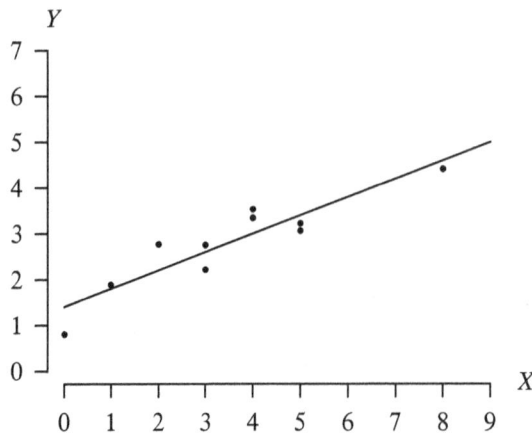

Figure 1.9: Scatterplot of simulated data pairs and fitted regression line (replication 1).

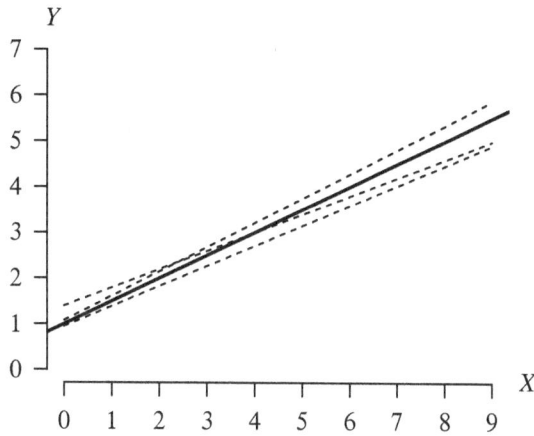

Figure 1.10: Population and fitted regression lines (replications 1–3).

the left and a histogram of the 5000 $\hat{\beta}_1$ values on the right. The `mfrow` (multiple frame by row) argument in `par` function sets up a 1×2 array of plots, and the `hist` function plots the histograms. Figure 1.11 contains the two histograms. The vertical axes have been suppressed because only the center and shape of the histogram is of interest.

```
par(mfrow = c(1, 2))
hist(beta0hat)
hist(beta1hat)
```

As predicted by Theorem 1.2, the histogram of the $\hat{\beta}_0$ values is centered around $\beta_0 = 1$ and the histogram of the $\hat{\beta}_1$ values is centered around $\beta_1 = 1/2$. Both histograms have a bell shape, indicating that the extreme values for the intercepts and slopes are less likely as you move further away from the population values. Although the error terms in the model are mutually independent $U(-1, 1)$ random variables, the summations

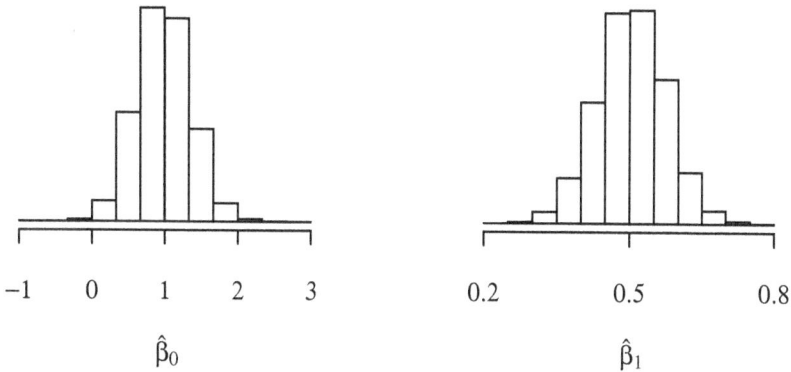

Figure 1.11: Histograms of estimated intercepts (left) and estimated slopes (right).

involved with the computation of $\hat{\beta}_0$ and $\hat{\beta}_1$ allow the central limit theorem to produce a histogram shape that is quite close to that of a normal probability density function.

The two histograms in Figure 1.11 do not indicate whether $\hat{\beta}_0$ and $\hat{\beta}_1$ are independent or dependent random variables. The *additional* R command

```
plot(beta0hat, beta1hat)
```

plots the 5000 $\left(\hat{\beta}_0, \hat{\beta}_1\right)$ pairs, which is displayed in Figure 1.12. The Monte Carlo simulation indicates that the estimated intercepts and slopes are negatively correlated. They tend to be on the opposite sides of their respective means. A larger-than-usual slope is likely to be associated with a smaller-than-usual intercept, and vice versa.

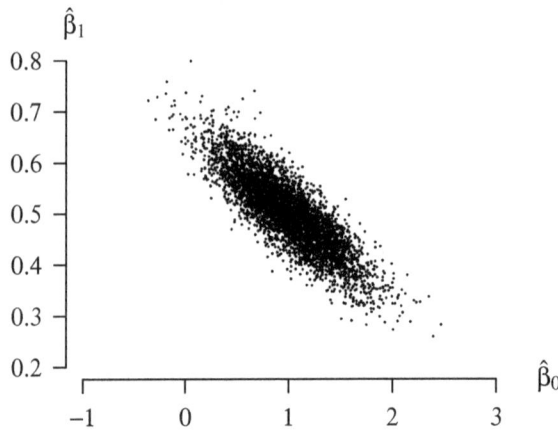

Figure 1.12: Estimated intercepts and slopes for 5000 Monte Carlo simulation replications.

The two key take-aways from this Monte Carlo experiment are:

- $\hat{\beta}_0$ and $\hat{\beta}_1$ being unbiased estimators of β_0 and β_1 via Theorem 1.2 is supported by the histograms in Figure 1.11, and

- $\hat{\beta}_0$ and $\hat{\beta}_1$ appear to be negatively correlated for this particular simple linear regression model by Figure 1.12.

1.5.2 $\hat{\beta}_0$ and $\hat{\beta}_1$ are Linear Combinations of $Y_1, Y_2, \ldots, Y_n$

Theorem 1.2, which states that $E\left[\hat{\beta}_0\right] = \beta_0$ and $E\left[\hat{\beta}_1\right] = \beta_1$, concerns the *accuracy* of the least squares estimators $\hat{\beta}_0$ and $\hat{\beta}_1$. These estimators are "on target" in the sense that their expected values equal their associated population values. The histograms in Figure 1.11 show that the estimators for β_0 and β_1 do not systematically deviate above or below their population values.

The *precision* of the estimators $\hat{\beta}_0$ and $\hat{\beta}_1$ is also of interest. This requires that we also compute their population variances. Before doing so, it is helpful to see that both of these point estimators can be written as linear combinations of the values of the dependent variables $Y_1, Y_2, \ldots, Y_n$.

It is not immediately apparent from the formula for the point estimator for the slope of the regression line $\hat{\beta}_1 = S_{XY}/S_{XX}$, but the estimator can be written as a linear combination of the dependent variables:

$$
\begin{aligned}
\hat{\beta}_1 &= \frac{S_{XY}}{S_{XX}} \\
&= \frac{\sum_{i=1}^{n} \left(X_i - \bar{X} \right) \left(Y_i - \bar{Y} \right)}{\sum_{i=1}^{n} \left(X_i - \bar{X} \right)^2} \\
&= \frac{\sum_{i=1}^{n} \left(X_i - \bar{X} \right) Y_i}{\sum_{i=1}^{n} \left(X_i - \bar{X} \right)^2}
\end{aligned}
$$

because $\bar{Y} \sum_{i=1}^{n} \left(X_i - \bar{X} \right) = \bar{Y} \left(n\bar{X} - n\bar{X} \right) = 0$. This formula indicates that the point estimator for the slope of the regression line is the linear combination

$$
\hat{\beta}_1 = a_1 Y_1 + a_2 Y_2 + \cdots + a_n Y_n,
$$

where

$$
a_i = \frac{X_i - \bar{X}}{\sum_{i=1}^{n} \left(X_i - \bar{X} \right)^2}
$$

for $i = 1, 2, \ldots, n$.

The coefficients $a_1, a_2, \ldots, a_n$ in the linear combination $\hat{\beta}_1 = a_1 Y_1 + a_2 Y_2 + \cdots + a_n Y_n$ satisfy three properties. First, $\sum_{i=1}^{n} a_i = 0$ because

$$
\sum_{i=1}^{n} a_i = \frac{1}{S_{XX}} \sum_{i=1}^{n} \left(X_i - \bar{X} \right) = \frac{n\bar{X} - n\bar{X}}{S_{XX}} = 0.
$$

Second, $\sum_{i=1}^{n} a_i X_i = 1$ because

$$
\sum_{i=1}^{n} a_i X_i = \frac{1}{S_{XX}} \sum_{i=1}^{n} \left(X_i - \bar{X} \right) X_i = \frac{1}{S_{XX}} \left[\sum_{i=1}^{n} X_i^2 - n\bar{X}^2 \right] = \frac{S_{XX}}{S_{XX}} = 1.
$$

Third, $\sum_{i=1}^{n} a_i^2 = 1/S_{XX}$ because

$$
\sum_{i=1}^{n} a_i^2 = \frac{1}{S_{XX}^2} \sum_{i=1}^{n} \left(X_i - \bar{X} \right)^2 = \frac{S_{XX}}{S_{XX}^2} = \frac{1}{S_{XX}}.
$$

These properties can be useful in deriving results associated with the simple linear regression model.

Likewise, the least squares point estimator for the intercept of the regression line is also a linear combination of the Y_i values:

$$
\begin{aligned}
\hat{\beta}_0 &= \bar{Y} - \hat{\beta}_1 \bar{X} \\
&= \frac{1}{n} \sum_{i=1}^{n} Y_i - \bar{X} \sum_{i=1}^{n} \frac{X_i - \bar{X}}{\sum_{i=1}^{n} \left(X_i - \bar{X} \right)^2} Y_i \\
&= \sum_{i=1}^{n} \left(\frac{1}{n} - \bar{X} \cdot \frac{X_i - \bar{X}}{\sum_{i=1}^{n} \left(X_i - \bar{X} \right)^2} \right) Y_i.
\end{aligned}
$$

This formula indicates that the point estimator for the intercept of the regression line can also be written as a linear combination:

$$\hat{\beta}_0 = c_1 Y_1 + c_2 Y_2 + \cdots + c_n Y_n,$$

where

$$c_i = \frac{1}{n} - \bar{X} \cdot \frac{X_i - \bar{X}}{\sum_{i=1}^{n} \left(X_i - \bar{X}\right)^2}$$

for $i = 1, 2, \ldots, n$. This derivation constitutes a proof of the following result.

Theorem 1.3 The least squares estimators of the parameters β_0 and β_1 in the simple linear regression model can be written as linear combinations of the dependent variables:

$$\hat{\beta}_0 = c_1 Y_1 + c_2 Y_2 + \cdots + c_n Y_n$$

and

$$\hat{\beta}_1 = a_1 Y_1 + a_2 Y_2 + \cdots + a_n Y_n,$$

where

$$c_i = \frac{1}{n} - \bar{X} \cdot \frac{X_i - \bar{X}}{S_{XX}} \qquad \text{and} \qquad a_i = \frac{X_i - \bar{X}}{S_{XX}}$$

for $i = 1, 2, \ldots, n$, and

$$\sum_{i=1}^{n} a_i = 0, \qquad \sum_{i=1}^{n} a_i X_i = 1, \qquad \text{and} \qquad \sum_{i=1}^{n} a_i^2 = \frac{1}{S_{XX}}.$$

These formulas will be illustrated for the small data set consisting of $n = 3$ data pairs.

Example 1.5 Consider again the $n = 3$ data pairs

$$(X_1, Y_1) = (6, 2), \qquad (X_2, Y_2) = (8, 9), \qquad \text{and} \qquad (X_3, Y_3) = (2, 2)$$

from Example 1.3. Recall that the independent variable X is Cheryl's number of sales per week. Each sale results in a random amount of revenue to the company. The dependent random variable Y is the associated total revenue from the sales that Cheryl completes for a particular week, in thousands of dollars. Find the least squares estimates of the intercept β_0 and slope β_1 for the simple linear regression model using the formulas that express the estimates as linear combinations of Y_1, Y_2, Y_3 from Theorem 1.3.

The sample mean of the independent variables is

$$\bar{X} = \frac{6 + 8 + 2}{3} = \frac{16}{3}.$$

The value of S_{XX} is

$$S_{XX} = \sum_{i=1}^{3} \left(X_i - \bar{X}\right)^2 = \left(6 - \frac{16}{3}\right)^2 + \left(8 - \frac{16}{3}\right)^2 + \left(2 - \frac{16}{3}\right)^2 = \frac{4}{9} + \frac{64}{9} + \frac{100}{9} = \frac{56}{3}.$$

The coefficients for the linear combination associated with $\hat{\beta}_1$ are

$$a_i = \frac{X_i - \bar{X}}{S_{XX}}$$

for $i = 1, 2, 3$, or

$$a_1 = \frac{6 - 16/3}{56/3} = \frac{1}{28}, \qquad a_2 = \frac{8 - 16/3}{56/3} = \frac{1}{7}, \qquad a_3 = \frac{2 - 16/3}{56/3} = -\frac{5}{28}.$$

You might want to check that the three properties associated with the coefficients a_1, a_2, and a_3 from Theorem 1.3, namely $a_1 + a_2 + a_3 = 0$, $a_1 X_1 + a_2 X_2 + a_3 X_3 = 1$, and $a_1^2 + a_2^2 + a_3^2 = 1/S_{XX}$, are all satisfied as expected. The least squares estimate of the slope of the regression line is

$$\hat{\beta}_1 = a_1 Y_1 + a_2 Y_2 + a_3 Y_3 = \frac{1}{28} \cdot 2 + \frac{1}{7} \cdot 9 - \frac{5}{28} \cdot 2 = \frac{1}{14} + \frac{9}{7} - \frac{5}{14} = 1.$$

The R code for performing these calculations is given below.

```
x = c(6, 8, 2)
y = c(2, 9, 2)
a = (x - mean(x)) / sum((x - mean(x)) ^ 2)
beta1hat = sum(a * y)
```

The coefficients for the linear combination associated with $\hat{\beta}_0$ are

$$c_i = \frac{1}{n} - \bar{X} \cdot \frac{X_i - \bar{X}}{S_{XX}} = \frac{1}{n} - \bar{X} \cdot a_i$$

for $i = 1, 2, 3$, or

$$c_1 = \frac{1}{3} - \frac{16}{3} \cdot \frac{1}{28} = \frac{1}{7}, \qquad c_2 = \frac{1}{3} - \frac{16}{3} \cdot \frac{1}{7} = -\frac{3}{7}, \qquad c_3 = \frac{1}{3} - \frac{16}{3} \cdot \frac{-5}{28} = \frac{9}{7}.$$

The least squares estimate of the intercept of the regression line is

$$\hat{\beta}_0 = c_1 Y_1 + c_2 Y_2 + c_3 Y_3 = \frac{1}{7} \cdot 2 - \frac{3}{7} \cdot 9 + \frac{9}{7} \cdot 2 = \frac{2}{7} - \frac{27}{7} + \frac{18}{7} = -1.$$

The R code for performing these calculations follows.

```
x = c(6, 8, 2)
y = c(2, 9, 2)
n = length(x)
c = 1 / n - mean(x) * (x - mean(x)) / sum((x - mean(x)) ^ 2)
beta0hat = sum(c * y)
```

In both cases the point estimates match the associated values calculated by the standard formulas for $\hat{\beta}_0$ and $\hat{\beta}_1$ from Theorem 1.1 that were used in Example 1.3, as expected.

1.5.3 Variance–Covariance Matrix of $\hat{\beta}_0$ and $\hat{\beta}_1$

Theorem 1.2 states that $\hat{\beta}_0$ and $\hat{\beta}_1$ are unbiased estimators of β_0 and β_1 because $E[\hat{\beta}_0] = \beta_0$ and $E[\hat{\beta}_1] = \beta_1$. This result concerns the *accuracy* of the least squares estimators, but does not address the *precision* of the least squares estimators. We now return to the question of assessing the precision of the point estimators. Being able to express the point estimators of the least squares estimators as linear combinations of the dependent variables as summarized in Theorem 1.3 will be very useful as we proceed. In order to assess the precision of $\hat{\beta}_0$ and $\hat{\beta}_1$, it is necessary to compute $V\left[\hat{\beta}_0\right]$ and $V\left[\hat{\beta}_1\right]$. More generally, we will compute the variance–covariance matrix of $\hat{\beta}_0$ and $\hat{\beta}_1$ in this subsection. Returning to the Monte Carlo simulation in Example 1.4, the magnitudes of the diagonal elements of the variance–covariance matrix reflect the spread of the histograms in Figure 1.11, and the off-diagonal elements of the variance–covariance matrix give the population covariance between $\hat{\beta}_0$ and $\hat{\beta}_1$ which is apparent in the simulation results displayed in Figure 1.12. The general form for the population covariance between $\hat{\beta}_0$ and $\hat{\beta}_1$ will indicate whether the negative sample covariance between $\hat{\beta}_0$ and $\hat{\beta}_1$ that was encountered in the Monte Carlo simulation was due to the particular values of the parameters in the simple linear regression model or whether the negative covariance is generally the case.

We begin with the lower-right-hand element of the variance–covariance matrix of $\hat{\beta}_0$ and $\hat{\beta}_1$. In the simple linear regression model

$$Y_i = \beta_0 + \beta_1 X_i + \varepsilon_i$$

for $i = 1, 2, \ldots, n$, the error terms $\varepsilon_1, \varepsilon_2, \ldots, \varepsilon_n$ are assumed to be mutually independent random variables. This implies that the dependent variables $Y_1, Y_2, \ldots, Y_n$ are also mutually independent random variables. Using the fact that $\hat{\beta}_1$ can be written as a linear combination of the dependent variables from Theorem 1.3, the population variance of $\hat{\beta}_1$ is

$$\begin{aligned} V\left[\hat{\beta}_1\right] &= V\left[a_1 Y_1 + a_2 Y_2 + \cdots + a_n Y_n\right] \\ &= \sum_{i=1}^n V\left[a_i Y_i\right] \\ &= \sum_{i=1}^n a_i^2 V\left[Y_i\right] \\ &= \left(\sum_{i=1}^n a_i^2\right)\sigma^2 \\ &= \frac{\sigma^2}{S_{XX}} \end{aligned}$$

because $\sum_{i=1}^n a_i^2 = 1/S_{XX}$ by Theorem 1.3. Although the experimenter typically has no control over σ^2, the experimenter may have control over selecting the values of $X_1, X_2, \ldots, X_n$ in some applications of simple linear regression. In order to make $V\left[\hat{\beta}_1\right]$ as small as possible, the experimenter should make S_{XX} as large as possible. Spreading the X_i values as much as possible gives the most stability to the estimated slope of the regression line. Simple linear regression modeling can still be performed when the X_i values are tightly clustered together, but the estimated slope will be less stable, and the scope of the model will be limited. As an extreme example of spreading the X_i values, consider clustering all of the X_i values at a left-most and a right-most extreme possible values for the independent variable. The good news is that this will give you the largest possible S_{XX} and the associated smallest possible $V\left[\hat{\beta}_1\right]$. The bad news is that you will not be able to assess linearity in this case because you have observed the dependent variable at only two values of the independent variable. A multitude of functions can model the average of the dependent variables at these two extreme values of the independent variable. So the usual practice is to select the X_i values in an approximately uniform fashion over as wide a range as possible. This gives the experimenter the opportunity to assess linearity and also achieves a large S_{XX}, resulting in an associated small $V\left[\hat{\beta}_1\right]$.

The next step is to calculate the upper-left-hand element of the variance–covariance matrix of $\hat{\beta}_0$ and $\hat{\beta}_1$. Before calculating the population variance of $\hat{\beta}_0$, it is necessary to establish that $\bar{Y}$ and $\hat{\beta}_1$ are uncorrelated. Since $Y_1, Y_2, \ldots, Y_n$ are mutually independent random variables, each with population variance $V[Y_i] = \sigma^2$, the population covariance between $\bar{Y}$ and $\hat{\beta}_1$ is

$$
\begin{aligned}
\text{Cov}\left(\bar{Y}, \hat{\beta}_1\right) &= \text{Cov}\left(\frac{Y_1}{n} + \frac{Y_2}{n} + \cdots + \frac{Y_n}{n}, a_1 Y_1 + a_2 Y_2 + \cdots + a_n Y_n\right) \\
&= \sum_{i=1}^{n} \sum_{j=1}^{n} \text{Cov}\left(\frac{Y_i}{n}, a_j Y_j\right) \\
&= \sum_{i=1}^{n} \text{Cov}\left(\frac{Y_i}{n}, a_i Y_i\right) \\
&= \sum_{i=1}^{n} \frac{a_i}{n} V[Y_i] \\
&= \frac{\sigma^2}{n} \sum_{i=1}^{n} a_i \\
&= 0
\end{aligned}
$$

because $\sum_{i=1}^{n} a_i = 0$ by Theorem 1.3. So $\bar{Y}$ and $\hat{\beta}_1$ are uncorrelated.

Based on the fact that the population covariance between $\bar{Y}$ and $\hat{\beta}_1$ is zero, the population variance of $\hat{\beta}_0$ is

$$
\begin{aligned}
V\left[\hat{\beta}_0\right] &= V\left[\bar{Y} - \hat{\beta}_1 \bar{X}\right] \\
&= V\left[\bar{Y}\right] + \bar{X}^2 V\left[\hat{\beta}_1\right] \\
&= \frac{\sigma^2}{n} + \frac{\bar{X}^2 \sigma^2}{S_{XX}} \\
&= \left[\frac{1}{n} + \frac{\bar{X}^2}{S_{XX}}\right] \sigma^2 \\
&= \left[\frac{\sum_{i=1}^{n}\left(X_i - \bar{X}\right)^2 + n\bar{X}^2}{n \sum_{i=1}^{n}\left(X_i - \bar{X}\right)^2}\right] \sigma^2 \\
&= \frac{\sum_{i=1}^{n} X_i^2}{n S_{XX}} \sigma^2.
\end{aligned}
$$

The last step is to calculate the off-diagonal elements of the variance–covariance matrix of $\hat{\beta}_0$ and $\hat{\beta}_1$. Since $\text{Cov}\left(\bar{Y}, \hat{\beta}_1\right) = 0$, the population covariance between $\hat{\beta}_0$ and $\hat{\beta}_1$ is

$$
\begin{aligned}
\text{Cov}\left(\hat{\beta}_0, \hat{\beta}_1\right) &= \text{Cov}\left(\bar{Y} - \hat{\beta}_1 \bar{X}, \hat{\beta}_1\right) \\
&= \text{Cov}\left(\bar{Y}, \hat{\beta}_1\right) - \text{Cov}\left(\hat{\beta}_1 \bar{X}, \hat{\beta}_1\right) \\
&= -\text{Cov}\left(\hat{\beta}_1 \bar{X}, \hat{\beta}_1\right) \\
&= -\bar{X} \text{Cov}\left(\hat{\beta}_1, \hat{\beta}_1\right) \\
&= -\bar{X} V\left[\hat{\beta}_1\right] \\
&= -\frac{\bar{X} \sigma^2}{S_{XX}}.
\end{aligned}
$$

All of the elements of the variance–covariance matrix have now been established, which constitutes a proof of the following theorem.

Theorem 1.4 The least squares estimators of the parameters β_0 and β_1 in the simple linear regression model have variance–covariance matrix

$$\begin{bmatrix} V[\hat{\beta}_0] & \text{Cov}(\hat{\beta}_0, \hat{\beta}_1) \\ \text{Cov}(\hat{\beta}_1, \hat{\beta}_0) & V[\hat{\beta}_1] \end{bmatrix} = \begin{bmatrix} \sum_{i=1}^n X_i^2 / (nS_{XX}) & -\bar{X}/S_{XX} \\ -\bar{X}/S_{XX} & 1/S_{XX} \end{bmatrix} \sigma^2.$$

There are two important observations that can be made from Theorem 1.4. First, the elements of the variance–covariance matrix of $\hat{\beta}_0$ and $\hat{\beta}_1$ are a function of only the X_i values and the typically unknown population error variance σ^2; the values of $Y_1, Y_2, \ldots, Y_n$ do not play a role. Recall from Definition 1.1 that the independent variable observations $X_1, X_2, \ldots, X_n$ are assumed to be observed without error. Second, since $S_{XX} > 0$ because at least two of the X_i values are distinct, the population covariance between $\hat{\beta}_0$ and $\hat{\beta}_1$ takes the opposite sign of $\bar{X}$. This provides an explanation of why $\hat{\beta}_0$ and $\hat{\beta}_1$ appeared to have negative covariance in the results of the 5000 simulated estimates plotted in Figure 1.12.

Example 1.6 Consider again the simple linear regression model

$$Y = \beta_0 + \beta_1 X + \varepsilon$$

from Example 1.4 in which

- the population intercept is $\beta_0 = 1$,
- the population slope is $\beta_1 = 1/2$, and
- the error term ε has a $U(-1, 1)$ distribution.

The error term distribution has population mean zero, so this model satisfies the conditions of a simple linear regression model. Find the variance–covariance matrix for the least squares estimators $\hat{\beta}_0$ and $\hat{\beta}_1$ associated with a single Monte Carlo replication of $n = 10$ data pairs. Assume that the X_i values are equally likely to be one of the integers $0, 1, 2, \ldots, 9$.

The R code that follows conducts a single replication of the Monte Carlo experiment. The results of this single replication were illustrated by the fitted regression line in Figure 1.9. Since the error terms are mutually independent $U(-1, 1)$ random variables and the population variance of a $U(a, b)$ random variable is $(b-a)^2/12$, the population variance of the error terms is $\sigma^2 = (1+1)^2/12 = 1/3$. Although the dependent variables are generated and stored in the vector y, they are not used in the calculation of the variance–covariance matrix.

```
n        = 10
beta0    = 1
beta1    = 1 / 2
sigma2   = 1 / 3
set.seed(100)

x = sample(0:9, n, replace = TRUE)
```

```
if (min(x) == max(x)) stop("All x values are equal")
y = beta0 + beta1 * x + runif(n, -1, 1)

sxx        = sum((x - mean(x)) ^ 2)
vcm        = matrix(nrow = 2, ncol = 2)
vcm[1, 1] = sum(x ^ 2) / (n * sxx)
vcm[1, 2] = vcm[2, 1] = - mean(x) / sxx
vcm[2, 2] = 1 / sxx
vcm        = vcm * sigma2
print(vcm)
```

The variance–covariance matrix for this single replication of the Monte Carlo simulation experiment, reported to four digits, is

$$\begin{bmatrix} V[\hat{\beta}_0] & \text{Cov}(\hat{\beta}_0, \hat{\beta}_1) \\ \text{Cov}(\hat{\beta}_1, \hat{\beta}_0) & V[\hat{\beta}_1] \end{bmatrix} = \begin{bmatrix} 0.1211 & -0.02509 \\ -0.02509 & 0.007168 \end{bmatrix}.$$

If additional Monte Carlo simulation replications were made, this matrix would vary from one replication to the next because the X_i values vary from one replication to the next. Taking the square roots of the diagonal elements yields

$$\sqrt{V[\hat{\beta}_0]} = 0.3481 \qquad \text{and} \qquad \sqrt{V[\hat{\beta}_1]} = 0.0847,$$

which are estimates of the standard deviation of the intercept and slope of the regression line, often referred to as the *standard errors* of the estimated parameters. These two standard deviations are roughly in line with the spread of the histograms generated from the 5000 simulation replications depicted in Figure 1.11. The negative values of the off-diagonal elements of the variance–covariance matrix are consistent with the plot of 5000 simulated $(\hat{\beta}_0, \hat{\beta}_1)$ values given in Figure 1.12.

So far we have found the expected values and the variance–covariance matrix of the least squares estimators $\hat{\beta}_0$ and $\hat{\beta}_1$. But there is a lingering doubt as to whether better point estimators for β_0 and β_1 exist. An example of such a better point estimator would be an unbiased estimator of β_0 with a smaller population variance than the least squares estimator of β_0. This lingering doubt will be addressed in the next subsection.

1.5.4   Gauss–Markov Theorem

Recall from Theorem 1.3 that the least squares estimators for the slope and intercept of the regression line were expressed as linear combinations of the dependent variables:

$$\hat{\beta}_1 = a_1 Y_1 + a_2 Y_2 + \cdots + a_n Y_n$$

and

$$\hat{\beta}_0 = c_1 Y_1 + c_2 Y_2 + \cdots + c_n Y_n.$$

But are these linear combinations the best possible linear combinations for estimating β_1 and β_0? The Gauss–Markov theorem is used to show that these estimators have the minimum variance of all possible unbiased estimators which are linear combinations of the dependent variables. These estimators are known as *Best Linear Unbiased Estimators*, typically abbreviated with the colorful acronym BLUE. The Venn diagram in Figure 1.13 might be helpful in categorizing the various types of estimators. The set L consists of all point estimators for the regression parameters β_0 and β_1 which can be expressed as linear combinations of the dependent variables $Y_1, Y_2, \ldots, Y_n$. The set U consists of all point estimators for the regression parameters β_0 and β_1 which are unbiased estimators of β_0 and β_1. The shaded intersection of L and U (that is, $L \cap U$) is all estimators which are both linear combinations of $Y_1, Y_2, \ldots, Y_n$ and unbiased. An example of an estimator of β_1 which is neither in L nor in U is Y_1^2. The Gauss–Markov theorem states that the least squares estimators have the smallest possible variance among all estimators in $L \cap U$.

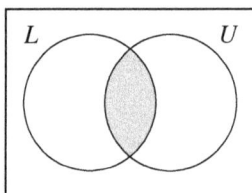

Figure 1.13: Venn diagram of sets L (linear combinations) and U (unbiased estimators).

Theorem 1.5 (Gauss–Markov theorem) The least squares estimators of β_0 and β_1 associated with a simple linear regression model have the smallest population variance among all unbiased estimators that can be expressed as a linear combination of the dependent variables.

Proof (partial proof) This proof will show that $\hat{\beta}_1$ has the smallest population variance among the class of all linear unbiased estimators for β_1. The proof for $\hat{\beta}_0$ is similar but left as an exercise for the reader. Let

$$\hat{\beta}_1 = a_1 Y_1 + a_2 Y_2 + \cdots + a_n Y_n$$

be the unbiased least squares estimator of the population slope β_1 from Theorem 1.3, where $a_i = \left(X_i - \bar{X} \right) / S_{XX}$ for $i = 1, 2, \ldots, n$. Consider another linear combination of the dependent variables which is also an unbiased estimator of β_1 that can be written as

$$\hat{\beta}_1' = k_1 Y_1 + k_2 Y_2 + \cdots + k_n Y_n$$

for some real-valued constants $k_1, k_2, \ldots, k_n$. Since $E[Y_i] = \beta_0 + \beta_1 X_i$, the expected value of $\hat{\beta}_1'$ is

$$
\begin{aligned}
E\left[\hat{\beta}_1' \right] &= E\left[\sum_{i=1}^{n} k_i Y_i \right] \\
&= \sum_{i=1}^{n} k_i E\left[Y_i \right] \\
&= \sum_{i=1}^{n} k_i \left(\beta_0 + \beta_1 X_i \right) \\
&= \beta_0 \sum_{i=1}^{n} k_i + \beta_1 \sum_{i=1}^{n} k_i X_i.
\end{aligned}
$$

Since $\hat{\beta}_1'$ is an unbiased estimator of β_1, $E\left[\hat{\beta}_1'\right] = \beta_1$. In order for this to be the case, the following conditions must hold:

$$\sum_{i=1}^n k_i = 0 \qquad \text{and} \qquad \sum_{i=1}^n k_i X_i = 1.$$

These two conditions will be used in the last step of the derivation that follows. Now let $k_i = a_i + d_i$, for $i = 1, 2, \ldots, n$. We want to find the d_i values that meet the two conditions given above and minimize $V\left[\hat{\beta}_1'\right]$, which is

$$V\left[\hat{\beta}_1'\right] = V\left[\sum_{i=1}^n k_i Y_i\right]$$

$$= \sum_{i=1}^n k_i^2 V\left[Y_i\right]$$

$$= \sum_{i=1}^n k_i^2 \sigma^2$$

$$= \sigma^2 \sum_{i=1}^n (a_i + d_i)^2$$

$$= \sigma^2 \left[\sum_{i=1}^n a_i^2 + \sum_{i=1}^n d_i^2 + 2\sum_{i=1}^n a_i d_i\right]$$

$$= V\left[\hat{\beta}_1\right] + \sigma^2 \sum_{i=1}^n d_i^2 + 2\sigma^2 \sum_{i=1}^n a_i d_i$$

$$= V\left[\hat{\beta}_1\right] + \sigma^2 \sum_{i=1}^n d_i^2 + 2\sigma^2 \sum_{i=1}^n a_i(k_i - a_i)$$

$$= V\left[\hat{\beta}_1\right] + \sigma^2 \sum_{i=1}^n d_i^2 + 2\sigma^2 \left(\sum_{i=1}^n a_i k_i - \sum_{i=1}^n a_i^2\right)$$

$$= V\left[\hat{\beta}_1\right] + \sigma^2 \sum_{i=1}^n d_i^2 + 2\sigma^2 \left(\sum_{i=1}^n k_i \cdot \frac{X_i - \bar{X}}{S_{XX}} - \frac{1}{S_{XX}}\right)$$

$$= V\left[\hat{\beta}_1\right] + \sigma^2 \sum_{i=1}^n d_i^2 + 2\sigma^2 \left(\frac{\sum_{i=1}^n k_i X_i - \bar{X}\sum_{i=1}^n k_i - 1}{S_{XX}}\right)$$

$$= V\left[\hat{\beta}_1\right] + \sigma^2 \sum_{i=1}^n d_i^2.$$

In order to minimize $V\left[\hat{\beta}_1'\right]$ the d_i values should be selected to minimize $\sum_{i=1}^n d_i^2$. This sum of squares is minimized when $d_1 = d_2 = \cdots = d_n = 0$. Therefore, the least squares estimator $\hat{\beta}_1$, with coefficients $k_i = a_i$ for $i = 1, 2, \ldots, n$, has the smallest variance among all unbiased estimators that can be written as linear combinations of $Y_1, Y_2, \ldots, Y_n$ and is therefore a best linear unbiased estimator. $\square$

The Gauss–Markov theorem indicates that the least squares estimators for β_0 and β_1 have minimal variance among all linear estimators. It does not indicate whether the least squares estimators for β_0 and β_1 have minimal variance among all estimators. The Gauss–Markov theorem extends to the case of multiple linear regression in which there are several independent variables. The least squares estimators are also the best linear unbiased estimators in this case.

To review the results that have been introduced so far, the simple linear regression model

$$Y = \beta_0 + \beta_1 X + \varepsilon$$

defines a linear statistical relationship between an independent variable X, observed without error, and a random dependent variable Y as given in Definition 1.1. The point estimators for β_1 and β_0 from n data pairs $(X_1, Y_1), (X_2, Y_2), \ldots, (X_n, Y_n)$ using the least squares criterion are

$$\hat{\beta}_1 = \frac{S_{XY}}{S_{XX}} \qquad \text{and} \qquad \hat{\beta}_0 = \bar{Y} - \hat{\beta}_1 \bar{X}$$

as given in Theorem 1.1. The least squares estimators are unbiased estimators of their associated parameters because

$$E\left[\hat{\beta}_1\right] = \beta_1 \qquad \text{and} \qquad E\left[\hat{\beta}_0\right] = \beta_0$$

as given in Theorem 1.2. The least squares estimators of β_0 and β_1 can be expressed as linear combinations of $Y_1, Y_2, \ldots, Y_n$ as

$$\hat{\beta}_0 = c_1 Y_1 + c_2 Y_2 + \cdots + c_n Y_n \qquad \text{and} \qquad \hat{\beta}_1 = a_1 Y_1 + a_2 Y_2 + \cdots + a_n Y_n,$$

with coefficients $c_1, c_2, \ldots, c_n$ and $a_1, a_2, \ldots, a_n$ given in Theorem 1.3. The variance–covariance matrix of $\hat{\beta}_0$ and $\hat{\beta}_1$ is

$$\begin{bmatrix} V[\hat{\beta}_0] & \mathrm{Cov}(\hat{\beta}_0, \hat{\beta}_1) \\ \mathrm{Cov}(\hat{\beta}_1, \hat{\beta}_0) & V[\hat{\beta}_1] \end{bmatrix} = \begin{bmatrix} \sum_{i=1}^{n} X_i^2 / (n S_{XX}) & -\bar{X}/S_{XX} \\ -\bar{X}/S_{XX} & 1/S_{XX} \end{bmatrix} \sigma^2$$

as given in Theorem 1.4. Finally, the Gauss–Markov theorem given in Theorem 1.5 states that the least squares estimators of β_0 and β_1 have the smallest population variance among all unbiased estimators that can be expressed as a linear combination of $Y_1, Y_2, \ldots, Y_n$.

The next section defines fitted values and residuals. Fitted values are the heights of the regression line associated with the observed values of the independent variable $X_1, X_2, \ldots, X_n$. The residuals are the vertical signed distances between the observed values of the dependent variable $Y_1, Y_2, \ldots, Y_n$ and the associated fitted values that fall on the regression line. Residuals play an analogous role to the error terms in the simple linear regression model.

1.6   Fitted Values and Residuals

The simple linear regression model

$$Y = \beta_0 + \beta_1 X + \varepsilon$$

was introduced in the previous section as a linear statistical model for describing the relationship between an independent variable X and a dependent variable Y. Taking the expected value of both sides of this equation yields

$$E[Y] = \beta_0 + \beta_1 X$$

because $E[\varepsilon] = 0$ and X is a fixed value assumed to be observed without error, which are two key assumptions in Definition 1.1. When the population intercept β_0 and the population slope β_1 are replaced by their associated least squares point estimators $\hat{\beta}_0$ and $\hat{\beta}_1$, the resulting estimated regression line is

$$\hat{Y} = \hat{\beta}_0 + \hat{\beta}_1 X.$$

This estimated regression line is typically plotted on a scatterplot that contains the data pairs (X_1, Y_1), (X_2, Y_2), ..., (X_n, Y_n). Seeing the data pairs and the least squares regression line on the same plot often makes the visual assessment of linearity easier. For any value X in which the simple linear regression model is valid, $\hat{Y}$ is the point estimator for the value of the dependent variable based on the data pairs and associated estimated regression line. This equation can be rewritten for the particular values of the independent variable collected as

$$\hat{Y}_i = \hat{\beta}_0 + \hat{\beta}_1 X_i$$

for $i = 1, 2, \ldots, n$. The value $\hat{Y}_i$ is known as the *fitted value* associated with data pair i, for $i = 1, 2,$ ..., n. When $\hat{Y}_i \neq Y_i$, which is almost always the case in applications, the fitted value does not fall on the estimated regression line; when $\hat{Y}_i = Y_i$, the fitted value falls on the estimated regression line. The next example illustrates the notion of fitted values for the sales data set.

Example 1.7 Consider the sales data set from Example 1.3 with just $n = 3$ data pairs:

$$(X_1, Y_1) = (6, 2), \qquad (X_2, Y_2) = (8, 9), \qquad (X_3, Y_3) = (2, 2).$$

Find the fitted values $\hat{Y}_1$, $\hat{Y}_2$, and $\hat{Y}_3$ associated with the least squares regression line.

From Examples 1.3 and 1.5, the point estimates for the population intercept and population slope are

$$\hat{\beta}_0 = -1 \qquad \text{and} \qquad \hat{\beta}_1 = 1.$$

Hence, the estimated regression line is $\hat{Y} = \hat{\beta}_0 + \hat{\beta}_1 X$, or

$$\hat{Y} = -1 + X,$$

which is plotted along with the scatterplot of the data pairs in Figure 1.14. So calculating the fitted values is just a matter of using the X_i values as arguments in the estimated regression line:

$$\hat{Y}_1 = -1 + X_1 = -1 + 6 = 5 \qquad \Rightarrow \qquad (X_1, \hat{Y}_1) = (6, 5)$$
$$\hat{Y}_2 = -1 + X_2 = -1 + 8 = 7 \qquad \Rightarrow \qquad (X_2, \hat{Y}_2) = (8, 7)$$
$$\hat{Y}_3 = -1 + X_3 = -1 + 2 = 1 \qquad \Rightarrow \qquad (X_3, \hat{Y}_3) = (2, 1).$$

The fitted values are also plotted as points that lie on the estimated regression line in Figure 1.14. Recall from the previous section that the fitted least squares line is the line which minimizes the sum of the squares of the lengths of the vertical dashed lines which connect the data pair with its associated fitted value. The fitted values are calculated and stored in a component named `fitted` in the list returned by the R `lm` function. The R code below confirms the fitted values calculated above by hand.

```
x = c(6, 8, 2)
y = c(2, 9, 2)
lm(y ~ x)$fitted
```

The spread of the data pair (X_i, Y_i) from the fitted regression line $\hat{Y} = \hat{\beta}_0 + \hat{\beta}_1 X$ is reflected in the vertical signed distance between the data pair (X_i, Y_i) and the associated fitted value $(X_i, \hat{Y}_i)$, These signed distances are known as the *residuals*, and are defined by

$$e_i = Y_i - \hat{Y}_i$$

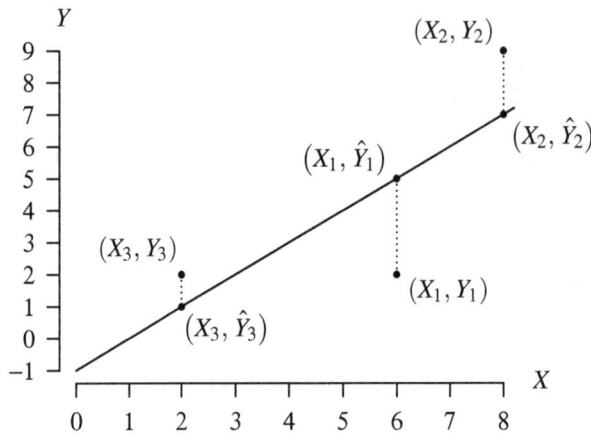

Figure 1.14: A scatterplot of the sales data pairs with the fitted values.

for $i = 1, 2, \ldots, n$. Data pairs that fall above the regression line correspond to positive residuals; data pairs that fall below the regression line correspond to negative residuals. The least squares approach used so far in estimating the intercept and slope of the regression line is a matter of finding the values of $\hat{\beta}_0$ and $\hat{\beta}_1$ which minimize the sum of the squares of the residuals. In other words, minimize

$$S = \sum_{i=1}^{n} e_i^2.$$

The fitted values and residuals are formally defined next.

Definition 1.2 Let $\hat{\beta}_0$ and $\hat{\beta}_1$ denote the least squares estimators of the parameters β_0 and β_1 in the simple linear regression model with data pairs $(X_1, Y_1), (X_2, Y_2), \ldots, (X_n, Y_n)$. The *fitted value* associated with the ith data pair (X_i, Y_i) is $\hat{Y}_i = \hat{\beta}_0 + \hat{\beta}_1 X_i$, for $i = 1, 2, \ldots, n$. The *residual* associated with ith data pair (X_i, Y_i) is $e_i = Y_i - \hat{Y}_i$, for $i = 1, 2, \ldots, n$.

Choosing to use the *vertical distance* between the observed value of the dependent variable and the regression line in the definition of the residual was based on the fact that the values of the independent variable $X_1, X_2, \ldots, X_n$ are assumed to be observed without error in Definition 1.1. The mathematics associated with simple linear regression changes substantially if both X and Y are considered to be random variables.

A subtle but important distinction should be drawn between the model error term ε_i for data pair i and the residual e_i for data pair i. The model error terms are defined by

$$\varepsilon_i = Y_i - (\beta_0 + \beta_1 X_i)$$

for $i = 1, 2, \ldots, n$, and represent the vertical distances between the observed dependent variable Y_i and the *true* (population) regression line $Y = \beta_0 + \beta_1 X$. The simple linear regression model assumes that $\varepsilon_1, \varepsilon_2, \ldots, \varepsilon_n$ are mutually independent random variables. In nearly all applications, however, β_0 and β_1 are unknown. This means that for a particular data set, these model error terms are also unknown. On the other hand, the residuals are defined by

$$e_i = Y_i - \hat{Y}_i = Y_i - \left(\hat{\beta}_0 + \hat{\beta}_1 X_i\right)$$

for $i = 1, 2, \ldots, n$, and represent the error for data pair i when compared to the *estimated* regression line $\hat{Y} = \hat{\beta}_0 + \hat{\beta}_1 X$, which is calculated from the n data pairs. Thus, $\hat{\varepsilon}_i = e_i$, for $i = 1, 2, \ldots, n$. The $e_1, e_2, \ldots, e_n$ values are *not* mutually independent random variables because they must sum to zero. (This will be proven subsequently in Theorem 1.6.) For a particular data set, these residuals are known. The residuals are calculated for the sales data next.

Example 1.8 Consider again the sales data set from Example 1.3 with $n = 3$ data pairs:

$$(X_1, Y_1) = (6, 2) \qquad (X_2, Y_2) = (8, 9) \qquad (X_3, Y_3) = (2, 2).$$

Calculate the residuals e_1, e_2, and e_3 associated with the least squares regression line and display them on a scatterplot that includes the regression line.

Table 1.2 contains the calculations required to calculate the residuals and their squares. The sum of the squared residuals for these data pairs is

$$S = \sum_{i=1}^{3} e_i^2 = (-3)^2 + 2^2 + 1^2 = 9 + 4 + 1 = 14.$$

This total is consistent with the sum of the areas of the squares from Figure 1.7. The data pairs were handpicked in this example to make the residuals all integers. This will not be the case in nearly all applications of simple linear regression. This value for S which is associated with the estimated regression line is the smallest possible value for the sum of squared residuals. Any other line will be associated with a larger sum of squared residuals.

Figure 1.15 shows the residuals e_1, e_2, and e_3 along with the data pairs and the estimated regression line. Unless all of the data pairs fall in a line (which would correspond to $S = 0$), there will always be one or more data values falling above the line and one or more data values falling below the line.

The values of the residuals are stored in a component named `residuals` in the list returned by the R `lm` function. The R code below calculates and displays the residuals that were calculated by hand and displayed in Table 1.2.

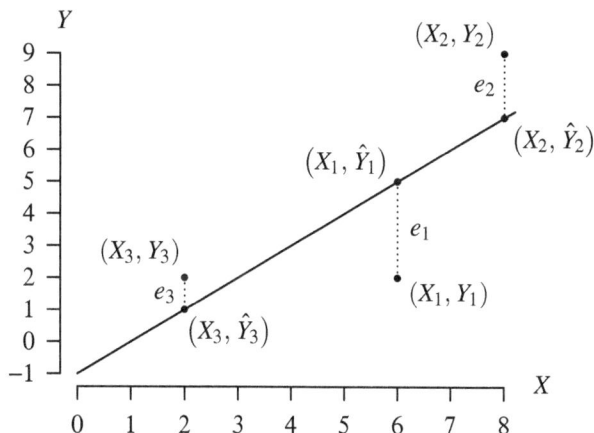

Figure 1.15: A scatterplot of the sales data pairs with the fitted values and residuals.

Observation number i	Number of sales X_i	Total revenue Y_i	Fitted value $\hat{Y}_i$	Residual $e_i = Y_i - \hat{Y}_i$	Squared residual e_i^2
1	6	2	5	-3	9
2	8	9	7	2	4
3	2	2	1	1	1
Sum	16	13	13	0	14

Table 1.2: Data pairs, fitted values, residuals, and squared residuals.

```
x = c(6, 8, 2)
y = c(2, 9, 2)
lm(y ~ x)$residuals
```

A close inspection of the entries in Table 1.2 reveals that there are some curious outcomes that occur, such as

$$\sum_{i=1}^{n} e_i = 0 \qquad \text{and} \qquad \sum_{i=1}^{n} Y_i = \sum_{i=1}^{n} \hat{Y}_i.$$

In other words, (a) the sum of the residuals is zero, and (b) the sum of the observed values of the dependent variable equals the sum of the fitted values. These were not just a matter of coincidence. The following theorem confirms that these relationships, along with a few other relationships, are true in general.

Theorem 1.6 Let $(X_1, Y_1), (X_2, Y_2), \ldots, (X_n, Y_n)$ be n data pairs associated with the simple linear regression model

$$Y = \beta_0 + \beta_1 X + \varepsilon.$$

Using the notation from Definition 1.2, the fitted values are $\hat{Y}_1, \hat{Y}_2, \ldots, \hat{Y}_n$ and the residuals are $e_1, e_2, \ldots, e_n$. Then

- $\displaystyle\sum_{i=1}^{n} e_i = 0$,

- $\displaystyle\sum_{i=1}^{n} Y_i = \sum_{i=1}^{n} \hat{Y}_i$,

- $\displaystyle\sum_{i=1}^{n} X_i e_i = 0$,

- $\displaystyle\sum_{i=1}^{n} \hat{Y}_i e_i = 0$,

- $\left(\bar{X}, \bar{Y}\right)$ is a point that lies on the estimated regression line.

Proof Each of the five results will be proven individually.

- Since $\hat{\beta}_0 = \bar{Y} - \hat{\beta}_1 \bar{X}$ from Theorem 1.1, the sum of the residuals is

$$\sum_{i=1}^{n} e_i = \sum_{i=1}^{n} \left(Y_i - \hat{Y}_i \right)$$

$$= \sum_{i=1}^{n} \left(Y_i - \hat{\beta}_0 - \hat{\beta}_1 X_i \right)$$

$$= \sum_{i=1}^{n} Y_i - n\hat{\beta}_0 - \hat{\beta}_1 \sum_{i=1}^{n} X_i$$

$$= \sum_{i=1}^{n} Y_i - \sum_{i=1}^{n} Y_i + \hat{\beta}_1 \sum_{i=1}^{n} X_i - \hat{\beta}_1 \sum_{i=1}^{n} X_i$$

$$= 0.$$

- Since $\hat{\beta}_0 = \bar{Y} - \hat{\beta}_1 \bar{X}$ from Theorem 1.1, the sum of the fitted values is

$$\sum_{i=1}^{n} \hat{Y}_i = \sum_{i=1}^{n} \left(\hat{\beta}_0 + \hat{\beta}_1 X_i \right)$$

$$= n\hat{\beta}_0 + \hat{\beta}_1 \sum_{i=1}^{n} X_i$$

$$= n\left(\bar{Y} - \hat{\beta}_1 \bar{X} \right) + \hat{\beta}_1 \sum_{i=1}^{n} X_i$$

$$= \sum_{i=1}^{n} Y_i.$$

Thus, the sum of the values of the dependent variable always equals the sum of the fitted values.

- The sum of the products of the independent variables and residuals is

$$\sum_{i=1}^{n} X_i e_i = \sum_{i=1}^{n} X_i \left(Y_i - \hat{Y}_i \right)$$

$$= \sum_{i=1}^{n} X_i Y_i - \sum_{i=1}^{n} X_i \hat{Y}_i$$

$$= \sum_{i=1}^{n} X_i Y_i - \sum_{i=1}^{n} X_i \left(\hat{\beta}_0 + \hat{\beta}_1 X_i \right)$$

$$= \sum_{i=1}^{n} X_i Y_i - \hat{\beta}_0 \sum_{i=1}^{n} X_i - \hat{\beta}_1 \sum_{i=1}^{n} X_i^2$$

$$= 0.$$

The final step uses the second normal equation from Theorem 1.1.

- Using the first and third result in this theorem, the sum of the products of the fitted values and residuals is

$$\sum_{i=1}^{n} \hat{Y}_i e_i = \sum_{i=1}^{n} \left(\hat{\beta}_0 + \hat{\beta}_1 X_i \right) e_i = \hat{\beta}_0 \sum_{i=1}^{n} e_i + \hat{\beta}_1 \sum_{i=1}^{n} X_i e_i = \hat{\beta}_0 \cdot 0 + \hat{\beta}_1 \cdot 0 = 0.$$

- The first normal equation from Theorem 1.1 is

$$n\hat{\beta}_0 + \hat{\beta}_1 \sum_{i=1}^{n} X_i = \sum_{i=1}^{n} Y_i.$$

Dividing both sides by n,

$$\hat{\beta}_0 + \hat{\beta}_1 \bar{X} = \bar{Y},$$

which indicates that the point $(\bar{X}, \bar{Y})$ lies on the estimated regression line.   □

These five results from Theorem 1.6 will be illustrated for the sales data in the example that follows.

Example 1.9 Calculate the quantities given in Theorem 1.6 for the $n = 3$ data pairs from the sales data set from Example 1.3:

$$(X_1, Y_1) = (6, 2) \qquad (X_2, Y_2) = (8, 9) \qquad (X_3, Y_3) = (2, 2).$$

From Examples 1.3 and 1.5, the point estimate for the intercept is $\hat{\beta}_0 = -1$ and the point estimate for the slope is $\hat{\beta}_1 = 1$. Table 1.3 contains the calculations necessary to illustrate the results given in Theorem 1.6. More specifically,

- $\sum_{i=1}^{3} e_i = 0,$

- $\sum_{i=1}^{3} Y_i = \sum_{i=1}^{3} \hat{Y}_i = 13,$

- $\sum_{i=1}^{3} X_i e_i = 0,$

- $\sum_{i=1}^{3} \hat{Y}_i e_i = 0.$

Finally, the point $(\bar{X}, \bar{Y}) = (16/3, 13/3)$ lies on the estimated regression line $\hat{Y} = -1 + X$, as illustrated in Figure 1.16.

i	X_i	Y_i	$\hat{Y}_i$	e_i	e_i^2	$X_i e_i$	$\hat{Y}_i e_i$
1	6	2	5	-3	9	-18	-15
2	8	9	7	2	4	16	14
3	2	2	1	1	1	2	1
Sum	16	13	13	0	14	0	0

Table 1.3: Calculation of quantities from Theorem 1.6.

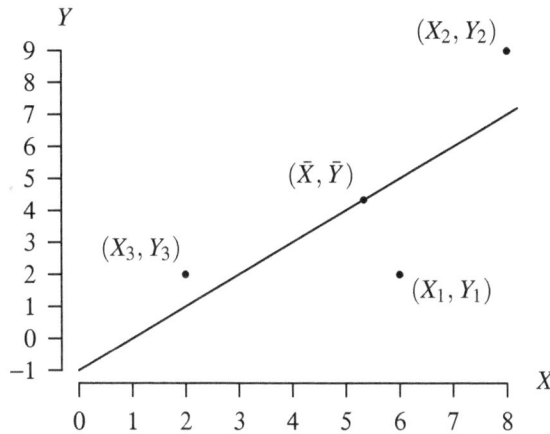

Figure 1.16: The point $(\bar{X}, \bar{Y})$ falls on the estimated regression line.

1.7 Estimating the Variance of the Error Terms

The emphasis so far has been focused on the estimation of the intercept and slope of the regression line. While $\hat{\beta}_0$ and $\hat{\beta}_1$ are the most critical parameters in most applications of a simple linear regression model, there is another parameter, the population variance of the error terms σ^2, which should also be estimated from the data pairs.

To establish a foundation for the estimation of σ^2, assume *for this paragraph only* that there is a univariate, rather than a bivariate, sample of values denoted by $X_1, X_2, \ldots, X_n$. These will not be fixed values observed without error as they were in regression modeling. It is assumed that these values constitute a random sample from a population that has finite population mean μ and finite population variance σ^2. The goal in this paragraph is to estimate σ^2 as a function of the data values. If the population mean μ is known (which is rare in practice), then an unbiased estimator of σ^2 is

$$\frac{1}{n} \sum_{i=1}^{n} (X_i - \mu)^2 \, .$$

If the first $n - 1$ deviations between the sample values and the population mean $X_1 - \mu$, $X_2 - \mu$, $\ldots$, $X_{n-1} - \mu$ were known, the final deviation, $X_n - \mu$, would be free to take on any value. It is in this sense that the sum of squares

$$\sum_{i=1}^{n} (X_i - \mu)^2$$

is said to have n "degrees of freedom." It is common practice in statistics to divide a sum of squares by its degrees of freedom to arrive at a point estimator. In this particular instance, dividing by n makes the point estimator an unbiased estimator of σ^2. The problem that arises more often in practice is to estimate σ^2 when μ is unknown. An unbiased estimator of σ^2 in this case is the sample variance

$$\frac{1}{n-1} \sum_{i=1}^{n} (X_i - \bar{X})^2 \, ,$$

which is typically denoted by S^2 by statisticians. There are three reasons why the term outside of the summation has $n - 1$ in the denominator. The first reason is that this is the appropriate term so that this estimator is an unbiased estimator of σ^2. This can be stated as $E\left[S^2\right] = \sigma^2$. The second reason is that one can't estimate the dispersion of a distribution from a single data value, so the sample variance is undefined when $n = 1$. The third reason is that the sum of squares has $n - 1$ degrees of freedom. One degree of freedom is lost because the sample mean $\bar{X}$ is used to estimate the population mean μ. If the first $n - 1$ deviations between the sample values and the sample mean $X_1 - \bar{X}, X_2 - \bar{X}, \ldots, X_{n-1} - \bar{X}$ were known, the final deviation, $X_n - \bar{X}$, could be calculated from the other $n - 1$ values because

$$\sum_{i=1}^{n} \left(X_i - \bar{X}\right) = \sum_{i=1}^{n} X_i - n\bar{X} = 0.$$

It is in this sense that the sum of squares

$$\sum_{i=1}^{n} \left(X_i - \bar{X}\right)^2$$

is said to have $n - 1$ degrees of freedom. This ends the discussion of degrees of freedom for a univariate data set.

We now return to the problem of estimating σ^2 in simple linear regression. The independent variables $X_1, X_2, \ldots, X_n$ are once again assumed to be fixed values observed without error as they have been throughout this chapter. Based on the fact that the error terms $\varepsilon_1, \varepsilon_2, \ldots, \varepsilon_n$ in the simple linear regression model are assumed to be mutually independent and identically distributed random variables, each with population mean 0 and finite population variance σ^2, the population variance of the error terms can be estimated with the unbiased estimator

$$\frac{1}{n} \sum_{i=1}^{n} \varepsilon_i^2 = \frac{1}{n} \sum_{i=1}^{n} (Y_i - \beta_0 - \beta_1 X_i)^2$$

if β_0 and β_1 were known. But in practice, the two parameters β_0 and β_1 are estimated from the data pairs $(X_1, Y_1), (X_2, Y_2), \ldots, (X_n, Y_n)$, so two degrees of freedom are lost and an appropriate point estimator for the population variance σ^2 is given by

$$\hat{\sigma}^2 = \frac{1}{n-2} \sum_{i=1}^{n} e_i^2 = \frac{1}{n-2} \sum_{i=1}^{n} \left(Y_i - \hat{\beta}_0 - \hat{\beta}_1 X_i\right)^2.$$

It is important that the population variance of the error terms σ^2 remain constant over the range of X values in which the simple linear regression model is appropriate. One tool for visually assessing this assumption is a scatterplot of the data pairs with the estimated regression line superimposed.

The point estimator for σ^2 when β_0 and β_1 are estimated from the data pairs involves the sum of squares of the residuals, and this is often abbreviated as *SSE*, for *sum of squares for error*:

$$SSE = \sum_{i=1}^{n} e_i^2,$$

which is also known as the *error sum of squares*, *residual sum of squares*, and *sum of squares due to error*. When this quantity is divided by its degrees of freedom, it is known as the *mean square error*, which is abbreviated by *MSE*:

$$\hat{\sigma}^2 = MSE = \frac{SSE}{n-2} = \frac{1}{n-2} \sum_{i=1}^{n} e_i^2.$$

Some good news is provided by the next result, which states that $MSE = \hat{\sigma}^2$ is an unbiased estimator of σ^2.

Theorem 1.7 If $e_1, e_2, \ldots, e_n$ are the residuals in a simple linear regression model, then the point estimator

$$\hat{\sigma}^2 = \frac{1}{n-2} \sum_{i=1}^{n} e_i^2$$

is an unbiased estimator of σ^2.

Proof The simple linear regression model is

$$Y_i = \beta_0 + \beta_1 X_i + \varepsilon_i$$

for $i = 1, 2, \ldots, n$. Summing both sides of this equation and dividing by n yields

$$\bar{Y} = \beta_0 + \beta_1 \bar{X} + \bar{\varepsilon}.$$

Taking the difference between the previous two equations results in

$$Y_i - \bar{Y} = \beta_1 \left(X_i - \bar{X} \right) + \varepsilon_i - \bar{\varepsilon} \tag{1}$$

for $i = 1, 2, \ldots, n$. The definition of the residual associated with data pair i is

$$e_i = Y_i - \hat{\beta}_0 - \hat{\beta}_1 X_i$$

for $i = 1, 2, \ldots, n$. Recognizing that the residuals sum to zero via Theorem 1.6, summing both sides of this equation, and dividing by n yields

$$0 = \bar{Y} - \hat{\beta}_0 - \hat{\beta}_1 \bar{X}.$$

Taking the difference between the previous two equations results in

$$e_i = Y_i - \bar{Y} - \hat{\beta}_1 \left(X_i - \bar{X} \right) \tag{2}$$

for $i = 1, 2, \ldots, n$. Substituting the right-hand side of equation (1) for $Y_i - \bar{Y}$ in equation (2) gives

$$e_i = \beta_1 \left(X_i - \bar{X} \right) + \varepsilon_i - \bar{\varepsilon} - \hat{\beta}_1 \left(X_i - \bar{X} \right)$$
$$= \left(\beta_1 - \hat{\beta}_1 \right) \left(X_i - \bar{X} \right) + \left(\varepsilon_i - \bar{\varepsilon} \right)$$

for $i = 1, 2, \ldots, n$. Squaring both sides of this equation and summing gives

$$\sum_{i=1}^{n} e_i^2 = \left(\hat{\beta}_1 - \beta_1 \right)^2 \sum_{i=1}^{n} \left(X_i - \bar{X} \right)^2 - 2 \left(\hat{\beta}_1 - \beta_1 \right) \sum_{i=1}^{n} \left(X_i - \bar{X} \right) \left(\varepsilon_i - \bar{\varepsilon} \right) + \sum_{i=1}^{n} \left(\varepsilon_i - \bar{\varepsilon} \right)^2 .$$

Taking into account that the X_i values are assumed to be fixed constants in a simple linear regression model, the expected value of both sides of this equation is

$$E \left[\sum_{i=1}^{n} e_i^2 \right] = E \left[\left(\hat{\beta}_1 - \beta_1 \right)^2 \right] \sum_{i=1}^{n} \left(X_i - \bar{X} \right)^2 - 2 E \left[\left(\hat{\beta}_1 - \beta_1 \right) \sum_{i=1}^{n} \left(X_i - \bar{X} \right) \left(\varepsilon_i - \bar{\varepsilon} \right) \right]$$

$$+ E\left[\sum_{i=1}^{n} (\varepsilon_i - \bar{\varepsilon})^2 \right]. \qquad (3)$$

There are three terms on the right-hand side of equation (3). Each term will be considered separately. The first term contains $E\left[(\hat{\beta}_1 - \beta_1)^2 \right]$, which is an expression for the population variance of $\hat{\beta}_1$ because $\hat{\beta}_1$ is an unbiased estimator for β_1 via Theorem 1.2. This population variance is the lower-right entry of the variance–covariance matrix given in Theorem 1.4. So the first term on the right-hand side of equation (3) reduces to

$$E\left[(\hat{\beta}_1 - \beta_1)^2 \right] \sum_{i=1}^{n} (X_i - \bar{X})^2 = V\left[\hat{\beta}_1 \right] \cdot S_{XX} = \frac{\sigma^2}{S_{XX}} \cdot S_{XX} = \sigma^2.$$

Before considering the second term on the right-hand side of equation (3), recall from Theorem 1.3 that $\hat{\beta}_1$ can be written as a linear combination of the observations of the dependent variable $Y_1, Y_2, \ldots, Y_n$ as $\hat{\beta}_1 = a_1 Y_1 + a_2 Y_2 + \cdots + a_n Y_n$, where $a_i = (X_i - \bar{X})/S_{XX}$ for $i = 1, 2, \ldots, n$. So an expression for the least squares point estimator of β_1 can be written as

$$\hat{\beta}_1 = \sum_{i=1}^{n} a_i Y_i$$

$$= \sum_{i=1}^{n} a_i (\beta_0 + \beta_1 X_i + \varepsilon_i)$$

$$= \beta_0 \sum_{i=1}^{n} a_i + \beta_1 \sum_{i=1}^{n} a_i X_i + \sum_{i=1}^{n} a_i \varepsilon_i$$

$$= \beta_1 + \sum_{i=1}^{n} a_i \varepsilon_i$$

via Theorem 1.3. Temporarily ignoring the -2 coefficient on the second term in equation (3) and using the fact that $\varepsilon_1, \varepsilon_2, \ldots, \varepsilon_n$ are mutually independent random variables with population mean zero and population variance σ^2, the expected value in the second term on the right-hand side of equation (3) is

$$E\left[(\hat{\beta}_1 - \beta_1) \sum_{i=1}^{n} (X_i - \bar{X})(\varepsilon_i - \bar{\varepsilon}) \right] = E\left[\left(\beta_1 + \sum_{i=1}^{n} a_i \varepsilon_i - \beta_1 \right) \sum_{i=1}^{n} (X_i - \bar{X})(\varepsilon_i - \bar{\varepsilon}) \right]$$

$$= E\left[\left(\sum_{i=1}^{n} a_i \varepsilon_i \right) \left(\sum_{i=1}^{n} (X_i - \bar{X}) \varepsilon_i - \bar{\varepsilon} \sum_{i=1}^{n} (X_i - \bar{X}) \right) \right]$$

$$= E\left[\frac{1}{S_{XX}} \left(\sum_{i=1}^{n} (X_i - \bar{X}) \varepsilon_i \right) \left(\sum_{i=1}^{n} (X_i - \bar{X}) \varepsilon_i \right) \right]$$

$$= \frac{1}{S_{XX}} E\left[\sum_{i=1}^{n} (X_i - \bar{X})^2 \varepsilon_i^2 + \sum\sum_{i \neq j} (X_i - \bar{X})(X_j - \bar{X}) \varepsilon_i \varepsilon_j \right]$$

$$= \frac{1}{S_{XX}} \sum_{i=1}^{n} (X_i - \bar{X})^2 E\left[\varepsilon_i^2 \right]$$

$$= \sigma^2.$$

Finally, consider the third term on the right-hand side of equation (3). Using (*i*) the shortcut formula for the population variance, (*ii*) the fact that the expected value operator E is a linear operator, (*iii*) the fact that the population variance of a sample mean comprised of mutually independent and identically distributed random variables is the ratio of the population variance to the sample size, and (*iv*) the fact that $E[\varepsilon_i] = 0$ for $i = 1, 2, \ldots, n$ and therefore $E[\bar{\varepsilon}] = 0$, the third term on the right-hand side of equation (3) is

$$
E\left[\sum_{i=1}^{n}(\varepsilon_i - \bar{\varepsilon})^2\right] = E\left[\sum_{i=1}^{n}\varepsilon_i^2 - 2\bar{\varepsilon}\sum_{i=1}^{n}\varepsilon_i + n\bar{\varepsilon}^2\right]
$$

$$
= E\left[\sum_{i=1}^{n}\varepsilon_i^2 - 2n\bar{\varepsilon}^2 + n\bar{\varepsilon}^2\right]
$$

$$
= E\left[\sum_{i=1}^{n}\varepsilon_i^2 - n\bar{\varepsilon}^2\right]
$$

$$
= \sum_{i=1}^{n}E[\varepsilon_i^2] - nE[\bar{\varepsilon}^2]
$$

$$
= \sum_{i=1}^{n}\left(V[\varepsilon_i] + E[\varepsilon_i]^2\right) - n\left(V[\bar{\varepsilon}] + E[\bar{\varepsilon}]^2\right)
$$

$$
= \sum_{i=1}^{n}\sigma^2 - n\cdot\frac{\sigma^2}{n}
$$

$$
= n\sigma^2 - \sigma^2
$$

$$
= (n-1)\sigma^2.
$$

Combining the three terms together, equation (3) becomes

$$
E\left[\sum_{i=1}^{n}e_i^2\right] = \sigma^2 - 2\sigma^2 + (n-1)\sigma^2 = (n-2)\sigma^2.
$$

Dividing both sides of this equation by $n-2$ indicates that the *MSE* is an unbiased estimator of σ^2:

$$
E\left[\frac{1}{n-2}\sum_{i=1}^{n}e_i^2\right] = \sigma^2. \qquad \Box
$$

To summarize, there are three parameters in a simple linear regression model: the population intercept β_0, the population slope β_1, and the population variance of the error terms σ^2. These parameters can be estimated from n data pairs $(X_1, Y_1), (X_2, Y_2), \ldots, (X_n, Y_n)$ by the least squares method. Theorem 1.2 indicates that the least squares point estimator $\hat{\beta}_0$ is an unbiased estimator of β_0 and the least squares point estimator $\hat{\beta}_1$ is an unbiased estimator of β_1. Theorem 1.7 indicates that the *MSE* is an unbiased estimator of σ^2. All three parameter estimators are on target on average. The next three examples illustrate the estimation of σ^2.

Example 1.10 Estimate the variance of the error terms σ^2 for the $n = 3$ data pairs from the sales data set in Example 1.3:

$$
(X_1, Y_1) = (6, 2), \qquad (X_2, Y_2) = (8, 9), \qquad (X_3, Y_3) = (2, 2).
$$

Using the calculations in Example 1.9, the estimated variance of the error terms is

$$\hat{\sigma}^2 = \frac{1}{n-2} \sum_{i=1}^{n} e_i^2 = \frac{1}{3-2}(9+4+1) = 14.$$

The R code to compute the value of the point estimate for σ^2 is given below.

```
x = c(6, 8, 2)
y = c(2, 9, 2)
n = length(x)
fit = lm(y ~ x)
sum(fit$residuals ^ 2) / (n - 2)
```

The magnitude of the point estimate of σ^2 is a reflection of whether the data points are tightly clustered about the estimated regression line (for small values of $\hat{\sigma}^2$) or whether the data points stray significantly from the estimated regression line (for large values of $\hat{\sigma}^2$). In the previous example involving the sales data pairs, there is significant vertical deviation between the data points and the associated fitted values, as seen in Figure 1.15. The next example illustrates the case in which the data pairs are tightly clustered about the regression line.

Example 1.11 Scottish physicist James Forbes wanted to devise a technique to estimate the altitude above sea level without transporting a fragile mercury barometer to the location of interest. He knew that the altitude could be computed from the barometric pressure, with lower barometric pressures corresponding to higher altitudes. He wanted to see if the boiling point of water could be used as a surrogate to determine the barometric pressure. In the 1840's and 1850's, he gathered $n = 17$ data pairs $(X_1, Y_1), (X_2, Y_2), \ldots, (X_{17}, Y_{17})$ from various locations at different altitudes in the Alps, where

> X_i: the boiling point of water in degrees Fahrenheit at location i, and
> Y_i: the adjusted barometric pressure in inches of mercury at location i,

for $i = 1, 2, \ldots, 17$. The data was published in an 1857 article in the *Transactions of the Royal Society of Edinburgh* titled "Further Experiments and Remarks on the Measurement of Heights and Boiling Point of Water." The $n = 17$ data pairs in Forbes' data set are shown in Table 1.4. Make a scatterplot of the data values to determine whether a simple linear regression model is appropriate. If it is an appropriate model, estimate the model parameters β_0, β_1, and σ^2.

A scatterplot of the data is plotted with the R commands given below.

```
x = c(194.5, 194.3, 197.9, 198.4, 199.4, 199.9, 200.9, 201.1, 201.4,
      201.3, 203.6, 204.6, 209.5, 208.6, 210.7, 211.9, 212.2)
y = c(20.79, 20.79, 22.40, 22.67, 23.15, 23.35, 23.89, 23.99, 24.02,
      24.01, 25.14, 26.57, 28.49, 27.76, 29.04, 29.88, 30.06)
plot(x, y)
```

The scatterplot is given in Figure 1.17. On the range of the independent variable X that was collected by Forbes, which is $194.3 \leq X \leq 212.2$, there appears to be a linear relationship between the boiling temperature X and the barometric pressure Y, so it

Boiling point	Barometric pressure
194.5	20.79
194.3	20.79
197.9	22.40
198.4	22.67
199.4	23.15
199.9	23.35
200.9	23.89
201.1	23.99
201.4	24.02
201.3	24.01
203.6	25.14
204.6	26.57
209.5	28.49
208.6	27.76
210.7	29.04
211.9	29.88
212.2	30.06

Table 1.4: Data pairs for Forbes' experiment.

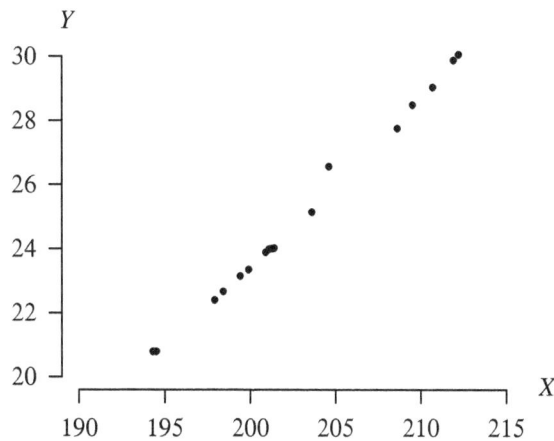

Figure 1.17: Scatterplot of the Forbes data.

is reasonable to proceed with fitting a simple linear regression model. The point that seems to stray slightly from the linear relationship, namely $(X_{12}, Y_{12}) = (204.6, 26.57)$, could be due to (*i*) random sampling variability, (*ii*) measurement error associated with the barometric pressure $Y_{12} = 26.57$, or (*iii*) measurement error associated with the boiling point $X_{12} = 204.6$ even though the simple linear regression model assumes that the boiling points are measured without error.

The R code below plots the fitted regression line on the scatterplot, which is shown in

Figure 1.18. Forbes' data pairs are in a built-in data frame named `forbes` that resides in the `MASS` package. The first column in `forbes` is named bp (for boiling point) and the second column is named `pres` (for barometric pressure).

```
library(MASS)
x = forbes$bp
y = forbes$pres
plot(x, y)
fit = lm(y ~ x)
abline(fit$coefficients)
```

Figure 1.18 confirms our conclusion about the linear relationship between X and Y from the scatterplot on the range of X values collected by Forbes. A simple linear regression model seems appropriate in this setting. The *additional* R commands that follow print the estimates for β_0, β_1, and σ^2 for Forbes' $n = 14$ data pairs.

```
n = length(x)
print(fit$coefficients)
print(sum(fit$residuals ^ 2) / (n - 2))
```

These yield the three unbiased point estimates for the simple linear regression model as

$$\hat{\beta}_0 = -81.0637 \qquad \hat{\beta}_1 = 0.5229 \qquad \hat{\sigma}^2 = 0.05421.$$

So the estimated regression line is

$$\hat{Y} = -81.0637 + 0.5229X.$$

Using the usual interpretation of the estimated intercept, when the boiling point of water is $0°$ Fahrenheit, the barometric pressure is estimated to be -81 inches of mercury. This is obviously an inappropriate conclusion and highlights the fact that this simple linear

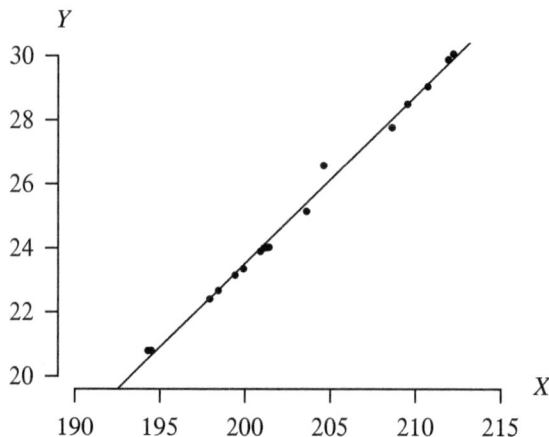

Figure 1.18: Scatterplot of the Forbes data with the estimated regression line.

regression model is only appropriate for a limited range of X values. The interpretation of $\hat{\beta}_1$, however, is meaningful. The barometric pressure increases by an estimated 0.5229 inches of mercury for every degree increase in the boiling point of water over the range of X values collected by Forbes. Finally, the estimated variance of the error terms, $\hat{\sigma}^2 = 0.05421$, is small (particularly relative to the estimated variance of the dependent variable observations, $S_{YY}/(n-1) = 9.12$, calculated with the *additional* R command var(forbes$pres)). This small estimated variance indicates that the data values are tightly clustered about the regression line. This is clearly the case in Figure 1.18.

The *additional* R command plot(fit$residuals) generates a plot of the residuals. Figure 1.19 shows the $n = 17$ residuals, along with a dashed horizontal line at a residual value of zero to show which observations fall above and below the regression line. (Notice that some of the X_i values are not in increasing order.) Six of the residuals are positive and 11 are negative. The reason that more residuals are negative is that the 12th data pair $(X_{12}, Y_{12}) = (204.6, 26.57)$ exerts a strong upwards "tug" on the fitted regression line, which is reflected in the plot of the residuals in Figure 1.19.

The non-symmetry in the values of $e_1, e_2, \ldots, e_{17}$ will also be reflected in a histogram of the residuals. Although $n = 17$ is a relatively small sample size for drawing a histogram and having a meaningful interpretation, one is displayed in Figure 1.20. This histogram can be generated with the *additional* R command hist(fit$residuals). The histogram reveals a bell-shaped distribution for the residuals, with a single extreme value in the right-hand tail associated with the residual $e_{12} = 0.65$. This is consistent with the plot of the residuals in Figure 1.19.

In conclusion, the regression analysis seems to indicate that Forbes' experiment was a success. The barometric pressure does appear to be a function of the boiling point of water, and furthermore, the relationship between the two variables appears to be reasonably linear on the range of data pairs collected by Forbes. For a particular boiling point X that falls within that range of X values, the barometric pressure can be estimated

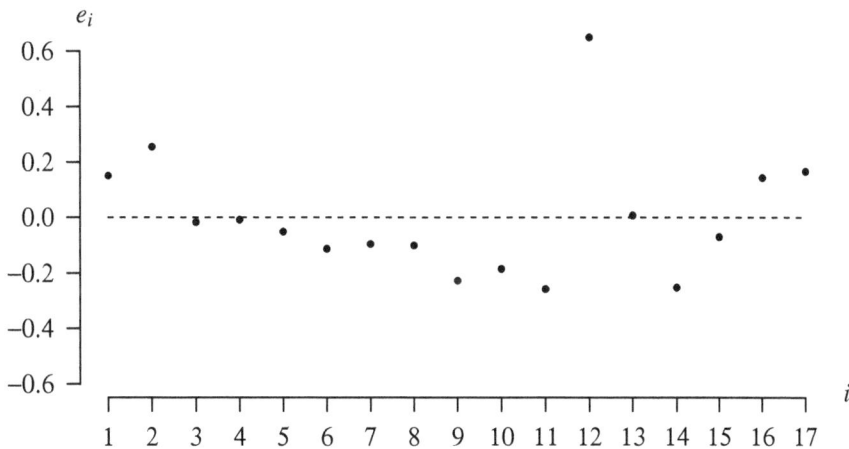

Figure 1.19: Residuals for the Forbes data.

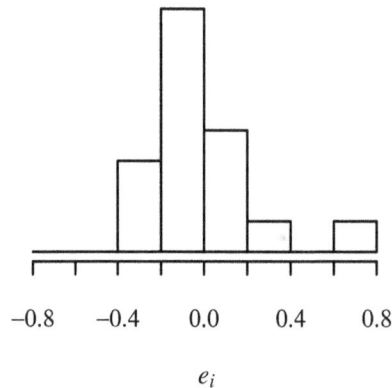

Figure 1.20: Histogram of the residuals for the Forbes data.

by

$$\hat{Y} = \hat{\beta}_0 + \hat{\beta}_1 X = -81.0637 + 0.5229X.$$

The altitude can, in turn, be estimated from the estimate of the barometric pressure provided by the regression analysis.

In the two previous examples, point estimates of the population variance of the error terms σ^2 were calculated. In the sales data example, the estimated error term variance $\hat{\sigma}^2 = 14$ indicated that the data pairs strayed a large distance from the estimated regression line, as illustrated in Figure 1.6. In the Forbes data set, the estimated error term variance $\hat{\sigma}^2 = 0.05421$ reflects data pairs that cluster closely to the estimated regression line, as illustrated in Figure 1.18. But these two examples involving individual data sets do not indicate anything about the *distribution* of $\hat{\sigma}^2$. The next example addresses this topic by extending the Monte Carlo simulation experiment from Example 1.4.

Example 1.12 Consider again the simple linear regression model

$$Y = \beta_0 + \beta_1 X + \varepsilon$$

from Example 1.4, where

- the population intercept is $\beta_0 = 1$,
- the population slope is $\beta_1 = 1/2$, and
- the error term ε has a $U(-1, 1)$ distribution.

The focus in this example will be on the estimation of the probability distribution of $\hat{\sigma}^2$. Recall that the error term distribution has population mean zero and finite population variance, so it satisfies the conditions of a simple linear regression model from Definition 1.1. Conduct a Monte Carlo simulation with 5000 replications that estimates the probability distribution of the estimated variance of the error terms $\hat{\sigma}^2$ for $n = 10$ data pairs. Assume that the X_i values are equally likely to be one of the integers $0, 1, 2, \ldots, 9$.

The R code below conducts 5000 replications of the Monte Carlo experiment. The simulated regression model is fit by the lm function and the results are stored in the list named `fit`. The component of the list named `fit$residuals` contains the residuals $e_1, e_2, \ldots, e_{10}$ for a particular simulation replication. The estimator of the variance of the error term in the simple linear regression model is given by the *MSE*:

$$\hat{\sigma}^2 = \frac{1}{n-2} \sum_{i=1}^{n} e_i^2,$$

which is an unbiased estimator estimator of σ^2 by Theorem 1.7. The code generates a histogram of the 5000 estimates of the variance of the error terms.

```
nrep    = 5000
n       = 10
beta0   = 1
beta1   = 1 / 2
sig2hat = numeric(nrep)
set.seed(100)
for (i in 1:nrep) {
  x = sample(0:9, n, replace = TRUE)
  if (min(x) == max(x)) stop("All x values are equal")
  y = beta0 + beta1 * x + runif(n, -1, 1)
  fit = lm(y ~ x)
  sig2hat[i] = sum(fit$residuals ^ 2) / (n - 2)
}
hist(sig2hat)
```

The histogram that is produced by this Monte Carlo simulation is given in Figure 1.21. The histogram is centered around the population variance of the error terms

$$\sigma^2 = \frac{(1+1)^2}{12} = \frac{4}{12} = \frac{1}{3}$$

because the population variance of the $U(a, b)$ distribution is

$$\sigma^2 = \frac{(b-a)^2}{12},$$

where $a = -1$ and $b = 1$. So the Monte Carlo simulation supports the fact that $\hat{\sigma}^2$ is an unbiased estimator of σ^2 via Theorem 1.7. Although the distribution of the probability density function is bell-shaped, a careful examination of the histogram indicates that the right-hand tail of the distribution appears to be slightly heavier than the left-hand tail of the distribution. The probability density function of $\hat{\sigma}^2$ is not symmetric. This nonsymmetry is a universal result which extends beyond this particular simple linear regression model. This should not be surprising because the support of $\hat{\sigma}^2$ is the positive real numbers, unlike the support of $\hat{\beta}_0$ and $\hat{\beta}_1$ whose support is the entire real number line.

So the conclusions of the Monte Carlo simulation experiment are that (a) Theorem 1.7 is supported because the histogram in Figure 1.21 is centered around σ^2, and (b) the probability density function of $\hat{\sigma}^2$ is nearly bell-shaped with a slight bit of nonsymmetry.

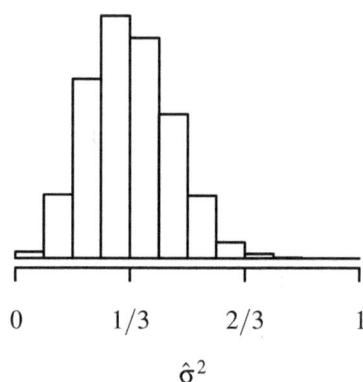

Figure 1.21: Histogram of the error term estimates for the Monte Carlo simulation.

Before leaving the topic of the estimation of σ^2 behind, consider the case of collecting just $n = 2$ data pairs (X_1, Y_1) and (X_2, Y_2), as illustrated in Figure 1.22. One of the assumptions associated with the observations in a simple linear regression model is that there are at least two distinct values of the independent variable observed. So when $n = 2$, it must be the case that $X_1 \neq X_2$. In this case, the least squares regression line will pass through the points (X_1, Y_1) and (X_2, Y_2). This means that the fitted values are identical to the data pairs, and hence, both residuals are zero. So the sum of squares for error is $SSE = e_1^2 + e_2^2 = 0$. But is an SSE of zero an appropriate estimate for the population variance of the spread of the values about the regression line? Can one conclude that this is really a deterministic relationship and any additional data pairs collected will fall on the fitted regression line? Certainly not, because it is not possible to draw that conclusion based on just two data pairs. A third data pair might fall on the regression line or fall significantly off of the regression line, as was the case with the sales data from Example 1.3. The unbiased estimator of σ^2 is undefined because of the $n - 2$ in the denominator of the formula for $\hat{\sigma}^2$, as it should be. Two data pairs are adequate

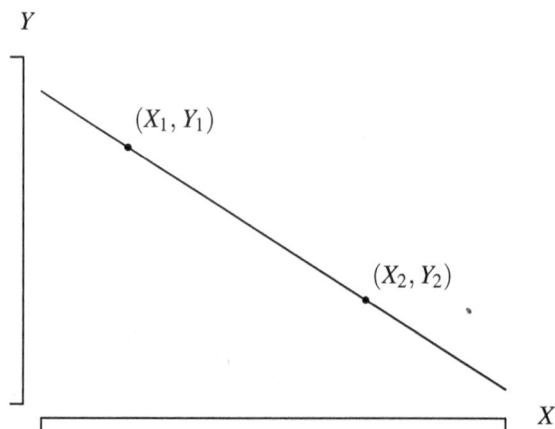

Figure 1.22: Scatterplot and estimated regression line for $n = 2$ data pairs.

for estimating the population slope and population intercept of the regression line, but they are not adequate for estimating σ^2. The mathematics and intuition are consistent in this setting.

1.8 Sums of Squares

Certain sums of squares play a key role in simple linear regression. This section considers three topics related to these sums of squares: (*a*) partitioning the total sum of squares, (*b*) defining and interpreting the coefficient of determination and the coefficient of correlation, and (*c*) displaying the sums of squares in an ANOVA table.

1.8.1 Partitioning the Total Sum of Squares

A topic that is closely related to fitted values and residuals is the partitioning of the total sum of squares. Figure 1.23 provides the geometric framework for the mathematical derivation provided next. There are only three points plotted in Figure 1.23. The first point plotted is (X_i, Y_i), which is a generic data pair. The other $n-1$ data pairs are not plotted in order to keep the figure uncluttered. The estimated regression line associated with the n data pairs, which happens to have a negative slope, is also plotted. The second point plotted is the fitted value $(X_i, \hat{Y}_i)$ associated with the ith data pair, which is located directly below data pair i and falls on the estimated regression line. The third point plotted is $(\bar{X}, \bar{Y})$, which, by Theorem 1.6, will always fall on the regression line.

Figure 1.23 provides a geometric proof of the relationship

$$Y_i - \bar{Y} = \hat{Y}_i - \bar{Y} + Y_i - \hat{Y}_i$$

for $i = 1, 2, \ldots, n$. The relationship can also be established algebraically by recognizing that the right-hand side of this equation can be determined by just adding and subtracting $\hat{Y}_i$ to the left-hand side of the equation. As will be stated and proved subsequently, squaring both sides of this equation and summing results in

$$\sum_{i=1}^{n} \left(Y_i - \bar{Y}\right)^2 = \sum_{i=1}^{n} \left(\hat{Y}_i - \bar{Y}\right)^2 + \sum_{i=1}^{n} \left(Y_i - \hat{Y}_i\right)^2.$$

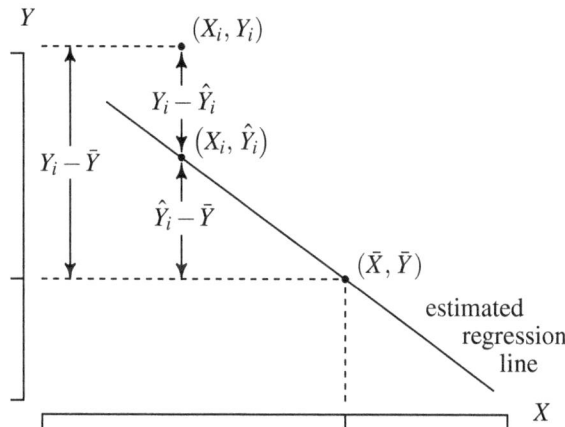

Figure 1.23: Partitioning the total sum of squares.

This equation involves three sums of squares that occur so often in regression analysis that they are given the abbreviations

$$SST = SSR + SSE,$$

where SST stands for total sum of squares, SSR stands for sum of squares for regression, and SSE stands for sum of squares for error. (The sum of squares for error has already been encountered in Theorem 1.7.) This equation expresses the total variation of the observed values of the dependent variable $Y_1, Y_2, \ldots, Y_n$ about their sample mean $\bar{Y}$ in SST as the sum of two sums of squares. The first term on the right-hand side, SSR, reflects the variation of the fitted values $\hat{Y}_1, \hat{Y}_2, \ldots, \hat{Y}_n$ about the sample mean $\bar{Y}$. The second term on the right-hand side, SSE, reflects the variation of the observed values $Y_1, Y_2, \ldots, Y_n$ about their associated fitted values $\hat{Y}_1, \hat{Y}_2, \ldots, \hat{Y}_n$. Since all three terms in this equation are sums of squares, all three terms are nonnegative. Notice that $SST/(n-1)$ is the sample variance of $Y_1, Y_2, \ldots, Y_n$.

The equation

$$SST = SSR + SSE$$

partitions SST into two pieces: SSR, which accounts for the total variability in $Y_1, Y_2, \ldots, Y_n$ that is accounted for by the regression line (that is, the linear relationship between X and Y), and SSE, which accounts for the remaining variability that is not associated with the regression line. This is why SSR measures the total variability in $Y_1, Y_2, \ldots, Y_n$ "explained" by the relationship between X and Y, whereas SSE measures the total variability in $Y_1, Y_2, \ldots, Y_n$ left "unexplained" by the relationship between X and Y. It is reasonable to think of SSR as measuring the "signal" associated with the linear relationship and SSE as measuring the "noise" associated with the linear relationship. The result is stated formally and proven next.

Theorem 1.8 Let $\hat{\beta}_0$ and $\hat{\beta}_1$ denote the least squares estimators of the parameters β_0 and β_1 in the simple linear regression model fitted to the data pairs $(X_1, Y_1), (X_2, Y_2), \ldots, (X_n, Y_n)$. Let $\hat{Y}_i = \hat{\beta}_0 + \hat{\beta}_1 X_i$ be the fitted value associated with data pair i, for $i = 1, 2, \ldots, n$. Let $\bar{Y}$ be the sample mean of $Y_1, Y_2, \ldots, Y_n$. Then

$$\sum_{i=1}^{n} \left(Y_i - \bar{Y}\right)^2 = \sum_{i=1}^{n} \left(\hat{Y}_i - \bar{Y}\right)^2 + \sum_{i=1}^{n} \left(Y_i - \hat{Y}_i\right)^2,$$

or, equivalently,

$$SST = SSR + SSE.$$

Proof Beginning with $Y_i - \bar{Y}$, adding and subtracting $\hat{Y}_i$ gives

$$Y_i - \bar{Y} = \hat{Y}_i - \bar{Y} + Y_i - \hat{Y}_i$$

for $i = 1, 2, \ldots, n$. Grouping the two terms on the right-hand side of this equation as $\left(\hat{Y}_i - \bar{Y}\right)$ and $\left(Y_i - \hat{Y}_i\right)$, squaring both sides of the equation, and summing gives

$$\sum_{i=1}^{n} \left(Y_i - \bar{Y}\right)^2 = \sum_{i=1}^{n} \left(\hat{Y}_i - \bar{Y}\right)^2 + 2 \sum_{i=1}^{n} \left(\hat{Y}_i - \bar{Y}\right)\left(Y_i - \hat{Y}_i\right) + \sum_{i=1}^{n} \left(Y_i - \hat{Y}_i\right)^2.$$

The middle summation on the right-hand side of this equation is zero because

$$2 \sum_{i=1}^{n} \left(\hat{Y}_i - \bar{Y}\right)\left(Y_i - \hat{Y}_i\right) = 2 \sum_{i=1}^{n} \hat{Y}_i\left(Y_i - \hat{Y}_i\right) - 2\bar{Y} \sum_{i=1}^{n} \left(Y_i - \hat{Y}_i\right) = 2 \sum_{i=1}^{n} \hat{Y}_i e_i - 2\bar{Y} \sum_{i=1}^{n} e_i = 0$$

by Theorem 1.6, which proves the result. $\square$

1.8.2   Coefficients of Determination and Correlation

There are two measures that are helpful in assessing the degree of the linear relationship between X and Y in a simple linear regression model. The *coefficient of determination* and the *coefficient of correlation* are defined next. The thinking behind the way that the coefficient of determination $R^2 = SSR/SST$ is defined is as follows. The value of SST reflects the variability in $Y_1, Y_2, \ldots, Y_n$ when the values of the associated independent variables $X_1, X_2, \ldots, X_n$ are ignored. The value of SSE reflects the variability in $Y_1, Y_2, \ldots, Y_n$ when a fitted regression model uses $X_1, X_2, \ldots, X_n$ as predictors. Their difference, $SSR = SST - SSE$, reflects the reduction in variability associated with using the regression model. The ratio SSR/SST captures the fraction of that reduction in variability.

Definition 1.3 Let SST, SSR, and SSE for a simple linear regression model be defined as in Theorem 1.8. The *coefficient of determination* is

$$R^2 = \frac{SSR}{SST} = \frac{SST - SSE}{SST} = 1 - \frac{SSE}{SST}$$

when $SST \neq 0$. The *coefficient of correlation* (a.k.a. the *sample correlation coefficient*) is

$$r = \pm\sqrt{R^2},$$

where the sign associated with r is positive (negative) when the slope of the estimated regression line is positive (negative).

The coefficient of determination R^2 is the fraction of the variation in $Y_1, Y_2, \ldots, Y_n$ about $\bar{Y}$ that is accounted for by the linear relationship between X and Y. Based on the result from Theorem 1.8, $SST = SSR + SSE$, the coefficient of determination must satisfy $0 \leq R^2 \leq 1$. Likewise, the coefficient of correlation must satisfy $-1 \leq r \leq 1$, which is true for all population and sample correlations.

Values of R^2 that are near 1 indicate that nearly all of the variation in $Y_1, Y_2, \ldots, Y_n$ about $\bar{Y}$ can be explained by the linear relationship between X and Y. This in turn implies that X is a useful predictor for Y. On the other hand, values of R^2 that are near 0 indicate that very little of the variation in $Y_1, Y_2, \ldots, Y_n$ about $\bar{Y}$ can be explained by the linear relationship between X and Y. This in turn implies that X is not a useful predictor for Y. It is in this sense that R^2 is a measure of the strength of the linear relationship between X and Y.

There are some important limitations associated with R^2 and r. First, it is important to remember that the linear relationship between X and Y might only be appropriate on a limited range of X values. Second, even a relatively large value of R^2 might not provide the precision necessary for a particular application. Third, regardless of the value of R^2, the scatterplot of the data pairs must always be inspected to see if a simple linear regression model is warranted. Both high and low values of R^2 can be associated with a strong *nonlinear* relationship between X and Y. Fourth, in the case in which the experimenter can control the values of $X_1, X_2, \ldots, X_n$, the magnitude of R^2 depends on the choices of the independent variables, which clouds its interpretation. Fifth, the usual interpretation of the coefficient of correlation r as an estimator of $\rho = \text{Cov}(X, Y)/(\sigma_X \sigma_Y)$ is only appropriate when X and Y are random variables, which is not the case in simple linear regression because X assumed to be observed without error.

It is a useful thought experiment to consider the scatterplots associated with the values of SST, SSR, and SSE at their extremes. These three extreme cases will be described in the next three paragraphs.

The first of these extreme cases is illustrated for $n = 7$ in Figure 1.24 in which

$$SSE = \sum_{i=1}^{n} \left(Y_i - \hat{Y}_i\right)^2 = \sum_{i=1}^{n} e_i^2 = 0.$$

The only way to achieve a sum of squares for error of zero is to have the data pairs (X_1, Y_1), (X_2, Y_2), ..., (X_n, Y_n) all fall on a line, which is the regression line. Using the result from Theorem 1.8 that $SST = SSR + SSE$, in this case $SST = SSR$, which implies that $R^2 = 1$. Therefore, *all* of the variation in $Y_1, Y_2, \ldots, Y_n$ is explained by the linear relationship between X and Y. In addition, $r = -1$ if the slope of the regression line is negative and $r = 1$ if the slope of the regression line is positive.

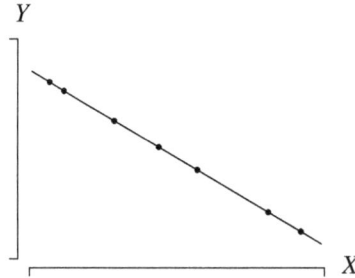

Figure 1.24: Data pairs with $SSE = 0$ and $\hat{\beta}_1 \neq 0$ (which implies that $SST = SSR$ and $R^2 = 1$).

The second of these extreme cases is illustrated for $n = 7$ in Figure 1.25 in which

$$SSR = \sum_{i=1}^{n} \left(\hat{Y}_i - \bar{Y}\right)^2 = 0.$$

The only way to achieve a sum of squares for regression of zero is to have an estimated regression line with slope zero. Using the result from Theorem 1.8 that $SST = SSR + SSE$, in this case $SST = SSE$, which implies that $R^2 = 0$. This means that *none* of the variation in $Y_1, Y_2, \ldots, Y_n$ is explained by the linear relationship between X and Y. In addition, $r = 0$.

The third of these extreme cases is illustrated for $n = 7$ in Figure 1.26 in which

$$SST = \sum_{i=1}^{n} \left(Y_i - \bar{Y}\right)^2 = 0.$$

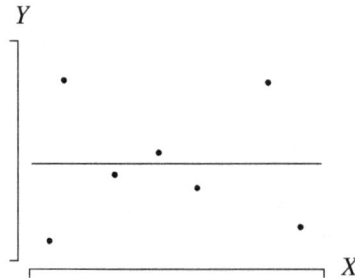

Figure 1.25: Data pairs with $SSR = 0$ (which implies that $SST = SSE$ and $R^2 = 0$).

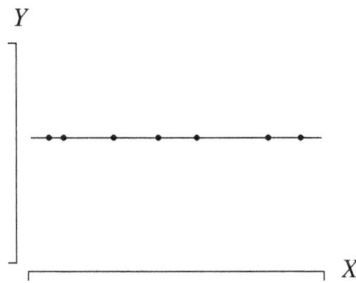

Figure 1.26: Data pairs with $SST = 0$ (which implies that $SSR = SSE = 0$ and R^2 is undefined).

The only way to achieve a total sum of squares of zero is to have an estimated regression line with slope zero and all points lying on the estimated regression line. Using the result from Theorem 1.8 that $SST = SSR + SSE$, in this case $SSR = SSE = 0$, and the coefficient of determination and coefficient of correlation are undefined because the denominator is zero.

Each of the sums of squares has an associated degrees of freedom. The total sum of squares

$$SST = \sum_{i=1}^{n} \left(Y_i - \bar{Y}\right)^2$$

has $n - 1$ degrees of freedom for either of two reasons: (1) one degree of freedom is lost because $\bar{Y}$ is used to estimate the population mean, and (2) the terms in the summation above are subject to the one constraint—they must sum to zero. The sum of squares for regression

$$SSR = \sum_{i=1}^{n} \left(\hat{Y}_i - \bar{Y}\right)^2$$

has 1 degree of freedom because each of the $\hat{Y}_i$ values is calculated from the same regression line which has two degrees of freedom, but is subject to the additional constraint $\sum_{i=1}^{n} \left(\hat{Y}_i - \bar{Y}\right) = 0$ by Theorem 1.6. The sum of squares for error

$$SSE = \sum_{i=1}^{n} \left(Y_i - \hat{Y}_i\right)^2$$

has $n - 2$ degrees of freedom for the reasons outlined just before Theorem 1.7.

An alternative definition for computing the coefficient of correlation r can save on computation time, as given in the following theorem.

Theorem 1.9 The coefficient of correlation r is

$$r = \hat{\beta}_1 \sqrt{\frac{S_{XX}}{S_{YY}}}.$$

Proof Recall from Definition 1.3 that the coefficient of correlation is

$$r = \pm \sqrt{\frac{SSR}{SST}},$$

where the sign associated with r is the same as the sign of $\hat{\beta}_1$. Since $\sum_{i=1}^{n} \hat{Y}_i = \sum_{i=1}^{n} Y_i$ by Theorem 1.6, this can be rewritten as

$$r = \pm \sqrt{\frac{SSR}{SST}}$$

$$= \pm \sqrt{\frac{\sum_{i=1}^{n} \left(\hat{Y}_i - \bar{Y}\right)^2}{S_{YY}}}$$

$$= \pm \sqrt{\frac{\sum_{i=1}^{n} \hat{Y}_i^2 - 2\bar{Y} \sum_{i=1}^{n} \hat{Y}_i + n\bar{Y}^2}{S_{YY}}}$$

$$= \pm \sqrt{\frac{\sum_{i=1}^{n} \hat{Y}_i^2 - 2\bar{Y} \sum_{i=1}^{n} Y_i + n\bar{Y}^2}{S_{YY}}}$$

$$= \pm \sqrt{\frac{\sum_{i=1}^{n} \hat{Y}_i^2 - n\bar{Y}^2}{S_{YY}}}$$

$$= \pm \sqrt{\frac{\sum_{i=1}^{n} \left(\hat{\beta}_0 + \hat{\beta}_1 X_i\right)^2 - n\bar{Y}^2}{S_{YY}}}$$

$$= \pm \sqrt{\frac{\sum_{i=1}^{n} \left(\bar{Y} - \hat{\beta}_1 \bar{X} + \hat{\beta}_1 X_i\right)^2 - n\bar{Y}^2}{S_{YY}}}$$

$$= \pm \sqrt{\frac{n\bar{Y}^2 + 2\bar{Y}\hat{\beta}_1 \sum_{i=1}^{n} (X_i - \bar{X}) + \hat{\beta}_1^2 \sum_{i=1}^{n} (X_i - \bar{X})^2 - n\bar{Y}^2}{S_{YY}}}$$

$$= \pm \sqrt{\frac{\hat{\beta}_1^2 \sum_{i=1}^{n} (X_i - \bar{X})^2}{S_{YY}}}$$

$$= \hat{\beta}_1 \sqrt{\frac{S_{XX}}{S_{YY}}},$$

which proves the theorem. $\square$

1.8.3 The ANOVA Table

The three sums of squares for the simple linear regression model and their associated degrees of freedom can be summarized in an analysis of variance (ANOVA) table. The four columns in the generic ANOVA table shown in Table 1.5 are (a) the source of variation, (b) the sum of squares, (c) the degrees of freedom, and (d) the mean square. The sums of squares and the degrees of freedom add to the values in the row labeled "Total". The mean square is the ratio of the sum of

Source	SS	df	MS
Regression	SSR	1	MSR
Error	SSE	$n-2$	MSE
Total	SST	$n-1$	

Table 1.5: Partial ANOVA table for simple linear regression.

squares to the associated degrees of freedom. The regression mean square is $MSR = SSR/1 = SSR$. The mean square error is $MSE = SSE/(n-2)$. The mean square entries do not add. Tradition dictates that the mean square associated with SST is not reported in an ANOVA table, but it does have meaning as the sample variance of $Y_1, Y_2, \ldots, Y_n$. More information on how the ANOVA table can be used for hypothesis testing concerning the population slope β_0 by adding a fifth column to the ANOVA table will be given in the next chapter.

Example 1.13 Consider the Forbes data set from Example 1.11 in which the independent variable X is the boiling point of water in degrees Fahrenheit and the dependent variable Y is the adjusted barometric pressure in inches of mercury. There are $n = 17$ data pairs collected from various locations. Calculate the three sums of squares (SST, SSR, and SSE), show that Theorem 1.8 is satisfied, calculate R^2 and r, and present the results in an ANOVA table.

The scatterplot with the estimated regression line superimposed from Example 1.11 is reproduced in Figure 1.27. The R commands below calculate the three sums of squares.

```
library(MASS)
x   = forbes$bp
y   = forbes$pres
fit = lm(y ~ x)
sst = sum((y - mean(y)) ^ 2)
ssr = sum((fit$fitted - mean(y)) ^ 2)
sse = sum(fit$residuals ^ 2)
print(c(sst, ssr, sse))
```

These commands result in the following values for the three sums of squares:

$$SST = 145.9378 \qquad SSR = 145.1246 \qquad SSE = 0.8131.$$

Ignoring the roundoff error in the fourth digit after the decimal point, these values sat-

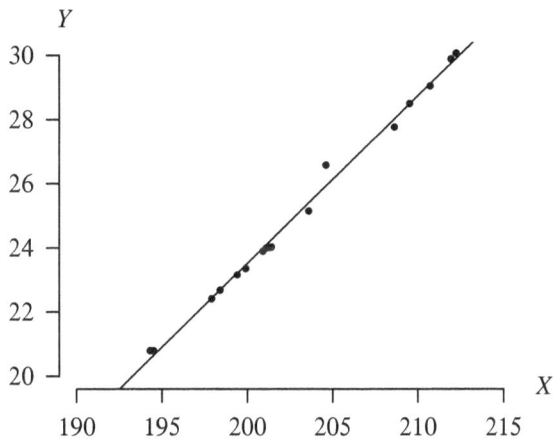

Figure 1.27: Scatterplot of the Forbes data with the estimated regression line.

isfy the result in Theorem 1.8:

$$SST = SSR + SSE.$$

The fact that *SSR* is more than two orders of magnitude greater than *SSE* indicates that there is much more of the total variation in $Y_1, Y_2, \ldots, Y_n$ that is explained by the relationship between X and Y than unexplained. This interpretation is consistent with the scatterplot and estimated regression line given in Figure 1.27.

The value of the coefficient of determination and the coefficient of correlation for this data set can be calculated by the *additional* R commands

```
R2 = ssr / sst
r  = sign(fit$coefficients[2]) * sqrt(R2)
print(c(R2, r))
```

via Definition 1.3 or

```
sxx = sum((x - mean(x)) ^ 2)
syy = sum((y - mean(y)) ^ 2)
R2  = ssr / sst
r   = fit$coefficients[2] * sqrt(sxx / syy)
print(c(R2, r))
```

via Theorem 1.9. Both code segments print the values

$$R^2 = 0.9944 \qquad \text{and} \qquad r = 0.9972.$$

The proper interpretation of R^2 is that 99.44% of the total variation in $Y_1, Y_2, \ldots, Y_n$ can be explained by the linear relationship between X and Y. This high percentage is consistent with the scatterplot and estimated regression line in Figure 1.27, which shows a nearly perfect linear relationship between boiling point of water and the barometric pressure, and data values that lie very close to the estimated regression line. Table 1.6 contains the sums of squares, degrees of freedom, and mean squares for the $n = 17$ data pairs collected by Forbes. This ANOVA table can be generated with the *additional* R command

```
anova(fit)
```

The anova function returns a data frame, and values in that data frame can be extracted using the $ extractor. The degrees of freedom for the sum of squares for error, for example, can be extracted with the R command

Source	SS	df	MS
Regression	145.1246	1	145.1246
Error	0.8131	15	0.0542
Total	145.9378	16	

Table 1.6: Partial ANOVA table for the Forbes data.

```
anova(fit)$Df[2]
```

The definitions and theorems that are associated with fitted values, residuals, estimating the population variance σ^2, partitioning the sums of squares, the coefficient of determination, the coefficient of correlation, and the ANOVA table are briefly reviewed here. The simple linear regression model

$$Y = \beta_0 + \beta_1 X + \varepsilon$$

from Definition 1.1 establishes a linear statistical relationship between an independent variable X and a dependent random variable Y. The error term ε has population mean 0 and finite population variance σ^2. The n data pairs collected are denoted by $(X_1, Y_1), (X_2, Y_2), \ldots, (X_n, Y_n)$. The fitted values $\hat{Y}_1, \hat{Y}_2, \ldots, \hat{Y}_n$ are the values on the estimated regression line associated with the independent variables $X_1, X_2, \ldots, X_n$:

$$\hat{Y}_i = \hat{\beta}_0 + \hat{\beta}_1 X_i$$

for $i = 1, 2, \ldots, n$, as established in Definition 1.2. The associated residuals are defined by

$$e_i = Y_i - \hat{Y}_i$$

for $i = 1, 2, \ldots, n$, as established in Definition 1.2. An unbiased estimator of the population variance of the error terms is

$$\hat{\sigma}^2 = \frac{1}{n-2} \sum_{i=1}^{n} e_i^2$$

as given in Theorem 1.7. The total sum of squares SST can be partitioned into the regression sum of squares SSR and the sum of squares for error SSE as

$$SST = SSR + SSE$$

or

$$\sum_{i=1}^{n} \left(Y_i - \bar{Y} \right)^2 = \sum_{i=1}^{n} \left(\hat{Y}_i - \bar{Y} \right)^2 + \sum_{i=1}^{n} \left(Y_i - \hat{Y}_i \right)^2$$

as given in Theorem 1.8. Two quantities that measure the linear association between X and Y are the coefficient of determination

$$R^2 = \frac{SSR}{SST},$$

which satisfies $0 \leq R^2 \leq 1$, and the coefficient of correlation

$$r = \pm \sqrt{R^2},$$

which satisfies $-1 \leq r \leq 1$ as defined in Definition 1.3. The coefficient of determination is the fraction of variation in $Y_1, Y_2, \ldots, Y_n$ that is explained by the linear relationship with X. The sums of squares are often presented in an ANOVA table, which includes columns for the source of variation, the sum of squares, the associated degrees of freedom, and the mean squares. An additional column will be added to the ANOVA table in the next chapter, when statistical inference in simple linear regression is introduced.

The point estimators for β_0, β_1, and σ^2 in the simple linear regression model have now all been established and many of their properties have been surveyed. But without additional assumptions, it is not possible to easily obtain interval estimators or perform hypothesis testing concerning these parameters. The next chapter addresses this issue.

1.9 Exercises

1.1 Establish a linear deterministic relationship between the independent variable X, the temperature in degrees Fahrenheit, and the dependent variable Y, the associated temperature in degrees Celsius.

1.2 Establish a nonlinear deterministic relationship between the independent variable X, the distance between two objects with fixed masses m_1 and m_2, and the dependent variable Y, the gravitational force acting between the two objects, using Newton's Law of Universal Gravitation.

1.3 For the following interpretations of the independent and dependent variables, predict whether the estimated slope $\hat{\beta}_1$ in a simple linear regression model will be positive or negative.

    (a) The independent variable X is a car's speed and the dependent variable Y is the car's stopping distance.

    (b) The independent variable X is a car's weight and the dependent variable Y is the car's fuel efficiency measured in miles per gallon.

    (c) The independent variable X is a husband's height and the dependent variable Y is the wife's height for a married couple.

    (d) The independent variable X is the average annual unemployment rate and the dependent variable Y is the annual GDP for a particular country.

1.4 For the simple linear regression model, show that solving the 2×2 set of linear normal equations

$$n\hat{\beta}_0 + \hat{\beta}_1 \sum_{i=1}^{n} X_i = \sum_{i=1}^{n} Y_i$$

$$\hat{\beta}_0 \sum_{i=1}^{n} X_i + \hat{\beta}_1 \sum_{i=1}^{n} X_i^2 = \sum_{i=1}^{n} X_i Y_i$$

for $\hat{\beta}_0$ and $\hat{\beta}_1$ gives the expressions for $\hat{\beta}_0$ and $\hat{\beta}_1$ given in Theorem 1.1.

1.5 Consider the simple linear regression model

$$Y = \beta_0 + \beta_1 X + \varepsilon,$$

where

- the population intercept is $\beta_0 = 1$,
- the population slope is $\beta_1 = 1/2$, and
- the error term ε has a $U(-1, 1)$ distribution.

Assume that $n = 10$ data pairs $(X_1, Y_1), (X_2, Y_2), \ldots, (X_{10}, Y_{10})$ are collected. The values of the independent variable X are equally likely to be one of the integers $0, 1, 2, \ldots, 9$, What are the minimum and maximum values that the estimated parameters $\hat{\beta}_0$ and $\hat{\beta}_1$ can assume?

1.6 For the values of the independent variables $X_1, X_2, \ldots, X_n$, show that

$$\sum_{i=1}^{n} \left(X_i - \bar{X} \right) = 0.$$

1.7 Write R commands to plot contours of the sum of squares for the sales data pairs

$$(X_1, Y_1) = (6, 2), \qquad (X_2, Y_2) = (8, 9), \qquad (X_3, Y_3) = (2, 2)$$

in the (β_0, β_1) plane.

1.8 The least squares criterion applied to a simple linear regression model minimizes

$$S = \sum_{i=1}^{n} (Y_i - \beta_0 - \beta_1 X_i)^2.$$

If instead the *least absolute deviation* criterion (also known as the minimum absolute deviation or MAD criterion) were applied to a simple linear regression model to minimize

$$S = \sum_{i=1}^{n} |Y_i - \beta_0 - \beta_1 X_i|,$$

what are the values of $\hat{\beta}_0$ and $\hat{\beta}_1$ for the sales data pairs

$$(X_1, Y_1) = (6, 2) \qquad (X_2, Y_2) = (8, 9) \qquad (X_3, Y_3) = (2, 2)?$$

1.9 Write a Monte Carlo simulation experiment that uses the same parameters as those in Example 1.4 (that is, $\beta_0 = 1$, $\beta_1 = 1/2$, $\varepsilon \sim U(-1, 1)$, $n = 10$) for 5000 replications, but this time selects the independent variable values to be equally likely integers from -5 and 5. Produce analogous figures to those of Figure 1.11 and Figure 1.12. Comment on your figures and how they relate to the variance–covariance matrix from Theorem 1.4.

1.10 For a simple linear regression model with $X_1 = 1$, $X_2 = 2$, $\ldots$, $X_n = n$ and $\sigma^2 = 1$, find the variance–covariance matrix of $\hat{\beta}_0$ and $\hat{\beta}_1$.

1.11 Use Theorems 1.2 and 1.4 to show that the least squares estimator of the intercept of the regression line β_0 in the simple linear regression model is a consistent estimator of β_0.

1.12 Example 1.6 calculates the variance–covariance matrix for a single replication of a Monte Carlo simulation experiment. Conduct this experiment for 5000 replications and report the average of the values in the variance–covariance matrix.

1.13 Let L be the set of all linear estimators of the slope β_1 in a simple linear regression model. Let U be the set of all unbiased estimators of the slope β_1 in a simple linear regression model. Give an example of an estimator of β_1 in $L \cap U'$.

1.14 Show that the fitted simple linear regression model

$$\hat{Y}_i = \hat{\beta}_0 + \hat{\beta}_1 X_i$$

for $i = 1, 2, \ldots, n$ can be written as

$$\hat{Y}_i - \bar{Y} = \hat{\beta}_1 (X_i - \bar{X}),$$

where $\hat{\beta}_0$ and $\hat{\beta}_1$ are the least squares estimators of β_0 and β_1 and $\bar{X}$ and $\bar{Y}$ are the sample means of the observed values of the independent and dependent variables.

1.15 Write a paragraph that argues why a fitted least squares regression line cannot pass through all data pairs except for one of the data pairs.

1.16 One of the most common error distributions used in simple linear regression is the normal distribution with population mean 0 and finite population variance σ^2, which has probability density function

$$f(x) = \frac{1}{\sqrt{2\pi}\sigma} e^{-x^2/(2\sigma^2)} \qquad -\infty < x < \infty.$$

An alternative error distribution is the Laplace distribution with probability density function

$$f(x) = \frac{1}{\sqrt{2}\sigma} e^{-\sqrt{2}|x-\mu|/\sigma} \qquad -\infty < x < \infty.$$

Since the error distribution must have expected value zero by assumption, this reduces to

$$f(x) = \frac{1}{\sqrt{2}\sigma} e^{-\sqrt{2}|x|/\sigma} \qquad -\infty < x < \infty.$$

As parameterized here, the Laplace distribution has population variance σ^2. Both of these distributions are symmetric and centered about zero.

(a) Plot the normal and Laplace error probability density functions on $-3 < x < 3$ and comment on any differences between the two error distributions. Use $\sigma = 1$ for the plots.

(b) Plot the normal and Laplace error probability density functions on $4 < x < 5$ and comment on any differences between the tails of the two error distributions.

(c) Fit both of these error distributions (that is, find $\hat\sigma^2$ for each distribution) for the `forbes` data set from the `MASS` package in R using the simple linear regression model.

1.17 Let the independent variable X be a car's speed and the dependent variable Y be the car's stopping distance, which are going to be modeled with a simple linear regression model. In which of the following scenarios do you expect to have a larger population variance of the error term?

(a) The data pairs $(X_1, Y_1), (X_2, Y_2), \ldots, (X_{20}, Y_{20})$ are $n = 20$ new cars that are all of the same make and model.

(b) The data pairs $(X_1, Y_1), (X_2, Y_2), \ldots, (X_{20}, Y_{20})$ are $n = 20$ new cars from $n = 20$ different car manufacturers.

1.18 Show that the sum of squares for regression in a simple linear regression model can be written as

$$SSR = \hat\beta_1 S_{XY}.$$

1.19 Show that the sum of squares for regression in a simple linear regression model can be written as

$$SSR = \hat\beta_1^2 S_{XX}.$$

1.20 Consider the data pairs in the `Formaldehyde` data set built into the base R language. Use the `help` function in R to determine the interpretation of the independent and dependent variables. Fit a simple linear regression model to the data pairs and interpret the meaning of $\hat\beta_0$, $\hat\beta_1$, and $\hat\sigma^2$. Also, calculate SST, SSR, and SSE for this data set.

1.21 Consider the data pairs collected by James Forbes that are given in the data frame `forbes` contained in the `MASS` package in R. The independent variable is the boiling point (in degrees Fahrenheit) and the dependent variable is the barometric pressure (in inches of mercury). For a simple linear regression model, calculate

- the fitted values,
- the residuals,
- the sum of squares for error, and
- the mean square error

without using the `lm` function. Then use the `lm` function to check the correctness of the values that you calculate.

1.22 This exercise investigates the effect of controllable values of $X_1, X_2, \ldots, X_n$ on the coefficient of determination R^2 in simple linear regression. Consider the simple linear regression model

$$Y = \beta_0 + \beta_1 X + \varepsilon,$$

where

- the population intercept is $\beta_0 = 1$,
- the population slope is $\beta_1 = 1/2$, and
- the error term ε has a $N(0, 1)$ distribution.

Conduct a Monte Carlo simulation with 40,000 replications that estimates the expected coefficient of determination for $n = 10$ data pairs under the following two ways of setting the values of $X_1, X_2, \ldots, X_{10}$.

(a) Let $X_i = i$ for $1, 2, \ldots, 10$.

(b) Let $X_1 = X_2 = \cdots = X_5 = 5$ and $X_6 = X_7 = \cdots = X_{10} = 6$.

1.23 Let S_X and S_Y be the sample standard deviations of the independent and dependent variables, respectively. Show that the following four definitions of the coefficient of correlation are equivalent.

(a) $r = \dfrac{1}{n-1} \sum_{i=1}^{n} \left(\dfrac{X_i - \bar{X}}{S_X} \right) \left(\dfrac{Y_i - \bar{Y}}{S_Y} \right)$

(b) $r = \pm \sqrt{\dfrac{SSR}{SSE}}$

(c) $r = \dfrac{S_{XY}}{\sqrt{S_{XX} S_{YY}}}$

(d) $r = \hat{\beta}_1 \sqrt{\dfrac{S_{XX}}{S_{YY}}}$

Chapter 2

Inference in Simple Linear Regression

The focus now shifts to statistical inference in the setting of a simple linear regression model applied to a data set containing the n data pairs $(X_1, Y_1), (X_2, Y_2), \ldots, (X_n, Y_n)$. The statistical inference typically takes the form of confidence intervals and hypothesis tests concerning the various parameters in the simple linear regression model. More specifically, the sections that follow concern statistical inference concerning σ^2, β_1, β_0, $E[Y_h]$, $Y_h^\star$, and joint statistical inference concerning β_0 and β_1.

2.1 Simple Linear Regression with Normal Error Terms

Drawing mathematically tractable statistical inferences concerning the parameters in a simple linear regression model is not possible with the current assumptions given in Definition 1.1. The problem lies in the vagueness of the assumptions about the error term. The assumption in a simple linear regression model is that the error term ε is a random variable with population mean 0 and finite population variance σ^2. The most common way of making this assumption more specific is to assume that the error term is *normally distributed* with population mean 0 and finite population variance σ^2. This will be stated formally in the following definition.

Definition 2.1 A *simple linear regression model with normal error terms* is given by

$$Y = \beta_0 + \beta_1 X + \varepsilon,$$

where X, Y, β_0, and β_1 are as in Definition 1.1 and $\varepsilon \sim N\left(0, \sigma^2\right)$.

Instead of just any probability distribution with a population mean of zero, we now specify that the error term should have a bell-shaped distribution centered about zero. Even though this is a more limiting assumption, it will allow us to establish exact confidence intervals and perform the associated hypothesis tests on the model parameters and other aspects of the model that might be of interest. Under this more restricted model, it is important to assure that the residuals (which estimate the error terms) do indeed have a bell-shaped distribution which has constant variance over the values of the independent variable in which the model is valid. Another way of stating Definition 2.1 is

$$Y \sim N\left(\beta_0 + \beta_1 X, \sigma^2\right).$$

Since this model is a special case of the simple linear regression model from Definition 1.1, all of the results from the previous chapter still apply to the simple linear regression model with normal error terms. As before, for the n data pairs $(X_1, Y_1), (X_2, Y_2), \ldots, (X_n, Y_n)$, the model becomes

$$Y_i = \beta_0 + \beta_1 X_i + \varepsilon_i$$

for $i = 1, 2, \ldots, n$, where $\varepsilon_1, \varepsilon_2, \ldots, \varepsilon_n$ are mutually independent and identically distributed $N\left(0, \sigma^2\right)$ random variables. The geometry associated with this model is shown in Figure 2.1. The *model* regression line (not the estimated regression line) $E[Y] = \beta_0 + \beta_1 X$ is shown with a negative slope. There are $n = 4$ data pairs collected from this simple linear regression model with normal error terms. The probability density function of each of the Y_i values, rotated clockwise by $90°$ highlights the fact that the population error distribution is normal with a population variance that does not change from one data pair to the next. The geometry illustrated here indicates how a simulation of a simple linear regression model with normal error terms is conducted. Once an X_i value has been established, a Y_i value is generated as $Y_i \sim N\left(\beta_0 + \beta_1 X_i, \sigma^2\right)$, for $i = 1, 2, \ldots, n$. A realization of four data pairs $(X_1, Y_1), (X_2, Y_2), (X_3, Y_3), (X_4, Y_4)$ is given by the points plotted in Figure 2.1. The *estimated* regression line $\hat{Y} = \hat{\beta}_0 + \hat{\beta}_1 X$ can be calculated from these four data pairs in the usual fashion.

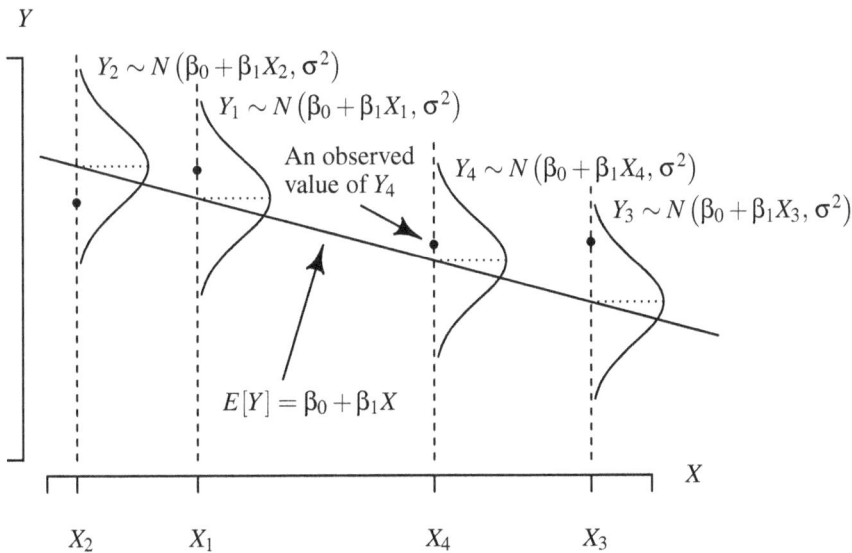

Figure 2.1: Simple linear regression model with normal error terms.

2.2 Maximum Likelihood Estimators

Since we have now specified a parametric distribution for the error terms, maximum likelihood estimation can be used to determine parameter estimates for β_0, β_1, and σ^2. As seen in the next result, the news is good. The maximum likelihood estimators for β_0 and β_1 are identical to the least squares estimators and the maximum likelihood estimator for σ^2 differs from the associated least squares estimator by a constant factor.

Theorem 2.1 Under the simple linear regression model with normal error terms, the maximum likelihood estimators for β_1, β_0, and σ^2 are

$$\hat{\beta}_1 = \frac{S_{XY}}{S_{XX}},$$

$$\hat{\beta}_0 = \bar{Y} - \hat{\beta}_1 \bar{X},$$

and

$$\hat{\sigma}^2 = \frac{1}{n} \sum_{i=1}^{n} e_i^2,$$

where

$$S_{XY} = \sum_{i=1}^{n} (X_i - \bar{X})(Y_i - \bar{Y}), \qquad S_{XX} = \sum_{i=1}^{n} (X_i - \bar{X})^2, \qquad \text{and} \qquad e_i = Y_i - \hat{\beta}_0 - \hat{\beta}_1 X_i$$

for $i = 1, 2, \ldots, n$, when $\sum_{i=1}^{n} e_i^2 > 0$.

Proof Since $Y_i \sim N\left(\beta_0 + \beta_1 X_i, \sigma^2\right)$ for $i = 1, 2, \ldots, n$ under the assumption of normally distributed errors, the likelihood function is

$$L\left(\beta_0, \beta_1, \sigma^2\right) = \prod_{i=1}^{n} \frac{1}{\sqrt{2\pi\sigma^2}} e^{-(Y_i - \beta_0 - \beta_1 X_i)^2/(2\sigma^2)}$$

$$= \left(2\pi\sigma^2\right)^{-n/2} e^{-\sum_{i=1}^{n}(Y_i - \beta_0 - \beta_1 X_i)^2/(2\sigma^2)}.$$

The log likelihood function is

$$\ln L\left(\beta_0, \beta_1, \sigma^2\right) = -\frac{n}{2} \ln\left(2\pi\sigma^2\right) - \frac{1}{2\sigma^2} \sum_{i=1}^{n} (Y_i - \beta_0 - \beta_1 X_i)^2.$$

The score vector consists of the partial derivatives of the log likelihood function with respect to the unknown parameters β_0, β_1, and σ^2. Its components are

$$\frac{\partial \ln L\left(\beta_0, \beta_1, \sigma^2\right)}{\partial \beta_0} = \frac{1}{\sigma^2} \sum_{i=1}^{n} (Y_i - \beta_0 - \beta_1 X_i),$$

$$\frac{\partial \ln L\left(\beta_0, \beta_1, \sigma^2\right)}{\partial \beta_1} = \frac{1}{\sigma^2} \sum_{i=1}^{n} (Y_i - \beta_0 - \beta_1 X_i) X_i,$$

$$\frac{\partial \ln L\left(\beta_0, \beta_1, \sigma^2\right)}{\partial \sigma^2} = -\frac{n}{2\sigma^2} + \frac{1}{2\sigma^4} \sum_{i=1}^{n} (Y_i - \beta_0 - \beta_1 X_i)^2.$$

Equating the first two elements of the score vector to zero, simplifying, and using the hat notation to denote the maximum likelihood estimators results in the normal equations

$$n\hat{\beta}_0 + \hat{\beta}_1 \sum_{i=1}^{n} X_i = \sum_{i=1}^{n} Y_i$$

$$\hat{\beta}_0 \sum_{i=1}^{n} X_i + \hat{\beta}_1 \sum_{i=1}^{n} X_i^2 = \sum_{i=1}^{n} X_i Y_i,$$

which are identical to those in Theorem 1.1. So the maximum likelihood estimators $\hat{\beta}_0$ and $\hat{\beta}_1$ are identical to the associated least squares estimators from Theorem 1.1. Equating the third element of the score vector to zero and solving for $\hat{\sigma}^2$ results in

$$\hat{\sigma}^2 = \frac{1}{n} \sum_{i=1}^{n} \left(Y_i - \hat{\beta}_0 - \hat{\beta}_1 X_i \right)^2 = \frac{1}{n} \sum_{i=1}^{n} e_i^2.$$

Next, determine whether these maximum likelihood estimators maximize the log likelihood function. The symmetric Hessian matrix $\mathbf{H}$ is the matrix of second partial derivatives with respect to the parameters β_0, β_1, and σ^2:

$$\mathbf{H} = \begin{bmatrix} \dfrac{\partial^2 \ln L\left(\beta_0, \beta_1, \sigma^2\right)}{\partial \beta_0^2} & \dfrac{\partial^2 \ln L\left(\beta_0, \beta_1, \sigma^2\right)}{\partial \beta_0 \partial \beta_1} & \dfrac{\partial^2 \ln L\left(\beta_0, \beta_1, \sigma^2\right)}{\partial \beta_0 \partial \sigma^2} \\[3mm] \dfrac{\partial^2 \ln L\left(\beta_0, \beta_1, \sigma^2\right)}{\partial \beta_1 \partial \beta_0} & \dfrac{\partial^2 \ln L\left(\beta_0, \beta_1, \sigma^2\right)}{\partial \beta_1^2} & \dfrac{\partial^2 \ln L\left(\beta_0, \beta_1, \sigma^2\right)}{\partial \beta_1 \partial \sigma^2} \\[3mm] \dfrac{\partial^2 \ln L\left(\beta_0, \beta_1, \sigma^2\right)}{\partial \sigma^2 \partial \beta_0} & \dfrac{\partial^2 \ln L\left(\beta_0, \beta_1, \sigma^2\right)}{\partial \sigma^2 \partial \beta_1} & \dfrac{\partial^2 \ln L\left(\beta_0, \beta_1, \sigma^2\right)}{\partial \left(\sigma^2\right)^2} \end{bmatrix}.$$

Taking the second partial derivatives results in the Hessian matrix

$$\mathbf{H} = \begin{bmatrix} -\dfrac{n}{\sigma^2} & -\dfrac{1}{\sigma^2} \sum_{i=1}^{n} X_i & -\dfrac{1}{\sigma^4} \sum_{i=1}^{n} (Y_i - \beta_0 - \beta_1 X_i) \\[3mm] -\dfrac{1}{\sigma^2} \sum_{i=1}^{n} X_i & -\dfrac{1}{\sigma^2} \sum_{i=1}^{n} X_i^2 & -\dfrac{1}{\sigma^4} \sum_{i=1}^{n} (Y_i - \beta_0 - \beta_1 X_i) X_i \\[3mm] -\dfrac{1}{\sigma^4} \sum_{i=1}^{n} (Y_i - \beta_0 - \beta_1 X_i) & -\dfrac{1}{\sigma^4} \sum_{i=1}^{n} (Y_i - \beta_0 - \beta_1 X_i) X_i & \dfrac{n}{2\sigma^4} - \dfrac{1}{\sigma^6} \sum_{i=1}^{n} (Y_i - \beta_0 - \beta_1 X_i)^2 \end{bmatrix}.$$

After some simplification, the Hessian matrix evaluated at the maximum likelihood estimators $\hat{\beta}_0$, $\hat{\beta}_1$, and $\hat{\sigma}^2$ is

$$\mathbf{H} = \begin{bmatrix} -\dfrac{n^2}{SSE} & -\dfrac{n \sum_{i=1}^{n} X_i}{SSE} & 0 \\[3mm] -\dfrac{n \sum_{i=1}^{n} X_i}{SSE} & -\dfrac{n \sum_{i=1}^{n} X_i^2}{SSE} & 0 \\[3mm] 0 & 0 & -\dfrac{n^3}{2(SSE)^2} \end{bmatrix}$$

using the value of the maximum likelihood estimator for σ^2

$$\hat{\sigma}^2 = \frac{SSE}{n} = \frac{1}{n} \sum_{i=1}^{n} e_i^2 = \frac{1}{n} \sum_{i=1}^{n} \left(Y_i - \hat{Y}_i \right)^2 = \frac{1}{n} \sum_{i=1}^{n} \left(Y_i - \hat{\beta}_0 - \hat{\beta}_1 X_i \right)^2,$$

two of the results from Theorem 1.6, and the definition of the sum of squares for error as $SSE = \sum_{i=1}^{n} e_i^2$ from Theorem 1.8. If $\mathbf{H}$ is a negative definite matrix when evaluated at the maximum likelihood estimators, then the maximum likelihood estimators maximize the likelihood function. In order to show that $\mathbf{H}$ is a negative definite matrix when evaluated at the maximum likelihood estimators $\hat{\beta}_0$, $\hat{\beta}_1$, and $\hat{\sigma}^2$, it is sufficient to show that the leading principal minors of $\mathbf{H}$ alternate in sign in the following fashion: negative, positive, negative. The first leading principal minor is the upper-left hand entry, which is negative when $SSE > 0$ by inspection. The second leading principal minor is the determinant of the upper-left-hand 2×2 submatrix of $\mathbf{H}$, which is

$$\frac{n^3 \sum_{i=1}^n X_i^2}{(SSE)^2} - \frac{n^2 \left(\sum_{i=1}^n X_i\right)^2}{(SSE)^2}$$

and is positive by the Cauchy–Schwartz inequality when $SSE > 0$. (For details, see the proof to Theorem 1.1.) The third leading principal minor is the determinant of $\mathbf{H}$. Taking advantage of the elements in $\mathbf{H}$ which are zero, the determinant of $\mathbf{H}$ is the lower-right element of $\mathbf{H}$ (which is negative when $SSE > 0$) multiplied by the second leading principal minor. Thus, the determinant of $\mathbf{H}$ is negative when evaluated at the maximum likelihood estimators. Since the leading principal minors of $\mathbf{H}$ are negative, positive, and negative when evaluated at the maximum likelihood estimators, $\mathbf{H}$ is a negative definite matrix. Hence, the maximum likelihood estimators $\hat{\beta}_0$, $\hat{\beta}_1$, and $\hat{\sigma}^2$ maximize the likelihood function.   □

The restriction that $SSE > 0$ in Theorem 2.1 is not a particularly restrictive assumption in practice. The only way to achieve a sum of squares for error of zero is to have all of the data pairs fall on a line. If this is indeed the case, then it is possible that a deterministic, rather than a statistical model, is appropriate.

The fact that the least squares estimators and maximum likelihood estimators for β_0 and β_1 are identical is welcome news. Since both techniques give the same values for $\hat{\beta}_0$ and $\hat{\beta}_1$, there is no lingering doubt as to which technique is appropriate for a particular modeling situation. But there is a slight difference between the estimators for σ^2. In the previous section, the sum of squares for error was divided by the appropriate degrees of freedom to arrive at the following unbiased estimator for σ^2:

$$\hat{\sigma}^2 = \frac{1}{n-2} \sum_{i=1}^n e_i^2,$$

or $\hat{\sigma}^2 = SSE/(n-2)$. On the other hand, the maximum likelihood estimator for σ^2 uses a similar formula, but with an n rather than an $n-2$ in the denominator. For large n, the difference is slight. But for small n, the difference can be significant. For the $n = 3$ sales data pairs first introduced in Example 1.3 with the variance of the error terms estimated in Example 1.10, for instance, the unbiased estimate of σ^2 is $\hat{\sigma}^2 = 14$, whereas the maximum likelihood estimate of σ^2 is $\hat{\sigma}^2 = 14/3$. The standard practice in regression analysis is to use the unbiased estimator. In general, maximum likelihood estimators are not guaranteed to be unbiased, although they are consistent and asymptotically efficient. For the simple linear regression model with normal error terms, the maximum likelihood estimators for the slope and intercept are unbiased, but the maximum likelihood estimator for σ^2 is biased.

In a more advanced course on regression, you will prove that the maximum likelihood estimators for the population intercept β_0 and the population slope β_1 (which are the same as the least squares estimators) are consistent, sufficient, and efficient. The property of consistency indicates that the estimators will converge to the associated population values as $n \to \infty$; symbolically,

$$\lim_{n \to \infty} P\left(\left|\hat{\beta}_0 - \beta_0\right| < \delta\right) = 1$$

and

$$\lim_{n \to \infty} P\left(\left|\hat{\beta}_1 - \beta_1\right| < \delta\right) = 1$$

for any $\delta > 0$. The property of sufficiency indicates that all of the information concerning the estimation of β_0 and β_1 is encapsulated in $\hat{\beta}_0$ and $\hat{\beta}_1$, respectively. The property of efficiency indicates that $\hat{\beta}_0$ and $\hat{\beta}_1$ have the smallest possible population variance among all unbiased estimators for β_0 and β_1, respectively.

2.3   Inference in Simple Linear Regression

Inference concerning the parameters in the simple linear regression model with normal error terms, which usually is performed in terms of constructing confidence intervals and performing hypothesis tests, is considered in this section. The following three subsections consider the sampling distributions of $\hat{\beta}_0$, $\hat{\beta}_1$, and $\hat{\sigma}^2$ under the simple linear regression model with normal error terms. We begin with σ^2.

2.3.1   Inference Concerning σ^2

Even though statistical inference concerning σ^2 typically has the least interest of the three parameters in simple linear regression, there is an important result concerning the probability distribution of SSE/σ^2 that is critical to the derivation of other results, so it is taken up first.

Theorem 2.2 Under the simple linear regression model with normal error terms,

$$\frac{SSE}{\sigma^2} \sim \chi^2(n-2),$$

and is independent of $\hat{\beta}_0$ and $\hat{\beta}_1$.

Proof (outline only) The proof of Theorem 1.7 includes the results

$$Y_i - \bar{Y} = \beta_1 \left(X_i - \bar{X} \right) + \varepsilon_i - \bar{\varepsilon} \tag{4}$$

for $i = 1, 2, \ldots, n$ and

$$\sum_{i=1}^{n} e_i^2 = \left(\hat{\beta}_1 - \beta_1 \right)^2 \sum_{i=1}^{n} \left(X_i - \bar{X} \right)^2 - 2\left(\hat{\beta}_1 - \beta_1 \right) \sum_{i=1}^{n} \left(X_i - \bar{X} \right) \left(\varepsilon_i - \bar{\varepsilon} \right) + \sum_{i=1}^{n} \left(\varepsilon_i - \bar{\varepsilon} \right)^2. \tag{5}$$

Using the fact that $\hat{\beta}_1 = S_{XY}/S_{XX}$ and equation (4),

$$\hat{\beta}_1 \sum_{i=1}^{n} (X_i - \bar{X})^2 = \frac{S_{XY}}{S_{XX}} \cdot S_{XX}$$

$$= S_{XY}$$

$$= \sum_{i=1}^{n} (X_i - \bar{X})(Y_i - \bar{Y})$$

$$= \sum_{i=1}^{n} (X_i - \bar{X})\left(\beta_1 (X_i - \bar{X}) + \varepsilon_i - \bar{\varepsilon} \right)$$

$$= \beta_1 \sum_{i=1}^{n} (X_i - \bar{X})^2 + \sum_{i=1}^{n} (X_i - \bar{X})(\varepsilon_i - \bar{\varepsilon}).$$

Solving for $\sum_{i=1}^{n} (X_i - \bar{X})(\varepsilon_i - \bar{\varepsilon})$ gives

$$\sum_{i=1}^{n} (X_i - \bar{X})(\varepsilon_i - \bar{\varepsilon}) = \left(\hat{\beta}_1 - \beta_1\right) \sum_{i=1}^{n} (X_i - \bar{X})^2.$$

Substituting this into equation (5) and rearranging gives

$$\sum_{i=1}^{n} e_i^2 = \sum_{i=1}^{n} (\varepsilon_i - \bar{\varepsilon})^2 + \left(\hat{\beta}_1 - \beta_1\right)^2 \sum_{i=1}^{n} (X_i - \bar{X})^2 - 2\left(\hat{\beta}_1 - \beta_1\right)^2 \sum_{i=1}^{n} (X_i - \bar{X})^2$$

or

$$SSE = \sum_{i=1}^{n} \varepsilon_i^2 - n\bar{\varepsilon}^2 - \left(\hat{\beta}_1 - \beta_1\right)^2 \sum_{i=1}^{n} (X_i - \bar{X})^2.$$

Dividing both sides of this equation by σ^2 and rearranging gives

$$\sum_{i=1}^{n} \left(\frac{\varepsilon_i}{\sigma}\right)^2 = \frac{SSE}{\sigma^2} + \left(\frac{\bar{\varepsilon}}{\sigma/\sqrt{n}}\right)^2 + \left(\frac{\hat{\beta}_1 - \beta_1}{\sigma/\sqrt{S_{XX}}}\right)^2. \tag{6}$$

Since $\varepsilon_1, \varepsilon_2, \ldots, \varepsilon_n$ are mutually independent and identically distributed error terms under the simple linear regression model with normal error terms, $\varepsilon_i \sim N\left(0, \sigma^2\right)$ for $i = 1, 2, \ldots, n$. A well-known result from probability theory concerning sample means indicates that $\bar{\varepsilon} \sim N\left(0, \sigma^2/n\right)$. Furthermore, since $\hat{\beta}_1 \sim N\left(\beta_1, \sigma^2/S_{XX}\right)$ via Theorem 2.4, it can be seen that the three random quantities in parentheses in equation (6) are normally distributed random variables which have been standardized by subtracting their population means and dividing by their population standard deviations. Thus, the three random quantities in parentheses are standard normal random variables. The square of a standard normal random variable has the chi-square distribution with one degree of freedom, so the last two terms on the right-hand side of equation (4) have the chi-square distribution with one degree of freedom. Also, since the sum of n mutually independent chi-square random variables also has the chi-square distribution with n degrees of freedom, the left-hand side of equation (6) is $\chi^2(n)$. Since the sum of mutually independent chi-square random variables also has the chi-square distribution (with degrees of freedom summing),

$$\frac{SSE}{\sigma^2} \sim \chi^2(n-2).$$

The part of this proof that is incomplete is proving that the four terms in equation (6) are mutually independent, which is left as an exercise.   □

    As an illustration of the use of Theorem 2.2, the derivation that follows develops an exact two-sided $100(1-\alpha)\%$ confidence interval for σ^2. Under the simple linear regression model with normal error terms, Theorem 2.2 states that

$$\frac{SSE}{\sigma^2} \sim \chi^2(n-2).$$

For some α value between 0 and 1, placing an area of $\alpha/2$ in each tail of the chi-square distribution with $n-2$ degrees of freedom gives

$$P\left(\chi^2_{n-2, 1-\alpha/2} < \frac{SSE}{\sigma^2} < \chi^2_{n-2, \alpha/2}\right) = 1 - \alpha,$$

where the second value in the subscripts corresponds to right-hand tail probabilities. Rearranging the inequality to isolate σ^2 in the center of the inequality gives an exact two-sided $100(1-\alpha)\%$ confidence interval for σ^2 as

$$\frac{SSE}{\chi^2_{n-2,\alpha/2}} < \sigma^2 < \frac{SSE}{\chi^2_{n-2,1-\alpha/2}}.$$

This derivation is a proof of the following theorem.

Theorem 2.3 Under the simple linear regression model with normal error terms,

$$\frac{SSE}{\chi^2_{n-2,\alpha/2}} < \sigma^2 < \frac{SSE}{\chi^2_{n-2,1-\alpha/2}}$$

is an exact two-sided $100(1-\alpha)\%$ confidence interval for σ^2.

Example 2.1 Consider again the $n = 17$ data pairs in the Forbes data set that was introduced in Example 1.11, where the independent variable X is the boiling point of water in degrees Fahrenheit and the dependent variable Y is the adjusted barometric pressure in inches of mercury. Give a point estimate and a 95% confidence interval for σ^2.

Using the calculations in Example 1.13, the unbiased point estimate for σ^2 is

$$\hat{\sigma}^2 = MSE = \frac{SSE}{n-2} = \frac{0.8131}{17-2} = 0.05421.$$

Since the scatterplot in Figure 1.18 showed that the data pairs fall very close to the estimated regression line, we expect a narrow confidence interval for σ^2. Using the formula from Theorem 2.3, an exact two-sided 95% confidence interval for σ^2 is

$$\frac{SSE}{\chi^2_{n-2,\alpha/2}} < \sigma^2 < \frac{SSE}{\chi^2_{n-2,1-\alpha/2}}$$

or

$$\frac{0.8131}{27.4884} < \sigma^2 < \frac{0.8131}{6.2621}$$

or

$$0.02958 < \sigma^2 < 0.1299.$$

This confidence interval is nonsymmetric about the point estimate because the chi-square distribution is not a symmetric probability distribution, and the quantiles appear in the denominators of the confidence interval formula. The width of the confidence interval is controlled by two factors: the sample size n and the vertical distances that the data pairs stray from the estimated regression line. Larger values of n result in narrower confidence intervals; data pairs that lie close to the regression line result in narrower confidence intervals.

The R code to compute the point and interval estimates follows. It uses the lm function to fit the simple linear regression model and the qchisq function to calculate the quantiles of the chi-square distribution. Notice that qchisq uses left-hand-tail probabilities whereas our formulas use right-hand-tail probabilities when computing quantiles.

```
library(MASS)
x     = forbes$bp
y     = forbes$pres
n     = length(x)
fit   = lm(y ~ x)
sse   = sum(fit$residuals ^ 2)
mse   = sse / (n - 2)
alpha = 0.05
lo    = sse / qchisq(1 - alpha / 2, n - 2)
hi    = sse / qchisq(alpha / 2, n - 2)
print(c(lo, mse, hi))
```

2.3.2   Inference Concerning β_1

In order to perform statistical inference concerning the population slope of the regression line β_1, it is first necessary to establish the sampling distribution of the estimator $\hat{\beta}_1$.

Since the error terms $\varepsilon_1, \varepsilon_2, \ldots, \varepsilon_n$ are mutually independent and identically distributed $N\left(0, \sigma^2\right)$ random variables under the simple linear regression model with normal error terms from Definition 2.1, the associated dependent variables $Y_1, Y_2, \ldots, Y_n$ are also mutually independent normally distributed random variables because $Y_i = \beta_0 + \beta_1 X_i + \varepsilon_i$ for $i = 1, 2, \ldots, n$. Furthermore, recall from Theorem 1.3 that $\hat{\beta}_1$ can be written as a linear combination of $Y_1, Y_2, \ldots, Y_n$ as

$$\hat{\beta}_1 = a_1 Y_1 + a_2 Y_2 + \cdots + a_n Y_n.$$

Since a linear combination of mutually independent normally distributed random variables is itself normally distributed, we can conclude that $\hat{\beta}_1$ is normally distributed.

Now that the normality of $\hat{\beta}_1$ has been established, the next step is to find the population mean and population variance of the point estimator $\hat{\beta}_1$, which will completely determine the distribution of $\hat{\beta}_1$. From Theorem 1.2 and Theorem 1.4, the population mean and the population variance of the point estimator $\hat{\beta}_1$ are

$$E\left[\hat{\beta}_1\right] = \beta_1 \qquad \text{and} \qquad V\left[\hat{\beta}_1\right] = \frac{\sigma^2}{S_{XX}}.$$

This establishes the result given in Theorem 2.4.

Theorem 2.4 Under the simple linear regression model with normal error terms,

$$\hat{\beta}_1 \sim N\left(\beta_1, \frac{\sigma^2}{S_{XX}}\right).$$

The usual method for conducting statistical inference on a test statistic that is normally distributed is to subtract the population mean and divide by the population standard deviation. A problem that arises here is that the population variance of $\hat{\beta}_1$ in Theorem 2.4 is not known for a particular set of n data pairs because σ^2 is not known. The population variance of $\hat{\beta}_1$, however, can be estimated by

$$\hat{V}\left[\hat{\beta}_1\right] = \frac{\hat{\sigma}^2}{S_{XX}},$$

where $\hat{\sigma}^2 = MSE = SSE/(n-2)$, which is a quantity that can be estimated from n data pairs. We can now use

$$\frac{\hat{\beta}_1 - \beta_1}{\sqrt{\hat{V}[\hat{\beta}_1]}}$$

as a pivotal quantity in the following result.

Theorem 2.5 Under the simple linear regression model with normal error terms,

$$\frac{\hat{\beta}_1 - \beta_1}{\sqrt{\hat{V}[\hat{\beta}_1]}} \sim t(n-2).$$

Proof This proof is based on the fact that the ratio of a standard normal random variable to the square root of an independent chi-square random variable divided by its degrees of freedom is a t random variable with the same number of degrees of freedom. In the particular setting here with n data pairs drawn from a simple linear regression model with normal error terms, this is

$$\frac{N(0, 1)}{\sqrt{\chi^2(n-2)/(n-2)}} \sim t(n-2),$$

where $N(0, 1)$ denotes a standard normal random variable, $\chi^2(n-2)$ denotes a chi-square random variable with $n-2$ degrees of freedom, and $t(n-2)$ denotes a t random variable with $n-2$ degrees of freedom. The normal and chi-square random variables are assumed to be independent. Begin by dividing the numerator and the denominator of the pivotal quantity by the square root of $V[\hat{\beta}_1] = \sigma^2/S_{XX}$:

$$\frac{\hat{\beta}_1 - \beta_1}{\sqrt{\hat{V}[\hat{\beta}_1]}} = \frac{\dfrac{\hat{\beta}_1 - \beta_1}{\sqrt{V[\hat{\beta}_1]}}}{\sqrt{\dfrac{\hat{V}[\hat{\beta}_1]}{V[\hat{\beta}_1]}}}. \tag{7}$$

Focus initially on the numerator of the right-hand side of equation (7). Because Theorem 2.4 states that $\hat{\beta}_1$ has a normal distribution with population mean β_1 and population standard deviation $\sqrt{V[\hat{\beta}_1]} = \sqrt{\sigma^2/S_{XX}}$, the numerator is a normal random variable minus its mean, divided by its standard deviation. Thus, the numerator of the right-hand side of equation (7) is a $N(0, 1)$ random variable. In other words,

$$\frac{\hat{\beta}_1 - \beta_1}{\sqrt{V[\hat{\beta}_1]}} \sim N(0, 1).$$

The focus now shifts to the denominator of the right-hand side of equation (7). Since $V[\hat{\beta}_1] = \sigma^2/S_{XX}$ is estimated by

$$\hat{V}[\hat{\beta}_1] = \frac{\hat{\sigma}^2}{S_{XX}} = \frac{SSE}{(n-2)S_{XX}},$$

the denominator of the right-hand side of equation (7) can be written as

$$\sqrt{\frac{\hat{V}\left[\hat{\beta}_1\right]}{V\left[\hat{\beta}_1\right]}} = \sqrt{\frac{\dfrac{SSE}{(n-2)S_{XX}}}{\dfrac{\sigma^2}{S_{XX}}}} = \sqrt{\frac{SSE}{(n-2)\sigma^2}} \sim \sqrt{\chi^2(n-2)/(n-2)}$$

because $SSE/\sigma^2 \sim \chi^2(n-2)$ and is independent of $\hat{\beta}_0$ and $\hat{\beta}_1$ by Theorem 2.2. Since the numerator of equation (7) is a standard normal random variable and the denominator is the square root of an independent chi-square random variable with $n-2$ degrees of freedom divided by its degrees of freedom, the pivotal quantity

$$\frac{\hat{\beta}_1 - \beta_1}{\sqrt{\hat{V}\left[\hat{\beta}_1\right]}} \sim t(n-2). \hspace{3cm} \square$$

Theorem 2.5 can be used to construct confidence intervals and perform hypothesis tests concerning β_1. In many applications, β_1 is the key parameter in the regression analysis because statistical evidence showing that it differs from zero indicates a linear relationship between X and Y if the assumptions associated with a simple linear regression model with normal error terms are met.

As an illustration, an exact two-sided $100(1-\alpha)\%$ confidence interval for β_1 is developed as follows. Theorem 2.5 states that

$$\frac{\hat{\beta}_1 - \beta_1}{\sqrt{\hat{V}\left[\hat{\beta}_1\right]}} \sim t(n-2).$$

For some α between 0 and 1, placing an area of $\alpha/2$ in each tail of the t distribution with $n-2$ degrees of freedom gives

$$P\left(-t_{n-2,\alpha/2} < \frac{\hat{\beta}_1 - \beta_1}{\sqrt{\hat{V}\left[\hat{\beta}_1\right]}} < t_{n-2,\alpha/2}\right) = 1 - \alpha,$$

where the second value in the subscripts corresponds to right-hand tail probabilities. Rearranging the inequality to isolate β_1 in the center of the inequality gives an exact two-sided $100(1-\alpha)\%$ confidence interval for β_1 as

$$\hat{\beta}_1 - t_{n-2,\alpha/2}\sqrt{\hat{V}\left[\hat{\beta}_1\right]} < \beta_1 < \hat{\beta}_1 + t_{n-2,\alpha/2}\sqrt{\hat{V}\left[\hat{\beta}_1\right]},$$

where

$$\hat{\beta}_1 = \frac{S_{XY}}{S_{XX}} \hspace{1cm} \text{and} \hspace{1cm} \hat{V}\left[\hat{\beta}_1\right] = \frac{\hat{\sigma}^2}{S_{XX}} = \frac{MSE}{S_{XX}}.$$

This constitutes a derivation of the following theorem.

Theorem 2.6 Under the simple linear regression model with normal error terms,

$$\hat{\beta}_1 - t_{n-2,\alpha/2}\sqrt{\hat{V}\left[\hat{\beta}_1\right]} < \beta_1 < \hat{\beta}_1 + t_{n-2,\alpha/2}\sqrt{\hat{V}\left[\hat{\beta}_1\right]}$$

is an exact two-sided $100(1-\alpha)\%$ confidence interval for β_1.

Example 2.2 Calculate a point estimate and an exact two-sided 95% confidence interval for the population slope β_1 for the Forbes data set from Example 1.11.

The unbiased point estimate for β_1 is

$$\hat{\beta}_1 = \frac{S_{XY}}{S_{XX}} = \frac{277.5}{530.8} = 0.5229.$$

The barometric pressure increases by an estimated 0.5229 inches of mercury for every degree increase in the boiling point of water over the range of values collected by Forbes. This value is reported to four-digit accuracy because that was the number of digits given in the data pairs. As was seen in the scatterplot in Figure 1.18, the $n = 17$ data pairs cluster tightly about the regression line, so we expect a fairly narrow confidence interval for β_1 even though the sample size is moderate. Using the formula from Theorem 2.6, an exact two-sided confidence interval for β_1 is

$$\hat{\beta}_1 - t_{n-2,\alpha/2}\sqrt{\frac{MSE}{S_{XX}}} < \beta_1 < \hat{\beta}_1 + t_{n-2,\alpha/2}\sqrt{\frac{MSE}{S_{XX}}}$$

or

$$0.5229 - 2.131\sqrt{\frac{0.05421}{530.8}} < \beta_1 < 0.5229 + 2.131\sqrt{\frac{0.05421}{530.8}}$$

or

$$0.5014 < \beta_1 < 0.5444.$$

Unlike the confidence interval for σ^2, this confidence interval is symmetric about the point estimate. The R code to compute the point and interval estimates is given below. The lm function fits the simple linear regression model and the qt function calculates the quantiles of the appropriate t distribution.

```
library(MASS)
x         = forbes$bp
y         = forbes$pres
n         = length(x)
xbar      = mean(x)
sxx       = sum((x - xbar) ^ 2)
fit       = lm(y ~ x)
beta1hat  = fit$coefficients[2]
sse       = sum(fit$residuals ^ 2)
mse       = sse / (n - 2)
alpha     = 0.05
lo        = beta1hat - qt(1 - alpha / 2, n - 2) * sqrt(mse / sxx)
hi        = beta1hat + qt(1 - alpha / 2, n - 2) * sqrt(mse / sxx)
print(c(lo, beta1hat, hi))
```

Statisticians perform these calculations so often that R has a built-in confint function to calculate the bounds of the confidence interval, as illustrated below. The first argument is the name of the fitted regression model, the second argument is the name of the parameter being estimated, and the third argument, which defaults to 0.95, is the confidence level.

```
library(MASS)
x   = forbes$bp
y   = forbes$pres
fit = lm(y ~ x)
confint(fit, "x", level = 0.95)
```

The hypothesis test concerning β_1 with the null hypothesis

$$H_0 : \beta_1 = \beta_1^\star$$

is based on the test statistic

$$\frac{\hat{\beta}_1 - \beta_1^\star}{\sqrt{\hat{V}\left[\hat{\beta}_1\right]}},$$

which has the t distribution with $n-2$ degrees of freedom under H_0 and the simple linear regression model with normal errors. The most common value for $\beta_1^\star$ in the null hypothesis is $\beta_1^\star = 0$, which tests whether the estimated slope of the regression line $\hat{\beta}_1$ differs significantly from zero. This type of hypothesis test concerning β_1 will be illustrated later in this chapter.

2.3.3   Inference Concerning β_0

In order to perform statistical inference concerning the population intercept of the regression line β_0, it is first necessary to establish the sampling distribution of $\hat{\beta}_0$.

Since the error terms $\varepsilon_1, \varepsilon_2, \ldots, \varepsilon_n$ are mutually independent and identically distributed $N\left(0, \sigma^2\right)$ random variables under the simple linear regression model with normal error terms from Definition 2.1, the associated dependent variables $Y_1, Y_2, \ldots, Y_n$ are also mutually independent normally distributed random variables. Furthermore, recall from Theorem 1.3 that $\hat{\beta}_0$ can be written as a linear combination of $Y_1, Y_2, \ldots, Y_n$ as

$$\hat{\beta}_0 = c_1 Y_1 + c_2 Y_2 + \cdots + c_n Y_n.$$

Since a linear combination of mutually independent normally distributed random variables is itself normally distributed, we can conclude that $\hat{\beta}_0$ is normally distributed.

Now that the normality of $\hat{\beta}_0$ has been established, the next step is to find the population mean and population variance of the point estimator $\hat{\beta}_0$, which will completely determine the distribution of $\hat{\beta}_0$. From Theorem 1.2 and Theorem 1.4, the population mean and the population variance of the point estimator $\hat{\beta}_0$ are

$$E\left[\hat{\beta}_0\right] = \beta_0 \qquad \text{and} \qquad V\left[\hat{\beta}_0\right] = \frac{\sigma^2 \sum_{i=1}^{n} X_i^2}{n S_{XX}}.$$

This establishes the result given in Theorem 2.7.

Theorem 2.7 Under the simple linear regression model with normal error terms,

$$\hat{\beta}_0 \sim N\left(\beta_0, \frac{\sigma^2 \sum_{i=1}^{n} X_i^2}{n S_{XX}}\right).$$

The usual method for conducting statistical inference on a test statistic that is normally distributed is to subtract the population mean and divide by the population standard deviation. A problem that arises here is that the population variance of $\hat{\beta}_0$ in Theorem 2.7 is not known for a particular set of n data pairs because σ^2 is not known. The population variance of $\hat{\beta}_0$, however, can be estimated by

$$\hat{V}\left[\hat{\beta}_0\right] = \frac{\hat{\sigma}^2 \sum_{i=1}^n X_i^2}{n S_{XX}},$$

where $\hat{\sigma}^2 = MSE = SSE/(n-2)$, which is a quantity that can be estimated from n data pairs. We can now use

$$\frac{\hat{\beta}_0 - \beta_0}{\sqrt{\hat{V}\left[\hat{\beta}_0\right]}}$$

as a pivotal quantity in the following result.

Theorem 2.8 Under the simple linear regression model with normal error terms,

$$\frac{\hat{\beta}_0 - \beta_0}{\sqrt{\hat{V}\left[\hat{\beta}_0\right]}} \sim t(n-2).$$

Proof This proof is based on the fact that the ratio of a standard normal random variable to the square root of an independent chi-square random variable divided by its degrees of freedom is a t random variable with the same number of degrees of freedom. In the particular setting here with n data pairs drawn from a simple linear regression model with normal error terms, this is

$$\frac{N(0,1)}{\sqrt{\chi^2(n-2)/(n-2)}} \sim t(n-2),$$

where $N(0,1)$ denotes a standard normal random variable, $\chi^2(n-2)$ denotes a chi-square random variable with $n-2$ degrees of freedom, and $t(n-2)$ denotes a t random variable with $n-2$ degrees of freedom. Begin by dividing the numerator and the denominator of the pivotal quantity by the square root of $V\left[\hat{\beta}_0\right] = \sigma^2 \sum_{i=1}^n X_i^2/(n S_{XX})$:

$$\frac{\hat{\beta}_0 - \beta_0}{\sqrt{\hat{V}\left[\hat{\beta}_0\right]}} = \frac{\dfrac{\hat{\beta}_0 - \beta_0}{\sqrt{V\left[\hat{\beta}_0\right]}}}{\sqrt{\dfrac{\hat{V}\left[\hat{\beta}_0\right]}{V\left[\hat{\beta}_0\right]}}}. \tag{8}$$

Focus initially on the numerator of the right-hand side of equation (8). Because Theorem 2.7 states that $\hat{\beta}_0$ has a normal distribution with population mean β_0 and population standard deviation $\sqrt{V\left[\hat{\beta}_0\right]} = \sqrt{\sigma^2 \sum_{i=1}^n X_i^2/(n S_{XX})}$, the numerator is a normal random variable minus its mean, divided by its standard deviation. Thus, the numerator of the right-hand side of equation (8) is a $N(0,1)$ random variable. In other words,

$$\frac{\hat{\beta}_0 - \beta_0}{\sqrt{V\left[\hat{\beta}_0\right]}} \sim N(0,1).$$

The focus now shifts to the denominator of the right-hand side of equation (8). Since $V\left[\hat{\beta}_0\right] = \sigma^2 \sum_{i=1}^n X_i^2/(nS_{XX})$ is estimated by

$$\hat{V}\left[\hat{\beta}_0\right] = \frac{\hat{\sigma}^2 \sum_{i=1}^n X_i^2}{nS_{XX}} = \frac{SSE \sum_{i=1}^n X_i^2}{n(n-2)S_{XX}},$$

the denominator of the right-hand side of equation (8) can be written as

$$\sqrt{\frac{\hat{V}\left[\hat{\beta}_0\right]}{V\left[\hat{\beta}_0\right]}} = \sqrt{\frac{\frac{SSE \sum_{i=1}^n X_i^2}{n(n-2)S_{XX}}}{\frac{\sigma^2 \sum_{i=1}^n X_i^2}{nS_{XX}}}} = \sqrt{\frac{SSE}{(n-2)\sigma^2}} \sim \sqrt{\chi^2(n-2)/(n-2)}$$

because $SSE/\sigma^2 \sim \chi^2(n-2)$ and is independent of $\hat{\beta}_0$ and $\hat{\beta}_1$ by Theorem 2.2. Since the numerator of equation (8) is a standard normal random variable and the denominator is the square root of an independent chi-square random variable with $n-2$ degrees of freedom divided by its degrees of freedom, the pivotal quantity

$$\frac{\hat{\beta}_0 - \beta_0}{\sqrt{\hat{V}\left[\hat{\beta}_0\right]}} \sim t(n-2). \qquad\qquad \Box$$

Theorem 2.8 can be used to construct confidence intervals and perform hypothesis tests concerning β_0. In many applications, there is an interest in whether β_0 is statistically different from 0. The results of this hypothesis test and the particular setting for the simple linear regression model indicate whether forcing a simple linear regression model through the origin is appropriate.

As an illustration, an exact two-sided $100(1-\alpha)\%$ confidence interval for β_0 is developed as follows. Theorem 2.8 states that

$$\frac{\hat{\beta}_0 - \beta_0}{\sqrt{\hat{V}\left[\hat{\beta}_0\right]}} \sim t(n-2).$$

For some α between 0 and 1, placing an area of $\alpha/2$ in each tail of the t distribution with $n-2$ degrees of freedom gives

$$P\left(-t_{n-2,\alpha/2} < \frac{\hat{\beta}_0 - \beta_0}{\sqrt{\hat{V}\left[\hat{\beta}_0\right]}} < t_{n-2,\alpha/2}\right) = 1 - \alpha,$$

where the second value in the subscripts corresponds to right-hand tail probabilities. Rearranging the inequality to isolate β_0 in the center of the inequality gives an exact two-sided $100(1-\alpha)\%$ confidence interval for β_0 as

$$\hat{\beta}_0 - t_{n-2,\alpha/2}\sqrt{\hat{V}\left[\hat{\beta}_0\right]} < \beta_0 < \hat{\beta}_0 + t_{n-2,\alpha/2}\sqrt{\hat{V}\left[\hat{\beta}_0\right]},$$

where

$$\hat{\beta}_0 = \bar{Y} - \hat{\beta}_1\bar{X} \qquad \text{and} \qquad \hat{V}\left[\hat{\beta}_0\right] = \frac{\hat{\sigma}^2 \sum_{i=1}^n X_i^2}{nS_{XX}} = \frac{MSE \sum_{i=1}^n X_i^2}{nS_{XX}}.$$

This constitutes a derivation of the following theorem.

Theorem 2.9 Under the simple linear regression model with normal error terms,

$$\hat{\beta}_0 - t_{n-2,\,\alpha/2}\sqrt{\hat{V}\left[\hat{\beta}_0\right]} < \beta_0 < \hat{\beta}_0 + t_{n-2,\,\alpha/2}\sqrt{\hat{V}\left[\hat{\beta}_0\right]}$$

is an exact two-sided $100(1-\alpha)\%$ confidence interval for β_0.

Example 2.3 Calculate a point estimate and an exact two-sided 95% confidence interval for β_0 for the Forbes data set from Example 1.11.

In this particular application, there is little meaning associated with the parameter β_0. Since the independent variable X is the boiling point of water in degrees Fahrenheit and the dependent variable Y is the associated barometric pressure, the intercept β_0 is interpreted as the barometric pressure when the boiling point is zero degrees Fahrenheit. Since the X_i values range from a minimum of 194.3 to a maximum of 212.2, a boiling point of zero degrees Fahrenheit is way outside of the scope of the model. Nevertheless, to illustrate the mechanics associated with the R code to compute the point and interval estimator, we proceed with the calculations. This also illustrates that just because we *can* perform a calculation does not mean that we *should*. The R code below uses the lm and confint functions to calculate the point and interval estimators for β_0. The first argument to confint is the fitted regression model and the second argument is the name of the parameter being estimated.

```
library(MASS)
x   = forbes$bp
y   = forbes$pres
fit = lm(y ~ x)
print(fit$coefficients[1])
confint(fit, "(Intercept)")
```

The unbiased point estimator of β_0 is displayed by R as

$$\hat{\beta}_0 = -81.06$$

and the exact two-sided 95% confidence interval for β_0 is

$$-85.44 < \beta_0 < -76.69.$$

The reason that the confidence intervals for σ^2 and β_1 are so narrow and this confidence interval is much wider is that $X = 0$ is so far out of the scope of the simple linear regression model with normal error terms. Typing just confint(fit) gives exact 95% confidence intervals for both β_0 and β_1. More realistic applications of statistical inference on β_0 are given later in this chapter.

The hypothesis test concerning β_0 with the null hypothesis

$$H_0 : \beta_0 = \beta_0^\star$$

is based on the test statistic

$$\frac{\hat{\beta}_0 - \beta_0^\star}{\sqrt{\hat{V}\left[\hat{\beta}_0\right]}},$$

which has the t distribution with $n-2$ degrees of freedom under H_0 and the simple linear regression model with normal errors. The most common value for $\beta_0^\star$ in the null hypothesis is $\beta_0^\star = 0$, which is for testing whether the estimated intercept of the regression line $\hat{\beta}_0$ differs significantly from zero. The p-value associated with this hypothesis test and the context associated with the meaning of X and Y might influence a modeler whether or not to fit a simple linear regression model which is forced through the origin.

2.3.4   Inference Concerning $E[Y_h]$

Many applications of simple linear regression require not only point and interval estimates for the regression parameters β_0, β_1, and σ^2, but also a point and interval estimate for the expected value of Y associated with a particular value of X. In this context, the simple linear regression model is being used to *forecast* the conditional expected value of Y from the data pairs. Denote the X-value of interest by X_h, which is a fixed constant that is observed without error within the scope of the simple linear regression model. The associated random Y-value is denoted by Y_h, which has conditional expected value $E[Y_h]$. This compact notation for the conditional expected value is adopted over the more precise $E[Y_h \mid X = X_h]$. If the parameters β_0 and β_1 are known, then the point estimator for $E[Y_h]$ is

$$\hat{Y}_h = \beta_0 + \beta_1 X_h,$$

which is simply the height of the *population* regression line at X_h. In nearly all applications, however, we estimate the parameters β_0, β_1, and σ^2 from the data pairs $(X_1, Y_1), (X_2, Y_2), \ldots, (X_n, Y_n)$. In this case, the point estimator for $E[Y_h]$ is

$$\hat{Y}_h = \hat{\beta}_0 + \hat{\beta}_1 X_h,$$

which is simply the height of the *estimated* regression line at X_h. When the data pairs (X_1, Y_1), $(X_2, Y_2), \ldots, (X_n, Y_n)$ are tightly clustered about the regression line, we expect a fairly precise point estimate for $E[Y_h]$. A more explicit notation for $\hat{Y}_h$ is $\hat{E}[Y_h \mid X = X_h]$ or $\hat{\mu}_{Y_h \mid X = X_h}$. We opt for the more compact $\hat{Y}_h$ and leave it to the reader to mentally convert this to the more explicit meaning.

The value of X_h might correspond to one of $X_1, X_2, \ldots, X_n$, or might correspond to another value of X. It is critical that X_h fall in the scope of the simple linear regression model. If X_h is less than $\min\{X_1, X_2, \ldots, X_n\}$ or greater than $\max\{X_1, X_2, \ldots, X_n\}$, then there should be some evidence, perhaps evidence based on data sets collected previously or evidence provided by experts in the subject matter, that the relationship between X and Y remains linear outside of the scope of the data pairs. Without evidence of this nature, one should not extrapolate beyond the scope of the simple linear regression model.

With the point estimator for $E[Y_h]$ established, we now seek a pivotal quantity which can be used to construct confidence intervals and perform hypothesis tests concerning $E[Y_h]$. We continue to assume that the simple linear regression model with normally distributed error terms is appropriate. Recall from Theorem 1.3 that $\hat{\beta}_0$ and $\hat{\beta}_1$ can be written as written as linear combinations of $Y_1, Y_2, \ldots, Y_n$:

$$\hat{\beta}_0 = c_1 Y_1 + c_2 Y_2 + \cdots + c_n Y_n$$

and

$$\hat{\beta}_1 = a_1 Y_1 + a_2 Y_2 + \cdots + a_n Y_n$$

for constants $c_1, c_2, \ldots, c_n$ and $a_1, a_2, \ldots, a_n$. Furthermore, $Y_1, Y_2, \ldots, Y_n$ are mutually independent random variables because $\varepsilon_1, \varepsilon_2, \ldots, \varepsilon_n$ are mutually independent random variables in the simple linear regression model $Y_i = \beta_0 + \beta_1 X_i + \varepsilon_i$ for $i = 1, 2, \ldots, n$. This implies that $\hat{Y}_h$ can be written as

$$\hat{Y}_h = \hat{\beta}_0 + \hat{\beta}_1 X_h = (c_1 + a_1 X_h) Y_1 + (c_2 + a_2 X_h) Y_2 + \cdots + (c_n + a_n X_h) Y_n.$$

Since a linear combination of mutually independent normally distributed random variables is itself normally distributed, $\hat{Y}_h$ is normally distributed under the simple linear regression model with normal error terms.

Now that the normality of $\hat{Y}_h$ has been established, we seek its population mean and population variance, which will completely define its probability distribution. Since $\hat{\beta}_0$ and $\hat{\beta}_1$ are unbiased estimators of β_0 and β_1, respectively, the population mean of $\hat{Y}_h$ is

$$E\left[\hat{Y}_h\right] = E\left[\hat{\beta}_0 + \hat{\beta}_1 X_h\right] = E\left[\hat{\beta}_0\right] + X_h E\left[\hat{\beta}_1\right] = \beta_0 + \beta_1 X_h$$

via Theorem 1.2. So the point estimator $\hat{Y}_h = \hat{\beta}_0 + \hat{\beta}_1 X_h$ is an unbiased estimator of $Y_h = \beta_0 + \beta_1 X_h$. Next, we calculate the population variance of $\hat{Y}_h$. Since $\bar{Y}$ and $\hat{\beta}_1$ are independent random variables (this was shown in the derivation prior to the establishment of the variance–covariance matrix of $\hat{\beta}_0$ and $\hat{\beta}_1$ in Theorem 1.4),

$$
\begin{aligned}
V\left[\hat{Y}_h\right] &= V\left[\hat{\beta}_0 + \hat{\beta}_1 X_h\right] \\
&= V\left[\bar{Y} - \hat{\beta}_1 \bar{X} + \hat{\beta}_1 X_h\right] \\
&= V\left[\bar{Y} + \hat{\beta}_1 \left(X_h - \bar{X}\right)\right] \\
&= V\left[\bar{Y}\right] + \left(X_h - \bar{X}\right)^2 V\left[\hat{\beta}_1\right] \\
&= \frac{\sigma^2}{n} + \left(X_h - \bar{X}\right)^2 \frac{\sigma^2}{S_{XX}} \\
&= \left[\frac{1}{n} + \frac{\left(X_h - \bar{X}\right)^2}{S_{XX}}\right] \sigma^2
\end{aligned}
$$

using the lower-right hand entry in the variance–covariance matrix for $\hat{\beta}_0$ and $\hat{\beta}_1$ from Theorem 1.4. This constitutes a derivation of the following result.

Theorem 2.10 Under the simple linear regression model with normal error terms,

$$\hat{Y}_h \sim N\left(\beta_0 + \beta_1 X_h, \ \left[\frac{1}{n} + \frac{\left(X_h - \bar{X}\right)^2}{S_{XX}}\right] \sigma^2\right),$$

where X_h is a fixed value of the independent variable within the scope of the simple linear regression model and $\hat{Y}_h = \hat{\beta}_0 + \hat{\beta}_1 X_h$.

The population variance of $\hat{Y}_h$ in Theorem 2.10 is of particular interest. If the experimenter has complete control over the choice of the values of the independent variables $X_1, X_2, \ldots, X_n$ in the data pairs $(X_1, Y_1), (X_2, Y_2), \ldots, (X_n, Y_n)$, the best choice is to (a) choose $X_1, X_2, \ldots, X_n$ so that S_{XX} is as large as possible (that is, spread the $X_1, X_2, \ldots, X_n$ out as much as possible), and (b) choose $X_1, X_2, \ldots, X_n$ such that $\bar{X}$ equals X_h. These choices for the values of the independent variables will result in the smallest possible population variance for $\hat{Y}_h$.

The geometry associated with the choice of the $X_1, X_2, \ldots, X_n$ values is illustrated in Figure 2.2. In each of the two scatterplots, there are $n = 24$ simulated data pairs drawn from simple linear regression models with normal error terms having identical population parameters β_0, β_1, and σ^2.

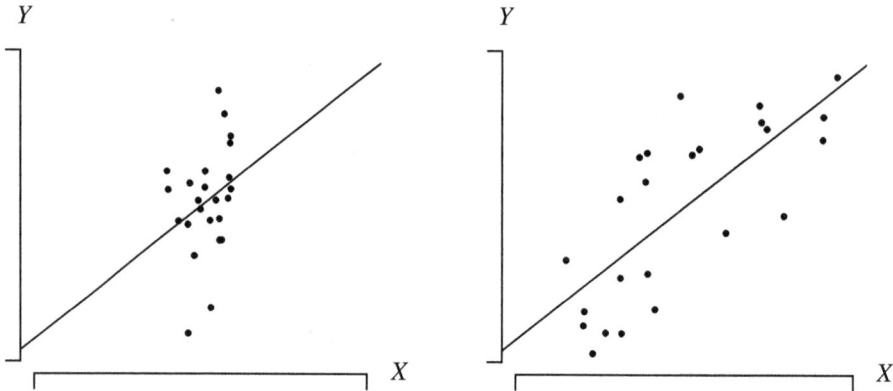

Figure 2.2: The effect of spreading $X_1, X_2, \ldots, X_n$.

Although they are not labeled, the axes on the two graphs have identical scales, and the two regression lines have nearly the same slope and intercept. The key difference between the two graphs is that the values of the independent variable are less spread out in the left-hand graph and more spread out in the right-hand graph. The spread of $X_1, X_2, \ldots, X_n$ results in three conclusions. First, the scope of the regression model is narrower in the graph on the left. Second, the estimation of β_1 is less stable when $X_1, X_2, \ldots, X_n$ are tightly clustered as in the graph on the left. Third, inference on $E[Y_h]$ will be less precise in the graph on the left because the variance of $\hat{Y}_h$ is larger via Theorem 2.10.

The development of a pivotal quantity for statistical inference concerning $E[Y_h]$ follows along the same line of reasoning as that for β_1 and β_0. We can't calculate the population variance of $\hat{Y}_h$ from n data pairs because the value of σ^2 is unknown, so it is estimated by

$$\hat{V}\left[\hat{Y}_h\right] = \left[\frac{1}{n} + \frac{(X_h - \bar{X})^2}{S_{XX}}\right] MSE,$$

where $\hat{\sigma}^2 = MSE = SSE/(n-2)$, which is a quantity that can be estimated from n data pairs. We can now use

$$\frac{\hat{Y}_h - E[Y_h]}{\sqrt{\hat{V}\left[\hat{Y}_h\right]}}$$

as a pivotal quantity in the following result.

Theorem 2.11 Under the simple linear regression model with normal error terms,

$$\frac{\hat{Y}_h - E[Y_h]}{\sqrt{\hat{V}\left[\hat{Y}_h\right]}} \sim t(n-2),$$

where X_h is a fixed value of the independent variable within the scope of the simple linear regression model and $\hat{Y}_h = \hat{\beta}_0 + \hat{\beta}_1 X_h$.

The proof of this result is analogous to the associated proofs for the pivotal quantities for inference concerning β_0 and β_1. This pivotal quantity can be used as a test statistic when conducting a hypothesis test concerning $E[Y_h]$. Proceeding in an analogous fashion to the development of the confidence intervals for β_1 and β_0, an exact two-sided $100(1-\alpha)\%$ confidence interval for $E[Y_h]$ is given next.

Theorem 2.12 Under the simple linear regression model with normal error terms,

$$\hat{Y}_h - t_{n-2,\alpha/2}\sqrt{\hat{V}\left[\hat{Y}_h\right]} < E[Y_h] < \hat{Y}_h + t_{n-2,\alpha/2}\sqrt{\hat{V}\left[\hat{Y}_h\right]},$$

is an exact two-sided $100(1-\alpha)\%$ confidence interval for $E[Y_h]$, where

$$\hat{Y}_h = \hat{\beta}_0 + \hat{\beta}_1 X_h \qquad \text{and} \qquad \hat{V}\left[\hat{Y}_h\right] = \left[\frac{1}{n} + \frac{\left(X_h - \bar{X}\right)^2}{S_{XX}}\right] MSE$$

and X_h is a fixed value of the independent variable within the scope of the simple linear regression model.

The calculation of an exact two-sided confidence interval for $E[Y_h]$ from a data set consisting of n data pairs $(X_1, Y_1), (X_2, Y_2), \ldots, (X_n, Y_n)$ using Theorem 2.12 will be illustrated in the next example.

Example 2.4 Calculate a point estimate and an exact two-sided 95% confidence interval for the expected barometric pressure $E[Y_h]$ associated with a boiling point of $X_h = 206$ degrees Fahrenheit for the Forbes data set from Example 1.11.

The R code below implements the formula in Theorem 2.12 for the Forbes data pairs. The lm function is used to fit the simple linear regression model. The estimated regression coefficients and the residuals are extracted from the fitted model in order to complete the computations.

```
library(MASS)
x     = forbes$bp
y     = forbes$pres
n     = length(x)
fit   = lm(y ~ x)
xh    = 206
yhat  = fit$coefficients[1] + fit$coefficients[2] * xh
sse   = sum(fit$residuals ^ 2)
mse   = sse / (n - 2)
sxx   = sum((x - mean(x)) ^ 2)
vhat  = (1 / n + (xh - mean(x)) ^ 2 / sxx) * mse
alpha = 0.05
crit  = qt(1 - alpha / 2, n - 2)
lo    = yhat - crit * sqrt(vhat)
hi    = yhat + crit * sqrt(vhat)
print(c(lo, yhat, hi))
```

This code returns the point estimator for the population mean barometric pressure $\hat{Y}_h = 26.65$ inches of mercury corresponding to the boiling point $X_h = 206$ degrees

Fahrenheit, which is associated with the exact two-sided 95% confidence interval

$$26.52 < E[Y_h] < 26.79.$$

We are 95% confident that the population mean barometric pressure lies between 26.52 and 26.79 inches of mercury when the boiling point is 206° Fahrenheit based on the $n = 17$ data pairs. Again, the confidence interval for $E[Y_h]$ is narrow because of the tight clustering of the data pairs around the estimated regression line. These calculations are also routinely performed by statisticians, so they can be performed with fewer lines of code by using the R built-in generic `predict` function. After the `predict` function recognizes the object `fit`, given as the first argument, as a fitted regression model, it internally calls the `predict.lm` function to calculate the point estimate and the interval estimate. The second argument to `predict` is a data frame that contains the value of X_h, which is $X_h = 206$ in this example. The `interval` argument is set to the character string `"confidence"` because a confidence interval is being requested. The default value for α is 0.05, which yields a two-sided 95% confidence interval for $E[Y_h]$, and can be altered with the `level` argument.

```
library(MASS)
x   = forbes$bp
y   = forbes$pres
fit = lm(y ~ x)
predict(fit, data.frame(x = 206), interval = "confidence")
```

The `predict` function displays the output given below.

```
        fit      lwr      upr
1 26.65211 26.51501 26.7892
```

These values match the point estimator $\hat{Y}_h = 26.65$ inches of mercury and the associated exact two-sided confidence interval $26.52 < E[Y_h] < 26.79$ generated by the previous code segment. Figure 2.3 contains a scatterplot of the data, the fitted regression line, and a (tiny) vertical line segment indicating the width of the confidence interval for $E[Y_h]$. This segment is symmetric about the fitted Y-value, which is the point estimator $\hat{Y}_h = 26.65$.

The previous example illustrated the steps required to calculate a point and interval estimate of $E[Y_h]$. The width of the confidence interval for $E[Y_h]$ is a function of

- n (a narrower confidence interval for larger values of n),

- α (a narrower confidence interval for smaller values of α),

- the dispersion of the data pairs about the regression line as measured by SSE (a narrower confidence interval for smaller values of SSE),

- the spread of the X values selected in the experiment as measured by S_{XX} (a narrower confidence interval for larger values of S_{XX}), and

- the proximity of X_h to $\bar{X}$ (a narrower confidence interval for X_h closer to $\bar{X}$).

Each of these conclusions concerning the width of the confidence interval is apparent in the formula for the confidence interval for $E[Y_h]$ given in Theorem 2.12. The next section considers the closely-related prediction interval associated with the introduction of a new data pair.

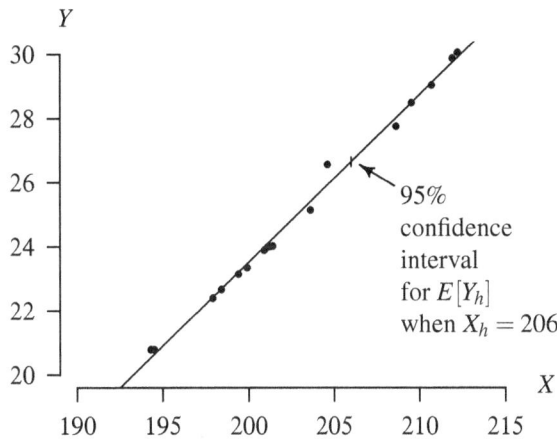

Figure 2.3: Predicted barometric pressure at a boiling point of $206°$ Fahrenheit.

2.3.5 Inference Concerning $Y_h^\star$

The previous section considered statistical inference on the mean response associated with a value X_h for the independent variable associated with the data pairs $(X_1, Y_1), (X_2, Y_2), \ldots, (X_n, Y_n)$. This section considers statistical inference associated with the introduction of a new data pair. The value of the independent variable for this new data pair is, as before, X_h, which is a fixed constant observed without error within the scope of the model. We would like to perform some type of statistical inference on the associated value of the dependent variable $Y_h^\star$. The star superscript is to denote that this is an additional data pair that is not one of the original n data pairs used to fit the simple linear regression model. There is a similar, but fundamentally different, analysis that must be used when we would like to consider the introduction of an *additional* data pair

$$(X_{n+1}, Y_{n+1}) = (X_h, Y_h^\star).$$

Three examples in which this type of analysis is appropriate are given below.

- A sociologist collects the $n = 50$ data pairs $(X_1, Y_1), (X_2, Y_2), \ldots, (X_{50}, Y_{50})$, where the independent variable X is the wife's height and the dependent variable Y is the husband's height for 50 married couples. These data pairs represent 50 couples surveyed by the sociologist. If the sociologist knows the height of a married woman who is not in the group of 50, what statistical inference can the sociologist make about her husband's height?

- An economist collects the $n = 50$ data pairs $(X_1, Y_1), (X_2, Y_2), \ldots, (X_{50}, Y_{50})$, where the independent variable is the average annual unemployment rate and the dependent variable is the annual gross domestic product (GDP) for a particular country. If these data pairs represent the last 50 years of data, and the economist knows the average annual unemployment rate for next year, what statistical inference can the economist perform on the random GDP for next year?

- An engineer collects the $n = 50$ data pairs $(X_1, Y_1), (X_2, Y_2), \ldots, (X_{50}, Y_{50})$, where the independent variable is the speed of a car and the dependent variable is the car's stopping distance for 50 different cars. If the engineer knows the speed of a 51st car to be tested, what statistical inference can the engineer perform on its random stopping distance?

The common thread that runs through the three examples is that there is a new data pair, $(X_{51}, Y_{51}) = (X_h, Y_h^\star)$, that is being introduced.

So we would like to predict the outcome for a new value of the dependent variable associated with the new value of the independent variable, namely X_h. As before, the value of X_h need not necessarily correspond to one of the $X_1, X_2, \ldots, X_n$ values, but needs to fall within the scope of the model unless there is some prevailing evidence to make statistical inference outside of the scope of the model. In order to help frame the issues associated with the case of a new data pair being introduced, the next paragraph considers the very rare case in which all of the parameters in the simple linear regression model are known.

Consider the simplest case in which all parameters are known in the simple linear regression model. In the previous section, Theorem 2.12 gave a *confidence interval* for $E[Y_h]$, which is a fixed constant. In this section, we desire a statistical interval for $Y_h^\star$, which is a random variable. Because of this fundamental difference in the nature of $E[Y_h]$ and $Y_h^\star$, the interval derived here for $Y_h^\star$ is a *prediction interval*. If all of the parameters in the regression model are known, Definition 2.1 indicates that

$$Y_h^\star \sim N\left(\beta_0 + \beta_1 X_h, \sigma^2\right).$$

Standardizing this normally distributed random variable,

$$\frac{Y_h^\star - (\beta_0 + \beta_1 X_h)}{\sigma} \sim N(0, 1).$$

The probability that this standard normal random variable lies between the $\alpha/2$ and $1 - \alpha/2$ fractiles of the standard normal distribution is

$$P\left(-z_{\alpha/2} < \frac{Y_h^\star - (\beta_0 + \beta_1 X_h)}{\sigma} < z_{\alpha/2}\right) = 1 - \alpha.$$

Some algebra on the inequality gives an exact two-sided $100(1 - \alpha)\%$ prediction interval for $Y_h^\star$ as

$$\beta_0 + \beta_1 X_h - z_{\alpha/2}\sigma < Y_h^\star < \beta_0 + \beta_1 X_h + z_{\alpha/2}\sigma.$$

Although this derivation is straightforward, the vast majority of regression applications do not have parameters which are known a priori, and we now pivot to the more practical question.

In the case in which the parameters in the simple linear regression model are unknown, they must be estimated from the n data pairs. The point estimator for $Y_h^\star$ is the same as the point estimator in the previous section:

$$\hat{Y}_h^\star = \hat{\beta}_0 + \hat{\beta}_1 X_h.$$

Handling the population variance of $Y_h^\star$ requires a little more finesse. In the case of the parameters being estimated from n data pairs, the population variance of $Y_h^\star$ comes from two sources:

- the population variance associated with a new observation of the dependent variable, and

- the population variance induced by estimating the intercept and slope of the fitted regression line from the n data pairs.

Since the new data pair is independent of the original n data pairs, the population variance of the prediction error is

$$V\left[Y_h^\star - \left(\hat{\beta}_0 + \hat{\beta}_1 X_h\right)\right] = V\left[Y_h^\star - \hat{Y}_h\right] = V\left[Y_h^\star\right] + V\left[\hat{Y}_h\right] = \sigma^2 + V\left[\hat{Y}_h\right].$$

Since $\hat{Y}_h$ is normally distributed via Theorem 2.10 and $Y_h^\star$ is independent of $\hat{Y}_h$ and is also normally distributed, we have the following result.

Theorem 2.13 Under the simple linear regression model with normal error terms,

$$Y_h^\star \sim N\left(\beta_0 + \beta_1 X_h, \left[1 + \frac{1}{n} + \frac{(X_h - \bar{X})^2}{S_{XX}}\right]\sigma^2\right),$$

where X_h is a fixed value of the independent variable within the scope of the simple linear regression model and $\hat{Y}_h^\star = \hat{\beta}_0 + \hat{\beta}_1 X_h$.

The population mean of the normal distribution in Theorem 2.13 is estimated by

$$\hat{Y}_h^\star = \hat{\beta}_0 + \hat{\beta}_1 X_h$$

and the population variance of the normal distribution is estimated by

$$\hat{V}\left[\hat{Y}_h^\star\right] = \left[1 + \frac{1}{n} + \frac{(X_h - \bar{X})^2}{S_{XX}}\right]\hat{\sigma}^2 = \left[1 + \frac{1}{n} + \frac{(X_h - \bar{X})^2}{S_{XX}}\right]MSE.$$

Using an analogous approach to the pivotal quantities in the previous sections, the following quantity can be used for statistical inference concerning $\hat{Y}_h^\star$.

Theorem 2.14 Under the simple linear regression model with normal error terms,

$$\frac{\hat{Y}_h^\star - E[\hat{Y}_h^\star]}{\sqrt{\hat{V}\left[\hat{Y}_h^\star\right]}} \sim t(n-2),$$

where X_h is a fixed value of the independent variable within the scope of the simple linear regression model and $\hat{Y}_h^\star = \hat{\beta}_0 + \hat{\beta}_1 X_h$.

The proof of this result is analogous to the associated proofs for the pivotal quantities for inference concerning β_0 and β_1. This pivotal quantity can be used as a test statistic when conducting a hypothesis test concerning $Y_h^\star$. Proceeding in an analogous fashion to the development of the confidence intervals for β_1 and β_0, an exact two-sided $100(1-\alpha)\%$ prediction interval for $Y_h^\star$ is given next.

Theorem 2.15 Under the simple linear regression model with normal error terms and parameters estimated from the data pairs $(X_1, Y_1), (X_2, Y_2), \ldots, (X_n, Y_n)$,

$$\hat{Y}_h^\star - t_{n-2,\alpha/2}\sqrt{\hat{V}\left[\hat{Y}_h^\star\right]} < Y_h^\star < \hat{Y}_h^\star + t_{n-2,\alpha/2}\sqrt{\hat{V}\left[\hat{Y}_h^\star\right]},$$

is an exact two-sided $100(1-\alpha)\%$ prediction interval for $Y_h^\star$, where

$$\hat{Y}_h^\star = \hat{\beta}_0 + \hat{\beta}_1 X_h \qquad \text{and} \qquad \hat{V}\left[\hat{Y}_h^\star\right] = \left[1 + \frac{1}{n} + \frac{(X_h - \bar{X})^2}{S_{XX}}\right]MSE$$

and X_h is a fixed value of the independent variable within the scope of the simple linear regression model.

Adding a 1 inside of the expression for $\hat{V}\left[\hat{Y}_h^\star\right]$ ensures that the prediction interval for $\hat{Y}_h$ will be wider than the associated confidence interval for $E[Y_h]$ from Theorem 2.12. In both results, the intervals are narrowest when X_h is near $\bar{X}$ and the observations of the independent variable are spread out so as to maximize S_{XX}.

Example 2.5 Calculate a point estimate and an exact two-sided 95% prediction interval for the barometric pressure $Y_h^\star$ associated with a new observation with a boiling point of $X_h = 206$ degrees Fahrenheit for the Forbes data set from Example 1.11.

A point estimate and an exact two-sided 95% prediction interval for the barometric pressure associated with a new data pair having boiling point $X_h = 206$ can be computed with the R predict function as shown below.

```
library(MASS)
x   = forbes$bp
y   = forbes$pres
fit = lm(y ~ x)
predict(fit, data.frame(x = 206), interval = "prediction")
```

The output from these statements is given below.

```
        fit       lwr       upr
1 26.65211  26.13726  27.16696
```

The point estimator for $Y_h^\star$ is $\hat{Y}_h^\star = 26.65$ (this value matches the point estimate from Example 2.4) and the 95% two-sided prediction interval returned is

$$26.14 < Y_h^\star < 27.17.$$

Figure 2.4 contains a scatterplot of the data, the fitted regression line, and a (not-as-tiny-as-before) vertical line segment indicating the width of the exact two-sided 95%

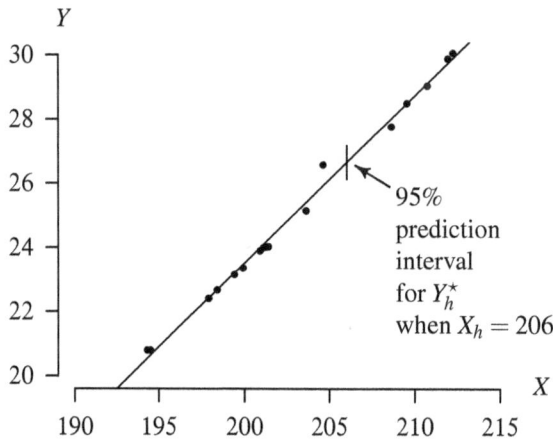

Figure 2.4: Prediction interval for a new data pair with boiling point of 206° F.

prediction interval for $Y_h^\star$. This segment is symmetric about the predicted Y-value, which is the point estimator $\hat{Y}_h^\star = 26.65$. In this particular setting, the 1 inside the expression for $\hat{V}\left[\hat{Y}_h^\star\right]$ resulted in a significantly wider 95% prediction interval than the associated 95% confidence interval.

A thought experiment that helps clarify the difference between the confidence interval for $E[Y_h]$ and the prediction interval for $\hat{Y}_h^\star$ is to consider the two intervals associated with $X_h = \bar{X}$. A careful inspection of the confidence interval given in Theorem 2.12 indicates that the width of the confidence interval for $E[Y_h]$ approaches zero as $n \to \infty$. Increasing the number of data pairs without bound results in perfect precision for the point estimator for the conditional expected value $\hat{Y}_h = \hat{\beta}_0 + \hat{\beta}_1 X_h$. On the other hand, a careful inspection of the prediction interval given in Theorem 2.15 indicates that the width of the prediction interval for $\hat{Y}_h^\star$ approaches a finite, nonzero value as $n \to \infty$. When a new observation associated with independent variable $X_h = \bar{X}$, the associated point estimator for the conditional expected value of the dependent variable $\hat{Y}_h^\star = \hat{\beta}_0 + \hat{\beta}_1 X_h$ has a population variance that approaches the MSE (which, in turn, approaches σ^2) as $n \to \infty$. It is not possible to predict the random response to a new data pair with perfect precision.

This section and the previous four sections have introduced various techniques for statistical inference in the setting of a simple linear regression model with normal error terms. Table 2.1 summarizes many of the key results from these sections. The first column gives the parameter of interest. The second column gives the pivotal quantity and its distribution. This pivotal quantity serves as the test statistic in a hypothesis test concerning the parameter of interest. The third column gives an exact two-sided $100(1-\alpha)\%$ confidence interval for the parameter of interest for the first four rows and an exact two-sided $100(1-\alpha)\%$ prediction interval for the parameter of interest for the last row.

parameter	pivotal quantity	exact two-sided $100(1-\alpha)\%$ statistical interval
σ^2	$\dfrac{SSE}{\sigma^2} \sim \chi^2(n-2)$	$\dfrac{SSE}{\chi^2_{n-2,\alpha/2}} < \sigma^2 < \dfrac{SSE}{\chi^2_{n-2,1-\alpha/2}}$
β_1	$\dfrac{\hat{\beta}_1 - \beta_1}{\sqrt{\hat{V}\left[\hat{\beta}_1\right]}} \sim t(n-2)$	$\hat{\beta}_1 \pm t_{n-2,\alpha/2}\sqrt{\dfrac{MSE}{S_{XX}}}$
β_0	$\dfrac{\hat{\beta}_0 - \beta_0}{\sqrt{\hat{V}\left[\hat{\beta}_0\right]}} \sim t(n-2)$	$\hat{\beta}_0 \pm t_{n-2,\alpha/2}\sqrt{\dfrac{MSE\sum_{i=1}^n X_i^2}{nS_{XX}}}$
$E[Y_h]$	$\dfrac{\hat{Y}_h - E[Y_h]}{\sqrt{\hat{V}\left[\hat{Y}_h\right]}} \sim t(n-2)$	$\hat{\beta}_0 + \hat{\beta}_1 X_h \pm t_{n-2,\alpha/2}\sqrt{\left(\dfrac{1}{n} + \dfrac{(X_h - \bar{X})^2}{S_{XX}}\right)MSE}$
$Y_h^\star$	$\dfrac{\hat{Y}_h^\star - E[\hat{Y}_h^\star]}{\sqrt{\hat{V}\left[\hat{Y}_h^\star\right]}} \sim t(n-2)$	$\hat{\beta}_0 + \hat{\beta}_1 X_h \pm t_{n-2,\alpha/2}\sqrt{\left(1 + \dfrac{1}{n} + \dfrac{(X_h - \bar{X})^2}{S_{XX}}\right)MSE}$

Table 2.1: Pivotal quantities and exact statistical intervals for a simple linear regression model.

2.3.6   Joint Inference Concerning β_0 and β_1

The exact two-sided $100(1-\alpha)\%$ confidence intervals for the intercept β_0 and slope β_1 in a simple linear regression model with normal error terms developed in Sections 2.3.2 and 2.3.3 might be combined to provide a joint confidence region on both parameters. Occasions arise in regression modeling in which joint inference on both β_0 and β_1 simultaneously is required. As a particular instance, recall from Examples 2.2 and 2.3 that the unbiased point estimators for β_0 and β_1 for the Forbes data set were

$$\hat{\beta}_0 = -81.06 \qquad \text{and} \qquad \hat{\beta}_1 = 0.5229$$

and the associated exact two-sided 95% confidence intervals for β_0 and β_1 calculated separately were

$$-85.44 < \beta_0 < -76.69 \qquad \text{and} \qquad 0.5014 < \beta_1 < 0.5444.$$

The union of these two confidence intervals is depicted by the rectangle in Figure 2.5. The point estimates for β_0 and β_1 are depicted by the point at the center of the rectangle. Does the union of the two confidence intervals depicted by the rectangle in Figure 2.5 constitute an exact 95% confidence region for β_0 and β_1? It does not. The problems associated with this rectangular-shaped confidence region are outlined in the next two paragraphs.

   If the two confidence intervals were constructed independently, the actual coverage associated with the confidence region would be $(0.95)(0.95) = 0.9025$. This would be a 90.25% confidence region. If the confidence intervals were constructed independently, then we could simply adjust the coverages of the individual confidence intervals for β_0 and β_1 to $\sqrt{0.95} \cong 0.9747$ in order to get an exact 95% confidence region for β_0 and β_1. But the two confidence intervals are constructed from the same data set, and, as seen by the off-diagonal elements in the variance–covariance matrix in Theorem 1.4, the covariance between $\hat{\beta}_0$ and $\hat{\beta}_1$ is zero only when $\bar{X} = 0$. This is seldom the case in practice. So while the rectangular region in Figure 2.5 is a confidence region, it is not one that we can easily find the associated actual coverage. Some help is provided by the Bonferroni inequality, which states that the actual coverage for the rectangular region is at least $1 - 2\alpha$, which in this setting is $1 - (2)(0.05) = 0.90$. Both confidence intervals contain the true value of β_0 and β_1

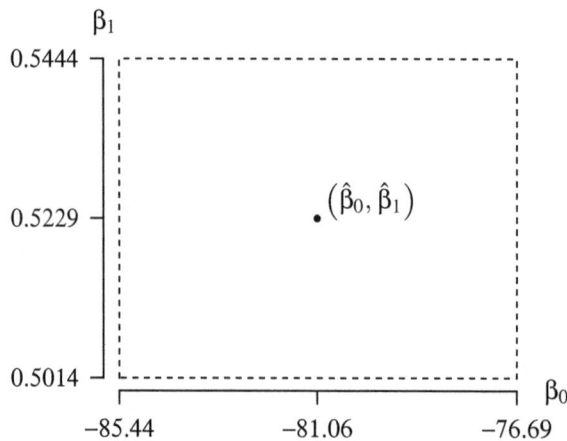

Figure 2.5: Confidence region for β_0 and β_1.

with at least 90% confidence, but this is all that can be stated concerning the actual coverage of the rectangular-shaped confidence region.

Since we know that the point estimators for β_0 and β_1 are only independent in the rare case of $\bar{X} = 0$ from Theorem 1.4, perhaps a rectangular-shaped confidence region is not appropriate. This is certainly the impression that one gets from the Monte Carlo simulation experiment conducted in Example 1.4. The problem here is that the point estimators $\hat{\beta}_0$ and $\hat{\beta}_1$ are typically dependent random variables, which means that a non-rectangular confidence region is appropriate. In an advanced class in regression, you will prove the following result, which is used to determine an exact $100(1-\alpha)\%$ confidence region for β_0 and β_1.

Theorem 2.16 Under the simple linear regression model with normal errors and parameters estimated from the data pairs $(X_1, Y_1), (X_2, Y_2), \ldots, (X_n, Y_n)$,

$$\frac{n-2}{2\sum_{i=1}^{n} e_i^2} \left[n\left(\hat{\beta}_0 - \beta_0\right)^2 + 2\left(\hat{\beta}_0 - \beta_0\right)\left(\hat{\beta}_1 - \beta_1\right)\sum_{i=1}^{n} X_i + \left(\hat{\beta}_1 - \beta_1\right)^2 \sum_{i=1}^{n} X_i^2 \right] \sim F(2, n-2).$$

Let $F_{2,n-2,\alpha}$ be the $1-\alpha$ percentile of the F distribution with 2 and $n-2$ degrees of freedom. Theorem 2.16 implies that

$$P\left(\frac{n-2}{2\sum_{i=1}^{n} e_i^2} \left[n\left(\hat{\beta}_0 - \beta_0\right)^2 + 2\left(\hat{\beta}_0 - \beta_0\right)\left(\hat{\beta}_1 - \beta_1\right)\sum_{i=1}^{n} X_i + \left(\hat{\beta}_1 - \beta_1\right)^2 \sum_{i=1}^{n} X_i^2 \right] \leq F_{2,n-2,\alpha} \right) = 1-\alpha.$$

This inequality can be used to construct an exact $100(1-\alpha)\%$ confidence region for β_0 and β_1.

Theorem 2.17 Under the simple linear regression model with normal error terms and parameters estimated from the data pairs $(X_1, Y_1), (X_2, Y_2), \ldots, (X_n, Y_n)$, all β_0 and β_1 values satisfying

$$\frac{n-2}{2\sum_{i=1}^{n} e_i^2} \left[n\left(\hat{\beta}_0 - \beta_0\right)^2 + 2\left(\hat{\beta}_0 - \beta_0\right)\left(\hat{\beta}_1 - \beta_1\right)\sum_{i=1}^{n} X_i + \left(\hat{\beta}_1 - \beta_1\right)^2 \sum_{i=1}^{n} X_i^2 \right] \leq F_{2,n-2,\alpha}$$

are an exact joint $100(1-\alpha)\%$ confidence region for β_0 and β_1.

The boundary of the confidence region in the (β_0, β_1) plane is an ellipse centered at $\left(\hat{\beta}_0, \hat{\beta}_1\right)$. The boundary is found by replacing the inequality in Theorem 2.17 with an equality. The tilt of the ellipse is a function of $\mathrm{Cov}\left(\hat{\beta}_0, \hat{\beta}_1\right)$, which is $-\bar{X}S_{XX}/\sigma^2$ by Theorem 1.4. Notice that $S_{XX} > 0$ and $\sigma^2 > 0$ under the simple linear regression model assumptions given in Definition 1.1. If $\bar{X} > 0$, then the covariance between the parameter estimates is negative, which implies that the error associated with the parameter estimates and their true values tends to be in the opposite direction. If $\hat{\beta}_0 > \beta_0$, for example, then it is more likely that $\hat{\beta}_1 < \beta_1$. This is the more common situation in practice. Conversely, if $\bar{X} < 0$, then $\mathrm{Cov}\left(\hat{\beta}_0, \hat{\beta}_1\right) > 0$, which implies that the error associated with the parameter estimates and their true values tends to be in the same direction.

The confidence region given in Theorem 2.17 can be plotted for the data pairs $(X_1, Y_1), (X_2, Y_2), \ldots, (X_n, Y_n)$ using numerical methods. Plotting the boundary of the confidence region in the (β_0, β_1) plane can be performed using a two-dimensional search for points on the boundary. Alternatively, a ray can be extended from $\left(\hat{\beta}_0, \hat{\beta}_1\right)$ at a particular angle, and a one-dimensional search can be conducted to find a point on the boundary. The details associated with plotting such a confidence region will be given in one of the examples in the next section.

2.4   The ANOVA Table

In most applications of simple linear regression, the slope of the regression line, β_1, is the most critical of the three parameters in the model. The most common statistical test that is performed in a simple linear regression application is testing whether the population slope β_1 is zero against the two-tailed alternative:

$$H_0 : \beta_1 = 0$$

versus

$$H_1 : \beta_1 \neq 0.$$

This choice of H_0 and H_1 is designed to determine whether the independent variable X is a statistically significant predictor of the dependent variable Y. Rejecting H_0 indicates that the independent variable is providing some predictive capability. Although this test can be conducted based on Theorem 2.5, a second test based on the F distribution is introduced in this section and its equivalency to the test based on the t distribution is established. Both tests are exact. In addition, the ANOVA table which was introduced in Section 1.8 will be expanded in this section to include an additional column.

Cochran's theorem, named after American statistician William Cochran (1909–1980), concerns writing sums of squares of independent and identically distributed $N(0, \sigma^2)$ random variables as the sum of positive semi-definite quadratic forms of these random variables. Applying his theorem to the simple linear regression model with normal error terms yields the following result.

Theorem 2.18 For the simple linear regression model with normal error terms,

- $SSR/\sigma^2 \sim \chi^2(1)$, and

- $SSE/\sigma^2 \sim \chi^2(n-2)$,

- SSR and SSE are independent

under H_0.

The second of the three results has already been seen in Theorem 2.2. The first and third results are necessary to derive the F test for the significance of the slope β_1, which is given in the following theorem.

Theorem 2.19 Under the simple linear regression model with normal error terms,

$$\frac{MSR}{MSE} \sim F(1, n-2)$$

under H_0.

Proof Since $SSR/\sigma^2 \sim \chi^2(1)$, $SSE/\sigma^2 \sim \chi^2(n-2)$, SSR and SSE are independent by Theorem 2.18, and the ratio of two independent chi-square random variables divided by their degrees of freedom has the F distribution, under H_0,

$$\frac{\dfrac{SSR/\sigma^2}{1}}{\dfrac{SSE/\sigma^2}{n-2}} = \frac{MSR}{MSE} \sim F(1, n-2). \qquad \square$$

The ANOVA table which was first introduced in Section 1.8 can be expanded to include an additional column on the right as shown in Table 2.2. Some computer packages will add yet another column on the right-hand side of the ANOVA table which contains the p-value associated with the F test.

Source	SS	df	MS	F
Regression	SSR	1	MSR	MSR/MSE
Error	SSE	$n-2$	MSE	
Total	SST	$n-1$		

Table 2.2: Basic ANOVA table for simple linear regression.

So the F test for the statistical significance of the slope parameter in the regression model with normal error terms begins by computing the test statistic $F = MSR/MSE$. If $F < F_{1,n-2,1-\alpha/2}$ or $F > F_{1,n-2,\alpha/2}$, then H_0 is rejected. The ANOVA table will be illustrated in one of the examples in the next section.

To show that the F-test developed here is equivalent to the same test based on the t distribution in Section 2.3.2,

$$F = \frac{MSR}{MSE} = \frac{\hat{\beta}_1^2 S_{XX}}{\hat{V}\left[\hat{\beta}_1\right] S_{XX}} = \frac{\hat{\beta}_1^2}{\hat{V}\left[\hat{\beta}_1\right]} = t^2,$$

because $MSR = SSR = \hat{\beta}_1^2 S_{XX}$ (which is an exercise from Chapter 1), where t is the test statistic for the hypothesis based on Theorem 2.5. Since the square of a t random variable has the F distribution with the appropriate degrees of freedom, the two tests are equivalent.

We do not pursue the F test any further because the test of significance for the slope of the regression line based on the F distribution is less flexible than that based on the t distribution from Section 2.3.2. The test based on the t distribution is superior because (a) it can adapt to one-tailed alternative hypotheses, and (b) it is capable of testing for slopes other than $\beta_1^\star = 0$. The primary purpose of introducing the F test here is to append the additional column to the right of the ANOVA table and provide an insightful link between regression, which is presented here, and experimental design, which relies heavily on ANOVA tables.

2.5 Examples

This section contains four examples that illustrate the implementation of the simple linear regression modeling techniques that have been developed so far.

Example 2.6 This first example is more of a cautionary tale than a real-world example. Francis Anscombe (1918–2001) was a British statistician who devised four sets, each consisting of $n = 11$ data pairs. These four data sets have come to be known as *Anscombe's quartet*, which are given in Table 2.3. Make scatterplots of the four data sets, along with the associated estimated regression lines.

Anscombe's quartet is contained in a data frame in R named `anscombe`. The R code below creates scatterplots of the four data sets in a 2×2 set of graphs using common horizontal and vertical scales. The four scatterplots and the associated regression lines are given in Figure 2.6.

Data set I		Data set II		Data set III		Data set IV	
X_i	Y_i	X_i	Y_i	X_i	Y_i	X_i	Y_i
10.0	8.04	10.0	9.14	10.0	7.46	8.0	6.58
8.0	6.95	8.0	8.14	8.0	6.77	8.0	5.76
13.0	7.58	13.0	8.74	13.0	12.74	8.0	7.71
9.0	8.81	9.0	8.77	9.0	7.11	8.0	8.84
11.0	8.33	11.0	9.26	11.0	7.81	8.0	8.47
14.0	9.96	14.0	8.10	14.0	8.84	8.0	7.04
6.0	7.24	6.0	6.13	6.0	6.08	8.0	5.25
4.0	4.26	4.0	3.10	4.0	5.39	19.0	12.50
12.0	10.84	12.0	9.13	12.0	8.15	8.0	5.56
7.0	4.82	7.0	7.26	7.0	6.42	8.0	7.91
5.0	5.68	5.0	4.74	5.0	5.73	8.0	6.89

Table 2.3: Anscombe's quartet.

```
par(mfrow = c(2, 2))
for (i in 1:4) {
  x = anscombe[ , i]
  y = anscombe[ , i + 4]
  plot(x, y, xlim = c(4, 19), ylim = c(3, 13), pch = 16)
  abline(lm(y ~ x))
}
```

Reading the plots row-wise, the first plot shows $n = 11$ data pairs could have come from a simple linear regression model with normal error terms. The residuals could possibly have emanated from a normal population with population mean zero and finite population variance σ^2. The second plot show that there is clearly a *relationship* between X and Y, but the relationship is *nonlinear* rather than *linear*. It appears that a quadratic model, rather than a linear model, best describes the relationship between X and Y. The third plot appears to contain an *outlier*, which might have been coded improperly. The fourth plot highlights the *leverage* that the far-right data pair exerts on the estimated regression line. Leverage points are those data pairs that exert more influence on the fitted model than others, typically by having a value of its independent variable which is distant from the values of the independent variable for other data pairs. The far-right point exerts that influence in the fourth plot. To summarize, only the first of the four data sets would be appropriate for a simple linear regression model. What if we bypassed the plotting of the scatterplots? If we did so and went directly to fitting the simple linear regression models, we would find that

$$\bar{X} = 9.0, \quad \bar{Y} = 7.5, \quad S_{XX} = 1001, \quad S_{YY} = 660, \quad \hat{\beta}_0 = 3.0, \quad \hat{\beta}_1 = 0.5, \quad r = 0.67$$

for all four data sets! (Some of these values are exactly the same for all four data sets and others match for two or three digits.) The estimated regression lines from Figure 2.6 are basically identical for all four data sets. Had we neglected to plot the data pairs in a scatterplot and proceeded directly to the regression analysis, we would conclude that the four data sets are basically identical. But the scatterplots show that this is clearly

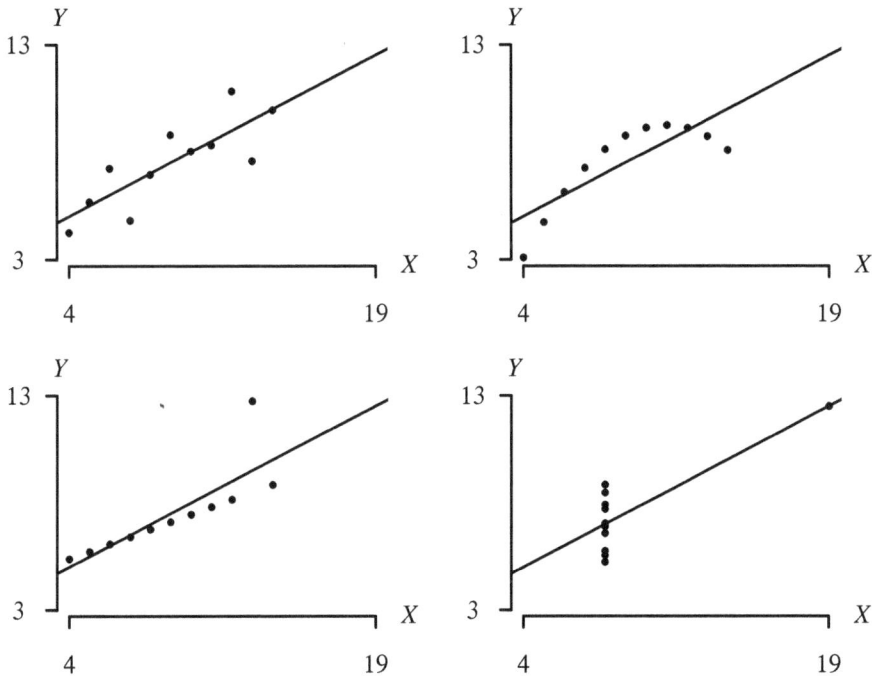

Figure 2.6: Scatterplots and estimated regression lines for Anscombe's quartet.

not the case. Only the first of the four data sets supports the simple linear regression model $Y = \beta_0 + \beta_1 X + \varepsilon$, with $N\left(0, \sigma^2\right)$ error terms.

The moral to the cautionary tale is to never bypass the critical step of making a scatterplot of the data pairs and visually assessing whether or not a simple linear regression model is appropriate. While it is easy to input your data into a statistical package and quickly get numerical estimates for the parameters, this can lead to adopting a statistical model which is inappropriate. In addition, if the visual assessment of the scatterplot leads you to believe that a simple linear regression model is feasible, this should be followed by a residual plot to assess the normality of the error terms.

The second example illustrates the assessment of the simple linear regression model, point estimation, and interval estimation for a large data set.

Example 2.7 A sociologist might be interested in the following question. Do taller-than-average women tend to date and eventually marry taller-than-average men? This question can be answered by collecting the heights of husband and wife pairs and examining the associated scatterplot to see if a regression model is appropriate. If it is appropriate, then a hypothesis test can be conducted to answer the question. The R data frame named `heights` contained in the R package `PBImisc` contains $n = 96$ pairs of heights (measured in centimeters) which will be used to answer the question. The first five and last five data pairs, ordered by the wife's height, are given in Table 2.4. We will use these data pairs of heights to assess the hypothesis by executing the following steps.

Wife's height	Husband's height
141	152
143	156
145	160
146	164
147	178
⋮	⋮
178	187
179	192
180	192
181	186
181	188

Table 2.4: Couple's heights ($n = 96$).

(a) Make a scatterplot of the data values and make an initial visual assessment of whether a simple linear regression model might be a reasonable approximation to the relationship between the husband's height and the wife's height.

(b) Fit a simple linear regression model and interpret the estimated slope and intercept of the regression line.

(c) Assess the adequacy of the model by making a plot of the residuals ordered by the values of the independent variable, plotting a histogram of the residuals, constructing a QQ plot, and performing a goodness-of-fit test for the normality of the residuals.

(d) Perform a hypothesis test with the null hypothesis

$$H_0 : \beta_1 = 0$$

versus the alternative hypothesis

$$H_1 : \beta_1 > 0$$

based on the data pairs, where β_1 is the slope of the regression line.

(e) Give a point estimator and a 95% confidence interval for $E[Y_h]$ when $X_h = 150$ centimeters.

(f) Give a point estimator and a 95% prediction interval for $\hat{Y}_h^\star$ when $X_h = 150$ centimeters.

This is an unusual data set in that it is not clear whether the husband's height or the wife's height should serve as the independent variable. Both spouses choose one another, so the analysis could be performed with either height serving as the independent variable. It could also be performed treating both heights as random variables. For the analysis performed here, we assume that the wife's height is a fixed value X and the husband's height is the random response Y.

(a) The R code below installs and loads the PBImisc package in R and generates a scatterplot using the plot function, which is displayed in Figure 2.7.

```
install.packages("PBImisc")
library(PBImisc)
x = heights$Wife
y = heights$Husband
plot(x, y)
```

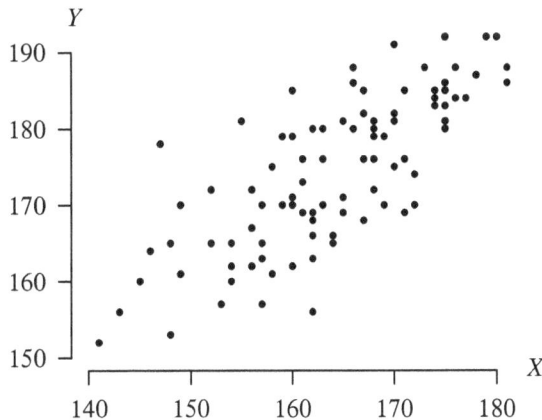

Figure 2.7: Scatterplot of wife's height X and husband's height Y for $n = 96$ data pairs.

Figure 2.7 appears to contain only 90 of the 96 points because there are six tied data pairs, such as $(165, 181)$, that occur in the data set. Some analysts prefer to *jitter* the tied data values slightly in order to avoid obscuring tied pairs. The paucity of points in the upper-left and lower-right corner of the scatterplot indicates that the two heights are positively correlated. There does not appear to be any systematic change in the variance of the data values moving from left to right, so it is reasonable to move forward with a simple linear regression model. Adding the line through the origin with slope 1 with the additional R command abline(c(0, 1)) reveals that one of the points, $(157, 157)$, has equal heights for the husband and wife, three points have a taller wife than her husband, and 92 of the points have a taller husband than his wife.

(b) The R statements below use the lm function to fit a simple linear regression model to the data pairs and the abline function to plot the associated regression line on the scatterplot.

```
library(PBImisc)
x = heights$Wife
y = heights$Husband
plot(x, y)
fit = lm(y ~ x)
abline(fit$coefficients)
```

The point estimates of the intercept and slope are

$$\hat{\beta}_0 = 37.8 \qquad \text{and} \qquad \hat{\beta}_1 = 0.833.$$

The interpretation of the slope is that the expected husband's height is 0.833 centimeter greater for each increase in the wife's height by one centimeter. The remaining question is whether this positive slope differs significantly from zero. The intercept, on the other hand, does not have a meaningful interpretation in this setting (a woman who is zero centimeters tall marries a man who has an average height of 37.8 centimeters). The intercept is way outside of the scope of the model and has no practical meaning here. Any conclusions drawn should be made within the range of collected heights of the women, which range from 141 to 181 centimeters. The fitted regression line is superimposed over the scatterplot in Figure 2.8.

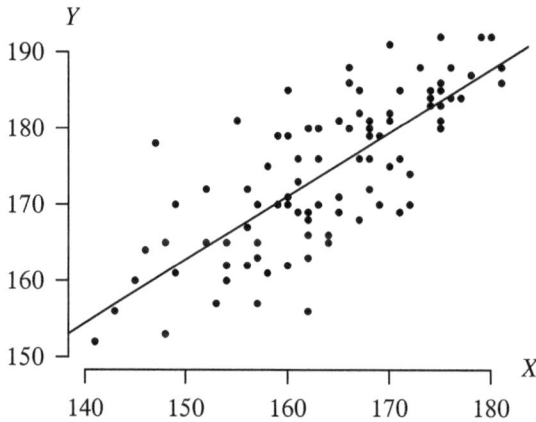

Figure 2.8: Fitted regression model for the $n = 96$ data pairs.

(c) Before conducting a hypothesis test concerning the slope, it is critical to assess the validity of the simple linear regression model by examining the residuals. The R code below orders the 96 data pairs by the wife's height, and plots the index on the horizontal axis and the associated residual on the vertical axis, which is displayed in Figure 2.9. Normally distributed error terms seem plausible from this graph, but there is some evidence that the early observations incur more variability on the high side. The fifth ordered residual plotted corresponds to the spectacular data pair $(147, 178)$, with a 31-centimeter difference between the two heights. Could these early extreme positive residuals correspond to shorter women having a greater array of options than taller women?

```
library(PBImisc)
x    = heights$Wife
y    = heights$Husband
i    = order(x)
x    = x[i]
y    = y[i]
```

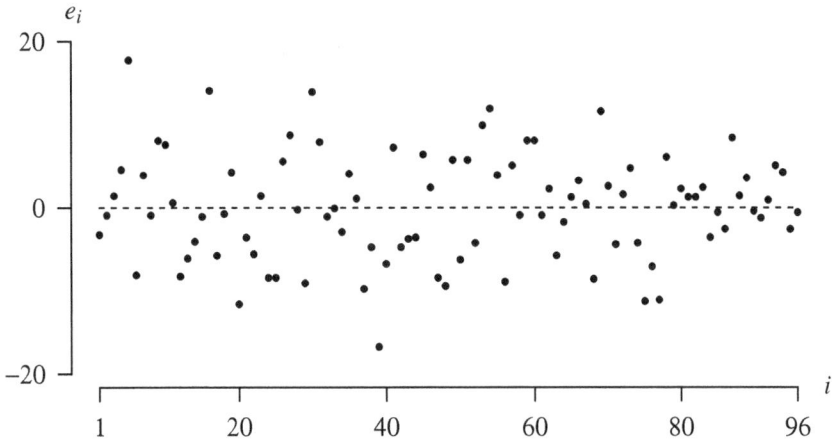

Figure 2.9: Residuals for the heights data.

```
n   = length(x)
fit = lm(y ~ x)
plot(1:n, fit$residuals)
```

If the residuals are approximately normally distributed, we can move forward with the statistical inference techniques associated with the simple linear regression model with normal error terms. The R code below uses the hist function to draw a histogram of the residuals, which is displayed in Figure 2.10. This reflects a population bell-shaped probability distribution for the error terms in the model.

```
fit = lm(Husband ~ Wife, data = PBImisc::heights)
hist(fit$residuals)
qqnorm(fit$residuals)
shapiro.test(fit$residuals)
```

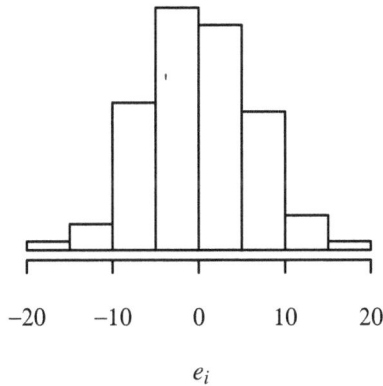

Figure 2.10: Histogram of the residuals for the heights data.

The histogram is useful for a preliminary visual assessment of the normality of
the residuals, but the partitioning of observations into cells can make conclusions
drawn from the histogram misleading. A second technique for visually assessing the normality of the residuals is to inspect a QQ (quantile–quantile) plot of
the residuals. A QQ plot displays the theoretical quantiles of the residuals on
the horizontal axis and the associated sample quantiles on the vertical axis. The
second-to-last line of the R code given above uses the qqnorm function to draw a
QQ plot of the residuals, which is displayed in Figure 2.11. To provide some detail
on the two most extreme points on this plot, the smallest residual is -16.7 which
corresponds to a wife who is 162 centimeters tall who is married to a husband
who is 156 centimeters tall. The theoretical quantile corresponds to a left-hand
tail probability for the standard normal distribution of $0.5/96$ (in general this left-
hand tail probability is $(i-0.5)/n$ for $i = 1, 2, \ldots, n$), which can be calculated in
R with qnorm(0.5 / 96), resulting in a theoretical quantile of -2.56. The point
$(-2.56, -16.7)$ is plotted in the lower-left-hand corner of Figure 2.11. Similarly,
the largest residual is 17.8, which corresponds to a wife who is 147 centimeters
tall who is married to a husband who is 178 centimeters tall. The theoretical
quantile corresponds to a left-hand tail probability for the standard normal distribution of $95.5/96$, which is calculated with qnorm(95.5 / 96), which gives a
theoretical quantile of 2.56. The point $(2.56, 17.8)$ is plotted in the upper-right-
hand corner of Figure 2.11. If the points on a QQ plot fall in an approximately
linear fashion, an analyst can conclude that the assumption of normality is reasonable. Before deciding whether the points fall close enough to a line in this
case with $n = 96$ values, you should make a dozen or so runs of the command
qqnorm(rnorm(96)) so your eye can assess how much deviation from linearity
occurs when the 96 values are truly from a normal distribution. In the case of Figure 2.11, the appropriate conclusion is that these residuals could have been drawn
from a normal population. Any slight departures from linearity on the QQ plot
can be attributed to random sampling variability. This conclusion is consistent
with the conclusion that was drawn from the histogram in Figure 2.10.

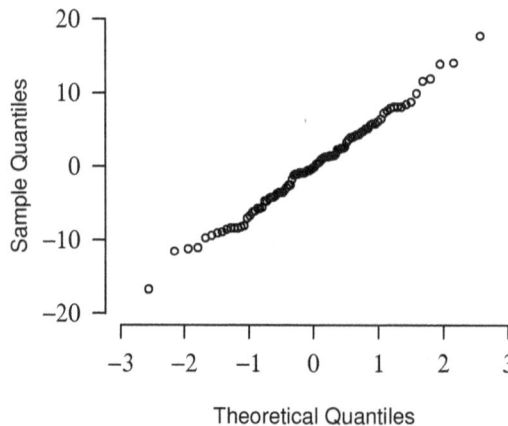

Figure 2.11: QQ normal plot of residuals for the $n = 96$ heights data pairs.

The two visual assessments drawn by examining the histogram and the QQ plot are subjective. A formal statistical goodness-of-fit test should be conducted to confirm the visual assessments. The final statement in the R code invokes the built-in `shapiro.test` function, which executes the Shapiro–Wilk test for normality. The Shapiro–Wilk test was chosen because it has been shown to have superior power over the Anderson–Darling, Kolmogorov–Smirnov, and Lilliefors goodness-of-fit tests. The details associated with the Shapiro–Wilk test can be found in any applied statistics textbook. The p-value for the Shapiro–Wilk test returned by the `shapiro.test` function is 0.953. The null hypothesis for the Shapiro–Wilk is that the residuals have been drawn with a normal population, so the high p-value indicates that we should fail to reject H_0, and we can move forward with using the simple linear regression model with normal error terms for the purposes of statistical inference. Although we have some slight misgivings about non-constant variability (shorter wives marrying husbands with possibly slightly greater variability than their taller counterparts), we will move forward with using the fitted simple linear regression model with normal error terms. All other aspects of the modeling assumptions are satisfied for these data pairs.

(d) Now that the simple linear regression with normal error terms has been established, we can proceed to addressing questions that require statistical inference techniques. Since the original question posed concerned whether the slope of the regression line had a statistically significant *positive* slope, the appropriate hypothesis test is

$$H_0 : \beta_1 = 0$$

versus the one-sided alternative hypothesis

$$H_1 : \beta_1 > 0.$$

The appropriate test statistic is based on Theorem 2.5 which states that

$$\frac{\hat{\beta}_1 - \beta_1}{\sqrt{MSE/S_{XX}}} \sim t(n-2).$$

under H_0, where $\beta_1 = 0$ in this setting. The R code below calculates the test statistic and associated p-value for the hypothesis test with a one-sided alternative.

```
library(PBImisc)
x         = heights$Wife
y         = heights$Husband
n         = length(x)
xbar      = mean(x)
sxx       = sum((x - xbar) ^ 2)
fit       = lm(y ~ x)
beta1hat  = fit$coefficients[2]
sse       = sum(fit$residuals ^ 2)
mse       = sse / (n - 2)
stderror  = sqrt(mse / sxx)
teststat  = beta1hat / stderror
p         = 1 - pt(teststat, n - 2)
```

The test statistic calculated by this code is $t = 11.5$, which corresponds to a p-value of approximately 0 for the one-sided alternative hypothesis.

Some keystrokes can be saved by using R's lm function to calculate the p-value for this test. The three R statements

```
library(PBImisc)
fit = lm(Husband ~ Wife, data = heights)
summary(fit)
```

generate the regression summary given below.

```
Call:
lm(formula = Husband ~ Wife, data = heights)

Residuals:
    Min       1Q   Median       3Q      Max
-16.7438  -4.2838  -0.1615   4.2562  17.7500

Coefficients:
             Estimate Std. Error t value Pr(>|t|)
(Intercept)  37.81005   11.93231   3.169  0.00207 **
heights$Wife  0.83292    0.07269  11.458  < 2e-16 ***
---
Signif. codes:  0 '***' 0.001 '**' 0.01 '*' 0.05 '.' 0.1 ' ' 1

Residual standard error: 6.468 on 94 degrees of freedom
Multiple R-squared:  0.5828,     Adjusted R-squared:  0.5783
F-statistic: 131.3 on 1 and 94 DF,  p-value: < 2.2e-16
```

The first section of the output from the call to the summary function echos the call that was made to the lm function. The second section gives the minimum, maximum, and the quartiles of the residuals. The third section concerns the co-efficients β_0 (on the first line) and β_1 (on the second line). Reading across the second line, (a) the column labeled Estimate contains the least squares estimate $\hat{\beta}_1 = 0.8329$, which was stored in beta1hat in the earlier R code, (b) the column labeled Standard Error contains $\sqrt{MSE/S_{XX}} = 0.07269$, which was stored in stderror in the earlier R code, (c) the column labeled t value contains the test statistic $t = \hat{\beta}_1/\sqrt{MSE/S_{XX}} = 11.46$, which was stored in teststat in the earlier R code, and (d) the column labeled Pr(>|t|) contains the p-value for the test, which was stored in p in the earlier R code, which was calculated using the pt function. The default for R is a two-sided alternative hypothesis, so the p-value given here should be halved in order to obtain the p-value for the hypothesis test with the one-sided alternative hypothesis. The three stars *** that follow the p-value indicate that the p-value is less than 0.001.

So the null hypothesis is rejected. There is *overwhelming* statistical evidence in these data pairs that the slope of the regression line is positive, which implies that height is a statistically significant factor in the selection of a spouse.

(e) The point estimator for the expected height of a husband married to a wife who is $x_h = 150$ centimeters tall is simply the fitted value. An exact 95% confidence interval for $E[Y_h]$ is given by Theorem 2.12. Since the details associated with the calculations are given in Example 2.4, we simply use the R `predict` function to calculate the point and interval estimates.

```
library(PBImisc)
x = heights$Wife
y = heights$Husband
fit = lm(y ~ x)
predict(fit, data.frame(x = 150), interval = "confidence")
```

The point estimate for the expected husband's height is $\hat{Y}_h = 162.7$ centimeters and the associated exact 95% confidence interval is

$$160.4 < E[Y_h] < 165.1.$$

We are 95% confident that the mean husband's height associated with a wife whose height is $X_h = 150$ is between 160.4 and 165.1 centimeters. This confidence interval is illustrated in Figure 2.12.

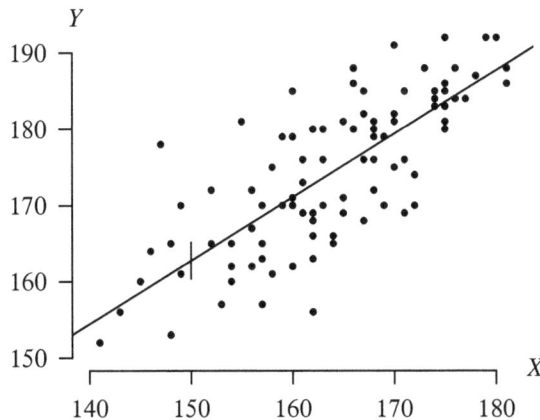

Figure 2.12: Point estimate and 95% confidence interval associated with $X_h = 150$.

(f) Now consider a 97th wife who is not part of the original $n = 96$ data pairs and is $x_h = 150$ centimeters tall. What conclusions can we draw concerning the height of her husband? The point estimator for his expected height is again just the fitted value. An exact 95% confidence interval for $Y_h^\star$ is given by Theorem 2.15. Since the details associated with the calculations are given in Example 2.5, we use the R `predict` function to calculate the point and interval estimates.

```
library(PBImisc)
x        = heights$Wife
y        = heights$Husband
fit = lm(y ~ x)
predict(fit, data.frame(x = 150), interval = "predict")
```

The point estimate for the expected husband's height is again $\hat{Y}_h = 162.7$ centimeters and the associated exact 95% prediction interval is

$$149.7 < Y_h^* < 175.8.$$

The probability that the husband's height associated with a wife whose height is $X_h = 150$ is between 149.7 and 175.8 centimeters is 0.95. This prediction interval is illustrated in Figure 2.13. As expected, it is significantly wider than the associated confidence interval.

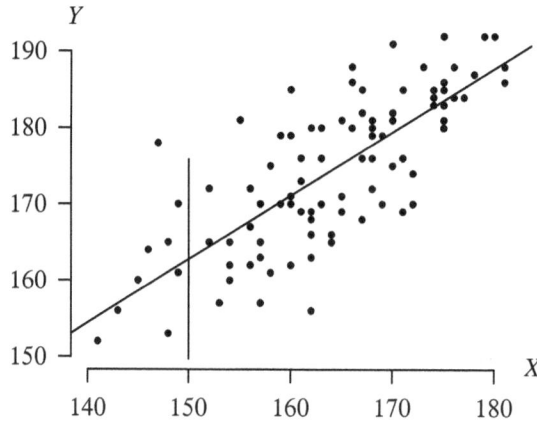

Figure 2.13: Point estimate and 95% prediction interval associated with $X_h = 150$.

The confidence interval and prediction interval associated with $X_h = 150$ could have been calculated for any X_h within the scope of the model. Figure 2.14 shows these confidence and prediction intervals for values of the independent variable (the wife's height) in the scope of the model $141 < X < 181$. The darker gray bands contain the confidence intervals; the lighter gray bands contain the prediction intervals. As indicated in Theorems 2.12 and 2.15, these intervals are narrowest at $\bar{X} = 163.9$ centimeters. The R code for generating a plot similar to that in Figure 2.14 is given below.

```
library(PBImisc)
x = heights$Wife
y = heights$Husband
n = length(x)
plot(NULL, xlim = c(140, 182), ylim = c(140, 202),
     xlab = "Wife's height (cm)", ylab = "Husband's height (cm)",
     axes = FALSE)
axis(side = 1, labels = TRUE, at = seq(140, 180, by = 10))
axis(side = 2, labels = TRUE, at = seq(140, 200, by = 10))
fit = lm(y ~ x)
x1  = 141:181
y1  = predict(fit, data.frame(x = x1), interval = "prediction")
polygon(c(x1, rev(x1)), c(y1[ , 2], rev(y1[ , 3])),
```

```
          col = "gray70", border = NA)
y2  = predict(fit, data.frame(x = x1), interval = "confidence")
polygon(c(x1, rev(x1)), c(y2[ , 2], rev(y2[ , 3])),
          col = "gray40", border = NA)
abline(fit$coefficients)
points(x, y, pch = 16, cex = 0.75)
```

The confidence and prediction intervals are calculated with the `predict` function, the confidence and prediction intervals are plotted with the `polygon` function, the regression line is plotted with the `abline` function, and finally the data pairs are plotted as solid points with the `points` function.

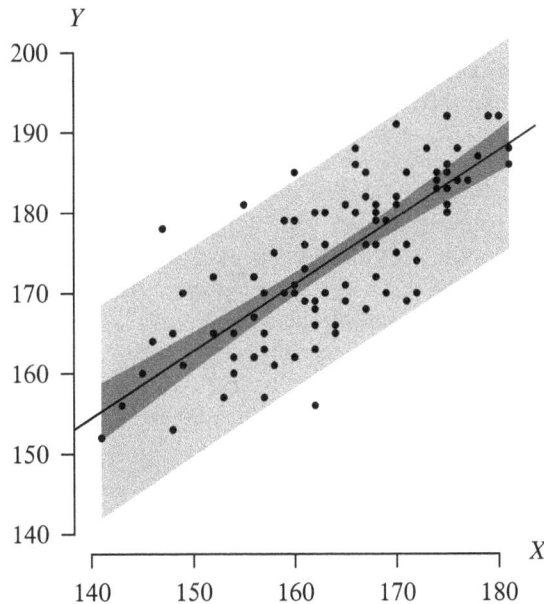

Figure 2.14: Scatterplot, regression line, and 95% confidence and prediction intervals.

The previous example might leave you wondering whether taller (and shorter) women marrying taller (and shorter) men, and having taller (and shorter) children might eventually result in a planet filled with people of more extreme heights. As first noticed by Sir Francis Galton in 1886 and usually known as "regression to the mean," this will probably not be the case. Consider the right-hand tail of the height distribution. A taller-than-average woman will indeed typically date and marry a taller-than-average man, but the husband's height, on average, will not fall as far out into the right-hand-tail of his height distribution as the wife's percentile in her height distribution. Some mathematics associated with the simple linear regression model backs this up. Recall from Definition 1.3 that the coefficient of correlation is

$$r = \pm\sqrt{\frac{SSR}{SSE}},$$

where the sign associated with r is the same as the sign of $\hat{\beta}_1$. Theorem 1.9 gave the alternate formula

$$r = \hat{\beta}_1 \sqrt{\frac{S_{XX}}{S_{YY}}}.$$

This can be rewritten as $\hat{\beta}_1 S_X = r S_Y$, where S_X is the sample standard deviation of $X_1, X_2, \ldots, X_n$ and S_Y is the sample standard deviation of $Y_1, Y_2, \ldots, Y_n$. The left-hand side of this equation represents the expected increase (or decrease) in the dependent variable for a one standard deviation increase in the independent variable. But since $|r| < 1$ in nearly all applications (the only exception is when all data pairs fall in a line), this standard deviation increase in X will result in less than a standard deviation increase in Y. In the previous example, where $r = 0.763$ was the correlation coefficient between the heights of the wives and their husbands, a standard deviation increase in the height of a wife results in a increase of just $0.763 S_Y$ increase in the height of her husband. Women do tend to marry taller men on average, but the height of their husbands, on average, are at a lesser percentile of the men's height distribution than the wife's height percentile.

The next example considers an automotive application of regression which uses speed as an independent variable and stopping distance as a dependent variable.

Example 2.8 R contains a built-in data frame named cars, which consists of $n = 50$ data pairs of speeds (which will be the independent variable) and associated stopping distances (which will be the dependent variable) for cars. The speed X is measured in miles per hour and the stopping distance Y is measured in feet. The data pairs were gathered in the 1920s, which accounts for the top speed of just 25 miles per hour. We would like to establish the relationship between X and Y. Common sense indicates that faster moving cars take a longer distance to stop, so we anticipate a positively correlated set of data pairs. Draw a scatterplot of the data pairs to determine if a simple linear regression model is appropriate, fit a simple linear regression model to the data pairs and assess the adequacy of the model.

The R code

```
x = cars$speed
y = cars$dist
plot(x, y, xlim = c(0, 25), ylim = c(0, 120))
```

generates the scatterplot in Figure 2.15. The xlim and ylim arguments on the plot function are used to include the origin in the scatterplot. The number of data pairs plotted on the scatterplot appears to be only 49 because the data pair $(13, 34)$ appears twice. As expected, Y increases as X increases. The relationship between X and Y is approximately linear, but some complicating factors cast doubt on a linear regression model. First, the relationship between X and Y should pass through the origin (stationary cars require zero feet to stop), but a fitted regression line will miss the origin by a significant margin. This might be evidence that a nonlinear relationship, such as a quadratic relationship, might provide a better fit than a linear relationship. Second, the population variance of the error terms, σ^2, might be increasing as the speeds increase. In spite of these misgivings, we will proceed forward and fit the simple linear regression model and assess whether it is an appropriate model. In the next chapter, other regression models will prove to provide a better fit to this data set.

The scatterplot with the fitted regression line for a simple linear regression model can be generated with the R commands below.

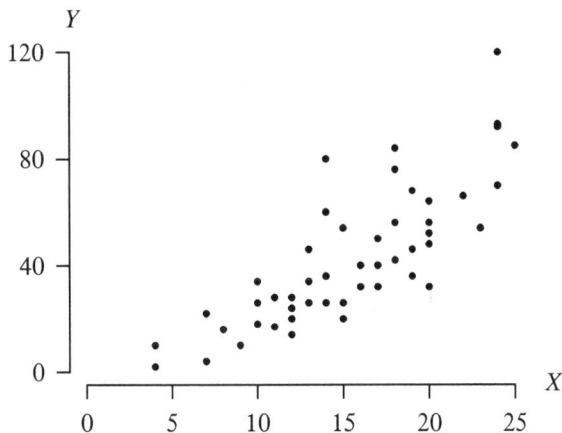

Figure 2.15: Scatterplot of speed X and stopping distance Y for the `cars` data.

```
x = cars$speed
y = cars$dist
plot(NULL, xlim = c(0, 25), ylim = c(0, 120))
polygon(c(0, 0, 25, 25), c(0, 120, 120, 0), col = "gray")
abline(v = seq(0, 25, by = 5), col = "white")
abline(h = seq(0, 120, by = 20), col = "white")
fit = lm(y ~ x)
abline(fit$coefficients)
points(cars, pch = 16)
```

This plot is given in Figure 2.16. The regression is performed by the `lm` function. Some extra features have been added to the plot to give it a slightly different look

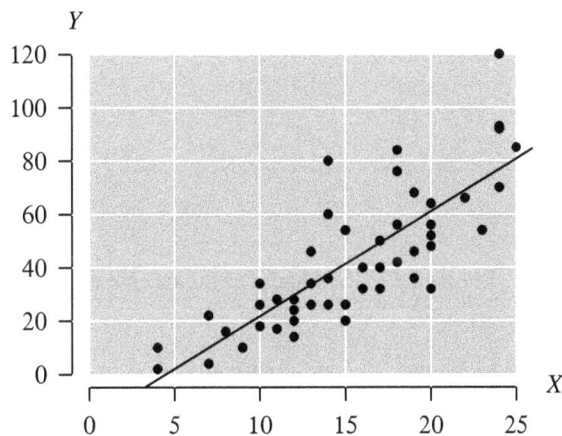

Figure 2.16: Fitted model of speed X and stopping distance Y for the `cars` data.

than previous scatterplots. The `polygon` function colors the plotting region gray. The `abline` function draws the vertical and horizontal white grid lines. Finally, a call to the `points` function with the plotting character parameter `pch` set to 16 plots the points as solid dots on top of the gray background. The intercept and slope of the least squares regression line are

$$\hat{\beta}_0 = -17.6 \qquad \text{and} \qquad \hat{\beta}_1 = 3.9.$$

These correspond to a minimized sum of squares of $SSE = 11,354$. As anticipated, the regression line falls below the origin. Having an estimated stopping distance of $\hat{\beta}_0 = -17.6$ feet for a stationary car makes the simple linear regression model less plausible. The slope of $\hat{\beta}_1 = 3.9$, indicates that there are about four extra feet of stopping distance for each additional mile per hour of speed.

We now investigate whether the residuals appear to be independent and identically distributed observations from a normal population. For those programmers who like succinct coding, a plot of the residuals can be generated with the single R command

```
plot(lm(dist ~ speed, data = cars)$residuals)
```

because the data pairs in the `cars` data frame are sorted by the speeds. The plot of the residuals is given in Figure 2.17. The misgivings that were identified from the scatterplot are also evident in the plot of the residuals in Figure 2.17. The first 21 data pairs (corresponding to the slower speeds and shorter stopping distances) seem to have less dispersion about the regression line than the subsequent 29 data pairs.

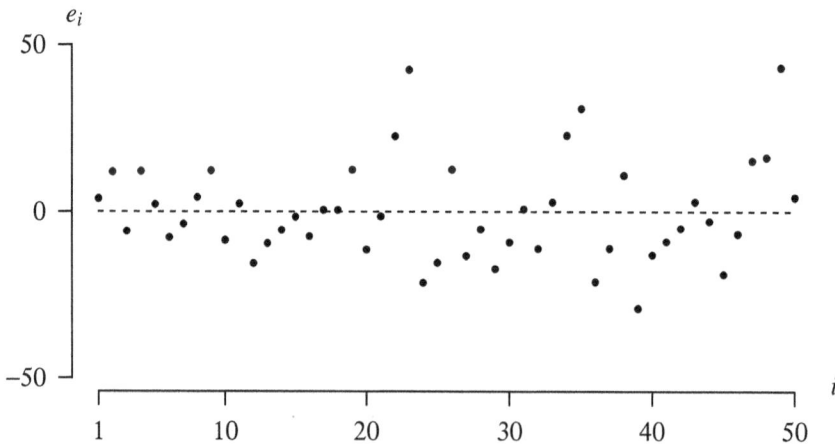

Figure 2.17: Residuals for the `cars` data.

Three tweaks applied to Figure 2.17 are commonly used in regression when analyzing the residuals. First, the residuals can be standardized by subtracting their sample mean and dividing by their estimated standard deviation: $(e_i - \bar{e})/\sqrt{MSE} = e_i/\sqrt{MSE}$, where $\sqrt{MSE}$ is an approximation to the standard deviation of $e_1, e_2, \ldots, e_n$. If the *standardized residuals* are independent and identically distributed realizations from a standard normal population, then approximately 95% of the standardized residuals will fall between -2 and 2. Second, the value of the independent variable is plotted on the horizontal axis rather than using the index of the observation. This ties the plot of the residuals more closely to the scatterplot. Third, the tied value at the data pair $(13, 34)$ has the associated standardized residuals jittered. Continuing with the theme of succinct coding, the R commands

```
res = lm(dist ~ speed, data = cars)$residuals
plot(cars$speed, res / sqrt(sum(res ^ 2) / (length(cars$speed) - 2)))
```

give the appropriate plot (without the jittering). The associated plot of standardized residuals that includes horizontal lines at 0 and ± 2 is given in Figure 2.18. The variance of the deviations from the regression line appear to be increasing as the speed increases, with a smaller spread for speeds between 4 and 13 miles per hour and a larger spread for speeds between 14 and 25 miles per hour. This change in dispersion is inconsistent with the assumption of constant variance of the error terms for a simple linear regression model in Definition 1.1.

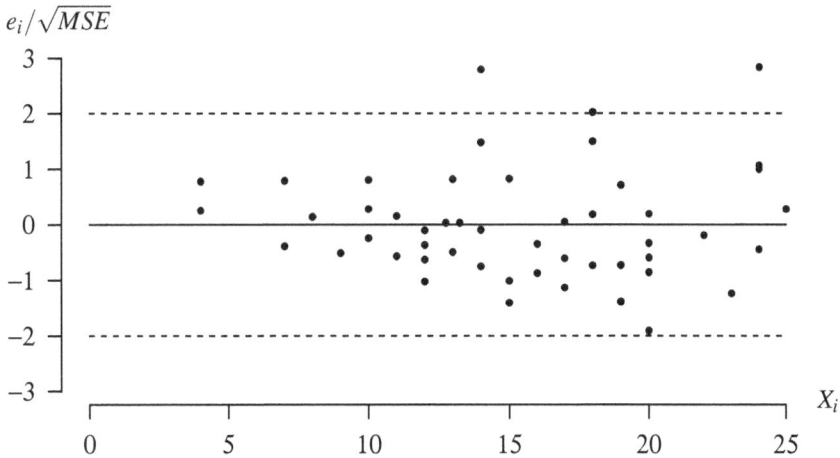

Figure 2.18: Standardized residuals for the `cars` data.

A histogram of the standardized residuals is generated with the *additional* R command

```
hist(res / sqrt(sum(res ^ 2) / (length(cars$speed) - 2)))
```

The histogram of the standardized residuals is given in Figure 2.19. The longer stopping distances in the right-hand tail of this histogram cast doubt on the assumption of normal error terms.

A QQ plot to assess the normality of the residuals is generated with the R command

```
qqnorm(lm(dist ~ speed, data = cars)$residuals)
```

The QQ plot will assume the same shape regardless of whether the residuals or standardized residuals are examined; the only difference will be in the scale used on the

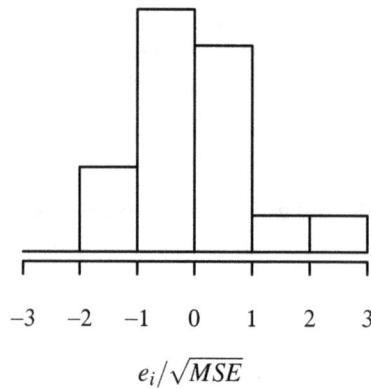

Figure 2.19: Histogram of the standardized residuals for the `cars` data.

vertical axis. The QQ plot is displayed in Figure 2.20. The QQ plot shows some significant departures from linearity. First, there is a large jump in the sequence of observations between the 37th ordered residual $e_{(37)} = 4.27$ and 38th ordered residual $e_{(38)} = 10.86$, which are the values plotted on the vertical axis. Second, the two largest residuals, $e_{(49)} = 42.53$ and $e_{(50)} = 43.20$, might indicate that the right-hand tail of the distribution of the error terms is not symmetric with the left-hand tail of the distribution.

The visual assessment that normally distributed error terms are not appropriate for these data pairs can be confirmed by conducting the Shapiro–Wilk test for normality. The R statement

```
shapiro.test(lm(dist ~ speed, data = cars)$residuals)
```

conducts the Shapiro–Wilk test on the residuals and returns a p-value of 0.0215. The null hypothesis that the error terms are normally distributed is rejected in this case,

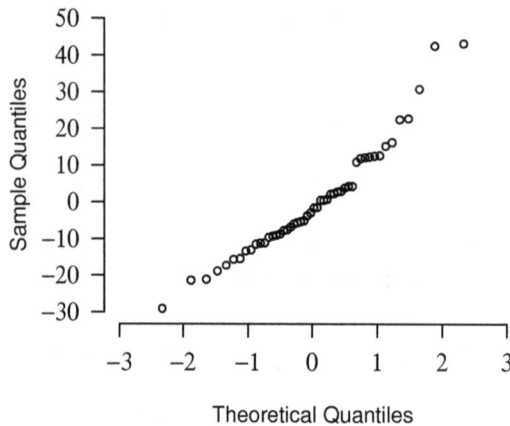

Figure 2.20: QQ normal plot of residuals for the $n = 50$ `cars` data pairs.

which confirms our visual assessments via the plot of the residuals, the histogram of the residuals, and the QQ plot for normality.

We have identified four misgivings with respect to using the simple linear regression model with normal error terms to model the relationship between the speed of a car and its stopping distance:

- the relationship between X and Y might be nonlinear,
- the variance of the error terms appears to be increasing in X (this is known to regression modelers as *heteroscadasticity*),
- the regression line does not pass near the origin as one would expect it would from the problem setting because a stationary vehicle does not require any distance to stop, and
- the non-normality of the errors as indicated by the plot of the residuals, the histogram of the residuals, the QQ plot for normality, and the Shapiro–Wilk test.

So we abandon using the use of a simple linear regression model to describe the relationship between speed and stopping distance. Although simple linear regression *can* be used to model the relationship between speed and stopping distance, it *should not* be used here because the model is not valid. This data set will be reexamined in the next chapter in an effort to establish a regression model that overcomes some of the difficulties described here.

The fourth and final example concerns a large data set of home sale prices and associated predictors. Real estate platforms, such as Zillow and Trulia, are able to assess home values using a variety of predictors, and one key predictor is illustrated in the final example.

Example 2.9 The `ames` data frame in the `modeldata` package in R contains 2930 rows and 74 columns of data concerning houses that sold in Ames, Iowa from 2006 to 2010. Each row in the data frame contains data on one particular home. Each column in the data frame contains data on one particular aspect of a home, such as the number of bedrooms, the acreage of the lot, whether the home has a pool, or the area of the living space measured in square feet. One of the primary factors that a real estate assessor uses to determine the value of a home is the number of square feet in the home. This example concerns the modeling of the selling price of a home in Ames as a function of the number of square feet of living space.

The following R code installs the `modeldata` package, loads the `modeldata` package into the current R session, extracts the living space column from the `ames` data frame and places it in the vector x, extracts the sales price column and places it in the vector y, and generates a scatterplot of the $n = 2930$ data pairs, which is displayed in Figure 2.21.

```
install.packages("modeldata")
library(modeldata)
x = ames$Gr_Liv_Area
y = ames$Sale_Price
plot(x, y)
```

As expected, larger homes sell for higher prices on average. The scatterplot clearly shows that a simple linear regression model is *not* appropriate for this data set because

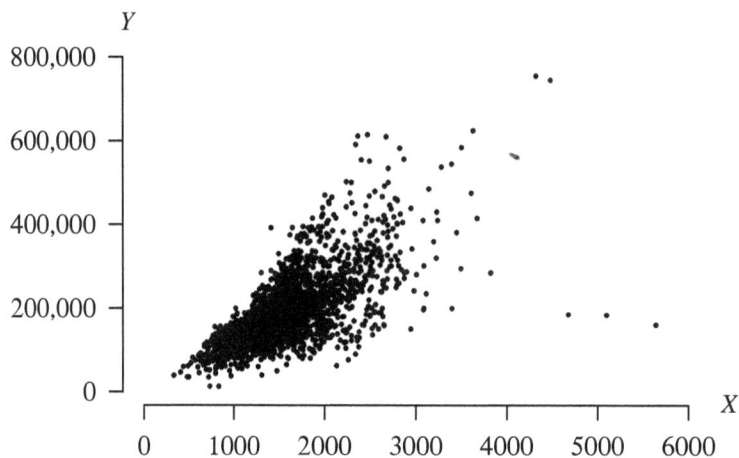

Figure 2.21: Scatterplot of living area X and sale price Y for the `ames` data.

the variance of the error terms is not constant over the various values of X. The variance of the error terms increases as the size of a home increases. In addition, the three large-but-relatively-inexpensive homes will exert significant leverage over a regression line. Although some remedial procedures to account for handling nonconstant variance of the error terms are given in the next chapter, we take the approach of restricting the home sizes to 2500 ft^2 to 3500 ft^2 in the hope that the simple linear regression assumptions will be satisfied on the restricted scope. The R code below generates the scatterplot and the associated regression line for the $n = 120$ homes satisfying $2500 \leq X \leq 3500$ displayed in Figure 2.22.

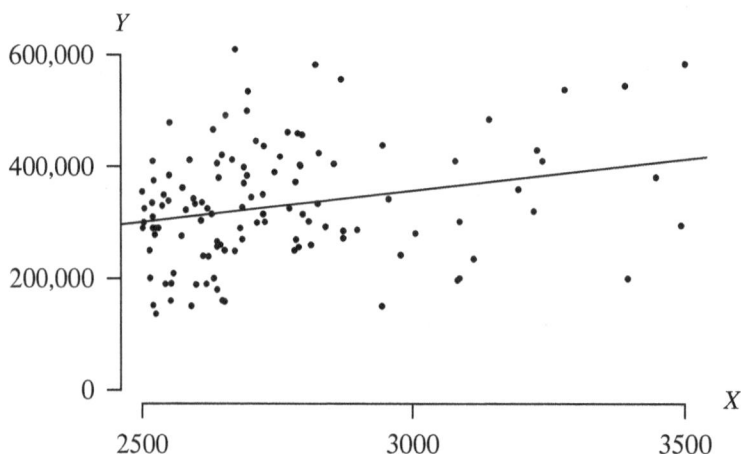

Figure 2.22: Scatterplot of living area X and sale price Y for the `ames` data, $2500 \leq X \leq 3500$.

```
library(modeldata)
i   = ames$Gr_Liv_Area >= 2500 & ames$Gr_Liv_Area <= 3500
x   = ames$Gr_Liv_Area[i]
y   = ames$Sale_Price[i]
fit = lm(y ~ x)
plot(x, y)
abline(fit$coefficients)
```

As was the case with the larger data set, there is a positive correlation between the size of the home and its sales price. Even though the observations are clustered more densely on the left-hand side of the scatterplot, the variance of the error terms does not seem to vary over this restricted scope of $2500 \leq X \leq 3500$. The least squares estimators of the intercept and slope of the regression line are

$$\hat{\beta}_0 = \$21,233 \qquad \text{and} \qquad \hat{\beta}_1 = \$112.$$

The price of a home increases by an average of $112 for every additional square foot in the home and the estimated price of a home with zero square feet is $21,223. In other words, the estimated value of the land is $21,223. The estimated value of the land will not be very precise because we have eliminated homes with less than 2500 square feet in the reduced data set, leaving the intercept way outside of the scope of the reduced simple linear regression model. We anticipate a particularly wide confidence interval for β_0 if we determine that the simple linear regression model is valid.

The next step is to assess the residuals for the $n = 120$ homes with square footage between 2500 and 3500 to determine whether a simple linear regression model with normal error terms is appropriate. The R code below (*a*) generates a plot of the standardized residuals, (*b*) generates a histogram of the standardized residuals (*c*) generates a QQ plot of the residuals, and (*d*) performs the Shapiro–Wilk test for normality of the residuals.

```
library(modeldata)
i   = ames$Gr_Liv_Area >= 2500 & ames$Gr_Liv_Area <= 3500
x   = ames$Gr_Liv_Area[i]
y   = ames$Sale_Price[i]
fit = lm(y ~ x)
plot(x, scale(fit$residuals))
hist(scale(fit$residuals))
qqnorm(fit$residuals)
shapiro.test(fit$residuals)
```

The plot of the standardized residuals is given in Figure 2.23. Although there are a few more homes that sell significantly above their predicted value than significantly below their predicted value (that is, outside of the dashed lines in Figure 2.23), normally distributed error terms seems plausible.

The histogram of the standardized residuals is given in Figure 2.24. The histogram is consistent with a bell-shaped distribution. The nonsymmetry between the heights of the two most central bars in the histogram might possibly be due to the particular binning

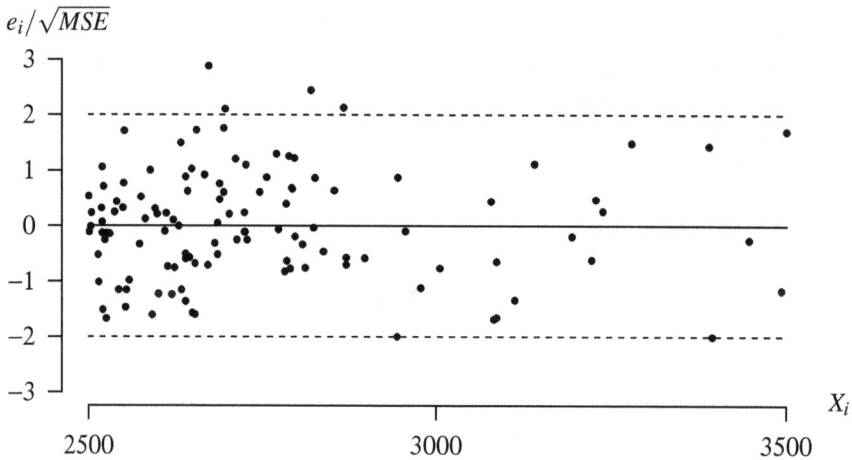

Figure 2.23: Standardized residuals for the ames data on $2500 \leq X \leq 3500$.

that was performed internally in R. The QQ plot tends to be a better graphic than the histogram to visually assess the normality of the residuals.

The QQ plot for the normality of the residuals is given in Figure 2.25. The graph of the sample and theoretical quantiles appears to be fairly close to linear, so we expect that the Shapiro–Wilk test will yield a fairly high p-value associated with normally distributed error terms.

The Shapiro–Wilk test for normality of the residuals yields a p-value of $p = 0.4339$. Since this p-value exceeds 0.05, we fail to reject the null hypothesis of normally distributed error terms. This analysis of the residuals is enough evidence for us to proceed with the simple linear regression model with normal error terms.

An ANOVA table can be generated in R with the anova function using the code below. This is an organized way to display the sum of squares and associated mean squares.

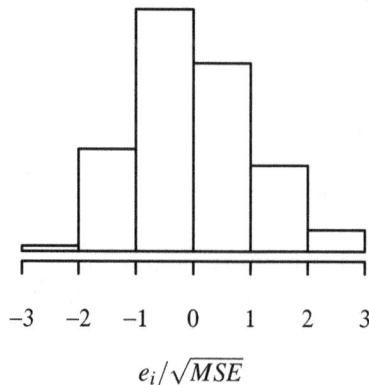

Figure 2.24: Histogram of the standardized residuals for the ames data.

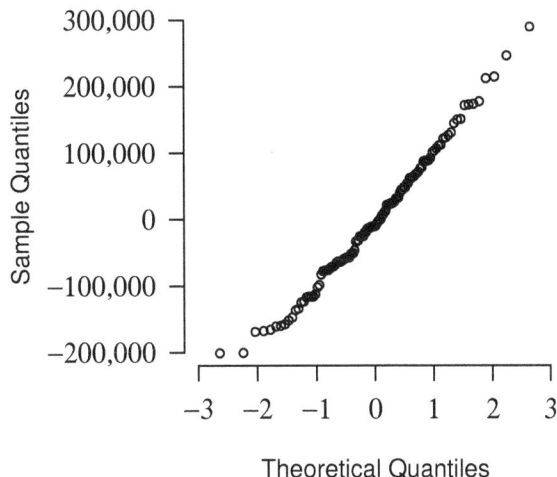

Figure 2.25: QQ normal plot of residuals for the $n = 120$ ames data pairs.

```
library(modeldata)
i   = ames$Gr_Liv_Area >= 2500 & ames$Gr_Liv_Area <= 3500
x   = ames$Gr_Liv_Area[i]
y   = ames$Sale_Price[i]
fit = lm(y ~ x)
anova(fit)
```

This code returns the output given below.

```
Analysis of Variance Table

Response: y
          Df     Sum Sq    Mean Sq F value    Pr(>F)
x          1 8.0882e+10 8.0882e+10  8.0016  0.005493 **
Residuals 118 1.1928e+12 1.0108e+10
---
Signif. codes:  0 '***' 0.001 '**' 0.01 '*' 0.05 '.' 0.1 ' ' 1
```

This output corresponds to the ANOVA table given in Table 2.5.

Source	SS	df	MS	F	p
Regression	$8.0882 \cdot 10^{10}$	1	$8.0882 \cdot 10^{10}$	8.0016	0.005493
Error	$1.1928 \cdot 10^{12}$	118	$1.0108 \cdot 10^{10}$		
Total	$1.2737 \cdot 10^{12}$	120			

Table 2.5: ANOVA table for the restricted ames housing data.

The test statistic $F = 8.002$ and associated p-value $p = 0.005$ indicate that the null hypothesis

$$H_0 : \beta_1 = 0$$

should be rejected in favor of

$$H_1 : \beta_1 \neq 0$$

for the restricted data pairs. The statistically significant positive slope of the regression line indicates that larger homes, on average, have higher selling prices on the range $2500 \leq X \leq 3500$, which is consistent with intuition. Based on this F test, we expect that a confidence interval for β_1 will not include $\beta_1 = 0$. The square root of the mean square error, which is $\hat{\sigma} = 100{,}540$, provides an estimate of the standard deviation of the error terms in the model. Using Theorem 2.3, an exact two-sided $100(1 - \alpha)\%$ confidence interval for σ is

$$\sqrt{\frac{SSE}{\chi^2_{n-2,\,\alpha/2}}} < \sigma < \sqrt{\frac{SSE}{\chi^2_{n-2,\,1-\alpha/2}}}.$$

So an exact two-sided 95% confidence interval for σ for the restricted $n = 120$ ames housing data pairs is

$$\sqrt{\frac{1{,}192{,}772{,}613{,}044}{149.96}} < \sigma < \sqrt{\frac{1{,}192{,}772{,}613{,}044}{89.83}}$$

or

$$89{,}186 < \sigma < 115{,}233.$$

The coefficient of determination and the coefficient of correlation can be calculated using the values from the ANOVA table. Using Definition 1.3, the coefficient of determination is

$$R^2 = \frac{SSR}{SST} = \frac{80{,}881{,}836{,}066}{1{,}273{,}654{,}449{,}110} = 0.0635$$

and the coefficient of correlation is

$$r = \sqrt{R^2} = 0.252.$$

So 6.35% of the variation in the selling price of a home is explained by the square footage of the home over the range $2500 \leq X \leq 3500$.

With the simple linear regression model with normal error terms established, we can compute confidence intervals for β_0 and β_1. The R code below uses the `confint` function to compute the lower and upper bounds of 95% confidence intervals for the intercept and slope of the regression line.

```
library(modeldata)
i       = ames$Gr_Liv_Area >= 2500 & ames$Gr_Liv_Area <= 3500
x       = ames$Gr_Liv_Area[i]
y       = ames$Sale_Price[i]
fit     = lm(y ~ x)
confint(fit, "x", level = 0.95)
confint(fit, "(Intercept)", level = 0.95)
```

The 95% confidence intervals are

$$-195{,}284 < \beta_0 < 237{,}749 \qquad \text{and} \qquad 33.57 < \beta_1 < 190.27.$$

The extraordinarily wide confidence interval for β_0 is due to the significant vertical distances between the data pairs and the regression line, and the large gap between the smallest value of the independent variable ($X = 2500$ square feet) and the value of the independent variable for an empty lot ($X = 0$ square feet). Including some of the other predictors of the sale price of a home in a regression model would narrow this confidence interval. Both confidence intervals would be narrowed with a larger number of data pairs.

In addition to individual confidence intervals for β_0 and β_1, a joint confidence region for both of the parameters can be generated. From Theorem 2.17, the boundary of the exact joint $100(1-\alpha)\%$ confidence region for β_0 and β_1 consists all β_0 and β_1 values satisfying

$$\frac{n-2}{2\sum_{i=1}^{n} e_i^2} \left[n\left(\hat{\beta}_0 - \beta_0\right)^2 + 2\left(\hat{\beta}_0 - \beta_0\right)\left(\hat{\beta}_1 - \beta_1\right)\sum_{i=1}^{n} X_i + \left(\hat{\beta}_1 - \beta_1\right)^2 \sum_{i=1}^{n} X_i^2 \right] = F_{2,n-2,\alpha},$$

where $F_{2,n-2,\alpha}$ is the $1-\alpha$ fractile of an F distribution with 2 and $n-2$ degrees of freedom. The boundary of the confidence region is an ellipse in the β_0 and β_1 plane. Plotting this ellipse requires the use of numerical methods and some significant coding, so it is easiest to use an R package that provides this capability. The `ellipse` function in the `ellipse` package is capable of plotting the ellipse. The R code below generates the confidence region plotted in Figure 2.26. The point estimators $\hat{\beta}_0$ and $\hat{\beta}_1$ are plotted as a point at the center of the ellipse. Dashed lines have been added at the confidence interval bounds for the individual confidence intervals for β_0 and β_1.

```
library(modeldata)
library(ellipse)
```

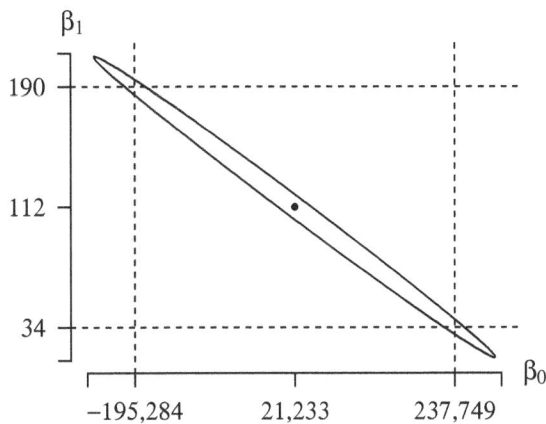

Figure 2.26: Exact 95% confidence region for β_0 and β_1 for the restricted `ames` data.

```
i   = ames$Gr_Liv_Area >= 2500 & ames$Gr_Liv_Area <= 3500
x   = ames$Gr_Liv_Area[i]
y   = ames$Sale_Price[i]
fit = lm(y ~ x)
plot(ellipse(fit))
```

Not surprisingly based on the off-diagonal elements of the variance–covariance matrix of $(\hat{\beta}_0, \hat{\beta}_1)$ from Theorem 1.4, there is a negative correlation between $\hat{\beta}_0$ and $\hat{\beta}_1$ which accounts for the tilt in the ellipse.

2.6 Exercises

2.1 True or false: An alternative way to express the simple linear regression model with normal error terms is

$$Y \sim N\left(\beta_0 + \beta_1 X, \sigma^2\right)$$

or

$$Y_i \sim N\left(\beta_0 + \beta_1 X_i, \sigma^2\right)$$

for $i = 1, 2, \ldots, n$.

2.2 Consider a simple linear regression model with normal error terms and population parameters $\beta_0 = 5$, $\beta_1 = 2$, and $\sigma = 2$. The independent variable assumes the values $x = 1, 2, \ldots, 10$, and $n = 10$ data pairs are collected, one for each potential value of the independent variable.

(a) Conduct a Monte Carlo simulation experiment which determines the shape of the marginal distribution of Y.

(b) How do you think the marginal distribution of Y will change as $\sigma \to 0$.

(c) How do you think the marginal distribution of Y will change as $\sigma \to \infty$.

2.3 Show that

$$\frac{SSE}{\sigma^2} \sim \chi^2(n-2).$$

2.4 For a simple linear regression model with normal error terms and known value of σ^2, give an exact two-sided $100(1-\alpha)\%$ confidence interval for β_1.

2.5 Fit the data pairs from the first of the Anscombe's quartet from Example 2.6 to the simple linear regression model with normal error terms and give point and exact two-sided 95% confidence intervals for the parameters β_0, β_1, and σ^2.

2.6 For what value of the independent variable is the confidence interval for the expected value of the associated dependent variable the narrowest?

2.7 For a simple linear regression model with normal error terms, known value of σ^2, and a fixed value X_h in the scope of the model, give an exact two-sided $100(1-\alpha)\%$ confidence interval for $E[Y_h] = \beta_0 + \beta_1 X_h$.

2.8 Conduct a Monte Carlo simulation that yields compelling numerical evidence that the confidence interval for $E[Y_h]$ from Theorem 2.12 is an *exact* confidence interval for the following parameter settings: $n = 10$, $\beta_0 = 1$, $\beta_1 = 1/2$, $\sigma^2 = 1$, $X_h = 3$, $\alpha = 0.05$, and $X_i = i$ for $i = 1, 2, \ldots, 10$.

2.9 Prove Theorem 2.11.

2.10 True or false: The width of a 95% confidence interval for $E[Y_h]$ shrinks to zero in the limit as $n \to \infty$.

2.11 True or false: The width of a 95% prediction interval for $Y_h^\star$ shrinks to zero in the limit as $n \to \infty$.

2.12 The R data frame named `longley` contains seven macroeconomical variables from the United States collected from 1947 to 1962. Use the number of people employed to predict the gross national product (GNP) measured in constant 1954 dollars. Assuming that the simple linear regression model with normal error terms is appropriate,

  (a) make a scatterplot of the $n = 16$ data pairs and superimpose the regression line,

  (b) make a plot of the standardized residuals,

  (c) make a QQ plot of the residuals,

  (d) conduct the Shapiro–Wilk test for normality of the residuals,

  (e) give a point estimate and an exact 95% confidence interval for the slope β_1,

  (f) give a point estimate and an exact 95% confidence interval for the intercept β_0,

  (g) give a point estimate and an exact 95% confidence interval for the mean value of the GNP, $E[Y_h]$, when $X_h = 65$ million people are employed, and

  (h) give a point estimate and an exact 95% prediction interval for the GNP, $Y_h^\star$, associated with a new data pair when $X_h = 65$ million people are employed.

2.13 Under the simple linear regression model with normal error terms and parameters estimated from the data pairs (X_1, Y_1), (X_2, Y_2), ..., (X_n, Y_n), the exact two-sided $100(1 - \alpha)\%$ prediction interval for $Y_h^\star$ given in Theorem 2.15 is appropriate for a *single* new observation associated with a fixed value of the independent variable X_h. What if there are m new observations? In this case, an exact two-sided $100(1 - \alpha)\%$ prediction interval for the mean response $Y_h^\star$ is

$$\hat{Y}_h^\star - t_{n-2,\alpha/2}\sqrt{\hat{V}\left[\hat{Y}_h^\star\right]} < Y_h^\star < \hat{Y}_h^\star + t_{n-2,\alpha/2}\sqrt{\hat{V}\left[\hat{Y}_h^\star\right]},$$

for $Y_h^\star$, where

$$\hat{Y}_h^\star = \hat{\beta}_0 + \hat{\beta}_1 X_h \qquad \text{and} \qquad \hat{V}\left[\hat{Y}_h^\star\right] = \left[\frac{1}{m} + \frac{1}{n} + \frac{(X_h - \bar{X})^2}{S_{XX}}\right] MSE$$

and X_h is a fixed value of the independent variable within the scope of the simple linear regression model. Find a 95% prediction interval for the `heights` data pairs (using the wife's height as the independent variable) from the `PBImisc` package from Example 2.7 with $m = 4$ and $X_h = 150$.

2.14   For the prediction interval for the population mean of the average of m new observations at a single setting of the independent variable X_h given in the previous question, what does the prediction interval collapse to in the limit as $m \to \infty$.

2.15   Consider the built-in data frame in R named `trees`, which contains data pairs of diameters (which will be the independent variable and is erroneously labeled `Girth` in the data frame) measured at 4 feet 6 inches above the ground and associated volumes (which will be the dependent variable) for $n = 31$ felled black cherry trees. Assuming that the simple linear regression model with normal error terms is appropriate, perform the following statistical inference procedures.

    (a) Plot the data pairs and the associated regression line.

    (b) Find a point estimate and an exact 95% confidence interval for β_1. Interpret the point estimate and the confidence interval.

    (c) Find a point estimate and an exact 95% confidence interval for the mean volume, $E[Y_h]$, when the diameter is $X_h = 20$ inches.

    (d) Find a point estimate and an exact 95% prediction interval for the volume, $Y_h^\star$, associated with a new data pair with a diameter of $X_h = 20$ inches.

    (e) Graph all values of the exact 95% confidence interval bounds for the expected volume for all diameters in the scope of the simple linear regression model. Also, graph all values of the exact 95% prediction interval bounds for the volume for a 32nd tree for all diameters in the scope of the simple linear regression model.

2.16   Plot a 95% confidence region for the data pairs in the `cars` data set under a simple linear regression model with normal error terms

    (a) using numerical methods, and

    (b) using the `ellipse` function from the `ellipse` package.

Include the maximum likelihood estimates for $\hat{\beta}_0$ and $\hat{\beta}_1$ and 95% confidence intervals for β_0 and β_1 on your plot.

2.17   Conduct a Monte Carlo simulation that provides convincing numerical evidence that the confidence region given in Theorem 2.17 is an exact confidence region for the following parameter settings: $n = 10$, $\beta_0 = 1$, $\beta_1 = 1/2$, $\sigma^2 = 1$, $\alpha = 0.05$, and $X_i = i$ for $i = 1, 2, \ldots, 10$.

2.18   Consider the simple linear regression model with normal error terms applied to the first set of $n = 11$ data pairs from Anscombe's quartet from Example 2.6. Show that the p-values are identical for testing

$$H_0 : \beta_1 = 0$$

versus

$$H_1 : \beta_1 \neq 0$$

using

    (a) the F test based on using the test statistic which is the ratio of MSR to MSE, and

    (b) the t test based on using the test statistic $\hat{\beta}_1 / \sqrt{MSE / S_{XX}}$.

2.19 Figures 1.24, 1.25, and 1.26 depict three examples of extreme cases for SSE, SSR, and SST for $n = 7$ data pairs. Assuming the simple linear regression model with normal error terms is an appropriate model,

(a) plot and label the potential points associated with the extreme cases when SSR is plotted on the horizontal axis and SSE is plotted on the vertical axis, and

(b) on this same graph, shade the area associated with rejecting H_0 at level of significance at $\alpha = 0.05$ for the statistical test

$$H_0 : \beta_1 = 0$$

versus

$$H_1 : \beta_1 \neq 0.$$

2.20 Plot a power function for the F test for testing

$$H_0 : \beta_1 = 0$$

versus

$$H_0 : \beta_1 \neq 0$$

for $n = 10$, $\beta_0 = 1$, $\sigma^2 = 1$, $\alpha = 0.05$, and $X_i = i$, for $i = 1, 2, \ldots, n$. You may use Monte Carlo simulation or the noncentral F distribution to generate the power function. Allow β_1 to vary from -1 to 1 in the plot.

2.21 Make plots of the standardized residuals for the four data sets from Anscombe's quartet given in Example 2.6.

2.22 The confidence interval for $E[Y_h]$ given in Theorem 2.12 is meaningful for a fixed value of the independent variable X_h. What if a *confidence band* that contains the entire regression line with a prescribed probability is desired. The Working–Hotelling $100(1-\alpha)\%$ confidence band for the regression line at any level X_h is given by

$$\hat{Y}_h - t_{n-2,\alpha/2}\sqrt{\hat{V}\left[\hat{Y}_h\right]} < E[Y_h] < \hat{Y}_h + t_{n-2,\alpha/2}\sqrt{\hat{V}\left[\hat{Y}_h\right]},$$

under the simple linear regression model with normal error terms, where

$$\hat{Y}_h = \hat{\beta}_0 + \hat{\beta}_1 X_h \qquad \text{and} \qquad \hat{V}\left[\hat{Y}_h\right] = 2F_{2,n-2,\alpha}MSE\left[\frac{1}{n} + \frac{\left(X_h - \bar{X}\right)^2}{S_{XX}}\right].$$

Plot a 95% confidence band for the `heights` data pairs from Example 2.7.

Chapter 3

Topics in Regression

The previous two chapters have provided a detailed introduction to the basic principles underlying simple linear regression. This chapter will cover some additional topics in regression, but not with the same detail as in the previous two chapters. Sometimes just a single example will illustrate a regression topic that deserves an entire chapter in a full-semester regression course. The topics considered in this chapter are forcing a regression line through the origin, diagnostics, remedial procedures, the matrix approach to simple linear regression, multiple linear regression, weighted least squares estimators, regression models with nonlinear terms, and logistic regression.

3.1 Regression Through the Origin

Applications occasionally arise in which it is of benefit to force a regression line to pass through the origin. To illustrate such applications, return to Examples 1.1 and 1.3 in which Bob and Cheryl each had the number of sales per week as an independent variable X. In both of the examples, $X = 0$ sales per week corresponds to $Y = 0$ commissions (for Bob) and $Y = 0$ revenue per week (from Cheryl's sales). In these settings it is sensible to force the regression line to pass through the origin; estimating a population intercept does not make sense. The resulting regression model does not contain the β_0 parameter. The simple linear regression model forced through the origin, sometimes abbreviated RTO for regression through the origin, is defined next.

Definition 3.1 A *simple linear regression model forced through the origin* is given by

$$Y = \beta_1 X + \varepsilon,$$

where

- X is the independent variable, assumed to be a fixed value observed without error,

- Y is the dependent variable, which is a continuous random variable,

- β_1 is the population slope of the regression line, which is an unknown constant, and

- ε is the error term, a random variable that accounts for the randomness in the relationship between X and Y, which has population mean zero and finite population variance σ^2.

The regression parameter β_1 can be estimated using least squares from the data pairs (X_i, Y_i) for $i = 1, 2, \ldots, n$.

Theorem 3.1 Let $(X_1, Y_1), (X_2, Y_2), \ldots, (X_n, Y_n)$ be n data pairs satisfying $\sum_{i=1}^{n} X_i^2 > 0$. The *least squares estimator* of β_1,

$$\hat{\beta}_1 = \frac{\sum_{i=1}^{n} X_i Y_i}{\sum_{i=1}^{n} X_i^2},$$

minimizes the sum of the squared deviations between Y_i and the associated fitted value $\hat{\beta}_1 X_i$ in the simple linear regression model forced through the origin.

Proof The sum of squared deviations between the observed values of the dependent variable and the associated fitted values is

$$S = \sum_{i=1}^{n} (Y_i - \beta_1 X_i)^2.$$

To minimize S with respect to β_1, take the derivative of S with respect to β_1:

$$\frac{dS}{d\beta_1} = -2 \sum_{i=1}^{n} X_i (Y_i - \beta_1 X_i) = 0$$

or

$$\sum_{i=1}^{n} X_i Y_i - \beta_1 \sum_{i=1}^{n} X_i^2 = 0.$$

This equation can be solved in closed-form for $\hat{\beta}_1$ as

$$\hat{\beta}_1 = \frac{\sum_{i=1}^{n} X_i Y_i}{\sum_{i=1}^{n} X_i^2}.$$

To show that the *least* squares estimator $\hat{\beta}_1$ minimizes S, take a second derivative of S:

$$\frac{d^2 S}{d\beta_1^2} = 2 \sum_{i=1}^{n} X_i^2.$$

Since $\sum_{i=1}^{n} X_i^2 > 0$, this second derivative, which is just twice a sum of squares, must be positive. Hence, S is minimized at $\hat{\beta}_1$. □

The next example conducts a hypothesis test to determine whether it is appropriate to drop the intercept term from the simple linear regression model based on the data pairs, and then proceeds to fit the reduced model.

Example 3.1 The R built-in data set `Formaldehyde` consists of the $n = 6$ data pairs given in Table 3.1. The independent variable `carb` is the carbohydrate level (ml) and the dependent variable `optden` is the optical density in a chemical experiment. Fit a simple linear regression to the model using the ordinary least squares estimates. If there is no statistically significant difference between the estimated intercept and zero, then fit a simple linear regression model forcing the regression line to pass through the origin to the data pairs.

carb	optden
0.1	0.086
0.3	0.269
0.5	0.446
0.6	0.538
0.7	0.626
0.9	0.782

Table 3.1: `Formaldehyde` data set from R.

The scatterplot given in Figure 3.1 shows a strong linear relationship between carbohydrates (measured in ml) and optical density (measured by the reading of the resulting purple color on a spectrophotometer) for the $n = 6$ data pairs. The nearly-perfect linear relationship provides overwhelming visual evidence that a simple linear regression model is appropriate for approximating the relationship between X and Y.

The R commands below fit the standard simple linear regression model (including an intercept) to the six data pairs.

```
fit = lm(optden ~ carb, data = Formaldehyde)
summary(fit)
```

The point estimates for the intercept and slope of the regression line are

$$\hat{\beta}_0 = 0.00509 \qquad \text{and} \qquad \hat{\beta}_1 = 0.876.$$

The call to `summary(fit)` indicates that there is no statistically significant difference between the point estimate for the intercept and 0. The p-value associated with the hypothesis test

$$H_0 : \beta_0 = 0$$

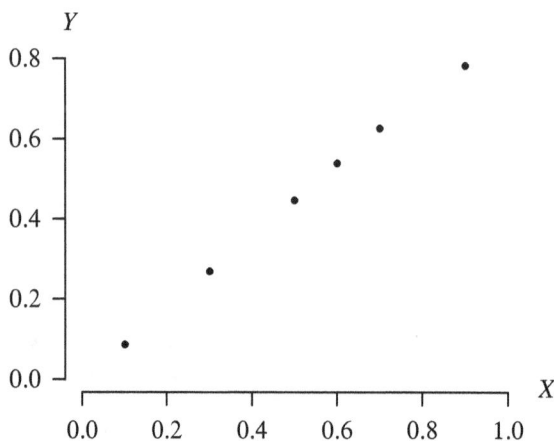

Figure 3.1: A scatterplot of the `Formaldehyde` data pairs.

versus

$$H_0 : \beta_0 \neq 0$$

is 0.55, which is statistical evidence that the intercept does not differ significantly from $\beta_0 = 0$. This p-value, perhaps along with some information about the chemical experiment itself, might cause the experimenter to consider the reduced model which is forced through the origin. This hypothesis test requires normally distributed error terms. The usual analysis of residuals to determine whether a simple linear regression model with normal error terms is appropriate in this setting will be abandoned here because of the small sample size. Histograms and statistical tests have diminished meaning with only $n = 6$ data pairs. The best we can do to assess the normality of the error terms is to use a graphical display such as a QQ plot.

Using Theorem 3.1, the least squares estimate for the slope of the regression line forced through the origin is

$$\hat{\beta}_1 = \frac{\sum_{i=1}^{n} X_i Y_i}{\sum_{i=1}^{n} X_i^2} = 0.884,$$

which can be calculated with the R statements given below.

```
x    = Formaldehyde$carb
y    = Formaldehyde$optden
beta = sum(x * y) / sum(x * x)
print(beta)
```

Not surprisingly, the slope of the regression line forced through the origin is very close to the slope of the regression line with the model that includes an intercept. The optical density increases by 0.884 for every unit increase in the carbohydrates. Figure 3.2 contains a scatterplot of the data pairs and the associated regression line forced through the origin. The model clearly provides an adequate approximation to the relationship between the independent variable X and the dependent variable Y over the scope of the model shown in Figure 3.2.

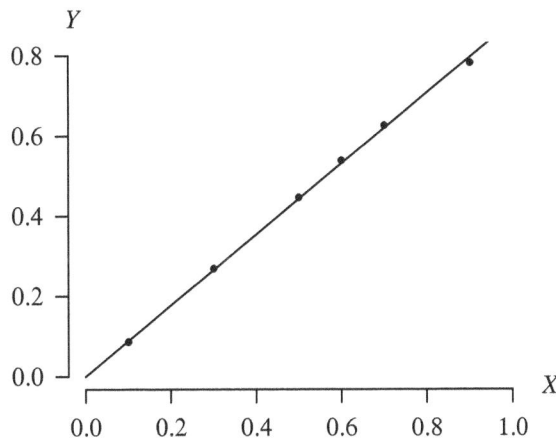

Figure 3.2: A scatterplot of the `Formaldehyde` data pairs with the fitted regression line.

These calculations can be performed in R by adding `-1` or `+0` to the `formula` argument in the `lm` function, which forces the regression line to pass through the origin.

```
fit2 = lm(optden ~ carb - 1, data = Formaldehyde)
fit2$coefficients
```

These R statements calculate the estimated slope of the regression line as $\hat{\beta}_1 = 0.884$.

Analogous theorems to those that were applied to simple linear regression with a population intercept parameter β_0 and a population slope parameter β_1 from Chapter 1 can also be derived associated with the simple linear regression model forced through the origin. In addition, the assumption of normal error terms from Chapter 2 can be added to the simple linear regression model forced through the origin, which allows for statistical inference (that is, constructing confidence intervals and performing hypothesis tests) concerning the population slope of the regression line β_1. For example, the *additional* R command

```
confint(fit2)
```

gives a very narrow 95% confidence interval for β_1 as

$$0.869 < \beta_1 < 0.899.$$

The narrowness of the confidence interval is a reflection of how close the points fall to the regression line in Figure 3.2.

The next example revisits the regression modeling of the stopping distance as a function of the speed of a car in the built-in `cars` data frame.

Example 3.2 Recall from Example 2.8 that X, the speed of a car in miles per hour, was used as an independent variable, and Y, the stopping distance in feet, was used as a dependent variable in a simple linear regression model. There are $n = 50$ data pairs in the `cars` data frame that is built into R. One critique of the simple linear regression model that was constructed for the data pairs in the built-in `cars` data frame from Example 2.8 was that the regression function did not pass through the origin (stationary cars require no stopping distance). Write R code to estimate the slope of the regression line through the origin and comment on the acceptability of this model.

The physics of the experiment indicates that stationary cars require no distance to stop, so forcing a regression line through the origin is appropriate in this setting. The R code below estimates the slope of the regression line that is forced to pass through the origin.

```
x   = cars$speed
y   = cars$dist
fit = lm(y ~ x - 1)
```

Figure 3.3 is a scatterplot of the data pairs (not jittered for ties) with the regression line superimposed. A car requires an additional distance of $\hat{\beta}_1 = 2.91$ feet to stop for every additional mile per hour in speed.

The *additional* R statements

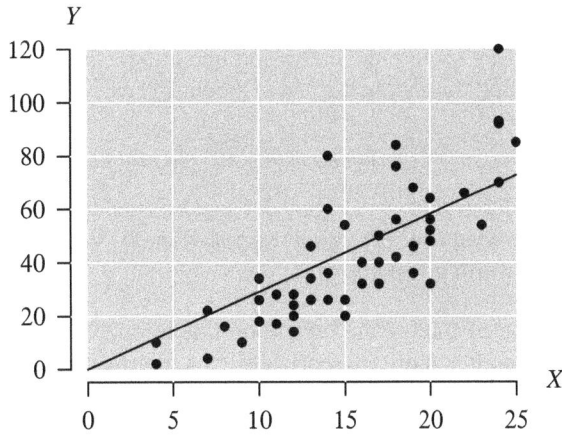

Figure 3.3: Fitted model $Y = \hat{\beta}_1 X$ of speed X and stopping distance Y for the `cars` data.

```
table(sign(fit$residuals))
sum(fit$residuals ^ 2)
```

reveal that 32 data pairs fall below the regression line and only 18 data pairs fall above the regression line. A plot of the standardized residuals can be generated with the R statements

```
res = lm(dist ~ speed - 1, data = cars)$residuals
plot(cars$speed, res / sqrt(sum(res ^ 2) / (length(cars$speed) - 2)))
```

and is given in Figure 3.4. The sum of squares increases from $SSE = 11{,}354$ as calculated in Example 2.8 for the full simple linear regression model to $SSE = 12{,}954$ by

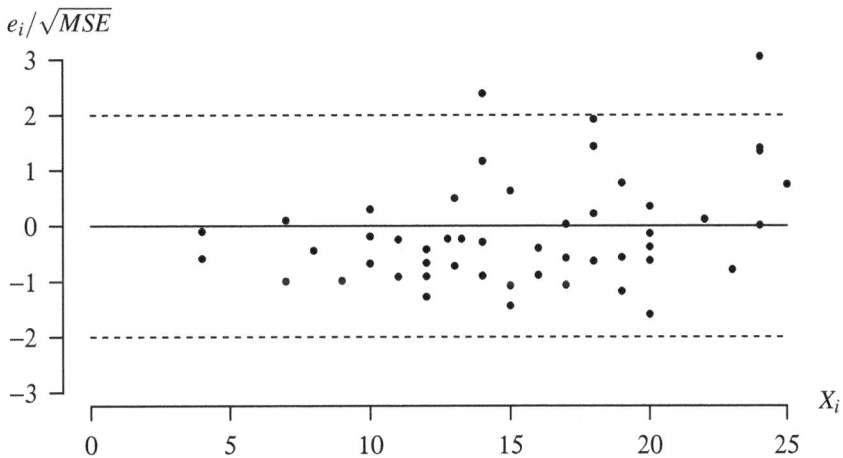

Figure 3.4: Standardized residuals for the `cars` data.

forcing the regression line through the origin. It is universally the case that SSE stays the same or increases by forcing the regression line to pass through the origin. Using the model as a subscript, this can be written symbolically as

$$SSE_{Y=\beta_0+\beta_1 X+\varepsilon} \le SSE_{Y=\beta_1 X+\varepsilon}.$$

The nonsymmetry of the residuals in Figure 3.4 suggests that the fitted linear regression function might not be adequate. Perhaps a regression model with higher-order terms or a nonlinear model is worth investigating.

This ends the discussion of forcing the regression line through the origin. Occasions arise in regression modeling in which it is more appropriate to fit a statistical model with fewer parameters. Some of the results from the full simple linear regression model generalize to simple linear regression forced through the origin. The point estimate for β_1, for example, is unbiased. Three examples of results that do *not* generalize are (a) the residuals do not necessarily sum to zero, (b) the regression line does not necessarily pass through the point $(\bar{X}, \bar{Y})$, and (c) it is possible that SSE can exceed the total sum of squares SST, which can result in a negative value of R^2.

3.2 Diagnostics

Diagnostic procedures are applied to fitted regression models to assess their conformity to the assumptions (for example, constant variance of the error terms) implicit in the simple linear regression model. We have already considered one such diagnostic procedure from the previous chapter, which is the examination of the residuals to assess their independence, constant variance, and normality. Two other diagnostic procedures will be examined here, which are the identification of data pairs known as *leverage points* and the identification of data pairs known as *influential points*. The subsequent section considers *remedial procedures*, which can be applied to a regression model that fails to satisfy one or more of the assumptions implicit in a regression model.

3.2.1 Leverage

Data pairs that have the ability to exert more influence on the regression line than other data pairs due to their independent variable values are known as *leverage points*. These data pairs should be given more scrutiny than the others because of the potential tug that they have on the regression line. More specifically, when the value of the independent variable is unusually far from $\bar{X}$ (either low or high), the data pair has the potential to exert more pull on the regression line than other points.

We begin developing the notion of leverage by expressing the predicted value of Y_i, denoted by $\hat{Y}_i$, as a function of Y_i. Using Theorems 1.1 and 1.3, the predicted value of Y_i is

$$\begin{aligned}
\hat{Y}_i &= \hat{\beta}_0 + \hat{\beta}_1 X_i \\
&= \bar{Y} - \hat{\beta}_1 \bar{X} + \hat{\beta}_1 X_i \\
&= \bar{Y} + \hat{\beta}_1 (X_i - \bar{X}) \\
&= \frac{1}{n} \sum_{j=1}^n Y_j + \sum_{j=1}^n a_j Y_j (X_i - \bar{X}) \\
&= \frac{1}{n} \sum_{j=1}^n Y_j + \sum_{j=1}^n \frac{X_j - \bar{X}}{S_{XX}} Y_j (X_i - \bar{X})
\end{aligned}$$

$$= \sum_{j=1}^{n} \left[\frac{1}{n} + \frac{(X_i - \bar{X})(X_j - \bar{X})}{S_{XX}} \right] Y_j$$

$$= \sum_{j=1}^{n} h_{ij} Y_j$$

for $i = 1, 2, \ldots, n$. The h_{ij} values form the elements of an $n \times n$ matrix $\mathbf{H}$, which is often referred to as the *hat matrix* or the *projection matrix*. The reason that this matrix is known as the projection matrix is that it provides a linear transformation from the observed values of the dependent variable to the associated fitted values. The diagonal elements of the hat matrix are known as the leverages of the data pairs, which are defined next.

Definition 3.2 The *leverage* of data pair (X_i, Y_i) in a simple linear regression model is

$$h_{ii} = \frac{1}{n} + \frac{(X_i - \bar{X})^2}{S_{XX}}$$

for $i = 1, 2, \ldots, n$.

The leverage is a measure of a data pair's potential to influence the regression line. Notice that the leverage is a function of the values of the independent variable $X_1, X_2, \ldots, X_n$ only; the heights of the data pairs do not play a role. Since the two denominators in the expression from Definition 3.2 are constants for a particular data set, only the numerator $(X_i - \bar{X})^2$ changes for each value of X_i. It reflects the distance between a particular X_i value and its associated sample mean. The leverage increases as the distance between X_i and $\bar{X}$ increases. There are several results concerning the leverages; one that concerns the average of the leverages is presented next.

Theorem 3.2 For data pairs $(X_1, Y_1), (X_2, Y_2), \ldots, (X_n, Y_n)$ in a simple linear regression model, the sample mean of the leverages is $2/n$.

Proof The sample mean of the leverages is

$$\frac{h_{11} + h_{22} + \cdots + h_{nn}}{n} = \frac{1}{n} \left[\frac{1}{n} + \frac{(X_1 - \bar{X})^2}{S_{XX}} + \frac{1}{n} + \frac{(X_2 - \bar{X})^2}{S_{XX}} + \cdots + \frac{1}{n} + \frac{(X_n - \bar{X})^2}{S_{XX}} \right]$$

$$= \frac{1}{n} \left[1 + \frac{(X_1 - \bar{X})^2 + (X_2 - \bar{X})^2 + \cdots + (X_n - \bar{X})^2}{S_{XX}} \right]$$

$$= \frac{1}{n} \left[1 + \frac{S_{XX}}{S_{XX}} \right]$$

$$= \frac{2}{n}. \qquad \square$$

To summarize what we know about the n leverages,

- the leverages are the diagonal elements of the hat matrix $\mathbf{H}$,

- all leverages are positive, with a minimum of $1/n$ (for $X_i = \bar{X}$) and a maximum of 1, and

- the sum of the leverages is 2, so the average of the leverages is $2/n$.

If all of the leverages are equal (this is always the case, for example, for $n = 2$ data pairs), then each leverage is $2/n$, which is the average from Theorem 3.2. We would like to establish a threshold at which a data pair has the ability to exert a significant influence over the regression line so that such points might be examined with additional scrutiny. Such data pairs are known as *leverage points*. Although not used universally, a common way to identify a leverage point is if the leverage h_{ii} is more than twice the average of the leverages. Symbolically, a point is designated a leverage point if

$$h_{ii} > \frac{4}{n}.$$

This threshold will be illustrated in the next example.

Example 3.3 To illustrate the identification of leverage points, we consider the first data set in Anscombe's quartet. For notational convenience, the $n = 11$ data pairs have been ordered by their independent variable values in Table 3.2. We will investigate the leverages associated with this data set and two other data sets with an extra data pair appended.

X_i	Y_i
4.0	4.26
5.0	5.68
6.0	7.24
7.0	4.82
8.0	6.95
9.0	8.81
10.0	8.04
11.0	8.33
12.0	10.84
13.0	7.58
14.0	9.96

Table 3.2: Data set I (sorted by X_i) in Anscombe's quartet.

The R code below calculates the $n = 11$ leverages using the formula from Definition 3.2.

```
x         = 4:14
xbar      = mean(x)
sxx       = sum((x - xbar) ^ 2)
n         = length(x)
leverages = 1 / n + (x - xbar) ^ 2 / sxx
```

Notice that the values of $Y_1, Y_2, \ldots, Y_{11}$ are not needed to compute the leverages. The leverages are displayed in Table 3.3. Not surprisingly, the leverages are symmetric about $\bar{X} = 9$ because the values of the independent variable are equally spaced. The leverage for $X_6 = 9$ is just $1/n = 1/11 \cong 0.09$, which is the first term in h_{ii} in Definition 3.2. None of the leverages exceeds the threshold value $4/n = 4/11 \cong 0.36$, so this data set does not contain any leverage points.

Calculating leverages is so common in regression analysis that R has two built-in functions that calculate leverages. The hat function calculates the leverages for Anscombe's first data set with the single statement

i	1	2	3	4	5	6	7	8	9	10	11
X_i	4	5	6	7	8	9	10	11	12	13	14
h_{ii}	0.32	0.24	0.17	0.13	0.10	0.09	0.10	0.13	0.17	0.24	0.32

Table 3.3: Leverages for data set I in Anscombe's quartet.

```
hat(4:14)
```

Alternatively, the `hatvalues` function with the fitted model as an argument can be used to calculate the leverages.

```
x = 4:14
y = c(4.26, 5.68, 7.24, 4.82, 6.95, 8.81, 8.04, 8.33, 10.84, 7.58, 9.96)
fit = lm(y ~ x)
hatvalues(fit)
```

The top graph in Figure 3.5 is a scatterplot of the data pairs and the associated regression line. From a cursory visual assessment, using a simple linear regression model to describe the relationship between X and Y seems reasonable for these data pairs. The leverages for the first three data pairs are identified on the graph. All three graphs in Figure 3.5 have the same horizontal and vertical scales for easier comparison.

The middle graph in Figure 3.5 includes all of the data values from the Anscombe's first data set, but adds the additional data pair $(19, 12.5)$, which was gleaned from Anscombe's fourth data set. The leverages are given in Table 3.4, with the leverage for the data pair $(19, 12.5)$ set in boldface because it has a leverage that exceeds $4/n = 4/12 \cong 0.33$. This data pair is a leverage point that warrants particular scrutiny. Although the data pair has the *ability* to exert unusual effect on the regression line, it is clear that the data point does not alter the regression line from where it was in the top graph. So although the new data pair is a leverage point (and is therefore circled in the middle graph), it does not contradict the existing trend from the other 11 points. In this sense, the leverage point provides some (scant) evidence that the scope of the model can be extended from $4 \leq X \leq 14$ to $4 \leq X \leq 19$.

The bottom graph in Figure 3.5 includes all of the data values from the Anscombe's first data set, but adds the additional data pair $(19, 4)$. Since the values of the independent variable have not changed, the leverages match those from Table 3.4. The leverage point $(19, 4)$ is circled on the graph. This leverage point exerts a significant downward tug on the right side of the regression line relative to the pattern established by the first 11 data pairs. A simple linear regression model is not appropriate in this case. There are several potential explanations for the deleterious effects of this leverage point.

i	1	2	3	4	5	6	7	8	9	10	11	12
X_i	4	5	6	7	8	9	10	11	12	13	14	19
h_{ii}	0.25	0.20	0.16	0.12	0.10	0.09	0.08	0.09	0.11	0.13	0.17	**0.50**

Table 3.4: Leverages for data set I in Anscombe's quartet with appended $X_{12} = 19$.

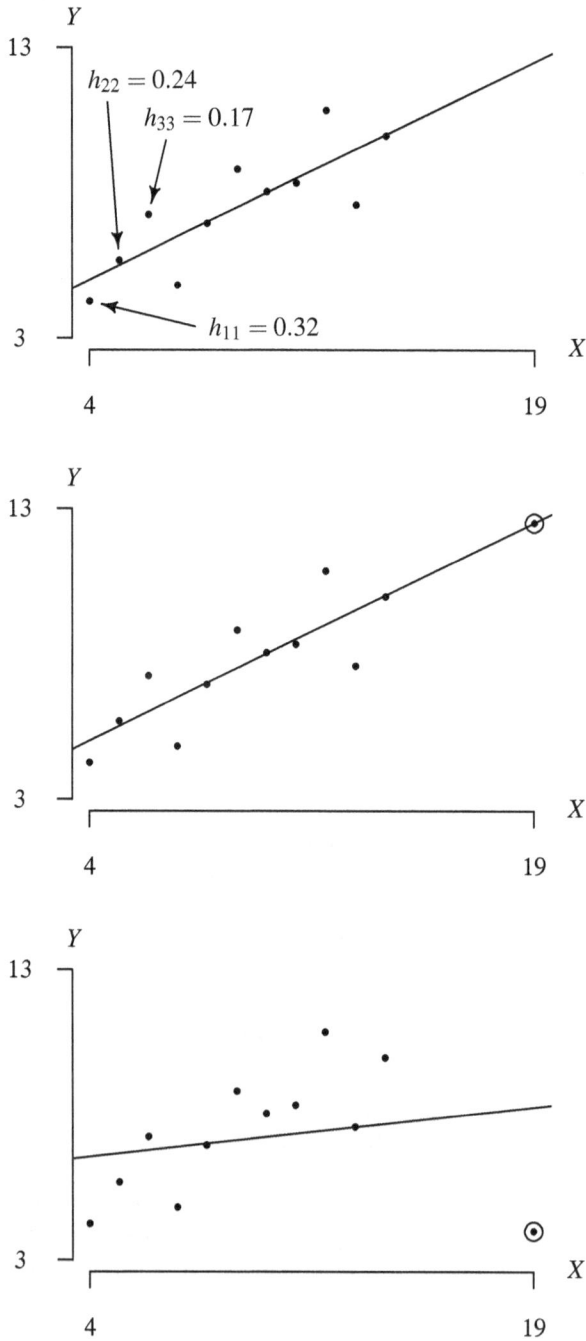

Figure 3.5: Fitted regression models and leverage points.

- The leverage point might have been incorrectly recorded.
- The leverage point might be fundamentally different than the others and does not belong in the data set.
- The leverage point might indicate that a nonlinear regression model is appropriate.
- The leverage point might signal that the scope of the model should be restricted to $4 \leq X \leq 14$, where a simple linear regression appears to be appropriate.
- The leverage point is legitimate and not fundamentally different than the others. It might just happen to be an extreme value. The linear model still might be appropriate, but more data pairs need to be collected to show that this is the case.

The previous example has indicated a fitted simple linear regression model is likely to pass close to a leverage point. Leverage points exert more tug on the regression line than those points whose independent variable value is closer to $\bar{X}$. The next illustration of identifying leverage points revisits the heights of couples from Example 2.7.

Example 3.4 Identify the leverage points for the $n = 96$ pairs of couples heights from Example 2.7.

The following R statements load the PBImisc package, set x to the heights of the wives, set y to the associated heights of the husbands, calculate the leverages using the hat function, store the indexes of those points whose leverage exceeds $4/n$ in the vector i, plot the data pairs using the plot function, plot the regression line using the abline function, and circle the leverage points using the symbols function.

```
library(PBImisc)
x         = heights$Wife
y         = heights$Husband
n         = length(x)
leverages = hat(x)
i         = leverages > 4 / n
m         = sum(i)
fit       = lm(y ~ x)
plot(x, y, pch = 16)
abline(fit$coefficients)
symbols(x[i], y[i], circles = rep(0.7, m), inches = FALSE, add = TRUE)
```

The resulting graph is displayed in Figure 3.6. There are a total of ten leverage points— seven on the left end of the scope of the model and three on the right end of the scope of the model. Examining each of the ten leverage points carefully, nine of the ten do not seem out of step with the rest of the data values. The leverage point $(147, 178)$, however, which corresponds to an unusually short wife marrying and fairly tall husband, is clearly a point that exerts a significant upward tug on the left side of the regression line. Assuming that the X and Y values were recorded correctly, there is no reason to remove this point from the data set. The impact of this point on the slope of the regression line is minimized by the large sample size.

Identifying leverage points is helpful for knowing which points to more carefully scrutinize. It is not appropriate to simply delete a leverage point because it falls far from the regression line. Leverage points can be helpful in highlighting an aspect of the model that was not originally considered

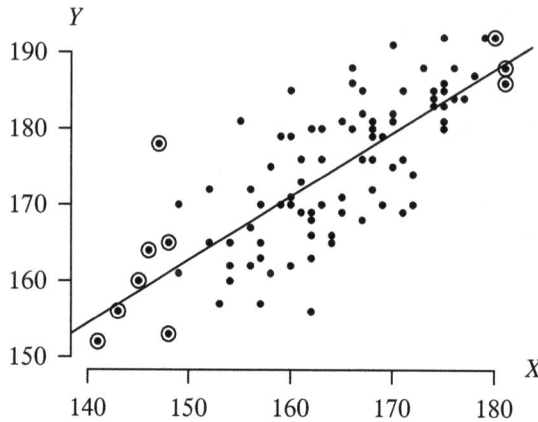

Figure 3.6: Fitted regression model and leverage points for the $n = 96$ data pairs.

relevant. The next subsection considers how to determine if a leverage point (or any other point) does produce a significant impact on $\hat{\beta}_0$ and $\hat{\beta}_1$.

3.2.2   Influential Points

Leverage points have the *potential* to produce large changes in the values of $\hat{\beta}_0$ and $\hat{\beta}_1$ when they are deleted. How can we determine whether a leverage point (or any other point) *does* have significant impact on the regression line? American statistician R. Dennis Cook suggested a quantity that measures the influence of each data pair on the regression line.

Definition 3.3 For a simple linear regression model, Cook's distances $D_1, D_2, \ldots, D_n$ associated with the n data pairs have the following three equivalent definitions.

- $D_i = \dfrac{\sum_{j=1}^{n} \left(\hat{Y}_j - \hat{Y}_{j(i)} \right)^2}{2 \cdot MSE}$,

- $D_i = \dfrac{n\left(\hat{\beta}_{0(i)} - \hat{\beta}_0 \right)^2 + 2\left(\hat{\beta}_{0(i)} - \hat{\beta}_0 \right)\left(\hat{\beta}_{1(i)} - \hat{\beta}_1 \right) \sum_{i=1}^{n} X_i + \left(\hat{\beta}_{1(i)} - \hat{\beta}_1 \right)^2 \sum_{i=1}^{n} X_i^2}{2 \cdot MSE}$,

- $D_i = \dfrac{e_i^2 h_{ii}}{2 \cdot MSE \left(1 - h_{ii} \right)^2}$,

where MSE is the mean square error (see Theorem 1.8), $\hat{Y}_{j(i)}$ is the fitted value of data pair j with data pair i removed, $\beta_{0(i)}$ is the estimated intercept of the regression line for the simple linear regression model with data pair i removed, $\beta_{1(i)}$ is the estimated slope of the regression line for the simple linear regression model with data pair i removed, and h_{ii} is the leverage of data pair i (see Definition 3.2), for $i = 1, 2, \ldots, n$.

The equivalence between the three very diverse formulas in Definition 3.3 is left as an exercise. The data pairs must not be collinear because MSE appears in the denominator of each formula. Each of the three formulas is helpful in developing intuition about Cook's distance, so each is illustrated in the following three examples.

Example 3.5 Use the first formula from Definition 3.3 to calculate the Cook's distances for the $n = 11$ data pairs in the Anscombe's first data set (sorted by the values of the independent variable), appended with the point $(X_{12}, Y_{12}) = (19, 4)$. This was the last data set encountered in Example 3.3.

The bottom graph in Figure 3.5 shows that the first 11 data pairs are consistent with an underlying linear model, but the 12th data pair is not consistent with this model. The first formula from Definition 3.3 is

$$D_i = \frac{\sum_{j=1}^{n} \left(\hat{Y}_j - \hat{Y}_{j(i)} \right)^2}{2 \cdot MSE}$$

for $i = 1, 2, \ldots, n$. Since the term $\hat{Y}_j - \hat{Y}_{j(i)}$ is a measure of the effect of dropping data pair i from the data set on the fitted value, larger values for D_i indicate that data pair i is more influential. Squaring $\hat{Y}_j - \hat{Y}_{j(i)}$ assures that the direction of the fitted value when data pair i is dropped makes a positive contribution to D_i. The R code below loops through the data points, excluding the data pairs one-by-one. Hence there will in general be a total of $n + 1$ simple linear regression models fitted when using the first formula for computing Cook's distance—one regression model for all data pairs included and n other regression models for dropping each data pair once.

```
x       = c(4:14, 19)
y       = c(4.26, 5.68, 7.24, 4.82, 6.95, 8.81, 8.04, 8.33, 10.84,
            7.58, 9.96, 4)
n       = length(x)
fit     = lm(y ~ x)
mse     = sum(fit$residuals ^ 2) / (n - 2)
fitted = fit$fitted.values
cooks   = numeric(n)
for (i in 1:n) {
  fit.exclude     = lm(y[-i] ~ x[-i])
  beta0           = fit.exclude$coefficients[1]
  beta1           = fit.exclude$coefficients[2]
  fitted.exclude = beta0 + beta1 * x
  cooks[i]        = sum((fitted - fitted.exclude) ^ 2) / (2 * mse)
}
print(cooks)
```

Several of the Cook's distances are given in Table 3.5. Consistent with the bottom graph in Figure 3.5, the 12th Cook's distance $D_{12} = 3.621$ is substantially larger than

i	1	2	3	4	$\cdots$	11	12
D_i	0.236	0.029	0.005	0.069	$\cdots$	0.128	3.621

Table 3.5: Cook's distances for Anscombe's data set I with $(X_{12}, Y_{12}) = (19, 4)$ appended.

the second-largest Cook's distance $D_1 = 0.236$. So the 12th data pair, (X_{12}, Y_{12}), is the most influential point. The first data pair, (X_1, Y_1), is the second most influential point. Notice that these are the two points with the highest leverage (see Table 3.4).

To show some of the geometry associated with the calculation of $D_1, D_2, \ldots, D_n$, Figure 3.7 shows the regression line

$$Y = \hat{\beta}_0 + \hat{\beta}_1 X = 6.09 + 0.114X$$

fitted to all $n = 12$ data pairs, which are indicated by solid points (●). This regression line corresponds to the fitted value at $X_{12} = 19$ of

$$\hat{Y}_{12} = 6.09 + (0.114)(19) = 8.25.$$

The other regression line,

$$Y = \hat{\beta}_{0(12)} + \hat{\beta}_{1(12)}X = 3.00 + 0.500X,$$

is the regression that excludes the influential 12th data pair $(X_{12}, Y_{12}) = (19, 4)$. This regression line corresponds to the fitted value at $X_{12} = 19$ of

$$\hat{Y}_{12(12)} = 3.00 + (0.500)(19) = 12.50.$$

The two fitted values are indicated by open points (○). So when calculating D_{12} using the first formula in Definition 3.3, one of the terms in the numerator is

$$\left(\hat{Y}_{12} - \hat{Y}_{12(12)}\right)^2 = (8.25 - 12.50)^2 = (-4.25)^2 = 18.07,$$

which makes a huge contribution to the numerator of D_{12}.

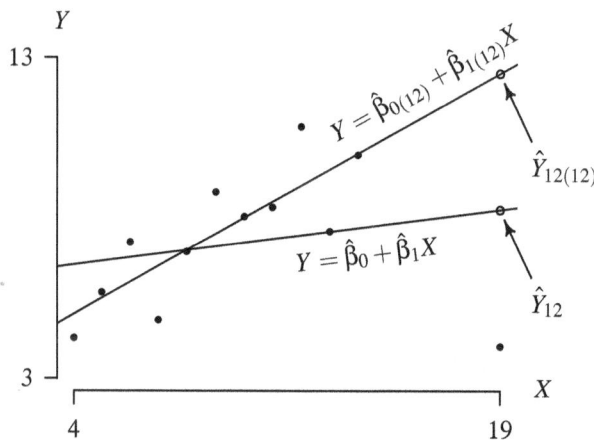

Figure 3.7: Calculating Cook's distances using fitted values.

The previous example has indicated that Cook's distance is a measure of the influence of each data pair based on the effect of removing each data pair sequentially, and measuring the associated impact on the fitted values. If the fitted values are not substantially altered by removing data pair i, then D_i will be small; if the fitted values are substantially altered by removing data pair i, then D_i will be large. This, however, does not explain why the denominator $2 \cdot MSE$ is in all four formulas in Definition 3.3. That will be addressed in the next example.

Example 3.6 Use the second formula from Definition 3.3 to calculate the Cook's distances for the $n = 11$ data pairs in the Anscombe's first data set (sorted by the values of the independent variable), appended with the point $(X_{12}, Y_{12}) = (19, 4)$.

The second formula for computing Cook's distance for data pair i from Definition 3.3 is

$$D_i = \frac{n(\hat{\beta}_{0(i)} - \hat{\beta}_0)^2 + 2(\hat{\beta}_{0(i)} - \hat{\beta}_0)(\hat{\beta}_{1(i)} - \hat{\beta}_1)\sum_{i=1}^{n} X_i + (\hat{\beta}_{1(i)} - \hat{\beta}_1)^2 \sum_{i=1}^{n} X_i^2}{2 \cdot MSE}$$

for $i = 1, 2, \ldots, n$. This formula emphasizes the change in the regression coefficients when data pair i is dropped. Figure 3.8 shows (a) the estimators $(\hat{\beta}_0, \hat{\beta}_1)$ for all $n = 12$ data pairs as a $+$, (b) the associated confidence regions for β_0 and β_1 at levels 0.25, 0.5, and 0.75, and (c) twelve points indicated by solid circles (•) giving the values of the slope and intercept when data pair i is dropped, for $i = 1, 2, \ldots, n$. Not surprisingly, the estimated slope and intercept when the 12th data point, $(X_{12}, Y_{12}) = (19, 4)$, is dropped, strays the furthest from $(\hat{\beta}_0, \hat{\beta}_1)$. The other 11 estimated slope and intercept pairs all fall within the 0.25 confidence region.

The connection with the confidence region for β_0 and β_1 in this case illuminates why the $2 \cdot MSE$ appears in the denominator of all of the formulas for D_i in Definition 3.3.

Compare the right-hand side of the second formula in Definition 3.3 with the expression in Theorem 2.16. They are identical except that β_0 is replaced by $\beta_{0(i)}$ and β_1 is replaced by $\beta_{1(i)}$. So under the assumption that the data pairs are drawn from a simple linear regression model, one would expect that D_i is approximately $F(2, n-2)$. Some suggest using the median of a $F(2, n-2)$ distribution as a threshold for classifying a data pair as an influential point. Another approach is to observe that the population mean and variance of an $F(2, n-2)$ random variable are

$$E[D_i] = \frac{n-2}{n-4} \quad \text{(for } n > 4) \qquad \text{and} \qquad V[D_i] = \frac{(n-2)^3}{(n-4)^2(n-6)} \quad \text{(for } n > 6).$$

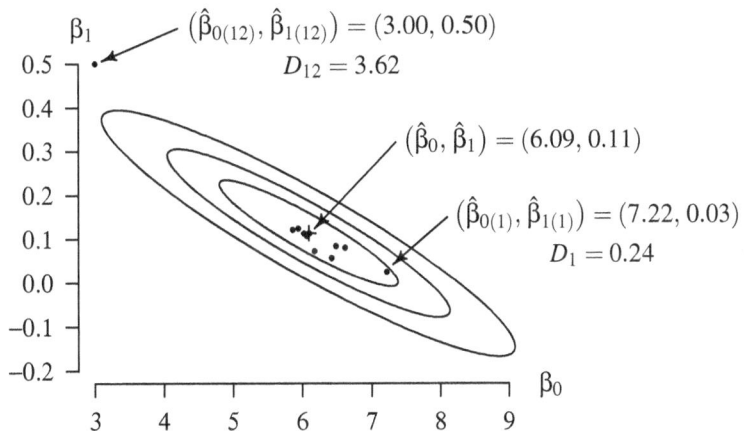

Figure 3.8: Calculating Cook's distances using the parameter estimates.

So in the limit as the number of data pairs increases,

$$\lim_{n \to \infty} E[D_i] = 1 \qquad \text{and} \qquad \lim_{n \to \infty} V[D_i] = 1.$$

It is for this reason that a threshold of 1 is used as a simple threshold for classifying a data point as influential based on Cook's distance. Regardless of whether the median of an $F(2,10)$ random variable (which is 0.743) or 1 is used as a threshold, the first 11 points are not deemed to be influential points, and the 12th point, $(19, 4)$, is deemed to be an influential point.

One weakness associated with the first two formulas for computing the Cook's distances in Definition 3.3 involves computation time. There are $n+1$ regression lines to estimate (one for all of the data pairs and then another n associated with dropping each of the data pairs). For large values of n, this can require significant computation time. The third formula is much faster, as illustrated next.

Example 3.7 Use the third formula from Definition 3.3 to calculate the Cook's distances for the $n = 96$ data pairs in the data set of heights of wives and husbands from Example 2.7.

The third formula for computing Cook's distance for data pair i from Definition 3.3 is

$$D_i = \frac{e_i^2 h_{ii}}{2 \cdot MSE \left(1 - h_{ii}\right)^2}$$

for $i = 1, 2, \ldots, n$. The advantage to using this formula over the other two formulas is that it only requires one regression line to be calculated, rather than $n+1$ regression lines in the other two formulas. This is a substantial time savings for large values of n. The R code below calculates Cook's distances for the heights data.

```
library(PBImisc)
x      = heights$Wife
y      = heights$Husband
n      = length(x)
fit    = lm(y ~ x)
mse    = sum(fit$residuals ^ 2) / (n - 2)
lev    = hat(x)
cooks  = fit$residuals ^ 2 * lev / (2 * mse * (1 - lev) ^ 2)
plot(cooks)
```

The $n = 96$ Cook's distances are plotted in Figure 3.9. The 12th data pair, which is $(X_{12}, Y_{12}) = (147, 178)$, has a spectacular Cook's distance of $D_{12} = 0.192$. Since this does not exceed the first threshold (which is the median of an F random variable with 2 and 94 degrees of freedom: 0.698) or the second threshold (which is 1 using the asymptotic result), we conclude that there are no influential points. Cook's distances are calculated so frequently in regression analysis that R includes a function named cooks.distance that calculates the Cook's distances, as illustrated below.

```
library(PBImisc)
x    = heights$Wife
```

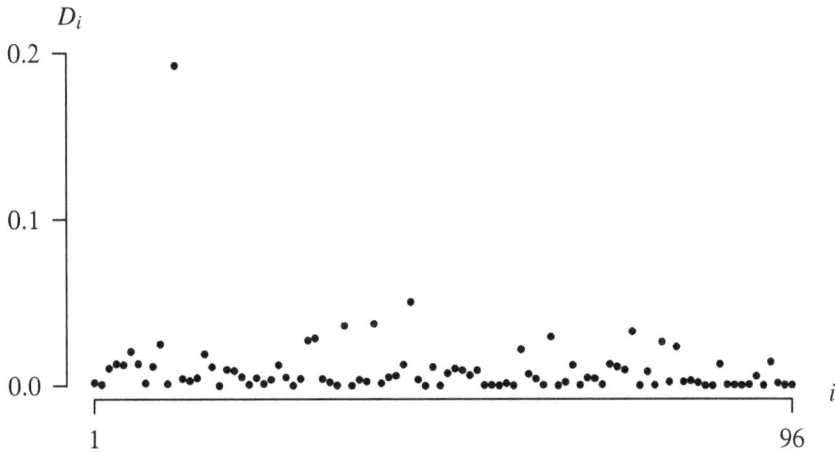

Figure 3.9: Cook's distances for the `heights` data.

```
y   = heights$Husband
fit = lm(y ~ x)
cooks.distance(fit)
```

Cook's distances are effective for identifying influential points. Once an influential point in a simple linear regression model has been identified, there are several possible next steps.

- The influential point might have been recorded or coded improperly; a typographical error has occurred. In most situations, this is easily remedied.

- The influential point has some unusual characteristic that is not present with the other data points that might account for it being deemed influential. Depending on the setting, the influential point can be removed and the regression model can be refitted without the influential point.

- The influential point might provide some evidence that an alternative regression model is appropriate. This might be a nonlinear regression model or a linear regression model with additional independent variables.

- The influential point might be at one of the extremes of the scope of the model. This might indicate that the scope of the model is too wide; narrowing the scope should be considered. It is often the case that a simple linear regression model is valid only over a rather limited scope. This might result in eliminating all data points outside of the narrowed scope and refitting the simple linear regression model.

- The high-leverage point is indeed within the scope of the model and was recorded correctly, but its extreme influence on the regression line is resulting in poor diagnostic measures. One approach here is to collect more data values, particularly at the extreme values of the independent variable within the scope of the model in order to mitigate the effect of the influential point.

3.3   Remedial Procedures

The diagnostic procedures presented in the previous section are designed to *identify* assumptions associated with the simple linear regression model that are not satisfied for a particular set of n data pairs. But these diagnostic procedures do not suggest *remedies* when model assumptions are not satisfied. This section considers remedial procedures.

Reasons that simple linear regression model with normal error terms can fail to satisfy the assumptions given in Definition 2.1 include

- the regression function is not linear,

- the regression model has not included an important independent variable,

- the error terms have a variance that varies with X,

- the error terms are not independent,

- the error terms are not normally distributed,

- the scope of the regression model is too wide,

- the scope of the regression model is too narrow, and

- an influential point has an unusually strong effect on the regression line.

Two common approaches to handling a regression model which violates one or more of the assumptions are (a) formulate and fit a regression model with nonlinear terms, and (b) transform the X-values or the Y-values (or both) in a fashion so that the simple linear regression assumptions are satisfied. Regression models with nonlinear terms will be considered in a subsequent section in this chapter; transformations will be considered here. Transformations will be illustrated in a single (long) example.

> **Example 3.8** A simple linear regression model with normal error terms for the speed of a car X (in miles per hour) versus the stopping distance Y (in feet) for the built-in R `cars` data set was abandoned in Example 2.8 for several reasons. A scatterplot (without jittering) with the associated regression line is displayed in Figure 3.10. The purpose of this example is to see whether a transformation can overcome the problems associated with
>
> - the relationship between X and Y appears to be slightly nonlinear,
> - the variance of the error terms appears to be increasing in X, and
> - the residuals do not appear to be normally distributed.

Rather than providing a complete inventory of all possible patterns and associated potential helpful transformations, four transformations will be illustrated here. This trial-and-error approach is not what is typically relied on in practice. There are some patterns associated with data pairs that tend to give clues as to which transformations will be effective.

The first transformation is $X' = X^2$. The R code below implements the transformation, generates a scatterplot of the transformed data pairs, and plots the associated regression line.

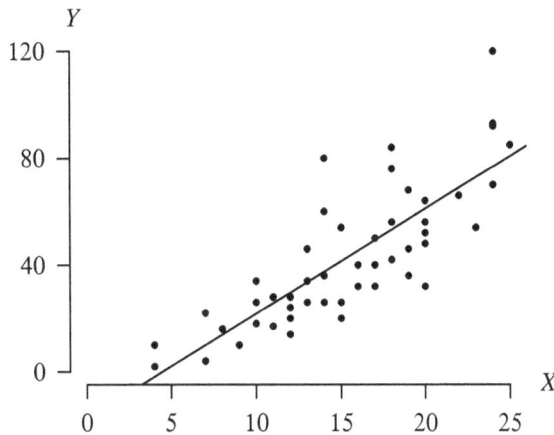

Figure 3.10: Scatterplot and regression line of speed X and stopping distance Y.

```
x = cars$speed ^ 2
y = cars$dist
plot(x, y)
abline(lm(y ~ x)$coefficients)
```

This scatterplot appears in the upper-left graph in Figure 3.11. Tick mark labels have been suppressed on these graphs because the interest is in gazing at the data pairs in order to determine whether the transformed data pairs conform to the simple linear regression model with normal error terms. For the transformation $X' = X^2$, little progress is made on the constant variance issue. The first 19 data pairs, which are associated with speeds from 4 to 13 miles per hour, seem to have a smaller variance in their stopping distances than the faster speeds. This transformation is deemed ineffective.

The second transformation is $Y' = \ln Y$. The R code below implements the transformation, generates a scatterplot of the transformed data pairs, and plots the associated regression line.

```
x = cars$speed
y = log(cars$dist)
plot(x, y)
abline(lm(y ~ x)$coefficients)
```

This scatterplot appears in the upper-right graph in Figure 3.11. The transformation $Y' = \ln Y$ also results in a nonconstant variance in the error terms; this time the variance in the stopping distances is greater for the slower speeds. So this transformation is also abandoned for lack of constant variance of the error terms.

The third transformation is $Y' = \sqrt{Y}$. The R code below implements the transformation, generates a scatterplot of the transformed data pairs, and plots the associated regression line.

```
x = cars$speed
```

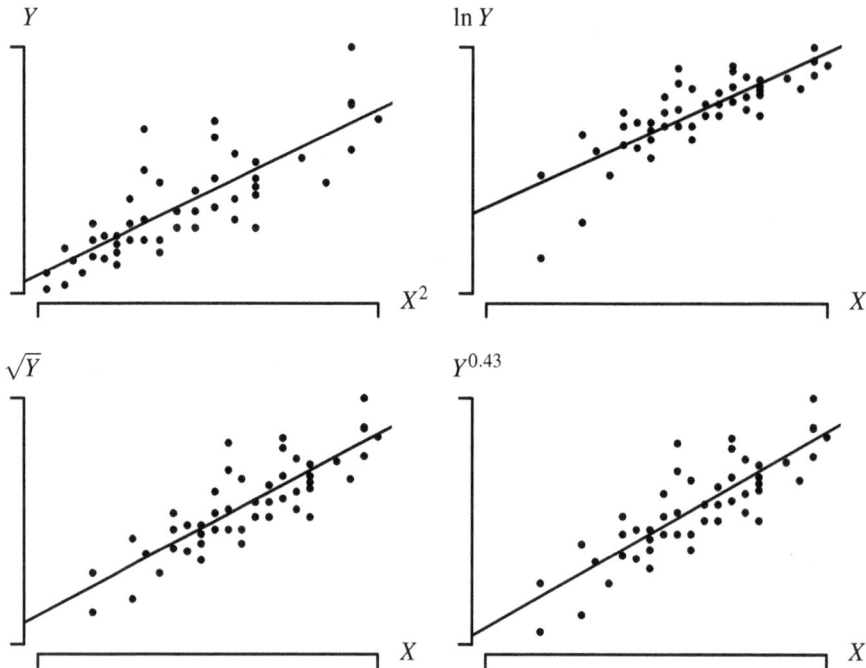

Figure 3.11: Scatterplots and estimated regression lines for transformed `cars` data.

```
y = sqrt(cars$dist)
plot(x, y)
abline(lm(y ~ x)$coefficients)
```

This scatterplot appears in the lower-left graph in Figure 3.11. The transformation $Y' = \sqrt{Y}$ is the first to show some promise for the use of the simple linear regression model with normal error terms. The variance of the error terms appears to be constant over the scope of the model. There is nothing magical, however, about the $1/2$ power in the transformation $Y' = \sqrt{Y} = Y^{1/2}$. Might the cube root be a superior transformation to the square root? This prompts a fourth transformation, which is $Y' = Y^\lambda$, and is known as the Box–Cox transformation, named after British statisticians George Box and David Cox. They suggested a similar transformation in 1964, which is

$$Y' = \frac{Y^\lambda - 1}{\lambda},$$

and the fitting of the λ parameter by maximum likelihood estimation can be performed by the `boxcox` function in the `MASS` package in R.

So the fourth transformation is $Y' = (Y^\lambda - 1)/\lambda$. The R code below calculates the maximum likelihood estimator of λ, implements the transformation, generates a scatterplot of the transformed data pairs, and plots the associated regression line. The `boxcox` function generates the log likelihood function for estimating λ, and the `which.max` function extracts the maximum likelihood estimator.

```
library(MASS)
x       = cars$speed
y       = cars$dist
bc      = boxcox(y ~ x, plotit = FALSE, lambda = seq(0, 1, by = 0.01))
lambda  = bc$x[which.max(bc$y)]
y       = (y ^ lambda - 1) / lambda
plot(x, y)
abline(lm(y ~ x)$coefficients)
```

The log likelihood function and an associated 95% confidence interval for λ is generated by setting the `plotit` argument to `FALSE` in the call to `boxcox`. This confidence interval includes $\lambda = 1/2$. The maximum likelihood estimator $\hat{\lambda} = 0.43$ falls between a square root and cube root transformation. This scatterplot appears in the lower-right graph in Figure 3.11, and is very similar to the square root transformation; either would work fine for this data set. Since the last two scatterplots and associated regression lines are nearly identical, we move forward with the transformation $Y' = \sqrt{Y}$. So the tentative fitted model is

$$E\left[\sqrt{Y}\right] = 1.28 + 0.322X$$

where the regression coefficients $\beta'_0 = 1.28$ and $\beta'_1 = 0.322$ are calculated with the R statement

```
lm(sqrt(cars$dist) ~ cars$speed)$coefficients
```

The next step is to assess the aptness of the model by examining the residuals. The four graphs (read row-wise) in Figure 3.12 are (*a*) the residuals associated with the transformed model $\sqrt{Y} = 1.28 + 0.322X$ plotted against their index, (*b*) the standardized residuals $e_i/\sqrt{MSE}$ associated with the transformed model plotted against the value of the independent variable X_i, (*c*) a histogram of the standardized residuals $e_i/\sqrt{MSE}$ for the transformed model, and (*d*) a QQ plot of the standardized residuals $e_i/\sqrt{MSE}$ for the transformed model with theoretical quantiles on the horizontal axis and sample quantiles on the vertical axis. Although there is some nonsymmetry in the histogram of the residuals (which might be due to the binning of the 50 data pairs), the residual plots and the QQ plot make the simple linear regression model with normal error terms for the transformed data pairs seem plausible. A roughly mound-shaped histogram is typically adequate for the normality assumption. Moving from the visual assessment to statistical tests, the R code

```
x = cars$speed
y = sqrt(cars$dist)
fit = lm(y ~ x)
shapiro.test(fit$residuals)
max(cooks.distance(fit))
```

gives a *p*-value for the Shapiro–Wilk test of $p = 0.314$. This is a big improvement over the *p*-value obtained in Example 2.8, which rejected normality with $p = 0.0215$. The transformation is effective. The largest Cook's distance is 0.134, which occurs at the 49th observation $(X_{49}, Y_{49}) = (24, 120)$. Returning to the 49th observation in

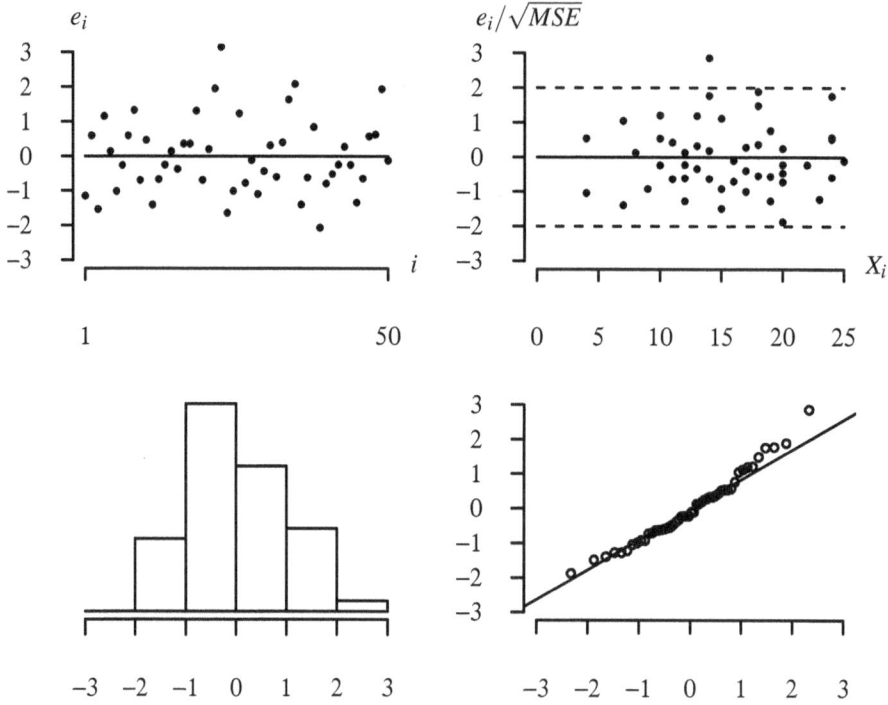

Figure 3.12: Visual assessment of the residuals of the transformed model.

Figure 3.12, we see that it achieves the largest Cook's distance because of its leverage, but does not appear to be inconsistent with the transformed model.

So the visual assessment and statistical tests lead us to believe that a simple linear regression model with normal error terms for the transformed data is appropriate. The fitted regression model is

$$E\left[\sqrt{Y}\right] = 1.28 + 0.322X.$$

All of the statistical inference techniques can now be applied to the transformed data. For example, confidence intervals for the β_0' and β_1' (the intercept and slope of the regression line for the transformed data) can be calculated with the R statements

```
x   = cars$speed
y   = sqrt(cars$dist)
fit = lm(y ~ x)
confint(fit)
```

which give the 95% confidence intervals

$$0.303 < \beta_0' < 2.25 \qquad \text{and} \qquad 0.263 < \beta_1' < 0.382.$$

Figure 3.13 displays all of the exact two-sided 95% confidence intervals for $E\left[\sqrt{Y_h}\right]$ and all of the exact two-sided 95% prediction intervals for $\sqrt{Y_h^*}$ for all values of X_h in

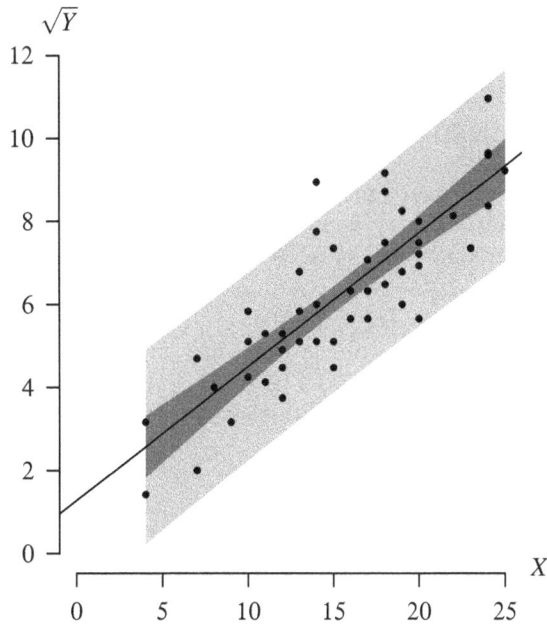

Figure 3.13: Transformed `cars` model 95% confidence and prediction intervals.

the scope of the regression model. For $X_h = 21$ miles per hour, for example, an exact two-sided 95% prediction interval for $\sqrt{Y_h^*}$ is

$$5.78 < \sqrt{Y_h^*} < 10.3,$$

which can be calculated with the R commands

```
x   = cars$speed
y   = sqrt(cars$dist)
fit = lm(y ~ x)
predict(fit, data.frame(x = 21), interval = "prediction")
```

So to translate this back to the original units, for a 51st car going $X_h = 21$ miles per hour, the expected stopping distance using the transformed model is

$$\hat{Y}_h = \left(1.28 + 0.322 \cdot 21\right)^2 = 64.8$$

feet, and an exact two-sided 95% prediction interval for the associated stopping distance is

$$33.5 < Y_h^* < 106.$$

The previous example took a trial-and-error approach to determining an appropriate transformation to apply to the raw data pairs in order to satisfy the assumptions implicit in a simple linear regression model with normal error terms. There are templates that can give a more systematic approach to determining these transformations.

There is a nice synergy between matrix algebra and regression, which will be presented in the next section.

3.4   Matrix Approach to Simple Linear Regression

So far, a purely algebraic approach has been taken to simple linear regression modeling. This section considers a matrix-based approach. There are (at least) four reasons to take this approach. First, the mathematical expressions are in many cases much more compact; summations from the algebraic approach are often equivalent to matrix multiplications. Second, matrix algebra can easily be implemented on a computer. Third, the matrix approach generalizes very easily to the multiple regression case in which there are several independent variables. Fourth, the matrix approach generalizes very easily to weighted least squares, which will be introduced in the next section.

We begin the matrix approach by defining certain critical matrices, which will be set in boldface. Let $\mathbf{X}$ be an $n \times 2$ matrix whose first column is all ones and whose second column contains the observed values of the independent variable, $\mathbf{Y}$ be an $n \times 1$ vector which holds the observed values of the dependent variable, $\boldsymbol{\beta}$ be a 2×1 vector which holds the population intercept and slope, and $\boldsymbol{\varepsilon}$ be an $n \times 1$ vector which holds the error terms:

$$\mathbf{X} = \begin{bmatrix} 1 & X_1 \\ 1 & X_2 \\ \vdots & \vdots \\ 1 & X_n \end{bmatrix}, \qquad \mathbf{Y} = \begin{bmatrix} Y_1 \\ Y_2 \\ \vdots \\ Y_n \end{bmatrix}, \qquad \boldsymbol{\beta} = \begin{bmatrix} \beta_0 \\ \beta_1 \end{bmatrix}, \qquad \text{and} \qquad \boldsymbol{\varepsilon} = \begin{bmatrix} \varepsilon_1 \\ \varepsilon_2 \\ \vdots \\ \varepsilon_n \end{bmatrix}.$$

The $\mathbf{X}$ matrix is known as the *design matrix*.

As before, the values of the independent variable (the second column of $\mathbf{X}$) are assumed to be fixed constants observed without error with at least two distinct values, the values of the dependent variable contained in $\mathbf{Y}$ are assumed to be continuous random responses, and the elements of the vector $\boldsymbol{\varepsilon}$ are assumed to be mutually independent random variables, each with population mean 0 and finite positive population variance σ^2. Stated another way, the expected value of $\boldsymbol{\varepsilon}$ is the zero vector and the variance–covariance matrix of $\boldsymbol{\varepsilon}$ is

$$\begin{bmatrix} \sigma^2 & 0 & \cdots & 0 \\ 0 & \sigma^2 & \cdots & 0 \\ \vdots & \vdots & \ddots & \vdots \\ 0 & 0 & \cdots & \sigma^2 \end{bmatrix}.$$

The simple linear regression model

$$Y_i = \beta_0 + \beta_1 X_i + \varepsilon_i$$

for $i = 1, 2, \ldots, n$, can be written more explicitly in terms of each observed data pair as

$$Y_1 = \beta_0 + \beta_1 X_1 + \varepsilon_1$$
$$Y_2 = \beta_0 + \beta_1 X_2 + \varepsilon_2$$
$$\vdots$$
$$Y_n = \beta_0 + \beta_1 X_n + \varepsilon_n$$

which, in matrix form, is

$$\begin{bmatrix} Y_1 \\ Y_2 \\ \vdots \\ Y_n \end{bmatrix} = \begin{bmatrix} 1 & X_1 \\ 1 & X_2 \\ \vdots & \vdots \\ 1 & X_n \end{bmatrix} \cdot \begin{bmatrix} \beta_0 \\ \beta_1 \end{bmatrix} + \begin{bmatrix} \varepsilon_1 \\ \varepsilon_2 \\ \vdots \\ \varepsilon_n \end{bmatrix}$$

or simply

$$\mathbf{Y} = \mathbf{X}\boldsymbol{\beta} + \boldsymbol{\varepsilon}.$$

This explains why the artificial column of ones appears as the first column of the $\mathbf{X}$ matrix; it is to account for the intercept term. To force a regression line through the origin, simply omit the column of ones in the $\mathbf{X}$ matrix. Taking the expected value of both sides of this equation results in

$$E[\mathbf{Y}] = \mathbf{X}\boldsymbol{\beta}$$

because $E[\varepsilon_i] = 0$, for $i = 1, 2, \ldots, n$, (that is, $E[\boldsymbol{\varepsilon}] = \mathbf{0}$). The left-hand side of this equation, $E[\mathbf{Y}]$, is an n-element column vector with elements $E[Y_1], E[Y_2], \ldots, E[Y_n]$. The sum of squares which is to be minimized to find the least squares estimators is

$$S = (\mathbf{Y} - \mathbf{X}\boldsymbol{\beta})' (\mathbf{Y} - \mathbf{X}\boldsymbol{\beta}).$$

With this notation established, the algebraic results concerning the simple linear regression model can be restated more compactly in terms of these matrices. The results have already been proved, so there is no need to prove them again when stated in matrix form. The $'$ superscript denotes transpose. It is a good exercise to perform the algebra necessary to see that the algebraic and matrix versions of these definitions and theorems match. The dimensions of the matrices should be checked for conformity.

- **Definition 1.1.** The *simple linear regression model* is

$$\mathbf{Y} = \mathbf{X}\boldsymbol{\beta} + \boldsymbol{\varepsilon},$$

where $E[\boldsymbol{\varepsilon}] = \mathbf{0}$, $V[\boldsymbol{\varepsilon}] = \sigma^2 \mathbf{I}$, and $\mathbf{I}$ is the $n \times n$ identity matrix.

- **Theorem 1.1.** The least squares estimators of $\boldsymbol{\beta}$, denoted by $\hat{\boldsymbol{\beta}} = \left(\hat{\beta}_0, \hat{\beta}_1 \right)'$, solve the normal equations

$$\mathbf{X}'\mathbf{X}\hat{\boldsymbol{\beta}} = \mathbf{X}'\mathbf{Y}.$$

The $\mathbf{X}$ matrix has rank 2 because there are at least two distinct X_i values. So $\mathbf{X}'\mathbf{X}$ is invertible and the normal equations have the unique solution

$$\hat{\boldsymbol{\beta}} = \left(\mathbf{X}'\mathbf{X} \right)^{-1} \mathbf{X}'\mathbf{Y},$$

by premultiplying both sides of the normal equations by $\left(\mathbf{X}'\mathbf{X} \right)^{-1}$.

- **Theorem 1.2.** The least squares estimator of $\boldsymbol{\beta}$ in a simple linear regression model is an unbiased estimator of $\boldsymbol{\beta}$ because

$$E\left[\hat{\boldsymbol{\beta}} \right] = \boldsymbol{\beta}.$$

- **Theorem 1.3.** The least squares estimators of $\boldsymbol{\beta}$ in the simple linear regression model can be written as linear combinations of the dependent variables:

$$\hat{\boldsymbol{\beta}} = \left(\mathbf{X}'\mathbf{X} \right)^{-1} \mathbf{X}'\mathbf{Y},$$

where the coefficients in the linear combinations are given by $\left(\mathbf{X}'\mathbf{X} \right)^{-1} \mathbf{X}'$.

- **Theorem 1.4.** The variance–covariance matrix of the least squares estimators of $\boldsymbol{\beta}$ is

$$\sigma^2 \left(\mathbf{X}'\mathbf{X} \right)^{-1}.$$

- **Theorem 1.5** (Gauss–Markov theorem). The least squares estimators of $\boldsymbol{\beta}$ in a simple linear regression model, $\hat{\boldsymbol{\beta}} = (\mathbf{X'X})^{-1}\mathbf{X'Y}$, have the smallest population variance amongst all linear unbiased estimators of $\boldsymbol{\beta}$.

- **Definition 1.2.** The vector of *fitted values* in a simple linear regression model is the $n \times 1$ column vector

$$\hat{\mathbf{Y}} = \mathbf{X}\hat{\boldsymbol{\beta}} = \mathbf{X}(\mathbf{X'X})^{-1}\mathbf{X'Y},$$

which is a linear combination of the dependent variables. The vector of *residuals* is the $n \times 1$ column vector

$$\begin{aligned}
\mathbf{e} &= \mathbf{Y} - \hat{\mathbf{Y}} \\
&= \mathbf{Y} - \mathbf{X}\hat{\boldsymbol{\beta}} \\
&= \mathbf{Y} - \mathbf{X}(\mathbf{X'X})^{-1}\mathbf{X'Y} \\
&= (\mathbf{I} - \mathbf{X}(\mathbf{X'X})^{-1}\mathbf{X'})\mathbf{Y},
\end{aligned}$$

which is also a linear combination of the dependent variables. The matrix $\mathbf{I}$ is the $n \times n$ identity matrix.

- **Theorem 1.6.** For the simple linear regression model with fitted values $\hat{\mathbf{Y}}$ and residuals $\mathbf{e}$,

  - $\mathbf{e'1} = 0$,
  - $\mathbf{Y'1} = \hat{\mathbf{Y}}'\mathbf{1}$
  - $\hat{\mathbf{Y}}'\mathbf{e} = 0$,

where $\mathbf{1}$ is an n-element column vector of ones.

- **Theorem 1.7.** An unbiased estimator of σ^2 in a simple linear regression model is

$$\hat{\sigma}^2 = MSE = \frac{\mathbf{e'e}}{n-2}.$$

- **Theorem 1.8.** The sums of squares can be partitioned in a simple linear regression model as $SST = SSR + SSE$ or

$$\left(\mathbf{Y} - \bar{\mathbf{Y}}\right)'\left(\mathbf{Y} - \bar{\mathbf{Y}}\right) = \left(\hat{\mathbf{Y}} - \bar{\mathbf{Y}}\right)'\left(\hat{\mathbf{Y}} - \bar{\mathbf{Y}}\right) + \left(\mathbf{Y} - \hat{\mathbf{Y}}\right)'\left(\mathbf{Y} - \hat{\mathbf{Y}}\right),$$

where $\bar{\mathbf{Y}}$ is an n-element column vector with identical elements which are each the sample mean of the values of the dependent variable.

- **Definition 1.3.** The *coefficient of determination* in a simple linear regression model is

$$R^2 = \frac{SSR}{SST} = \frac{\left(\hat{\mathbf{Y}} - \bar{\mathbf{Y}}\right)'\left(\hat{\mathbf{Y}} - \bar{\mathbf{Y}}\right)}{\left(\mathbf{Y} - \bar{\mathbf{Y}}\right)'\left(\mathbf{Y} - \bar{\mathbf{Y}}\right)},$$

when $\left(\mathbf{Y} - \bar{\mathbf{Y}}\right)'\left(\mathbf{Y} - \bar{\mathbf{Y}}\right) \neq 0$. The *coefficient of correlation* is

$$r = \pm\sqrt{R^2},$$

where the sign associated with r is positive when $\hat{\beta}_1 \geq 0$ and negative when $\hat{\beta}_1 < 0$.

- **Definition 2.1.** The *simple linear regression model with normal error terms* is

$$\mathbf{Y} = \mathbf{X}\boldsymbol{\beta} + \boldsymbol{\varepsilon},$$

where $\boldsymbol{\varepsilon} \sim N\left(\mathbf{0}, \sigma^2 \mathbf{I}\right)$.

- **Theorem 2.1.** For the simple linear regression model with normal error terms, the maximum likelihood estimators of $\boldsymbol{\beta}$ are

$$\hat{\boldsymbol{\beta}} = (\mathbf{X}'\mathbf{X})^{-1}\mathbf{X}'\mathbf{Y}$$

and the maximum likelihood estimator of σ^2 is

$$\hat{\sigma}^2 = \frac{1}{n}\left(\mathbf{Y} - \mathbf{X}\hat{\boldsymbol{\beta}}\right)'\left(\mathbf{Y} - \mathbf{X}\hat{\boldsymbol{\beta}}\right).$$

Since the vector of error terms $\boldsymbol{\varepsilon}$ consists of independent and identically distributed normal random variables, $\mathbf{Y} = \mathbf{X}\boldsymbol{\beta} + \boldsymbol{\varepsilon}$ is a vector of independent and identically distributed normal random variables, and the linear transformation $\hat{\boldsymbol{\beta}} = (\mathbf{X}'\mathbf{X})^{-1}\mathbf{X}'\mathbf{Y}$ has normally distributed elements.

- **Theorem 2.2.** For the simple linear regression model with normal error terms,

$$\frac{\mathbf{e}'\mathbf{e}}{\sigma^2} \sim \chi^2(n-2),$$

and is independent of $\hat{\boldsymbol{\beta}}$.

- **Theorem 2.3.** For the simple linear regression model with normal error terms, an exact two-sided $100(1-\alpha)\%$ confidence interval for σ^2 is

$$\frac{\mathbf{e}'\mathbf{e}}{\chi^2_{n-2,\alpha/2}} < \sigma^2 < \frac{\mathbf{e}'\mathbf{e}}{\chi^2_{n-2,1-\alpha/2}}.$$

- **Theorems 2.4 and 2.7.** For the simple linear regression model with normal error terms,

$$\hat{\boldsymbol{\beta}} \sim N\left(\boldsymbol{\beta}, \sigma^2 (\mathbf{X}'\mathbf{X})^{-1}\right).$$

- **Theorem 2.12.** For the simple linear regression model with normal error terms, an exact two-sided $100(1-\alpha)\%$ confidence interval for $E[Y_h]$ for a given value of the independent variable X_h is

$$\mathbf{X}'_h\hat{\boldsymbol{\beta}} - t_{n-2,\alpha/2}\sqrt{\hat{\sigma}^2 \mathbf{X}'_h(\mathbf{X}'\mathbf{X})^{-1}\mathbf{X}_h} < E[Y_h] < \mathbf{X}'_h\hat{\boldsymbol{\beta}} + t_{n-2,\alpha/2}\sqrt{\hat{\sigma}^2 \mathbf{X}'_h(\mathbf{X}'\mathbf{X})^{-1}\mathbf{X}_h},$$

where $\mathbf{X}_h = (1, X_h)'$ and $\hat{\sigma}^2 = MSE$.

- **Theorem 2.15.** For the simple linear regression model with normal error terms, an exact two-sided $100(1-\alpha)\%$ prediction interval for $Y_h^{\star}$ for a given value of the independent variable X_h is

$$\mathbf{X}'_h\hat{\boldsymbol{\beta}} - t_{n-2,\alpha/2}\sqrt{\hat{\sigma}^2\left(1 + \mathbf{X}'_h\left(\mathbf{X}'\mathbf{X}\right)^{-1}\mathbf{X}_h\right)} < Y_h^{\star} < \mathbf{X}'_h\hat{\boldsymbol{\beta}} + t_{n-2,\alpha/2}\sqrt{\hat{\sigma}^2\left(1 + \mathbf{X}'_h\left(\mathbf{X}'\mathbf{X}\right)^{-1}\mathbf{X}_h\right)},$$

where $\mathbf{X}_h = (1, X_h)'$ and $\hat{\sigma}^2 = MSE$.

- **Theorem 2.16.** Under the simple linear regression model with normal error terms and parameters estimated from the data pairs $(X_1, Y_1), (X_2, Y_2), \ldots, (X_n, Y_n)$,

$$\frac{(\hat{\boldsymbol{\beta}} - \boldsymbol{\beta})' \mathbf{X}'\mathbf{X}(\hat{\boldsymbol{\beta}} - \boldsymbol{\beta})}{2 \cdot MSE} \sim F(2, n - 2).$$

- **Theorem 2.17.** Under the simple linear regression model with normal error terms and parameters estimated from the data pairs $(X_1, Y_1), (X_2, Y_2), \ldots, (X_n, Y_n)$, the values of β_0 and β_1 satisfying

$$\frac{(\hat{\boldsymbol{\beta}} - \boldsymbol{\beta})' \mathbf{X}'\mathbf{X}(\hat{\boldsymbol{\beta}} - \boldsymbol{\beta})}{2 \cdot MSE} \le F_{2, n-2, \alpha}$$

  form an exact joint $100(1 - \alpha)\%$ confidence region for β_0 and β_1.

- **Definition 3.2.** Under the simple linear regression model, the *hat matrix* is

$$\mathbf{H} = \mathbf{X} \left(\mathbf{X}'\mathbf{X}\right)^{-1} \mathbf{X}'.$$

  The diagonal elements of the hat matrix are the *leverages*. The matrix equation

$$\hat{\mathbf{Y}} = \mathbf{H}\mathbf{Y}$$

  indicates that $\mathbf{H}$ transforms $\mathbf{Y}$ to $\hat{\mathbf{Y}}$. The hat matrix is symmetric (that is, $\mathbf{H} = \mathbf{H}'$) and idempotent (that is, $\mathbf{H}\mathbf{H} = \mathbf{H}$).

The matrix approach applied to a simple linear regression model is illustrated for a small sample size next.

Example 3.9 Consider again the sales data set from Example 1.3. Let the independent variable X be the *number* of sales per week that Cheryl completes. Each sale results in a random amount of revenue to the company that can be attributed to Cheryl. Let the dependent random variable Y be the associated total revenue to the company from the sales attributed to Cheryl for that week, in thousands of dollars. The data pairs for the past $n = 3$ weeks are

$$(X_1, Y_1) = (6, 2), \qquad (X_2, Y_2) = (8, 9), \qquad \text{and} \qquad (X_3, Y_3) = (2, 2).$$

Use the matrix approach to simple linear regression to define the matrices $\mathbf{X}, \mathbf{Y}, \boldsymbol{\beta}$, and $\boldsymbol{\varepsilon}$. Calculate the least squares estimates of the population intercept β_0 and population slope β_1, the fitted values, the hat matrix, the residuals, the unbiased estimate of the variance of the error terms, SST, SSR, SSE, R^2, r, an exact 95% confidence interval for $E[Y_h]$ when $X_h = 5$ weekly sales, and an exact 95% prediction interval for $Y_h^{\star}$ when $X_h = 5$ weekly sales using the matrix approach to simple linear regression.

The $\mathbf{X}, \mathbf{Y}, \boldsymbol{\beta}$, and $\boldsymbol{\varepsilon}$ matrices associated with the $n = 3$ data pairs are

$$\mathbf{X} = \begin{bmatrix} 1 & 6 \\ 1 & 8 \\ 1 & 2 \end{bmatrix}, \qquad \mathbf{Y} = \begin{bmatrix} 2 \\ 9 \\ 2 \end{bmatrix}, \qquad \boldsymbol{\beta} = \begin{bmatrix} \beta_0 \\ \beta_1 \end{bmatrix}, \qquad \text{and} \qquad \boldsymbol{\varepsilon} = \begin{bmatrix} \varepsilon_1 \\ \varepsilon_2 \\ \varepsilon_3 \end{bmatrix}.$$

The R code below uses the matrix approach to simple linear regression to calculate the estimate of the intercept $\hat{\beta}_0$, the estimate of the slope $\hat{\beta}_1$, the fitted values $\hat{\mathbf{Y}}$, the hat matrix $\mathbf{H}$, the residuals $\mathbf{e}$, and the estimate of the population variance of the error terms $\hat{\sigma}^2$. *SST*, *SSR*, *SSE*, R^2, r, an exact 95% confidence interval for $E[Y_h]$ when $X_h = 5$, and an exact 95% prediction interval for $Y_h^\ast$ when $X_h = 5$ using the matrix approach to simple linear regression. The t function computes a matrix transpose, the diag function creates an identity matrix, and the solve function computes the inverse of $X'X$. The matrix multiplication operator is %*%.

```
x       = c(6, 8, 2)
y       = c(2, 9, 2)
x       = cbind(1, x)
beta    = solve(t(x) %*% x) %*% t(x) %*% y
yhat    = x %*% beta
H       = x %*% solve(t(x) %*% x) %*% t(x)
n       = length(y)
e       = (diag(n) - H) %*% y
sighat  = (t(e) %*% e) / (n - 2)
ybar    = rep(mean(y), n)
sst     = t(y - ybar) %*% (y - ybar)
ssr     = t(yhat - ybar) %*% (yhat - ybar)
sse     = t(y - yhat) %*% (y - yhat)
R2      = ssr / sst
r       = sign(beta[2]) * sqrt(R2)
alpha   = 0.05
conf1   = c(sum(e ^ 2) / qchisq(1 - alpha / 2, n - 2),
             sum(e ^ 2) / qchisq(alpha / 2, n - 2))
xh      = matrix(c(1, 5), 2, 1)
half2   = qt(1 - alpha / 2, n - 2) *
             sqrt(sse / (n - 2) * t(xh) %*% solve(t(x) %*% x) %*% xh)
conf2   = c(t(xh) %*% beta - half2, t(xh) %*% beta + half2)
half3   = qt(1 - alpha / 2, n - 2) *
             sqrt(sse / (n - 2) * (1 + t(xh) %*% solve(t(x) %*% x) %*% xh))
conf3   = c(t(xh) %*% beta - half3, t(xh) %*% beta + half3)
```

The output of this code is given in the equations that follow. The least squares estimators of the intercept and slope of the regression line are

$$\hat{\boldsymbol{\beta}} = \left(\mathbf{X}'\mathbf{X}\right)^{-1}\mathbf{X}'\mathbf{Y} = \begin{bmatrix} 3 & 16 \\ 16 & 104 \end{bmatrix}^{-1} \begin{bmatrix} 1 & 1 & 1 \\ 6 & 8 & 2 \end{bmatrix} \begin{bmatrix} 2 \\ 9 \\ 2 \end{bmatrix} = \begin{bmatrix} -1 \\ 1 \end{bmatrix}.$$

The fitted values are

$$\hat{\mathbf{Y}} = \mathbf{X}\hat{\boldsymbol{\beta}} = \begin{bmatrix} 1 & 6 \\ 1 & 8 \\ 1 & 2 \end{bmatrix} \begin{bmatrix} -1 \\ 1 \end{bmatrix} = \begin{bmatrix} 5 \\ 7 \\ 1 \end{bmatrix}.$$

The 3×3 hat matrix $\mathbf{H}$ is

$$\mathbf{H} = \mathbf{X}\left(\mathbf{X}'\mathbf{X}\right)^{-1}\mathbf{X}' = \begin{bmatrix} 1 & 6 \\ 1 & 8 \\ 1 & 2 \end{bmatrix} \begin{bmatrix} 3 & 16 \\ 16 & 104 \end{bmatrix}^{-1} \begin{bmatrix} 1 & 1 & 1 \\ 6 & 8 & 2 \end{bmatrix} = \begin{bmatrix} 5/14 & 3/7 & 3/14 \\ 3/7 & 5/7 & -1/7 \\ 3/14 & -1/7 & 13/14 \end{bmatrix}.$$

The diagonal elements of the hat matrix are the leverages h_{11}, h_{22}, h_{33}. The vector of residuals is

$$
\mathbf{e} = (\mathbf{I} - \mathbf{H})\mathbf{Y} = \begin{bmatrix} 9/14 & -3/7 & -3/14 \\ -3/7 & 2/7 & 1/7 \\ -3/14 & 1/7 & 1/14 \end{bmatrix} \begin{bmatrix} 2 \\ 9 \\ 2 \end{bmatrix} = \begin{bmatrix} -3 \\ 2 \\ 1 \end{bmatrix}.
$$

The fitted values and the residuals computed here are consistent with the geometry shown in Figure 1.15 from Example 1.8. The unbiased estimate of the population variance of the error terms is

$$
\hat{\sigma}^2 = MSE = \frac{\mathbf{e}'\mathbf{e}}{n-2} = \frac{1}{3-2} \begin{bmatrix} -3 & 2 & 1 \end{bmatrix} \begin{bmatrix} -3 \\ 2 \\ 1 \end{bmatrix} = 14.
$$

The sums of squares can be partitioned as $SST = SSR + SSE$ using

$$
\left(\mathbf{Y} - \bar{\mathbf{Y}}\right)'\left(\mathbf{Y} - \bar{\mathbf{Y}}\right) = \left(\hat{\mathbf{Y}} - \bar{\mathbf{Y}}\right)'\left(\hat{\mathbf{Y}} - \bar{\mathbf{Y}}\right) + \left(\mathbf{Y} - \hat{\mathbf{Y}}\right)'\left(\mathbf{Y} - \hat{\mathbf{Y}}\right),
$$

where $\bar{\mathbf{Y}}$ is an n-element column vector with identical elements which are each the sample mean of the values of the dependent variable. For the $n = 3$ data pairs, this becomes

$$
\left(-\frac{7}{3}\right)^2 + \left(\frac{14}{3}\right)^2 + \left(-\frac{7}{3}\right)^2 = \left(\frac{2}{3}\right)^2 + \left(\frac{8}{3}\right)^2 + \left(-\frac{10}{3}\right)^2 + (-3)^2 + 2^2 + 1^2
$$

or

$$
\frac{98}{3} = \frac{56}{3} + 14.
$$

Figure 3.14 show the geometry associated with $SST = SSR + SSE$ for the three data pairs. The sum of the areas of the three squares in the top graph is SST; the sum of the areas of the three squares in the middle graph is SSR; the sum of the areas of the three squares in the bottom graph is SSE.

The coefficient of determination and the correlation coefficient in a simple linear regression model are

$$
R^2 = \frac{SSR}{SST} = \frac{\left(\hat{\mathbf{Y}} - \bar{\mathbf{Y}}\right)'\left(\hat{\mathbf{Y}} - \bar{\mathbf{Y}}\right)}{\left(\mathbf{Y} - \bar{\mathbf{Y}}\right)'\left(\mathbf{Y} - \bar{\mathbf{Y}}\right)} = \frac{56/3}{98/3} = \frac{4}{7} = 0.57 \qquad \text{and} \qquad r = 0.76.
$$

The three intervals are

$$
2.8 < \sigma^2 < 14000,
$$

$$
-24 < E[Y_h] < 32,
$$

and

$$
-51 < Y_h^\star < 59.
$$

The intervals are unusually wide because there are only $n = 3$ data pairs which have significant deviation from the regression line. Notice that these results match those obtained earlier by algebraic methods and by using the $\texttt{lm}$ (linear model) function as given in Examples 1.3, 1.7, 1.8, and 1.10.

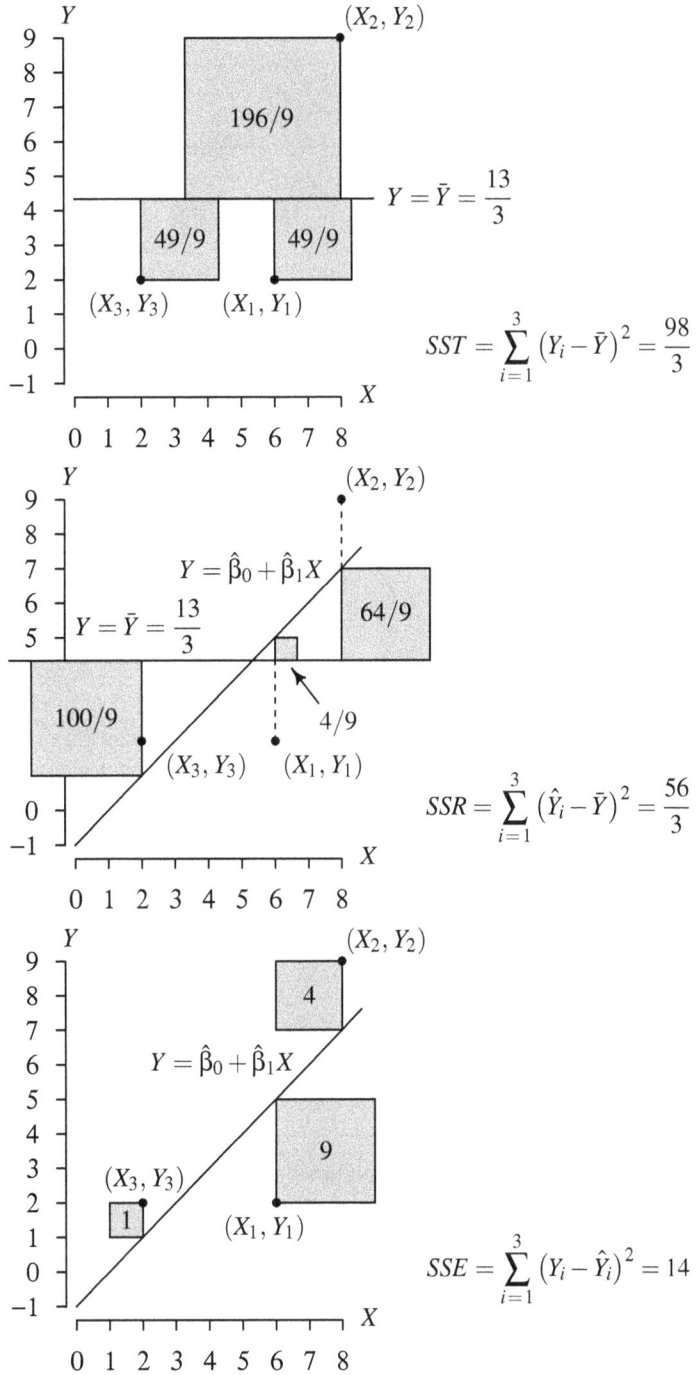

Figure 3.14: Geometry associated with $SST = SSR + SSE$ for the sales data.

Theorem 2.2 stated that under the simple linear regression model with normal errors,

$$\frac{SSE}{\sigma^2} \sim \chi^2(n-2).$$

An outline of the proof of Theorem 2.2 was given in Chapter 2 in purely algebraic terms. An outline of the proof to the result using the matrix approach to simple linear regression is given here to contrast the difference between the two approaches.

> **Proof** (Outline only; matrix approach) As given in the matrix version of Definition 1.2, the vector of *fitted values* in a simple linear regression model is the $n \times 1$ column vector
>
> $$\hat{\mathbf{Y}} = \mathbf{X}(\mathbf{X}'\mathbf{X})^{-1}\mathbf{X}'\mathbf{Y}.$$

The sum of squares for error in matrix form is

$$
\begin{aligned}
SSE &= \left(\mathbf{Y} - \hat{\mathbf{Y}}\right)'\left(\mathbf{Y} - \hat{\mathbf{Y}}\right) \\
&= \left[\mathbf{Y} - \mathbf{X}(\mathbf{X}'\mathbf{X})^{-1}\mathbf{X}'\mathbf{Y}\right]'\left[\mathbf{Y} - \mathbf{X}(\mathbf{X}'\mathbf{X})^{-1}\mathbf{X}'\mathbf{Y}\right] \\
&= \left[\mathbf{Y}' - \mathbf{Y}'\mathbf{X}''\left((\mathbf{X}'\mathbf{X})'\right)^{-1}\mathbf{X}'\right]\left[\mathbf{Y} - \mathbf{X}(\mathbf{X}'\mathbf{X})^{-1}\mathbf{X}'\mathbf{Y}\right] \\
&= \mathbf{Y}'\left[\mathbf{I} - \mathbf{X}(\mathbf{X}'\mathbf{X})^{-1}\mathbf{X}'\right]\left[\mathbf{I} - \mathbf{X}(\mathbf{X}'\mathbf{X})^{-1}\mathbf{X}'\right]\mathbf{Y}.
\end{aligned}
$$

Let $\mathbf{R} = \mathbf{I} - \mathbf{X}(\mathbf{X}'\mathbf{X})^{-1}\mathbf{X}'$, where I is the $n \times n$ identity matrix. This matrix plays a critical role in the proof. The matrix $\mathbf{R}$ is symmetric because

$$\mathbf{R}' = \left[\mathbf{I} - \mathbf{X}(\mathbf{X}'\mathbf{X})^{-1}\mathbf{X}'\right]' = \mathbf{I}' - \mathbf{X}''\left((\mathbf{X}'\mathbf{X})'\right)^{-1}\mathbf{X}' = \mathbf{I} - \mathbf{X}(\mathbf{X}'\mathbf{X})^{-1} = \mathbf{R}.$$

The matrix $\mathbf{R}$ is idempotent because

$$
\begin{aligned}
\mathbf{R}^2 &= \left[\mathbf{I} - \mathbf{X}(\mathbf{X}'\mathbf{X})^{-1}\mathbf{X}'\right]\left[\mathbf{I} - \mathbf{X}(\mathbf{X}'\mathbf{X})^{-1}\mathbf{X}'\right] \\
&= \mathbf{I}^2 - 2\mathbf{X}(\mathbf{X}'\mathbf{X})^{-1}\mathbf{X}' + \mathbf{X}(\mathbf{X}'\mathbf{X})^{-1}\mathbf{X}'\mathbf{X}(\mathbf{X}'\mathbf{X})^{-1}\mathbf{X}' \\
&= \mathbf{I} - \mathbf{X}(\mathbf{X}'\mathbf{X})^{-1}\mathbf{X}' \\
&= \mathbf{R}.
\end{aligned}
$$

Since $\mathbf{R}$ is a symmetric idempotent matrix, it is a projection matrix. This has two implications. First, the rank of $\mathbf{R}$ equals the trace of $\mathbf{R}$, which in this case is $n-2$. Second, all eigenvalues of $\mathbf{R}$ are either zero or one, and in this setting, there are $n-2$ ones and 2 zeros. The rest of the proof proceeds as follows. Since $\mathbf{R}$ is symmetric matrix it can be orthogonally diagonalized as $\mathbf{R} = \mathbf{UDU}'$, where $\mathbf{U}$ is an orthogonal matrix and $\mathbf{D}$ is a diagonal matrix with $n-2$ ones and 2 zeros on the diagonal. The assumed normality of the error terms in the model results in normally distributed residuals, which can be simplified to yield $SSE/\sigma^2 \sim \chi^2(n-2)$. □

The matrix approach gives an alternative way of computing measures of interest in a simple linear regression. Using matrices also allows the following two helpful extensions to simple linear regression.

- Removing the first column of the **X** matrix that consists entirely of ones corresponds to forcing a regression line through the origin.

- Adding additional columns to the **X** matrix corresponds to including additional independent variables to the regression model, which is known as *multiple linear regression*. This is the topic of the next section.

3.5   Multiple Linear Regression

Multiple linear regression can often be applied when there are several independent variables (or predictors) $X_1, X_2, \ldots, X_p$ which can be used to explain a continuous dependent (or response) variable Y. Three examples are listed below.

- The asking price of a home Y is a function of

  - the number of square feet in the home,

  - the number of bedrooms, and

  - acreage of the land associated with the home.

- The annual amount of money a person donates to charity Y is a function of

  - the nationality of the person,

  - the annual income of the person,

  - the net worth of the person,

  - the religious affiliation of the person,

  - the age of the person, and

  - the gender of the person.

- The stopping distance of a car Y is a function of

  - the speed of the car,

  - the weight of the car, and

  - the type of brakes installed on the car.

One way to formulate a multiple linear regression model is to treat the left-hand side of the model as an expected value:

$$E[Y] = \beta_0 + \beta_1 X_1 + \beta_2 X_2 + \cdots + \beta_p X_p.$$

Since $E[Y]$ denotes a *conditional expectation* of Y given the values of the p independent variables $X_1, X_2, \ldots, X_n$, a more careful way to write this model is

$$E[Y \,|\, X_1, X_2, \ldots, X_n] = \beta_0 + \beta_1 X_1 + \beta_2 X_2 + \cdots + \beta_p X_p.$$

So far, there has been no consideration of the probability distribution of the error terms, and that is addressed in the formal definition of a multiple linear regression model given next.

Definition 3.4 A *multiple linear regression model* is given by

$$Y = \beta_0 + \beta_1 X_1 + \beta_2 X_2 + \cdots + \beta_p X_p + \varepsilon,$$

where

- $X_1, X_2, \ldots, X_p$ are the independent variables, assumed to be a fixed values observed without error,

- Y is the dependent variable, which is a continuous random variable,

- β_0 is the population intercept of the regression plane, an unknown constant parameter,

- $\beta_1, \beta_2, \ldots, \beta_p$ are unknown constant parameters which control the inclination of the regression plane, and

- ε is the error term, a continuous random variable with population mean zero and positive, finite population variance σ^2 that accounts for the randomness in the relationship between $X_1, X_2, \ldots, X_p$ and Y.

To estimate the parameters in a multiple linear regression model, we collect n observations which each consist of the p independent variables and the associated dependent variable. In most applications, $p > n$. Occasions arise (often in biostatistical applications) in which $p < n$. The formulation of the simple linear regression model with notation included for the n observations is

$$Y_i = \beta_0 + \beta_1 X_{i1} + \beta_2 X_{i2} + \cdots + \beta_p X_{ip} + \varepsilon_i$$

for $i = 1, 2, \ldots, n$. So X_{ij} denotes the value of the jth independent variable collected on the ith observational unit. In the real estate example given at the beginning of this section, X_{83} is the value of the third independent variable (acreage) collected on the 8th home collected by the analyst. The associated asking price of the 8th home is Y_8.

Figure 3.15 shows a portion of the *population* regression plane $E[Y] = \beta_0 + \beta_1 X_1 + \beta_2 X_2$ for a multiple linear regression model with $p = 2$ independent variables X_1 and X_2. The plane extends outward from the portion shown in Figure 3.15. The regression parameters β_0, β_1, and β_2 are fixed constants. The intercept β_0 is positive in Figure 3.15 because the plane strikes the Y-axis above the origin. Based on the inclination of the population regression plane relative to the X_1- and X_2-axes it is clear that $\beta_1 < 0$ and $\beta_2 > 0$. To avoid clutter and highlight the geometry and notation, only the ith data triple (X_{i1}, X_{i2}, Y_i) and the associated error term ε_i are shown in the figure.

Figure 3.16 shows a portion of the *estimated* regression plane $Y = \hat{\beta}_0 + \hat{\beta}_1 X_1 + \hat{\beta}_2 X_2$ for a multiple linear regression model with $p = 2$ independent variables X_1 and X_2. The estimated regression parameters $\hat{\beta}_0$, $\hat{\beta}_1$, and $\hat{\beta}_2$ are random variables which are estimated from n data triples (X_{11}, X_{12}, Y_1), (X_{21}, X_{22}, Y_2), $\ldots$, (X_{n1}, X_{n2}, Y_n). The estimated regression parameters are random variables because the dependent variable values $Y_1, Y_2, \ldots, Y_n$ are random variables. The estimated intercept $\hat{\beta}_0$ is positive in Figure 3.16 because the plane strikes the Y-axis above the origin. Based on the inclination of the estimated regression plane relative to the X_1- and X_2-axes it is clear that $\hat{\beta}_1 < 0$ and $\hat{\beta}_2 > 0$. To avoid clutter and highlight the geometry and notation, just the ith data triple (X_{i1}, X_{i2}, Y_i), the associated fitted value $(X_{i1}, X_{i2}, \hat{Y}_i)$, and the associated residual e_i are shown in the figure.

When there are $p > 2$ independent variables, the estimated regression model is a *hyperplane* in $\mathcal{R}^{p+1}$. Residual i is the distance $e_i = Y_i - \hat{Y}_i$, for $i = 1, 2, \ldots, n$.

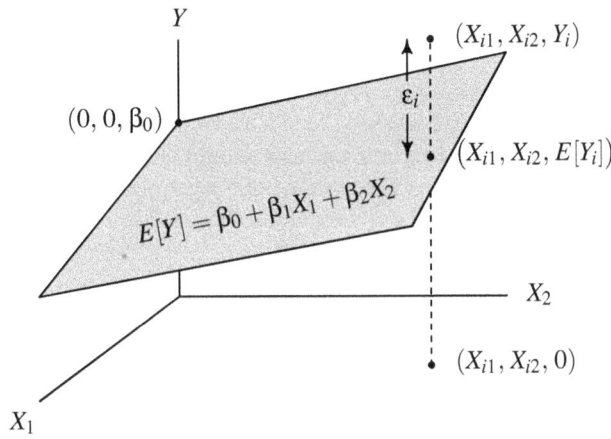

Figure 3.15: Population regression plane and a sample point.

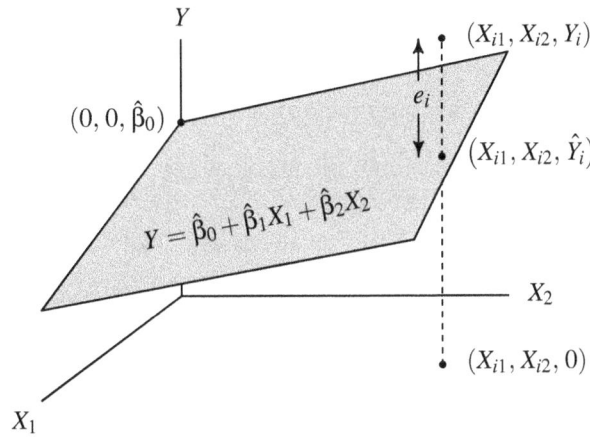

Figure 3.16: Estimated regression plane and a sample point.

When the error terms are assumed to be normally distributed, this model is known as the *multiple linear regression model with normal error terms*. This additional assumption allows for statistical inference concerning parameters and predicted values in a similar manner to that described in Chapter 2.

The multiple linear regression model can also be expressed in terms of matrices. Relative to the simple linear regression model, additional columns are appended to the $\mathbf{X}$ matrix, and the $\boldsymbol{\beta}$ vector is expanded to include the parameters associated with the additional parameters:

$$
\mathbf{X} = \begin{bmatrix} 1 & X_{11} & X_{12} & \cdots & X_{1p} \\ 1 & X_{21} & X_{22} & \cdots & X_{2p} \\ \vdots & \vdots & \vdots & \ddots & \vdots \\ 1 & X_{n1} & X_{n2} & \cdots & X_{np} \end{bmatrix}, \quad \mathbf{Y} = \begin{bmatrix} Y_1 \\ Y_2 \\ \vdots \\ Y_n \end{bmatrix}, \quad \boldsymbol{\beta} = \begin{bmatrix} \beta_0 \\ \beta_1 \\ \vdots \\ \beta_p \end{bmatrix}, \quad \text{and} \quad \boldsymbol{\varepsilon} = \begin{bmatrix} \varepsilon_1 \\ \varepsilon_2 \\ \vdots \\ \varepsilon_n \end{bmatrix}.
$$

The vectors $\mathbf{Y}$ and $\boldsymbol{\varepsilon}$ remain unchanged from the simple linear regression formulation. The first row of $\mathbf{X}$ corresponds to the values of the independent variables collected on the first observational unit, the second row of $\mathbf{X}$ corresponds to the values of the independent variables collected on the second observational unit, etc. As was the case in simple linear regression, $\mathbf{X}$ is known as the *design matrix*.

The good news about the matrix approach to multiple linear regression is that the definitions and results from simple linear regression only require some minor tweaking in order to generalize to multiple regression. Several of these definitions and results are given below. In many cases, it is just a matter of replacing the word "simple" with the word "multiple" or updating the degrees of freedom to account for the p independent variables. It is assumed that the $\mathbf{X}$ matrix has rank $p+1$ (that is, a *full rank* matrix), which means that the columns of $\mathbf{X}$ are linearly independent.

- The *multiple linear regression model* is

$$\mathbf{Y} = \mathbf{X}\boldsymbol{\beta} + \boldsymbol{\varepsilon},$$

  where $E[\boldsymbol{\varepsilon}] = \mathbf{0}$, $V[\boldsymbol{\varepsilon}] = \sigma^2 \mathbf{I}$, and $\mathbf{I}$ is the $n \times n$ identity matrix.

- The least squares estimators of $\boldsymbol{\beta}$, denoted by $\hat{\boldsymbol{\beta}} = (\hat{\beta}_0, \hat{\beta}_1, \ldots, \hat{\beta}_p)'$, solve the normal equations

$$\mathbf{X}'\mathbf{X}\hat{\boldsymbol{\beta}} = \mathbf{X}'\mathbf{Y}.$$

  Since $\mathbf{X}$ has full rank, $\mathbf{X}'\mathbf{X}$ is invertible and the normal equations have the unique solution

$$\hat{\boldsymbol{\beta}} = (\mathbf{X}'\mathbf{X})^{-1}\mathbf{X}'\mathbf{Y},$$

  by premultiplying both sides of the normal equations by $(\mathbf{X}'\mathbf{X})^{-1}$.

- The least squares estimator of $\boldsymbol{\beta}$ in a multiple linear regression model is an unbiased estimator of $\boldsymbol{\beta}$ because

$$E[\hat{\boldsymbol{\beta}}] = \boldsymbol{\beta}.$$

- The least squares estimators of $\boldsymbol{\beta}$ in the multiple linear regression model can be written as linear combinations of the dependent variables:

$$\hat{\boldsymbol{\beta}} = (\mathbf{X}'\mathbf{X})^{-1}\mathbf{X}'\mathbf{Y},$$

  where the coefficients in the linear combinations are given by $(\mathbf{X}'\mathbf{X})^{-1}\mathbf{X}'$.

- The variance–covariance matrix of the least squares estimators of $\boldsymbol{\beta}$ is

$$\sigma^2 (\mathbf{X}'\mathbf{X})^{-1}.$$

- (Gauss–Markov theorem) The least squares estimators of $\boldsymbol{\beta}$ in a multiple linear regression model, $\hat{\boldsymbol{\beta}} = (\mathbf{X}'\mathbf{X})^{-1}\mathbf{X}'\mathbf{Y}$, have the smallest population variance amongst all linear unbiased estimators of $\boldsymbol{\beta}$.

- The vector of *fitted values* in a multiple linear regression model is the $n \times 1$ column vector

$$\hat{\mathbf{Y}} = \mathbf{X}\hat{\boldsymbol{\beta}} = \mathbf{X}(\mathbf{X}'\mathbf{X})^{-1}\mathbf{X}'\mathbf{Y},$$

which is a linear combination of the dependent variables. The vector of *residuals* is the $n \times 1$ column vector

$$\begin{aligned} \mathbf{e} &= \mathbf{Y} - \hat{\mathbf{Y}} \\ &= \mathbf{Y} - \mathbf{X}\hat{\boldsymbol{\beta}} \\ &= \mathbf{Y} - \mathbf{X}(\mathbf{X}'\mathbf{X})^{-1}\mathbf{X}'\mathbf{Y} \\ &= \left(\mathbf{I} - \mathbf{X}(\mathbf{X}'\mathbf{X})^{-1}\mathbf{X}'\right)\mathbf{Y}, \end{aligned}$$

which is also a linear combination of the dependent variables. The matrix $\mathbf{I}$ is the $n \times n$ identity matrix.

- The *multiple linear regression model with normal error terms* is

$$\mathbf{Y} = \mathbf{X}\boldsymbol{\beta} + \boldsymbol{\varepsilon},$$

where $\boldsymbol{\varepsilon} \sim N\left(\mathbf{0}, \sigma^2\mathbf{I}\right)$.

- For the multiple linear regression model with normal error terms, the maximum likelihood estimators of $\boldsymbol{\beta}$ are

$$\hat{\boldsymbol{\beta}} = (\mathbf{X}'\mathbf{X})^{-1}\mathbf{X}'\mathbf{Y}$$

and the maximum likelihood estimator of σ^2 is

$$\hat{\sigma}^2 = \frac{1}{n}\left(\mathbf{Y} - \mathbf{X}\hat{\boldsymbol{\beta}}\right)'\left(\mathbf{Y} - \mathbf{X}\hat{\boldsymbol{\beta}}\right).$$

Since the vector of error terms $\boldsymbol{\varepsilon}$ consists of independent and identically distributed normal random variables, $\mathbf{Y} = \mathbf{X}\boldsymbol{\beta} + \boldsymbol{\varepsilon}$ is a vector of independent and identically distributed normal random variables. Since $\hat{\boldsymbol{\beta}}$ is a linear transformation of Y, $\hat{\boldsymbol{\beta}} \sim N\left(\boldsymbol{\beta}, \sigma^2(\mathbf{X}'\mathbf{X})^{-1}\right)$.

- Under the multiple linear regression model, the $n \times n$ *hat matrix* is

$$\mathbf{H} = \mathbf{X}\left(\mathbf{X}'\mathbf{X}\right)^{-1}\mathbf{X}'.$$

The diagonal elements of the hat matrix are the *leverages*. The matrix equation

$$\hat{\mathbf{Y}} = \mathbf{H}\mathbf{Y}$$

indicates that $\mathbf{H}$ transforms $\mathbf{Y}$ to $\hat{\mathbf{Y}}$. The hat matrix is symmetric (that is, $\mathbf{H} = \mathbf{H}'$) and idempotent (that is, $\mathbf{H}\mathbf{H} = \mathbf{H}$). The trace of the hat matrix is $\sum_{i=1}^{n} h_{ii} = p + 1$.

The example of multiple linear regression that follows considers $p = 2$ predictors of the sales price of a home.

Example 3.10 In Example 2.9, the sales price, Y, of homes sold in Ames, Iowa between 2006 and 2010 with between 2500 and 3500 square feet were fitted to a simple linear regression model with the square footage as an independent variable X. There were $n = 120$ homes in the data frame that fit this criteria. In that analysis, the value of the land was estimated to be \$21,233 (although this was outside of the scope of the simple linear regression model), and the price of the home increased by an average of \$112 with each additional square foot of indoor space. Fit a multiple linear regression model with normal error terms to the same data set using two independent variables, X_1, the square footage of indoor space, and X_2, the square footage of the lot. The dependent variable is again the sales price Y.

The multiple regression model in this setting is

$$Y = \beta_0 + \beta_1 X_1 + \beta_2 X_2 + \varepsilon,$$

where $\varepsilon \sim N\left(\mu, \sigma_Z^2\right)$. The R code below estimates the regression parameters β_0, β_1, and β_2. The regression function

$$E[Y] = \beta_0 + \beta_1 X_1 + \beta_2 X_2,$$

is a plane in $\mathcal{R}^3$. The values of β_1 and β_2 control the tilt of the regression plane, and the value of β_0 is the intercept of the regression plane with the $E[Y]$ axis. The regression plane will be fitted in two fashions in R: the matrix approach to multiple linear regression and the built-in `lm` function. The R code below defines the **X** and **Y** matrices, and then uses the formula

$$\hat{\beta} = \left(\mathbf{X}'\mathbf{X}\right)^{-1}\mathbf{X}'\mathbf{Y}$$

to calculate the estimates of the regression coefficients.

```
library(modeldata)
i       = ames$Gr_Liv_Area >= 2500 & ames$Gr_Liv_Area <= 3500
sqft    = ames$Gr_Liv_Area[i]
lotarea = ames$Lot_Area[i]
X       = cbind(1, sqft, lotarea)
Y       = ames$Sale_Price[i]
beta    = solve(t(X) %*% X) %*% t(X) %*% Y
```

These R statements return the least squares regression parameter estimates $\hat{\beta}_0 = 26{,}515$, $\hat{\beta}_1 = 96.88$, and $\hat{\beta}_2 = 2.65$. The intercept is not meaningful in this setting because it is associated with a home with 0 square feet and no land. This situation does not make sense nor does it fall in the scope of the model. The naive interpretation of the other regression coefficients in the fitted model are (a) the sales price of a home increases by an average of \$96.88 for each additional square foot in the home, and (b) the sales price of the home increases by \$2.65 for each additional square foot in the lot size. The interpretation of the estimated regression coefficients is more nuanced in the case of multiple independent variables because those independent variables are often correlated. So reporting that "the value of $\hat{\beta}_1 = 96.88$ means that the sales price of the house increases by an average of \$96.88 for each additional square foot of interior space with the lot size fixed" is not quite accurate because the interior space and lot size might be correlated. Larger homes might be built on larger lots, for example. Regression analysts acknowledge possible correlations between the independent variables by just stating "the sales price increases by an average of \$96.88 for each additional square foot of interior space, adjusted for lot size" when interpreting $\hat{\beta}_1$. Likewise, "the sales price increases by an average of \$2.65 for each additional square foot of lot size, adjusted for interior square footage" when interpreting $\hat{\beta}_2$.

A second way to calculate the estimated regression coefficients is to use R's built-in `lm` function.

```
library(modeldata)
i       = ames$Gr_Liv_Area >= 2500 & ames$Gr_Liv_Area <= 3500
sqft    = ames$Gr_Liv_Area[i]
lotarea = ames$Lot_Area[i]
price   = ames$Sale_Price[i]
fit     = lm(price ~ sqft + lotarea)
summary(fit)
```

The call to the summary function prints the following output concerning the fitted multiple linear regression model.

```
Call:
lm(formula = price ~ sqft + lotarea)

Residuals:
    Min       1Q   Median       3Q      Max
-226718   -61645    -5756    62774   288215

Coefficients:
              Estimate Std. Error t value Pr(>|t|)
(Intercept) 2.652e+04  1.087e+05   0.244   0.8077
sqft        9.688e+01  4.043e+01   2.396   0.0181 *
lotarea     2.645e+00  1.660e+00   1.593   0.1138
---
Signif. codes:  0 '***' 0.001 '**' 0.01 '*' 0.05 '.' 0.1 ' ' 1

Residual standard error: 99890 on 117 degrees of freedom
Multiple R-squared:  0.08339,   Adjusted R-squared:  0.06772
F-statistic: 5.322 on 2 and 117 DF,  p-value: 0.006134
```

The estimated regression coefficients match those that were calculated using the matrix approach to multiple linear regression. The right-hand column of p-values tells us that the size of a home is a statistically significant predictor of the sales price of a home, but the lot size is not a statistically significant predictor of the sales price of a home.

A multiple linear regression model can easily be adapted to include nonlinear terms. A multiple regression model with two independent variables X_1 and X_2, for example, with a linear relationship between X_1 and Y and a quadratic relationship between X_2 and Y which includes an intercept term is

$$Y = \beta_0 + \beta_1 X_1 + \beta_2 X_2 + \beta_3 X_2^2 + \varepsilon.$$

Using the R lm function to estimate the coefficients will be illustrated in Section 3.7.

Multiple linear regression has many more modeling issues that arise than simple linear regression. The subsections that follow consider the following topics within multiple regression: (a) handling categorical independent variables which fall in categories rather than quantitative values, (b) handling the case in which independent variables have interactive effects, (c) extending the ANOVA table to multiple independent variables, (d) calculation of the coefficient of determination for multiple linear regression, and an adjustment that can be made to reduce its bias, (e) the effect of multicollinearity among the independent variables, and (f) algorithms for model selection.

3.5.1   Categorical Independent Variables

Some regression models include independent variables which are not naturally quantitative, but are rather categorical. These categorical independent variables require some special treatment in order to be included in a multiple linear regression model. The cases in which a categorical independent variable falls in one of two categories will be considered separately from the case in which a categorical independent variable falls in one of more than two categories.

    Categorical independent variable which falls in one of two categories. Consider a multiple linear regression model with $p = 2$ independent variables, X_1, which is age, and X_2, which is gender. The dependent variable is the annual salary Y. So the multiple linear regression model is

$$Y = \beta_0 + \beta_1 X_1 + \beta_2 X_2 + \varepsilon.$$

Regression models assume that the independent variables are quantitative rather than categorical like gender. One solution to this problem is to code the gender as 0 for female and 1 for male. The independent variable X_2 in this case is known as a *dummy variable* or an *indicator variable*. As a particular instance, consider $n = 6$ data points consisting of three women (ages 26, 71, and 34) and three men (ages 44, 65, and 21). In this case the design matrix is

$$\mathbf{X} = \begin{bmatrix} 1 & 26 & 0 \\ 1 & 71 & 0 \\ 1 & 34 & 0 \\ 1 & 44 & 1 \\ 1 & 65 & 1 \\ 1 & 21 & 1 \end{bmatrix}.$$

The elements of the six-element column vector $\mathbf{Y}$ are the associated salaries. The value of $\hat{\beta}_0$ is not meaningful here. Not only is it outside of the scope of the model, its interpretation as the annual salary of a newborn baby girl doesn't fit with societal norms. Newborn baby girls seldom earn annual salaries. The value of $\hat{\beta}_1$ indicates the increase in annual salary for each additional year in age, adjusted for gender. Since salaries tend to rise over time, we anticipate that $\hat{\beta}_1$ will be positive. The value of $\hat{\beta}_2$ indicates the change in salary associated being male rather than female, adjusted for age. If $\hat{\beta}_2$ is significantly greater than zero, then men's salaries are significantly higher than women's salaries, adjusted for age; if $\hat{\beta}_2$ is significantly less than zero, then women's salaries are significantly higher than men's salaries, adjusted for age. The choice of using an indicator of 0 for women and 1 for men was arbitrary. See if you can predict what would happen if instead we used 0 for men and 1 for women.

    Categorical independent variable which falls in one of more than two categories. Let's extend the regression model to predict the annual salary to include another categorical variable: political affiliation. This categorical variable will have three levels: Republican, Democrat, and Independent. The third category includes anyone who is not affiliated with the two main political parties in the United States. Although it might be tempting to just let $X_3 = 1$ denote a Republican, $X_3 = 2$ denote a Democrat, and $X_3 = 3$ denote an Independent, this will likely produce erroneous results for two reasons. First, using the ordering $X_3 = 1$, $X_3 = 2$, and $X_3 = 3$ implies an ordering of the salaries associated with individuals from the three different political affiliations for $\beta_3 > 0$, or the opposite ordering of the salaries associated with individuals from the three political affiliations for $\beta_3 < 0$. This ordering might not be the correct ordering. Second, leaving a gap of 1 between each of the values of X_3 indicates that there is a known and equal salary gap between individuals from the ordered different political affiliations. The usual way to account for a categorical independent variable which can take on c values is to define $c - 1$ independent indicator variables. In the case of political affiliation, the independent variables X_3 and X_4 can be defined as

$$X_3 = \begin{cases} 0 & \text{not a Republican} \\ 1 & \text{Republican} \end{cases}$$

and

$$X_4 = \begin{cases} 0 & \text{not a Democrat} \\ 1 & \text{Democrat.} \end{cases}$$

So now the multiple linear regression model with $p = 4$ independent variables is

$$Y = \beta_0 + \beta_1 X_1 + \beta_2 X_2 + \beta_3 X_3 + \beta_4 X_4 + \varepsilon.$$

In this fashion, the expected value of an Independent's salary is given by

$$E[Y] = \beta_0 + \beta_1 X_1 + \beta_2 X_2,$$

the expected value of an Republican's salary is given by

$$E[Y] = \beta_0 + \beta_1 X_1 + \beta_2 X_2 + \beta_3 X_3,$$

and the expected value of a Democrat's salary is given by

$$E[Y] = \beta_0 + \beta_1 X_1 + \beta_2 X_2 + \beta_4 X_4.$$

With this arrangement of the levels of the categorical variable representing the political affiliation, there is no predicted ordering of salaries by the three political affiliations nor are the gaps between the affiliations necessarily equal.

As a particular instance, consider $n = 6$ data points with three women (a 26-year-old Independent, a 71-year-old Democrat, and a 34-year-old Republican) and three men (a 44-year-old Independent, a 65-year-old Democrat, and a 21-year-old Republican) in the study. The appropriate design matrix is

$$\mathbf{X} = \begin{bmatrix} 1 & 26 & 0 & 0 & 0 \\ 1 & 71 & 0 & 0 & 1 \\ 1 & 34 & 0 & 1 & 0 \\ 1 & 44 & 1 & 0 & 0 \\ 1 & 65 & 1 & 0 & 1 \\ 1 & 21 & 1 & 1 & 0 \end{bmatrix}.$$

The value of $\hat{\beta}_3$ is the estimated difference between the mean annual salary of an Independent and a Republican, adjusted for age and gender. The value of $\hat{\beta}_3$ is the estimated difference between the mean annual salary of an Independent and a Democrat, adjusted for age and gender. This example has been for illustrative purposes only. Estimating five parameters $\beta_0, \beta_1, \ldots, \beta_4$ from just six data values will almost certainly not provide strong statistical evidence concerning the effect of age, gender, and political affiliation on salary. Furthermore, many other important factors, such as years of education, years on the job, and type of work, have not been included in this regression model.

3.5.2 Interaction Terms

The multiple linear regression model

$$Y = \beta_0 + \beta_1 X_1 + \beta_2 X_2 + \cdots + \beta_p X_p + \varepsilon$$

assumes a linear relationship between each independent variable and Y and the slope associated with an independent variable is identical at all values of the other independent variables within the scope of the multiple linear regression model. This relationship is illustrated for some selected data points of smaller homes from the Ames, Iowa housing data set from Examples 2.9 and 3.10. In this case, X_1 is the interior square footage, X_2 is an indicator variable reflecting the lot size,

$$X_2 = \begin{cases} 0 & \text{lot size is less than or equal to 10,000 square feet} \\ 1 & \text{lot size is greater than 10,000 square feet,} \end{cases}$$

and Y is the sales price. The multiple linear regression model with the $p = 2$ independent variables is

$$Y = \beta_0 + \beta_1 X_1 + \beta_2 X_2 + \varepsilon.$$

Figure 3.17 shows a scatterplot of the interior square footage and sales price of homes on smaller lots ($X_2 = 0$ as open points) and larger lots ($X_2 = 1$ as solid points). The values of $\hat{\beta}_0$, $\hat{\beta}_1$, and $\hat{\beta}_2$ are indicated on the graph. The estimated intercept $\hat{\beta}_0 = 21{,}473$, although slightly outside of the scope of the model, gives the estimated sales price of a small lot containing no dwelling as \$21,473. The estimated regression coefficient $\hat{\beta}_1 = 31.33$ indicates that the sales price of a home increases by an estimated \$31.33 for each additional interior square foot, adjusted for lot size. The estimated regression coefficient $\hat{\beta}_2 = 35{,}693$ indicates that homes on larger lots cost \$35,693 more, on average, than homes on smaller lots, adjusted for interior square feet. Notice that this formulation of the multiple linear regression model forces the slopes of the two lines in Figure 3.17 to be identical, regardless of the value of X_2.

But is the assumption of equal slopes of the two lines in Figure 3.17 justified? Separate simple linear regression models are fitted to the homes built on smaller and larger lots, and the results are plotted in Figure 3.18. The lines do not appear to be parallel in this case, indicating that a more complex regression model is warranted. There appears, in this case, to be an *interaction effect*

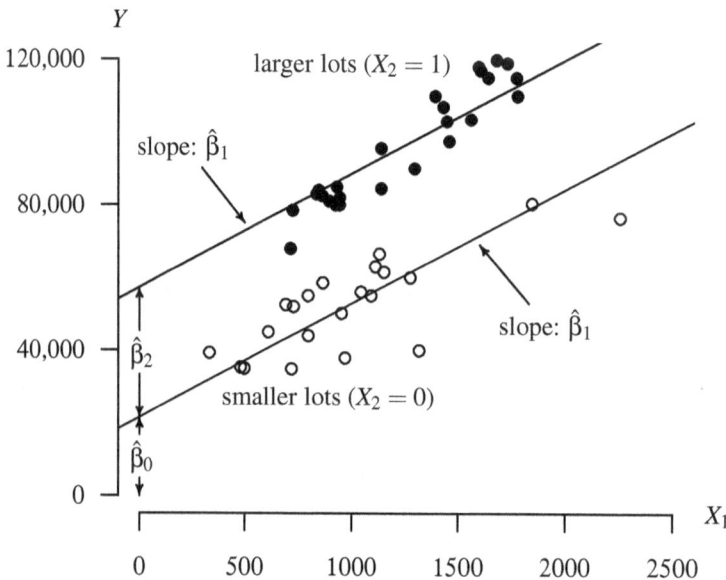

Figure 3.17: Fitted multiple linear regression model $Y = \beta_0 + \beta_1 X_1 + \beta_2 X_2 + \varepsilon$.

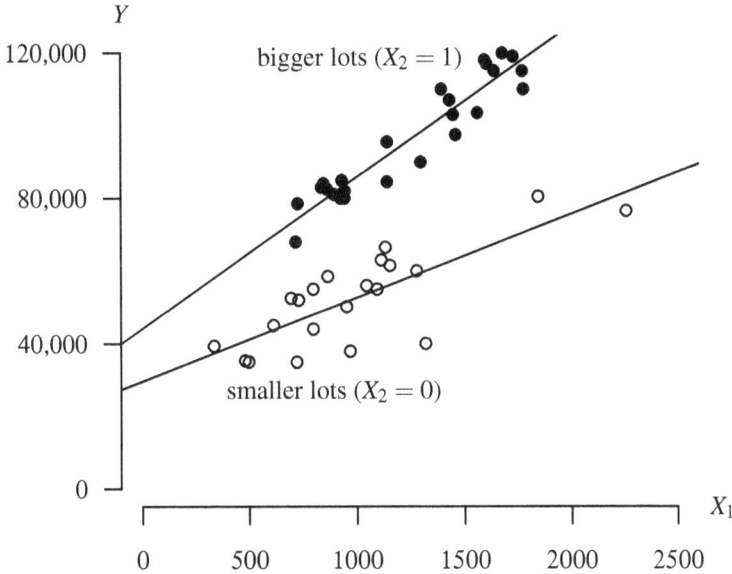

Figure 3.18: Fitted simple linear regression models $Y = \beta_0 + \beta_1 X_1 + \varepsilon$.

between X_1 and X_2. This means that the effect of one independent variable (X_1, for example, the interior size) on Y is altered based on the value of another independent variable (X_2, the lot size indicator).

Regression analysts account for this interaction by including cross-product terms in the regression model. In this Ames housing data set example, the regression model with an interaction term is

$$Y = \beta_0 + \beta_1 X_1 + \beta_2 X_2 + \beta_3 X_1 X_2 + \varepsilon.$$

If the regression parameter $\hat{\beta}_3$ differs statistically from 0, then the inclusion of the interaction term is warranted. Notice that when $X_2 = 0$ (smaller lots), the model reduces to

$$Y = \beta_0 + \beta_1 X_1 + \varepsilon,$$

which is a simple linear regression model with intercept parameter β_0 and slope parameter β_1. On the other hand, when $X_2 = 1$ (larger lots), the model reduces to

$$Y = \beta_0 + \beta_1 X_1 + \beta_2 + \beta_3 X_1 + \varepsilon$$

or

$$Y = \beta_0 + \beta_2 + (\beta_1 + \beta_3) X_1 + \varepsilon$$

which is a simple linear regression model with intercept parameter $\beta_0 + \beta_2$ and slope parameter $\beta_1 + \beta_3$. It is in this fashion that the two non-parallel lines depicted in Figure 3.18 can be estimated in a single regression model. Not surprisingly, it requires four parameters, β_0, β_1, β_2, and β_3, to do so. The multiple linear regression model with an interaction term can be fitted using the lm function in R by simply replacing the usual + in the formula with *. All four parameters are statistically significant at the 0.05 level in this case, so the inclusion of an interaction term is warranted.

3.5.3 The ANOVA Table

The degrees of freedom for the sums of squares in multiple linear regression are modified because of the additional parameters estimated relative to those given in the ANOVA table from Table 2.2 for simple linear regression. The ANOVA table for a multiple linear regression model with p independent variables and normal error terms is given in Table 3.6. Formulas for the sums of squares

Source	SS	df	MS	F
Regression	SSR	p	MSR	MSR/MSE
Error	SSE	$n-p-1$	MSE	
Total	SST	$n-1$		

Table 3.6: Basic ANOVA table for multiple linear regression.

using the matrix formulation for multiple linear regression are $SST = SSR + SSE$, which is

$$\left(\mathbf{Y} - \bar{\mathbf{Y}}\right)'\left(\mathbf{Y} - \bar{\mathbf{Y}}\right) = \left(\hat{\mathbf{Y}} - \bar{\mathbf{Y}}\right)'\left(\hat{\mathbf{Y}} - \bar{\mathbf{Y}}\right) + \left(\mathbf{Y} - \hat{\mathbf{Y}}\right)'\left(\mathbf{Y} - \hat{\mathbf{Y}}\right),$$

where $\bar{\mathbf{Y}}$ is an n-element column vector with identical elements which are each the sample mean of the values of the dependent variable. Equivalently,

$$SST = \mathbf{Y}'\mathbf{Y} - \mathbf{Y}'\mathbf{J}\mathbf{Y}/n, \qquad SSR = \hat{\boldsymbol{\beta}}'\mathbf{X}'\mathbf{Y} - \mathbf{Y}'\mathbf{J}\mathbf{Y}/n, \qquad SSE = \mathbf{Y}'\mathbf{Y} - \hat{\boldsymbol{\beta}}'\mathbf{X}'\mathbf{Y},$$

where $\mathbf{J}$ is an $n \times n$ matrix with all elements being equal to 1. The mean square error for regression is $MSR = SSR/p$, the mean square error is $MSE = SSE/(n-p-1)$, and the test statistic $F = MSR/MSE$ can be used for testing

$$H_0 : \beta_1 = \beta_2 = \cdots = \beta_p = 0$$

versus

$$H_1 : \text{not all } \beta_1, \beta_2, \ldots, \beta_p \text{ equal } 0$$

where F has an $F(p, n-p-1)$ distribution under H_0. The anova function in R can be used to generate an ANOVA table associated with a multiple linear regression model fitted by the lm function. For the Ames, Iowa housing data from Example 3.10 which used $p = 2$ independent variables (interior square footage and lot size), the R summary function returns the test statistic $F = 5.322$, which is associated with a p-value of $p = 0.006$ based on the F distribution with $p = 2$ and $n - p - 1 = 120 - 2 - 1 = 117$ degrees of freedom. There is strong statistical evidence that one or both of the coefficients $\hat{\beta}_1$ and $\hat{\beta}_2$ is statistically different from zero. One or both of the independent variables is effective in predicting the sales price.

3.5.4 Adjusted Coefficient of Determination

The coefficient of determination for a multiple linear regression model is defined as

$$R^2 = \frac{SSR}{SST} = \frac{SST - SSE}{SST} = 1 - \frac{SSE}{SST},$$

and it measures the fraction of variation in $Y_1, Y_2, \ldots, Y_n$ about $\bar{Y}$ that is accounted for by the linear relationship between the independent variables $X_1, X_2, \ldots, X_p$ and Y. As before $0 \leq R^2 \leq 1$, and the extreme cases are associated with $\hat{\beta}_1 = \hat{\beta}_2 = \cdots = \hat{\beta}_p = 0$ (for $R^2 = 0$) and all Y-values falling in the estimated regression hyperplane (for $R^2 = 1$).

Now consider a multiple linear regression model with p independent variables $X_1, X_2, \ldots, X_p$. What is the effect on *SST* and *SSE* of adding another independent variable, X_{p+1}, to the model? Adding another independent variable does not affect *SST* because it depends only on $Y_1, Y_2, \ldots, Y_n$. The value of *SSE* cannot increase with the addition of the new independent variable because either (*a*) *SSE* will remain the same if $\hat{\beta}_{p+1} = 0$, or (*b*) *SSE* will decrease if $\hat{\beta}_{p+1} \neq 0$. The impact on R^2 is that it must stay the same or increase for every additional independent variable that is added to the model.

It is for this reason that R^2 tends to be a biased estimator of the fraction of variation in $Y_1, Y_2, \ldots, Y_n$ accounted for by the independent variables. Some regression software (including R) calculate an *adjusted coefficient of variation* by dividing the sums of squares by their associated degrees of freedom

$$R^2_{\text{adj}} = 1 - \frac{SSE/(n-p-1)}{SST/(n-1)}.$$

Both values are reported in the call to the `summary` function with the Ames, Iowa housing data in Example 3.10 as

$$R^2 = 0.08339 \qquad \text{and} \qquad R^2_{\text{adj}} = 0.06772.$$

3.5.5 Multicollinearity

In many settings, the values of the independent variables are correlated. In the housing data set from Example 3.10, for example, the independent variables X_1 (interior square footage) and X_2 (lot size) are probably positively correlated. Intuition suggests that larger homes are built on larger lots, on average. In the extreme case, what if homes in Ames were required by some bizarre municipal code to all be single story homes with the square footage of the lot always exactly four times the square footage of the interior of the home? In this case, $X_2 = 4X_1$, so knowing the value of either X_1 or X_2 allows you to know the value of the other. Intuitively, one of the two independent variables is superfluous. When this is the case, the design matrix $\mathbf{X}$ has two columns which are multiples of one another, so these columns are linearly dependent and the matrix does not have full rank. This implies that the matrix $\mathbf{X}'\mathbf{X}$ (which is used in computing the estimates of the regression coefficients) is singular, so it does not have an inverse. In this case, the usual formula for the regression coefficients,

$$\hat{\boldsymbol{\beta}} = \left(\mathbf{X}'\mathbf{X}\right)^{-1}\mathbf{X}'\mathbf{Y},$$

is undefined because the matrix $\mathbf{X}'\mathbf{X}$ does not have an inverse. In the case in which $X_2 = 4X_1$, all pairs of the independent variables fall on a line, so it is impossible to know the proper tilt of the fitted regression plane in $\mathcal{R}^3$. There are many planes that minimize the sum of squared errors.

Multicollinearity is the condition associated with independent variables that are highly correlated among themselves in a multiple regression model. More specifically, multicollinearity occurs when two or more of the independent variables have a high correlation. This can appear as an approximately linear relationship between two of the independent variables. Multicollinearity is a condition associated with the design matrix $\mathbf{X}$ rather than the values of the dependent variable $\mathbf{Y}$ or the model $\mathbf{Y} = \mathbf{X}\boldsymbol{\beta} + \boldsymbol{\varepsilon}$. In cases in which multicollinearity exists, the matrix $\mathbf{X}'\mathbf{X}$ has an inverse, but it is ill-conditioned and subject to slight variations in the data or is unstable because of large differences in the magnitudes of the various values of the independent variables. One of the key practical issues when multicollinearity is present is that an estimated regression coefficient for a particular independent variable depends on whether the other independent variables are included or left out of the model.

So multicollinearity has been loosely defined as high correlation among the independent variables. There is redundancy to the information contained in the independent variables. The next paragraphs describe how to detect multicollinearity, its consequences, and some remedies.

Although the hypothetical perfect correlation between the interior space and the lot size of a home from Ames, Iowa described previously occurs seldom in practice, highly correlated independent variables can result in some unusual behavior of regression coefficients as a regression model is constructed. Some signs that multicollinearity might be present in a multiple linear regression model include the following.

- Large values of the estimated standard deviations of the regression coefficients.

- Including or not including an independent variable in the model results in large changes to the estimated regression coefficients.

- An estimated regression coefficient that is statistically significant when the associated independent variable is considered alone, but becomes insignificant when one or more other independent variables are added to the model.

- An estimated regression coefficient with a sign that is inconsistent with expected sign or inconsistent with previous similar data sets.

- The pairwise sample correlation among the independent variables is high. The cor function in R can be used to assess the correlation among independent variables. The R statement

```
cor(swiss)
```

for example, calculates the correlation matrix for the columns of the built-in data frame named swiss. The off-diagonal elements of this matrix range from -0.69 to 0.70, indicating that multicollinearity is present.

All of the criteria listed above are informal. A more formal way to determine whether multicollinearity is present is to introduce a statistic which reflects multicollinearity. The estimate of the variance of $\hat{\beta}_j$ can be written as

$$\hat{V}\left[\hat{\beta}_j\right] = \frac{1}{1-R_j^2}\left[\frac{MSE}{\sum_{i=1}^n (X_{ij}-\bar{X}_j)^2}\right],$$

where $\bar{X}_j = \sum_{i=1}^n X_{ij}$, $MSE = SSE/(n-p-1)$ for the full multiple regression model, and R_j^2 is the coefficient of determination obtained by conducting a multiple linear regression with X_j as the dependent variable and the other $p-1$ X-values as the independent variables, for $j = 1, 2, \ldots, p$. The coefficient on the right-hand side of this equation,

$$VIF_j = \frac{1}{1-R_j^2},$$

is known as a *variance inflation factor* for independent variable j, for $j = 1, 2, \ldots, p$. In the extreme case when $R_j^2 = 0$, the associated variance inflation factor is $VIF_j = 1$. This corresponds to the case in which X_j is not linearly related to the other independent variables. As R_j^2 increases, VIF_j also increases, corresponding to increased correlation between the independent variables. When the largest of the VIF_j values exceeds the threshold value of 10, one can conclude that the multicollinearity is present among the independent variables.

The R code below calculates the variance inflation factors for the data values in the `swiss` data frame, where the independent variables

- X_1, the percentage of males involved in agriculture as an occupation,

- X_2, the percentage of draftees receiving the highest make on an army examination,

- X_3, the percentage of draftees with education beyond the primary school,

- X_4, the percentage of Catholics, and

- X_5, the percentage of live births who live less than one year,

are used to predict Y, a common standardized fertility measure, from the $n = 47$ French-speaking provinces of Switzerland in about the year 1888. The R code below computes the variance inflation factors for the $p = 5$ independent variables.

```
swiss = as.matrix(swiss)
p    = 5
y    = swiss[ , 1]
n    = length(y)
x    = cbind(1, swiss[ , 2:(p + 1)])
for (i in 2:(p + 1)) {
  yy      = x[ , i]
  xx      = x[ , -i]
  beta    = solve(t(xx) %*% xx) %*% t(xx) %*% yy
  fitted  = xx %*% beta
  resid   = yy - fitted
  sse     = sum(resid ^ 2)
  m       = mean(yy)
  sst     = sum((yy - m) ^ 2)
  r2      = 1 - sse / sst
  vif     = 1 / (1 - r2)
  print(vif)
}
```

The variance inflation factors for the $p = 5$ independent variables are

$$VIF_1 = 2.28, VIF_2 = 3.68, VIF_3 = 2.77, VIF_4 = 1.94, VIF_5 = 1.11.$$

Since none of these five values exceeds 10, we can conclude that the multicollinearity that exists in the independent variables is not strong enough to cause concern. (Some regression analysts use 5 as a threshold rather than 10.) Some keystrokes can be saved by using the `vif` function from the `car` package on a multiple linear regression model fitted by the lm function.

```
library(car)
fit = lm(Fertility ~ Agriculture + Examination + Education +
         Catholic + Infant.Mortality, data = swiss)
summary(fit)
vif(fit)
```

One popular remedy for multicollinearity is known as *ridge regression*, which is a parameter estimation technique that abandons the requirement of unbiased parameter estimates. The approach taken with ridge regression is to choose estimates for the regression parameters that are biased, but have a smaller variance than the ordinary least squares estimates. The goal is to generate parameter estimates with tolerable bias but smaller variance. The typical approach used in statistics to overcome this bias/variability trade-off is to use the estimates that minimize the mean square errors. Assuming that the X and Y values have been centered, we can dispense with the need for an intercept term in the multiple regression model. Rather than minimizing the usual sum of squared errors

$$S = \sum_{i=1}^{n} (Y_i - \beta_1 X_{i1} - \beta_2 X_{i2} - \cdots - \beta_p X_{ip})^2,$$

ridge regression minimizes

$$S_R = \sum_{i=1}^{n} (Y_i - \beta_1 X_{i1} - \beta_2 X_{i2} - \cdots - \beta_p X_{ip})^2 + \lambda \sum_{j=1}^{p} \beta_j^2.$$

There are now two terms in the modified sum of squares. The second term in S_R is known as the *penalty term*. The new parameter λ is known as the *penalty parameter*. When $\lambda = 0$, S_R reduces to the ordinary least squares case and achieves a value SSE at the ordinary least squares estimators. As λ increases, the estimators converge to $\hat{\beta}_1 = \hat{\beta}_2 = \cdots = \beta_p = 0$. We desire a λ value that introduces some bias into the parameter estimates, but also have a reduced variance.

The geometry associated with ridge regression for $p = 2$ independent variables X_1 and X_2 in a multiple linear regression model is illustrated in Figure 3.19. The ellipses are level surfaces of the first term in S_R. The center of the ellipses is the ordinary least squares estimators of $(\beta_1, \beta_2) = (\hat{\beta}_1, \hat{\beta}_2)$, which are the values that minimize the first term of S_R. The circles centered at the origin are level surfaces of the second term in S_R. The ridge regression estimators for β_1 and β_2 will occur at the intersection of one of elliptical and circular contours. In Figure 3.19 the two outermost level surfaces intersect at a point, which is a value of the ridge regression estimates of β_1 and β_2 which correspond to one particular value of the penalty parameter λ. The point at which this intersection

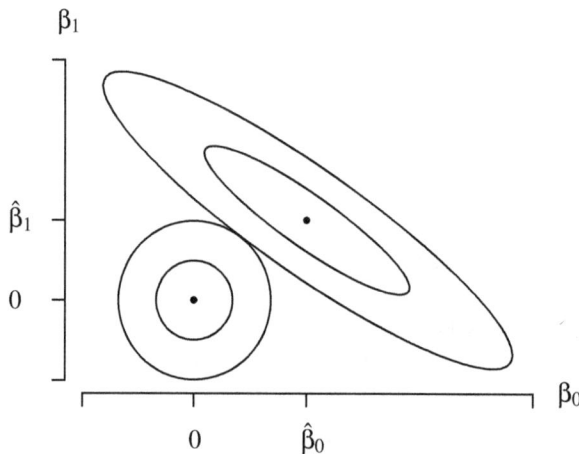

Figure 3.19: Ridge regression geometry for $p = 2$ independent variables.

occurs is a function of the penalty parameter λ. In higher dimensions, the circles become spheres and the ellipses become ellipsoids.

Determining the value of the penalty parameter is critical in ridge regression, but its choice depends on the regression model and associated data set. A common technique for determining an optimal value for λ is known as k-fold cross-validation. There are several functions in R which can perform ridge regression: the `lm.ridge` function from the `MASS` package, the `linearRidge` function from the `ridge` package, and the `glmnet` function from the `glmnet` package. Ridge regression is related to the lasso (least absolute shrinkage and selection operator) estimator and elastic net regularization, two other popular parameter estimation techniques that are often applied for large values of p.

Is there a way to completely avoid multicollinearity? In some settings, the answer is yes. When the values of the independent variables are chosen so that they are uncorrelated, the regression coefficients associated with a simple linear regression model of each independent variable separately match the regression coefficients of any model involving more independent variables. This fact provides a strong argument for a designed experiment which can result in uncorrelated independent variables whenever the setting of the regression problem make this possible.

3.5.6 Model Selection

It is common in regression modeling to have a large number of potential independent variables that might adequately predict the dependent variable Y that need to be sifted through in order to decide whether each should be included or excluded from the regression model. If there are p potential independent variables in the multiple linear regression model

$$Y = \beta_0 + \beta_1 X_1 + \beta_2 X_2 + \cdots + \beta_p X_p + \varepsilon$$

then there are 2^p possible regression models (always including an intercept term and not considering interaction terms or nonlinear terms) because each independent variable will either be included or not included in the regression model. Since the number of regression models to fit can be daunting, even for moderate values of p, we desire an algorithm for selecting the appropriate independent variables to include in the model. *Forward stepwise regression* is one such automatic search procedure used to select the independent variables to include in a multiple linear regression model. The procedure begins with the null model $Y = \beta_0 + \varepsilon$ and progressively adds independent variables to the model that are deemed to be statistically significant. In the initial step, p simple linear regression models are fit for each potential independent variable. The independent variable with the smallest p-value falling below a prescribed threshold (commonly, $\alpha = 0.05$) associated with the t-test described in Section 2.3.2 is added to the model. In the second step, $p - 1$ multiple linear regression models with two independent variables are fitted using the previously selected independent variable and each of the other potential independent variables. The independent variable with the smallest p-value is added to the model. This process continues until no more independent variables meet the criteria. This is the multiple linear regression model selected by forward stepwise regression. Several other variants of forward stepwise regression and other model selection algorithms are outlined below.

- Foreward stepwise regression often includes a test to determine whether independent variables that have previously been added to the model have p-values that exceed the threshold and should consequently be removed from the model.

- Backward stepwise regression starts by including all p independent variables in the regression model and eliminates the independent variable with the largest p-value on each step.

Unfortunately, there is no guarantee that forward stepwise regression and backward stepwise regression will result in the same final regression model.

- Once this statistically significant independent variables have been identified, a similar stepwise procedure can be executed to test for statistically significant interaction terms.

- A similar stepwise procedure can be executed to test for the significance of nonlinear terms in the regression model.

- With increased computer speeds and a moderate value of p, the number of independent variables, it is possible to fit all 2^p possible regression models and compare them to determine an appropriate final regression model.

- Comparing potential regression models using p-values is not universal. The Akaike Information Criterion (AIC) is a measure which extracts a penalty for each additional parameter in a model in an effort to avoid overfitting.

In summary, selecting a multiple linear regression model is not easy. The skills required to select a model include the ability to (a) detect and remedy multicollinearity, (b) assess evidence of interaction effects between independent variables and include them in the model when appropriate, (c) assess evidence of nonlinear relationships between some or all of the independent variables and the dependent variable and include appropriate terms in the model, (d) execute the appropriate multidimensional diagnostic procedures (outlined in the simple linear regression case in Section 3.2) and execute the appropriate remedial procedures (outlined in the simple linear regression case in Section 3.3) when model assumptions are violated, and (e) assess the normality of the residuals.

3.6   Weighted Least Squares

The three approaches to estimating the parameters in a simple linear regression model that we have encountered thus far,

- the algebraic approach,

- the matrix approach,

- using the R lm (linear model) function,

all have the same assumptions regarding the independent variable, the dependent variable, and the model $Y = \beta_0 + \beta_1 X + \varepsilon$. In all three approaches, the error terms are assumed to be mutually independent random variables, each with population mean 0 and population variance–covariance matrix $V[\varepsilon] = \sigma_Z^2 I$, where I is the $n \times n$ identity matrix. This means that $V[\varepsilon_i] = \sigma_Z^2$, for $i = 1, 2, \ldots, n$. There is also an implicit assumption that each of the data pairs (X_i, Y_i) are each given equal weight in the regression.

Settings occasionally arise in which some data values should be given different weights. There might be evidence that some of the Y_i values have more precision than others. Weights can be placed on each of the data pairs to account for this difference in precision. This leads to a *weighted least squares* approach to estimating the coefficients in a regression model.

In the standard simple linear regression model, the assumption

$$V[\varepsilon_i] = \sigma_Z^2,$$

for $i = 1, 2, \ldots, n$, means that the variance of the dependent variable from the regression line is equal for all of the n data pairs, regardless of the value of the independent variable. In *weighted least squares* modeling, the positive weights $w_1, w_2, \ldots, w_n$ are determined so that

$$V[\varepsilon_i] = \sigma_Z^2 / w_i$$

for $i = 1, 2, \ldots, n$, which means that certain data pairs have more precision than other data pairs. The weights are fixed constants. There is no requirement that the weights sum to one. Data pairs with larger weights are assumed to have a lower variability to their error terms. This allows for a population variance that changes from one data pair to another.

As an illustration, the values of the dependent variable Y might be sample means at the various values of the independent variable X. Furthermore, if the sample sizes associated with the sample means are known and unequal, then we would like to assign higher weights to the data pairs associated with larger sample sizes. If n_i is the sample size for data pair i, for $i = 1, 2, \ldots, n$, then the appropriate weight for data pair i is $w_i = n_i$ so that

$$V[\varepsilon_i] = \sigma_Z^2 / n_i$$

for $i = 1, 2, \ldots, n$.

So rather than minimizing the sum of squares

$$S = \sum_{i=1}^{n} (Y_i - \beta_0 - \beta_1 X_i)^2$$

as was the case in the standard simple linear regression model, weighted least squares minimizes the weighted sum of squares

$$S = \sum_{i=1}^{n} w_i (Y_i - \beta_0 - \beta_1 X_i)^2.$$

Notice that this reduces to the ordinary sum of squares when $w_1 = w_2 = \cdots = w_n = 1$. As before, calculus can be used to minimize S with respect to β_0 and β_1 to arrive at the least squares estimators $\hat{\beta}_0$ and $\hat{\beta}_1$. The partial derivatives of S with respect to β_0 and β_1 are

$$\frac{\partial S}{\partial \beta_0} = -2 \sum_{i=1}^{n} w_i (Y_i - \beta_0 - \beta_1 X_i) = 0$$

and

$$\frac{\partial S}{\partial \beta_1} = -2 \sum_{i=1}^{n} w_i X_i (Y_i - \beta_0 - \beta_1 X_i) = 0.$$

These can be simplified to give the normal equations

$$\beta_0 \sum_{i=1}^{n} w_i + \beta_1 \sum_{i=1}^{n} w_i X_i = \sum_{i=1}^{n} w_i Y_i$$

and

$$\beta_0 \sum_{i=1}^{n} w_i X_i + \beta_1 w_i X_i^2 = \sum_{i=1}^{n} w_i X_i Y_i.$$

The normal equations are a system of two linear equations in the two unknowns β_0 and β_1, given the data pairs $(X_1, Y_1), (X_2, Y_2), \ldots, (X_n, Y_n)$ and the weights $w_1, w_2, \ldots, w_n$. The normal equations can be solved to yield the weighted least squares estimators. This derivation constitutes a proof of the following theorem.

Theorem 3.3 Let $(X_1, Y_1), (X_2, Y_2), \ldots, (X_n, Y_n)$ be n data pairs with at least two distinct X_i values. Let $w_1, w_2, \ldots, w_n$ be the weights associated with the data pairs. The *weighted least squares estimators* of β_0 and β_1 in the simple linear regression model are the solution to the simultaneous *normal equations*

$$\beta_0 \sum_{i=1}^{n} w_i + \beta_1 \sum_{i=1}^{n} w_i X_i = \sum_{i=1}^{n} w_i Y_i$$

$$\beta_0 \sum_{i=1}^{n} w_i X_i + \beta_1 w_i X_i^2 = \sum_{i=1}^{n} w_i X_i Y_i.$$

and are given by

$$\hat{\beta}_1 = \frac{\sum_{i=1}^{n} w_i (X_i - \bar{X}_w)(Y_i - \bar{Y}_w)}{\sum_{i=1}^{n} w_i (X_i - \bar{X}_w)^2}$$

and

$$\hat{\beta}_0 = \bar{Y}_w - \hat{\beta}_1 \bar{X}_w,$$

where $\bar{X}_w$ and $\bar{Y}_w$ are the weighted sample means

$$\bar{X}_w = \frac{\sum_{i=1}^{n} w_i X_i}{\sum_{i=1}^{n} w_i} \qquad \text{and} \qquad \bar{Y}_w = \frac{\sum_{i=1}^{n} w_i Y_i}{\sum_{i=1}^{n} w_i}.$$

The matrix approach can also be applied to weighted least squares. Define the $\mathbf{X}$, $\mathbf{Y}$, $\boldsymbol{\beta}$ and $\boldsymbol{\varepsilon}$ matrices as in Section 3.4:

$$\mathbf{X} = \begin{bmatrix} 1 & X_1 \\ 1 & X_2 \\ \vdots & \vdots \\ 1 & X_n \end{bmatrix}, \qquad \mathbf{Y} = \begin{bmatrix} Y_1 \\ Y_2 \\ \vdots \\ Y_n \end{bmatrix}, \qquad \boldsymbol{\beta} = \begin{bmatrix} \beta_0 \\ \beta_1 \end{bmatrix}, \qquad \text{and} \qquad \boldsymbol{\varepsilon} = \begin{bmatrix} \varepsilon_1 \\ \varepsilon_2 \\ \vdots \\ \varepsilon_n \end{bmatrix}.$$

In addition, assume that the matrix $\mathbf{W}$ is a diagonal matrix with the weights $w_1, w_2, \ldots, w_n$ on the diagonal:

$$\mathbf{W} = \begin{bmatrix} w_1 & 0 & \cdots & 0 \\ 0 & w_2 & \cdots & 0 \\ \vdots & \vdots & \ddots & \vdots \\ 0 & 0 & \cdots & w_n \end{bmatrix}.$$

In this case, the normal equations can be written in matrix form as

$$\mathbf{X}'\mathbf{W}\mathbf{X}\boldsymbol{\beta} = \mathbf{X}'\mathbf{W}\mathbf{Y}.$$

Pre-multiplying both sides of this equation by $(\mathbf{X}'\mathbf{W}\mathbf{X})^{-1}$ gives the least squares estimators for the regression parameters in matrix form as

$$\hat{\boldsymbol{\beta}} = (\mathbf{X}'\mathbf{W}\mathbf{X})^{-1}\mathbf{X}'\mathbf{W}\mathbf{Y}.$$

As before, the fitted values can also be written in matrix form as

$$\hat{\mathbf{Y}} = \mathbf{X}\hat{\boldsymbol{\beta}}$$

or

$$\hat{\mathbf{Y}} = \mathbf{X}\left(\mathbf{X}'\mathbf{W}\mathbf{X}\right)^{-1}\mathbf{X}'\mathbf{W}\mathbf{Y}.$$

The residuals $e_i = Y_i - \hat{Y}_i$ for $i = 1, 2, \ldots, n$, can also be written in matrix form as

$$\begin{aligned}
\mathbf{e} &= \mathbf{Y} - \hat{\mathbf{Y}} \\
&= \mathbf{Y} - \mathbf{X}\hat{\boldsymbol{\beta}} \\
&= \mathbf{Y} - \mathbf{X}\left(\mathbf{X}'\mathbf{W}\mathbf{X}\right)^{-1}\mathbf{X}'\mathbf{W}\mathbf{Y} \\
&= \left(\mathbf{I} - \mathbf{X}\left(\mathbf{X}'\mathbf{W}\mathbf{X}\right)^{-1}\mathbf{X}'\mathbf{W}\right)\mathbf{Y},
\end{aligned}$$

where $\mathbf{e}$ is the column vector of residuals $\mathbf{e} = (e_1, e_2, \ldots, e_n)'$. These matrix results are summarized in the following theorem.

Theorem 3.4 Let $\mathbf{X}$, $\mathbf{Y}$, $\boldsymbol{\beta}$, and $\boldsymbol{\varepsilon}$ be the matrices associated with a simple linear regression model with weights $w_1, w_2, \ldots, w_n$ associated with the data pairs (X_1, Y_1), (X_2, Y_2), $\ldots$, (X_n, Y_n). Let $\mathbf{W}$ be an $n \times n$ diagonal matrix with the weights on the diagonal elements. The least squares estimators of β_0 and β_1 are

$$\hat{\boldsymbol{\beta}} = \left(\mathbf{X}'\mathbf{W}\mathbf{X}\right)^{-1}\mathbf{X}'\mathbf{W}\mathbf{Y}.$$

The fitted values are

$$\hat{\mathbf{Y}} = \mathbf{X}\hat{\boldsymbol{\beta}} = \mathbf{X}\left(\mathbf{X}'\mathbf{W}\mathbf{X}\right)^{-1}\mathbf{X}'\mathbf{W}\mathbf{Y}.$$

The residuals are

$$\mathbf{e} = \left(\mathbf{I} - \mathbf{X}\left(\mathbf{X}'\mathbf{W}\mathbf{X}\right)^{-1}\mathbf{X}'\mathbf{W}\right)\mathbf{Y}.$$

The algebraic approach, matrix approach, and R approach to weighted least squares problem will be illustrated in the next example. Establishing the weights $w_1, w_2, \ldots, w_n$ can be a nontrivial problem, and differs depending on the setting in which the weighted regression model is employed.

Example 3.11 In reliability, *current status data* is generated by testing a randomly selected group of items with varying ages from a population at a particular fixed time in order to determine whether or not each item has failed or is operating at its particular age. Items were selected at ages 100, 200, 300, and 400 hours to see if they are operating. In this case, the independent variable X is the age, measured in hours, at which an item is tested. Each item tested is deemed to be either operating or failed. Table 3.7 contains the results of the test. Notice that 100 items were tested at ages $X_1 = 100$ and $X_2 = 200$, but only 10 items were tested at ages $X_3 = 300$ and $X_4 = 400$. The dependent variable in this setting is the fraction of items that survive to a particular age. The sample size at each testing age is denoted by n_i, $i = 1, 2, 3, 4$. So a total of $n_1 + n_2 + n_3 + n_4 = 220$ items were tested. The number of items that are operating at each testing age is denoted by S_i, $i = 1, 2, 3, 4$. The fraction of items that are operating at each testing age, which is the dependent variable in the regression, is denoted by Y_i, $i = 1, 2, 3, 4$. Notice that the fraction surviving is not necessarily decreasing from one time to the next because of random sampling variability. The small sample sizes at

Time (hours)	$X_1 = 100$	$X_2 = 200$	$X_3 = 300$	$X_4 = 400$
Sample size	$n_1 = 100$	$n_2 = 100$	$n_3 = 10$	$n_4 = 10$
Number surviving	$S_1 = 50$	$S_2 = 25$	$S_3 = 4$	$S_4 = 3$
Fraction surviving	$Y_1 = 0.5$	$Y_2 = 0.25$	$Y_3 = 0.4$	$Y_4 = 0.3$

Table 3.7: Current status data test results.

times $X_3 = 300$ and $X_4 = 400$ magnify this problem with the data set. The goal here is to establish a regression function that will adequately smooth the data values in order to estimate the survivor function for the items at any time.

Assume for now that the standard (non-weighted) least squares approach using the $n = 4$ data pairs

$$(100, 0.5), \qquad (200, 0.25), \qquad (300, 0.4), \qquad \text{and} \qquad (400, 0.3)$$

is taken to this problem. The R code below fits the simple linear regression model to the data.

```
x   = c(100, 200, 300, 400)
n   = c(100, 100,  10,  10)
s   = c( 50,  25,   4,   3)
y   = s / n
fit = lm(y ~ x)
fit$coefficients
```

The regression line in this case has intercept $\hat{\beta}_0 = 0.475$ and slope $\hat{\beta}_1 = -0.00045$. The survival probability of a brand-new item is estimated to be 0.475, and the survival probability decreases by 0.00045 for every hour that passes. The unimpressive survival probability of 0.475 for a new item is outside of the scope of the simple linear regression model, so its interpretation is not meaningful.

But using the standard simple linear regression approach is not appropriate here. The first two data pairs, both of which involved testing 100 items, should be weighted more heavily that the last two data pairs, which only involved testing 10 items. Determining the appropriate weights, however, is nontrivial.

Assume that the test results for each item are mutually independent Bernoulli trials. The number of items that survive a test at one particular time (that is, S_i using the notation from Table 3.7) is a binomial random variable with parameters n_i and p_i, where p_i is the population probability that item i is operating at time X_i. The population variance of the dependent variable $Y_i = S_i / n_i$ is

$$V[\hat{p}_i] = V[Y_i] = V\left[\frac{S_i}{n_i}\right] = \frac{1}{n_i^2} V[S_i] = \frac{n_i p_i(1 - p_i)}{n_i^2} = \frac{p_i(1 - p_i)}{n_i},$$

for $i = 1, 2, 3, 4$. Using the point estimate for p_i on the right-hand side of this expression results in the following estimated variances for the four dependent variables:

$$\hat{V}[Y_1] = \frac{\frac{50}{100}\left(1 - \frac{50}{100}\right)}{100} = \frac{1}{400}, \qquad\qquad \hat{V}[Y_2] = \frac{\frac{25}{100}\left(1 - \frac{25}{100}\right)}{100} = \frac{3}{1600},$$

$$\hat{V}[Y_3] = \frac{\frac{4}{10}\left(1-\frac{4}{10}\right)}{10} = \frac{24}{1000}, \qquad \hat{V}[Y_4] = \frac{\frac{3}{10}\left(1-\frac{3}{10}\right)}{10} = \frac{21}{1000}.$$

Not surprisingly, the first two variance estimates are about an order of magnitude smaller than the second two variance estimates because of the differences in the sample sizes. This approach will have problems if one of the testing times has all successes ($S_i = n_i$) or all failures ($S_i = 0$).

Since the weights w_i appear in the denominator of the expression $V[\varepsilon_i] = \sigma_Z^2/w_i$, the reciprocals of these variance estimates will be used as the weights in the weighted least squares regression:

$$w_1 = \frac{400}{1}, \qquad w_2 = \frac{1600}{3}, \qquad w_3 = \frac{1000}{24}, \qquad w_4 = \frac{1000}{21}.$$

The regression coefficients will be calculated in three ways, all of which yield identical results: the algebraic approach, the matrix approach, and using the lm function.

First, the algebraic approach for calculating the slope and intercept of the regression line using weighted least squares uses the following R statements. These are an implementation of Theorem 3.3.

```
x     = c(100, 200, 300, 400)
n     = c(100, 100,  10,  10)
s     = c( 50,  25,   4,   3)
y     = s / n
w     = n / (y * (1 - y))
meanx = sum(w * x) / sum(w)
meany = sum(w * y) / sum(w)
slope = sum(w * (x - meanx) * (y - meany)) / (sum(w * (x - meanx) ^ 2))
inter = meany - slope * meanx
print(c(inter, slope))
```

The weighted mean of the X values is

$$\bar{X}_w = \frac{\sum_{i=1}^n w_i X_i}{\sum_{i=1}^n w_i} = 174.2725.$$

Notice that this is slightly lower than the unweighted mean of the x values, which is $(100+200+300+400)/4 = 250$ hours. This is due to the larger sample sizes at testing times 100 and 200, resulting in larger weights for these values. The weighted mean of the Y values is

$$\bar{Y}_w = \frac{\sum_{i=1}^n w_i Y_i}{\sum_{i=1}^n w_i} = 0.3562.$$

The estimates for the slope and intercept of the regression line for weighted least squares is

$$\hat{\beta}_1 = \frac{\sum_{i=1}^n w_i(X_i - \bar{X}_w)(Y_i - \bar{Y}_w)}{\sum_{i=1}^n w_i(X_i - \bar{X}_w)^2} = -0.001081$$

and

$$\hat{\beta}_0 = \bar{Y}_w - \hat{\beta}_1\bar{X}_w = 0.5447.$$

The interpretation of these estimates is that the estimated probability of survival at time 0 is 0.5447 and the probability of survival decreases by 0.001081 with every hour that passes.

Second, using the matrix approach, the **X**, **Y**, and **W** matrices associated with this data set are

$$
\mathbf{X} = \begin{bmatrix} 1 & 100 \\ 1 & 200 \\ 1 & 300 \\ 1 & 400 \end{bmatrix}, \quad \mathbf{Y} = \begin{bmatrix} 0.50 \\ 0.25 \\ 0.40 \\ 0.30 \end{bmatrix}, \quad \text{and} \quad \mathbf{W} = \begin{bmatrix} \frac{400}{1} & 0 & 0 & 0 \\ 0 & \frac{1600}{3} & 0 & 0 \\ 0 & 0 & \frac{1000}{24} & 0 \\ 0 & 0 & 0 & \frac{1000}{21} \end{bmatrix}.
$$

The R code below uses the matrix approach to simple linear regression with weights to calculate the estimated slope $\hat{\beta}_0$ and intercept $\hat{\beta}_1$, the fitted values $\hat{\mathbf{Y}}$, and the residuals **e** for the current status data set using Theorem 3.4. The R solve function is used to compute the inverse of $\mathbf{X'X}$.

```
options(digits = 4)
x      = c(100, 200, 300, 400)
n      = c(100, 100,  10,  10)
s      = c( 50,  25,   4,   3)
y      = s / n
w      = n / (y * (1 - y))
w      = diag(w)
x      = cbind(1, x)
beta   = solve(t(x) %*% w %*% x) %*% t(x) %*% w %*% y
fitted = x %*% beta
e      = y - fitted
```

The results of these calculations are given below. The point estimators of the slope and intercept are

$$
\hat{\boldsymbol{\beta}} = (\mathbf{X'WX})^{-1}\mathbf{X'WY} = \begin{bmatrix} 0.5447 \\ -0.001081 \end{bmatrix}.
$$

The fitted values are

$$
\hat{\mathbf{Y}} = \mathbf{X}\hat{\boldsymbol{\beta}} = \begin{bmatrix} 0.4365 \\ 0.3284 \\ 0.2203 \\ 0.1121 \end{bmatrix}.
$$

The residuals are

$$
\mathbf{e} = \mathbf{Y} - \hat{\mathbf{Y}} = \begin{bmatrix} 0.0635 \\ -0.0784 \\ 0.1797 \\ 0.1879 \end{bmatrix}.
$$

Third, the built-in function lm can be used for weighted least squares by using the weights argument. The R code below calculates the estimates of the regression coefficients, the fitted values, and the residuals.

```
x      = c(100, 200, 300, 400)
n      = c(100, 100,  10,  10)
```

```
s    = c( 50,   25,   4,    3)
y    = s / n
w    = n / (y * (1 - y))
fitw = lm(y ~ x, weights = w)
print(fitw$coefficients)
print(fitw$fitted.values)
print(fitw$residuals)
print(weighted.residuals(fitw))
```

The three approaches all yield the same results. The regression line associated with ordinary least squares and weighted least squares can be compared graphically. The R code below plots the four data pairs and the associated ordinary least squares and weighted least squares regression lines.

```
x    = c(100, 200, 300, 400)
n    = c(100, 100,  10,  10)
s    = c( 50,  25,   4,   3)
y    = s / n
fit  = lm(y ~ x)
w    = n / (y * (1 - y))
fitw = lm(y ~ x, weights = w)
plot(x, y)
abline(fit$coefficients)
abline(fitw$coefficients)
```

Figure 3.20 contains the resulting plot, which shows the ordinary least squares line with equal weighting to the four data values and the weighted least squares line with much more weight to the first two data pairs and much less weight to the last two data pairs. Extra circles have been added to the two data pairs associated with the larger sample sizes with larger weights in Figure 3.20. The effect of the larger weights on the first

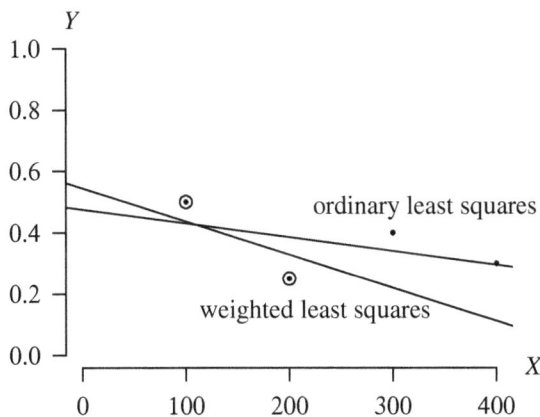

Figure 3.20: Current status data ordinary and weighted least squares fits.

two data pairs is apparent in the weighted least squares regression line. The rightmost two data pairs exert significantly less tug on the weighted least squares regression line because of their smaller weights.

Using simple linear regression in the previous example, either weighted or unweighted, might not be the best approach. The dependent variable Y is the probability that an item of age X is functioning. This dependent variable must lie between 0 and 1, but the regression line could potentially fall outside of that range within the scope of the model. Two potential remedies are given in the next two sections: using a regression model with nonlinear terms such as X^2 or X^3, or a survivor function of a lifetime model rather than a line, or a nonlinear model known as a *logistic regression model*, whose dependent variable necessarily lies between 0 and 1.

3.7   Regression Models with Nonlinear Terms

Regression models with nonlinear terms arise frequently in regression modeling. One simple example is polynomial regression. A quadratic regression model, for example, is

$$Y = \beta_0 + \beta_1 X + \beta_2 X^2 + \varepsilon,$$

where β_0, β_1, and β_2 are the regression coefficients, and ε is a white noise term. This model is still linear in β_0, β_1, and β_2. One way to think about this model is to consider X and X^2 to be the $p = 2$ independent variables in a multiple regression model. The next example fits a quadratic model to the data pairs in which the independent variable X is the speed of an automobile and the dependent variable Y is its stopping distance.

Example 3.12 Consider the $n = 50$ data pairs from Example 2.8 which give the speed (in miles per hour) as X and the stopping distance (in feet) as Y. These data pairs are built into the base R language in the data frame named `cars`, where the `speed` column contains the values of X and the `dist` column contains the values of Y. Fit a quadratic regression model forced through the origin to the data pairs.

Since the quadratic regression model is being forced through the origin in order to account for the fact that stationary cars ($X = 0$) do not require any distance ($Y = 0$) to stop, the quadratic regression model is

$$Y = \beta_1 X + \beta_2 X^2 + \varepsilon,$$

where $\varepsilon \sim WN\left(0, \sigma_Z^2\right)$. R is capable of fitting nonlinear models to data. The I (inhibit interpretation) function allows the modeling of some function of a particular independent variable. For the data pairs in the `cars` data frame, a quadratic regression model that is forced through the origin can be fit with `lm` function.

```
fit = lm(dist ~ speed + I(speed ^ 2) - 1, data = cars)
```

The -1 part of the formula forces the regression function to pass through the origin. The output generated by the `summary(fit)` statement is given below.

```
Call:
lm(formula = dist ~ speed + I(speed^2) - 1, data = cars)
```

```
Residuals:
    Min      1Q  Median      3Q     Max
-28.836  -9.071  -3.152   4.570  44.986
```

```
Coefficients:
            Estimate Std. Error t value Pr(>|t|)
speed        1.23903    0.55997   2.213  0.03171 *
I(speed^2)   0.09014    0.02939   3.067  0.00355 **
---
Signif. codes:  0 '***' 0.001 '**' 0.01 '*' 0.05 '.' 0.1 ' ' 1
```

```
Residual standard error: 15.02 on 48 degrees of freedom
Multiple R-squared:  0.9133,    Adjusted R-squared:  0.9097
F-statistic: 252.8 on 2 and 48 DF,  p-value: < 2.2e-16
```

The fitted quadratic regression model that is forced to pass through the origin is

$$Y = 1.24X + 0.0901X^2,$$

where X is speed and Y is stopping distance. Notice that $\hat{\beta}_2 = 0.0901 > 0$, which means that a graph of the fitted regression function—a parabola that passes through the origin—is concave up. Since the p-value associated with the linear term is $p = 0.032$ and the p-value associated with the quadratic term is $p = 0.0036$, both of the regression coefficients are statistically significant. The R commands

```
plot(cars, xlim = c(0, 25), pch = 16, las = 1)
fit = lm(dist ~ speed + I(speed ^ 2) - 1, data = cars)
x   = 0:25
y   = fit$coefficients[1] * x + fit$coefficients[2] * x ^ 2
lines(x, y)
```

plot the fitted model over the scatterplot. This graph appears in Figure 3.21.

How do we compare the simple linear regression model to the quadratic regression model forced through the origin? Both have two parameters, but which one of the models is a better approximation to the data pairs? One way to compare the two models is with the sum of squared residuals for each of the models, which are computed with the R commands

```
sum(lm(dist ~ speed, data = cars)$residuals ^ 2)
sum(lm(dist ~ speed + I(speed ^ 2) - 1, data = cars)$residuals ^ 2)
```

The simple linear regression model has a sum of squared residuals of $S = 11{,}353.52$, and the quadratic regression model forced through the origin has a sum of squared residuals of $S = 10{,}831.12$. Using the quadratic regression model forced through the origin reduces the sum of squared residuals by 522.4. Higher-order polynomials can be fit using the lm function in a similar manner. As was the case in multiple regression, adding more terms generally results in a reduction in the sum of squared residuals.

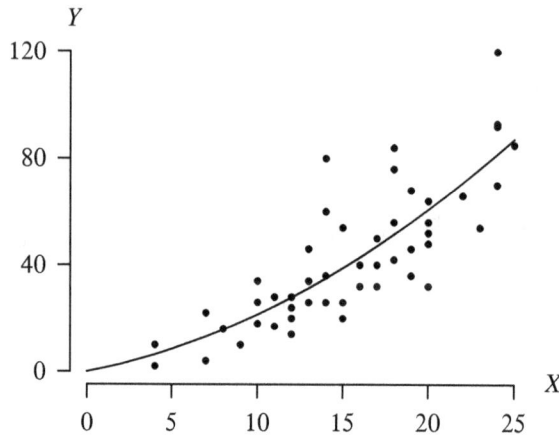

Figure 3.21: Scatterplot and quadratic fit of speed X and stopping distance Y.

Nonlinear regression modeling is not limited to just polynomial regression models. The next two examples fit the same data set concerning the national debt in the United States between 1970 and 2020 to a nonlinear regression model using two fundamentally different approaches. The first approach is to transform the nonlinear regression model to a linear regression model and then apply the standard techniques for parameter estimation to the transformed model. The second approach is to use numerical methods to minimize the sum of squares in the usual least squares fashion described previously.

Example 3.13 The national debt of the United States, in trillions of dollars, between 1970 and 2020 is given in Table 3.8. These values are not adjusted for inflation. Fit an exponential regression model to the national debt of the United States, where X is the year and Y is the debt, by transforming an exponential regression model to a linear model.

Year	Debt
1970	0.37
1975	0.53
1980	0.91
1985	1.82
1990	3.23
1995	4.97
2000	5.67
2005	7.93
2010	13.56
2015	18.15
2020	27.75

Table 3.8: United States national debt, 1970–2020.

The scatterplot in Figure 3.22 shows that a simple linear regression model is not appro-

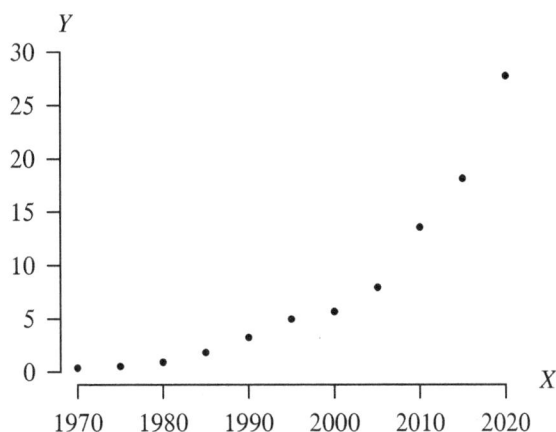

Figure 3.22: Scatterplot of the year X and the national debt Y.

priate for these data pairs. A regression model that reflects the exponential growth rate in the debt is warranted. Both savings and debt tend to grow exponentially, so an exponential regression model is a reasonable initial model to investigate. Consider fitting the regression model

$$Y = e^{\beta_0 + \beta_1 X + \varepsilon}$$

to the data set, where X is the year, Y is the debt, ε is an error term, and β_0 and β_1 are unknown regression parameters to be estimated from the data pairs. This model can be transformed to a linear model by taking the natural logarithm of both sides of the model:

$$\ln Y = \beta_0 + \beta_1 X + \varepsilon.$$

This model is now in the form of a simple linear regression with independent variable X and dependent variable $\ln Y$. The intercept of the fitted model is β_0 and the slope of the fitted model is β_1. So a graph that contains X on the horizontal axis and $\ln Y$ on the vertical axis should be approximately linear if this transformation approach is appropriate. Such a graph is given in Figure 3.23, which is much closer to linear than the raw data points. It is apparent that some work on debt reduction occurred in the late 1990s, resulting in a slight bit of nonlinearity. We will proceed with fitting the transformed model. The R code below follows a similar pattern to the earlier examples, but this time the `formula` used in the call to the `lm` function is `log(debt) ~ year`. The `curve` function is used to add the fitted regression function to the scatterplot.

```
year = seq(1970, 2020, by = 5)
debt = c(.37, .53, .91, 1.82, 3.23, 4.97, 5.67, 7.93, 13.56, 18.15, 27.75)
fit  = lm(log(debt) ~ year)              # fit an exponential model
b0   = coef(fit)[1]                      # estimated beta0 value
b1   = coef(fit)[2]                      # estimated beta1 value
plot(year, debt, las = 1, pch = 16)      # scatterplot of data pairs
curve(exp(b0 + b1 * x), add = TRUE)      # plot fitted model
```

The fitted model is displayed in Figure 3.24. The values of the estimated parameters are $\hat{\beta}_0 = -170.4$ and $\hat{\beta}_1 = 0.08606$.

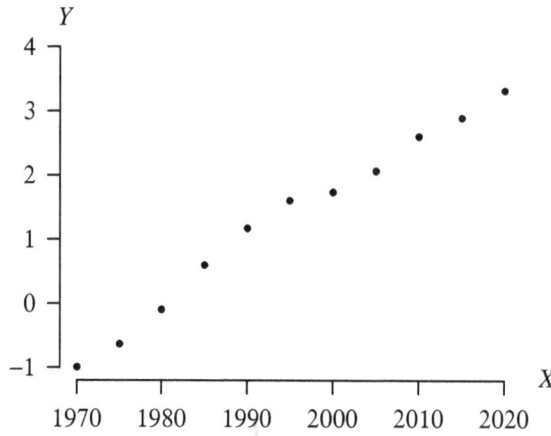

Figure 3.23: Scatterplot of the year X and the logarithm of the national debt Y.

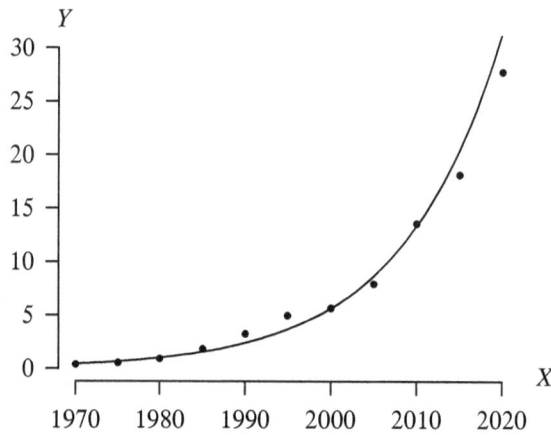

Figure 3.24: Scatterplot and exponential fit of year X and debt Y.

There is a second approach to fitting an exponential regression model to the national debt data pairs that follows the standard approach to least squares estimation, which is given next.

Example 3.14 Fit an additive exponential regression model to the United States national debt data pairs from Example 3.13.

The second approach to fitting an exponential regression model to the debt data pairs is to use the additive model

$$Y = e^{\beta_0 + \beta_1 X} + \varepsilon.$$

Using the traditional least squares approach, the sum of squares

$$S = \sum_{i=1}^{n} \left(Y_i - e^{\beta_0 + \beta_1 X_i} \right)^2$$

is minimized with respect to β_0 and β_1, yielding the associated least squares estimators $\hat{\beta}_0$ and $\hat{\beta}_1$. The estimators cannot be expressed in closed form, so numerical methods must be used to estimate β_0 and β_1. A nonlinear least squares R function named `nls` can be used to estimate the parameters. Here is a first attempt at fitting the model.

```
year = seq(1970, 2020, by = 5)
debt = c(.37, .53, .91, 1.82, 3.23, 4.97, 5.67, 7.93, 13.56, 18.15, 27.75)
fit  = nls(debt ~ exp(b0 + b1 * year))          # fit exponential model
```

This code returns an error message indicating that the `nls` function was unable to estimate the parameters. What went wrong? The way that the model has been formulated, the parameter e^{β_0} represents the United States national debt in the year 0. This is why we had the parameter estimate $e^{\hat{\beta}_0} = e^{-170.4} = 10^{-74}$ from the transformation approach in Example 3.13. The `nls` function attempts to do a search over all values of β_0 and β_1 to minimize the sum of squares. Finding the value of $\hat{\beta}_0$ is like finding a needle in a haystack. We need to give the `nls` function some help. We will give `nls` some starting values in a list named `start` to make the internal search performed by the `nls` function easier. The initial values for $\hat{\beta}_0$ and $\hat{\beta}_1$ will be the estimates for β_0 and β_1 from the transformation approach from the previous example.

```
year = seq(1970, 2020, by = 5)
debt = c(.37, .53, .91, 1.82, 3.23, 4.97, 5.67, 7.93, 13.56, 18.15, 27.75)
fit  = nls(debt ~ exp(b0 + b1 * year), start = list(b0 = -170, b1 = 0.09))
b0   = coef(fit)[1]                       # fitted beta0 value
b1   = coef(fit)[2]                       # fitted beta1 value
plot(year, debt, pch = 16)                # scatterplot of data pairs
curve(exp(b0 + b1 * x), add = TRUE)       # plot fitted model
```

The estimated parameters are $\hat{\beta}_0 = -151.7$ and $\hat{\beta}_1 = 0.07676$. Thus, the fitted nonlinear regression model is

$$E[Y] = e^{\hat{\beta}_0 + \hat{\beta}_1 X}.$$

The fitted exponential regression model is displayed in Figure 3.25. The two different exponential regression models can be compared by computing the sums of squares for the two models, which can be computed by the *additional* R command

```
sum((debt - exp(b0 + b1 * year)) ^ 2)          # calculate sum of squares
```

The sum of squares for fitting the exponential regression model using the transformation technique is 22.7 and the sum of squares for the nonlinear least squares is 3.1. So consistent with Figures 3.24 and 3.25, the nonlinear least squares approach provides a better fit to the data pairs.

One drawback that emerged from the survival function estimation example from the previous section (involving current status data) is that fitting a regression line results in a survival probability that can be negative or greater than one when extrapolated outside of the range of the independent variable in the data pairs. In addition, the estimated probability of survival at time zero for both the ordinary simple linear regression model and the weighted simple linear regression model seemed low. Typically, a brand-new item is not defective. A nonlinear regression function is an attractive alternative model in this particular setting. The next example combines a nonlinear regression model and weighted least squares estimators to provide an improved regression model.

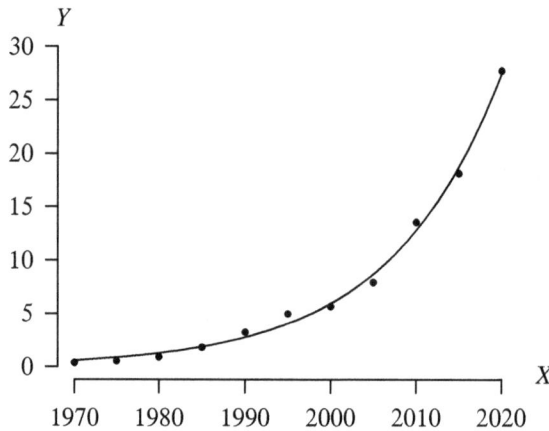

Figure 3.25: Scatterplot and exponential fit of year X and debt Y.

Example 3.15 Consider again the estimate of the probability of survival from the current status data given in Example 3.11. A simple nonlinear model that might be appropriate for the current status data set is to assume that the lifetime of the item under consideration follows the exponential(λ) distribution. The survivor function for an exponential random variable T with positive failure rate λ is

$$S(t) = P(T \geq t) = e^{-\lambda t} \qquad t > 0,$$

where t is the failure time in hours.

(a) Fit this nonlinear regression model using ordinary least squares.

(b) Fit this nonlinear regression model using weighted least squares.

(c) Compare the two fitted regression models.

(a) There are two ways to proceed with this regression problem. The first is to minimize the squared deviations

$$S = \sum_{i=1}^{n} \left(Y_i - e^{-\lambda X_i} \right)^2$$

with respect to λ to arrive at an appropriate regression parameter estimator. Equivalently, the least squares estimator of λ is

$$\hat{\lambda} = \underset{\lambda}{\operatorname{argmin}} \sum_{i=1}^{n} \left(Y_i - e^{-\lambda X_i} \right)^2.$$

This is the usual least squares approach. The second is to perform algebraic manipulations to the model in order to "linearize" the model so that the theory associated with the simple linear regression model can be implemented. The second approach is considered here. Treating this as a regression problem with X as time and Y as the survival probability results in the multiplicative nonlinear regression model

$$Y = e^{-\lambda X}\varepsilon.$$

Taking the natural logarithm of both sides of this model results in

$$\ln Y = -\lambda X + \varepsilon$$

or

$$-\ln Y = \lambda X + \varepsilon.$$

(Notice that when the error distribution is symmetric, which is often the case, the last step is justified.) This can be thought of as a linear regression problem with X as the independent variable and $-\ln Y$ as the dependent variable. There is no intercept in this model, so it can be treated as forcing the regression line through the origin and the single regression parameter λ corresponds to the slope of the regression line.

The R code below uses unweighted least squares to estimate the slope λ using the algebraic approach that forces the regression line through the origin using the techniques from Section 3.1.

```
x      = c(100, 200, 300, 400)
n      = c(100, 100,  10,  10)
s      = c( 50,  25,   4,   3)
y      = s / n
logy   = -log(y)
lamhat = sum(x * logy) / sum(x * x)
```

The R code using the matrix approach is identical to the algebraic approach in this case. Likewise, the regression parameter λ can be estimated using the lm function with the - 1 parameter to force the regression through the origin via the code below.

```
x    = c(100, 200, 300, 400)
n    = c(100, 100,  10,  10)
s    = c( 50,  25,   4,   3)
y    = s / n
logy = -log(y)
lm(logy ~ x - 1)$coefficients
```

Using any of these approaches to estimating λ, the estimate for the failure rate is

$$\hat{\lambda} = 0.003677$$

failures per hour.

(b) For the current status data set, it is sensible to incorporate the weights that are based on the various sample sizes into the regression model.

The algebraic and the matrix approach to the nonlinear weighted least squares model, which will be a regression model forced through the origin, have identical R code, which is given below.

```
x    = c(100, 200, 300, 400)
n    = c(100, 100,  10,  10)
s    = c( 50,  25,   4,   3)
y    = s / n
w    = n / (y * (1 - y))
logy = -log(y)
sum(x * w * logy) / sum(x * w * x)
```

The R code using the lm function to estimate the parameter λ is given below.

```
x      = c(100, 200, 300, 400)
n      = c(100, 100,  10,  10)
s      = c( 50,  25,   4,   3)
y      = s / n
w      = n / (y * (1 - y))
logy   = -log(y)
lamhat = lm(logy ~ x - 1, weights = w)$coefficients
```

Regardless of which approach is taken, the least squares estimate for the failure rate is

$$\hat{\lambda} = 0.005721$$

failures per hour, which is slightly higher than the estimated failure rate in the ordinary least squares approach.

(c) The two approaches (ordinary least squares and weighted least squares) for the nonlinear regression model can be compared graphically by plotting the two estimated survivor functions associated with the two fitted models. The R code below generates that plot. The estimated failure rate in the case of ordinary nonlinear least squares is stored in lambda.ols. The estimated failure rate in the case of weighted nonlinear least squares is stored in lambda.wls.

```
x          = c(100, 200, 300, 400)
n          = c(100, 100,  10,  10)
s          = c( 50,  25,   4,   3)
y          = s / n
logy       = -log(y)
lambda.ols = lm(logy ~ x - 1)$coefficients
w          = n / (y * (1 - y))
lambda.wls = lm(logy ~ x - 1, weights = w)$coefficients
xx         = 0:400
plot(x, y, xlim = c(0, 400), ylim = c(0, 1))
lines(xx, exp(-lambda.ols * xx))
lines(xx, exp(-lambda.wls * xx))
```

Figure 3.26 contains the graph. The ordinary least squares fit with $\hat{\lambda} = 0.003677$ gives equal weight to the four data pairs; the weighted least squares fit with $\hat{\lambda} = 0.005721$ gives significantly more weight to the first two data pairs. The two data pairs with the larger sample sizes are again circled in the figure. The weighted least squares model indicates that there is a higher estimated failure rate when increased weight is placed on the first two data pairs.

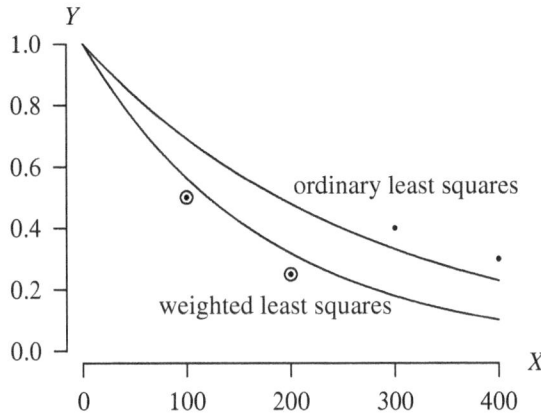

Figure 3.26: Current status data ordinary and weighted least squares fits for the exponential model.

3.8   Logistic Regression

Logistic regression is appropriate when the dependent variable Y can assume one of two values: zero and one. This is sometimes known as a *binary* or *dichotomous* response variable. For now, to keep the mathematics and interpretations simple, assume that there is a single predictor X. This is known as a *simple logistic regression model*, and is a special type of nonlinear regression model. Including multiple independent variables in a logistic regression model is a straightforward extension. For dichotomous data, instead of predicting 0 or 1, we predict the probability of getting a 1 [that is, $P(Y = 1)$]. So we need a regression model that predicts values of the interval $[0, 1]$.

The following example will be used throughout this section to motivate the need for a special model to accommodate a binary dependent variable, and to illustrate the techniques for the estimation of the model parameters.

Example 3.16 As an example to motivate the application of simple logistic regression, consider the $n = 948$ field goal attempts in the National Football League during the 2003 season. Let the independent variable X be the length of the field goal attempt (in yards) and the dependent variable Y be the outcome (0 for failure and 1 for success). The scatterplot (without jittering for ties) of the data values is shown in Figure 3.27, along with the associated least squares regression line with estimated intercept $\hat{\beta}_0 = 1.35$ and slope $\hat{\beta}_1 = -0.015$. The regression line is heading in the correct direction because longer field goals are less likely to be successful. Simple linear regression is clearly not an appropriate statistical model in this setting because it predicts probabilities outside of the interval $[0, 1]$. Even if predictions greater than 1 are set to 1 and negative predictions are set to zero, the model predicts that all 20-yard field goal attempts will be successful, and, at the other extreme, it predicts that the probability of kicking an 85-yard field goal is 0.06. This is inconsistent with the fact that the longest field goal ever in the NFL was a 66-yard field goal by Justin Tucker of the Baltimore Ravens on September 26, 2021. Obviously we can build a better regression model.

One of the initial considerations in developing a statistical model for the outcome of a field goal as a function of the length of the field goal attempt is to find a function that will only assume values

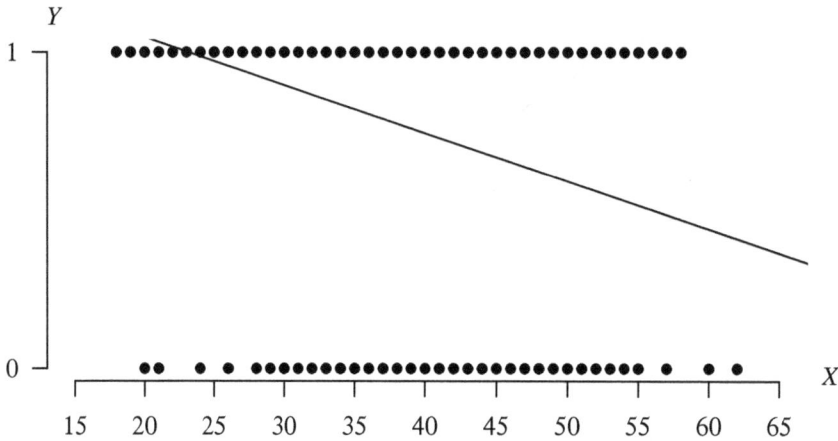

Figure 3.27: Scatterplot of field goal outcomes vs. yards with regression line.

between 0 and 1. A diagram that gives some guidance with regard to this function is to batch the data into 5-year increments. So the bins are all field goals that fall in the ranges 20 ± 2, 25 ± 2, ..., 60 ± 2. This window is long enough so that the random sampling variability associated with nearby attempts is damped considerably, and yet short enough so that outcome patterns as a function of yardage are still apparent. The R code below batches the independent variable into the 5-yard increments and plots the estimated probability of success for attempts in each batch at its midpoint. This estimated probability is just the fraction of successful field goals within a particular range. Furthermore, the area of each point plotted is proportional to the number of attempts in that particular bin. For example, there were 79 attempts in the first bin (18–22 yards) and only 4 attempts in the last bin (58–62 yards). The R code below reads a data set off of the web that contains the results of $n = 948$ NFL field goal attempts during 2003. The data consists of columns that give the length of the field goal attempt and the outcome, failure ($Y = 0$) or success ($Y = 1$). The R code rounds each length to the nearest 5 yards, and plots the midpoint of the rounded field goal lengths versus the estimated probability of success.

```
df = read.table("http://users.stat.ufl.edu/~winner/data/fieldgoal.dat")
yards = df[, 1]
outcome = df[, 2]
plot(NA, xlim = c(15, 65), ylim = c(0, 1))
yards = floor((yards + 2) / 5) * 5
for (i in 1:9) points(5 * i + 15, mean(outcome[yards / 5 - 3 == i]),
                      pch = 16, cex = 0.12 * sqrt(table(yards / 5 - 3)[i]))
```

While the performance of NFL field goal kickers varies from one kicker to the next, these points give us an idea of what we would like for a smooth regression function in this setting.

The results are shown in Figure 3.28. It is clear that the estimated probability of making a field goal decreases as the length of the field goal attempt increases, as one would expect. There is a strong relationship between the length of the field goal attempt and the probability of success. Our goal is to fit a nonlinear regression function to the raw data values that smooths the random sampling variability and can be used for the purpose of prediction.

$\hat{P}(\text{success})$

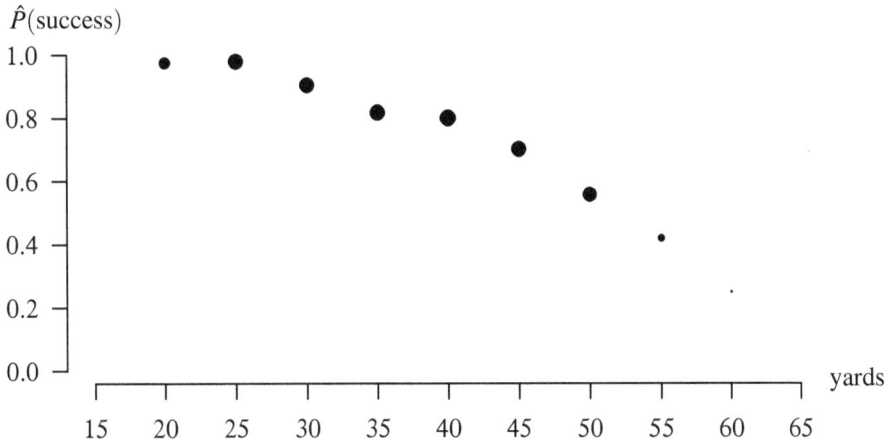

Figure 3.28: Field goal outcomes vs. yards in 5-yard increments.

When the dependent variable only takes on the values zero and one, the usual mean response function for the simple linear regression model

$$Y = \beta_0 + \beta_1 X + \varepsilon$$

is

$$E[Y] = \beta_0 + \beta_1 X,$$

where $E[Y]$ denotes the conditional expected value of Y given a particular setting of the independent variable X. This mean response function does not limit the values of Y to just zero and one. With normally distributed error terms, this model would allow for Y values which could be less than 0 or greater than 1.

In logistic regression, this type of curve, regardless of whether it begins near one and ends near zero or it begins near zero and ends near one, is known as a *sigmoidal* response function. A natural choice for the sigmoidal response function is a cumulative distribution function associated with a random variable, or its complement (the survivor function). Three popular probability distributions whose cumulative distribution functions are used in logistic regression are the standard logistic distribution (also commonly called the *logit* model), the standard normal distribution (also commonly called the *probit* model), and the standard extreme value distribution (also commonly called the complimentary *log-log* model). These are described in the next paragraph.

The standard logistic distribution has probability density function

$$f(x) = \frac{e^x}{(1+e^x)^2} \qquad -\infty < x < \infty$$

and cumulative distribution function

$$F(x) = \frac{e^x}{1+e^x} \qquad -\infty < x < \infty.$$

The probability density function is symmetric about the population mean $E[X] = 0$ and has population variance $V[X] = \pi^2/3$. The standard normal distribution has probability density function

$$f(x) = \frac{1}{\sqrt{2\pi}} e^{-x^2/2} \qquad -\infty < x < \infty$$

and cumulative distribution function

$$F(x) = \int_{-\infty}^{x} f(w)\,dw \qquad -\infty < x < \infty.$$

The probability density function is also symmetric about the population mean $E[X] = 0$ and has population variance $V[X] = 1$. The probability density function for the standard logistic distribution is similar in shape (that is, bell-shape) to that for the standard normal distribution, but has heavier tails. The symmetry of the probability density functions for the standard logistic distribution and the standard normal distribution limits the shape of the associated cumulative distribution function. A nonsymmetric distribution often provides a better fit. This leads to a search for a probability distribution with a nonsymmetric probability density function. One such probability distribution is the extreme value distribution. The standard extreme value distribution has probability density function

$$f(x) = e^{x - e^x} \qquad -\infty < x < \infty$$

and cumulative distribution function

$$F(x) = 1 - e^{-e^x} \qquad -\infty < x < \infty.$$

The population mean and the population variance are not mathematically tractable, but the numeric values, to ten digits, are

$$E[X] = -0.5772156649 \qquad \text{and} \qquad V[X] = 1.644934067.$$

The probability density function is not symmetric about the mean.

The R code below plots these three probability density functions on the same set of axes. The standard normal probability density function is taken directly from the formulas in the previous paragraph. The probability density functions for the standard logistic distribution and the standard extreme value distribution have been standardized (by subtracting their population mean and dividing by the population standard deviation) so that all three probability density functions can be viewed on an equal footing. The plot emphasizes the shape of the various probability density functions.

```
x = seq(-3, 3, by = 0.01)
k = pi / sqrt(3)
y = k * exp(k * x) / (1 + exp(k * x)) ^ 2
plot(x, y, type = "l", xlim = c(-3, 3), ylim = c(0, 0.5))
lines(x, dnorm(x))
mu = -0.5772156649
sig = sqrt(1.644934067)
y = sig * exp(mu + sig * x - exp(mu + sig * x))
lines(x, y)
```

The results are displayed in Figure 3.29. All three probability distributions have support on the entire real number line $-\infty < x < \infty$, although the graph only includes the values within three standard deviation units from the population mean. As expected, the probability density functions for the standard normal distribution and the standardized version of the standard logistic distribution are symmetric and bell-shaped. The probability density function of the standardized version of the standard extreme value distribution is nonsymmetric. The R code below plots the cumulative distribution function associated with the standardized version of the standard logistic distribution.

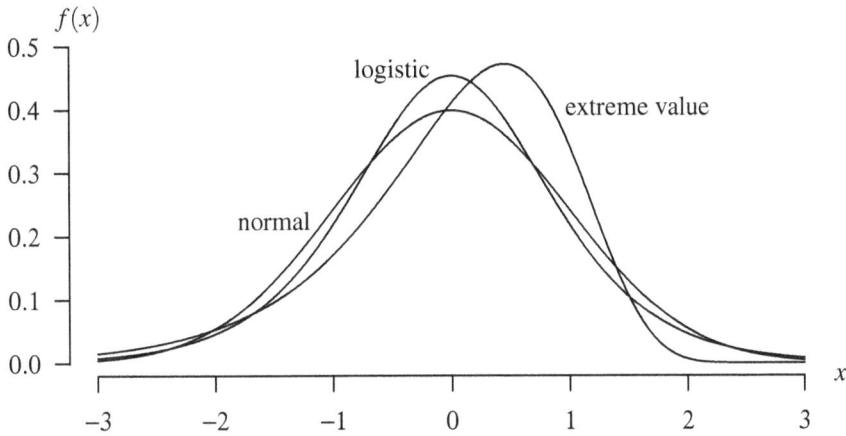

Figure 3.29: Standardized logistic, normal, and extreme value probability density functions.

```
x = seq(-3, 3, by = 0.01)
k = pi / sqrt(3)
y = exp(k * x) / (1 + exp(k * x))
plot(x, y, type = "l", xlim = c(-3, 3), ylim = c(0, 1))
```

The cumulative distribution function $F(x) = P(X \leq x)$ is graphed in Figure 3.30. This cumulative distribution function is monotone increasing and satisfies $\lim_{x \to -\infty} F(x) = 0$ and $\lim_{x \to \infty} F(x) = 1$. Notice that a plot of $F(-x)$ gives the complement of the cumulative distribution function. In other words, $S(x) = 1 - F(x) = P(X \geq x)$. This function is monotone decreasing and satisfies $\lim_{x \to -\infty} S(x) = 1$ and $\lim_{x \to \infty} S(x) = 0$. This function is known in survival analysis as the *survivor function*.

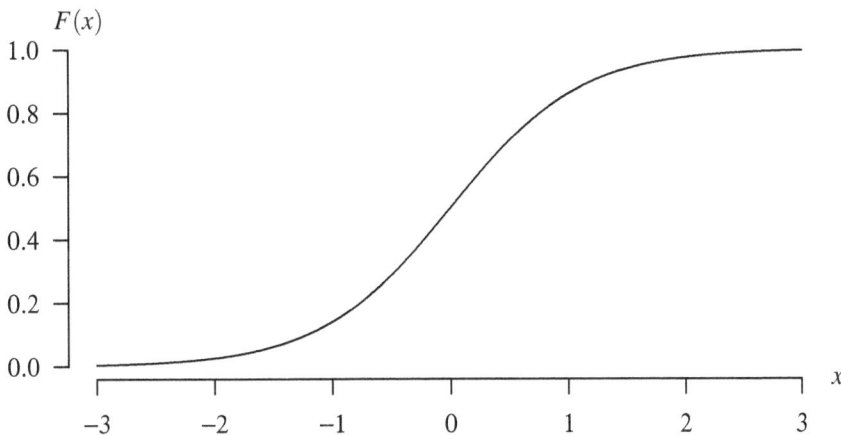

Figure 3.30: Standardized version of the standard logistic cumulative distribution function.

Now that cumulative distribution functions and their complements have been identified as a reasonable way to estimate the probability of success for the field goal data, we would like to establish a mechanism for incorporating the value of the predictor X into the probability model. The emphasis here will be on using the cumulative distribution function for the logistic distribution, since that seems to be the most commonly used in logistic regression.

The usual form of the mean response function for simple linear regression is

$$E[Y] = \beta_0 + \beta_1 X.$$

But in the case of a binary outcome, the constraint

$$0 \le E[Y] \le 1$$

must be imposed. This is done naturally using the cumulative distribution functions and their complements for the various probability distributions described earlier. Let $\pi(X)$ be the mean response function for a regression model with a binary response. Using the cumulative distribution function for the logistic distribution, the mean response function is

$$\pi(X) = E[Y] = \frac{e^{\beta_0 + \beta_1 X}}{1 + e^{\beta_0 + \beta_1 X}}.$$

Since the random variable Y can only assume the values 0 and 1 for a particular value of X, it is a Bernoulli random variable with probability of success $\pi(X)$. Since the expected value and the probability that a Bernoulli random variable assumes the value 1 are equal, the mean response function can also be expressed as

$$\pi(X) = P(Y = 1) = \frac{e^{\beta_0 + \beta_1 X}}{1 + e^{\beta_0 + \beta_1 X}},$$

where $P(Y = 1)$ is the probability that the dependent variable Y equals 1 for a particular fixed setting of the independent variable X. The parameters β_0 and β_1 assume the following roles.

- The sign of β_1 controls whether the mean response function is monotone increasing or decreasing. Table 3.9 shows the direction of the relationship associated with the sign of β_1. The statistical significance of the point estimator of β_1 depends on its magnitude.

- The magnitude of β_1 controls the steepness of the mean response function, with larger magnitudes corresponding to steeper mean response functions.

- The value of β_0 controls the location of the mean response function on the X-axis.

A graph that illustrates the effect of varying values of β_1 for the fixed value of $\beta_0 = 0$ on the mean response function $\pi(X)$ is given in Figure 3.31. As expected, the mean response function $\pi(X)$

Condition	$\lim\limits_{X \to -\infty} \pi(X)$	$\lim\limits_{X \to \infty} \pi(X)$
$\beta_1 < 0$	1	0
$\beta_1 > 0$	0	1
$\beta_1 = 0$	$e^{\beta_0}/(1+e^{\beta_0})$	$e^{\beta_0}/(1+e^{\beta_0})$

Table 3.9: Direction of monotonicity of $\pi(X)$.

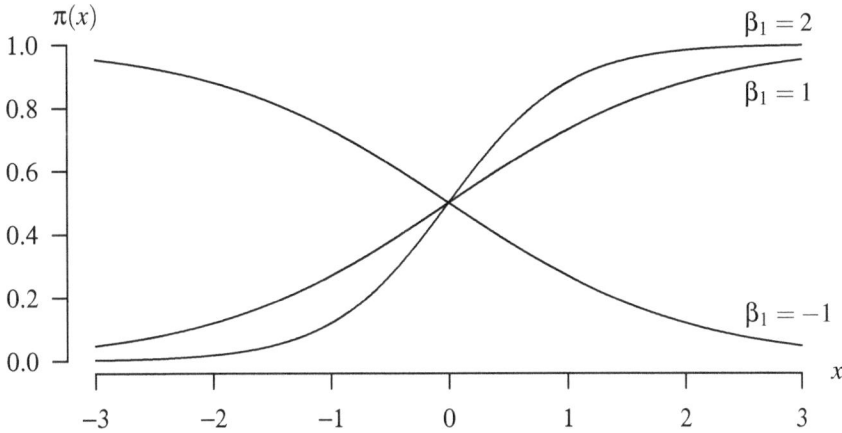

Figure 3.31: Mean response functions for $\beta_0 = 0$ and various β_1 values.

is monotone decreasing for $\beta_1 < 1$ and monotone increasing for $\beta_1 > 1$. The mean response function is steeper as the magnitude of β_1 increases.

A graph that illustrates the effect of varying values of β_0 for the fixed value of $\beta_1 = 1$ on the mean response function $\pi(X)$ is given in Figure 3.32. As expected, the mean response function $\pi(X)$ is monotone increasing in all cases because $\beta_1 > 1$. The effect of varying β_0 is to shift the mean response functions horizontally. The rationale behind the horizontal shift can be seen by writing the mean response function with $\beta_1 = 1$ as

$$\pi(X) = \frac{e^{\beta_0 + X}}{1 + e^{\beta_0 + X}}.$$

So the effect of increasing β_0 in this case is to shift the mean response function to the right (for

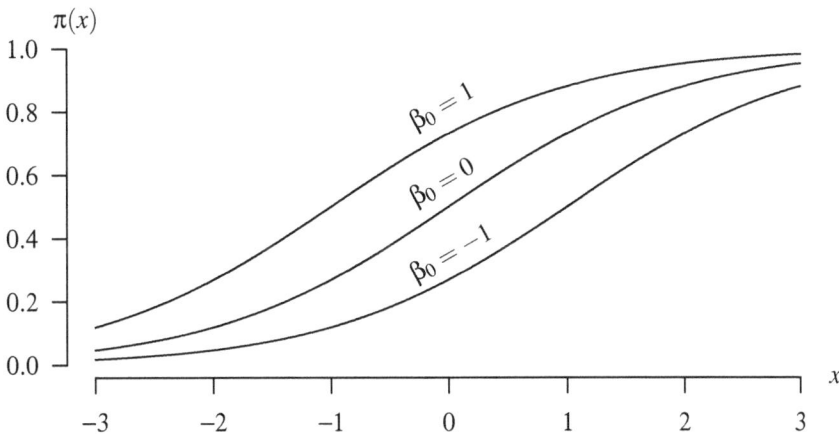

Figure 3.32: Mean response functions for $\beta_1 = 1$ and various β_0 values.

$\beta_1 < 1$) or to the left (for $\beta_1 > 1$) relative to the $\pi(X)$ curve associated with $\beta_0 = 0$.

To summarize, the sign of β_1 controls the direction of the monotonicity of $\pi(X)$, the magnitude of β_1 controls the steepness of $\pi(X)$, and β_0 controls the location of $\pi(X)$ along the X-axis.

We now consider the estimation of the parameters β_0 and β_1 from a data set consisting of the n data pairs $(X_1, Y_1), (X_2, Y_2), \ldots, (X_n, Y_n)$. The first components $X_1, X_2, \ldots, X_n$ are real numbers and the second components $Y_1, Y_2, \ldots, Y_n$ assume only the values 0 and 1. Since

$$P(Y = 1) = \pi(X) \qquad \text{and} \qquad P(Y = 0) = 1 - \pi(X)$$

the contribution to the likelihood function of the data pair (X_i, Y_i) is

$$\pi(X_i)^{Y_i} [1 - \pi(X_i)]^{1-Y_i}$$

for $i = 1, 2, \ldots, n$. When $Y_i = 0$, the contribution to the likelihood function is $1 - \pi(X_i)$, which is $P(Y_i = 0)$, where $P(Y_i = 0)$ is the probability that $Y_i = 0$ for the particular setting of the independent variable at X_i. When $Y_i = 1$, the contribution to the likelihood function is $\pi(X_i)$, which is $P(Y_i = 1)$. Since X_i is assumed to be observed without error, Y_i is a random binary response, and the responses are assumed to be mutually independent random variables, the likelihood function is

$$L(\beta_0, \beta_1) = \prod_{i=1}^{n} \pi(X_i)^{Y_i} [1 - \pi(X_i)]^{1-Y_i}.$$

The log likelihood function is

$$\ln L(\beta_0, \beta_1) = \sum_{i=1}^{n} Y_i \ln [\pi(X_i)] + (1 - Y_i) \ln [1 - \pi(X_i)].$$

This can be written in terms of β_0 and β_1 as

$$\ln L(\beta_0, \beta_1) = \sum_{i=1}^{n} Y_i \left[\beta_0 + \beta_1 X_i - \ln \left(1 + e^{\beta_0 + \beta_1 X_i} \right) \right] - (1 - Y_i) \ln \left(1 + e^{\beta_0 + \beta_1 X_i} \right)$$

or

$$\ln L(\beta_0, \beta_1) = \sum_{i=1}^{n} Y_i (\beta_0 + \beta_1 X_i) - \ln \left(1 + e^{\beta_0 + \beta_1 X_i} \right).$$

The likelihood function and the log likelihood function are maximized at the same values of β_0 and β_1 because the natural logarithm is a monotonic transformation. The score vector is comprised of the partial derivatives of the log likelihood function with respect to β_0 and β_1:

$$\frac{\partial \ln L(\beta_0, \beta_1)}{\partial \beta_0} = \sum_{i=1}^{n} \left(Y_i - \frac{e^{\beta_0 + \beta_1 X_i}}{1 + e^{\beta_0 + \beta_1 X_i}} \right)$$

and

$$\frac{\partial \ln L(\beta_0, \beta_1)}{\partial \beta_1} = \sum_{i=1}^{n} \left(X_i Y_i - \frac{X_i e^{\beta_0 + \beta_1 X_i}}{1 + e^{\beta_0 + \beta_1 X_i}} \right).$$

When these two equations are equated to zero, there is no closed form solution for $\hat{\beta}_0$ and $\hat{\beta}_1$, so numerical methods must be relied on to calculate these point estimates. The second derivatives of the log likelihood function after simplification are

$$\frac{\partial^2 \ln L(\beta_0, \beta_1)}{\partial \beta_0^2} = - \sum_{i=1}^{n} \frac{e^{\beta_0 + \beta_1 X_i}}{\left(1 + e^{\beta_0 + \beta_1 X_i} \right)^2},$$

$$\frac{\partial^2 \ln L\left(\beta_0, \beta_1\right)}{\partial \beta_0 \partial \beta_1} = -\sum_{i=1}^{n} \frac{X_i e^{\beta_0 + \beta_1 X_i}}{\left(1 + e^{\beta_0 + \beta_1 X_i}\right)^2},$$

and

$$\frac{\partial^2 \ln L\left(\beta_0, \beta_1\right)}{\partial \beta_1^2} = -\sum_{i=1}^{n} \frac{X_i^2 e^{\beta_0 + \beta_1 X_i}}{\left(1 + e^{\beta_0 + \beta_1 X_i}\right)^2}.$$

The Fisher information matrix is the matrix of expected values of these partial derivatives:

$$I\left(\beta_0, \beta_1\right) = \begin{pmatrix} E\left[\dfrac{-\partial^2 \ln L(\beta_0, \beta_1)}{\partial \beta_0^2}\right] & E\left[\dfrac{-\partial^2 \ln L(\beta_0, \beta_1)}{\partial \beta_0 \beta_1}\right] \\ E\left[\dfrac{-\partial^2 \ln L(\beta_0, \beta_1)}{\partial \beta_1 \beta_0}\right] & E\left[\dfrac{-\partial^2 \ln L(\beta_0, \beta_1)}{\partial \beta_1^2}\right] \end{pmatrix}.$$

The expected values in this matrix can be determined because they do not contain any random variables. Their values cannot be calculated, however, because the values of the parameters β_0 and β_1 are unknown. The observed information matrix

$$O\left(\hat{\beta}_0, \hat{\beta}_1\right) = \begin{pmatrix} \dfrac{-\partial^2 \ln L(\beta_0, \beta_1)}{\partial \beta_0^2} & \dfrac{-\partial^2 \ln L(\beta_0, \beta_1)}{\partial \beta_0 \beta_1} \\ \dfrac{-\partial^2 \ln L(\beta_0, \beta_1)}{\partial \beta_1 \beta_0} & \dfrac{-\partial^2 \ln L(\beta_0, \beta_1)}{\partial \beta_1^2} \end{pmatrix}_{\beta_0 = \hat{\beta}_0, \, \beta_1 = \hat{\beta}_1}$$

can be estimated from data values once the maximum likelihood estimators are computed. This matrix is the variance–covariance matrix of the score vector and its inverse is the asymptotic variance–covariance matrix of the maximum likelihood estimators. The square roots of the diagonal elements of this inverse matrix provide estimates of the standard errors of the maximum likelihood estimates.

The NFL field goal data set has a large sample size ($n = 948$) and a strong statistical relationship between the length of the field goal attempt and the probability of success. The R code below again uses the `optim` function to calculate the parameter estimates. The first argument to `optim` are initial parameter estimates. The second argument to `optim` is the function to be *minimized*, so the negative of the log likelihood function is given as the second argument. Once the maximum likelihood estimates are calculated, the observed information matrix, standard errors, z-statistics, and associated p-values are calculated.

```
df = read.table("http://users.stat.ufl.edu/~winner/data/fieldgoal.dat")
yards = df[, 1]
outcome = df[, 2]
logl = function(parameters) {
  beta0 = parameters[1]
  beta1 = parameters[2]
  sum(-outcome * (beta0 + beta1 * yards) + log(1 + exp(beta0 + beta1 * yards)))
}
fit         = optim(c(0, -1), logl)
beta0hat    = fit$par[1]
beta1hat    = fit$par[2]
oim         = matrix(0, 2, 2)
oim[1, 1]   = sum(exp(beta0hat + beta1hat * yards) /
                 (1 + exp(beta0hat + beta1hat * yards)) ^ 2)
```

```
oim[1, 2]    = sum(yards * exp(beta0hat + beta1hat * yards) /
                 (1 + exp(beta0hat + beta1hat * yards)) ^ 2)
oim[2, 1]    = oim[1, 2]
oim[2, 2]    = sum(yards * yards * exp(beta0hat + beta1hat * yards) /
                 (1 + exp(beta0hat + beta1hat * yards)) ^ 2)
print(oim)
se.beta0hat = sqrt(solve(oim)[1, 1])
se.beta1hat = sqrt(solve(oim)[2, 2])
z0           = beta0hat / se.beta0hat
z1           = beta1hat / se.beta1hat
p0           = 2 * (1 - pnorm(abs(z0)))
p1           = 2 * (1 - pnorm(abs(z1)))
print(c(beta0hat, se.beta0hat, z0, p0))
print(c(beta1hat, se.beta1hat, z1, p1))
```

The results of the code are summarized in Table 3.10. The values of $\hat{\beta}_0$ and $\hat{\beta}_1$ are both statistically significant with p-values near zero. The observed information matrix for the NFL field goal data set

i	$\hat{\beta}_i$	$\hat{\sigma}_{\hat{\beta}_i}$	z	p
0	5.69	0.451	12.6	0.00
1	-0.110	0.0106	-10.4	0.00

Table 3.10: Summary statistics for NFL field goal data.

is

$$O\left(\hat{\beta}_0, \hat{\beta}_1\right) = \begin{pmatrix} 130.83 & 5470.26 \\ 5470.26 & 237,653.57 \end{pmatrix}.$$

These values can be compared to the values obtained using the `glm` (generalized linear model) function:

```
df = read.table("http://users.stat.ufl.edu/~winner/data/fieldgoal.dat")
yards = df[, 1]
outcome = df[, 2]
fit = glm(outcome ~ yards, family = binomial(link = logit))
summary(fit)
```

The results match those given in Table 3.10. When the `link` parameter within the `binomial` family is set to `logit`, the cumulative distribution function (or its complement) for the standard logistic distribution is employed. When the `link` parameter is set to `probit`, the cumulative distribution function (or its complement) for the standard normal distribution is employed. The logit and probit choices force the sigmoidal function to be symmetric, so that it approaches 0 and 1 at the same rate. When the `link` parameter is set to `cloglog`, the cumulative distribution function (or its complement) for the standard extreme value distribution is employed. It approaches 0 and 1 at the different rates.

When the following R statements are added to the code that generated Figure 3.28, the fitted mean response function $\hat{\pi}(X)$ is added to the graph.

```
x = seq(15, 65, by = 0.1)
beta0hat = 5.6942693
```

```
beta1hat = -0.1098488
y = exp(beta0hat + beta1hat * x) / (1 + exp(beta0hat + beta1hat * x))
lines(x, y)
```

The graph is shown in Figure 3.33. The estimated mean response function is monotone decreasing because $\hat{\beta}_1 < 0$. Furthermore, the mean response curve does an adequate job of modeling the probability of success as the points lie very close to the estimated mean response function. The estimated

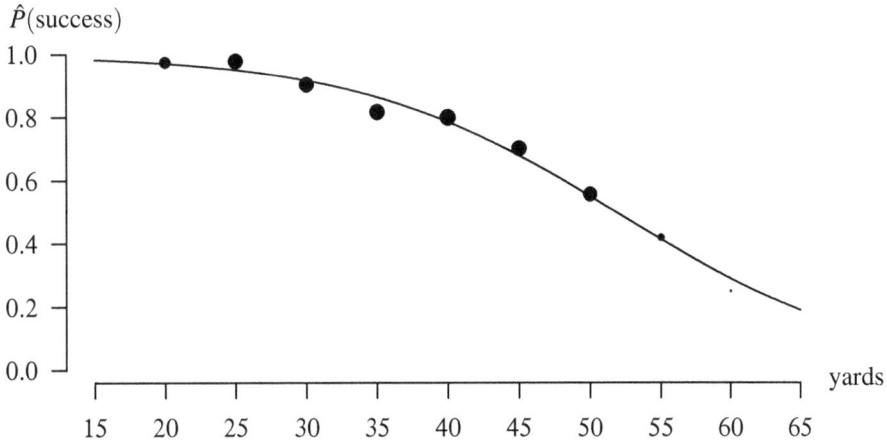

Figure 3.33: Field goal outcomes and estimated mean response function.

mean response function can be used for prediction. The estimated probability that a 38-yard field goal attempt is successful is

$$\hat{\pi}(38) = \frac{e^{5.6942693 - 0.1098488(38)}}{1 + e^{5.6942693 - 0.1098488(38)}} = 0.82.$$

This value can be generated with the `predict` function in R with the *additional* statements

```
linear = predict(fit, newdata = data.frame(yards = 38))
exp(linear) / (1 + exp(linear))
```

Some keystrokes can be saved by using the `type = "response"` argument in the call to `predict`.

```
predict(fit, newdata = data.frame(yards = 38), type = "response")
```

The limitations of a symmetric mean response function also become apparent in this case. The estimated probability that a 71-yard field goal attempt is successful is

$$\hat{\pi}(71) = \frac{e^{5.6942693 - 0.1098488(71)}}{1 + e^{5.6942693 - 0.1098488(71)}} = 0.11,$$

even though the NFL field goal record from 2021 is 66 yards. This is clearly a case of extrapolating beyond the range of the data, which is discouraged. The meaningful range of $\hat{\pi}(X)$ is over the scope of the model $18 \leq X \leq 62$, whose endpoints are the shortest and longest field goal attempt during the 2003 season. The symmetric nature of the logistic distribution makes the $\hat{\pi}(X)$ values associated with X-values greater than 62 yards higher than are meaningful.

Confidence intervals for the parameters in a logistic regression model can be calculated with the `confint` and `confint.default` functions. These confidence intervals give a measure of the precision of the point estimates. The R code below calculates the 95% confidence intervals for the parameters using the `confint` and `confint.default` functions for the NFL field goal data.

```
df = read.table("http://users.stat.ufl.edu/~winner/data/fieldgoal.dat")
yards = df[, 1]
outcome = df[, 2]
fit = glm(outcome ~ yards, family = binomial(link = logit))
confint(fit)
confint.default(fit)
```

The first set of confidence intervals that are returned via `confint` use the profiled log likelihood function to return the confidence intervals given in the output below. The default is a 95% confidence interval.

```
              2.5 %        97.5 %
(Intercept)   4.8435441    6.61425072
yards        -0.1312492   -0.08970744
```

To three significant digits, these 95% confidence intervals are

$$4.84 < \beta_0 < 6.61 \qquad \text{and} \qquad -0.131 < \beta_1 < -0.0897.$$

The second set of confidence intervals that are returned via `confint.default` are based on the asymptotic normality of the maximum likelihood estimators. The call to `confint.default` returns the confidence intervals given in the output below.

```
              2.5 %        97.5 %
(Intercept)   4.8137433    6.58201706
yards        -0.1306527   -0.08916745
```

To three significant digits, these 95% confidence intervals are

$$4.82 < \beta_0 < 6.58 \qquad \text{and} \qquad -0.131 < \beta_1 < -0.0892.$$

Alternatively, the 95% confidence interval for β_1 can be calculated by using the `qnorm` function to calculate the appropriate quantile from the standard normal distribution.

```
df = read.table("http://users.stat.ufl.edu/~winner/data/fieldgoal.dat")
yards = df[, 1]
outcome = df[, 2]
fit = glm(outcome ~ yards, family = binomial(logit))
coef(fit)[2] + c(-1, 1) * qnorm(0.975) * summary(fit)$coefficients[2, 2]
```

The 95% confidence interval for β_1 that is returned matches that returned by `confint.default`. The confidence intervals based on the asymptotic normality of the maximum likelihood estimator from `confint.default` will be symmetric about the maximum likelihood estimators, but the confidence interval based on the profiled log likelihood function from `confint` will not be symmetric about the maximum likelihood estimators. The confidence intervals given here are somewhat narrow because of the large sample size of $n = 948$ for the NFL field goal data.

The last topic is the interpretation of the point estimators for the coefficients. This interpretation is much more difficult than the interpretation of the coefficients in a standard simple linear regression model. The next paragraph defines the odds and the log odds. The subsequent paragraph relates the log odds to the logistic regression model.

Consider an event which occurs with probability 0.9. The probability that the event will not occur is 0.1. The odds are defined as the ratio of the probability that the event will occur to the probability that the event will not occur. In this case that ratio is 9, so the odds are often referred to as 9 to 1. Table 3.11 gives several probability values and associated odds for several probability values.

Probability	Odds
0.2	0.25
0.5	1
0.6	1.5
0.75	3
0.8	4
0.9	9
0.99	99

Table 3.11: Probability and odds.

The R code below generates a graph of the odds on the vertical axis versus the probability on the horizontal axis.

```
prob = seq(0, 0.9, by = 0.01)
odds = prob / (1 - prob)
plot(prob, odds, type = "l", xlim = c(0, 1), ylim = c(0, 9))
```

Figure 3.34 shows the transformation from probability to odds, which reveals a monotone increasing function. Probabilities fall on the interval $[0, 1]$; odds fall on the interval $[0, \infty)$. The natural loga-

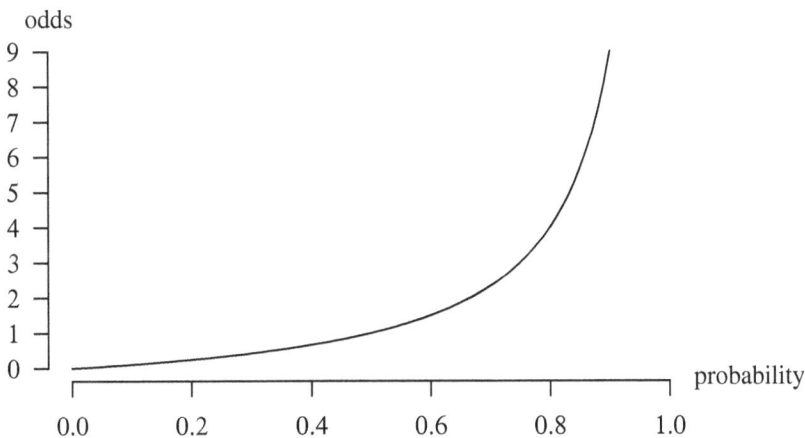

Figure 3.34: Odds versus probability.

rithm of the odds is the function, known as the log odds, which is a transformation of the probability p in the following fashion:

$$\ln\left(\frac{p}{1-p}\right).$$

This is a transformation from $[0, 1]$ to $(-\infty, \infty)$. Table 3.12 extends the previous table by including a column for the log odds. Notice that a probability of $1/2$ corresponds to a log odds of 0 and the symmetry of the log odds associated with the probabilities 0.2 and 0.8. The R code below graphs

Probability	Odds	Log Odds
0.2	0.25	-1.3863
0.5	1	0
0.6	1.5	0.4055
0.75	3	1.0986
0.8	4	1.3863
0.9	9	2.1972
0.99	99	4.5951

Table 3.12: Probability, odds, and log odds.

the log odds versus the probability.

```
prob    = seq(0.045, 0.955, by = 0.001)
odds    = prob / (1 - prob)
logodds = log(odds)
plot(prob, logodds, type = "l", xlim = c(0, 1), ylim = c(-3, 3))
```

The associated graph is shown in Figure 3.35. The shape of the log odds is a transformed version of the mean response functions seen earlier. The purpose of defining the log odds is to convert from probability, which has a restricted range between 0 and 1, and the log odds, which has an unrestricted range.

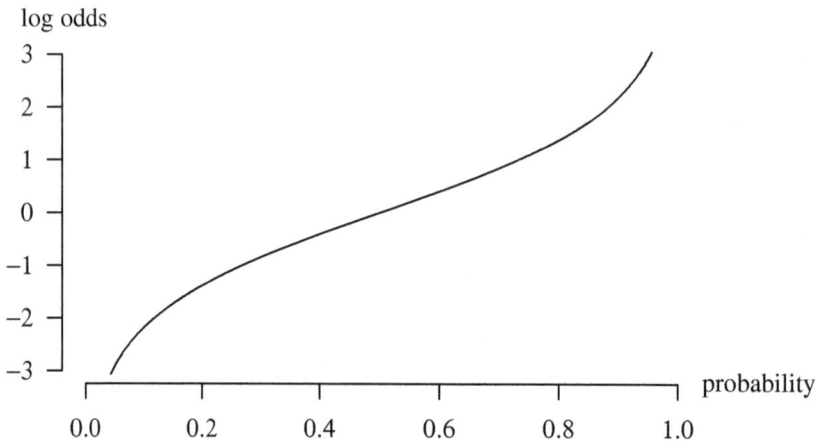

Figure 3.35: Log odds versus probability.

Now back to logistic regression and the interpretation of the estimated coefficients. Recall that for a simple logistic regression problem, the mean response function is

$$\pi(x) = E[Y \mid X = x] = P(Y = 1 \mid X = x) = \frac{e^{\beta_0 + \beta_1 x}}{1 + e^{\beta_0 + \beta_1 x}},$$

where x is the independent variable and Y is the response variable. The *logit transformation* of $\pi(x)$ is

$$\ln\left[\frac{\pi(x)}{1 - \pi(x)}\right] = \ln\left[e^{\beta_0 + \beta_1 x}\right] = \beta_0 + \beta_1 x.$$

Since $\pi(x)$ is a probability, the expression on the left-hand side of this equation is a log odds.

Now consider the NFL data. From the earlier work, the estimated intercept provided by the R glm function is $\hat{\beta}_0 = 5.6979$ and the estimated coefficient associated with the length of the field goal attempt in yards is $\hat{\beta}_1 = -0.1099$. The estimated intercept is the log odds of a kicker making a field goal from a (theoretical) zero yards, which has no meaningful interpretation in this setting. The value of $\hat{\beta}_1 = -0.1099$ is the change in the log odds for a one-yard change in the length of the field goal attempt. Additionally, the quantity

$$e^{\hat{\beta}_1} = e^{-0.1099} = 0.8959$$

is the multiplier that gives the change in the odds for a one-unit change in the independent variable. We expect to see a 10.4% decrease in the odds associated with the probability of success for a field goal attempt for every additional yard added to the field goal attempt. This value and an associated 95% confidence interval can be generated with the *additional* R statement

```
exp(cbind(oddsratio = coef(fit), confint(fit)))
```

The analysis of the NFL data given here is a composite of all kickers in the NFL during 2003. Individual kickers within the NFL will have their own logistic regression curve.

With this background concerning simple logistic regression in place, it is straightforward to extend this to more complicated modeling situations. Additional topics in logistic regression include constructing a confidence interval for a predicted value, the calculation of deviance residuals, including multiple independent variables in a logistic regression model, model assessment, and interpreting estimated coefficients for interaction terms.

3.9 Exercises

3.1 Write a paragraph that describes why the sum of squares for error associated with the simple linear regression model $Y = \beta_0 + \beta_1 X + \varepsilon$ will always be less than or equal to the sum of squares for error associated with the simple linear regression model forced through the origin $Y = \beta_1 X + \varepsilon$ for the same data pairs $(X_1, Y_1), (X_2, Y_2), \ldots, (X_n, Y_n)$.

3.2 Under what condition(s) does the regression line forced through the origin have the same sum of squares for error as the simple linear regression for the full model $Y = \beta_0 + \beta_1 X + \varepsilon$ for the same data pairs $(X_1, Y_1), (X_2, Y_2), \ldots, (X_n, Y_n)$.

3.3 Consider the simple linear regression model forced through the origin

$$Y = \beta_1 X + \varepsilon.$$

Show that the least squares estimator $\hat{\beta}_1$ is an unbiased estimator of β_1.

3.4   Consider the simple linear regression model forced through the origin

$$Y = \beta_1 X + \varepsilon.$$

Find $V\left[\hat{\beta}_1\right]$.

3.5   Consider the simple linear regression model forced through the origin with normal error terms,

$$Y = \beta_1 X + \varepsilon,$$

where $\varepsilon \sim N\left(0, \sigma^2\right)$.

(a) Find the maximum likelihood estimators of β_1 and σ^2.

(b) Show that the maximum likelihood estimators maximize the log likelihood function.

3.6   Give an example of $n = 2$ data pairs corresponding to the case in which a simple linear regression line forced through the origin contains the point $(\bar{X}, \bar{Y})$.

3.7   Give an example of $n = 2$ data pairs corresponding to the case in which a simple linear regression line forced through the origin does not contain the point $(\bar{X}, \bar{Y})$.

3.8   Consider the simple linear regression model forced through the origin with normal error terms

$$Y = \beta_1 X + \varepsilon,$$

with *known* parameters β_1 and σ^2. Find an exact two-sided $100(1 - \alpha)\%$ confidence interval for β_1 from n data pairs $(X_1, Y_1), (X_2, Y_2), \ldots, (X_n, Y_n)$.

3.9   Consider the simple linear regression model forced through the origin with normal error terms,

$$Y = \beta_1 X + \varepsilon,$$

with unknown parameters β_1 and σ^2. Show that the R statement

```
confint(lm(Formaldehyde$optden ~ Formaldehyde$carb - 1))
```

uses the formula

$$\hat{\beta}_1 - t_{n-1, \alpha/2} \sqrt{\frac{SSE}{(n-1)\sum_{i=1}^{n} X_i^2}} < \beta_1 < \hat{\beta}_1 + t_{n-1, \alpha/2} \sqrt{\frac{SSE}{(n-1)\sum_{i=1}^{n} X_i^2}}$$

to calculate the 95% two-sided confidence interval for β_1 for the data pairs in the built-in R data frame `Formaldehyde`. Notice that the degrees of freedom are one more than the associated degrees of freedom for the full simple linear regression model.

3.10   Consider the simple linear regression model forced through the origin with normal error terms,

$$Y = \beta_1 X + \varepsilon,$$

with unknown parameters β_1 and σ^2. Conduct a Monte Carlo simulation experiment to provide convincing numerical evidence that the two-sided $100(1 - \alpha)\%$ confidence interval

$$\hat{\beta}_1 - t_{n-1, \alpha/2} \sqrt{\frac{SSE}{(n-1)\sum_{i=1}^{n} X_i^2}} < \beta_1 < \hat{\beta}_1 + t_{n-1, \alpha/2} \sqrt{\frac{SSE}{(n-1)\sum_{i=1}^{n} X_i^2}}$$

is an *exact* confidence interval for β_1 for the following parameter settings: $n = 3$, $\alpha = 0.05$, $\beta_1 = 2$, $X_1 = 1$, $X_2 = 2$, $X_3 = 3$, and $\sigma^2 = 1$.

3.11 The Brown–Forsythe test can be used to determine whether the error terms have constant variance. In particular, it tests for equality of the variances of the error terms in two subsets of the data values. The test is analogous to a t-test. The test is robust with respect to departures from normality of the error terms. The data pairs are partitioned by a threshold value of X which is not one of the $X_1, X_2, \ldots, X_n$ values. Let n_1 be the number of data pairs with X-values less than the threshold value and n_2 be the number of data pairs with X-values greater than the threshold value so that $n = n_1 + n_2$. In addition, let

- e_{i1} be residual i for group 1,
- e_{i2} be residual i for group 2,
- $\tilde{e}_1$ be the sample median of the group 1 residuals,
- $\tilde{e}_2$ be the sample median of the group 2 residuals,
- $d_{i1} = |e_{i1} - \tilde{e}_1|$,
- $d_{i2} = |e_{i2} - \tilde{e}_2|$,
- $\bar{d}_1 = (1/n_1) \sum_{i=1}^{n_1} d_{i1}$, and
- $\bar{d}_2 = (1/n_2) \sum_{i=1}^{n_2} d_{i2}$.

The test statistic for the Brown–Forsythe test is

$$ t = \frac{\bar{d}_1 - \bar{d}_2}{s\sqrt{1/n_1 + 1/n_2}}, $$

where s^2 is the pooled sample variance

$$ s^2 = \frac{\sum_{i=1}^{n_1} \left(d_{i1} - \bar{d}_1\right)^2 + \sum_{i=1}^{n_2} \left(d_{i2} - \bar{d}_2\right)^2}{n - 2}. $$

The test statistic is approximately $t(n-2)$ when the population variances of the error terms in the two groups are equal n_1 and n_2 are large enough so that the dependency between the residuals is not too large. Write R code to compute the p-value for the Brown–Forsythe test for the `cars` data set using `speed` as the independent variable and `dist` as the dependent variable with a threshold value of 13.5 miles per hour.

3.12 Find the leverages for $n = 2$ data pairs in a simple linear regression model.

3.13 For a simple linear regression model with $X_i = i$, for $i = 1, 2, \ldots, n$, derive a formula for the leverage of the ith data pair.

3.14 Write R functions named `cooks.distance1`, `cooks.distance2`, and `cooks.distance3`, which calculate the Cook's distances for each of the n data pairs associated with the simple linear regression model

$$ Y = \beta_0 + \beta_1 X + \varepsilon $$

using the three formulas from Definition 3.3. The arguments for these three functions are the vector **x**, which contains the n values of the independent variable, and the vector **y**, which contains the n values of the dependent variable. Test your functions on the `Formaldehyde` data set which is built into R, with `carb` as the independent variable and `optden` as the dependent variable.

3.15   Make a scatterplot (with associated regression line) of the $n = 11$ data pairs in the third data set in Anscombe's quartet with the R commands

```
x = anscombe[ , 3]
y = anscombe[ , 7]
plot(x, y, xlim = c(4, 19), ylim = c(3, 13), pch = 16)
abline(lm(y ~ x))
```

Without doing any calculations,

(a) circle the point(s) with the largest leverage, and

(b) circle the point(s) with the largest Cook's distance.

3.16   What is the smallest and largest possible leverage?

3.17   Show that leverage is scale invariant. In other words, show that the leverages remain unchanged when the scale of the independent variable changes (for example, from centimeters to meters).

3.18   Use Monte Carlo simulation to estimate the probability that all of the Cook's distances are less than 1 for a simple linear regression model with normal error terms and the following parameter settings: $\beta_0 = 1$, $\beta_1 = 1/2$, $\sigma = 1$, $n = 10$, and $X_i = i$ for $i = 1, 2, \ldots, n$. Is this probability affected by changes is σ or n?

3.19   Use Monte Carlo simulation to draw empirical cumulative distribution functions of Cook's distances D_1, D_2, D_3, D_4, and D_5 for a simple linear regression model with the following parameter settings: $\beta_0 = 1$, $\beta_1 = 1/2$, $\sigma = 1$, $n = 10$, and $X_i = i$ for $i = 1, 2, \ldots, n$.

3.20   Consider a simple linear regression model with the independent variable X and the dependent variable Y having the same units (for example, centimeters). If the same linear transformation is applied to both X and Y so as to change their units (for example, from centimeters to meters), show that the Cook's distances remain unchanged.

3.21   Show that the row sums of the hat matrix are all equal to 1 for data pairs (X_1, Y_1), (X_2, Y_2), $\ldots$, (X_n, Y_n) in a simple linear regression model.

3.22   Perform a Monte Carlo simulation to provide convincing numerical evidence that

$$\frac{(\hat{\boldsymbol{\beta}} - \boldsymbol{\beta})' \mathbf{X}' \mathbf{X} (\hat{\boldsymbol{\beta}} - \boldsymbol{\beta})}{2 \cdot \text{MSE}} \sim F(2, n-2)$$

for a simple linear regression model with normal error terms of your choice. This result is used to establish a $100(1 - \alpha)\%$ confidence region for β_0 and β_1.

3.23   Show that the residuals $e_i = Y_i - \hat{Y}_i$ for $i = 1, 2, \ldots, n$, can be written in terms of the hat matrix $\mathbf{H}$ as

$$\mathbf{e} = (\mathbf{I} - \mathbf{H})\mathbf{Y}.$$

3.24   For the simple linear regression model with normal error terms, the variance–covariance matrix of $\hat{\boldsymbol{\beta}}$ is

$$\sigma^2 \left(\mathbf{X}' \mathbf{X} \right)^{-1}.$$

For data pairs (X_1, Y_1), (X_2, Y_2), $\ldots$, (X_n, Y_n), give an estimator for this matrix.

3.25 For the simple linear regression model, show that

$$\mathbf{X}_h' \left(\mathbf{X}'\mathbf{X} \right)^{-1} \mathbf{X}_h = \frac{1}{n} + \frac{(X_h - \bar{X})^2}{S_{XX}}.$$

3.26 For a simple linear regression model, show that the matrix equation

$$\mathbf{X}'\mathbf{X}\hat{\boldsymbol{\beta}} = \mathbf{X}'\mathbf{Y},$$

where

$$\mathbf{X} = \begin{bmatrix} 1 & X_1 \\ 1 & X_2 \\ \vdots & \vdots \\ 1 & X_n \end{bmatrix}, \qquad \mathbf{Y} = \begin{bmatrix} Y_1 \\ Y_2 \\ \vdots \\ Y_n \end{bmatrix}, \qquad \text{and} \qquad \hat{\boldsymbol{\beta}} = \begin{bmatrix} \hat{\beta}_0 \\ \hat{\beta}_1 \end{bmatrix},$$

corresponds to the normal equations given in Theorem 1.1 as

$$n\hat{\beta}_0 + \hat{\beta}_1 \sum_{i=1}^{n} X_i = \sum_{i=1}^{n} Y_i$$

$$\hat{\beta}_0 \sum_{i=1}^{n} X_i + \hat{\beta}_1 \sum_{i=1}^{n} X_i^2 = \sum_{i=1}^{n} X_i Y_i.$$

3.27 A multiple linear regression model is used to determine the relationship between the sales price of a home Y as a function of the two predictor variables: X_1, the number of square feet in the home, and X_2, the distance from downtown in miles. The fitted model is

$$Y = 170,024 + 133X_1 - 14,123X_2.$$

One home sells for \$314,159. Find the predicted sales price for a second home, which is the same size as the first but is ten miles further away from downtown that the first home.

3.28 The R built-in data frame named `swiss` contains a standardized fertility measure and five socio-economic indicators for 47 French-speaking provinces in Switzerland from about 1888.

(a) Using a forward stepwise regression with threshold $\alpha = 0.05$, determine a multiple linear regression model with a dependent variable Y, the standardized fertility measure, and the five associated potential independent variables.

(b) Using a backward stepwise regression with threshold $\alpha = 0.05$, determine a multiple linear regression model with a dependent variable Y, the standardized fertility measure, and the five associated potential independent variables.

(c) For one of the two final multiple linear regression models determined in parts (a) and (b), test the statistical significance of all possible interaction terms.

3.29 Show that when the independent variables X_1 and X_2 in a multiple linear regression model are uncorrelated, the estimator for $\hat{\beta}_1$ is the same for both the simple linear regression model involving just X_1 and Y and the multiple linear regression model involving X_1, X_2, and Y.

3.30 Consider a simple linear regression model that uses the weighted least squares estimation. When all of the weights $w_1, w_2, \ldots, w_n$ are equal, show that the weighted least squares normal equations reduce to the associated unweighted least squares normal equations.

3.31   "I first believed I was dreaming ... but it is absolutely certain and exact that the ratio which exists between the period times of any two planets is precisely the ratio of the 3/2th power of the mean distance" was the reaction of Johannes Kepler upon discovering the relationship

$$y = \beta x^{3/2}$$

as translated from *Harmonies of the World* by Kepler in 1619, where x is the distance between a planet and the sun and y is the period. Using the data from the Wikipedia webpage titled *Kepler's Laws of Planetary Motion*, the data values for the $n = 8$ planets are given below.

Planet	Semi-major axis (AU) x	Period (days) y
Mercury	0.38710	87.9693
Venus	0.72333	224.7008
Earth	1	365.2564
Mars	1.52366	686.9796
Jupiter	5.20336	4332.8201
Saturn	9.53707	10,775.599
Uranus	19.1913	30,687.153
Neptune	30.0690	60,190.03

The semi-major axes values are measured in Astronomical Units (AU).

(a) Make an appropriate scatterplot to visually assess whether a regression model is appropriate.

(b) Find the least squares point estimate for β.

(c) Perhaps fit a least squares model in another fashion.

(d) Interpret the value for $\hat{\beta}$.

(e) Find a 95% confidence interval for β.

3.32   Fit the quadratic regression function forced through the origin

$$Y = \beta_1 X^2 + \varepsilon,$$

to the data pairs in the `cars` data frame in R, where X is the speed of the car in miles per hour and Y is the stopping distance in feet.

3.33   Using an extreme value distribution as a link function, fit a regression function to the 2003 NFL field goal data from Section 3.8 and use the fitted model to predict that probability of success on a 38-yard field goal attempt.

Part II

SURVIVAL ANALYSIS

Chapter 4

Probability Models in Survival Analysis

One of the central aspects of survival analysis is the investigation of the probability distribution of a random variable T which has nonnegative support. In some settings, there are covariates that influence the probability distribution of T. In addition, the data collected on the random variable T is often right censored, which means that only a lower bound is available on the value of T. So there is a bit of a mental adjustment that needs to be made from probability theory, where X is usually used to denote a random variable, to survival analysis, where T is used to denote a random variable that can only assume nonnegative values. The choice of T is made because the random variable of interest is typically *time*. This chapter introduces probability models for T.

Four lifetime distribution representations that are commonly used to define the probability distribution of a random variable T are introduced in this chapter: the *survivor function*, the *probability density function*, the *hazard function*, and the *cumulative hazard function*. These four representations apply to both continuous (for example, the lifetime of a light bulb) and discrete (for example, the lifetime of the landing gear on an airplane) lifetimes. The survival time distribution of a drill bit, an automobile, a cat, and a recession are vastly different. One would certainly not want to use the same failure time distribution with identical parameters to model these diverse lifetimes. This chapter surveys two probability distributions (the *exponential* distribution as an example of a one-parameter distribution and the *Weibull* distribution as an example of a two-parameter distribution) that are commonly used to model lifetimes. The exponential distribution is central to survival analysis just as the normal distribution is central to classical statistics. After sections that survey other lifetime distributions and moment ratio diagrams, the *Cox proportional hazards model* is introduced. The proportional hazards model is appropriate for incorporating a vector of covariates that influence survival (for example, the turning speed and feed rate for a drill bit) into a lifetime model.

4.1 Lifetime Distribution Representations

The application areas associated with the probability distribution of the nonnegative random variable T are quite wide.

- In reliability engineering, T is typically the lifetime of a component or a system of components. Examples include the lifetime of a light bulb or the lifetime of a tennis racket.

- In biostatistics, T is typically the survival time of a patient. To be more specific, this might be the survival time of a patient after a particular type of surgery. More generally, the lifetime T could be the time between the end of radiation treatment for a particular cancer and the time the cancer recurs. In other words, T is the remission time.

- In actuarial science, T is often the lifetime of an insured individual in the life insurance industry. On the casualty and property side of actuarial science, T is often the lifetime of a structure or a vehicle.

- In sociology, T can model the duration of a strike, the duration of a marriage, or the duration of a business partnership. More generally, T might model the social distance between two strangers having a conversation.

- In economics, T can be the time between recessions or the absolute change in a stock market index from one year to the next.

- In systems engineering, T could be the length of time that it takes to screen a passenger at an airport. The time that a customer spends in a slow-moving queue before exiting the queue is another nonnegative random variable that might be of interest.

- In public policy, T could be the response time by emergency vehicles to a reported building fire. Alternatively, T could be the time for a released inmate to return to prison in a recidivism application within the criminal justice system.

- In library science, T could be the time that a book is checked out. The time between an interlibrary loan request and its fulfillment is another nonnegative random variable of interest to librarians.

- In meteorology, T could be the time between the formation of a tropical storm and the time it makes landfall. The time that a severe hurricane spends as a Category 5 hurricane is another nonnegative random variable of interest to meteorologists.

- In chemistry, T could be the length of time required to complete a chemical reaction. A chemist could also use T to denote the bond length between two atoms.

The long list given above is intended to highlight that survival analysis is a field that has a very wide range of applications. Although the letter T has been selected because it most often represents time, there are many applications in which it represents something other than time (for example, social distance or bond length).

When T represents time, T can be thought of as the time between two events. For this reason, this part of survival analysis is often referred to as *time-to-event* modeling. The time of purchase and the time of failure, for example, might be the two events for a manufactured product. Since the applications of survival analysis are wide, we will use the generic terms "failure" of an "item" when referring to the second of the two events.

This section introduces four functions that define the probability distribution of a continuous, nonnegative random variable T, the lifetime of an item. The four representations presented in this chapter are not the only ways to define the distribution of T. Other methods include the moment generating function $E\left[e^{sT}\right]$, the characteristic function $E\left[e^{isT}\right]$, the Mellin transform $E\left[T^s\right]$, the mean residual life function $E[T - t \mid T \geq t]$, and the reversed failure rate $f(t)/F(t)$. The four representations used here have been chosen because of their intuitive appeal, usefulness in problem solving, and popularity in the literature.

4.1.1 Survivor Function

The first lifetime distribution representation is the *survivor function* $S(t)$. The survivor function is the probability that an item is functioning at any time t.

Definition 4.1 The *survivor function* for a nonnegative random variable T is

$$S(t) = P(T \geq t) \qquad t \geq 0,$$

where $S(t) = 1$ for all $t < 0$.

A survivor function is also known as the reliability function [because $S(t)$ is the reliability of an item at time t] and the complementary cumulative distribution function [because $S(t) = 1 - F(t)$ for continuous random variables, where $F(t) = P(T \leq t)$ is the cumulative distribution function]. All survivor functions must satisfy three conditions:

$$S(0) = 1 \qquad \lim_{t \to \infty} S(t) = 0 \qquad S(t) \text{ is nonincreasing.}$$

There are two interpretations of the survivor function. First, $S(t)$ is the probability that an individual item is functioning at time t. Second, if there is a large population of items with identically distributed lifetimes, $S(t)$ is the expected fraction of the population that is functioning at time t.

The survivor function is useful for comparing the survival patterns of several populations of items. The graph in Figure 4.1 shows survivor functions $S_1(t)$ and $S_2(t)$, where $S_1(t)$ corresponds to population 1 and $S_2(t)$ corresponds to population 2. Since $S_1(t) \geq S_2(t)$ for all t values, it can be concluded that the items in population 1 are superior to those in population 2 with regard to survival.

The conditional survivor function, $S_{T \mid T \geq a}(t)$, is the survivor function of an item that is functioning at time a:

$$S_{T \mid T \geq a}(t) = \frac{P(T \geq t \text{ and } T \geq a)}{P(T \geq a)} = \frac{P(T \geq t)}{P(T \geq a)} = \frac{S(t)}{S(a)} \qquad t \geq a.$$

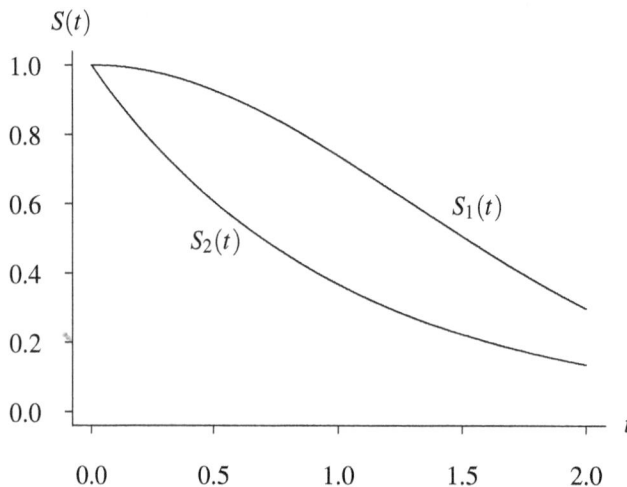

Figure 4.1: Two survivor functions.

Figure 4.2 shows the original survivor function $S(t)$ and the conditional survivor function $S_{T \mid T \geq a}(t)$ when $a = 0.5$. Since the conditional survivor function is rescaled by the factor $S(a)$, it has the same shape as the remaining portion of the original survivor function. The conditional survivor function is useful for comparing the survival experience of a group of items that has survived to time a. Examples include manufactured items surviving a burn-in test and cancer patients surviving 5 years after diagnosis and treatment. The conditional survivor function is of particular interest to actuaries. If a 37-year-old woman, for example, is purchasing a one-year term life insurance policy, an estimate of $S_{T \mid T \geq 37}(38)$ is required to determine an appropriate premium for the policy.

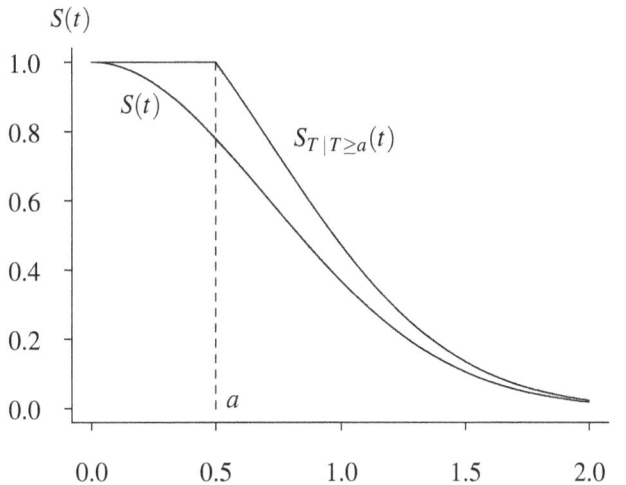

Figure 4.2: Conditional survivor function.

4.1.2 Probability Density Function

The second lifetime distribution representation is the familiar *probability density function*, which is defined as the negative of the derivative of the survivor function.

Definition 4.2 The *probability density function* of the nonnegative random variable T is

$$f(t) = -S'(t) \qquad t \geq 0,$$

where $S(t)$ is the survivor function and its derivative exists.

The probability density function has the probabilistic interpretation

$$f(t)\Delta t \cong P(t \leq T \leq t + \Delta t)$$

for small Δt values. Although the probability density function is not as effective as the survivor function for comparing the survival patterns of two populations, a graph of $f(t)$ indicates the likelihood of failure for any t. The probability of failure between times a and b is calculated by an integral:

$$P(a \leq T \leq b) = \int_a^b f(t)\,dt.$$

All probability density functions for lifetimes must satisfy two conditions:

$$\int_0^\infty f(t)\,dt = 1 \qquad\qquad f(t) \geq 0 \text{ for all } t \geq 0.$$

It is assumed that $f(t) = 0$ for all $t < 0$, which is consistent with our assumption that the random variable T is nonnegative. This assumption excludes distributions with negative support, such as the normal distribution. The probability density function shown in Figure 4.3 illustrates the relationship between the cumulative distribution function $F(t)$ and the survivor function $S(t)$ for a continuous lifetime. The area under $f(t)$ to the left of the arbitrary time t_0 is $F(t_0)$; the area under $f(t)$ to the right of t_0 is $S(t_0)$.

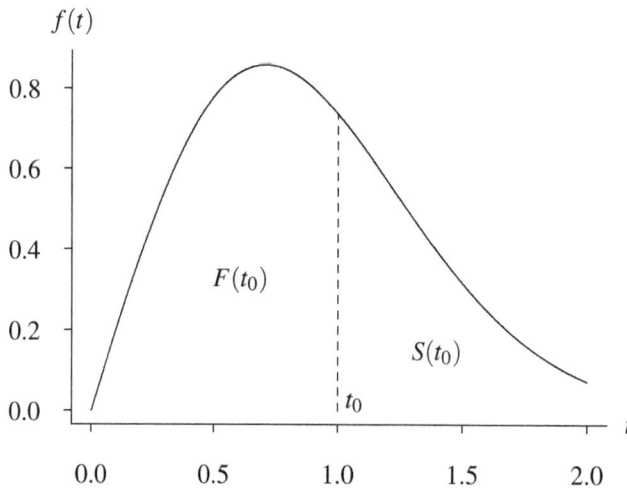

Figure 4.3: Relationship between the survivor and cumulative distribution functions.

4.1.3 Hazard Function

The *hazard function*, $h(t)$, is perhaps the most popular of the four representations for survival analysis due to its intuitive interpretation as the amount of *risk* associated with an item at time t. A second reason for its popularity is its usefulness in comparing the way risks change over time for several populations of items by plotting their hazard functions on a single axis. A third reason is that the hazard function is a special case of the intensity function for a nonhomogeneous Poisson process, which will be introduced in a subsequent chapter. A hazard function models the occurrence of one event, a failure, whereas the intensity function models the occurrence of a sequence of events over time. The hazard function goes by several aliases: in reliability it is also known as the hazard rate or failure rate; in actuarial science it is known as the force of mortality or force of decrement; in point process and extreme value theory it is known as the rate or intensity function; in vital statistics it is known as the age-specific death rate; and in economics its reciprocal is known as Mill's ratio.

    The hazard function can be derived using conditional probability. First, consider the probability of failure between t and $t + \Delta t$:

$$P(t \leq T \leq t + \Delta t) = \int_t^{t+\Delta t} f(\tau)d\tau = S(t) - S(t + \Delta t).$$

Conditioning on the event that the item is working at time t yields

$$P(t \leq T \leq t + \Delta t \,|\, T \geq t) = \frac{P(t \leq T \leq t + \Delta t)}{P(T \geq t)} = \frac{S(t) - S(t + \Delta t)}{S(t)}.$$

If this conditional probability is averaged over the interval $[t, \, t + \Delta t]$ by dividing by Δt, an average rate of failure is obtained:

$$\frac{S(t) - S(t + \Delta t)}{S(t)\Delta t}.$$

As $\Delta t \to 0$, this becomes the instantaneous failure rate, which is the hazard function

$$
\begin{aligned}
h(t) &= \lim_{\Delta t \to 0} \frac{S(t) - S(t + \Delta t)}{S(t)\Delta t} \\
&= -\frac{S'(t)}{S(t)} \\
&= \frac{f(t)}{S(t)} \qquad t \geq 0
\end{aligned}
$$

using the definition of the derivative from calculus. This forms the basis for the following definition.

Definition 4.3 The *hazard function* for a nonnegative random variable T is

$$h(t) = \frac{f(t)}{S(t)} \qquad t \geq 0,$$

where $f(t)$ is the probability density function and $S(t)$ is the survivor function.

Thus, the hazard function is the ratio of the probability density function to the survivor function. Using the previous derivation, a probabilistic interpretation of the hazard function is

$$h(t)\Delta t \cong P(t \leq T \leq t + \Delta t \,|\, T \geq t)$$

for small Δt values, which is a conditional version of the interpretation for the probability density function. All hazard functions must satisfy two conditions:

$$\int_0^\infty h(t)\,dt = \infty \qquad\qquad h(t) \geq 0 \text{ for all } t \geq 0.$$

Example 4.1 Consider the Weibull distribution defined by the survivor function

$$S(t) = e^{-(\lambda t)^\kappa} \qquad t \geq 0,$$

with positive scale parameter λ and positive shape parameter κ. Find the hazard function.

By differentiating the survivor function with respect to t and negating, the probability density function is

$$f(t) = \lambda \kappa (\lambda t)^{\kappa - 1} e^{-(\lambda t)^\kappa} \qquad t \geq 0,$$

so the hazard function is

$$h(t) = \frac{f(t)}{S(t)} = \lambda \kappa (\lambda t)^{\kappa - 1} \qquad t \geq 0.$$

Figure 4.4 illustrates the shape of the hazard function for the Weibull distribution with $\lambda = 1$ and three κ values. The hazard function is constant when $\kappa = 1$, increasing when $\kappa > 1$, and decreasing when $\kappa < 1$.

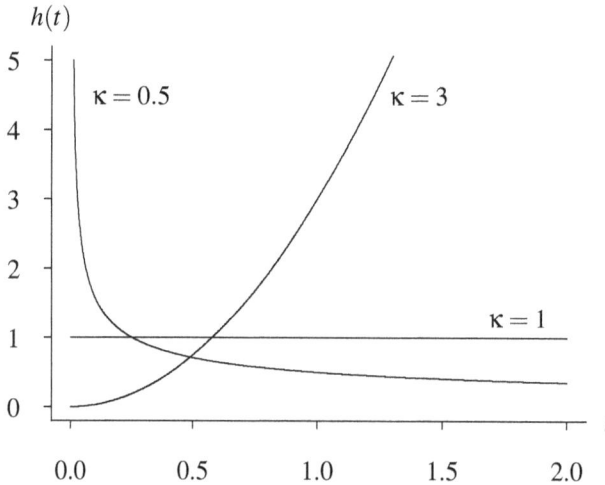

Figure 4.4: Hazard functions for the Weibull distribution.

The *units* on a hazard function are typically given in failures per unit time. In Example 4.1, if $\lambda = 0.01$, $\kappa = 1$, and time is measured in hours, then $h(t) = 0.01$ *failures per hour*. Manufactured items are often so reliable that to avoid hazard functions such as $h(t) = 0.00000128$ failures per hour the units are changed so that the hazard function may be expressed as $h(t) = 1.28$ failures per 10^6 hours. Another way to avoid writing too many leading zeroes is to change the units to years, where one year equals 8760 hours.

The shape of the hazard function indicates how an item ages. The intuitive interpretation of $h(t)$ as the amount of *risk* an item is subject to at time t implies that when the hazard function is larger the item is under greater risk of failure, and when the hazard function is smaller the item is under less risk of failure. The three hazard functions plotted in Figure 4.5 correspond to an increasing hazard function (labeled IFR for increasing failure rate), a decreasing hazard function (labeled DFR for decreasing failure rate), and a bathtub-shaped hazard function (labeled BT for bathtub-shaped failure rate).

The increasing hazard function is probably the most common situation of the three depicted in Figure 4.5. In this case, items are more likely to fail as time passes. In other words, items wear out or degrade with time. This is almost certainly the case with mechanical items that undergo wear or fatigue. It can also be the case in certain biomedical experiments. Let T, for example, be the time until a tumor appears after the injection of a substance into a laboratory animal. If the substance makes the tumor more likely to appear as time passes, then the hazard function associated with T is increasing. This leads to the formal definition of the IFR class. Notice the loose use of the term

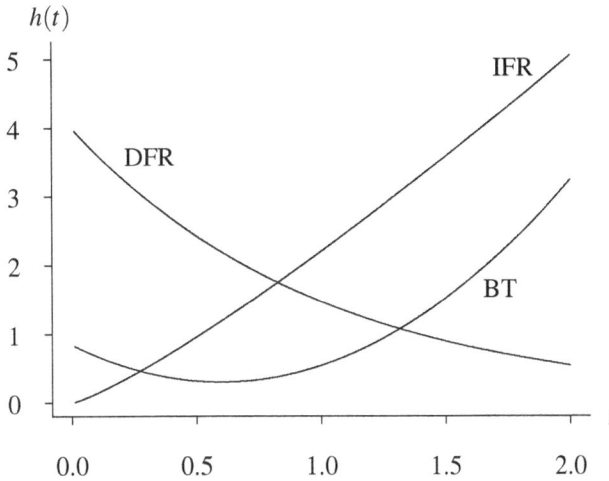

Figure 4.5: Common hazard function shapes.

increasing in the definition of the IFR class (because IFR distributions have *nondecreasing* hazard functions).

Definition 4.4 The distribution of a nonnegative random variable T is in the IFR (increasing failure rate) class if $h(t)$ is a nondecreasing function of t.

The second situation depicted in Figure 4.5, the decreasing hazard function, is less common. In this case, the item is less likely to fail as time passes. Items with this type of hazard function improve with time. Some metals work-harden through use and thus have increased strength as time passes. Another situation for which a decreasing hazard function might be appropriate for modeling is in working bugs out of computer programs. Bugs are more likely to appear initially, but the likelihood of them appearing decreases as time passes. This leads to the formal definition of the DFR class.

Definition 4.5 The distribution of a nonnegative random variable T is in the DFR (decreasing failure rate) class if $h(t)$ is a nonincreasing function of t.

The loose use of the term *increasing* in the definition of the IFR class and the term *decreasing* in the definition of the DFR class allows a distribution with a constant hazard function, the exponential distribution, to serve as a boundary between the two classes. The exponential distribution's hazard function $h(t) = \lambda$ for $t \geq 0$, is both nondecreasing and nonincreasing, so it belongs to both the IFR and DFR classes. As shown in the Venn diagram in Figure 4.6, this definition of IFR and DFR classifies all lifetime distributions into one of four sets: a constant hazard function (that is, the exponential distribution, which is the intersection of the IFR and DFR classes), strictly increasing hazard functions, strictly decreasing hazard functions, and other hazard functions (such as bathtub-shaped hazard functions).

The third situation depicted in Figure 4.5, a bathtub-shaped hazard function, occurs when the hazard function decreases initially and then increases as items age. Items improve initially and then degrade as time passes. One situation in which the bathtub-shaped hazard function arises is in the

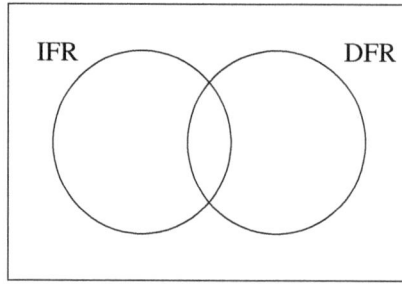

Figure 4.6: Venn diagram for IFR and DFR distribution classes.

lifetimes of manufactured items. Often manufacturing, design, or component defects cause early failures. The period in which these failures occur is sometimes called the *burn-in* period. If failure is particularly catastrophic, this part of the lifetime will often be consumed by the manufacturer in a controlled environment. The time value during which early failures have been eliminated may be valuable to a producer who is determining an appropriate warranty period. Once items pass through this early part of their lifetime, they have a fairly constant hazard function, and failures are equally likely to occur at any point in time. Finally, as items continue to age, the hazard function increases, resulting in *wear-out* failures. The three paragraphs that follow give examples of applications of the bathtub-shaped hazard function.

The bathtub-shaped hazard function can be envisioned for calculators; the burn-in period corresponds to the first few weeks of use when manufacturing, design, or component defects arise. Wear-out failures occur after a few years of use when the buttons are about ready to fall off. Failures due to calculators being dropped occur throughout the life of a calculator. If these failures are equally likely at any time, the hazard function will be increased by a constant that reflects the probability of dropping the calculator for all time values.

The bathtub-shaped hazard function also arises in the lifetimes of people. In this case, the early failures are known as *infant mortality* deaths and occur during the first few years of life. After this time, the hazard function has a very gentle increase through the teenage years and into adulthood. Finally, *old age* deaths occur during the later years of life. The magnitude of the hazard function depends on factors such as the standard of living and medical services available. Also, occupation (for example, flower arranger versus stunt man) and lifestyle (for example, eating habits, sleeping habits, smoking habits, stress level) affect the lifetime distribution of a person. The hazard function is used in actuarial science; the appropriate premium for a life insurance policy is based on probabilities associated with the lifetime distribution. The lowest life insurance premiums are usually for children who have survived the infant mortality part of their lifetimes.

There are dozens of other lifetime distribution classes beyond just the IFR and DFR classes. These include IFRA (increasing failure rate on average), DFRA (decreasing failure rate on average), IMRL (increasing mean residual life), and DMRL (decreasing mean residual life).

Care must be taken to differentiate between the hazard function for a *population* and the hazard function for an individual *item* under consideration. To use human lifetimes as an illustration, consider the following question: do two healthy 11-year-old boys living in the same town necessarily have the same hazard function? The answer is no. The reason is that all people are born with genetic predispositions that will influence their risk as they age. So, although a hazard function exists for all 11-year-old boys living in that particular town, it is an aggregate hazard function representing the population, and individual boys may be at increased or decreased risk. This is why life insurance companies typically require a medical exam to determine whether an individual is at higher risk than the rest of the population. The common assumption in most probabilistic models and statistical analyses is that of mutually independent and identically distributed random variables, which in this case are lifetimes. This assumption is not always valid in survival analysis applications because items are often manufactured in diverse conditions (for example, different temperatures or raw materials).

4.1.4 Cumulative Hazard Function

The fourth lifetime distribution representation, the *cumulative hazard function*, is defined as the integral of the hazard function.

Definition 4.6 The *hazard function* for a nonnegative random variable T is

$$H(t) = \int_0^t h(\tau)\,d\tau \qquad t \geq 0,$$

where $h(t)$ is the hazard function.

Whereas the hazard function reflects the risk pattern associated with an item over time, the cumulative hazard function gives the accumulated risk at time t. Similar to the way a cumulative distribution function accumulates probability, the cumulative hazard function $H(t)$ accumulates the risk from time 0 to time t. All cumulative hazard functions must satisfy three conditions:

$$H(0) = 0 \qquad \lim_{t \to \infty} H(t) = \infty \qquad H(t) \text{ is nondecreasing.}$$

The cumulative hazard function is valuable for random variate generation in Monte Carlo simulation, implementing certain procedures in statistical inference, and defining certain distribution classes (for example, the IFRA class).

The four lifetime distribution representations presented here are equivalent in the sense that each completely specifies a lifetime distribution. In addition, any one lifetime distribution representation implies the other three. Algebra and calculus can be used to find one lifetime distribution representation given that another is known. For example, if the survivor function is known, the cumulative hazard function can be determined by

$$H(t) = \int_0^t h(\tau)\,d\tau = \int_0^t \frac{f(\tau)}{S(\tau)}\,d\tau = -\ln S(t),$$

where ln is the natural logarithm (log base e). The from–to matrix in Table 4.1 shows that any of the three other lifetime distribution representations (given by the columns) can be found if one of the representations (given by the rows) is known. It is assumed that the support of the lifetime T is $[0, \infty)$ in Table 4.1.

Example 4.2 Given $h(t) = 18t$ for $t \geq 0$, find $f(t)$.

Using the $\big(h(t), f(t)\big)$ element of the from–to matrix in Table 4.1,

$$f(t) = h(t)e^{-\int_0^t h(\tau)\,d\tau}$$

$$= 18t e^{-\int_0^t 18\tau\,d\tau}$$

$$= 18t e^{-9t^2} \qquad t \geq 0,$$

which is a special case of the Weibull distribution with $\lambda = 3$ and $\kappa = 2$.

	$f(t)$	$S(t)$	$h(t)$	$H(t)$
$f(t)$	•	$\int_t^\infty f(\tau)d\tau$	$\dfrac{f(t)}{\int_t^\infty f(\tau)d\tau}$	$-\ln\left[\int_t^\infty f(\tau)d\tau\right]$
$S(t)$	$-S'(t)$	•	$\dfrac{-S'(t)}{S(t)}$	$-\ln S(t)$
$h(t)$	$h(t)\,e^{-\int_0^t h(\tau)d\tau}$	$e^{-\int_0^t h(\tau)d\tau}$	•	$\int_0^t h(\tau)d\tau$
$H(t)$	$H'(t)\,e^{-H(t)}$	$e^{-H(t)}$	$H'(t)$	•

Table 4.1: Lifetime distribution representation relationships.

4.2 Exponential Distribution

Just as the normal distribution plays a pivotal role in classical statistics because of the central limit theorem, the exponential distribution plays a pivotal role in survival analysis because it is the only continuous distribution with a constant hazard function. The exponential distribution has a single positive scale parameter λ, often called the *failure rate* by reliability engineers.

Definition 4.7 The four lifetime distribution representations associated with a random variable T having the exponential distribution with positive rate parameter λ are

$$S(t) = e^{-\lambda t} \qquad f(t) = \lambda e^{-\lambda t} \qquad h(t) = \lambda \qquad H(t) = \lambda t$$

for $t \geq 0$. Symbolically, this is written as $T \sim \text{exponential}(\lambda)$.

The four lifetime distribution representations are plotted in Figure 4.7 for $\lambda = 1$ and $\lambda = 2$. Two-parameter distributions, which are more complex but can model a wider variety of situations, are presented in subsequent sections.

The centrality, tractability, and importance of the exponential distribution make it a key probability distribution to know well. In that light, this section surveys several probabilistic properties

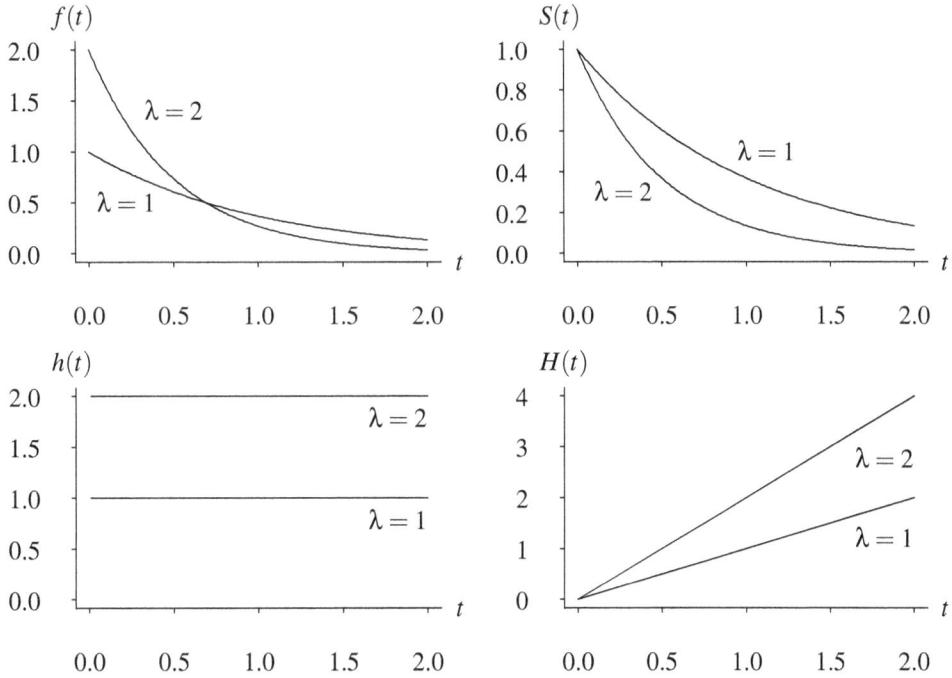

Figure 4.7: Lifetime distribution representations for the exponential distribution.

of the exponential distribution that are useful in understanding how it is unique and when it should be applied. In all the properties, it is assumed that the nonnegative lifetime T has the exponential distribution with parameter λ.

Theorem 4.1 (memoryless property) If $T \sim$ exponential(λ), then

$$P(T \geq t) = P(T \geq t + s \,|\, T \geq s) \qquad t \geq 0; \; s \geq 0.$$

Proof The probability of surviving to time $t + s$ conditioned on survival to time s is

$$
\begin{aligned}
P(T \geq t + s \,|\, T \geq s) &= \frac{P(T \geq t + s)}{P(T \geq s)} \\
&= \frac{S(t + s)}{S(s)} \\
&= \frac{e^{-\lambda(t+s)}}{e^{-\lambda s}} \\
&= e^{-\lambda t} \\
&= S(t) \\
&= P(T \geq t)
\end{aligned}
$$

for all $t \geq 0$ and $s \geq 0$. $\qquad\qquad\square$

As shown in Figure 4.8 for $\lambda = 1$ and $s = 0.5$, this result indicates that the conditional survivor function for the lifetime of an item that has survived to time s is identical to the survivor function for the lifetime of a brand new item. This used-as-good-as-new assumption is very strong. Consider, for example, whether the exponential distribution should be used to model the lifetime of a candle with an expected burning time of 5 hours. If several candles are sampled and burned, we could imagine a bell-shaped histogram for candle lifetimes, centered around 5 hours. The exponential lifetime model is certainly *not* appropriate in this case, because a candle that has burned for 4 hours does not have the same remaining lifetime distribution as that of a brand new candle. The exponential distribution would only be appropriate for candle lifetimes if the remaining lifetime of a used candle is identical to the lifetime of a new candle. An electrical component for which the exponential lifetime assumption might be justified is a fuse. A fuse is designed to fail when there is a power surge that causes the fuse to fail, resulting in a blown fuse which must be replaced. Assuming that the fuse does not undergo any weakening or degradation over time and that power surges that cause failure occur at a constant rate over time, the exponential lifetime assumption is appropriate, and a used fuse that has not failed is as good as a new one in terms of longevity.

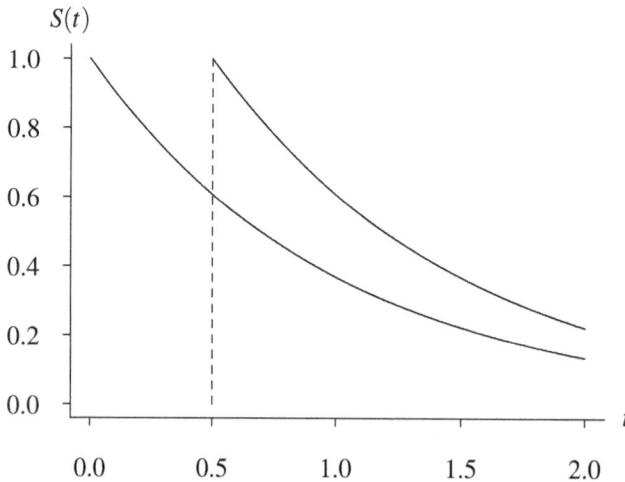

Figure 4.8: The memoryless property of the exponential distribution.

The exponential distribution should be applied judiciously because the memoryless property restricts its applicability. It is often misapplied for the sake of simplicity because the statistical techniques for the exponential distribution are particularly tractable, or because small sample sizes do not support more than a one-parameter distribution.

The exponential distribution is the only continuous distribution with the memoryless property. The exponential distribution is the only continuous lifetime distribution for which the conditional lifetime distribution of a used item is identical to the original lifetime distribution. The only discrete distribution with the memoryless property is the geometric distribution.

Theorem 4.2 If T is a continuous nonnegative random variable with cumulative hazard function $H(t)$, then $H(T) \sim$ exponential(1).

Proof The survivor function for λT is

$$P(\lambda T \geq t) = P(T \geq t/\lambda)$$
$$= e^{-\lambda(t/\lambda)}$$
$$= e^{-t} \qquad t \geq 0,$$

so λT has survivor function e^{-t}, which is exponential(1). $\qquad\qquad$ □

This property is mathematically equivalent to the probability integral transformation, which states that $F(T) \sim U(0, 1)$, resulting in the inverse-cdf technique for generating random variates for Monte Carlo simulation: $T \leftarrow F^{-1}(U)$, where $U \sim U(0, 1)$. Using Theorem 4.2, random lifetime variates are generated by

$$T \leftarrow H^{-1}\big(-\ln(1-U)\big)$$

because $-\ln(1-U)$ is a unit exponential random variate. Random lifetimes generated in this fashion are generated by the *cumulative hazard function technique*.

Example 4.3 Assuming that the failure time of an item has the Weibull distribution with survivor function

$$S(t) = e^{-(\lambda t)^\kappa} \qquad t \geq 0$$

for positive scale parameter λ and positive shape parameter κ, find an equation to convert $U(0, 1)$ random numbers to Weibull random variates.

The cumulative hazard function for the Weibull distribution is

$$H(t) = -\ln S(t) = (\lambda t)^\kappa \qquad t \geq 0,$$

which has inverse

$$H^{-1}(y) = \frac{y^{1/\kappa}}{\lambda} \qquad y \geq 0.$$

Weibull random variates can be generated by

$$T \leftarrow \frac{1}{\lambda} \left[-\ln(1-U)\right]^{1/\kappa},$$

where U is uniformly distributed between 0 and 1.

Figure 4.9 illustrates the geometry associated with generating a variate from the cumulative hazard function. The value of $-\ln(1-U)$, the unit exponential random variate, is indicated on the vertical axis, and the corresponding random variate T is indicated on the horizontal axis.

The next result gives a general expression for the sth moment of an exponential random variable.

Theorem 4.3 If $T \sim$ exponential(λ), then

$$E[T^s] = \frac{\Gamma(s+1)}{\lambda^s} \qquad s > -1,$$

where $\Gamma(\alpha) = \displaystyle\int_0^\infty x^{\alpha-1} e^{-x} dx.$

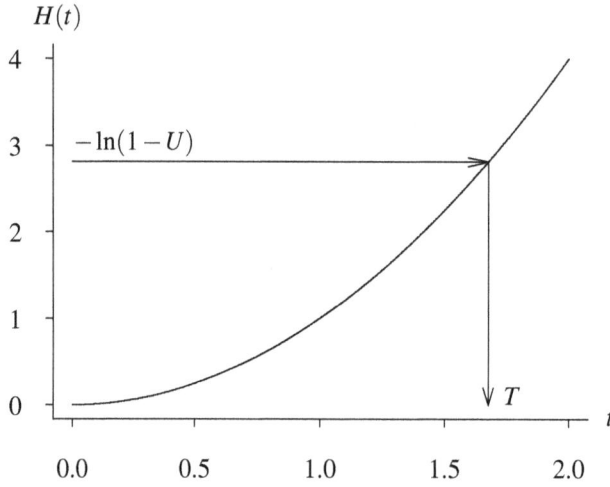

Figure 4.9: Generating a random variate by the inverse cumulative hazard function technique.

Proof Using the substitution $x = \lambda t$, the expected value of T^s is

$$E[T^s] = \int_0^\infty t^s \lambda e^{-\lambda t}\, dt$$
$$= \lambda^{-s} \int_0^\infty x^s e^{-x}\, dx$$
$$= \lambda^{-s} \Gamma(s+1) \qquad s > -1. \qquad \square$$

When s is a nonnegative integer, this expression reduces to $E[T^s] = s!/\lambda^s$. By setting $s = 1, 2, 3$, and 4, the population mean, variance, coefficient of variation, skewness, and kurtosis can be obtained:

$$E[T] = \frac{1}{\lambda} \qquad V[T] = \frac{1}{\lambda^2} \qquad \gamma = 1 \qquad \gamma_3 = 2 \qquad \gamma_4 = 9.$$

Since the coefficient of variation of an exponential random variable is 1, a quick check for exponentiality for a data set is to see if the ratio of the sample standard deviation to the sample mean is approximately 1. The histogram of the sample values should also have the appropriate shape.

Theorem 4.4 (self-reproducing) If $T_1, T_2, \ldots, T_n$ are mutually independent random variables, $T_i \sim$ exponential(λ_i), for $i = 1, 2, \ldots, n$, and $T = \min\{T_1, T_2, \ldots, T_n\}$, then

$$T \sim \text{exponential}\left(\sum_{i=1}^n \lambda_i\right).$$

Proof The survivor function for T is

$$S_T(t) = P(T \geq t)$$

$$
\begin{aligned}
&= P(\min\{T_1, T_2, \ldots, T_n\} \geq t) \\
&= P(T_1 \geq t, T_2 \geq t, \ldots, T_n \geq t) \\
&= P(T_1 \geq t) P(T_2 \geq t) \ldots P(T_n \geq t) \\
&= e^{-\lambda_1 t} e^{-\lambda_2 t} \ldots e^{-\lambda_n t} \\
&= e^{-\sum_{i=1}^{n} \lambda_i t} \qquad\qquad t \geq 0.
\end{aligned}
$$

Therefore, $T \sim$ exponential $\left(\sum_{i=1}^{n} \lambda_i\right)$. $\qquad\qquad \square$

This result indicates that the minimum of n exponential random lifetimes also has the exponential distribution. This is important in two applications. First, if n components, each with mutually independent exponential times to failure, are arranged in a series system, then the distribution of the system failure time is also exponential with a failure rate equal to the sum of the component failure rates. Second, when there are several mutually independent, exponentially distributed *causes* of failure competing for the lifetime of an item (for example, failing by open or short circuit for an electronic item or death by various diseases for a human being), then the lifetime can be modeled as the minimum of the individual lifetimes from each cause of failure.

Theorem 4.5 If $T_1, T_2, \ldots, T_n$ are mutually independent and identically distributed exponential(λ) random variables, then

$$
2\lambda \sum_{i=1}^{n} T_i \sim \chi^2(2n),
$$

where $\chi^2(2n)$ denotes the chi-square distribution with $2n$ degrees of freedom.

Proof Since $T_1, T_2, \ldots, T_n$ are mutually independent and identically distributed exponential(λ) random variables,

$$
\sum_{i=1}^{n} T_i \sim \text{Erlang}(\lambda, n).
$$

Furthermore

$$
\lambda \sum_{i=1}^{n} T_i \sim \text{Erlang}(1, n),
$$

which implies that

$$
2\lambda \sum_{i=1}^{n} T_i \sim \chi^2(2n). \qquad\qquad \square
$$

This property is useful for determining a confidence interval for λ based on a data set of n mutually independent exponential(λ) lifetimes. With probability $1 - \alpha$,

$$
\chi^2_{2n,\, 1-\alpha/2} < 2\lambda \sum_{i=1}^{n} T_i < \chi^2_{2n,\, \alpha/2},
$$

where the left- and right-hand sides of this inequality are the $\alpha/2$ and $1 - \alpha/2$ fractiles of the chi-square distribution with $2n$ degrees of freedom. This notation is illustrated in Figure 4.10, with the three areas under the probability density function of the chi-square random variable plotted on the

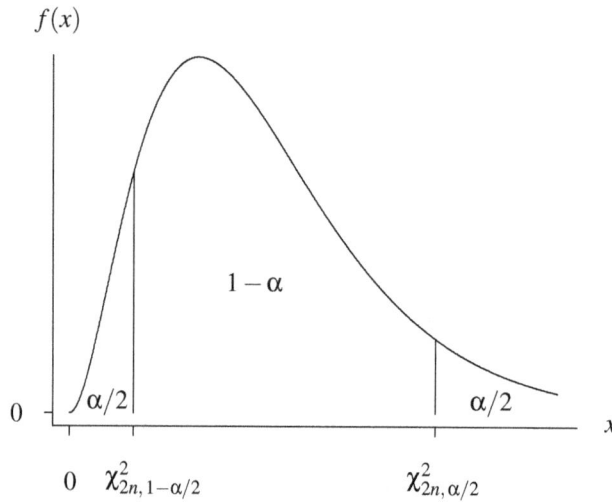

Figure 4.10: Fractiles of the chi-square distribution with $2n$ degrees of freedom.

graph. Rearranging this expression yields an exact $100(1 - \alpha)\%$ two-sided confidence interval for λ:

$$\frac{\chi^2_{2n,\,1-\alpha/2}}{2\sum\limits_{i=1}^{n} T_i} < \lambda < \frac{\chi^2_{2n,\,\alpha/2}}{2\sum\limits_{i=1}^{n} T_i}.$$

Theorem 4.6 If $T \sim$ exponential(λ), then $\lceil T \rceil \sim$ Geometric $\left(1 - e^{-\lambda}\right)$.

Proof Let $N = \lceil T \rceil$. The ceiling function applied to the continuous random variable T means that the random variable N is discrete. Since the support of T is $t \geq 0$, the support of N is $n = 1, 2, \dots$. The probability mass function of N is

$$f_N(n) = P(N = n) = P(n-1 \leq T < n) = \int_{n-1}^{n} \lambda e^{-\lambda t}\, dt = \left[-e^{-\lambda t}\right]_{n-1}^{n} = e^{-(n-1)\lambda} - e^{-n\lambda}$$

for $n = 1, 2, \dots$. Equivalently,

$$f_N(n) = e^{-(n-1)\lambda}\left(1 - e^{-\lambda}\right) \qquad n = 1, 2, \dots,$$

which can be recognized as a Geometric $\left(1 - e^{-\lambda}\right)$ random variable with support beginning at 1. $\qquad\square$

This property involves the only two probability distributions with the memoryless property. The ceiling function returns the next highest integer associated with the continuous failure time T. A modeling situation in which this property might be of interest occurs when an item with an exponential(λ) failure time distribution is placed on test at noon on a particular day. (The item could be a manufactured item such as a light bulb in the reliability setting or a subject such as a laboratory animal with cancer in a biostatistics setting.) Rather than continuously monitoring the item in order to observe its failure time T, you instead check the item for failure each subsequent day at noon. If time is measured in days, then the day number in which you observe failure is $\lceil T \rceil$, which the property indicates has a geometric distribution. Data collected in this fashion is known as *current status data*. A time to failure is known as *interval censored* when only a lower bound and upper bound are known on a failure time, as is the case in the scenario presented here.

The exponential distribution, for which the item under study does not age in a probabilistic sense, is the simplest of the lifetime models. Three are many other important properties of the exponential distribution in addition to those presented in this section. The two-parameter Weibull distribution, which includes the exponential distribution as a special case, is presented next. It is more flexible for modeling, although more complex mathematically.

4.3 Weibull Distribution

The exponential distribution is limited in applicability because of the memoryless property. The assumption that a lifetime has a constant failure rate is often too restrictive or inappropriate. Mechanical items, for instance, typically degrade over time and hence their lifetimes are more likely to follow a probability distribution with a strictly increasing hazard function. The Weibull distribution, named after Swedish mathematician Waloddi Weibull, is a generalization of the exponential distribution that is appropriate for modeling lifetimes having constant, strictly increasing, or strictly decreasing hazard functions.

Definition 4.8 The four lifetime distribution representations associated with a random variable T having the Weibull distribution with positive scale parameter λ and positive shape parameter κ are

$$S(t) = e^{-(\lambda t)^{\kappa}} \qquad f(t) = \kappa \lambda^{\kappa} t^{\kappa-1} e^{-(\lambda t)^{\kappa}} \qquad h(t) = \kappa \lambda^{\kappa} t^{\kappa-1} \qquad H(t) = (\lambda t)^{\kappa}$$

for $t \geq 0$. Symbolically, this is written as $T \sim \text{Weibull}(\lambda, \kappa)$.

The first four lifetime distribution representations for the Weibull(λ, κ) distribution are for $t \geq 0$, where $\lambda > 0$ and $\kappa > 0$ are the scale and shape parameters of the distribution. The hazard function approaches zero from infinity for $\kappa < 1$, is constant for $\kappa = 1$, the exponential case, and increases from zero for $\kappa > 1$. One other special case occurs when $\kappa = 2$, commonly known as the *Rayleigh distribution*, which has a linear hazard function with slope $2\lambda^2$. When $3 < \kappa < 4$, the probability density function resembles that of a normal probability density function, and the mode and median of the distribution are equal when $\kappa \cong 3.26$. The R code for plotting these lifetime distribution representations for $\lambda = 1$ and $\kappa = 0.5, 1, 2, 3$ is given below. The by argument in the call to the seq function controls the spacing between the t values plotted. The matplot function plots several functions on a single plot simultaneously.

```
par(mfrow = c(2, 2))
kappa = c(0.5, 1, 2, 3)
t     = seq(0, 1.5, by = 0.05)
f     = cbind(dweibull(t, kappa[1]), dweibull(t, kappa[2]),
              dweibull(t, kappa[3]), dweibull(t, kappa[4]))
matplot(t, f, type = "l")
S     = cbind(1 - pweibull(t, kappa[1]), 1 - pweibull(t, kappa[2]),
              1 - pweibull(t, kappa[3]), 1 - pweibull(t, kappa[4]))
matplot(t, S, type = "l")
```

```
h      = f / S
matplot(t, h, type = "l")
H      = -log(S)
matplot(t, H, type = "l")
```

These four functions are plotted in Figure 4.11 for $\lambda = 1$ and $\kappa = 0.5, 1, 2, 3$.

The *characteristic life* of the Weibull distribution is a special fractile defined by $t_c = 1/\lambda$. All Weibull survivor functions pass through the point $(1/\lambda, 1/e)$, regardless of the value of κ, as shown in Figure 4.11 for $\lambda = 1$. Also, since $H(t) = -\ln S(t)$, all Weibull cumulative hazard functions pass through the point $(1/\lambda, 1)$, regardless of the value of κ.

There are several ways to *parameterize* the Weibull distribution. The previous two paragraphs introduced one such parameterization with a scale parameter λ and a shape parameter κ. Another common way to parameterize the Weibull distribution is with the survivor function

$$S(t) = e^{-(t/\eta)^\beta} \qquad t \geq 0,$$

where η is a positive scale parameter and β is a positive shape parameter. This is the parameterization used in R. Comparing the two survivor functions, it is clear that the two shape parameters κ and β play identical roles, and the two scale parameters λ and η are reciprocals. Both parameterizations correspond to the Weibull distribution, but some careful bookkeeping is necessary to account for the different roles of the various parameters. The version of the Weibull distribution with parameters λ and κ will be used consistently throughout this book.

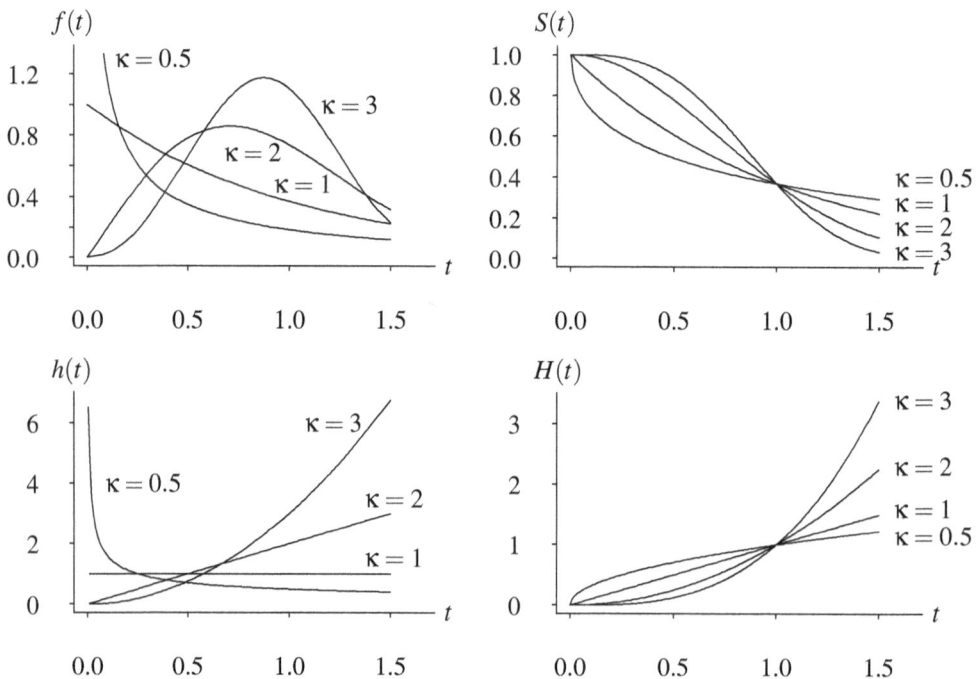

Figure 4.11: Lifetime distribution representations for the Weibull distribution for $\lambda = 1$.

Moments for the Weibull distribution are not as mathematically tractable as those for the exponential distribution. Using the substitution $u = (\lambda t)^\kappa$, the rth central moment about the origin is

$$
\begin{aligned}
E[T^r] &= \int_0^\infty t^r f(t)\, dt \\
&= \int_0^\infty \kappa \lambda^\kappa t^{\kappa+r-1} e^{-(\lambda t)^\kappa}\, dt \\
&= \frac{1}{\lambda^r} \int_0^\infty u^{r/\kappa} e^{-u}\, du \\
&= \frac{1}{\lambda^r} \Gamma\left(1 + \frac{r}{\kappa}\right) \\
&= \frac{r}{\kappa \lambda^r} \Gamma\left(\frac{r}{\kappa}\right)
\end{aligned}
$$

for $r = 1, 2, \ldots$. Using this expression for $E[T^r]$ with $r = 1$ and $r = 2$ and the shortcut formula for the population variance, the population mean and variance of a Weibull(λ, κ) random variable are

$$
E[T] = \frac{1}{\lambda} \Gamma\left(1 + \frac{1}{\kappa}\right) = \frac{1}{\lambda \kappa} \Gamma\left(\frac{1}{\kappa}\right)
$$

and

$$
V[T] = \frac{1}{\lambda^2}\left\{ \Gamma\left(1 + \frac{2}{\kappa}\right) - \left[\Gamma\left(1 + \frac{1}{\kappa}\right)\right]^2 \right\} = \frac{1}{\lambda^2}\left\{ \frac{2}{\kappa}\Gamma\left(\frac{2}{\kappa}\right) - \left[\frac{1}{\kappa}\Gamma\left(\frac{1}{\kappa}\right)\right]^2 \right\}.
$$

The associated coefficient of variation is

$$
\gamma = \frac{\sigma}{\mu} = \frac{\left\{ \frac{2}{\kappa}\Gamma\left(\frac{2}{\kappa}\right) - \left[\frac{1}{\kappa}\Gamma\left(\frac{1}{\kappa}\right)\right]^2 \right\}^{1/2}}{\frac{1}{\kappa}\Gamma\left(\frac{1}{\kappa}\right)}.
$$

Using this expression for $E[T^r]$ with $r = 3$ and $r = 4$ yields the population skewness and kurtosis:

$$
\gamma_3 = \left\{ \frac{2}{\kappa}\Gamma\left(\frac{2}{\kappa}\right) - \left[\frac{1}{\kappa}\Gamma\left(\frac{1}{\kappa}\right)\right]^2 \right\}^{-3/2} \left\{ \frac{3}{\kappa}\Gamma\left(\frac{3}{\kappa}\right) - \frac{6}{\kappa^2}\Gamma\left(\frac{1}{\kappa}\right)\Gamma\left(\frac{2}{\kappa}\right) + 2\left[\frac{1}{\kappa}\Gamma\left(\frac{1}{\kappa}\right)\right]^3 \right\},
$$

$$
\begin{aligned}
\gamma_4 = &\left\{ \frac{2}{\kappa}\Gamma\left(\frac{2}{\kappa}\right) - \left[\frac{1}{\kappa}\Gamma\left(\frac{1}{\kappa}\right)\right]^2 \right\}^{-2} \left\{ \frac{4}{\kappa}\Gamma\left(\frac{4}{\kappa}\right) - \frac{12}{\kappa^2}\Gamma\left(\frac{1}{\kappa}\right)\Gamma\left(\frac{3}{\kappa}\right) \right. \\
&\left. + \frac{12}{\kappa^3}\left[\Gamma\left(\frac{1}{\kappa}\right)\right]^2 \Gamma\left(\frac{2}{\kappa}\right) - \frac{3}{\kappa^4}\left[\Gamma\left(\frac{1}{\kappa}\right)\right]^4 \right\}.
\end{aligned}
$$

The next example applies the formulas developed thus far for the Weibull distribution to the lifetime of a spring.

Example 4.4 The lifetime of a certain type of spring used continuously under known operating conditions has the Weibull distribution with $\lambda = 0.0014$ and $\kappa = 1.28$, where time is measured in hours. (Estimating the parameters for the Weibull distribution from a data set is introduced in the next chapter; the parameters are assumed to be known constants in this example.)

- Find the population mean time to failure.
- Find the probability that a new spring will operate for 400 hours.
- Find the probability that a spring that has operated for 200 hours without failure will operate another 400 hours.

The population mean time to failure is

$$\mu = E[T] = \frac{1}{(0.0014)(1.28)} \Gamma\left(\frac{1}{1.28}\right) \cong 661.8 \text{ hours.}$$

The probability that a new spring will operate for 400 hours is

$$S(400) = e^{-[(0.0014)(400)]^{1.28}} \cong 0.6222.$$

To calculate the conditional probability that a used spring lasts another 400 hours requires a conditional survivor function. The conditional survivor function for a spring that has operated for 200 hours is

$$S_{T\,|\,T\geq 200}(t) = \frac{S(t)}{S(200)} = \frac{e^{-(0.0014t)^{1.28}}}{e^{-[(0.0014)(200)]^{1.28}}} \qquad t \geq 200.$$

So the conditional probability that a spring that has operated for 200 hours lasts another 400 hours is $S_{T\,|\,T\geq 200}(600) \cong 0.5469$, as illustrated in Figure 4.12. It is not surprising that this conditional survival probability is slightly lower than the probability that a new spring survives 400 hours. Since the shape parameter $\kappa = 1.28$ is greater than 1, the spring's lifetime is in the IFR class, which means that the spring degrades over time.

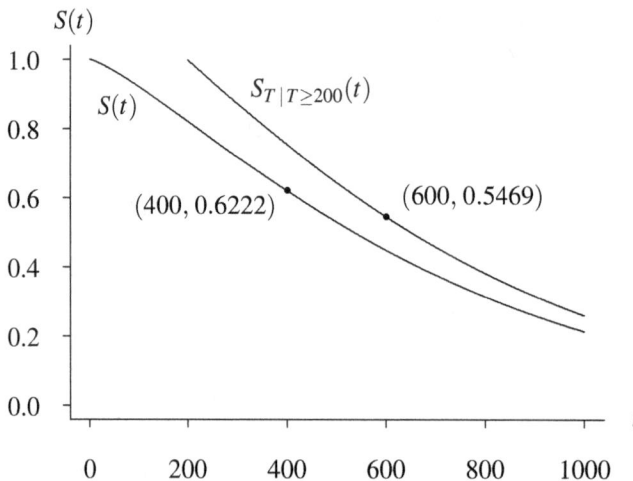

Figure 4.12: The original and conditional survivor functions.

Finding fractiles of the Weibull distribution requires only a few steps of algebra. The pth fractile of a Weibull(λ, κ) random variable, denoted by t_p, can be found by equating the cumulative distribution of T to p and solving for t_p:

$$p = 1 - e^{-(\lambda t_p)^\kappa} \qquad \Rightarrow \qquad t_p = \frac{1}{\lambda}\left[-\ln(1-p)\right]^{1/\kappa}.$$

These fractiles can be useful for establishing warranty periods or setting burn-in periods for manufactured items.

The Weibull distribution has the self-reproducing property. If $T_1, T_2, \ldots, T_n$ are mutually independent component lifetimes having the Weibull distribution with the same shape parameters, then the minimum of these values has the Weibull distribution. More specifically, if $T_i \sim$ Weibull(λ_i, κ) for $i = 1, 2, \ldots, n$, then min $\{T_1, T_2, \ldots, T_n\} \sim$ Weibull $\left((\sum_{i=1}^n \lambda_i^\kappa)^{1/\kappa}, \kappa\right)$.

4.4   Other Lifetime Distributions

Although the exponential and Weibull distributions are popular lifetime models, they are limited in their modeling capability. For example, if it were determined that an item had a bathtub-shaped hazard function, none of these three models would be appropriate unless a piecewise model over time segments of the lifetime were used. Several other models that may be used to describe the distribution of a continuous lifetime T are surveyed in this section.

The distributions for the nonnegative random variable T described here have three types of parameters: location parameters, denoted by a, b, and μ; scale parameters, denoted by λ and σ; and shape parameters, denoted by κ, γ, and δ. All distributions have support on $[0, \infty)$ except for the uniform and Pareto distributions.

The one-parameter lifetime models that are summarized in this section are the exponential(λ) and Muth(κ) distributions. The two-parameter lifetime models summarized here are the Weibull(λ, κ), gamma(λ, κ), uniform(a, b), log normal(μ, σ), log logistic(λ, κ), inverse Gaussian(λ, μ), exponential power(λ, κ), Pareto(λ, κ), and Gompertz(δ, κ) distributions. The three-parameter lifetime models summarized here are the Makeham(δ, κ, γ), IDB(δ, κ, γ), and generalized Pareto(δ, κ, γ) distributions. The n-parameter lifetime models summarized here are the hypoexponential$(\lambda_1, \lambda_2, \ldots, \lambda_n)$ and hyperexponential$(\lambda_1, \lambda_2, \ldots, \lambda_n)$ distributions.

The shapes of the lifetime distribution representations, particularly the hazard function, are useful in determining the appropriate distribution to use to model a lifetime. One-, two-, three-, and n-parameter lifetime distributions are described consecutively in the following paragraphs.

4.4.1   Some One-Parameter Lifetime Models

The one-parameter lifetime distributions defined here have their $f(t)$, $S(t)$, $h(t)$, and $H(t)$ functions given in Table 4.2. The simplest lifetime distribution is the *exponential distribution*, with a positive scale parameter λ. As indicated in Section 4.2, it is the only continuous distribution with a constant failure rate.

Muth developed a distribution with a single shape parameter κ $(0 < \kappa \le 1)$. The *Muth distribution* is asymptotically equivalent to the unit exponential distribution as $\kappa \to 0$ and has a hazard function that increases from $h(0) = 1 - \kappa$, for all κ.

4.4.2   Some Two-Parameter Lifetime Models

The two-parameter lifetime distributions defined here have their lifetime distribution representations given in Table 4.3. As outlined in Section 4.3, the *Weibull distribution*, having positive scale parameter λ and positive shape parameter κ, is one of the most popular two-parameter lifetime models

Distribution	$f(t)$	$S(t)$	$h(t)$	$H(t)$	Parameters
Exponential	$\lambda e^{-\lambda t}$	$e^{-\lambda t}$	λ	λt	$\lambda > 0$
Muth	$(e^{\kappa t} - \kappa)e^{\left[-\frac{1}{\kappa}e^{\kappa t} + \kappa t + \frac{1}{\kappa}\right]}$	$e^{\left[-\frac{1}{\kappa}e^{\kappa t} + \kappa t + \frac{1}{\kappa}\right]}$	$e^{\kappa t} - \kappa$	$\frac{1}{\kappa}e^{\kappa t} - \kappa t - \frac{1}{\kappa}$	$0 < \kappa \le 1$

Table 4.2: One-parameter univariate lifetime distributions.

used in survival analysis. The Weibull distribution includes the exponential distribution as a special case when $\kappa = 1$, and the hazard function increases from zero to infinity when $\kappa > 1$ and decreases from infinity to zero when $\kappa < 1$.

The *gamma distribution*, has positive scale parameter λ and positive shape parameter κ. As with

Distribution	$f(t)$	$S(t)$	$h(t)$	$H(t)$	Parameters
Weibull	$\kappa\lambda^{\kappa}t^{\kappa-1}e^{-(\lambda t)^{\kappa}}$	$e^{-(\lambda t)^{\kappa}}$	$\kappa\lambda^{\kappa}t^{\kappa-1}$	$(\lambda t)^{\kappa}$	$\lambda > 0;\ \kappa > 0$
Gamma	$\dfrac{\lambda(\lambda t)^{\kappa-1}e^{-\lambda t}}{\Gamma(\kappa)}$	$1 - I(\kappa, \lambda t)$	$\dfrac{\lambda(\lambda t)^{\kappa-1}e^{-\lambda t}}{\Gamma(\kappa)[1-I(\kappa, \lambda t)]}$	$-\ln[1-I(\kappa, \lambda t)]$	$\lambda > 0;\ \kappa > 0$
Uniform	$\dfrac{1}{b-a}$	$\dfrac{b-t}{b-a}$	$\dfrac{1}{b-t}$	$-\ln\left(\dfrac{b-t}{b-a}\right)$	$a \le t \le b;\ 0 \le a < b$
Log normal	$\dfrac{1}{\sigma t\sqrt{2\pi}}e^{-(\ln t - \mu)^2/2\sigma^2}$	$\int_t^{\infty} f(\tau)d\tau$	$\dfrac{f(t)}{S(t)}$	$-\ln S(t)$	$-\infty < \mu < \infty;\ \sigma > 0$
Log logistic	$\dfrac{\lambda\kappa(\lambda t)^{\kappa-1}}{[1+(\lambda t)^{\kappa}]^2}$	$\dfrac{1}{1+(\lambda t)^{\kappa}}$	$\dfrac{\lambda\kappa(\lambda t)^{\kappa-1}}{1+(\lambda t)^{\kappa}}$	$\ln[1+(\lambda t)^{\kappa}]$	$\lambda > 0;\ \kappa > 0$
Inverse Gaussian	$\sqrt{\dfrac{\lambda}{2\pi t^3}}e^{-\lambda(t-\mu)^2/2\mu^2 t}$	$\int_t^{\infty} f(\tau)d\tau$	$\dfrac{f(t)}{S(t)}$	$-\ln S(t)$	$\lambda > 0;\ \mu > 0$
Exponential Power	$\lambda\kappa t^{\kappa-1}e^{1-e^{\lambda t^{\kappa}}+\lambda t^{\kappa}}$	$e^{1-e^{\lambda t^{\kappa}}}$	$e^{\lambda t^{\kappa}}\lambda\kappa t^{\kappa-1}$	$e^{\lambda t^{\kappa}} - 1$	$\lambda > 0;\ \kappa > 0$
Pareto	$\dfrac{\kappa\lambda^{\kappa}}{t^{\kappa+1}}$	$\left(\dfrac{\lambda}{t}\right)^{\kappa}$	$\dfrac{\kappa}{t}$	$\kappa\ln\left(\dfrac{t}{\lambda}\right)$	$t \ge \lambda;\ \lambda > 0;\ \kappa > 0$
Gompertz	$\delta\kappa^t e^{-\delta(\kappa^t-1)/\ln\kappa}$	$e^{-\delta(\kappa^t-1)/\ln\kappa}$	$\delta\kappa^t$	$\dfrac{\delta(\kappa^t-1)}{\ln\kappa}$	$\kappa > 1;\ \delta > 0$

Table 4.3: Two-parameter univariate lifetime distributions.

the Weibull distribution, the gamma distribution includes the exponential distribution as a special case when $\kappa = 1$. The hazard function increases from zero to λ when $\kappa > 1$, decreases from infinity to λ when $\kappa < 1$.

The *uniform distribution* is a simple two-parameter model. The main application of the uniform distribution in survival analysis is to approximate lifetime distributions over relatively small intervals. The uniform distribution has support on $[a, b]$ with location parameters a and b, where $0 \leq a < b$. The hazard function increases from $h(a) = 1/(b-a)$ to infinity. When $a = 0$ and $b = 1$, the uniform distribution can be used to generate random variates for Monte Carlo simulation by inversion of the cumulative distribution function based on the probability integral transformation.

The *log normal distribution* has a hazard function shape that places it in the UBT [upside-down bathtub-shaped, or hump-shaped, where $h(t)$ increases initially and then decreases] class. It is parameterized by μ and σ because the logarithm of a log normal random variable is a normal random variable with population mean μ and standard deviation σ. One historical reason that the log normal distribution has been less popular than the Weibull distribution is that its survivor function is not closed form. This is important for estimating parameters for right-censored data sets, although widespread algorithms and computer routines can overcome this issue. The survivor function for a log normal random variable is

$$S(t) = 1 - \Phi\left(\frac{\ln t - \mu}{\sigma}\right) \qquad t \geq 0,$$

where Φ is the cumulative distribution function of a standard normal random variable.

The *log logistic distribution* has positive scale parameter λ and positive shape parameter κ. The hazard function is decreasing when $\kappa \leq 1$ and is UBT for $\kappa > 1$. As with the exponential and Weibull distributions, its survivor function can be inverted in closed form, so log logistic variates can easily be generated by inversion for Monte Carlo simulation. The log logistic distribution is widely used in biomedical applications.

The *inverse Gaussian distribution* has a positive parameter μ and positive scale parameter λ. Similar to the log normal distribution, the inverse Gaussian distribution is also in the UBT class. The survivor function is not closed form, but can be written in terms of the cumulative distribution function of a standard normal random variable. The population mean of the inverse Gaussian distribution is μ and the population variance is μ^3/λ, so the parameter μ is not a true location parameter because it does more than just shift the location of the distribution.

The *exponential power distribution* has a positive scale parameter λ and a positive shape parameter κ. The exponential power distribution has two properties that make it unique. First, the hazard function increases exponentially in t, whereas the Weibull hazard function increases in a polynomial fashion. Second, the exponential power distribution is one of the few two-parameter distributions that has a hazard function that can assume a bathtub shape. The hazard function achieves a minimum at $t = [(1-\kappa)/(\lambda\kappa)]^{1/\kappa}$ when $\kappa < 1$. For $\kappa > 1$, the hazard function increases from zero to infinity, and for $\kappa = 1$ the hazard function increases from λ. The distribution has a characteristic life of $(1/\lambda)^{1/\kappa}$. The exponential power distribution's survivor function, which is

$$S(t) = e^{1 - e^{\lambda t^\kappa}} \qquad t \geq 0,$$

can be inverted in closed form, so random variates can easily be generated by inversion.

Pareto devised a probability distribution with support on $t \geq \lambda$, where κ is a positive shape parameter and λ is a positive scale parameter. The hazard function for the *Pareto distribution* decreases to zero from $h(\lambda) = \kappa/\lambda$.

The *Gompertz distribution* is a lifetime model that has been used to model adult lifetimes in actuarial applications. This distribution has positive shape parameters δ and κ. Gompertz assumed that Mill's ratio, the reciprocal of the hazard function, measures human resistance to death. He assumed this resistance decreases over time at a rate proportional to itself; that is,

$$\frac{d}{dt}\left[\frac{1}{h(t)}\right] = \kappa\left[\frac{1}{h(t)}\right],$$

where κ is a constant. The solution to this separable differential equation is $h(t) = \delta e^{ct}$, where $e^c = \kappa$. The hazard function increases from $h(0) = \delta$.

4.4.3 Some Three-Parameter Lifetime Models

The three-parameter lifetime distributions defined here have their $f(t)$, $S(t)$, $h(t)$, and $H(t)$ functions given in Table 4.4. The *Makeham distribution* has three positive shape parameters and is a generalization of the Gompertz distribution with γ included in the hazard function. Whereas the Gompertz distribution has been used to model lifetimes in terms of death from natural causes, the Makeham distribution takes into account the possibility of accidental deaths by including the extra parameter. The hazard function increases from $\delta + \gamma$.

The *IDB (increasing, decreasing, bathtub) distribution* is a three-parameter model with a hazard function that can exhibit increasing ($\delta \geq \gamma\kappa$), decreasing ($\delta = 0$), and bathtub shapes ($0 < \delta < \gamma\kappa$). The distribution has shape parameters $\delta \geq 0$, $\kappa \geq 0$, and $\gamma \geq 0$. Special cases of the IDB distribution are the Rayleigh distribution when $\gamma = 0$ and the exponential distribution when $\delta = \kappa = 0$ and $\gamma > 0$.

The *generalized Pareto distribution* is another three-parameter distribution with shape parameters δ, κ, and γ. It is able to achieve an increasing hazard function when $\kappa < 0$, a decreasing hazard function when $\kappa > 0$, and a constant hazard function when $\kappa = 0$. For all parameter values, $h(0) = \gamma + \kappa/\delta$ and $\lim_{t \to \infty} h(t) = \gamma$. The special cases of $\gamma = 0$ and $\kappa = -\delta\gamma$ result in the hazard functions

$$h(t) = \frac{\kappa}{t+\delta} \qquad \text{and} \qquad h(t) = \frac{\gamma t}{t+\delta}$$

for $t \geq 0$.

Distribution	$f(t)$	$S(t)$	$h(t)$	$H(t)$	Parameters
Makeham	$(\gamma+\delta\kappa^t)e^{-\gamma t-\delta(\kappa^t-1)/\ln\kappa}$	$e^{-\gamma t-\delta(\kappa^t-1)/\ln\kappa}$	$\gamma+\delta\kappa^t$	$\gamma t + \dfrac{\delta(\kappa^t-1)}{\ln\kappa}$	$\delta\geq 0;\ \kappa>1;$ $\gamma>0$
IDB	$\dfrac{(1+\kappa t)\delta t+\gamma}{(1+\kappa t)^{\gamma/\kappa+1}}e^{-\delta t^2/2}$	$(1+\kappa t)^{-\gamma/\kappa}e^{-\delta t^2/2}$	$\delta t+\dfrac{\gamma}{1+\kappa t}$	$\dfrac{\delta}{2}t^2+\dfrac{\gamma}{\kappa}\ln(1+\kappa t)$	$\delta\geq 0;\ \kappa\geq 0;$ $\gamma\geq 0$
Generalized Pareto	$\left(\gamma+\dfrac{\kappa}{t+\delta}\right)\left(1+\dfrac{t}{\delta}\right)^{-\kappa}e^{-\gamma t}$	$\left(1+\dfrac{t}{\delta}\right)^{-\kappa}e^{-\gamma t}$	$\gamma+\dfrac{\kappa}{t+\delta}$	$\gamma t+\kappa\ln\left(1+\dfrac{t}{\delta}\right)$	$\delta> 0;\ \gamma\geq 0;$ $\kappa\geq -\delta\gamma$

Table 4.4: Three-parameter univariate lifetime distributions.

4.4.4   Some n-Parameter Lifetime Models

Two n-parameter distributions are related to the exponential distribution. The first is the *hypoexponential distribution*. If $T_i \sim$ exponential(λ_i) for $i = 1, 2, \ldots, n$, then $T = T_1 + T_2 + \cdots + T_n$ has the hypoexponential distribution. The hypoexponential distribution collapses to the Erlang distribution with parameters λ and n when $\lambda = \lambda_1 = \lambda_2 = \cdots = \lambda_n$. The hypoexponential distribution is in the IFR class for all values of its parameters.

A second n-parameter distribution is the *hyperexponential distribution*. If $T_i \sim$ exponential(λ_i) for $i = 1, 2, \ldots, n$, and T has probability density function

$$f_T(t) = p_1 f_{T_1}(t) + p_2 f_{T_2}(t) + \cdots + p_n f_{T_n}(t),$$

where $p_1 + p_2 + \cdots + p_n = 1$ and $p_i > 0$ for $i = 1, 2, \ldots, n$, then T has the hyperexponential distribution. This lifetime distribution is a mixture of exponential distributions. The hyperexponential distribution collapses to the exponential distribution with failure rate λ when $\lambda = \lambda_1 = \lambda_2 = \cdots = \lambda_n$. The hyperexponential distribution is in the DFR class for all values of its parameters.

4.4.5   Summary

Figure 4.13 shows how these univariate lifetime distributions are related to one another. Each oval represents one lifetime distribution, listing its name, parameter(s), and support. Solid arrows connecting the distributions denote special cases and transformations. An example of a special case is the arrow pointing from the Weibull distribution to the exponential distribution with the label $\kappa = 1$. An example of a transformation is the arrow pointing from the exponential distribution to the chi-square distribution with the label $2\lambda \sum_{i=1}^{n} T_i$ (iid). This result is given in Theorem 4.5. Another example of a transformation is the self-loop on the exponential distribution, where the minimum of independent exponential random variables is also exponential. This result is given in Theorem 4.4. Dashed arrows denote limiting distributions, which typically arise as one of the parameters approaches 0 or infinity. An example of a limiting distribution is the arrow pointing from the gamma distribution to the normal distribution with the label $\kappa \to \infty$. The limiting distribution of a gamma random variable converges to the normal distribution as its shape parameter increases.

Table 4.5 contains a summary of the distribution classes to which the distributions belong. Double lines are used to separate the distributions by the number of parameters. For each class to which a distribution belongs, the corresponding set of parameter values is specified. The distribution classes that are considered are IFR, DFR, BT, and UBT.

4.5   Moment Ratio Diagrams

The lifetime distributions introduced in this chapter have been presented in a serial fashion without much attention being directed toward looking at all of them simultaneously. Isolating their presentation in this fashion is unfortunate; it would be of benefit to view all of these distributions simultaneously. One way to view these probability distributions simultaneously is to place them on a graph of their moments. These graphs are often known as *moment-ratio diagrams*. A moment-ratio diagram is the locus of pairs of standardized moments for a particular probability distribution plotted on a single set of axes. Moment-ratio diagrams are useful for (1) quantifying the "distance" or "proximity" between univariate probability distributions based on their second, third, and fourth moments, (2) illustrating the limiting behavior of probability distributions, (3) highlighting the versatility of a particular probability distribution based on the range of values that the moments can assume, and (4) generating a list of potential probability models based on a data set.

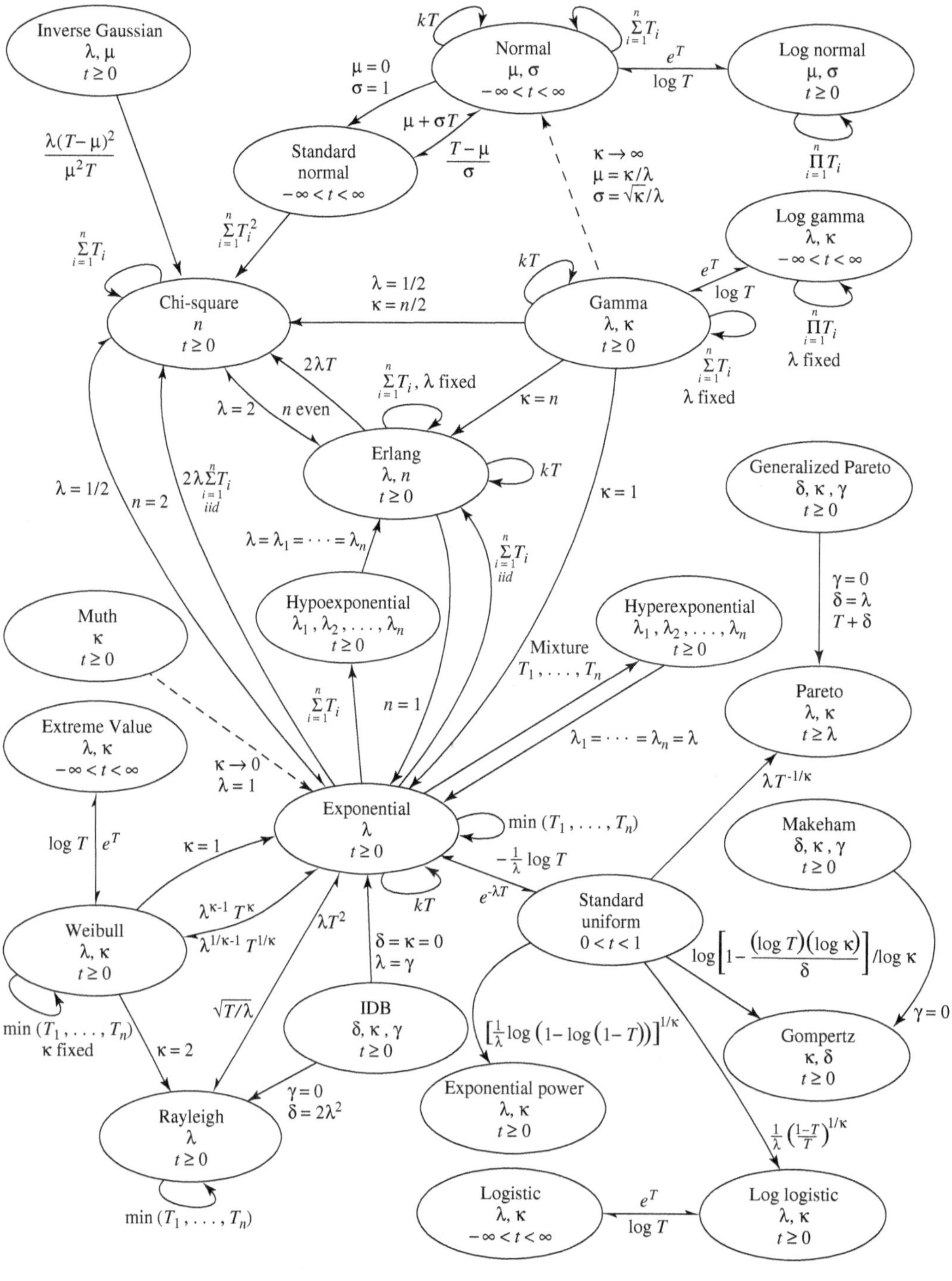

Figure 4.13: Relationships among continuous univariate lifetime distributions.

Distribution	IFR	DFR	BT	UBT
Exponential	YES$_{\text{all } \lambda}$	YES$_{\text{all } \lambda}$	NO	NO
Muth	YES$_{\text{all } \kappa}$	NO	NO	NO
Weibull	YES$_{\kappa \geq 1}$	YES$_{\kappa \leq 1}$	NO	NO
Gamma	YES$_{\kappa \geq 1}$	YES$_{\kappa \leq 1}$	NO	NO
Uniform	YES$_{\text{all } a \text{ and } b}$	NO	NO	NO
Log normal	NO	NO	NO	YES$_{\text{all } \mu \text{ and } \sigma}$
Log logistic	NO	YES$_{\kappa \leq 1}$	NO	YES$_{\kappa > 1}$
Inverse Gaussian	NO	NO	NO	YES$_{\text{all } \lambda \text{ and } \mu}$
Exponential Power	YES$_{\kappa \geq 1}$	NO	YES$_{\kappa < 1}$	NO
Pareto	NO	YES$_{\text{all } \kappa}$	NO	NO
Gompertz	YES$_{\text{all } \delta \text{ and } \kappa}$	NO	NO	NO
Makeham	YES$_{\text{all } \delta \text{ and } \kappa}$	NO	NO	NO
IDB	YES$_{\delta \geq \gamma \kappa}$	YES$_{\delta = 0}$	YES$_{0 < \delta < \gamma \kappa}$	NO
Generalized Pareto	YES$_{\kappa \leq 0}$	YES$_{\kappa \geq 0}$	NO	NO
Hypoexponential	YES$_{\text{all } \lambda_1, \lambda_2, \ldots, \lambda_n}$	YES$_{n=1}$	NO	NO
Hyperexponential	YES$_{\lambda_1 = \lambda_2 = \cdots = \lambda_n}$	YES$_{\text{all } \lambda_1, \lambda_2, \ldots, \lambda_n}$	NO	NO

Table 4.5: Distribution classes.

4.5.1 Skewness vs. Coefficient of Variation

As one illustration of a moment-ratio diagram, Figure 4.14 contains a plot of the population skewness

$$\gamma_3 = E\left[\left(\frac{T-\mu}{\sigma}\right)^3\right]$$

on the vertical axis, versus the population coefficient of variation

$$\gamma = \frac{\sigma}{\mu}$$

on the horizontal axis for several of the lifetime distributions introduced in this chapter, where μ and σ are the population mean and standard deviation of the random variable T. Some features of this moment-ratio diagram are listed below.

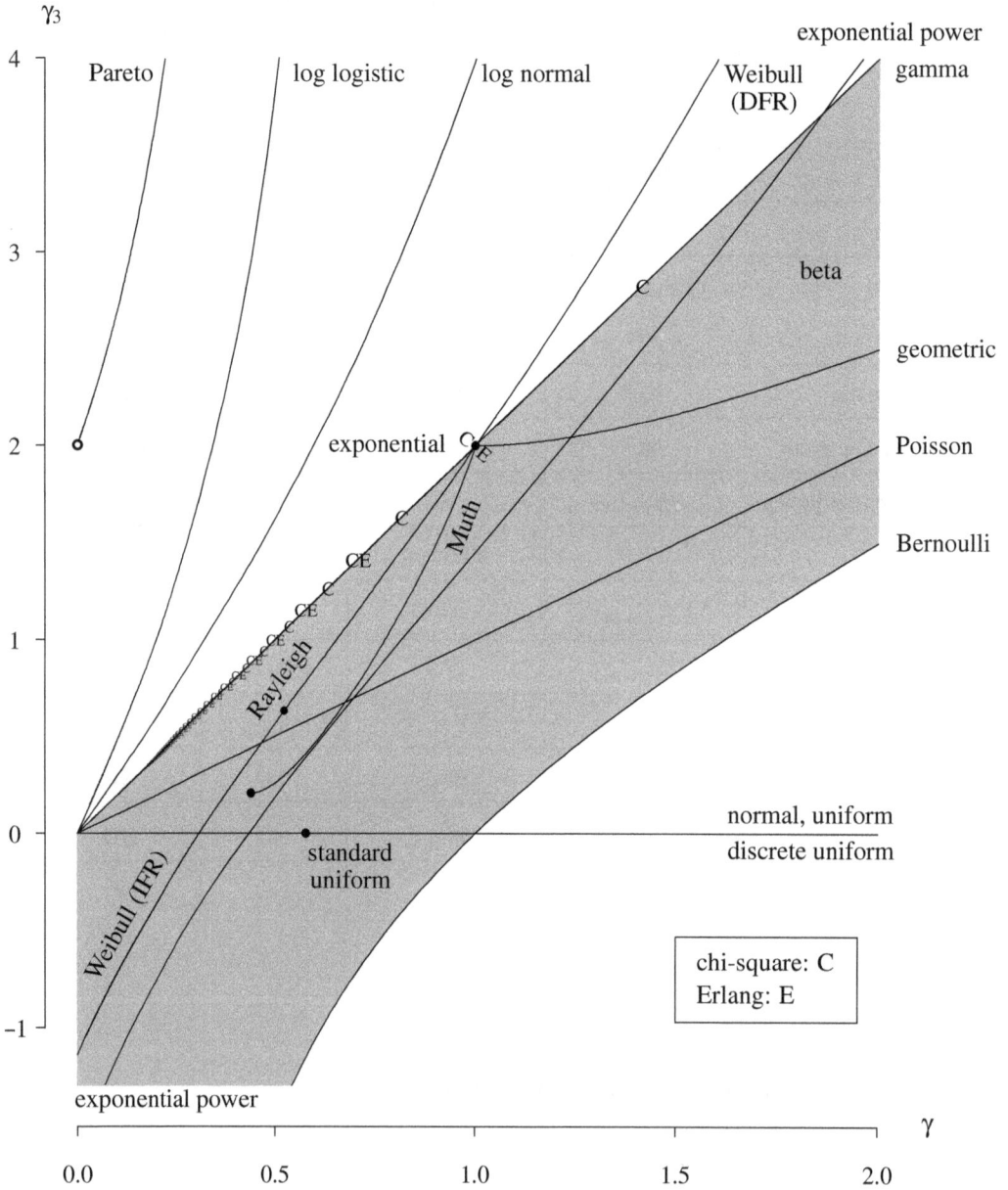

Figure 4.14: Skewness γ_3 versus coefficient of variation γ.

- The locus of points associated with the various probability distribution consist of either a single point (for example, the Rayleigh distribution), a curve (for example, the log logistic distribution), or a region (for example, the beta distribution).

- There are two gathering points: the exponential distribution at $(\gamma, \gamma_3) = (1, 2)$ and a degenerate distribution at $(\gamma, \gamma_3) = (0, 0)$.

- The Poisson distribution, with $\gamma_3 = \gamma$, and the gamma distribution, with $\gamma_3 = 2\gamma$, have linear relationships between γ and γ_3.

- The limiting values of the beta distribution region are the line associated with the gamma distribution $\gamma_3 = 2\gamma$ and the curve associated with the Bernoulli distribution $\gamma_3 = \gamma - 1/\gamma$.

- Symmetric distributions, such as the $N(\mu, \sigma^2)$, $U(a, b)$, and discrete uniform distributions, all have population skewness $\gamma_3 = 0$.

- The curves associated with the gamma and Weibull distributions intersect at the exponential distribution, which is associated with shape parameter $\kappa = 1$.

- The open point associated with the Pareto distribution gives the limiting distribution as $\kappa \to \infty$. The values of γ and γ_3 are defined for $\kappa > 3$.

- The chi-square distribution, indicated by a C for various values of its degrees of freedom, and the Erlang distribution, indicated by an E for various values of its integer shape parameter, coincide when the degrees of freedom for the chi-square distribution are even. This accounts for the alternating pattern of C and CE labels along the line for the gamma distribution.

4.5.2 Kurtosis vs. Skewness

A second moment-ratio diagram, which is given in Figure 4.15 is a plot of the population kurtosis

$$\gamma_4 = E\left[\left(\frac{T-\mu}{\sigma}\right)^4\right]$$

on the vertical axis, versus the population skewness

$$\gamma_3 = E\left[\left(\frac{T-\mu}{\sigma}\right)^3\right]$$

on the horizontal axis for several lifetime distributions introduced in this chapter. (Some authors prefer to work with the *excess population kurtosis* $\gamma_4 - 3$.) Although it uses higher-order moments, it is considered the more classic moment-ratio diagram because the distributions plotted consist of points, curves, and regions that are independent of location and scale parameters. The population skewness scale can sometimes be replaced by the squared skewness, resulting in what is known as a *Cullen and Frey graph*. Figure 4.15 contains a moment-ratio diagram for the population skewness versus the population kurtosis, plotted upside down per tradition.

The locus of (γ_3, γ_4) values that a distribution occupies in Figure 4.15 typically depends on the number of shape parameters. The Rayleigh(λ) distribution, for example, with just a scale parameter, occupies just the single point because it has no shape parameters. The gamma distribution, on the other hand, occupies the curve $\gamma_3 = 3\gamma_2^2 + 3$ because it has one shape parameter. Finally, the beta distribution occupies a region because it has two shape parameters. Some further features of this moment-ratio diagram are listed below.

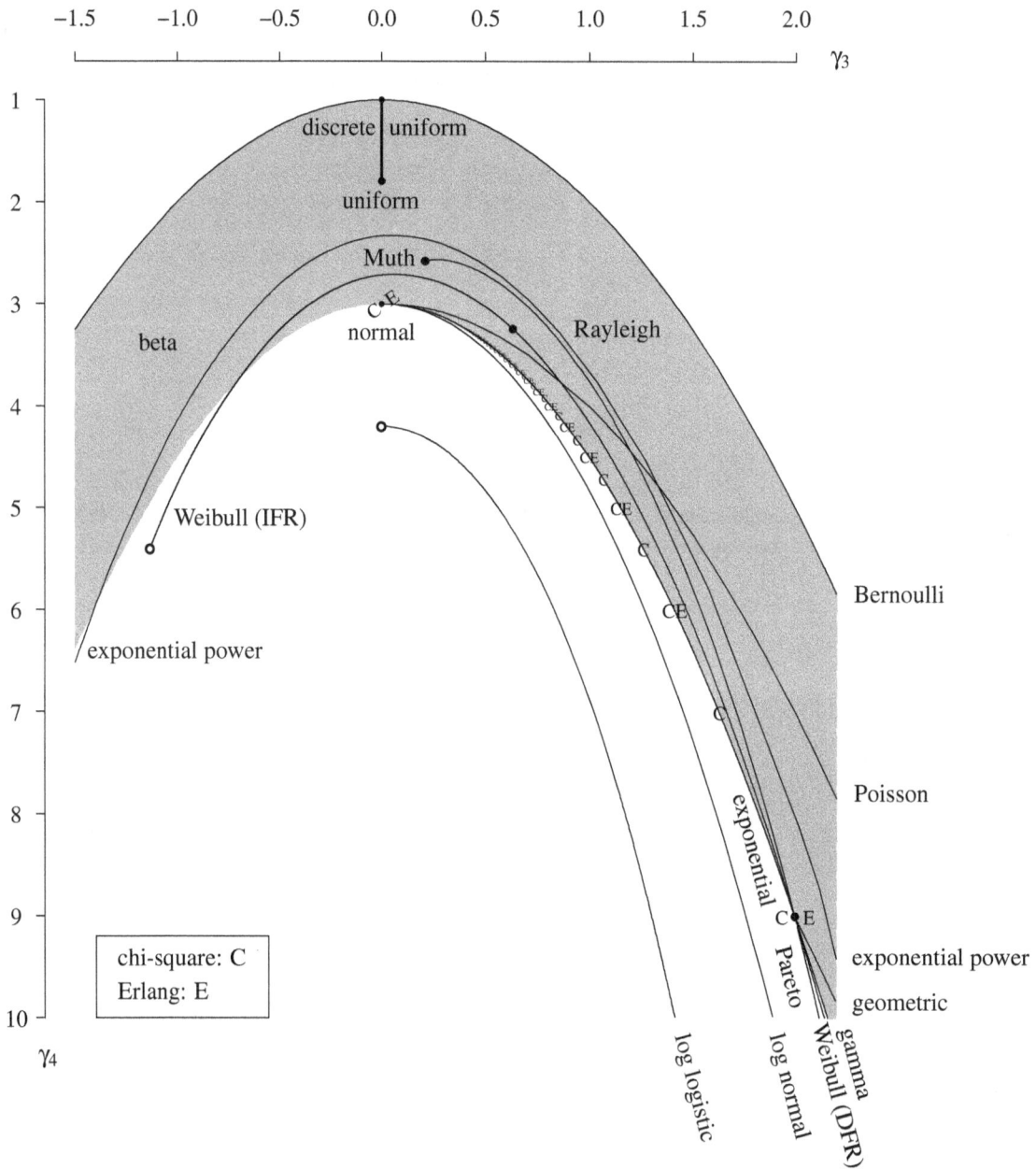

Figure 4.15: Population kurtosis γ_4 versus population skewness γ_3.

- There are two gathering points: the exponential distribution at $(\gamma_3, \gamma_4) = (2, 9)$ and the normal distribution at $(\gamma_3, \gamma_4) = (0, 3)$. This is further evidence of the centrality of these two distributions in probability theory: the exponential distribution plays a pivotal role in stochastic processes (for example, queueing theory and survival analysis) and the normal distribution plays a pivotal role in classical statistics. Stochastic processes can be thought of as probability over time; statistics can be thought of as probability applied to data.

- Once again, the beta distribution covers the largest amount of territory in Figure 4.15, high-lighting its versatility as a probability model.

- The smallest possible value of the population kurtosis is $\gamma_4 = 1$, which is achieved by the Bernoulli distribution with $p = 1/2$. This distribution is equivalent to the discrete uniform distribution with parameters 0 and 1.

- The curves associated with the gamma and Weibull distributions again intersect at the exponential distribution at $(\gamma_3, \gamma_4) = (2, 9)$, which corresponds to the shape parameter $\kappa = 1$.

- The discrete uniform distribution is plotted as a solid line between its limits as a two-mass value at $(\gamma_3, \gamma_4) = (0, 1)$ and its limiting distribution (as the number of mass values increases) at $(\gamma_3, \gamma_4) = (0, 1.8)$. The locus of points is actually a series of points along this line.

A statistician can plot the *sample skewness* and the *sample kurtosis*

$$\frac{1}{n} \sum_{i=1}^{n} \left(\frac{T_i - \bar{T}}{S} \right)^3 \qquad \text{and} \qquad \frac{1}{n} \sum_{i=1}^{n} \left(\frac{T_i - \bar{T}}{S} \right)^4$$

for a set of data values $T_1, T_2, \ldots, T_n$, sample mean $\bar{T}$, and sample standard deviation S on Figure 4.15 for a particular data set. Doing so gives a sense for which of the distributions might be candidate probability models for the implied population distribution. The moment-ratio diagram allows a statistician to compare several candidate distributions simultaneously in terms of their moments.

4.6 Proportional Hazards Model

The *proportional hazards* model is appropriate for including a vector of covariates (for example, the turning speed and feed rate for a drill bit) in a lifetime model. It is often known as the *Cox proportional hazards model* because it was devised by British statistician Sir David Cox in 1972. A *covariate*—often called an *explanatory variable*—is a variable that influences the survival time of the item under consideration. Covariates might account for the fact that the population is not truly homogeneous, or they might account for treatments imposed on the population.

The $q \times 1$ vector $z = (z_1, z_2, \ldots, z_q)'$ contains q covariates associated with a particular item. These covariates might be treatments, stresses, intrinsic properties of items, or exogenous (environmental) variables. The simplest case is the two-population situation modeled by a single ($q = 1$) binary covariate z, where $z = 0$ typically corresponds to the control group and $z = 1$ typically corresponds to the treatment group. A second, slightly more complicated example arises when a single covariate assumes a continuous value (for example, dosage in a medical setting or turning speed in a manufacturing setting). The objective in an analysis of this type might be to find the dosage or turning speed that minimizes risks or costs, respectively. Other possibilities for the elements of z include cumulative load applied, time-varying stresses, and environmental factors.

The covariates increase or decrease the hazard function in the proportional hazards model. This model was originally developed for medical settings in which covariates are usually patient characteristics such as age, gender, cholesterol level, or blood pressure. The models are often used to determine which covariate has the most significant impact on survival or to compare the survival patterns for different treatments (for example, chemotherapy versus surgery for cancer) by factoring out the impact of the covariates.

One issue of immediate interest is how to link the covariates to a lifetime distribution. One approach is to define one lifetime model when $z = 0$ (often called the *baseline* distribution) and other models when $z \neq 0$. One problem that arises with this approach is that there might be dozens or even thousands of possible values associated with $z \neq 0$, and a separate lifetime model would need to be defined for each of these vectors. The more practical approach is to define a single lifetime model which is appropriate for all values of z in order to simplify the modeling.

The baseline distribution corresponds to having all the covariates equal to zero. In a reliability setting, this is typically the normal operating conditions for the item. Other covariate vectors are often used for accelerated environmental conditions. In a biomedical setting, the baseline is typically the control group that receives either no treatment or the standard treatment for a particular disease. The covariates are linked to the lifetime by the link function $\psi(z)$, which typically satisfies $\psi(0) = 1$ and $\psi(z) > 0$ for all z. When a link function satisfies these conditions, then $z = 0$ implies that $S_0(t) \equiv S(t)$. The most general case is to let $\psi(z)$ be any function of the covariates.

Definition 4.9 Let the $q \times 1$ vector $z = (z_1, z_2, \ldots, z_q)'$ denote q covariates associated with the lifetime of an item. The *proportional hazards* model can be defined by

$$h(t) = \psi(z)h_0(t) \qquad t \geq 0,$$

where $h_0(t)$ is a baseline hazard function and $\psi(z) > 0$ is a link function.

The covariates increase the hazard function when $\psi(z) > 1$ or decrease the hazard function when $\psi(z) < 1$. A popular choice is the log-linear link function $\psi(z) = e^{\beta'z}$, where $\beta = (\beta_1, \beta_2, \ldots, \beta_q)'$ is a $q \times 1$ vector of regression coefficients corresponding to the q covariates. The log-linear link function satisfies $\psi(z) > 0$ for all vectors z and β. Other, less popular choices for the link function are $\psi(z) = \beta'z$ and $\psi(z) = (\beta'z)^{-1}$. Both alternative choices suffer from the limitation that $\psi(z) < 0$ for some values of β and z, resulting in a constrained optimization problem when the models are fitted to data. The left-hand side of this model is often written as $h(t; z)$ because survival is now a function of both time and the covariate vector z.

Regression modeling tools, such as indicator variables, modeling of interaction terms, modeling of nonlinear relationships between variables, and stepwise selection of significant covariates, can all be used here in the same fashion as in regression modeling covered earlier in the text. Estimation of the regression coefficients $\beta_1, \beta_2, \ldots, \beta_q$ and the baseline distribution parameters from a data set consisting of times to failure and associated covariates is introduced in the next chapter. The proportional hazards model has a unique feature that allows estimation of the regression parameters (the β vector) without knowledge of the baseline distribution.

Reliability engineers often use accelerated conditions to induce failures. These conditions include voltage, current, pressure, impact, and humidity. The results from the fitted proportional hazards model can then be extrapolated back to the standard operating conditions by adjusting the values of the covariates. The accelerated levels of the covariates must be chosen carefully based on sound engineering judgment and previous experience in order to assure that failure modes that would not occur in the standard operating conditions are not induced by the accelerated testing environment.

The other lifetime distribution representations can be determined for the proportional hazards model. For example, the cumulative hazard function for a random variable T with covariates z is

$$
\begin{aligned}
H(t) &= \int_0^t h(\tau)\,d\tau \\
&= \int_0^t \psi(z)h_0(\tau)\,d\tau \\
&= \psi(z)\int_0^t h_0(\tau)\,d\tau \\
&= \psi(z)H_0(t) \qquad t \geq 0.
\end{aligned}
$$

Similarly,

$$
\begin{aligned}
S(t) &= e^{-H(t)} \\
&= e^{-\psi(z)H_0(t)} \\
&= \left(e^{-H_0(t)}\right)^{\psi(z)} \\
&= \left[S_0(t)\right]^{\psi(z)} \qquad t \geq 0.
\end{aligned}
$$

Finally,

$$
\begin{aligned}
f(t) &= S(t)h(t) \\
&= \left[S_0(t)\right]^{\psi(z)}\psi(z)h_0(t) \\
&= \left[S_0(t)\right]^{\psi(z)-1}\psi(z)S_0(t)h_0(t) \\
&= \left[S_0(t)\right]^{\psi(z)-1}\psi(z)f_0(t) \qquad t \geq 0.
\end{aligned}
$$

The notation has been simplified in the three expressions above; these functions are more accurately expressed as $H(t, z)$, $S(t, z)$, and $f(t, z)$. Table 4.6 summarizes the various lifetime distribution representations for the proportional hazards models. This table allows a modeler to determine any of the four lifetime distribution representations for either model once the baseline distribution and link function are specified, as illustrated in the next example.

Representation	Proportional Hazards
$S(t)$	$\left[S_0(t)\right]^{\psi(z)}$
$f(t)$	$f_0(t)\psi(z)\left[S_0(t)\right]^{\psi(z)-1}$
$h(t)$	$\psi(z)h_0(t)$
$H(t)$	$\psi(z)H_0(t)$

Table 4.6: Lifetime distribution representations for the proportional hazards model.

Example 4.5 Consider the case of a Weibull baseline function in a proportional hazards model. Find the hazard function, survivor function, and the mean time to failure for an item having covariate vector **z**.

The baseline hazard function is Weibull with parameters λ and κ:

$$h_0(t) = \kappa \lambda^\kappa t^{\kappa-1} \qquad t \geq 0.$$

So the hazard function for an item with covariates z is

$$h(t) = \psi(z) h_0(t) = \psi(z) \kappa \lambda^\kappa t^{\kappa-1} \qquad t \geq 0.$$

Using Table 4.6, the appropriate formula for determining the survivor function is

$$S(t) = [S_0(t)]^{\psi(z)} \qquad t \geq 0.$$

Using the usual baseline survivor function for the Weibull distribution,

$$S(t) = \left[e^{-(\lambda t)^\kappa} \right]^{\psi(z)} = e^{-(\lambda t)^\kappa \psi(z)} \qquad t \geq 0.$$

This survivor function can be recognized as that of a Weibull lifetime with scale parameter $\lambda \psi(z)^{1/\kappa}$ and shape parameter κ. The population mean time to failure for an item with covariate vector z is

$$E[T] = \frac{1}{\lambda \psi(z)^{1/\kappa} \kappa} \Gamma\left(\frac{1}{\kappa}\right).$$

As before, the notation has been simplified. It is certainly more accurate to write this as $E[T \,|\, z]$.

This chapter has contained a brief introduction to probability models for univariate lifetime distributions, both without and with associated covariates. These models are appropriate for a nonnegative random variable T with applications in reliability, biostatistics, actuarial science, economics, sociology, etc. The distribution of T can be defined by one of five lifetime distribution representations: the survivor function, the probability density function, the hazard function, or the cumulative hazard function. The exponential distribution is a key central lifetime distribution because it is the only continuous distribution having both a constant hazard function and the memoryless property. The Weibull distribution is a two-parameter lifetime distribution that includes the exponential distribution as a special case when its shape parameter κ is equal to 1. The Cox proportional hazards model provides one way to incorporate a vector of covariates z into a lifetime model. This model contains a link function $\psi(z)$ which links the values of the covariates to the failure time distribution. The next chapter introduces statistical methods that can be applied to lifetime data.

4.7 Exercises

4.1 Let $t^* > 0$ be the mode value for a continuous lifetime T. Show that $h'(t^*) = [h(t^*)]^2$.

4.2 The probability that an item will survive a 1000-hour mission is 0.4. If the item is operating 800 hours into the mission, the probability of surviving the remaining 200 hours of the mission is 0.85. What is the probability that the item survives the initial 800 hours of the mission?

4.3 The hazard function shown below is for a continuous random variable measured in hours.

    (a) Find $S(4)$.

    (b) Find $S(10)$.

    (c) Find $f(10)$.

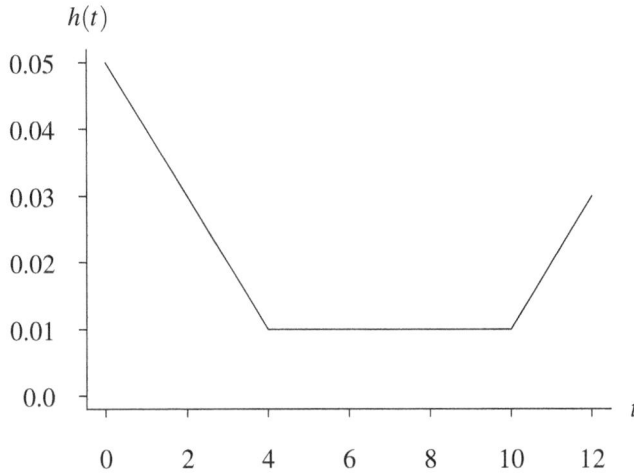

4.4 Draw the survivor function corresponding to the probability density function illustrated below. Use a straight edge whenever the function is linear. The rectangles and triangle on the probability density function all have area $1/3$.

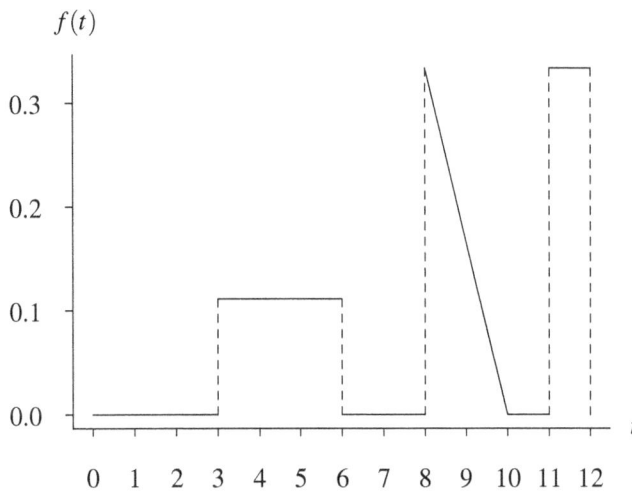

4.5 Consider the hazard function

$$h(t) = \alpha + e^{\beta t} \qquad t \geq 0.$$

What conditions must the parameters α and β meet for $h(t)$ to be a legitimate hazard function for a random lifetime T?

4.6   Jordan has designed a new screwdriver. Its lifetime, measured in years, follows a distribution with survivor function

$$S(t) = \frac{1}{1 + (e^{\lambda t} - 1)^\kappa} \qquad t \geq 0,$$

where λ and κ are positive parameters. If $\kappa = 1/2$ and $\lambda = 1/7$, find the probability that a screwdriver that is still functioning after 5 years of use will last another 3 years.

4.7   Let the time to failure of a bar code reader have survivor function

$$S(t) = \frac{2}{1 + e^{\theta t}} \qquad t \geq 0,$$

where $\theta > 0$. If Ellen places n bar code readers on test simultaneously at time $t = 0$, find the expected number that fail by time t_0.

4.8   The lifetime of a motor, in years, is a continuous random variable with probability density function

$$f(t) = 3(t + 3)^{-2} \qquad t \geq 0.$$

Find the warranty period so that 5% of the motors will fail during the warranty period on average.

4.9   Carrie purchases a hammer whose lifetime T is a random variable with survivor function

$$S(t) = e^{-(\lambda t)^\kappa} \qquad t \geq 0,$$

where λ is a positive scale parameter and κ is a positive shape parameter. Assuming that $\lambda = 0.001$ and $\kappa = 2$, find $P(T > 80 \,|\, T > 50)$.

4.10   Let the lifetime T have hazard function

$$h(t) = \begin{cases} \lambda & 0 < t < 1 \\ \lambda t & t \geq 1 \end{cases}$$

for positive parameter λ. Find the associated survivor function $S(t)$.

4.11   Consider a random lifetime T with survivor function

$$S(t) = \begin{cases} 1 & t \leq 3 \\ 1 - (t - 3)/(t - 2) & t > 3. \end{cases}$$

Give a variate generation algorithm for this probability distribution.

4.12   Show that

$$E[T] = \int_0^\infty t f(t) \, dt$$

can also be found by

$$E[T] = \int_0^\infty S(t) \, dt$$

for any continuous random variable T with nonnegative support and a finite population mean whenever $\lim_{t \to \infty} t S(t) = 0$.

4.13 Consider the random variable T with hazard function

$$h(t) = \begin{cases} 1 & 0 \le t \le 1 \\ t & t > 1. \end{cases}$$

(a) Find $S(t)$.

(c) Find the median of T.

(d) Find the 95th percentile of T.

4.14 Let T be a random variable with hazard function

$$h(t) = \kappa + t \qquad t \ge 0,$$

where κ is a positive parameter.

(a) Use numerical methods to determine a value of κ such that $E[T] = 1/5$.

(b) Conduct a Monte Carlo simulation experiment that supports the value of κ determined in part (a).

4.15 The random variable T has hazard function

$$h(t) = 1 + 2t \qquad t \ge 0.$$

Find $E[T]$.

4.16 An insurance company issues a 30-month warranty on an automobile transmission. Lisa has purchased a 30-month warranty on her transmission and has not made a claim during the 30 months. She would like to purchase a 12-month extension to the warranty. This insurance company will pay a fixed $5000 for a new transmission on the extended warranty if her transmission fails within the next 12 months. Let the continuous random variable T denote the lifetime of Lisa's transmission, measured in months, from the date of the purchase of the automobile. The survivor function of T is $S(t)$. Find an expression for the expected payment that the insurance company will make on Lisa's extended warranty.

4.17 Find the population skewness and kurtosis for an exponential(λ) random variable.

4.18 Which of the following lifetimes is the best candidate for being well approximated by an exponential distribution?

(a) The burning time for a particular type of candle.

(b) The duration of a woman's pregnancy associated with her newborn baby.

(c) The duration of a strike.

(d) The duration of a men's haircut by John at John's barbershop.

(e) The breaking strength of a particular type of yarn.

4.19 Tami purchases a battery whose lifetime T is exponential(λ), for some fixed, positive failure rate λ. The battery is placed in storage on the interval $[0, t_0]$, where t_0 is a fixed, positive constant time value, then monitored continuously for failure thereafter. Thus, the time to detect failure is $X = \max\{T, t_0\}$. Find $E[X]$.

4.20   Consider the continuous random variables T_1 and T_2, each with positive support. Let $S_1(t)$ denote the survivor function of T_1 and $S_2(t)$ denote the survivor function of T_2. When $S_1(t) \geq S_2(t)$ for all values of $t \geq 0$, T_1 is said to "stochastically dominate" T_2. This is one way of showing the superiority of the probability distribution of T_1 over the probability distribution of T_2. One measure of the difference between T_1 and T_2 is the Kolmogorov metric

$$D = \sup_t |S_1(t) - S_2(t)|.$$

Calculate the value of D for the independent random variables $T_1 \sim$ exponential(λ_1) and $T_2 \sim$ exponential(λ_2), where $\lambda_1 < \lambda_2$.

4.21   Rosie purchases a light bulb whose lifetime follows an exponential distribution. If the mean lifetime is one year longer than the median lifetime, find the value of the mean lifetime.

4.22   Marian purchases 30 sixty-watt light bulbs, each having a lifetime which is exponentially distribution with a mean of 1000 hours. If she places the 30 bulbs on a life test without replacement upon failure, find the probability that 10 or fewer of these light bulbs survive to 1200 hours.

4.23   An automobile insurance policy is structured as follows. For claims below \$1000, the policyholder bears the entire cost of the claim. For claims between \$1000 and \$2000, the policyholder bears the first \$1000 of the claim and the policy pays any additional amount. For claims over \$2000, the policyholder bears the first \$1000 of the claim and the policy pays \$1000 plus half of the amount that exceeds \$2000. The distribution of a claim T has the exponential distribution with mean \$3000. Find the cumulative distribution function of the reimbursement amount that the insurance company pays on a claim.

4.24   Let $T_1, T_2, \ldots, T_5$ be independent and identically distributed exponential(λ) random variables.

(a) Find the probability density function of the second order statistic $T_{(2)}$.

(b) Find $P\left(T_{(2)} \leq 1/\lambda\right)$.

4.25   Let T_1, T_2, T_3, be independent and identically distributed exponential(λ) random variables. Find the 96th percentile of the random variables:

(a) $3 \min\{T_1, T_2, T_3\}$,

(b) $T_1 + T_2 - T_3$.

4.26   Troy is a bicyclist. His bike has a front wheel with eight spokes. The wheel is in the failed state when two consecutive spokes have failed. The initial lifetime distributions of the eight spokes are independent and identically distributed exponential random variables with failure rates of one failure per year. When a spoke fails, the failure rate of the two adjacent operating spokes doubles. Conduct a Monte Carlo simulation experiment to estimate the population mean time to wheel failure (considering only the spokes) to two-digit accuracy.

4.27   For the Weibull random lifetime T, show that

$$P\left(T < \frac{1}{\lambda}\right) = 1 - \frac{1}{e},$$

regardless of the value of κ.

4.28 Find the population skewness and kurtosis of a Weibull(λ, κ) random variable.

4.29 Find the value of the shape parameter in the Weibull distribution associated with a population skewness of zero.

4.30 Drea is using a graphics design software package that has a Weibull time to failure. Find the mode of the time to failure distribution. For which parameter values is this value the mode? Find the probability that the software package is still functioning at the mode value.

4.31 Let $T \sim$ Weibull(λ, κ).

(a) Find expressions for the mean, median, and mode of T. (*Hint*: they might not all be closed-form.)

(b) Find parameter values associated with the following three cases: the median and mode of the distribution are equal; the mean and median of the distribution are equal; the mean and mode of the distribution are equal.

4.32 Katherine designs a scanner and desires a one-month reliability of 0.8. She finds that the failure time of the scanner has a Weibull distribution with parameters $\lambda = 8.33$ and $\kappa = 0.334$, with time measured in months. Unfortunately, she finds that the one-month reliability is

$$S(1) = e^{-8.33^{0.334}} = 0.13,$$

which is clearly unacceptable. Fortunately, this Weibull distribution has a decreasing failure rate, so she knows that if she burns in the scanners, she can increase their one-month reliability. How long should she burn in the scanners to achieve a one-month reliability of 0.8 for scanners that survive the test? What fraction of the scanners placed on the burn-in test will fail during the test?

4.33 Statistical applications involving the Weibull distribution can benefit from reparameterizing the distribution. One such reparameterization replaces the scale parameter λ with a particular fractile of the distribution. More specifically, let p_0 be a prescribed constant satisfying $0 < p_0 < 1$. Denote the associated fractile of the Weibull distribution as t_{p_0}. Perform the necessary algebra to write the survivor function of the reparameterized Weibull distribution in terms of the parameters t_{p_0} and κ.

4.34 Alex purchases a laptop computer with a lifetime T, in years, which has a Weibull distribution with $\lambda = 0.2$ and $\kappa = 2$. The laptop computer can be purchased for \$600. The manufacturer of the laptop provides a full refund if the laptop fails within the first year after purchase, a one-third refund if the laptop fails during the second year after purchase, and no refund if the laptop fails thereafter. What is the expected refund on a laptop?

4.35 Steve takes three generators to a work site. He will use the three generators in a cold standby system to provide electrical power. The lifetimes of the generators are exponentially distributed with mean 1000 hours. Find the variance of the total amount of time that electrical power can be supplied by the generators.

4.36 Find $E[T^r]$ for $r = 1, 2, \ldots$ for a log logistic random variable.

4.37 Find the population skewness and kurtosis of a log logistic random variable.

4.38   Lindsay purchases a jack hammer. Let the lifetime of an item be defined by a special case of the log logistic distribution with survivor function

$$S(t) = \frac{1}{1 + \lambda t} \qquad t \geq 0,$$

where λ is a positive scale parameter. If the item has been operating for a time units, find

   (a) the probability it will last another r time units,

   (b) the expected remaining time to failure.

4.39   Consider the random variable X having the *logistic* distribution with location parameter η, positive scale parameter $\rho > 0$, and probability density function

$$f(x) = \frac{e^{(x-\eta)/\rho}}{\rho \left(1 + e^{(x-\eta)/\rho}\right)^2} \qquad -\infty < x < \infty.$$

Show that e^X has the log logistic distribution.

4.40   Derive $\lim_{t \to \infty} h(t)$ for the log normal distribution.

4.41   Many life insurance companies offer a "last-to-die" policy for couples or business partners that pays out when the second of the two individuals dies. These policies are often purchased to pay tax liabilities on small businesses. Assume that David, age 40, and his wife Laura, age 35, celebrate their mutual birthday by purchasing a one-year, \$100,000 term last-to-die policy. Find, to the nearest penny, the revenue-neutral premium (that is, where the premium equals the expected payout). For simplicity, assume that

   • all new-born baby boys have Weibull random lifetimes with $\lambda = 1/65$ and $\kappa = 3/2$,

   • all new-born baby girls have exponential power random lifetimes with $\lambda = 1/12$ and $\kappa = 1/2$,

   • health care, lifestyle, environmental factors, etc. remain constant throughout David and Laura's lifetimes,

   • there is no overhead or profit associated with the premium,

   • the prevailing interest rate during the next year is 0%, and

   • their two lifetimes are independent.

4.42   Meghan purchases a book stand for a rare book, which has lifetime T. If T has the log logistic distribution, $S(1) = 1/5$ and $S(3) = 1/37$, find $S(2)$.

4.43   Joanna purchases a food truck whose lifetime is a continuous random variable T with the *power distribution*, having probability density function

$$f(t) = \frac{\beta t^{\beta-1}}{\alpha^\beta} \qquad 0 < t < \alpha,$$

where α is a positive scale parameter and β is a positive shape parameter. Find the median of T.

4.44 Summer and Brigid are conducting a study concerning the random time T required to reshelve a book after it has been returned to a library. The time between the return of a book and the time it is reshelved has a special case of the *extreme value distribution* with survivor function

$$S(t) = 1 - e^{-e^{-t}} \qquad -\infty < t < \infty.$$

For real constants a and b satisfying $a < b$, find $P(a < T < b)$.

4.45 Daneen is modeling the lifetimes of light bulbs (in years) with the proportional hazards model with $q = 2$ covariates: wattage (z_1) and operating temperature in degrees Fahrenheit (z_2). The baseline distribution is exponential with a failure rate of 1.1 failures per year and the log-linear form of the link function $\psi(\mathbf{z})$ is used. If previous data has shown that the associated regression coefficients are $\beta_1 = 0.003$ and $\beta_2 = 0.004$, what is the expected time to failure of a 60-watt bulb operating in a constant $72°F$ environment?

4.46 In a *log logistic regression model* with a single covariate z, the lifetime T can be expressed as

$$T = e^{\beta_0 + \beta_1 z + \theta Y},$$

where β_0 and β_1 are regression parameters, $\theta > 0$ is a parameter of the model, and Y has probability density function

$$f_Y(y) = \frac{e^y}{(1 + e^y)^2} \qquad -\infty < y < \infty.$$

  (a) Find the survivor function of T for one particular value of the covariate z; that is, find $S_{T|Z=z}(t \,|\, Z = z)$.

  (b) The *odds ratio*
$$\frac{1 - S_{T|Z=z}(t \,|\, Z = z)}{S_{T|Z=z}(t \,|\, Z = z)}$$

  gives the odds that an item fails by time t for one particular value of the covariate z. Calculate the odds ratio for the log logistic regression model.

  (c) Consider two different items with covariates z_1 and z_2. Prove that the quotient of their odds ratios is independent of t for any time $t > 0$.

4.47 Consider the baseline hazard function

$$h_0(t) = \begin{cases} 1 & 0 \le t < 1 \\ t & t \ge 1. \end{cases}$$

In a proportional hazards model, find the probability that an item with covariates $\mathbf{z}$ and link function $\psi(\mathbf{z})$ survives to time t.

4.48 A proportional hazards model is applied to a lifetime that has a single binary covariate z with regression coefficient β, link function $\psi(z) = e^{\beta z}$, and Weibull baseline hazard function. Find

  (a) the survivor function for the time to failure,

  (b) the mean time to failure when $z = 0$,

  (c) the mean time to failure when $z = 1$.

4.49   Ali purchases a freezer with a lifetime that is well approximated by the proportional hazards model with $q = 2$ covariates: external temperature z_1 (measured in degrees Celsius) and humidity z_2. Assume that the Weibull baseline distribution and a log-linear link function are used.

     (a) What would you expect the sign (positive or negative) of $\hat{\beta}_1$ to be if a large sample of failure times and associated covariates was collected? Explain your reasoning.

     (b) Find the probability that such a component survives to time t for any covariate vector z and regression coefficients $\boldsymbol{\beta}$.

4.50   Write a few sentences describing the suitability of the link functions

$$\psi(z) = 1 + \beta_1 z_1 + \beta_2 z_2 \qquad \text{and} \qquad \psi(z) = (\beta_1 z_1 + \beta_2 z_2)^2$$

for a proportional hazards model with $q = 2$ covariates.

4.51   Consider the Cox proportional hazards model

$$h(t) = \psi(z) h_0(t) \qquad\qquad t \geq 0,$$

with $q = 2$ covariates, z_1 and z_2, which includes an interaction term between the covariates. The link function assumes the log linear form:

$$\psi(z) = e^{\beta_1 z_1 + \beta_2 z_2 + \beta_3 z_1 z_2}.$$

Find the ratio of the hazard function for covariates z_1 and z_2 to the hazard function for covariates z_1 and $z_2 + 5$.

Chapter 5

Statistical Methods in Survival Analysis

The previous chapter introduced probability models that are frequently used in survival analysis. This chapter introduces the associated statistical methods.

The focus in this chapter is the use of maximum likelihood for parameter estimation and inference. Likelihood theory is illustrated in the first section. The matrix of the expected values of the opposite of the second partial derivatives of the log likelihood function is known as the *Fisher information matrix* and its statistical analog, the *observed information matrix*, is useful for determining confidence intervals for parameters. Asymptotic properties of the likelihood function, which are associated with large sample sizes, are reviewed in the second section. One distinctive feature of lifetime data is the presence of *censoring*, which occurs when only an upper or lower bound on the lifetime is known. Statistical methods for handling censored data values are introduced in the third section. The focus is on right censoring, where only a lower bound on the failure time is known. These methods are applied to the exponential distribution and the Weibull distribution in the next two sections. Finally, the last section indicates how to fit the proportional hazards model to a data set consisting of lifetimes with associated covariates.

5.1 Likelihood Theory

There are always merits in obtaining raw data (that is, exact individual failure times), as opposed to grouped data (counts of the number of failures over prescribed time intervals). Given raw data, we can always construct grouped data, but the converse is typically not true; therefore, we limit discussion in this chapter to the raw data case.

The random variable T has denoted a random lifetime in previous chapter. So it is natural to use $T_1, T_2, \ldots, T_n$ to denote a *random sample* of n such lifetimes, where n is the number of items on test. When specific values are given for realizations of such lifetimes, which is typically the case from this point forward, they are denoted by $t_1, t_2, \ldots, t_n$. In other words, $t_1, t_2, \ldots, t_n$ are the experimental values of the mutually independent and identically distributed random variables $T_1, T_2, \ldots, T_n$. The associated ordered observations, or order statistics, are denoted by $t_{(1)}, t_{(2)}, \ldots, t_{(n)}$.

The Greek letter θ is often used to denote a generic unknown parameter. We will refer to $\hat{\theta}$ in the abstract as a *point estimator*; when $\hat{\theta}$ assumes a specific numeric value, it will be referred to as a *point estimate*. The probability distribution of a statistic is referred to as a *sampling distribution*.

Assume that there is a single unknown parameter θ in the probability model for T. Assume further that the data values $t_1, t_2, \ldots, t_n$ are mutually independent and identically distributed random variables. The joint probability density function of the data values is the product of the marginal probability density functions of the individual observations:

$$L(t_1, t_2, \ldots, t_n, \theta) = \prod_{i=1}^{n} f(t_i; \theta).$$

This function is the *likelihood function*. In order to simplify the notation, the likelihood function is often written as simply

$$L(\theta) = \prod_{i=1}^{n} f(t_i).$$

The maximum likelihood estimator of θ, which is denoted by $\hat{\theta}$, is the value of θ that maximizes $L(\theta)$.

The next example reviews the associated notions of the log likelihood function, score vector, maximum likelihood estimator, Fisher information matrix, and observed information matrix for a two-parameter lifetime model. We assume for now that there are no censored observations in the data set; all of the failure times are observed.

Example 5.1 Let $t_1, t_2, \ldots, t_n$ be a random sample from an *inverse Gaussian* (Wald) population having unknown positive parameters λ and μ, where μ is the population mean. The probability density function of the inverse Gaussian distribution is

$$f(t) = \sqrt{\frac{\lambda}{2\pi}}\, t^{-3/2} e^{-\lambda(t-\mu)^2/(2\mu^2 t)} \qquad t > 0.$$

Find the likelihood function, log likelihood function, score vector, maximum likelihood estimator, Fisher information matrix, and observed information matrix.

The likelihood function is

$$L(t, \lambda, \mu) = \prod_{i=1}^{n} \sqrt{\frac{\lambda}{2\pi}}\, t_i^{-3/2} e^{-\lambda(t_i-\mu)^2/(2\mu^2 t_i)}$$

$$= \lambda^{n/2}(2\pi)^{-n/2}\left[\prod_{i=1}^{n} t_i\right]^{-3/2} e^{-\lambda/(2\mu^2)\sum_{i=1}^{n}(t_i-\mu)^2/t_i},$$

where $t = (t_1, t_2, \ldots, t_n)$. The likelihood function and any monotonic transformation of the likelihood function are maximized at the same value. Since the calculus and algebra is often easier when working with the logarithm of the likelihood function, we do so in this setting. The log likelihood function is

$$\ln L(t, \lambda, \mu) = \frac{n}{2}\ln \lambda - \frac{n}{2}\ln(2\pi) - \frac{3}{2}\sum_{i=1}^{n}\ln t_i - \frac{\lambda}{2\mu^2}\sum_{i=1}^{n}\frac{(t_i-\mu)^2}{t_i}.$$

The two-component score vector, $U(\lambda, \mu)$, consists of the partial derivatives with respect to the two unknown parameters:

$$\frac{\partial \ln L(t, \lambda, \mu)}{\partial \lambda} = \frac{n}{2\lambda} - \frac{1}{2\mu^2}\sum_{i=1}^{n}\frac{(t_i-\mu)^2}{t_i}$$

and

$$\frac{\partial \ln L(t, \lambda, \mu)}{\partial \mu} = \frac{\lambda}{\mu^3} \left[\sum_{i=1}^{n} t_i - n\mu \right].$$

When the second equation is equated to zero, the maximum likelihood estimator $\hat{\mu}$ is determined. Then using $\hat{\mu}$ as an argument in the first equation and solving for $\hat{\lambda}$ results in the maximum likelihood estimators

$$\hat{\lambda} = \left[\frac{1}{n} \sum_{i=1}^{n} \frac{1}{t_i} - \frac{n}{\sum_{i=1}^{n} t_i} \right]^{-1} \qquad \text{and} \qquad \hat{\mu} = \frac{1}{n} \sum_{i=1}^{n} t_i.$$

The second partial derivatives of the log likelihood function are

$$\frac{\partial^2 \ln L(t, \lambda, \mu)}{\partial \lambda^2} = -\frac{n}{2\lambda^2}$$

$$\frac{\partial^2 \ln L(t, \lambda, \mu)}{\partial \lambda \partial \mu} = \frac{1}{\mu^3} \sum_{i=1}^{n} t_i - \frac{n}{\mu^2}$$

$$\frac{\partial^2 \ln L(t, \lambda, \mu)}{\partial \mu^2} = -\frac{3\lambda}{\mu^4} \sum_{i=1}^{n} t_i + \frac{2n\lambda}{\mu^3}.$$

Since $E[T] = \mu$ for the inverse Gaussian distribution, the Fisher information matrix consists of the expected values of the negatives of these derivatives:

$$I(\lambda, \mu) = \begin{pmatrix} E\left[\dfrac{-\partial^2 \ln L(t, \lambda, \mu)}{\partial \lambda^2} \right] & E\left[\dfrac{-\partial^2 \ln L(t, \lambda, \mu)}{\partial \lambda \partial \mu} \right] \\ E\left[\dfrac{-\partial^2 \ln L(t, \lambda, \mu)}{\partial \mu \partial \lambda} \right] & E\left[\dfrac{-\partial^2 \ln L(t, \lambda, \mu)}{\partial \mu^2} \right] \end{pmatrix} = \begin{pmatrix} \dfrac{n}{2\lambda^2} & 0 \\ 0 & \dfrac{n\lambda}{\mu^3} \end{pmatrix}.$$

The Fisher information matrix is the variance–covariance matrix of the score vector. The off-diagonal elements being zero for the inverse Gaussian distribution implies that the elements of the score vector are uncorrelated. Although this example has simple closed-form expressions for the Fisher information matrix, it is more often the case that the elements of the Fisher information matrix are not closed form. The observed information matrix can be calculated for all distributions; it uses the maximum likelihood estimates:

$$O(\hat{\lambda}, \hat{\mu}) = \begin{pmatrix} \dfrac{-\partial^2 \ln L(t, \lambda, \mu)}{\partial \lambda^2} & \dfrac{-\partial^2 \ln L(t, \lambda, \mu)}{\partial \lambda \partial \mu} \\ \dfrac{-\partial^2 \ln L(t, \lambda, \mu)}{\partial \mu \partial \lambda} & \dfrac{-\partial^2 \ln L(t, \lambda, \mu)}{\partial \mu^2} \end{pmatrix}_{\lambda = \hat{\lambda}, \mu = \hat{\mu}} = \begin{pmatrix} \dfrac{n}{2\hat{\lambda}^2} & 0 \\ 0 & \dfrac{n\hat{\lambda}}{\hat{\mu}^3} \end{pmatrix}.$$

In some cases, it is possible to find the exact distribution of a pivotal quantity which results in exact statistical inference (that is, constructing exact confidence intervals and performing exact hypothesis tests). It is more often the case that exact statistical inference is not possible, and asymptotic properties associated with the likelihood function must be relied on for approximate inference. The next section reviews some asymptotic properties that arise in likelihood theory. When a large data set of lifetimes is available, these properties often lead to approximate statistical methods of inference.

5.2   Asymptotic Properties

When the number of items on test n is large, there are some asymptotic results concerning the likelihood function that are useful for constructing confidence intervals and performing hypothesis tests associated with a vector of p unknown parameters $\boldsymbol{\theta} = (\theta_1, \theta_2, \ldots, \theta_p)'$. As indicated in the example in the last section, the $p \times 1$ score vector $\boldsymbol{U}(\boldsymbol{\theta})$ has elements

$$U_i(\boldsymbol{\theta}) = \frac{\partial \ln L(\boldsymbol{t}, \boldsymbol{\theta})}{\partial \theta_i} = \frac{\partial}{\partial \theta_i} \sum_{j=1}^{n} \ln f(t_j, \boldsymbol{\theta})$$

for $i = 1, 2, \ldots, p$. Therefore, each element of the score vector is a sum of mutually independent random variables, and, when n is large, the elements of $\boldsymbol{U}(\boldsymbol{\theta})$ are asymptotically normally distributed by the central limit theorem. More specifically, the score vector $\boldsymbol{U}(\boldsymbol{\theta})$ is asymptotically normal with population mean $\boldsymbol{0}$ and variance–covariance matrix $I(\boldsymbol{\theta})$, where $I(\boldsymbol{\theta})$ is the Fisher information matrix. This means that when the true value for the parameter vector is $\boldsymbol{\theta}_0$ then

$$\boldsymbol{U}'(\boldsymbol{\theta}_0) I(\boldsymbol{\theta}_0)^{-1} \boldsymbol{U}(\boldsymbol{\theta}_0)$$

is asymptotically chi-square with p degrees of freedom. This can be used to determine confidence intervals and perform hypothesis tests with respect to $\boldsymbol{\theta}$.

  The maximum likelihood estimator for the parameter vector $\hat{\boldsymbol{\theta}}$ can also be used for confidence intervals and hypothesis testing. Since $\hat{\boldsymbol{\theta}}$ is asymptotically normal with population mean $\boldsymbol{\theta}$ and variance–covariance matrix $I^{-1}(\boldsymbol{\theta})$, when $\boldsymbol{\theta} = \boldsymbol{\theta}_0$,

$$\left(\hat{\boldsymbol{\theta}} - \boldsymbol{\theta}_0\right)' I(\boldsymbol{\theta}_0) \left(\hat{\boldsymbol{\theta}} - \boldsymbol{\theta}_0\right)$$

is also asymptotically chi-square with p degrees of freedom. Two statistics that are asymptotically equivalent to this statistic that can be used to estimate the value of the chi-square random variable are

$$\left(\hat{\boldsymbol{\theta}} - \boldsymbol{\theta}_0\right)' I\left(\hat{\boldsymbol{\theta}}\right) \left(\hat{\boldsymbol{\theta}} - \boldsymbol{\theta}_0\right)$$

and

$$\left(\hat{\boldsymbol{\theta}} - \boldsymbol{\theta}_0\right)' O\left(\hat{\boldsymbol{\theta}}\right) \left(\hat{\boldsymbol{\theta}} - \boldsymbol{\theta}_0\right).$$

  A third asymptotic result involves the likelihood ratio statistic

$$-2\left[\ln L(\boldsymbol{\theta}) - \ln L(\hat{\boldsymbol{\theta}})\right] = -2 \ln \left[\frac{L(\boldsymbol{\theta})}{L(\hat{\boldsymbol{\theta}})}\right],$$

which is asymptotically chi-square with p degrees of freedom. The conditions necessary for these asymptotic properties to apply are cited at the end of the chapter.

  These three asymptotic results are summarized in the result below, where the a above the $\sim$ is shorthand for "asymptotically distributed."

Theorem 5.1 Let $t_1, t_2, \ldots, t_n$ be mutually independent and identically distributed lifetimes from a population distribution with p unknown parameters $\boldsymbol{\theta} = (\theta_1, \theta_2, \ldots, \theta_p)'$. Then

$$\boldsymbol{U}'(\boldsymbol{\theta}_0) I(\boldsymbol{\theta}_0)^{-1} \boldsymbol{U}(\boldsymbol{\theta}_0) \overset{a}{\sim} \chi^2(p), \quad \left(\hat{\boldsymbol{\theta}} - \boldsymbol{\theta}_0\right)' O\left(\hat{\boldsymbol{\theta}}\right) \left(\hat{\boldsymbol{\theta}} - \boldsymbol{\theta}_0\right) \overset{a}{\sim} \chi^2(p), \quad -2 \ln \left[\frac{L(\boldsymbol{\theta})}{L(\hat{\boldsymbol{\theta}})}\right] \overset{a}{\sim} \chi^2(p).$$

Example 5.2 Let $t_1, t_2, \ldots, t_n$ be a random sample from a population with probability density function

$$f(t) = \frac{1}{\sqrt{2\pi t^3}} \, e^{-(t-\mu)^2/(2\mu^2 t)} \qquad t > 0,$$

where μ is a positive unknown parameter, which is the population mean. This population distribution is a special case of the two-parameter inverse Gaussian distribution. Use one of the asymptotic results from Theorem 5.1 to construct an asymptotically exact two-sided $100(1-\alpha)\%$ confidence interval for μ.

The first step is to find the maximum likelihood estimator of μ. The likelihood function is

$$L(t, \mu) = \prod_{i=1}^{n} \left(2\pi t_i^3\right)^{-1/2} e^{-(t_i-\mu)^2/(2\mu^2 t_i)}$$

$$= (2\pi)^{-n/2} \left[\prod_{i=1}^{n} t_i\right]^{-3/2} e^{-\sum_{i=1}^{n}(t_i-\mu)^2/(2\mu^2 t_i)}.$$

The log likelihood function is

$$\ln L(t, \mu) = -\frac{n}{2}\ln(2\pi) - \frac{3}{2}\sum_{i=1}^{n}\ln t_i - \frac{1}{2\mu^2}\sum_{i=1}^{n}\frac{(t_i-\mu)^2}{t_i}.$$

The score is the derivative of the log likelihood function with respect to μ, which, after simplification, is

$$\frac{\partial \ln L(t, \mu)}{\partial \mu} = \frac{1}{\mu^3}\left[\sum_{i=1}^{n} t_i - n\mu\right].$$

When this equation is equated to zero, the maximum likelihood estimator for μ is

$$\hat{\mu} = \frac{1}{n}\sum_{i=1}^{n} t_i,$$

which is the sample mean. The second partial derivative of the log likelihood function is

$$\frac{\partial^2 \ln L(t, \mu)}{\partial \mu^2} = -\frac{3}{\mu^4}\sum_{i=1}^{n} t_i + \frac{2n}{\mu^3},$$

which is negative at the maximum likelihood estimator, so the maximum likelihood estimator maximizes the log likelihood function. The next step is to find the 1×1 Fisher information matrix. Using the second partial derivative of the log likelihood function, the Fisher information matrix is

$$I(\mu) = E\left[-\frac{\partial^2 \ln L(t, \mu)}{\partial \mu^2}\right] = E\left[\frac{3}{\mu^4}\sum_{i=1}^{n} t_i - \frac{2n}{\mu^3}\right] = \frac{3n\mu}{\mu^4} - \frac{2n}{\mu^3} = \frac{n}{\mu^3}$$

because $E[X] = \mu$ for this population distribution. The 1×1 observed information is

$$O(\hat{\mu}) = \left[-\frac{\partial^2 \ln L(t, \mu)}{\partial \mu^2}\right]_{\mu=\hat{\mu}} = \frac{n}{\hat{\mu}^3}.$$

In order to construct an asymptotically exact two-sided $100(1-\alpha)\%$ confidence interval for μ, recall that $\hat{\mu}$ is asymptotically normal with population mean μ and variance–covariance matrix $I^{-1}(\mu)$. In other words,

$$\hat{\mu} \overset{a}{\sim} N\left(\mu, I^{-1}(\mu)\right).$$

For large values of n, we replace the Fisher information matrix with the observed information matrix:

$$\hat{\mu} \overset{a}{\sim} N\left(\mu, O^{-1}(\hat{\mu})\right)$$

or

$$\hat{\mu} \overset{a}{\sim} N\left(\mu, \frac{\hat{\mu}^3}{n}\right).$$

This random variable can be standardized by subtracting its population mean and dividing by its population standard deviation:

$$\frac{\hat{\mu}-\mu}{\sqrt{\hat{\mu}^3/n}} \overset{a}{\sim} N(0,1).$$

So the probability that this random variable falls between $-z_{\alpha/2}$ and $z_{\alpha/2}$ for large n is

$$\lim_{n\to\infty} P\left(-z_{\alpha/2} < \frac{\hat{\mu}-\mu}{\sqrt{\hat{\mu}^3/n}} < z_{\alpha/2}\right) = 1-\alpha.$$

where $z_{\alpha/2}$ is the $1-\alpha/2$ quantile of the standard normal distribution. Rearranging the inequality

$$-z_{\alpha/2} < \frac{\hat{\mu}-\mu}{\sqrt{\hat{\mu}^3/n}} < z_{\alpha/2}$$

so that μ is in the center of the inequality yields the asymptotically exact two-sided $100(1-\alpha)\%$ confidence interval

$$\hat{\mu} - z_{\alpha/2}\frac{\hat{\mu}^{3/2}}{\sqrt{n}} < \mu < \hat{\mu} + z_{\alpha/2}\frac{\hat{\mu}^{3/2}}{\sqrt{n}},$$

where $\hat{\mu}$ is the sample mean of the observed data values. The actual coverage of confidence intervals developed in this fashion typically approaches $1-\alpha$ as the number of items on test n increases.

All of the statistical methods developed thus far have assumed that we are able to observe all n of the items on test fail. The associated lifetimes are denoted by $t_1, t_2, \ldots, t_n$. Although this is ideal and might be the case in some settings, a short testing time or items with long lifetimes might result in some items that survive the test. The lifetimes of the items which do not fail during the test are known as *right-censored* observations. The lifetimes of these items are not observed, but are known to exceed the time at which the test is concluded. If a decision concerning the acceptability of the items must be made with some of the items still operating at the end of the test, then a statistical model must be formulated to account for the unobserved lifetimes of these items. The next section introduces the important topic of censoring, which is pervasive in survival analysis.

5.3   Censoring

Censoring occurs in lifetime data sets when only an upper or lower bound on the lifetime is known. Censoring occurs frequently in lifetime data sets because it is often impossible or impractical to observe the lifetimes of all the items on test. A data set for which all failure times are known is called a *complete data set*. Figure 5.1 illustrates a complete data set of $n = 5$ items placed on test simultaneously at time $t = 0$, where the $\times$'s denote failure times. Consider the two endpoints of each of the horizontal segments. It is critical to provide an unambiguous definition of the time origin (for example, the time a product is purchased or the time a cancer is diagnosed). Likewise, failure must be defined in an unambiguous fashion. This is easier to define for a light bulb or a fuse than for a ball bearing or a sock. Outside of a reliability setting, a data set of lifetimes is often generically referred to as *time-to-event* data, corresponding to the time between the time origin and the event of interest. A *censored observation* occurs when only a bound is known on the time of failure. If a data set contains one or more censored observations, it is called a *censored data set*.

The most common type of censoring is known as *right censoring*. In a right-censored data set, one or more items have only a lower bound known on their lifetime. The term *sample size* is now vague. From this point forward, we use n to denote the *number of items on test* and use r to denote the *number of observed failures*. In an industrial life testing situation, for example, $n = 12$ cell phones are put on a continuous, rigorous life test on January 1, and $r = 3$ of the cell phones have failed by December 31. These failed cell phones are discarded upon failure. The remaining $n - r = 12 - 3 = 9$ cell phones that are still operating on December 31 have lifetimes that exceed 365 days, and are therefore right-censored observations. Right censoring is not limited to just reliability applications. In a medical study in which T is the survival time after the diagnosis of a particular type of cancer, for example, a patient can either (*a*) still be alive at the end of a study, (*b*) die of a cause other than the particular type of cancer, constituting a right-censored observation, or (*c*) lose contact with the study (for example, if they leave town), constituting a right-censored observation.

Three special cases of right censoring are common in survival analysis. The first is Type II or *order statistic* censoring. As shown in Figure 5.2, this corresponds to terminating a study upon one of the ordered failures. The diagram corresponds to a set of $n = 5$ items placed on a test simultaneously at time $t = 0$. The test is terminated when $r = 3$ failures are observed. Time advances from left to right in Figure 5.2 and the failure of the first item (corresponding to the third ordered observed failure) terminates the test. The lifetimes of the third and fourth items are right censored. Observed failure times are indicated by an $\times$ and right-censoring times are indicated by a $\bigcirc$. In Type II censoring, the time to complete the test is random.

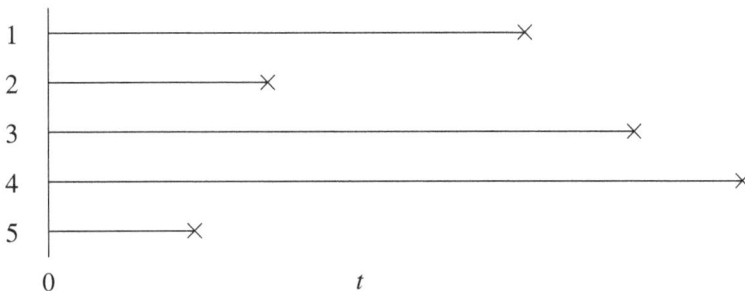

Figure 5.1: Complete data set with $n = 5$.

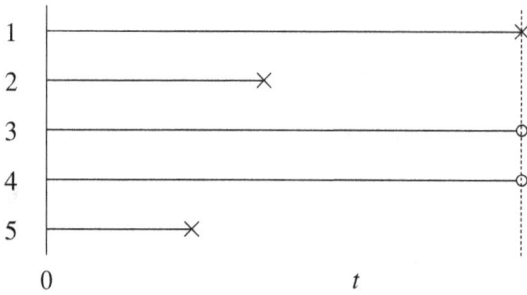

Figure 5.2: Type II right-censored data set with $n = 5$ and $r = 3$.

The second special case is Type I or *time* censoring. As shown in Figure 5.3, this corresponds to terminating the study at a particular time. The diagram shows a set of $n = 5$ items placed on a test simultaneously at $t = 0$ that is terminated at the time indicated by the dotted vertical line. For the realization illustrated in Figure 5.3, there are $r = 4$ observed failures. In Type I censoring, the number of failures r is random.

Finally, *random censoring* occurs when individual items are withdrawn from the test at any time during the study. Figure 5.4 illustrates a realization of a randomly right-censored life test with $n = 5$ items on test and $r = 2$ observed failures. It is usually assumed that the failure times and the censoring times are mutually independent random variables and that the probability distribution of the censoring times does not involve any unknown parameters from the failure time distribution. In other words, in a randomly censored data set, items cannot be more or less likely to be censored because they are at unusually high or low risk of failure.

Although other types of censoring exist, such as *left censoring* and *interval censoring*, the focus of this chapter will be on right censoring because it is the most common type of censoring. In the case of right censoring, the ratio r/n is the fraction of items which are observed to fail. When r/n is close to one, the data set is referred to as a *lightly censored* data set; when r/n is close to zero, the data set is referred to as a *heavily censored* data set. In the reliability setting, many data sets are heavily censored because the items have long lifetimes. In the biomedical setting, certain cancers have long remission times, resulting in heavily censored data sets.

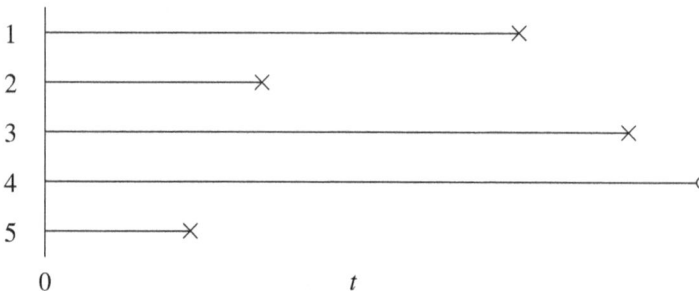

Figure 5.3: Type I right-censored data set with $n = 5$ and $r = 4$.

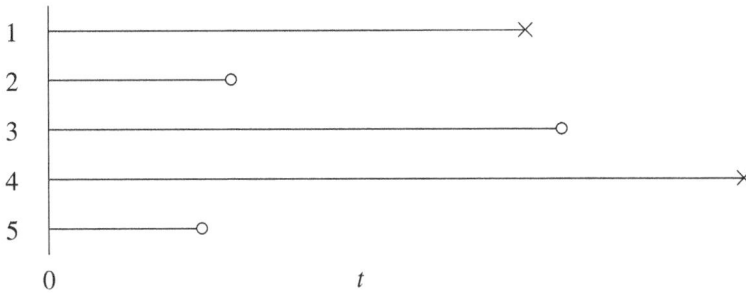

Figure 5.4: Randomly right-censored data set with $n = 5$ and $r = 2$.

Of the following three approaches to handling the problem of censoring, only one is both valid and practical. The first approach is to ignore all the censored values and to perform analysis only on those items that were observed to fail. Although this simplifies the mathematics involved, it is not a valid approach. If, for example, this approach is used on a right-censored data set, the analyst is discarding the right-censored values, and these are typically the items that have survived the longest. In this case, the analyst arrives at an overly pessimistic result concerning the lifetime distribution because the best items (that is, the right-censored observations) have been excluded from the analysis. A second approach is to wait for all the right-censored observations to fail. Although this approach is valid statistically, it is not practical. In an industrial setting, waiting for the last light bulb to burn out or the last machine to fail may take so long that the product being tested will not get to market in time. In a medical setting, waiting for the last patient to die from a particular disease may take decades. For these reasons, the proper approach is to handle censored observations probabilistically, including the censored values in the likelihood function.

The likelihood function for a censored data set can be written in several different equivalent forms. Let $t_1, t_2, \ldots, t_n$ be mutually independent observations denoting lifetimes sampled randomly from a population. The corresponding right-censoring times are denoted by $c_1, c_2, \ldots, c_n$. The t_i and c_i values are assumed to be independent, for $i = 1, 2, \ldots, n$. In the case of Type I right censoring, $c_1 = c_2 = \cdots = c_n = c$. The set U contains the indexes of the items that are observed to fail during the test (that is, the uncensored observations):

$$U = \{i \,|\, t_i \leq c_i\}.$$

The set C contains the indexes of the items whose failure time exceeds the corresponding censoring time (that is, those that are right censored):

$$C = \{i \,|\, t_i > c_i\}.$$

This notation, along with an important notion known as *alignment*, are illustrated in the next example.

Example 5.3 Consider the case of $n = 5$ items placed on test as indicated in Figure 5.5. Find the sets U and C.

Observe that the right-censored data set depicted in Figure 5.5, unlike the previous right-censored data sets with $n = 5$ items on test, does not have all of the items starting on test at time $t = 0$. This is quite common in practice. A software engineer, for example,

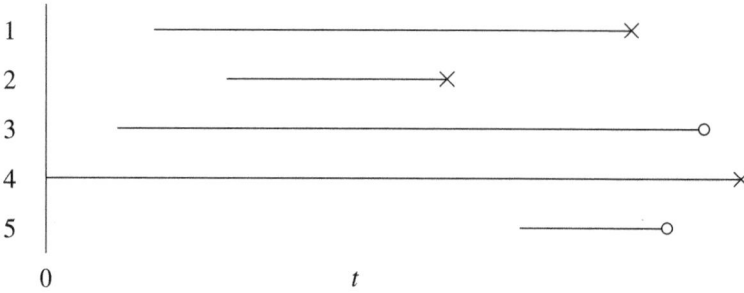

Figure 5.5: Randomly right-censored data set.

cannot get all customers to purchase a computer program at the same time; a medical researcher evaluating the time between first and second heart attacks cannot get all of the patients in the study to have their first heart attack at the same time; a casualty actuary cannot get all customers to purchase motorcycle insurance at the same time. In all cases, it is necessary to shift each data value back to a common origin. As long as there are not any changes to the items over the time window of observation, aligning the data values in this fashion is appropriate. Figure 5.6 displays the aligned data set. In this particular case, the first, second, and fourth items were observed to fail, and the failure times for the third and fifth items were right-censored. Therefore, the sets U and C are

$$U = \{1, 2, 4\} \qquad \text{and} \qquad C = \{3, 5\}.$$

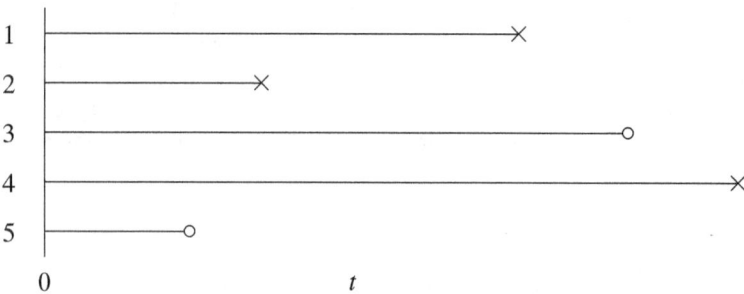

Figure 5.6: Aligned randomly right-censored data set.

The usual form for right-censored lifetime data is given by the pairs (x_i, δ_i), where $x_i = \min\{t_i, c_i\}$ and δ_i is a censoring indicator variable:

$$\delta_i = \begin{cases} 0 & t_i > c_i \\ 1 & t_i \leq c_i \end{cases}$$

for $i = 1, 2, \ldots, n$. The (x_i, δ_i) pairs can be reconstructed from the (t_i, c_i) pairs and vice versa. Hence, δ_i is 1 if the failure of item i is observed and 0 if the failure of item i is right censored, and

x_i is the failure time (when $\delta_i = 1$) or the censoring time (when $\delta_i = 0$). For the vector of unknown parameters $\boldsymbol{\theta} = (\theta_1, \theta_2, \ldots, \theta_p)'$, ignoring a constant factor, the likelihood function is

$$L(\boldsymbol{x}, \boldsymbol{\theta}) = \prod_{i=1}^{n} f(x_i, \boldsymbol{\theta})^{\delta_i} S(x_i, \boldsymbol{\theta})^{1-\delta_i} = \prod_{i \in U} f(t_i, \boldsymbol{\theta}) \prod_{i \in C} S(c_i, \boldsymbol{\theta})$$

where $S(c_i, \boldsymbol{\theta})$ is the survivor function of the population distribution with parameters $\boldsymbol{\theta}$ evaluated at censoring time c_i, $i \in C$. The reason that the survivor function is the appropriate term in the likelihood function for a right-censored observation is that $S(c_i, \boldsymbol{\theta})$ is the probability that item i survives to c_i. The log likelihood function is

$$\ln L(\boldsymbol{x}, \boldsymbol{\theta}) = \sum_{i \in U} \ln f(t_i, \boldsymbol{\theta}) + \sum_{i \in C} \ln S(c_i, \boldsymbol{\theta}),$$

or

$$\ln L(\boldsymbol{x}, \boldsymbol{\theta}) = \sum_{i \in U} \ln f(x_i, \boldsymbol{\theta}) + \sum_{i \in C} \ln S(x_i, \boldsymbol{\theta}).$$

Since the probability density function is the product of the hazard function and the survivor function, the log likelihood function can be simplified to

$$\ln L(\boldsymbol{x}, \boldsymbol{\theta}) = \sum_{i \in U} \ln h(x_i, \boldsymbol{\theta}) + \sum_{i \in U} \ln S(x_i, \boldsymbol{\theta}) + \sum_{i \in C} \ln S(x_i, \boldsymbol{\theta})$$

or

$$\ln L(\boldsymbol{x}, \boldsymbol{\theta}) = \sum_{i \in U} \ln h(x_i, \boldsymbol{\theta}) + \sum_{i=1}^{n} \ln S(x_i, \boldsymbol{\theta}),$$

where the second summation now includes all n items on test. Finally, to write the log likelihood in terms of the hazard and cumulative hazard functions only,

$$\ln L(\boldsymbol{x}, \boldsymbol{\theta}) = \sum_{i \in U} \ln h(x_i, \boldsymbol{\theta}) - \sum_{i=1}^{n} H(x_i, \boldsymbol{\theta}),$$

since $H(t) = -\ln S(t)$. The choice of which of these three expressions for the log likelihood to use for a particular distribution depends on the particular forms of $S(t)$, $f(t)$, $h(t)$, and $H(t)$. In other words, one of the distribution representations may possess a mathematical form that is advantageous over the others.

The next example will use the last version of the log likelihood function to find a maximum likelihood estimator and an asymptotically exact confidence interval for an unknown parameter.

Example 5.4 Consider a life test with n items on test with random right censoring and $r \geq 1$ observed failures. Assume that previous tests on these same items informs us that lifetimes of the items are drawn from a Rayleigh population with positive unknown parameter λ. Find the maximum likelihood estimator and construct an asymptotically exact two-sided $100(1 - \alpha)\%$ confidence interval for λ.

The survivor function for the Rayleigh distribution is

$$S(t) = e^{-(\lambda t)^2} \qquad t \geq 0.$$

The associated cumulative hazard function and hazard function are

$$H(t) = -\ln S(t) = (\lambda t)^2 \qquad t \geq 0$$

and
$$h(t) = H'(t) = 2\lambda^2 t \qquad\qquad t \geq 0.$$

In the case of random right censoring, the log likelihood function is

$$\ln L(\boldsymbol{x}, \lambda) = \sum_{i \in U} \ln h(x_i, \lambda) - \sum_{i=1}^{n} H(x_i, \lambda)$$

$$= \sum_{i \in U} \ln\left(2\lambda^2 x_i\right) - \sum_{i=1}^{n} (\lambda x_i)^2$$

$$= r \ln 2 + 2r \ln \lambda + \sum_{i \in U} \ln x_i - \lambda^2 \sum_{i=1}^{n} x_i^2,$$

where r is the number of observed failures. The single-element score vector can be found by differentiating the log likelihood function with respect to λ:

$$\frac{\partial \ln L(\boldsymbol{x}, \lambda)}{\partial \lambda} = \frac{2r}{\lambda} - 2\lambda \sum_{i=1}^{n} x_i^2.$$

Equating the score to zero and solving for λ yields the maximum likelihood estimator

$$\hat{\lambda} = \sqrt{\frac{r}{\sum_{i=1}^{n} x_i^2}}.$$

The second derivative of the log likelihood function is

$$\frac{\partial^2 \ln L(\boldsymbol{x}, \lambda)}{\partial \lambda^2} = -\frac{2r}{\lambda^2} - 2\sum_{i=1}^{n} x_i^2.$$

As an aside, the 1×1 Fisher information matrix

$$I(\lambda) = E\left[-\frac{\partial^2 \ln L(\boldsymbol{x}, \lambda)}{\partial \lambda^2}\right] = E\left[\frac{2r}{\lambda^2} + 2\sum_{i=1}^{n} x_i^2\right]$$

cannot be calculated without knowing the probability distribution of the censoring times. The observed information matrix, however, can be calculated as

$$O(\hat{\lambda}) = \left[-\frac{\partial^2 \ln L(\boldsymbol{x}, \lambda)}{\partial \lambda^2}\right]_{\lambda = \hat{\lambda}} = \frac{2r}{\hat{\lambda}^2} + 2\sum_{i=1}^{n} x_i^2 = 4\sum_{i=1}^{n} x_i^2.$$

For large values of n, we know that

$$\hat{\lambda} \stackrel{a}{\sim} N\left(\lambda, O^{-1}(\hat{\lambda})\right)$$

or

$$\hat{\lambda} \stackrel{a}{\sim} N\left(\lambda, \left(4\sum_{i=1}^{n} x_i^2\right)^{-1}\right).$$

Standardizing by subtracting the population mean and dividing by the population standard deviation of $\hat{\lambda}$ gives

$$\frac{\hat{\lambda} - \lambda}{\left(4\sum_{i=1}^{n} x_i^2\right)^{-1/2}} \stackrel{a}{\sim} N(0, 1),$$

which implies that

$$\lim_{n \to \infty} P\left(-z_{\alpha/2} < \frac{\hat{\lambda} - \lambda}{\left(4\sum_{i=1}^{n} x_i^2\right)^{-1/2}} < z_{\alpha/2}\right) = 1 - \alpha.$$

Performing the algebra required to isolate λ in the center of the inequality results in an asymptotically exact two-sided $100(1 - \alpha)\%$ confidence interval for λ:

$$\hat{\lambda} - z_{\alpha/2}\left(4\sum_{i=1}^{n} x_i^2\right)^{-1/2} < \lambda < \hat{\lambda} + z_{\alpha/2}\left(4\sum_{i=1}^{n} x_i^2\right)^{-1/2}.$$

This confidence interval narrows as $\sum_{i=1}^{n} x_i^2$ increases. So placing a large number of items on test with a lightly censored data set with r/n close to one will result in a narrow confidence interval for λ.

To provide a numerical illustration, assume that the $n = 5$ items on a randomly right-censored life test with $r = 3$ observed failures illustrated in Figure 5.6 are

$$1.3, 0.6, 1.6^*, 1.9, 0.4^*,$$

where the superscript $*$ denotes a right-censored observation. For this data set,

$$\sum_{i=1}^{n} x_i^2 = 1.3^2 + 0.6^2 + 1.6^2 + 1.9^2 + 0.4^2 = 1.69 + 0.36 + 2.56 + 3.61 + 0.16 = 8.38.$$

The maximum likelihood estimate of λ is

$$\hat{\lambda} = \sqrt{\frac{r}{\sum_{i=1}^{n} x_i^2}} = \sqrt{\frac{3}{8.38}} = 0.598.$$

An asymptotically exact two-sided 95% confidence interval for λ is

$$0.598 - 1.96\,(4 \cdot 8.38)^{-1/2} < \lambda < 0.598 + 1.96\,(4 \cdot 8.38)^{-1/2}$$

or

$$0.260 < \lambda < 0.937.$$

To summarize the material introduced so far in this chapter, point estimators are statistics calculated from a data set to estimate an unknown parameter. Confidence intervals reflect the precision of a point estimator. The most common technique for determining a point estimator for an unknown parameter is maximum likelihood estimation, which involves finding the parameter value(s) that make the observed data values the most likely. The maximum likelihood estimators are usually found by using calculus to maximize the log likelihood function. Most population lifetime distributions do not have exact confidence intervals for unknown parameters, so the asymptotic properties of the likelihood function can be used to generate approximate confidence intervals for unknown parameters. Finally, many data sets in reliability are censored, which means that only a bound is known on the lifetime for one or more of the data values. The most common censoring mechanism is known as right censoring, where only a lower bound on the lifetime is known. The number of items on test is denoted by n and the number of observed failures is denoted by r.

The next section applies the techniques developed so far in this chapter to the exponential distribution.

5.4   Exponential Distribution

The exponential distribution is popular due to its tractability for parameter estimation and inference. The exponential distribution can be parameterized by either its population rate λ or its population mean $\mu = 1/\lambda$. Using the rate to parameterize the distribution, the survivor, density, hazard, and cumulative hazard functions are

$$S(t, \lambda) = e^{-\lambda t} \qquad f(t, \lambda) = \lambda e^{-\lambda t} \qquad h(t, \lambda) = \lambda \qquad H(t, \lambda) = \lambda t$$

for $t \geq 0$. Note that the unknown parameter λ has been added as an argument in these lifetime distribution representations because it is now also an argument in the likelihood function and is estimated from data.

All the analysis in this and subsequent sections assumes that a *random sample* of n items from a population has been placed on a test and subjected to typical environmental conditions. Equivalently, $t_1, t_2, \ldots, t_n$ are independent and identically distributed random lifetimes from a particular population distribution (exponential in this section). As with all statistical inference, care must be taken to ensure that a random sample of lifetimes is collected. Consequently, random numbers should be used to determine which n items to place on test. In a reliability setting, laboratory conditions should adequately mimic field conditions. Only representative items should be placed on test because items manufactured using a previous design may have a different failure pattern than those with the current design. This is more difficult in a biomedical setting because of inherent differences between patients.

Four classes of data sets (complete, Type II right censored, Type I right censored, and randomly right censored) are considered in separate subsections. In all cases, n is the number of items placed on test and r is the number of observed failures.

5.4.1   Complete Data Sets

A complete data set is typically the easiest to analyze because extensive analytical work exists for finding point and interval estimators for parameters. Also, by testing each item to failure, we have equal confidence in the fitted model in both the left-hand and right-hand tails of the distribution. A heavily right-censored data set, on the other hand, might fit well in the left-hand tail of the distribution where failures were observed, but we have less confidence in the right-hand tail of the distribution where there were few or no failures.

A complete data set consists of failure times $t_1, t_2, \ldots, t_n$. Although lowercase letters are used to denote the failure times here to be consistent with the notation for censoring times, the failure times are nonnegative random variables. The likelihood function can be written as a product of the probability density functions evaluated at the failure times:

$$L(\lambda) = \prod_{i=1}^{n} f(t_i, \lambda).$$

Note that the t argument has been left out of the likelihood expression for compactness. Using the last expression for the log likelihood function (adapted for a complete data set) from Section 5.3,

$$\ln L(\lambda) = \sum_{i=1}^{n} \left[\ln h(t_i, \lambda) - H(t_i, \lambda) \right].$$

For the exponential distribution, this is

$$\ln L(\lambda) = \sum_{i=1}^{n} \left[\ln \lambda - \lambda t_i \right] = n \ln \lambda - \lambda \sum_{i=1}^{n} t_i.$$

To determine the maximum likelihood estimator for λ, the single-element score vector

$$U(\lambda) = \frac{\partial \ln L(\lambda)}{\partial \lambda} = \frac{n}{\lambda} - \sum_{i=1}^{n} t_i,$$

also known as the *score statistic*, is equated to zero and solved for λ, yielding

$$\hat{\lambda} = \frac{n}{\sum_{i=1}^{n} t_i},$$

where the denominator is often referred to as the *total time on test*. Not surprisingly, the maximum likelihood estimator $\hat{\lambda}$ is the reciprocal of the sample mean.

Theorem 5.2 Let $t_1, t_2, \ldots, t_n$ be the observed values of n mutually independent and identically distributed exponential(λ) random variables. The maximum likelihood estimator of λ is

$$\hat{\lambda} = \frac{n}{\sum_{i=1}^{n} t_i}.$$

Example 5.5 A complete data set of $n = 23$ ball bearing failure times associated with testing the endurance of deep-groove ball bearings has been extensively studied. The failure times measured in 10^6 revolutions, ordered for readability, are

17.88	28.92	33.00	41.52	42.12	45.60	48.48	51.84	51.96
54.12	55.56	67.80	68.64	68.64	68.88	84.12	93.12	98.64
		105.12	105.84	127.92	128.04	173.40.		

Notice that there is a single tied value of 68.64 million revolutions. Fit the exponential distribution to the $n = 23$ ball bearing failure times.

For this particular data set, the total time on test is $\sum_{i=1}^{n} t_i = 1661.16$ million revolutions, yielding the maximum likelihood estimate

$$\hat{\lambda} = \frac{n}{\sum_{i=1}^{n} t_i} = \frac{23}{1661.16} = 0.01385$$

failure per 10^6 revolutions. The number of significant digits reported in the point estimate matches the number of digits in the data set. The value of the log likelihood function at the maximum likelihood estimate is $\ln L(\hat{\lambda}) = -121.435$, which will be used later in this chapter to compare the exponential and Weibull fits to this data set.

Figure 5.7 displays a graph of the *empirical survivor function*, which takes a downward step of $1/n = 1/23$ at each data value, along with the fitted exponential survivor function $S(t) = e^{-\hat{\lambda}t}$. Empirical and fitted distributions are traditionally compared by plotting the two survivor functions on the same set of axes because the probability density function and hazard function suffer from the drawback of requiring the data to be divided into cells to plot the empirical distribution. It is apparent from this figure that the exponential distribution is a very poor fit. This particular data set was chosen for this example to illustrate one of the shortcomings of using the exponential distribution to model any data set without assessing the adequacy of the fit. Extreme caution must

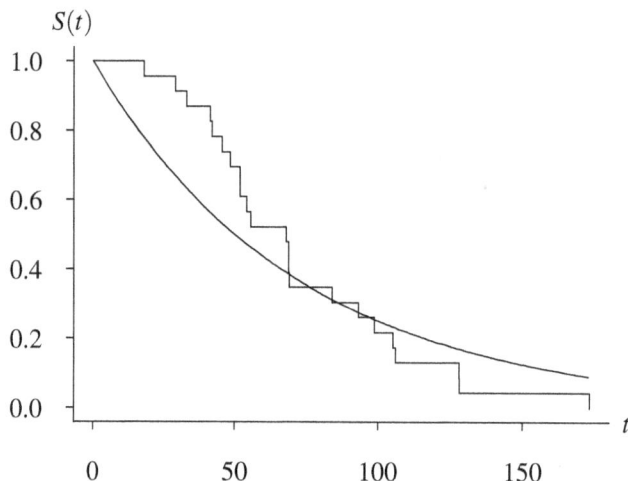

Figure 5.7: Empirical and exponential fitted survivor functions for the ball bearing data set.

be exercised when using the exponential distribution since, as indicated in Figure 5.7, the exponential distribution is not an adequate probability model for this data set.

There are two clues that the exponential distribution would perform poorly in this setting. First, we neglected to plot a histogram of the ball bearing failure times prior to fitting the exponential distribution. The histogram in Figure 5.8 indicates a nonzero mode to the population probability density function, implying that the exponential distribution is probably not going to be an adequate model. Second, knowing the physics of failure can be helpful in this case. Ball bearings typically fail by wearing out. When a ball bearing's diameter falls outside of a prescribed range, it is considered to be failed. This indicates that the hazard function for a ball bearing will probably increase over time, so a distribution with a monotone increasing hazard function from the IFR class

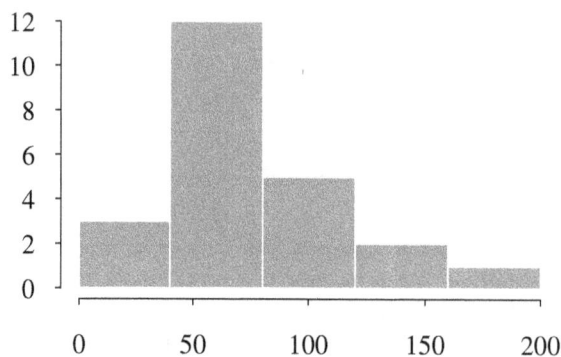

Figure 5.8: Histogram of the ball bearing failure times.

would be a better choice than the exponential distribution. As shown in the next section, the Weibull distribution provides a much better approximation to this particular data set. Since the exponential distribution can be fitted to any data set that has at least one observed failure, the adequacy of the model must always be assessed. The point and interval estimators associated with the exponential distribution are legitimate only if the data set is a random sample drawn from an exponential population. That is almost certainly *not* the case for this particular data set.

Information matrices. To find the information matrix associated with a complete data set from an exponential(λ) population, the derivative of the score statistic is required:

$$\frac{\partial^2 \ln L(\lambda)}{\partial \lambda^2} = -\frac{n}{\lambda^2}.$$

Taking the expected value of the negative of this quantity yields the 1×1 Fisher information matrix

$$I(\lambda) = E\left[\frac{-\partial^2 \ln L(\lambda)}{\partial \lambda^2}\right] = E\left[\frac{n}{\lambda^2}\right] = \frac{n}{\lambda^2}.$$

If the maximum likelihood estimator $\hat{\lambda}$ is used as an argument in the negative of the second partial derivative of the log likelihood function, the 1×1 observed information matrix is obtained:

$$O(\hat{\lambda}) = \left[\frac{-\partial^2 \ln L(\lambda)}{\partial \lambda^2}\right]_{\lambda = \hat{\lambda}} = \frac{n}{\hat{\lambda}^2} = \frac{\left(\sum_{i=1}^{n} t_i\right)^2}{n}.$$

Confidence interval for λ. Asymptotic confidence intervals for λ based on the likelihood ratio statistic or the observed information matrix are unnecessary for a complete data set because the sampling distribution of $\sum_{i=1}^{n} t_i$ is tractable. In particular, from Theorem 4.5,

$$2\lambda \sum_{i=1}^{n} t_i = \frac{2n\lambda}{\hat{\lambda}}$$

has the chi-square distribution with $2n$ degrees of freedom. Therefore, with probability $1 - \alpha$,

$$\chi^2_{2n,\, 1-\alpha/2} < \frac{2n\lambda}{\hat{\lambda}} < \chi^2_{2n,\, \alpha/2},$$

where $\chi^2_{2n,\, p}$ is the $(1 - p)$th fractile of the chi-square distribution with $2n$ degrees of freedom. Performing the algebra required to isolate λ in the middle of the inequality yields an exact two-sided $100(1 - \alpha)\%$ confidence interval for λ.

Theorem 5.3 Let $t_1, t_2, \ldots, t_n$ be the observed values of n mutually independent and identically distributed exponential(λ) random variables. Let $\hat{\lambda}$ denote the maximum likelihood estimator of λ. An exact two-sided $100(1 - \alpha)\%$ confidence interval for λ is

$$\frac{\hat{\lambda}\chi^2_{2n,\, 1-\alpha/2}}{2n} < \lambda < \frac{\hat{\lambda}\chi^2_{2n,\, \alpha/2}}{2n}.$$

A well-known example of a randomly right-censored data set is drawn from the biostatistical literature. The focus here is on determining an estimate of the remission rate of a complete data set of remission times for patients in a control group.

Example 5.6 A clinical trial is conducted to determine the effect of an experimental drug named 6–mercaptopurine (6–MP) on leukemia remission times. A sample of $n = 21$ leukemia patients is treated with 6–MP, and the remission times are recorded. There are $r = 9$ individuals for whom the remission time is observed, and the remission times for the remaining 12 individuals are randomly censored on the right. Letting an asterisk denote a right-censored observation, the remission times (in weeks) are

$$
\begin{array}{ccccccccccc}
6 & 6 & 6 & 6^* & 7 & 9^* & 10 & 10^* & 11^* & 13 & 16 \\
17^* & 19^* & 20^* & 22 & 23 & 25^* & 32^* & 32^* & 34^* & 35^*.
\end{array}
$$

In addition, 21 other leukemia patients are not given the drug, and they serve as a control group. For this group there is no censoring and the remission times are

$$
\begin{array}{ccccccccccc}
1 & 1 & 2 & 2 & 3 & 4 & 4 & 5 & 5 & 8 & 8 \\
8 & 8 & 11 & 11 & 12 & 12 & 15 & 17 & 22 & 23.
\end{array}
$$

This data set illustrates the simplest possible use of a covariate for modeling: a single binary covariate indicating the group to which each data value belongs. Fit the exponential distribution to the $n = 21$ remission times in the control group of the 6–MP clinical trial. Give a point estimator and a 95% confidence interval for λ.

Having learned our lesson from the previous example, we begin by drawing a histogram of the remission times, which is displayed in Figure 5.9. The shape of the histogram reveals significant random sampling variability which can be attributed to the small ($n = 21$) number patients in the control group. Modeling the remission times with a probability distribution that has a mode of zero seems reasonable based on the shape of the histogram, so we will proceed with fitting the exponential distribution.

The total time on test is

$$
\sum_{i=1}^{21} t_i = 182
$$

weeks. The maximum likelihood estimate is

$$
\hat{\lambda} = \frac{n}{\sum_{i=1}^{n} t_i} = \frac{21}{182} = 0.12
$$

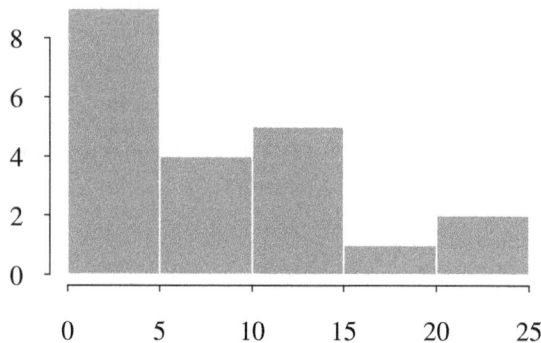

Figure 5.9: Histogram of the leukemia remission times.

remission per week. Figure 5.10 shows the empirical survivor function, which takes a downward step of $1/n = 1/21$ at each data point, along with the survivor function for the fitted exponential distribution. In spite of the discrete nature of the data, the excessive number of ties, and the fact that the number of patients in the control group is rather small, the exponential distribution does a reasonable job of approximating the empirical survivor function.

The observed information matrix is

$$O(\hat{\lambda}) = \left[\frac{-\partial^2 \ln L(\lambda)}{\partial \lambda^2} \right]_{\lambda=\hat{\lambda}} = \frac{(\sum_{i=1}^n t_i)^2}{n} = \frac{182^2}{21} = 1577.$$

Since the data set is complete, an exact two-sided 95% confidence interval for the failure rate of the distribution can be determined. Since $\chi^2_{42, 0.975} = 26.0$ and $\chi^2_{42, 0.025} = 61.8$, the formula for the confidence interval

$$\frac{\hat{\lambda} \chi^2_{2n, 1-\alpha/2}}{2n} < \lambda < \frac{\hat{\lambda} \chi^2_{2n, \alpha/2}}{2n}$$

becomes

$$\frac{(0.12)(26.0)}{42} < \lambda < \frac{(0.12)(61.8)}{42}$$

or

$$0.071 < \lambda < 0.17.$$

The involvement of the non-symmetric chi-square distribution in this confidence interval means that the interval is not symmetric about the maximum likelihood estimate. For this and subsequent examples, intermediate calculations involving numeric quantities, such as critical values or total time on test values, are performed to as much precision as possible, then final values are reported using only significant digits.

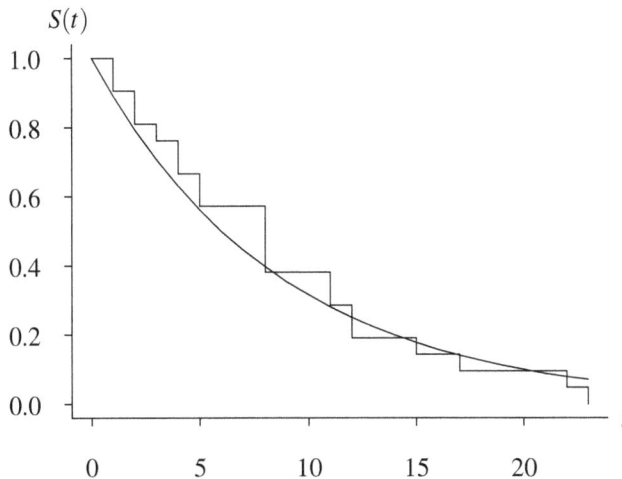

Figure 5.10: Empirical and exponential fitted survivor functions for the 6–MP control group.

The R code given below calculates the maximum likelihood estimator $\hat{\lambda}$, calculates the endpoints of the exact two-sided confidence interval for λ, and conducts the Kolmogorov–Smirnov goodness-of-fit test. $p = 0.55$.

```
x  = c(1, 1, 2, 2, 3, 4, 4, 5, 5, 8, 8, 8, 8, 11, 11, 12, 12, 15,
       17, 22, 23)
n  = length(x)
l  = n / sum(x)
lo = l * qchisq(0.025, 2 * n) / (2 * n)
hi = l * qchisq(0.975, 2 * n) / (2 * n)
p  = ks.test(x, "pexp", l, exact = FALSE)$p.value
print(c(lo, l, hi, p))
```

Since the p-value for the Kolmogorov–Smirnov test is $p = 0.55$, there is not sufficient evidence in the data to reject the null hypothesis that the data values were drawn from an exponential population. This conclusion is consistent with the empirical and exponential fitted survivor functions in Figure 5.10. The exponential distribution provides a reasonable approximation to the leukemia remission times.

The importance of assessing model adequacy applies to all fitted distributions—not just the exponential distribution. Furthermore, if a modeler knows the failure physics (for example, fatigue crack growth) underlying a process, then an appropriate probability model that is consistent with the failure physics should be chosen.

So far we have fitted the exponential distribution to two complete data sets: the ball bearing failure times from Example 5.5 and the 6–MP remission times for the control group from Example 5.6. We visually assessed the two fits in Figures 5.7 and 5.10 by comparing the empirical survivor function, which takes a downward step of $1/n$ at each data value, with the fitted survivor function $S(t) = e^{-\hat{\lambda}t}$ and concluded that the exponential distribution did a very poor job of approximating the ball bearing failure times and a (barely) adequate job of approximating the remission times of the patients in the control group of the 6–MP clinical trial. This visual assessment was subjective and was followed by a formal goodness-of-fit test in order to draw these conclusions for the 6–MP remission times.

Confidence intervals for measures other than λ. It is possible to find point and interval estimators for measures other than λ by using the invariance property for maximum likelihood estimators and by rearranging the confidence interval formula. Define

$$L = \frac{\hat{\lambda}\chi^2_{2n,\,1-\alpha/2}}{2n} \qquad \text{and} \qquad U = \frac{\hat{\lambda}\chi^2_{2n,\,\alpha/2}}{2n}$$

as the lower and upper bounds on the exact two-sided $100(1-\alpha)\%$ confidence interval for λ. If the measure of interest is $\mu = 1/\lambda$, for example, then the point estimator is the sample mean $\hat{\mu} = \frac{1}{n}\sum_{i=1}^{n} t_i$. Rearranging the confidence interval

$$L < \lambda < U$$

by taking reciprocals yields the exact two-sided $100(1-\alpha)\%$ confidence interval for μ:

$$\frac{1}{U} < \mu < \frac{1}{L}.$$

As a second example, consider the probability of survival to a fixed time t, $S(t) = e^{-\lambda t}$. By the invariance property of maximum likelihood estimators, the maximum likelihood estimator for the survivor function at time t is

$$\hat{S}(t) = e^{-\hat{\lambda}t}.$$

A confidence interval for $S(t)$, on the other hand, can be found by rearranging the confidence interval

$$L < \lambda < U$$

in the following fashion:

$$-U < -\lambda < -L$$
$$e^{-Ut} < e^{-\lambda t} < e^{-Lt}$$
$$e^{-Ut} < S(t) < e^{-Lt}.$$

These formulas for point and interval estimates for quantities other than λ are illustrated next for a complete data set that is assumed to be drawn from an exponential population.

Example 5.7 Assuming that the exponential distribution is an appropriate model for the remission times in the control group of the 6–MP clinical trial, find point estimators and exact two-sided 95% confidence intervals for the mean remission time and the probability that a patient in the control group has a remission time that exceeds 10 weeks.

The point estimators in this case are

$$\hat{\mu} = \frac{1}{n}\sum_{i=1}^{n} t_i = \frac{182}{21} = 8.7$$

weeks and

$$\hat{S}(t) = e^{-\hat{\lambda}t},$$

which is $\hat{S}(10) = e^{-(0.12)(10)} = 0.32$. The values of L and U for the exact two-sided 95% confidence interval for λ from the previous example are $L = 0.071$ and $U = 0.17$. Finding a confidence interval for the population mean requires taking reciprocals of these limits:

$$\frac{1}{0.17} < \mu < \frac{1}{0.071}$$
$$5.9 < \mu < 14.$$

An exact two-sided 95% confidence interval for $S(100)$ using the formula

$$e^{-Ut} < S(t) < e^{-Lt}$$

results in

$$e^{-(0.17)(10)} < S(100) < e^{-(0.071)(10)}$$

or

$$0.18 < S(10) < 0.49.$$

Although the manipulation of the confidence interval for λ is performed here in the case of a complete data set, these techniques may also be applied to any of the right-censoring mechanisms to be described in the next three subsections.

5.4.2   Type II Censored Data Sets

A life test of n items that is terminated when r failures have occurred produces a Type II right-censored data set. The previous subsection is a special case of Type II censoring when $r = n$. As before, assume that the failure times are $t_1, t_2, \ldots, t_n$, the test is terminated upon the rth ordered failure, the censoring times are $c_1 = c_2 = \cdots = c_n = t_{(r)}$ for all items, and $x_i = \min\{t_i, c_i\}$ for $i = 1, 2, \ldots, n$.

Since $h(t, \lambda) = \lambda$ and $H(t, \lambda) = \lambda t$ for $t \geq 0$, the log likelihood function is

$$\ln L(\lambda) = \sum_{i \in U} \ln h(x_i, \lambda) - \sum_{i=1}^{n} H(x_i, \lambda) = r \ln \lambda - \lambda \sum_{i=1}^{n} x_i$$

because there are r observed failures. The expression

$$\sum_{i=1}^{n} x_i = \sum_{i \in U} t_i + \sum_{i \in C} c_i = \sum_{i=1}^{r} t_{(i)} + (n - r) t_{(r)},$$

where $t_{(1)} < t_{(2)} < \cdots < t_{(r)}$ are the order statistics of the observed failure times, is the total time on test. It represents the total accumulated time that the n items accrue while on test.

To determine the maximum likelihood estimator, the log likelihood function is differentiated with respect to λ,

$$U(\lambda) = \frac{\partial \ln L(\lambda)}{\partial \lambda} = \frac{r}{\lambda} - \sum_{i=1}^{n} x_i$$

and is equated to zero, yielding the maximum likelihood estimator.

Theorem 5.4 Let $t_1, t_2, \ldots, t_n$ be the observed values of n mutually independent and identically distributed exponential(λ) random variables. The associated test is terminated at time $t_{(r)}$ (Type II right censoring) for $r \geq 1$. The censoring times are $c_1 = c_2 = \cdots = c_n = t_{(r)}$ for all items, and $x_i = \min\{t_i, c_i\}$ for $i = 1, 2, \ldots, n$. The maximum likelihood estimator of λ is

$$\hat{\lambda} = \frac{r}{\sum_{i=1}^{n} x_i}.$$

So the maximum likelihood estimator of the failure rate is the ratio of the number of observed failures to the total time on test. The second partial derivative of the log likelihood function is

$$\frac{\partial^2 \ln L(\lambda)}{\partial \lambda^2} = -\frac{r}{\lambda^2},$$

so the information matrix is

$$I(\lambda) = E\left[\frac{-\partial^2 \ln L(\lambda)}{\partial \lambda^2}\right] = \frac{r}{\lambda^2},$$

and the observed information matrix is

$$O(\hat{\lambda}) = \left[\frac{-\partial^2 \ln L(\lambda)}{\partial \lambda^2}\right]_{\lambda = \hat{\lambda}} = \frac{r}{\hat{\lambda}^2} = \frac{\left(\sum_{i=1}^{n} x_i\right)^2}{r}.$$

Exact confidence intervals and hypothesis tests concerning λ can be derived by using the result

$$2\lambda \sum_{i=1}^{n} x_i = \frac{2r\lambda}{\hat{\lambda}} \sim \chi^2(2r),$$

where $\chi^2(2r)$ is the chi-square distribution with $2r$ degrees of freedom. This result can be proved in a similar fashion to Theorem 4.5 of the exponential distribution from Section 4.2. Using this fact, an exact two-sided confidence interval for λ can be constructed in a similar fashion to that for a complete data set. It can be stated with probability $1 - \alpha$ that

$$\chi^2_{2r,\,1-\alpha/2} < \frac{2r\lambda}{\hat{\lambda}} < \chi^2_{2r,\,\alpha/2}.$$

Rearranging terms yields an exact two-sided $100(1 - \alpha)\%$ confidence interval for the failure rate λ.

Theorem 5.5 Let $t_1, t_2, \ldots, t_n$ be the observed values of n mutually independent and identically distributed exponential(λ) random variables. The associated test is terminated at time $t_{(r)}$ (Type II right censoring) for $r \geq 1$. The censoring times are $c_1 = c_2 = \cdots = c_n = t_{(r)}$ for all items, and $x_i = \min\{t_i, c_i\}$ for $i = 1, 2, \ldots, n$. An exact two-sided $100(1 - \alpha)\%$ confidence interval for the failure rate λ is

$$\frac{\hat{\lambda}\chi^2_{2r,\,1-\alpha/2}}{2r} < \lambda < \frac{\hat{\lambda}\chi^2_{2r,\,\alpha/2}}{2r}.$$

Example 5.8 A Type II right-censored data set of $n = 15$ automotive a/c switches has been collected. The test was terminated when the fifth failure occurred. The $r = 5$ ordered observed failure times measured in number of cycles are

$$t_{(1)} = 1410, \qquad t_{(2)} = 1872, \qquad t_{(3)} = 3138, \qquad t_{(4)} = 4218, \qquad t_{(5)} = 6971.$$

The remaining 10 automotive a/c switches are right-censored at 6971 cycles. Any parametric model that is fitted to this data set is only considered valid from 0 to 6971 cycles unless there is some evidence (perhaps from previous test results) that indicates that the parametric model is valid beyond 6971 cycles. Fit the exponential distribution to this data set and give point and interval estimates for the failure rate and the mean time to failure.

A diagram that can be helpful in visualizing lifetime data sets is given in Figure 5.11. The top five horizontal lines ending with $\times$ denote the $r = 5$ observed failures and the bottom $n - r = 15 - 5 = 10$ horizontal lines ending with $\circ$ denote the right-censored observations at 6971 cycles. Each of the right-censored observations will have an unseen $\times$ somewhere to the right of the censoring time indicated by the $\circ$. It is a worthwhile thought experiment to imagine where those $\times$s might occur for each right-censored observation in this particular data set. Once you have visualized the approximate positions of the ten right-censored failure times, try to guess the approximate population mean of the 15 failure times.

For this particular data set, the total time on test is $\sum_{i=1}^{n} x_i = 87{,}319$ cycles, yielding a maximum likelihood estimate

$$\hat{\lambda} = \frac{r}{\sum_{i=1}^{n} x_i} = \frac{5}{87{,}319} = 0.00005726$$

failure per cycle. Equivalently, the maximum likelihood estimate of the population mean time to failure is

$$\hat{\mu} = \frac{\sum_{i=1}^{n} x_i}{r} = \frac{87{,}319}{5} = 17{,}464$$

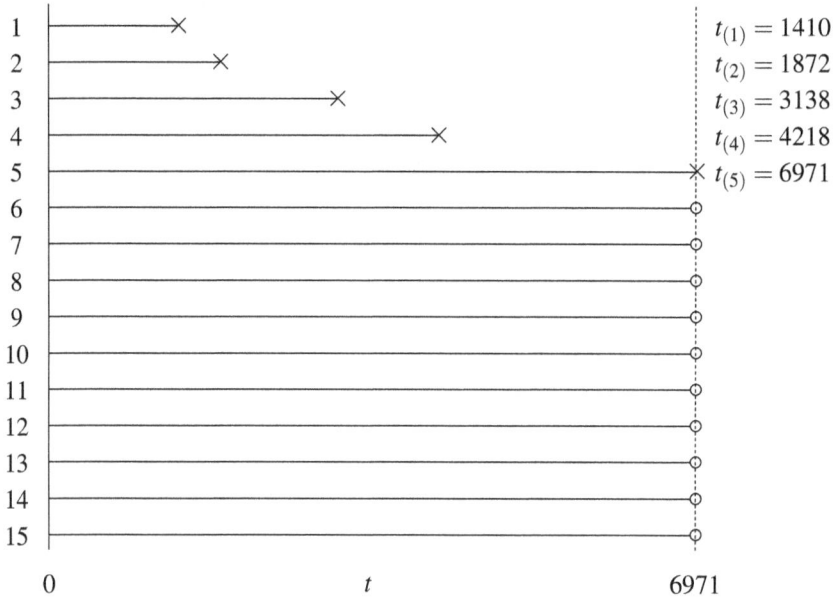

Figure 5.11: Automotive switches failure and censoring times with $n = 5$ and $r = 3$.

cycles. Notice that the estimated mean time to failure exceeds the largest observed failure time, $t_{(5)} = 6971$. As long as there is evidence, perhaps from previous testing on identical or similar automotive switches, to support the exponential failure time distribution, this estimate of the population mean time to failure is meaningful. Figure 5.12

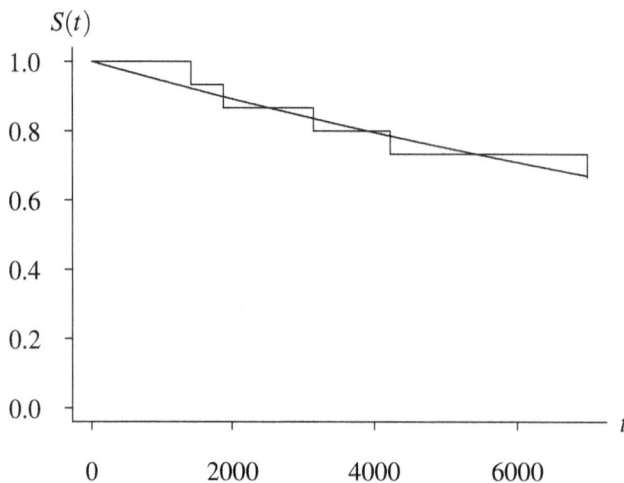

Figure 5.12: Empirical and exponential fitted survivor functions for the a/c switch data set.

shows the empirical survivor function (which takes downward steps of $1/n = 1/15$ at each of the five observed failure times) and the associated fitted exponential survivor function. In this case, the exponential distribution appears to adequately model the lifetimes through 6971 cycles. The confidence intervals given below are only exact when the data values are drawn from an exponential population, so assessing the fit is a crucial part of data analysis. Assessing the adequacy of the fit is more difficult for a right-censored data set because it is impossible to determine what the lifetime distribution looks like after the last observed failure time (6971 cycles for this data set) unless previous test results support the exponential model.

The observed information matrix based on using the failure rate as the unknown parameter is

$$O(\hat{\lambda}) = \left[\frac{-\partial^2 \ln L(\lambda)}{\partial \lambda^2}\right]_{\lambda=\hat{\lambda}} = \frac{(\sum_{i=1}^n x_i)^2}{r} = \frac{(87{,}319)^2}{5} = 1{,}525{,}000{,}000.$$

Since the data set is Type II right censored, an exact two-sided 95% confidence interval for the failure rate of the distribution can be determined. Using the chi-square critical values, $\chi^2_{10,0.975} = 3.247$ and $\chi^2_{10,0.025} = 20.49$, the formula for the confidence interval

$$\frac{\hat{\lambda}\chi^2_{2r,1-\alpha/2}}{2r} < \lambda < \frac{\hat{\lambda}\chi^2_{2r,\alpha/2}}{2r}$$

becomes

$$\frac{(0.00005726)(3.247)}{10} < \lambda < \frac{(0.00005726)(20.49)}{10}$$

or

$$0.00001859 < \lambda < 0.0001173.$$

Taking reciprocals, this is equivalent to an exact two-sided 95% confidence interval for the population mean number of cycles to failure of

$$8526 < \mu < 53{,}785.$$

Not surprisingly, with only $r = 5$ observed failures, this is a rather wide confidence interval for μ, and hence there is not as much precision as in the case of the 6–MP control group data, in which there were $n = r = 21$ observed remission times. The R code below calculates the point estimates for λ and μ and the associated exact two-sided 95% confidence intervals.

```
n       = 15
r       = 5
x       = c(1410, 1872, 3138, 4218, 6971, rep(6971, n - r))
lam     = r / sum(x)
mu      = 1 / lam
lam.lo  = lam * qchisq(0.025, 2 * r) / (2 * r)
lam.hi  = lam * qchisq(0.975, 2 * r) / (2 * r)
mu.lo   = 1 / lam.hi
mu.hi   = 1 / lam.lo
```

Hypothesis testing, which is the rough equivalent of interval estimation, is also possible in the case of Type II censoring because the sampling distribution of $2\lambda \sum_{i=1}^{n} x_i$ is tractable. Some aspects of hypothesis testing in the setting of Type II censoring, such as the alternative hypothesis, one- and two-tailed tests, and p-values are illustrated in the next example. The example shows how a life test can be used to check a manufacturer's claimed mean time to failure.

Example 5.9 The producer of the automotive switches tested in the previous example claims that the population mean time to failure of their switches is $\mu = 100{,}000$ cycles. Is there enough evidence in the data set of 15 switches placed on test to conclude that the population mean time to failure is less than 100,000 cycles? Assume that the automotive switch lifetimes are exponentially distributed.

The producer's claim is certainly suspect because the maximum likelihood estimator for the population mean time to failure is only $\hat{\mu} = 17{,}464$ from the previous example. The null and alternative hypotheses for the hypothesis test are

$$H_0 : \mu = 100{,}000$$
$$H_1 : \mu < 100{,}000$$

or, equivalently,

$$H_0 : \lambda = 0.00001$$
$$H_1 : \lambda > 0.00001$$

in terms of the failure rate. So the hypothesis test being conducted here is to determine whether there is statistically significant evidence in the data set to conclude that the population mean time to failure of the switches is less then 100,000 cycles. Since small values of $\sum_{i=1}^{n} x_i$ lead to rejecting H_0, the attained level of significance (p-value) is

$$p = P\left(\sum_{i=1}^{n} x_i < 87{,}319 \,\middle|\, \lambda = 0.00001 \right).$$

Since $2\lambda \sum_{i=1}^{n} x_i \sim \chi^2(2r)$, the p-value, when H_0 is true, is

$$p = P\left((2)(0.00001) \sum_{i=1}^{n} x_i < (2)(0.00001)(87{,}319) \right)$$
$$= P\left(\chi^2(10) < 1.746 \right)$$
$$= 0.002.$$

This p-value can be calculated with the following R statements.

```
n = 15
r = 5
x = c(1410, 1872, 3138, 4218, 6971, rep(6971, n - r))
pchisq(2 * 0.00001 * sum(x), 2 * r)
```

Although the number of observed failures is small, there is adequate evidence from this data set to conclude that the population mean number of cycles to failure is less than 100,000 (for example, the null hypothesis can be rejected at significance levels $\alpha = 0.10, 0.05,$ and 0.01). We conclude that the manufacturer is probably exaggerating the magnitude of the population mean time to failure based on this hypothesis test.

The fact that the distribution of $2\lambda \sum_{i=1}^{n} x_i = 2r\lambda/\hat{\lambda}$ is independent of n implies that $\hat{\lambda}$ has the same precision in a test of r items tested until all have failed as that for a test of n items tested until r items have failed. So the justification for obtaining a Type II censored data set over a complete data set is time savings. The additional costs associated with this time savings are the additional $n - r$ test stands and the additional $n - r$ items to place on test.

If a limited number of test stands are available for testing, the only way to speed up the test is to perform a test with replacement in which failed items are immediately replaced with new items. This will decrease the expected time to complete the test, which is terminated when r of the items fail. The sequence of failures in this case is a Poisson process with rate $n\lambda$.

Although the inference for Type II censoring is tractable, the unfortunate consequence is that the time to complete the test is a random variable. Constraints on the time to run a life test may make a Type I censored data set more practical.

5.4.3 Type I Censored Data Sets

The analysis for Type I censored data sets is similar to that for the Type II censoring case. The test is terminated at time c. The censoring times for each item on test are the same: $c_1 = c_2 = \cdots = c_n = c$. The number of observed failures, r, is a random variable. The total time on test in this case is

$$\sum_{i=1}^{n} x_i = \sum_{i \in U} t_i + \sum_{i \in C} c_i = \sum_{i=1}^{r} t_{(i)} + (n - r)c.$$

As before, the log likelihood function is

$$\ln L(\lambda) = \sum_{i \in U} \ln h(x_i, \lambda) - \sum_{i=1}^{n} H(x_i, \lambda) = r \ln \lambda - \lambda \sum_{i=1}^{n} x_i,$$

and the score statistic is

$$U(\lambda) = \frac{r}{\lambda} - \sum_{i=1}^{n} x_i.$$

The maximum likelihood estimator for $r > 0$ is

$$\hat{\lambda} = \frac{r}{\sum_{i=1}^{n} x_i},$$

the information matrix is

$$I(\lambda) = \frac{r}{\lambda^2},$$

and the observed information matrix is

$$O(\hat{\lambda}) = \frac{r}{\hat{\lambda}^2}.$$

The functional form of the maximum likelihood estimator is identical to the Type II censoring case. For identical values of r, Type I censoring has a larger total time on test $\sum_{i=1}^{n} x_i$ than the corresponding Type II censoring case because a Type I test ends *between* failures r and $r + 1$. Thus the expected value of $\hat{\lambda}$ is smaller for Type I censoring than for Type II censoring. One problem that arises with Type I censoring is that the sampling distribution of $\sum_{i=1}^{n} x_i$ is no longer tractable, so an exact confidence interval for λ has not been established. Although many more complicated methods exist, one of the best approximation methods is to assume that $2\lambda \sum_{i=1}^{n} x_i$ has the chi-square distribution with $2r + 1$ degrees of freedom. This approximation, illustrated in Figure 5.13, is based on the fact

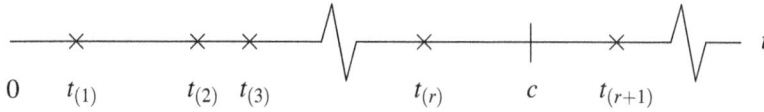

Figure 5.13: Approximation technique for confidence intervals for Type I censoring.

that if $c = t_{(r)}$, then $2\lambda \sum_{i=1}^{n} x_i \sim \chi^2(2r)$, and if $c = t_{(r+1)}$, then $2\lambda \sum_{i=1}^{n} x_i \sim \chi^2(2r+2)$. Since c is between $t_{(r)}$ and $t_{(r+1)}$, $2\lambda \sum_{i=1}^{n} x_i$ will be approximately chi-square with $2r+1$ degrees of freedom. This constitutes a proof, after a little algebra, of the following result.

Theorem 5.6 Let $t_1, t_2, \ldots, t_n$ be the observed values of n mutually independent and identically distributed exponential(λ) random variables. The associated test is terminated at time c (Type I right censoring) for some positive real number c. The censoring times are $c_1 = c_2 = \cdots = c_n = c$ for all items, and $x_i = \min\{t_i, c_i\}$ for $i = 1, 2, \ldots, n$. The maximum likelihood estimator for λ is

$$\hat{\lambda} = \frac{r}{\sum_{i=1}^{n} x_i}$$

and an approximate two-sided $100(1-\alpha)\%$ confidence interval for the failure rate λ is

$$\frac{\hat{\lambda} \chi^2_{2r+1, 1-\alpha/2}}{2r} < \lambda < \frac{\hat{\lambda} \chi^2_{2r+1, \alpha/2}}{2r}.$$

Example 5.10 A life test of $n = 100$ light bulbs is run for $c = 5000$ hours. Failed items are not replaced upon failure in this Type I right censored data set. If the total time on test is $\sum_{i=1}^{n} x_i = 384{,}968$ hours, and $r = 32$ failures are observed, find a point and interval estimator for the failure rate.

It is impossible to check to see whether the exponential distribution is an appropriate model for the light bulb failure times from the problem statement because the actual failure times are not given. Assuming that the exponential model is appropriate, the maximum likelihood estimate for the failure rate is

$$\hat{\lambda} = \frac{r}{\sum_{i=1}^{n} x_i} = \frac{32}{384{,}968} = 0.0000831$$

failure per hour, or, equivalently, the maximum likelihood estimate for the population mean time to failure is its reciprocal, 12,030 hours. To obtain an approximate 95% confidence interval for the failure rate, the chi-square critical values for $2r+1 = 65$ degrees of freedom must be determined. These critical values are $\chi^2_{65, 0.975} = 44.60$ and $\chi^2_{65, 0.025} = 89.18$. The approximate two-sided $100(1-\alpha)\%$ confidence interval for λ

$$\frac{\hat{\lambda} \chi^2_{2r+1, 1-\alpha/2}}{2r} < \lambda < \frac{\hat{\lambda} \chi^2_{2r+1, \alpha/2}}{2r}$$

becomes

$$\frac{(0.0000831)(44.60)}{64} < \lambda < \frac{(0.0000831)(89.18)}{64}$$

or
$$0.0000579 < \lambda < 0.000116.$$

Taking reciprocals, this is equivalent to an approximate 95% confidence interval for the population mean number of cycles to failure of

$$8630 < \mu < 17{,}260.$$

The R statements below compute the point and interval estimates.

```
n      = 100
r      = 32
ttt    = 384968
lam    = r / ttt
mu     = 1 / lam
lam.lo = lam * qchisq(0.025, 2 * r + 1) / (2 * r)
lam.hi = lam * qchisq(0.975, 2 * r + 1) / (2 * r)
mu.lo  = 1 / lam.hi
mu.hi  = 1 / lam.lo
```

5.4.4 Randomly Censored Data Sets

Many of the examples that have the random censoring mechanism for which the failure times $t_1, t_2, \ldots, t_n$ and the censoring times $c_1, c_2, \ldots, c_n$ are independent random variables are from biostatistics. Random censoring occurs frequently in biostatistics because it is not always possible to control the time patients enter and exit the study. The log likelihood function, score statistic, information matrix, and observed information matrix are the same as in the Type I censoring case. The total time on test is now simply

$$\sum_{i=1}^{n} x_i = \sum_{i \in U} t_i + \sum_{i \in C} c_i.$$

The sampling distribution of $\sum_{i=1}^{n} x_i$ is more complicated in this case, so asymptotic properties must be relied on to determine approximate confidence intervals for λ. In the example that follows, three different approximation procedures for determining a confidence interval for λ are illustrated.

The first technique is based on an approximation to a result that holds exactly in the Type II censoring case: $2\lambda \sum_{i=1}^{n} x_i \sim \chi^2(2r)$. The second technique is based on the likelihood ratio statistic, where $-2[\ln L(\lambda) - \ln L(\hat{\lambda})]$ is asymptotically chi-square with 1 degree of freedom. The third technique is based on the fact that the maximum likelihood estimator $\hat{\lambda}$ is asymptotically normal with population mean λ and a population variance that is the inverse of the observed information matrix. Since this third technique results in a symmetric confidence interval, it should only be used with large sample sizes.

Example 5.11 Find the maximum likelihood estimate and three approximate 95% confidence intervals for the remission rate λ for the treatment group (those who received the drug 6–MP) in the leukemia study described in Example 5.6.

For this data set, there are $n = 21$ individuals on test and $r = 9$ observed failures. A "failure" for this data set is the end of a remission period. The total time on test for this data set is $\sum_{i=1}^{n} x_i = 359$ weeks. The log likelihood function is

$$\ln L(\lambda) = r \ln \lambda - \lambda \sum_{i=1}^{n} x_i = 9 \ln \lambda - 359\lambda.$$

As shown by the vertical dashed line in Figure 5.14, this function is maximized at

$$\hat{\lambda} = \frac{r}{\sum_{i=1}^{n} x_i} = \frac{9}{359} = 0.0251$$

remission per week. The maximum likelihood estimate of the expected remission time is $\hat{\mu} = 359/9 = 39.9$ weeks. The value of the log likelihood function at the maximum likelihood estimate is $\ln L(\hat{\lambda}) = -42.17$, as indicated by the horizontal dashed line in Figure 5.14. The observed information matrix is

$$O(\hat{\lambda}) = \frac{\left(\sum_{i=1}^{n} x_i\right)^2}{r} = \frac{(359)^2}{9} = 14{,}320.$$

The three approximation techniques for determining a confidence interval for λ are outlined next. Under the assumption that $2\lambda \sum_{i=1}^{n} x_i$ is approximately chi-square with $2r$ degrees of freedom (this is satisfied exactly in the Type II censoring case), an approximate two-sided $100(1-\alpha)\%$ confidence interval for λ is

$$\frac{\hat{\lambda} \chi^2_{2r,1-\alpha/2}}{2r} < \lambda < \frac{\hat{\lambda} \chi^2_{2r,\alpha/2}}{2r},$$

which, for the 6–MP treatment group remission times with $\alpha = 0.05$, is

$$\frac{(9)(8.23)}{(359)(18)} < \lambda < \frac{(9)(31.53)}{(359)(18)}$$

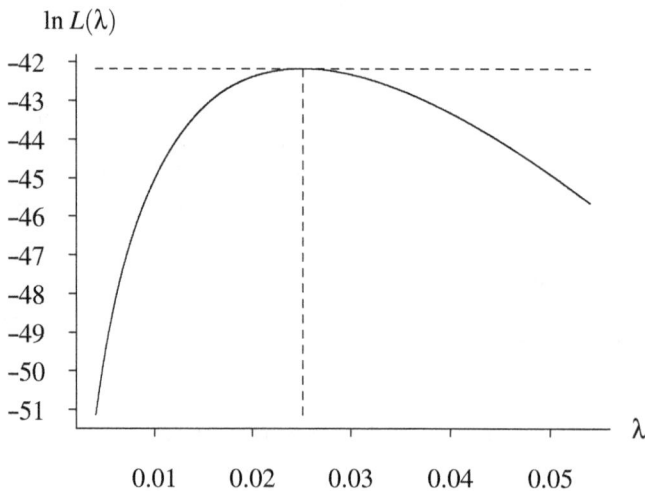

Figure 5.14: Log likelihood function for the 6–MP treatment group.

because $\chi^2_{18,0.975} = 8.23$ and $\chi^2_{18,0.025} = 31.53$, or

$$0.0115 < \lambda < 0.0439.$$

The second approximate confidence interval for λ is based on the likelihood ratio statistic, $-2[\ln L(\lambda) - \ln L(\hat{\lambda})]$, which is asymptotically chi-square with 1 degree of freedom. Thus, with probability $1 - \alpha$, the inequality

$$-2\left[\ln L(\lambda) - \ln L(\hat{\lambda})\right] < \chi^2_{1,\alpha}$$

is approximately satisfied. For the 6–MP remission times in the treatment group and $\alpha = 0.05$, this can be rearranged as

$$\ln L(\lambda) > \ln L(\hat{\lambda}) - \frac{3.84}{2}$$

because $\chi^2_{1,0.05} = 3.84$, or

$$\ln L(\lambda) > -42.17 - \frac{3.84}{2}.$$

As shown by the horizontal dashed lines in Figure 5.15, this corresponds to all values of λ for which the log likelihood function is within $3.84/2 = 1.92$ units of its largest value. The inequality reduces to

$$9\ln\lambda - 359\lambda > -42.17 - 1.92,$$

which can be solved numerically to determine the endpoints. Many computer languages have an equation solver that can determine the two λ values satisfying

$$9\ln\lambda - 359\lambda = -42.17 - 1.92 = -44.09.$$

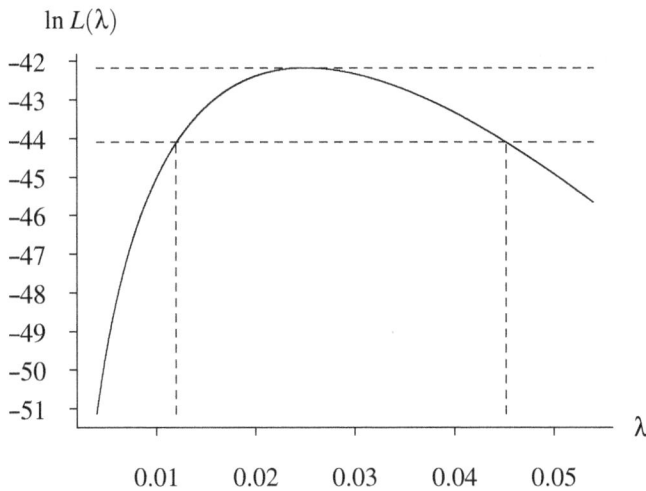

Figure 5.15: Log likelihood function and 95% confidence limits for λ for the 6–MP treatment group.

In this particular example, the approximate two-sided confidence interval for λ is

$$0.0120 < \lambda < 0.0452,$$

which is shifted slightly to the right of the previous confidence interval. The lower and upper bounds for this confidence interval are indicated by the vertical dashed lines in Figure 5.15.

The final confidence interval for λ is based on the fact that the sampling distribution of $\hat{\lambda}$ is asymptotically normal with population mean λ and population variance $I(\lambda)^{-1}$. Replacing $I(\lambda)$ by the observed information matrix $O(\hat{\lambda})$, with approximate probability $1 - \alpha$,

$$-z_{\alpha/2} < \frac{\hat{\lambda} - \lambda}{O(\hat{\lambda})^{-1/2}} < z_{\alpha/2},$$

where $z_{\alpha/2}$ is the $1 - \alpha/2$ fractile of the standard normal distribution. This is equivalent to

$$\hat{\lambda} - z_{\alpha/2} O(\hat{\lambda})^{-1/2} < \lambda < \hat{\lambda} + z_{\alpha/2} O(\hat{\lambda})^{-1/2}.$$

For the 6–MP treatment group remission times, an approximate two-sided 95% confidence interval for λ is

$$\frac{9}{359} - (1.96)(14,320)^{-1/2} < \lambda < \frac{9}{359} + (1.96)(14,320)^{-1/2}$$

or

$$0.0087 < \lambda < 0.0414,$$

which has smaller bounds than the previous two interval estimators.

To summarize the conclusions of this long example, the maximum likelihood estimate of the failure rate is

$$\hat{\lambda} = 0.0251$$

remission per week, which corresponds to an estimated mean remission time of 39.9 weeks. The three approximate two-sided 95% confidence intervals for λ are given in the second column of Table 5.1. Taking reciprocals, the third column contains the associated approximate two-sided 95% confidence intervals for the population mean remission time μ. The confidence intervals for λ associated with the first two techniques are not symmetric about the maximum likelihood estimator because they are based on the non-symmetric chi-square distribution. Since there are only $n = 21$ patients in the clinical trial and only $r = 9$ observed remission times, we have more faith in the actual coverage of the first two confidence interval techniques. This conclusion would need to be confirmed by a Monte Carlo simulation experiment.

Basis for confidence interval	Confidence interval for λ	Confidence interval for μ
Type II censoring approximate result	$0.0115 < \lambda < 0.0439$	$22.8 < \mu < 87.2$
Likelihood ratio statistic	$0.0120 < \lambda < 0.0452$	$22.1 < \mu < 83.0$
Asymptotic normality of the MLE	$0.0087 < \lambda < 0.0414$	$24.1 < \mu < 115.1$

Table 5.1: Approximate 95% confidence intervals for λ and μ for the 6–MP treatment group.

To summarize, the maximum likelihood estimator for the failure rate λ in the random censoring case is the same as in the complete, Type II and Type I censoring cases:

$$\hat{\lambda} = \frac{r}{\sum_{i=1}^{n} x_i}.$$

Three approximate confidence intervals for λ are based on (a) an exact result from Type II censoring, (b) the asymptotic distribution of the likelihood ratio statistic, and (c) the asymptotic normality of the maximum likelihood estimator. The confidence interval based on the asymptotic normality of the maximum likelihood estimator is symmetric and is therefore recommended only in the case of a large number of items on test.

5.5 Weibull Distribution

The Weibull distribution is typically more appropriate for modeling the lifetimes of items with a strictly increasing or decreasing hazard function, such as mechanical items. Rather than looking at each censoring mechanism (for example, no censoring, Type II censoring, Type I censoring) individually, we proceed directly to the general case of random censoring.

Maximum likelihood estimators. As before, let $t_1, t_2, \ldots, t_n$ be the failure times, $c_1, c_2, \ldots, c_n$ be the associated censoring times, and $x_i = \min\{t_i, c_i\}$ for $i = 1, 2, \ldots, n$. The Weibull distribution has hazard and cumulative hazard functions

$$h(t, \lambda, \kappa) = \kappa \lambda (\lambda t)^{\kappa-1} \qquad t \geq 0$$

and

$$H(t, \lambda, \kappa) = (\lambda t)^{\kappa} \qquad t \geq 0.$$

When there are r observed failures, the log likelihood function is

$$\ln L(\lambda, \kappa) = \sum_{i \in U} \ln h(x_i, \lambda, \kappa) - \sum_{i=1}^{n} H(x_i, \lambda, \kappa)$$

$$= \sum_{i \in U} \left(\ln \kappa + \kappa \ln \lambda + (\kappa - 1) \ln x_i \right) - \sum_{i=1}^{n} (\lambda x_i)^{\kappa}$$

$$= r \ln \kappa + \kappa r \ln \lambda + (\kappa - 1) \sum_{i \in U} \ln x_i - \lambda^{\kappa} \sum_{i=1}^{n} x_i^{\kappa},$$

and the 2×1 score vector has elements

$$U_1(\lambda, \kappa) = \frac{\partial \ln L(\lambda, \kappa)}{\partial \lambda} = \frac{\kappa r}{\lambda} - \kappa \lambda^{\kappa-1} \sum_{i=1}^{n} x_i^{\kappa}$$

and

$$U_2(\lambda, \kappa) = \frac{\partial \ln L(\lambda, \kappa)}{\partial \kappa} = \frac{r}{\kappa} + r \ln \lambda + \sum_{i \in U} \ln x_i - \sum_{i=1}^{n} (\lambda x_i)^{\kappa} \ln(\lambda x_i).$$

When these equations are set equal to zero, the simultaneous equations have no closed-form solution for $\hat{\lambda}$ and $\hat{\kappa}$:

$$\frac{\kappa r}{\lambda} - \kappa \lambda^{\kappa-1} \sum_{i=1}^{n} x_i^{\kappa} = 0,$$

$$\frac{r}{\kappa} + r \ln \lambda + \sum_{i \in U} \ln x_i - \sum_{i=1}^{n} (\lambda x_i)^{\kappa} \ln(\lambda x_i) = 0.$$

One piece of good fortune, however, to avoid solving a 2×2 set of nonlinear equations, is that the first equation can be solved for λ in terms of κ as

$$\lambda = \left(\frac{r}{\sum_{i=1}^{n} x_i^{\kappa}} \right)^{1/\kappa}.$$

Notice that λ reduces to the maximum likelihood estimator for the exponential distribution when $\kappa = 1$. Using this expression for λ in terms of κ in the second element of the score vector yields a single, albeit more complicated, expression with κ as the only unknown. After applying some algebra, this equation reduces to

$$g(\kappa) = \frac{r}{\kappa} + \sum_{i \in U} \ln x_i - \frac{r \sum_{i=1}^{n} x_i^{\kappa} \ln x_i}{\sum_{i=1}^{n} x_i^{\kappa}} = 0,$$

which must be solved iteratively. One technique that can be used to solve this equation is the Newton–Raphson procedure, which uses

$$\kappa_{j+1} = \kappa_j - \frac{g(\kappa_j)}{g'(\kappa_j)},$$

where κ_0 is an initial estimator. The iterative procedure can be repeated until the desired accuracy for κ is achieved; that is, $|\kappa_{j+1} - \kappa_j| < \varepsilon$, for some small positive real number ε. When the accuracy is achieved, the maximum likelihood estimator $\hat{\kappa}$ is used to calculate $\hat{\lambda} = \left(r / \sum_{i=1}^{n} x_i^{\hat{\kappa}} \right)^{1/\hat{\kappa}}$. The derivative of $g(\kappa)$ reduces to

$$g'(\kappa) = -\frac{r}{\kappa^2} - \frac{r}{\left(\sum_{i=1}^{n} x_i^{\kappa} \right)^2} \left[\left(\sum_{i=1}^{n} x_i^{\kappa} \right) \left(\sum_{i=1}^{n} (\ln x_i)^2 x_i^{\kappa} \right) - \left(\sum_{i=1}^{n} x_i^{\kappa} \ln x_i \right)^2 \right].$$

Determining an initial estimator κ_0 is not trivial. When there are no censored observations, Menon's initial estimator for κ_0 is

$$\kappa_0 = \left\{ \frac{6}{(n-1)\pi^2} \left[\sum_{i=1}^{n} (\ln t_i)^2 - \frac{\left(\sum_{i=1}^{n} \ln t_i \right)^2}{n} \right] \right\}^{-1/2}.$$

Least squares estimation can be used in the case of a right-censored data set. The Newton–Raphson procedure can fail to converge to the maximum likelihood estimators. A bisection algorithm or fixed point algorithm often provides more reliable convergence.

Fisher and observed information matrices. The 2×2 Fisher and observed information matrices are based on the following partial derivatives:

$$\frac{-\partial^2 \ln L(\lambda, \kappa)}{\partial \lambda^2} = \frac{\kappa r}{\lambda^2} + \kappa(\kappa - 1) \lambda^{\kappa - 2} \sum_{i=1}^{n} x_i^{\kappa},$$

$$\frac{-\partial^2 \ln L(\lambda, \kappa)}{\partial \lambda \partial \kappa} = -\frac{r}{\lambda} + \left[\left(\kappa \lambda^{\kappa - 1} \right) \left(\sum_{i=1}^{n} x_i^{\kappa} \ln x_i \right) + \left(\sum_{i=1}^{n} x_i^{\kappa} \right) \left(\kappa \lambda^{\kappa - 1} \ln \lambda + \lambda^{\kappa - 1} \right) \right]$$

$$= -\frac{r}{\lambda} + \lambda^{\kappa-1}\left[\kappa\sum_{i=1}^{n}x_i^{\kappa}\ln x_i + (1+\kappa\ln\lambda)\sum_{i=1}^{n}x_i^{\kappa}\right],$$

$$\frac{-\partial^2\ln L(\lambda,\kappa)}{\partial\kappa^2} = \frac{r}{\kappa^2} + \sum_{i=1}^{n}(\lambda x_i)^{\kappa}(\ln \lambda x_i)^2.$$

The expected values of these quantities are not tractable, so the Fisher information matrix does not have closed-form elements. The observed information matrix, however, can be determined by using $\hat{\lambda}$ and $\hat{\kappa}$ as arguments in these expressions.

Example 5.12 Example 5.5 showed that the exponential distribution was a poor approximation to the ball bearing data lifetimes. The histogram in Figure 5.8 indicated that a probability distribution with a nonzero mode and an increasing hazard function might provide a better fit. Fit the Weibull distribution to the ball bearing lifetimes and assess the fit.

The maximum likelihood estimates, using the Newton–Raphson technique described previously, are $\hat{\lambda} = 0.0122$ and $\hat{\kappa} = 2.10$. Figure 5.16 shows the empirical survivor function, along with the fitted exponential and Weibull survivor functions. It is clear that the Weibull distribution is superior to the exponential distribution in fitting the ball bearing failure times because it is capable of modeling wear out. The log likelihood function evaluated at the maximum likelihood estimators is $\ln L(\hat{\lambda}, \hat{\kappa}) = -113.691$. The log likelihood function is shown in Figure 5.17.

The observed information matrix is

$$O(\hat{\lambda}, \hat{\kappa}) = \begin{bmatrix} 681{,}000 & 875 \\ 875 & 10.4 \end{bmatrix},$$

revealing a positive correlation between the elements of the score vector. Using the fact that the likelihood ratio statistic, $-2\left[\ln L(\lambda, \kappa) - \ln L(\hat{\lambda}, \hat{\kappa})\right]$, is asymptotically $\chi^2(2)$,

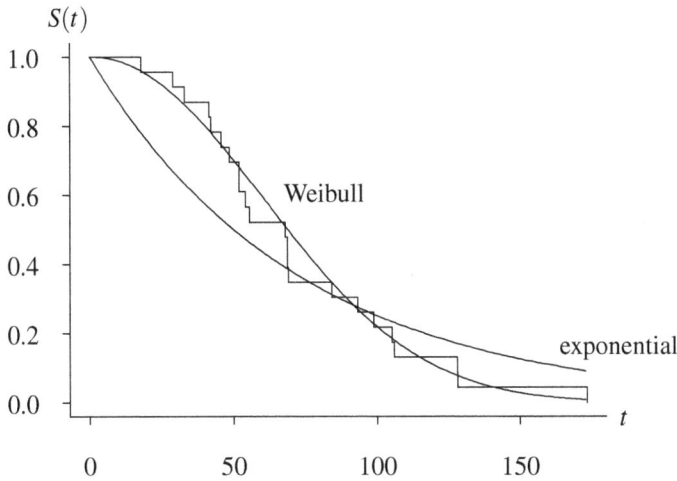

Figure 5.16: Exponential and Weibull fits to the ball bearing data.

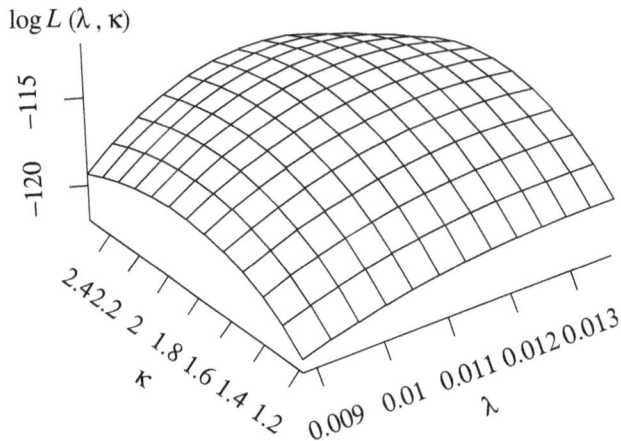

Figure 5.17: Log likelihood function for the ball bearing data.

an approximate 95% confidence region for the parameters is all λ and κ satisfying

$$-2[\ln L(\lambda, \kappa) + 113.691] < 5.99,$$

since $\chi^2_{2, 0.05} = 5.99$. The 95% confidence region is shown in Figure 5.18, and, not surprisingly, the line $\kappa = 1$ is not interior to the region. This indicates that the exponential distribution is not an appropriate model for this particular data set. This is yet more statistical evidence that the ball bearings are wearing out. Note that the boundary of this region is a level surface of the log likelihood function shown in Figure 5.17 that is cut $5.99/2$ units below the maximum of the log likelihood function.

The R code to generate the confidence region is given below. The `crplot` function contained in the `conf` package calculates the maximum likelihood estimates $\hat{\lambda}$ and $\hat{\kappa}$ and plots the 95% confidence region. The first argument to `crplot` contains the data values, the second argument contains α, and the third argument contains the name of the population distribution. Setting the `pts` argument to FALSE means the points along the boundary are connected by lines; setting the `origin` argument to TRUE means the origin is included in the plot; setting the `info` argument to TRUE means the maximum likelihood estimates and boundary points in the confidence region can easily be retrieved.

```
library(conf)
bb = c(17.88, 28.92, 33.00, 41.52, 42.12, 45.60, 48.48, 51.84,
       51.96, 54.12, 55.56, 67.80, 68.64, 68.64, 68.88, 84.12,
       93.12, 98.64, 105.12, 105.84, 127.92, 128.04, 173.40)
crplot(bb, 0.05, "weibull", pts = FALSE, origin = TRUE, info = TRUE)
```

As further evidence that the Weibull distribution is a significantly better model than the exponential, the likelihood ratio statistic can be used to determine whether κ is significant. Evaluating the log likelihood values at the maximum likelihood estimators

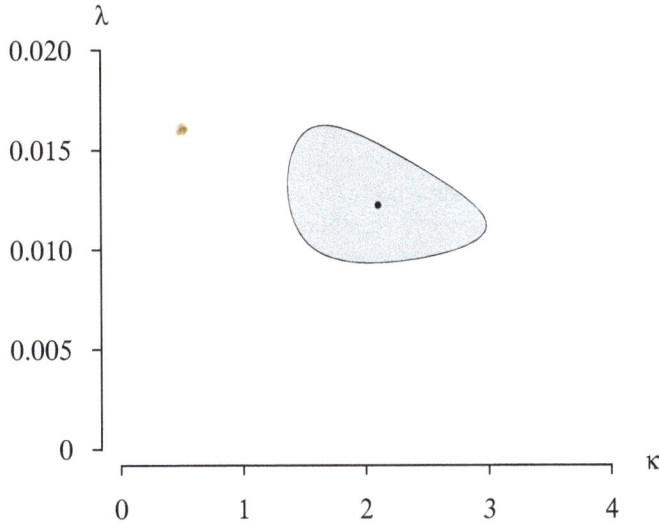

Figure 5.18: Confidence region ($\alpha = 0.05$) for λ and κ for the ball bearing data.

in the Weibull and exponential fits, the likelihood ratio statistic is

$$-2\left[\ln L(\hat{\lambda}) - \ln L(\hat{\lambda}, \hat{\kappa})\right] = -2[-121.435 + 113.691] = 15.488.$$

This value shows that there is a statistically significant difference between κ and 1 when it is compared with the critical value $\chi^2_{1,0.05} = 3.84$.

If we are still uncertain as to whether κ is significantly different from 1, the standard errors of the distribution of the parameter estimators can be computed by determining the inverse of the observed information matrix

$$O^{-1}(\hat{\lambda}, \hat{\kappa}) = \begin{bmatrix} 0.00000165 & -0.000139 \\ -0.000139 & 0.108 \end{bmatrix}.$$

This matrix is an estimate of the variance–covariance matrix for the parameter estimates $\hat{\lambda}$ and $\hat{\kappa}$. The standard errors of the parameter estimates are the square roots of the diagonal elements

$$\hat{\sigma}_{\hat{\lambda}} = 0.00128 \qquad \hat{\sigma}_{\hat{\kappa}} = 0.329.$$

Thus, an asymptotic 95% confidence interval for κ is

$$2.10 - (1.96)(0.329) < \kappa < 2.10 + (1.96)(0.329)$$

or

$$1.46 < \kappa < 2.74,$$

since $z_{0.025} = 1.96$. Since this confidence interval does not contain 1, the parameter κ is statistically significant. Three different techniques have all drawn the same conclusion: the ball bearings are wearing out because there is a statistically significant difference between $\hat{\kappa} = 2.10$ and $\kappa = 1$.

5.6   Proportional Hazards Model

Parameter estimation for the proportional hazards model, which was introduced in Section 4.6, is considered in this section. Since there is now a vector of covariates in addition to a failure or censoring time for each item on test, special notation must be established to accommodate the covariates. The proportional hazards model has the unique feature that the baseline distribution need not be defined in order to estimate the regression coefficients associated with the covariates.

A lifetime model that incorporates a vector of covariates $z = (z_1, z_2, \ldots, z_q)'$ models the impact of the covariates on survival. The reason for including this vector may be to determine which covariates significantly affect survival, to determine the distribution of the lifetime for a particular setting of the covariates, or to fit a more complicated distribution from a small data set, as opposed to fitting separate distributions for each level of the covariates.

The proportional hazards model was defined in Section 4.6 by

$$h(t, z) = \psi(z)h_0(t),$$

for $t \geq 0$, where $h_0(t)$ is a baseline hazard function. The covariates increase the hazard function when $\psi(z) > 1$ or decrease the hazard function when $\psi(z) < 1$. The goal of this section is to develop techniques for estimating the $q \times 1$ vector of regression coefficients β from a data set consisting of n items on test and r observed failure times.

The notation used to describe a data set in a lifetime model involving covariates will borrow some notation established earlier in this chapter, but also establish some new notation. As before, n is the number of items on test and r is the number of observed failures. The failure time of the ith item on test, t_i, is either observed or right censored at time c_i, for $i = 1, 2, \ldots, n$. As before, let $x_i = \min\{t_i, c_i\}$ and δ_i be a censoring indicator variable (1 for an observed failure and 0 for a right-censored value), for $i = 1, 2, \ldots, n$. In addition, a $q \times 1$ vector of covariates $z_i = (z_{i1}, z_{i2}, \ldots, z_{iq})'$ is collected for each item on test, for $i = 1, 2, \ldots, n$. Thus, z_{ij} is the value of covariate j for item i, for $i = 1, 2, \ldots, n$ and $j = 1, 2, \ldots, q$. This formulation of the problem can be stated in matrix form as

$$x = \begin{bmatrix} x_1 \\ x_2 \\ \vdots \\ x_n \end{bmatrix} \qquad \delta = \begin{bmatrix} \delta_1 \\ \delta_2 \\ \vdots \\ \delta_n \end{bmatrix} \qquad \text{and} \qquad Z = \begin{bmatrix} z_{11} & z_{12} & \cdots & z_{1q} \\ z_{21} & z_{22} & \cdots & z_{2q} \\ \vdots & \vdots & \ddots & \vdots \\ z_{n1} & z_{n2} & \cdots & z_{nq} \end{bmatrix}.$$

Each row in the Z matrix consists of the values of the q covariates collected on a particular item. The matrix approach is useful because complicated systems of equations can be expressed compactly and operations on data sets can be performed efficiently by a computer. For parameter estimation, the survivor, density, hazard, and cumulative hazard functions now have the extra arguments z and β associated with them:

$$S(t, z, \theta, \beta) \qquad f(t, z, \theta, \beta) \qquad h(t, z, \theta, \beta) \qquad H(t, z, \theta, \beta),$$

for $t \geq 0$, where the vector $\theta = (\theta_1, \theta_2, \ldots, \theta_p)'$ consists of the p unknown parameters associated with the baseline distribution, which must be estimated along with the regression coefficients β.

Parameter estimation for the proportional hazards model can be divided into two cases. The first case is when the baseline distribution is known. This case applies when previous test results have indicated that a particular functional form of the baseline distribution is appropriate. The second case is when the baseline distribution is unknown. This is almost certainly the case when looking at a data set of lifetimes and covariates for the first time without any guidance with respect to an appropriate baseline distribution.

5.6.1 Known Baseline Distribution

When the baseline distribution is known, the parameter estimation procedure follows along the same lines as in the previous sections. The hazard function and cumulative hazard function in the proportional hazards model are

$$h(t, z, \theta, \beta) = \psi(z)h_0(t)$$

and

$$H(t, z, \theta, \beta) = \psi(z)H_0(t)$$

for $t \geq 0$, where θ is a $p \times 1$ vector of unknown parameters associated with the baseline distribution. For simplicity and mathematical tractability, only the log linear form of the link function, which is $\psi(z) = e^{\beta' z}$, is considered here. This assumption is not necessary for some of the derivations, so many of the results apply to a wider range of link functions. When the log linear form of the link function is assumed, the hazard function and cumulative hazard function become

$$h(t, z, \theta, \beta) = e^{\beta' z}h_0(t)$$

and

$$H(t, z, \theta, \beta) = e^{\beta' z}H_0(t)$$

for $t \geq 0$, where θ is a $p \times 1$ vector of unknown parameters associated with the baseline distribution. The log likelihood function is

$$\ln L(\theta, \beta) = \sum_{i \in U} \ln h(x_i, z_i, \theta, \beta) - \sum_{i=1}^{n} H(x_i, z_i, \theta, \beta)$$

$$= \sum_{i \in U} \left[\beta' z_i + \ln h_0(x_i) \right] - \sum_{i=1}^{n} e^{\beta' z_i} H_0(x_i).$$

This expression can be differentiated with respect to all the unknown parameters to arrive at the score vector, which is then equated to zero and solved numerically to arrive at the maximum likelihood estimates.

Two observations with respect to this model formulation are important. First, the maximum likelihood estimates for θ and β for most of the models in this section cannot be expressed in closed form (as was the case for the exponential distribution in Section 5.4), so numerical methods typically need to be used to find the values of the estimates. Second, the choice of whether to use a model of dependence or to examine each population separately is dependent on the number of unique covariate vectors z and the number of items on test, n. If, for example, n is large and there is only a single binary covariate (that is, only two unique covariate vectors, $z_1 = 0$ and $z_1 = 1$), it is probably wiser to analyze each of the two populations separately by the techniques described earlier.

Although numerical methods are required to find $\hat{\theta}$ and $\hat{\beta}$ in general, there are closed-form expressions in a very narrow case that satisfies the following conditions.

- The log linear link function $\psi(z) = e^{\beta' z}$ is used to incorporate the vector of covariates z into the lifetime model.

- The baseline distribution is exponential(λ), which means that the baseline hazard function is $h_0(t) = \lambda$ and the baseline cumulative hazard function is $H_0(t) = \lambda t$ for $t \geq 0$.

Under these assumptions, the general form for the hazard function in the proportional hazards model

$$h(t, z, \theta, \beta) = \psi(z)h_0(t)$$

reduces to the special case

$$h(t, z, \lambda, \boldsymbol{\beta}) = \lambda e^{\boldsymbol{\beta}' z}.$$

for $t \geq 0$. It is often more convenient notationally to define an additional covariate, $z_0 = 1$, for all n items on test. This allows the baseline parameter $\lambda = e^{\beta_0 z_0}$ to be included in the vector of regression coefficients, rather than being considered separately. The baseline hazard function is effectively absorbed into the link function. In this case, the hazard function can be expressed as

$$h(t, z, \boldsymbol{\beta}) = e^{\boldsymbol{\beta}' z}$$

for $t \geq 0$, where $\boldsymbol{\beta} = (\beta_0, \beta_1, \ldots, \beta_q)'$ and $z = (z_0, z_1, \ldots, z_q)'$. The corresponding cumulative hazard function is

$$H(t, z, \boldsymbol{\beta}) = t e^{\boldsymbol{\beta}' z}$$

for $t \geq 0$. Using this parameterization, the log likelihood function is

$$\ln L(\boldsymbol{\beta}) = \sum_{i \in U} \ln h(x_i, z_i, \boldsymbol{\beta}) - \sum_{i=1}^{n} H(x_i, z_i, \boldsymbol{\beta})$$

$$= \sum_{i \in U} \boldsymbol{\beta}' z_i - \sum_{i=1}^{n} x_i e^{\boldsymbol{\beta}' z_i}.$$

Differentiating this expression with respect to β_j yields the elements of the score vector

$$\frac{\partial \ln L(\boldsymbol{\beta})}{\partial \beta_j} = \sum_{i \in U} z_{ij} - \sum_{i=1}^{n} x_i z_{ij} e^{\boldsymbol{\beta}' z_i}$$

for $j = 0, 1, \ldots, q$. When the elements of the score vector are equated to zero, the resulting set of $q + 1$ nonlinear equations in $\boldsymbol{\beta}$ must be solved numerically in the general case. There is a closed-form solution for this set of simultaneous equations when there is a single binary covariate, often referred to as the two-sample case.

To find the observed information matrix and the Fisher information matrix, a second partial derivative of the log likelihood function is required:

$$\frac{\partial^2 \ln L(\boldsymbol{\beta})}{\partial \beta_j \partial \beta_k} = - \sum_{i=1}^{n} x_i z_{ij} z_{ik} e^{\boldsymbol{\beta}' z_i}$$

for $j = 0, 1, \ldots, q$ and $k = 0, 1, \ldots, q$. The observed information matrix can be determined by using the maximum likelihood estimate $\hat{\boldsymbol{\beta}}$ as an argument in this second partial derivative. Thus, the (j, k) element of the observed information matrix is

$$\left[-\frac{\partial^2 \ln L(\boldsymbol{\beta})}{\partial \beta_j \partial \beta_k} \right]_{\boldsymbol{\beta} = \hat{\boldsymbol{\beta}}} = \sum_{i=1}^{n} x_i z_{ij} z_{ik} e^{\hat{\boldsymbol{\beta}}' z_i}$$

for $j = 0, 1, \ldots, q$ and $k = 0, 1, \ldots, q$. For computational purposes, this can be expressed in matrix form as

$$O(\hat{\boldsymbol{\beta}}) = Z' \hat{B} Z,$$

where $\hat{B}$ is an $n \times n$ diagonal matrix whose elements are $x_1 e^{\hat{\boldsymbol{\beta}}' z_1}, x_2 e^{\hat{\boldsymbol{\beta}}' z_2}, \ldots, x_n e^{\hat{\boldsymbol{\beta}}' z_n}$. The Fisher information matrix is more difficult to calculate because it involves the expected value of the second partial derivative:

$$E\left[-\frac{\partial^2 \ln L(\boldsymbol{\beta})}{\partial \beta_j \partial \beta_k} \right] = \sum_{i=1}^{n} z_{ij} z_{ik} e^{\boldsymbol{\beta}' z_i} E[x_i]$$

for $j = 0, 1, \ldots, q$ and $k = 0, 1, \ldots, q$. Determining the value of $E[x_i]$ will be considered separately in the paragraphs that follow for uncensored ($r = n$) and censored ($r < n$) data sets.

For a complete data set, $E[x_i] = E[t_i]$, for $i = 1, 2, \ldots, n$, because there is no censoring. Since the population mean of the exponential distribution is the reciprocal of the failure rate and the ith item on test has failure rate $e^{\boldsymbol{\beta}' \mathbf{z}_i}$, $E[x_i] = e^{-\boldsymbol{\beta}' \mathbf{z}_i}$. Returning to the Fisher information matrix, the (j, k) element is

$$E\left[-\frac{\partial^2 \ln L(\boldsymbol{\beta})}{\partial \beta_j \partial \beta_k}\right] = \sum_{i=1}^{n} z_{ij} z_{ik} e^{\boldsymbol{\beta}' \mathbf{z}_i} e^{-\boldsymbol{\beta}' \mathbf{z}_i} = \sum_{i=1}^{n} z_{ij} z_{ik}$$

for $j = 0, 1, \ldots, q$ and $k = 0, 1, \ldots, q$. This result for the Fisher information matrix has a particularly tractable matrix representation

$$I(\boldsymbol{\beta}) = \mathbf{Z}' \mathbf{Z},$$

which is a function of the matrix of covariates only.

For a censored data set, the expression for $E[x_i]$ is a bit more complicated. Since the failure rate for the ith item on test is $e^{\boldsymbol{\beta}' \mathbf{z}_i}$,

$$E[x_i] = E\left[\min\{t_i, c_i\}\right]$$
$$= \int_0^{c_i} t_i f_{T_i}(t_i) dt_i + c_i P[t_i \geq c_i]$$
$$= \int_0^{c_i} t_i e^{\boldsymbol{\beta}' \mathbf{z}_i} e^{-e^{\boldsymbol{\beta}' \mathbf{z}_i} t_i} dt_i + c_i e^{-e^{\boldsymbol{\beta}' \mathbf{z}_i} c_i}$$
$$= e^{-\boldsymbol{\beta}' \mathbf{z}_i} \left(1 - e^{-e^{\boldsymbol{\beta}' \mathbf{z}_i} c_i}\right)$$

for $i = 1, 2, \ldots, n$, by using integration by parts. This means that the (j, k) element of the Fisher information matrix is

$$E\left[-\frac{\partial^2 \ln L(\boldsymbol{\beta})}{\partial \beta_j \partial \beta_k}\right] = \sum_{i=1}^{n} z_{ij} z_{ik} e^{\boldsymbol{\beta}' \mathbf{z}_i} e^{-\boldsymbol{\beta}' \mathbf{z}_i} \left[1 - e^{-e^{\boldsymbol{\beta}' \mathbf{z}_i} c_i}\right] = \sum_{i=1}^{n} z_{ij} z_{ik} (1 - \gamma_i),$$

where $\gamma_i = e^{-e^{\boldsymbol{\beta}' \mathbf{z}_i} c_i}$ is the probability that the ith item on test is censored, for $i = 1, 2, \ldots, n$. The potential censoring time for the ith item on test, c_i, must be known for each item in order to compute the Fisher information matrix, which is not always the case in practice. Letting $\boldsymbol{\Gamma}$ be a diagonal matrix with elements $\gamma_1, \gamma_2, \ldots, \gamma_n$, the Fisher information matrix can be written in matrix form as

$$I(\boldsymbol{\beta}) = \mathbf{Z}'(\mathbf{I} - \boldsymbol{\Gamma})\mathbf{Z},$$

which is independent of the failure times.

Before ending the discussion on the exponential baseline distribution, the two-sample case, where a binary covariate z_1 is used to differentiate between the control ($z_1 = 0$) and treatment ($z_1 = 1$) cases, is considered. This case is of interest because the maximum likelihood estimates can be expressed in closed form. The notation for the two-sample case is summarized in Table 5.2. As before, $z_0 = 1$ is included in the vector of covariates to account for the baseline distribution. The set of two nonlinear equations for finding the estimates of $\boldsymbol{\beta} = (\beta_0, \beta_1)'$ obtained by setting the score vector equal to $\mathbf{0}$ is

$$\sum_{i \in U} z_{i0} - \sum_{i=1}^{n} x_i z_{i0} e^{\beta_0 z_{i0} + \beta_1 z_{i1}} = 0,$$

$$\sum_{i \in U} z_{i1} - \sum_{i=1}^{n} x_i z_{i1} e^{\beta_0 z_{i0} + \beta_1 z_{i1}} = 0.$$

Let $r_0 > 0$ be the number of observed failures in the control group ($z_1 = 0$), and let $r_1 > 0$ be the number of observed failures in the treatment group ($z_1 = 1$). Since $z_0 = 1$ for all items on test, the equations reduce to

$$r_0 + r_1 - \sum_{i=1}^{n} x_i e^{\beta_0 + \beta_1 z_{i1}} = 0,$$

$$r_1 - \sum_{i=1}^{n} x_i z_{i1} e^{\beta_0 + \beta_1 z_{i1}} = 0.$$

These equations can be further simplified by partitioning the summations based on the value of z_1:

$$r_0 + r_1 - \sum_{i \mid z_{i1}=0} x_i e^{\beta_0} - \sum_{i \mid z_{i1}=1} x_i e^{\beta_0 + \beta_1} = 0,$$

$$r_1 - \sum_{i \mid z_{i1}=1} x_i e^{\beta_0 + \beta_1} = 0.$$

Letting $\lambda_0 = e^{\beta_0}$ be the failure rate in the control group ($z_1 = 0$) and letting $\lambda_1 = e^{\beta_0 + \beta_1}$ be the failure rate in the treatment group ($z_1 = 1$), the equations become

$$r_0 + r_1 - \lambda_0 \sum_{i \mid z_{i1}=0} x_i - \lambda_1 \sum_{i \mid z_{i1}=1} x_i = 0,$$

$$r_1 - \lambda_1 \sum_{i \mid z_{i1}=1} x_i = 0.$$

When these equations are solved simultaneously, the maximum likelihood estimates for λ_0 and λ_1 are the same as those for the exponential distribution with two separate populations:

$$\hat{\lambda}_0 = \frac{r_0}{\sum_{i \mid z_{i1}=0} x_i} \qquad \text{and} \qquad \hat{\lambda}_1 = \frac{r_1}{\sum_{i \mid z_{i1}=1} x_i}.$$

These estimators are the ratio of the number of observed failures to the total time on test within the two groups.

Example 5.13 The patients in the 6–MP drug experiment described in Example 5.6 are broken down into a control group that did not receive the drug ($z_1 = 0$) and a treatment group that did receive the drug ($z_1 = 1$). The remission times, in weeks, for the 21 patients in the control group are

$$\begin{array}{ccccccccccc} 1 & 1 & 2 & 2 & 3 & 4 & 4 & 5 & 5 & 8 & 8 \\ 8 & 8 & 11 & 11 & 12 & 12 & 15 & 17 & 22 & 23. \end{array}$$

	Control Group	Treatment Group
Number of failures	r_0	r_1
Baseline covariate z_0	1	1
Binary covariate z_1	0	1

Table 5.2: Single binary covariate proportional hazards model notation.

The remission times for the 21 patients that received the drug are

$$
\begin{array}{ccccccccccc}
6 & 6 & 6 & 6^* & 7 & 9^* & 10 & 10^* & 11^* & 13 & 16 \\
17^* & 19^* & 20^* & 22 & 23 & 25^* & 32^* & 32^* & 34^* & 35^*.
\end{array}
$$

There are a total of $n = 42$ patients in the clinical trial, and there are a total of $r = 30$ observed cancer recurrences, $r_0 = 21$ of which are in the control group and $r_1 = 9$ of which are in the treatment group. The values of x, δ, and Z are given in Figure 5.19; the control group values have been arbitrarily placed first in the x vector. Note that for this analysis the *order* of the observations in the x vector is irrelevant. For tied values, the censored values have been placed last.

The maximum likelihood estimates for the failure rates for the two populations are

$$
\hat{\lambda}_0 = \frac{r_0}{\sum_{i \mid z_{i1}=0} x_i} = \frac{21}{182} = 0.115 \qquad \text{and} \qquad \hat{\lambda}_1 = \frac{r_1}{\sum_{i \mid z_{i1}=1} x_i} = \frac{9}{359} = 0.0251
$$

or, equivalently, in terms of the estimated mean remission times, the expected remission times of the control and treatment groups are estimated to be $\frac{182}{21} = 8.67$ weeks and $\frac{359}{9} = 39.9$ weeks, respectively. These estimates can be easily converted to the coefficients in the proportional hazards model:

$$
\hat{\beta}_0 = \ln\left[\frac{21}{182}\right] = -2.16 \qquad \text{and} \qquad \hat{\beta}_1 = \ln\left[\frac{(9)(182)}{(359)(21)}\right] = -1.53.
$$

Confidence intervals can be determined separately for the two populations because the remission times in each are assumed to be exponentially distributed. Using the techniques from Section 5.4, an exact two-sided 95% confidence interval for λ_0 is

$$
\frac{(0.115)(26.00)}{42} < \lambda_0 < \frac{(0.115)(61.78)}{42}
$$

$$
0.0714 < \lambda_0 < 0.170
$$

based on the chi-square distribution with 42 degrees of freedom. An approximate two-sided 95% confidence interval for λ_1 is

$$
\frac{(0.0251)(8.23)}{18} < \lambda_1 < \frac{(0.0251)(31.53)}{18}
$$

$$
0.0115 < \lambda_1 < 0.0439
$$

based on the chi-square distribution with 18 degrees of freedom. The first confidence interval is exact because the control group contains no censored observations, and the second confidence interval is approximate because the treatment group has randomly censored observations. Since these confidence intervals do not overlap, it can be concluded that 6–MP is effective in increasing remission times. If the side effects from 6–MP are minor, it should be prescribed to all leukemia patients.

Since exact confidence intervals apply only to the two-sample case with an exponential baseline distribution and Type II censoring, asymptotic intervals will also be calculated here to illustrate how they are developed in the general case. The Fisher information matrix cannot be calculated for this data set because the observed remission times do

$$
x = \begin{bmatrix} 1 \\ 1 \\ 2 \\ 2 \\ 3 \\ 4 \\ 4 \\ 5 \\ 5 \\ 8 \\ 8 \\ 8 \\ 8 \\ 11 \\ 11 \\ 12 \\ 12 \\ 15 \\ 17 \\ 22 \\ 23 \\ 6 \\ 6 \\ 6 \\ 6 \\ 7 \\ 9 \\ 10 \\ 10 \\ 11 \\ 13 \\ 16 \\ 17 \\ 19 \\ 20 \\ 22 \\ 23 \\ 25 \\ 32 \\ 32 \\ 34 \\ 35 \end{bmatrix}
\quad
\delta = \begin{bmatrix} 1 \\ 1 \\ 1 \\ 1 \\ 1 \\ 1 \\ 1 \\ 1 \\ 1 \\ 1 \\ 1 \\ 1 \\ 1 \\ 1 \\ 1 \\ 1 \\ 1 \\ 1 \\ 1 \\ 1 \\ 1 \\ 1 \\ 1 \\ 1 \\ 0 \\ 1 \\ 0 \\ 1 \\ 0 \\ 0 \\ 1 \\ 1 \\ 0 \\ 0 \\ 0 \\ 1 \\ 1 \\ 0 \\ 0 \\ 0 \\ 0 \\ 0 \end{bmatrix}
\quad
Z = \begin{bmatrix} 1 & 0 \\ 1 & 1 \\ 1 & 1 \end{bmatrix}
$$

Figure 5.19: Data values for the 6–MP experiment with a single binary covariate.

not have corresponding known censoring times. The observed information matrix, on the other hand, is easily calculated using the matrix formulation

$$O(\hat{\boldsymbol{\beta}}) = \boldsymbol{Z}'\hat{\boldsymbol{B}}\boldsymbol{Z} = \begin{bmatrix} 30 & 9 \\ 9 & 9 \end{bmatrix},$$

where $\hat{\boldsymbol{B}}$ is a 42×42 diagonal matrix with elements $x_1 e^{\hat{\boldsymbol{\beta}}' z_1}$, $x_2 e^{\hat{\boldsymbol{\beta}}' z_2}$, $\ldots$, $x_{42} e^{\hat{\boldsymbol{\beta}}' z_{42}}$. Since the determinant of this matrix is $(30)(9) - 9^2 = 189$, it has inverse

$$O^{-1}(\hat{\boldsymbol{\beta}}) = \begin{bmatrix} 9/189 & -9/189 \\ -9/189 & 30/189 \end{bmatrix},$$

which estimates the variance–covariance matrix of the maximum likelihood estimates. The off-diagonal elements of $O^{-1}(\hat{\boldsymbol{\beta}})$ indicate a negative correlation between $\hat{\beta}_0$ and $\hat{\beta}_1$. The square roots of the diagonal elements yield asymptotic estimates for the standard deviation of the regression parameter estimates. Thus, the asymptotic estimated standard deviation of the estimate for β_0 is

$$\sqrt{\hat{V}\left[\hat{\beta}_0\right]} = \sqrt{\frac{9}{189}} = 0.218,$$

and the asymptotic estimated standard deviation of the estimate for β_1 is

$$\sqrt{\hat{V}\left[\hat{\beta}_1\right]} = \sqrt{\frac{30}{189}} = 0.398.$$

These values can be used in the usual fashion to obtain asymptotically valid confidence intervals and perform hypothesis testing with respect to the regression parameter estimates. Note that $\hat{\beta}_1 = -1.53$ is more than three standard deviation units away from 0, supporting the conclusion that there is a statistically significant difference between the patients who take 6–MP versus those that do not with respect to their remission times. Since the sign of $\hat{\beta}_1$ is negative, the drug prolongs the remission times. More specifically, since the proportional hazards model is being used, a patient taking the 6–MP drug will have a hazard function that is estimated to be $e^{\hat{\beta}_1} = e^{-1.53} = 0.217$ times that of a patient who does not take the drug.

   Parameter estimation for single binary covariate is ideal in the sense that the parameter estimates can be expressed in closed form. The next subsection considers the more common situation in which the baseline distribution is unknown.

5.6.2 Unknown Baseline Distribution

In many applications, the baseline distribution is not known. Furthermore, the modeler may not be interested in the baseline distribution, rather only in the influence of the covariates on survival. A technique has been developed for the proportional hazards model that allows the coefficient vector $\boldsymbol{\beta}$ to be estimated without knowledge of the parametric form of the baseline distribution. This type of analysis might be appropriate when the modeler wants to detect which covariates are significant, to determine which covariate is the most significant, or to analyze interactions among covariates. This technique is characteristic of nonparametric methods because it is impossible to misspecify the baseline distribution.

The focus of this estimation technique is on the *indexes* of the components on test, as will be seen in the derivation to follow. Since this procedure is very different from all previous point estimation derivations, an example will be carried through the derivation to illustrate the notation and the method. The purpose in this small example is to determine whether light bulb wattage influences light bulb survival. This introduction to parameter estimation will alternate between the small example and the general case. In this example and the derivation, it is initially assumed that there is no censoring and there are no tied observations.

Example 5.14 A set of $n = 3$ light bulbs are placed on test. The first and second bulbs are 100-watt bulbs and the third bulb is a 60-watt bulb. A single ($q = 1$) covariate z_1 assumes the value 0 for a 60-watt bulb and 1 for a 100-watt bulb. The purpose of the test is to determine if the wattage has any influence on the survival distribution of the bulbs. The baseline distribution is unknown and unspecified, so there is only one parameter in the proportional hazards model, the regression coefficient β_1, that needs to be estimated. This small data set is used for illustrative purposes only, and we would obviously need to collect more than three data points to detect any statistically significant difference between the two wattages. Let $t_1 = 80$, $t_2 = 20$, and $t_3 = 50$ denote the lifetimes of the three bulbs. From the notation developed earlier in this chapter,

$$
x = \begin{bmatrix} 80 \\ 20 \\ 50 \end{bmatrix} \qquad \delta = \begin{bmatrix} 1 \\ 1 \\ 1 \end{bmatrix} \qquad Z = \begin{bmatrix} 1 \\ 1 \\ 0 \end{bmatrix}.
$$

The order statistics are $t_{(1)} = 20$, $t_{(2)} = 50$, and $t_{(3)} = 80$. Figure 5.20 illustrates the definitions made thus far. Recall that the first subscript on z_{ij} is the bulb number and the second subscript is the covariate number. The *risk set* $R(t)$, parameterized by the

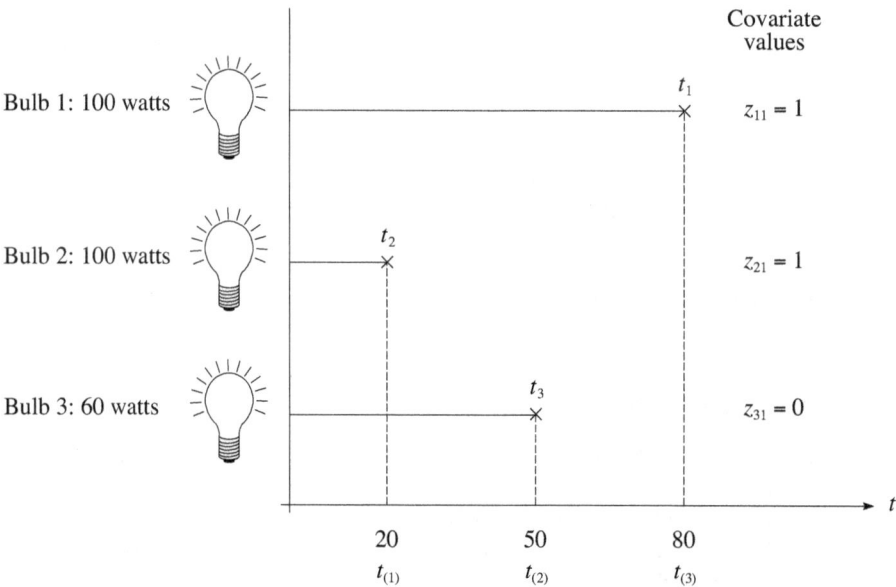

Figure 5.20: Proportional hazards parameter estimation notation.

failure times, is defined as the set of indexes of bulbs at risk just prior to time t. In this case

$$R(t_{(1)}) = R(t_2) = R(20) = \{1, 2, 3\}$$

since all bulbs are at risk just prior to $t_{(1)}$. At time $t_{(2)}$, the risk set is

$$R(t_{(2)}) = R(t_3) = R(50) = \{1, 3\}$$

since bulbs 1 and 3 are at risk just prior to $t_{(2)}$. Finally, at time $t_{(3)}$, the risk set is

$$R(t_{(3)}) = R(t_1) = R(80) = \{1\}$$

since only bulb 1 is still on test just prior to $t_{(3)}$. Similar to the concept of a pointer array from computer science, a *rank vector* r is used here to simplify the notation. The ith element of the rank vector is the index of the item that fails at $t_{(i)}$, for $i = 1, 2, 3$. For this particular data set,

$$r = \begin{bmatrix} 2 \\ 3 \\ 1 \end{bmatrix}$$

because bulb 2 fails first, bulb 3 fails next, and bulb 1 fails last. The failure times for each bulb can therefore be determined from the order statistics and the rank vector.

The notation defined in the example is easily extended from three items on test with a single binary covariate to the general case. Let $t_1, t_2, \ldots, t_n$ be n distinct lifetimes. Each lifetime t_i has an associated $q \times 1$ vector of covariates z_i, for $i = 1, 2, \ldots, n$. The ith order statistic is given by $t_{(i)}$, and the risk set $R(t_{(i)})$ is the set of indexes of all items that are at risk just prior to $t_{(i)}$, for $i = 1, 2, \ldots, n$. The ith element of the rank vector $r = (r_1, r_2, \ldots, r_n)'$ is the index of the item that fails at time $t_{(i)}$, for $i = 1, 2, \ldots, n$. The observed failure times and their associated indexes are equivalent to the observed order statistics and the associated rank vector. Now that the new notation has been defined, the emphasis transitions to determining the probability that a particular permutation of the indexes appears in the rank vector.

Example 5.15 We now return to the light bulb life test described in Example 5.14. The joint probability distribution of the elements of the rank vector, denoted by $f(r_1, r_2, r_3)$, is now considered for the data set containing $n = 3$ observations. In this case, there are $3! = 6$ possible permutations of the ranks of the observations:

$$\begin{bmatrix} 1 \\ 2 \\ 3 \end{bmatrix} \quad \begin{bmatrix} 1 \\ 3 \\ 2 \end{bmatrix} \quad \begin{bmatrix} 2 \\ 1 \\ 3 \end{bmatrix} \quad \begin{bmatrix} 2 \\ 3 \\ 1 \end{bmatrix} \quad \begin{bmatrix} 3 \\ 1 \\ 2 \end{bmatrix} \quad \begin{bmatrix} 3 \\ 2 \\ 1 \end{bmatrix}.$$

If the wattage of the light bulb had no influence on the survival time, then clearly $f(r_1, r_2, r_3) = \frac{1}{6}$ for all six permutations because all three items are drawn from a homogeneous population with respect to survival. Switching to the non-equally-likely case, the probability mass function for the rank vector will be determined by finding the conditional probabilities associated with the ranks. For example, assume that a failure has just occurred at time $t_{(2)} = 50$, and the history up to time 50, which is bulb 2 failed at time $t_{(1)} = 20$, is known. The bulb that fails at time 50 is either bulb 1 or bulb 3. For small Δt, the conditional probability that the bulb failing at time 50 is bulb 1 is

$$P\left(r_2 = 1 \,\middle|\, t_{(1)} = 20, t_{(2)} = 50, r_1 = 2\right) = \frac{P(\text{bulb 1 fails at time 50})}{P(\text{one item from } R(t_{(2)}) \text{ fails at time 50})}$$

$$= \frac{h(50, z_{11})\Delta t}{h(50, z_{11})\Delta t + h(50, z_{31})\Delta t}$$

$$= \frac{h(50, z_{11})}{h(50, z_{11}) + h(50, z_{31})}$$

$$= \frac{\psi(z_{11})h_0(50)}{\psi(z_{11})h_0(50) + \psi(z_{31})h_0(50)}$$

$$= \frac{\psi(z_{11})}{\psi(z_{11}) + \psi(z_{31})}$$

$$= \frac{e^{\beta_1 z_{11}}}{e^{\beta_1 z_{11}} + e^{\beta_1 z_{31}}}$$

$$= \frac{e^{\beta_1}}{e^{\beta_1} + 1}$$

because the first bulb is 100 watts ($z_{11} = 1$) and the third bulb is 60 watts ($z_{31} = 0$). Note that the baseline hazard function has dropped out of this expression, so this probability will be the same regardless of the choice of $h_0(t)$. Also, the first two order statistics, $t_{(1)}$ and $t_{(2)}$, were not used in the calculation of this conditional probability. By similar reasoning, the conditional probability that the 60-watt bulb is the second to fail is

$$P\left(r_2 = 3 \,\middle|\, t_{(1)} = 20, t_{(2)} = 50, r_1 = 2\right) = \frac{1}{e^{\beta_1} + 1}.$$

In the example, as well as in the general case, the conditional probability expression does not involve the failure times, making it possible to shorten $P(r_j = i \,|\, t_{(1)}, t_{(2)}, \ldots, t_{(j)}, r_1, r_2, \ldots, r_{j-1})$ to just $P(r_j = i \,|\, r_1, r_2, \ldots, r_{j-1})$. The probability that the jth element of the rank vector will be equal to i, given $t_{(j)}$ and the failure history up to $t_{(j)}$, is

$$P\left(r_j = i \,\middle|\, r_1, r_2, \ldots, r_{j-1}\right) = \frac{h(t_{(j)}, z_i)\Delta t}{\sum_{k \in R(t_{(j)})} h(t_{(j)}, z_k)\Delta t}$$

$$= \frac{h(t_{(j)}, z_i)}{\sum_{k \in R(t_{(j)})} h(t_{(j)}, z_k)}$$

$$= \frac{\psi(z_i)h_0(t_{(j)})}{\sum_{k \in R(t_{(j)})} \psi(z_k)h_0(t_{(j)})}$$

$$= \frac{\psi(z_i)}{\sum_{k \in R(t_{(j)})} \psi(z_k)}$$

$$= \frac{e^{\beta' z_i}}{\sum_{k \in R(t_{(j)})} e^{\beta' z_k}}.$$

Example 5.16 We continue with the light bulb life test with $n = 3$ bulbs on test from Examples 5.14 and 5.15. It is now a simple task to use this conditional probability to determine the probability mass function for the indexes. For the three light bulbs, this probability mass function is

$$f(r_1, r_2, r_3) = f(r_3 \,|\, r_1, r_2)f(r_1, r_2)$$
$$= f(r_3 \,|\, r_1, r_2)f(r_2 \,|\, r_1)f(r_1)$$
$$= f(r_1)f(r_2 \,|\, r_1)f(r_3 \,|\, r_1, r_2)$$

over all six permutations of the rank vector. Since the sequence that was observed for the rank vector was $r = (2, 3, 1)'$, this becomes

$$f(2, 3, 1) = f(2)f(3 \mid 2)f(1 \mid 2, 3)$$

$$= \frac{\psi(z_{21})}{\psi(z_{11}) + \psi(z_{21}) + \psi(z_{31})} \cdot \frac{\psi(z_{31})}{\psi(z_{11}) + \psi(z_{31})} \cdot \frac{\psi(z_{11})}{\psi(z_{11})}$$

$$= \frac{e^{\beta_1 z_{21}}}{e^{\beta_1 z_{11}} + e^{\beta_1 z_{21}} + e^{\beta_1 z_{31}}} \cdot \frac{e^{\beta_1 z_{31}}}{e^{\beta_1 z_{11}} + e^{\beta_1 z_{31}}}$$

$$= \frac{e^{\beta_1}}{e^{\beta_1} + e^{\beta_1} + 1} \cdot \frac{1}{e^{\beta_1} + 1}$$

$$= \frac{e^{\beta_1}}{\left(2e^{\beta_1} + 1\right)\left(e^{\beta_1} + 1\right)}.$$

Treating this expression as a likelihood function $L(\beta_1)$, the problem reduces to determining the β_1 value that maximizes the log likelihood function

$$\ln L(\beta_1) = \beta_1 - \ln\left(2e^{\beta_1} + 1\right) - \ln\left(e^{\beta_1} + 1\right).$$

The score statistic is

$$\frac{\partial \ln L(\beta_1)}{\partial \beta_1} = 1 - \frac{2e^{\beta_1}}{2e^{\beta_1} + 1} - \frac{e^{\beta_1}}{e^{\beta_1} + 1}.$$

Setting the score statistic to zero and solving for the maximum likelihood estimate, $\hat{\beta}_1 = (-\ln 2)/2 = -0.347$. Since $\hat{\beta}_1 < 0$, there is lower risk for the 100-watt bulbs than for 60-watt bulbs. More specifically, the hazard function for 100-watt light bulbs is $e^{\hat{\beta}_1} = \sqrt{2}/2 = e^{-0.347} = 0.707$ times that of the baseline hazard function for 60-watt bulbs, regardless of what baseline distribution is considered. To see if this regression coefficient is statistically significant involves calculating the negative of the derivative of the score:

$$-\frac{\partial^2 \ln L(\beta_1)}{\partial \beta_1^2} = \frac{2e^{\beta_1}}{\left(2e^{\beta_1} + 1\right)^2} + \frac{e^{\beta_1}}{\left(e^{\beta_1} + 1\right)^2}.$$

When this expression is evaluated at $\beta_1 = \hat{\beta}_1 = -0.347$, the 1×1 observed information matrix is 0.485, so the asymptotic estimate of the variance of $\hat{\beta}_1$ is $1/0.485 = 2.06$, and the asymptotic estimate of the standard deviation of $\hat{\beta}_1$ is $\sqrt{2.06} = 1.44$. Since $\hat{\beta}_1$ is only a fraction of a standard deviation away from 0, z_1 is not statistically significant. This result is not surprising considering the small number of light bulbs placed on the life test. Note that these values are only asymptotically correct and are obviously poor approximations when $n = 3$. The p-value for testing $H_0 : \beta_1 = 0$ versus $H_1 : \beta_1 \neq 0$ is 0.809, indicating that there is no statistical evidence that wattage influences the longevity of a light bulb for this tiny data set. In addition, only the order of the failure times and not their numerical values were used to find $\hat{\beta}_1$. This means, for example, that the failure time of the third bulb, t_3, could have fallen anywhere on the interval $(20, 80)$, and the estimate would have been the same because the order of the observed failure times was not changed.

The R code below confirms the calculations given above. The coxph function, which is part of the survival package, is used to calculate the estimated regression coefficient

$\hat{\beta}_1 = -0.347$, which is stored in b, the 1×1 observed information matrix, which is stored in v, and the p-value for the hypothesis test, which is stored in p.

```
library(survival)
failtimes = c(80, 20, 50)
censor    = c(1, 1, 1)
z         = c(1, 1, 0)
bulbs.fit = coxph(Surv(failtimes, censor) ~ z)
b         = bulbs.fit$coef
v         = bulbs.fit$var
p         = 2 * pnorm(b / sqrt(v))
```

The procedure for estimating β_1 can be generalized from the example without any significant difficulties. The probability mass function for the indexes, or the likelihood function for $\boldsymbol{\beta}$, is now

$$L(\boldsymbol{\beta}) = f(r_1)f(r_2 \mid r_1) \ldots f(r_n \mid r_1, r_2, \ldots, r_{n-1})$$

$$= \prod_{j=1}^{n} \frac{\psi(z_{r_j})}{\sum_{k \in R(t_{(j)})} \psi(z_k)}$$

$$= \prod_{j=1}^{n} \frac{e^{\boldsymbol{\beta}' z_{r_j}}}{\sum_{k \in R(t_{(j)})} e^{\boldsymbol{\beta}' z_k}}.$$

The log likelihood is

$$\ln L(\boldsymbol{\beta}) = \sum_{j=1}^{n} \left[\boldsymbol{\beta}' z_{r_j} - \ln \sum_{k \in R(t_{(j)})} e^{\boldsymbol{\beta}' z_k} \right].$$

The score vector has sth component

$$\frac{\partial \ln L(\boldsymbol{\beta})}{\partial \beta_s} = \sum_{j=1}^{n} \left[z_{sr_j} - \frac{\sum_{k \in R(t_{(j)})} z_{sk} e^{\boldsymbol{\beta}' z_k}}{\sum_{k \in R(t_{(j)})} e^{\boldsymbol{\beta}' z_k}} \right]$$

for $s = 1, 2, \ldots, q$. The vector of maximum likelihood estimators $\hat{\boldsymbol{\beta}}$ is obtained when the elements of the score vector are equated to zero and solved via numerical methods. To determine an estimate for the variance of $\hat{\boldsymbol{\beta}}$, the score vector must be differentiated to calculate the observed information matrix. The diagonal elements of the inverse of the observed information matrix are asymptotically valid estimates of the variance of $\hat{\boldsymbol{\beta}}$.

There are two approaches to handle right censoring that do not significantly complicate the derivation presented thus far. The first approach is to assume that right censoring occurs immediately after a failure occurs when a failure time and right-censoring time coincide. This assumption is valid for a Type II censored data set, but will involve an approximation for more general right-censoring schemes. In this case the rank vector is shortened to only r elements, corresponding to the indexes of the observed failure times $t_{(1)}, t_{(2)}, \ldots, t_{(r)}$. The likelihood function is

$$L(\boldsymbol{\beta}) = \prod_{j=1}^{r} \frac{\psi(z_{r_j})}{\sum_{k \in R(t_{(j)})} \psi(z_k)} = \prod_{j=1}^{r} \frac{e^{\boldsymbol{\beta}' z_{r_j}}}{\sum_{k \in R(t_{(j)})} e^{\boldsymbol{\beta}' z_k}}.$$

The log likelihood function is

$$\ln L(\boldsymbol{\beta}) = \sum_{j=1}^{r} \left[\boldsymbol{\beta}' \boldsymbol{z}_{r_j} - \ln \sum_{k \in R(t_{(j)})} e^{\boldsymbol{\beta}' \boldsymbol{z}_k} \right].$$

The score vector has sth component

$$\frac{\partial \ln L(\boldsymbol{\beta})}{\partial \beta_s} = \sum_{j=1}^{r} \left[z_{srj} - \frac{\sum_{k \in R(t_{(j)})} z_{sk} e^{\boldsymbol{\beta}' \boldsymbol{z}_k}}{\sum_{k \in R(t_{(j)})} e^{\boldsymbol{\beta}' \boldsymbol{z}_k}} \right]$$

for $s = 1, 2, \ldots, q$. Using the quotient rule, the derivative of the score vector is

$$\frac{\partial^2 \ln L(\boldsymbol{\beta})}{\partial \beta_s \partial \beta_t} = -\sum_{j=1}^{r} \frac{\left(\sum_{k \in R(t_{(j)})} e^{\boldsymbol{\beta}' \boldsymbol{z}_k} \right) \left(\sum_{k \in R(t_{(j)})} z_{sk} z_{tk} e^{\boldsymbol{\beta}' \boldsymbol{z}_k} \right) - \left(\sum_{k \in R(t_{(j)})} z_{sk} e^{\boldsymbol{\beta}' \boldsymbol{z}_k} \right) \left(\sum_{k \in R(t_{(j)})} z_{tk} e^{\boldsymbol{\beta}' \boldsymbol{z}_k} \right)}{\left(\sum_{k \in R(t_{(j)})} e^{\boldsymbol{\beta}' \boldsymbol{z}_k} \right)^2}$$

for $s = 1, 2, \ldots, q$ and $t = 1, 2, \ldots, q$. The elements of the observed information matrix are obtained by using the maximum likelihood estimates as arguments in the negative of this expression.

The second approach to right censoring is to write the likelihood function as the sum of all likelihoods for complete data sets that are consistent with the censoring pattern. Fortunately, this second approach yields the same likelihood function as the first approach, as illustrated by the following example.

Example 5.17 In the previous example, the data set consisted of three observed failure times: 80, 20, 50. Now, if the situation changes so that the lifetime of the third light bulb is right censored at time 50, the data set is 80, 20, 50*, as is illustrated in Figure 5.21. Using the first approach to right censoring, the observed rank vector is now $r = (2, 1)'$, and the likelihood function is

$$L(\beta_1) = \prod_{j=1}^{2} \frac{\psi(z_{r_j})}{\sum_{k \in R(t_{(j)})} \psi(z_k)}$$

$$= \frac{\psi(z_{21})}{\psi(z_{11}) + \psi(z_{21}) + \psi(z_{31})} \cdot \frac{\psi(z_{11})}{\psi(z_{11})}$$

$$= \frac{\psi(z_{21})}{\psi(z_{11}) + \psi(z_{21}) + \psi(z_{31})}.$$

For the second approach to right censoring, there are two possibilities for the rank vector if there was no censoring: if the third bulb failed before time 80, the observed rank vector would be $r = (2, 3, 1)'$; if the third bulb failed after time 80, the observed rank vector would be $r = (2, 1, 3)'$. In the first case, the likelihood function would be that from the previous example:

$$\frac{\psi(z_{21})}{\psi(z_{11}) + \psi(z_{21}) + \psi(z_{31})} \cdot \frac{\psi(z_{31})}{\psi(z_{11}) + \psi(z_{31})} \cdot \frac{\psi(z_{11})}{\psi(z_{11})}.$$

In the second case, the likelihood function would be

$$\frac{\psi(z_{21})}{\psi(z_{11}) + \psi(z_{21}) + \psi(z_{31})} \cdot \frac{\psi(z_{11})}{\psi(z_{11}) + \psi(z_{31})} \cdot \frac{\psi(z_{31})}{\psi(z_{31})}.$$

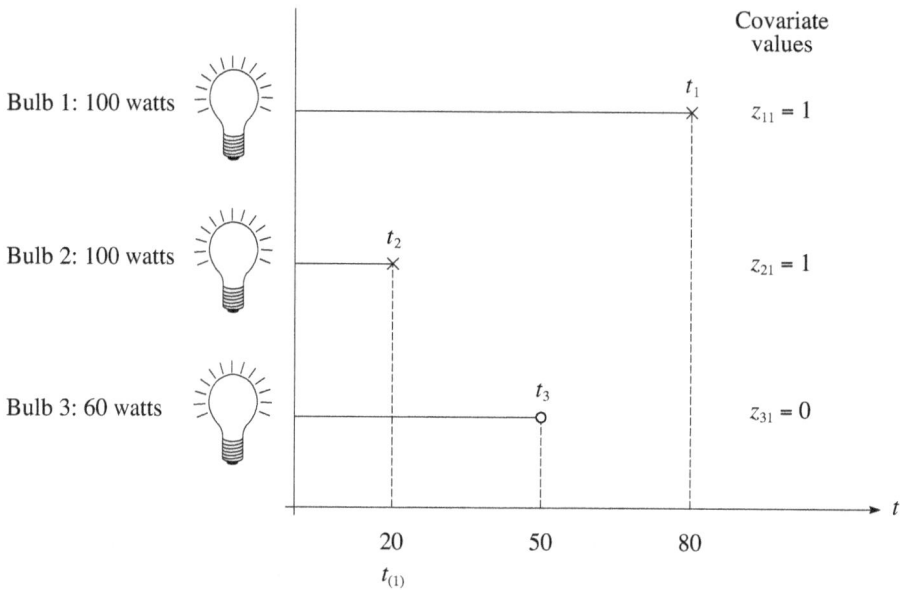

Figure 5.21: Proportional hazards model with censoring.

The sum of these two likelihood functions is

$$\frac{\psi(z_{21})}{\psi(z_{11}) + \psi(z_{21}) + \psi(z_{31})},$$

which is the same result as in the first approach to handling right censoring.

Tied lifetimes are typically handled by an approximation. When there are several failures at the same time value, each is assumed to contribute the same term to the likelihood function. Consequently, all the items with tied failure times are included in the risk set at the time of the tied observation. This approximation works well when there are not many tied observations in the data set and has been implemented in many software packages that estimate the vector of regression coefficients $\boldsymbol{\beta}$.

Example 5.18 Fit the Cox proportional hazards model via maximum likelihood to the remission times in the 6–MP clinical trial with a single binary covariate z_1 for the control ($z_1 = 0$) and treatment ($z_1 = 1$) groups. The data values are given in Example 5.6.

Using numerical methods, the maximum likelihood estimate for the single regression parameter is $\hat{\beta}_1 = -1.51$. The log likelihood function attains a value of -86.38 at the maximum likelihood value, the observed information matrix has a single value 5.962, and the inverse of the observed information matrix is $1/5.962 = 0.168$. This means that an asymptotic estimate of the standard deviation of the maximum likelihood estimate is

$$\sqrt{\hat{V}\left[\hat{\beta}_1\right]} = \sqrt{0.168} = 0.41,$$

which indicates that the maximum likelihood estimate is $1.51/0.41 = 3.7$ standard deviations units away from 0. It can be concluded, with a p-value less than 0.001, that the 6–MP drug is effective in increasing the remission times for leukemia patients, assuming that the proportional hazards model is appropriate here. Regardless of the baseline hazard function $h_0(t)$ chosen, the hazard function in the treatment case is $e^{\hat{\beta}_1} = e^{-1.51} = 0.221$ times that of the baseline hazard function for all time values. Note that no work has been done here to assess model adequacy, and all these conclusions have been based on the fact that the proportional hazards model adequately describes the distribution of the remission time with the single binary covariate.

The R code below uses the `coxph` function in the `survival` package to compute $\hat{\beta}_1$ and a p-value for the appropriate hypothesis test.

```
library(survival)
x1 = c(1, 1, 2, 2, 3, 4, 4, 5, 5, 8, 8, 8, 8, 11, 11, 12,
       12, 15, 17, 22, 23)
d1 = rep(1, length(x2))
z1 = rep(0, length(x2))
x2 = c(6, 6, 6, 6, 7, 9, 10, 10, 11, 13, 16, 17, 19, 20, 22, 23,
       25, 32, 32, 34, 35)
d2 = c(1, 1, 1, 0, 1, 0, 1, 0, 0, 1, 1, 0, 0, 0, 1, 1, 0, 0, 0, 0, 0)
z2 = rep(1, length(x1))
x  = c(x1, x2)
d  = c(d1, d2)
z  = c(z1, z2)
ph = coxph(Surv(x, d) ~ z)
summary(ph)
```

The data set is contained in the `gehan` data frame in the `MASS` package, so the Cox proportional hazards model can also be fitted with the statements

```
library(survival)
library(MASS)
summary(coxph(Surv(time, cens) ~ treat, data = gehan))
```

which reverses the roles of the treatment and control groups, resulting in the reversal of the sign of $\hat{\beta}_1$.

The last example moves from the single binary covariate case to the case in which there are $q > 1$ covariates which can assume discrete and continuous values. The survival analysis application comes from sociology, and the analyst is attempting to determine which of the covariates significantly influences survival.

Example 5.19 The proportional hazards model has been used in diverse applications. Recidivism considers the probability that an inmate will return to prison in the future after release. Recidivism can be predicted using survival models. Several factors related to inmate background that could affect an inmate's adjustment to society are potential screening variables. North Carolina collected recidivism data on $n = 1540$ prisoners in 1978. The lifetime of interest here is the time of release until the time of return to prison. Obviously, not all inmates will return to prison, so a more complicated *split model*, for which some of the lifetimes are assumed to be infinite, may also be used. In addition, there is significant right censoring in the data set. The purpose of the study is to assess the impact of the $q = 15$ covariates. The covariates $z_1, z_2, \ldots, z_{15}$ are time served, age, number of prior convictions, number of rule violations in prison, education, race, gender, alcohol problems, drug problems, marital status, probationary period, participation in a work release program, type of crime, crime against person, and crime against property. Many of these covariates are coded as indicator variables. Table 5.3 presents the estimates of the regression coefficients and their standard deviations in order of their significance. The column labeled *Covariate* gives a short description of the covariate considered. The next two columns give the regression coefficient estimator and an asymptotic estimate of its standard deviation. The column labeled $\hat{\beta}/\sqrt{\hat{V}\left[\hat{\beta}\right]}$ gives a test statistic for testing $H_0 : \beta_i = 0$ versus $H_1 : \beta_i \neq 0$, for $i = 1, 2, \ldots, 15$. The column labeled *p-value* indicates the attained significance of the covariates. A value less than $\alpha = 0.05$ indicates that a covariate is a statistically significant indicator of recidivism. Ten of the fifteen covariates are statistically significant. This example includes indicator variables (such as gender) and can easily be extended to include other regression modeling tools such as nonlinear and interaction terms in the regression model.

Name	Covariate	$\hat{\beta}$	$\sqrt{\hat{V}\left[\hat{\beta}\right]}$	$\dfrac{\hat{\beta}}{\sqrt{\hat{V}\left[\hat{\beta}\right]}}$	p-value	Significant
z_2	AGE	-3.3420	0.5195	-6.4328	0.0000	•
z_3	PRIORS	0.8355	0.1371	6.0957	0.0000	•
z_1	TSERVD	1.1666	0.1957	5.9616	0.0000	•
z_6	WHITE	-0.4444	0.0876	-5.0701	0.0000	•
z_8	ALCHY	0.4285	0.1043	4.1103	0.0000	•
z_{13}	FELON	-0.5782	0.1633	-3.5412	0.0002	•
z_9	JUNKY	0.2819	0.0970	2.9058	0.0018	•
z_7	MALE	0.6745	0.2423	2.7834	0.0027	•
z_{15}	PROPTY	0.3894	0.1578	2.4678	0.0068	•
z_4	RULE	3.0788	1.6890	1.8229	0.0342	•
z_{10}	MARRIED	-0.1532	0.1077	-1.4227	0.0774	
z_5	SCHOOL	-0.2507	0.1933	-1.2966	0.0974	
z_{12}	WORKREL	0.0865	0.0902	0.9587	0.1688	
z_{14}	PERSON	0.0737	0.2425	0.3039	0.3806	
z_{11}	SUPER	-0.0088	0.0966	-0.0914	0.4636	

Table 5.3: North Carolina recidivism model.

This chapter has contained a brief introduction to some of the statistical methods that are used in survival analysis. The key modeling features that indicate the use of survival analysis are (a) a population lifetime distribution with nonnegative support, (b) appreciable dispersion, (c) possibly right-censored data values, (d) possibly a vector of covariates which might influence the lifetime distribution. The exponential, Weibull, and Cox proportional hazards model were fitted to complete and right-censored data sets in this chapter.

5.7 Exercises

5.1   Consider a large batch of light bulbs whose lifetimes are known to have exponential(1) life-times. Gina knows that the population distribution is exponential, but she does not know the value of the population mean. She estimates the population mean lifetime of the light bulbs by averaging n observed lifetimes from bulbs chosen at random from the batch. Find the smallest value of n that assures, with probability of at least 0.95, that the sample mean is within 0.2 of the population mean

   (a) exactly,

   (b) approximately, using the central limit theorem.

5.2   Libby is a statistician for a light bulb company. She knows that the lifetimes of the 60-watt bulbs that her company manufactures are exponentially distributed with population mean 1500 hours. She conducts a life test in which 39 of their 60-watt bulbs are placed on life test until they fail and the average of the failure times is recorded. Find the probability that the sample mean exceeds 1600 hours using

   (a) the central limit theorem,

   (b) the exact distribution of the sample mean.

5.3   Let $t_1, t_2, \ldots, t_n$ be a random sample from an exponential(λ) population, where λ is a positive unknown failure rate parameter. Find an unbiasing constant c_n so that $c_n t_{(1)}$ is an unbiased estimator of $1/\lambda$, where $t_{(1)} = \min\{t_1, t_2, \ldots, t_n\}$ is the first order statistic. *Hint*: the unbiasing constant c_n is a function of the number of items on test n.

5.4   Debbie purchases a laptop computer with a random lifetime T whose probability distribution is a special case of the log logistic distribution with survivor function

$$S(t) = \frac{1}{1 + \lambda t} \qquad t > 0,$$

where λ is a positive unknown scale parameter. From just a single observation of the lifetime of her laptop computer, find an exact two-sided 90% confidence interval for λ.

5.5   Let $t_1, t_2, \ldots, t_n$ be a random sample from an exponential population with mean θ, where θ is a positive unknown parameter. An exact two-sided 90% confidence interval for θ is

$$27 < \theta < 55.$$

Carol is not concerned about large values of θ. Only small values of θ are of concern. What is an exact one-sided 95% confidence interval of the form $\theta > k$, for some constant k?

5.6   If $t_1, t_2, \ldots, t_n$ are n mutually independent observations from a log normal distribution with probability density function

$$f(t) = \frac{1}{\sqrt{2\pi}\sigma t} e^{-\frac{1}{2}\left(\frac{\ln t - \mu}{\sigma}\right)^2} \qquad t \geq 0$$

for $\sigma > 0$ and $-\infty < \mu < \infty$, find the maximum likelihood estimators of μ and σ and exact two-sided $100(1 - \alpha)\%$ confidence intervals for μ and σ in terms of $t_1, t_2, \ldots, t_n$.

5.7 Let $t_1, t_2, \ldots, t_7$ be a random sample of the lifetimes of $n = 7$ items on test drawn from an exponential population with positive unknown mean θ.

  (a) Find an exact two-sided 90% confidence interval for the median by finding a pivotal quantity based on the sample median $t_{(4)}$.

  (b) Give an exact two-sided 90% confidence interval for the median for the $n = 7$ rat survival times in the treatment group from Efron and Tibshirani (1993, page 11):

$$16, 23, 38, 94, 99, 141, 197.$$

  (c) Conduct a Monte Carlo simulation experiment to provide convincing numerical evidence that the exact two-sided 90% confidence interval for the median is indeed an exact two-sided 90% confidence interval for an exponential population when θ is arbitrarily set to 1.

5.8 This chapter has emphasized confidence intervals. Another type of statistical interval is known as a *prediction interval*, which contains a future value of an observation with a prescribed probability. Let $t_1, t_2, \ldots, t_n$ be a random sample from an exponential population with a positive unknown mean θ. Conduct a Monte Carlo simulation experiment that provides convincing numerical evidence that the $100(1 - \alpha)\%$ prediction interval for t_{n+1}

$$\frac{\bar{t}}{F_{2n, 2, \alpha/2}} < t_{n+1} < \frac{\bar{t}}{F_{2n, 2, 1-\alpha/2}}$$

is an *exact* prediction interval for the arbitrary parameter settings $n = 11$, $\alpha = 0.05$, and $\theta = 19$.

5.9 Let T_1, T_2, T_3 be mutually independent random variables such that T_i is exponentially distributed with mean $i\theta$, for $i = 1, 2, 3$, where θ is a positive unknown parameter.

  (a) Find the maximum likelihood estimator $\hat{\theta}$.

  (b) Find the probability density function of the maximum likelihood estimator $\hat{\theta}$.

  (c) Is $\hat{\theta}$ an unbiased estimator of θ?

  (d) Find an exact two-sided $100(1 - \alpha)\%$ confidence interval for θ.

  (e) Perform a Monte Carlo simulation experiment to evaluate the coverage of the confidence interval for $\theta = 10$ and $\alpha = 0.1$.

5.10 Let $t_1, t_2, \ldots, t_n$ be a random sample from a population with probability density function

$$f(t) = \frac{\theta}{t^{\theta+1}} \qquad t \geq 1,$$

where θ is a positive unknown parameter.

  (a) Find the maximum likelihood estimator of θ.

  (b) Use the invariance property to find the maximum likelihood estimator of the median of the distribution.

5.11 Let $t_1, t_2, \ldots, t_n$ be a random sample from a population with probability density function

$$f(t) = \sqrt{\frac{\lambda}{2\pi t^3}} \, e^{-\lambda(t-1)^2/(2t)} \qquad t > 0,$$

where λ is a positive unknown parameter. This distribution is known as the *standard Wald distribution* which is a special case of the inverse Gaussian distribution Find the maximum likelihood estimator of λ.

5.12 Let $T_1, T_2, \ldots, T_n$ be mutually independent and identically distributed random variables from a population having probability density function

$$f(t) = 7e^{-7(t-\theta)} \qquad t \geq \theta.$$

Find the limiting distribution of $n\left(T_{(1)} - \theta\right)$. Support this limiting distribution by conducting a Monte Carlo simulation experiment.

5.13 Let $t_1, t_2, \ldots, t_n$ be a random sample from a population with probability density function

$$f(t) = \frac{\theta}{(1+t)^{\theta+1}} \qquad t \geq 0,$$

where θ is a positive unknown parameter. Calculate an asymptotically exact two-sided $100(1-\alpha)\%$ confidence interval for θ based on the asymptotic normality of the maximum likelihood estimator.

5.14 Let $t_1, t_2, \ldots, t_n$ be a random sample from a population with probability density function

$$f(t) = \sqrt{\frac{1}{2\pi t^3}} \, e^{-(t-\theta)^2/(2t\theta^2)} \qquad t > 0,$$

where θ is a positive unknown parameter. This population distribution is a special case of the inverse Gaussian distribution. Calculate an asymptotically exact two-sided $100(1-\alpha)\%$ confidence interval for θ based on the asymptotic normality of the maximum likelihood estimator. *Hint*: the expected value of T is $E[T] = \theta$.

5.15 Let $t_1, t_2, \ldots, t_n$ be a random sample from a population with probability density function

$$f(t) = \frac{\theta}{(1+\theta t)^2} \qquad t \geq 0,$$

where θ is a positive unknown parameter. This is a special case of the log logistic distribution.

(a) Find the maximum likelihood estimator of θ. *Hint*: The maximum likelihood estimator cannot be expressed in closed form.

(b) Find the maximum likelihood estimate of θ for the $n = 7$ rat survival times (in days) of the treatment group from Efron and Tibshirani (1993, page 11):

$$16, 23, 38, 94, 99, 141, 197.$$

(c) Find an asymptotically exact two-sided 95% confidence interval for θ based on the likelihood ratio statistic for the rat survival times from part (b).

5.16 If n items from an exponential population with failure rate λ are placed on a life test that is terminated after r failures have occurred, show that

$$V\left[\sum_{i=1}^{n} x_i\right] = \frac{r}{\lambda^2},$$

where $x_i = \min\{t_i, c_i\}$, t_i is the time to failure of the ith item, and c_i is the right-censoring time for the ith item, $i = 1, 2, \ldots, n$.

5.17 If n items from an exponential population with failure rate λ are placed on a life test that is terminated after r failures have occurred, find the expected time to complete the test if

(a) failed items are not replaced,

(b) failed items are immediately replaced with new items.

5.18 Find the score, maximum likelihood estimator, and Fisher information matrix for a Type II censored random sample from a population with

$$f(t) = \frac{\theta}{t^{\theta+1}} \qquad t \geq 1,$$

where θ is a positive unknown parameter.

5.19 The lifetimes of studio light bulbs, measured in days, is exponentially distributed with an unknown failure rate λ. James places n studio light bulbs on test at noon on one day and subsequently checks for failed bulbs at noon on subsequent days until all bulbs have failed. Let $r_1, r_2, \ldots, r_k$ be the number of observed bulb failures, some of which may be zero, on the k days that the bulbs are inspected. Find the maximum likelihood estimator for λ. Also, give the maximum likelihood estimate for the data values $r_1 = 8$, $r_2 = 5$, $r_3 = 2$, $r_5 = 1$, and all other r_i values equal zero.

5.20 James's friend Alexandra decides to simplify matters from the previous question by assuming that all failures that occur during any interval occur at midnight. What is Alexandra's maximum likelihood estimator for λ as a function of n and $r_1, r_2, \ldots, r_k$?

5.21 Dre conducts a life test on n items from an exponential population with mean θ. He observes only the value of a single order statistic $t_{(k)}$, where k is known. So $k-1$ lifetimes are left censored at $t_{(k)}$, one lifetime is observed at $t_{(k)}$, and $n-k$ lifetimes are right censored at $t_{(k)}$.

(a) What is the score statistic for estimating θ?

(b) What is the maximum likelihood estimator for θ when $n = 30$, $k = 11$, and $t_{(11)} = 15.5$?

5.22 Consider a Type II right-censored life test with n items on test and $r = 1$ failure is observed at time $t_{(1)}$. Assume that the items placed on the life test have lifetimes that are well described by a Rayleigh(λ) population.

(a) What is the maximum likelihood estimator for λ?

(b) What is an exact confidence interval for λ?

(c) What is the expected width of the confidence interval from part (b)?

(d) Verify the coverage and expected width of the exact confidence interval for $\lambda = 2$, $n = 7$, and $\alpha = 0.05$ via Monte Carlo simulation.

5.23 A randomly right-censored data set is collected from a population with hazard function

$$h(t) = \theta(1+t) \qquad t \geq 0,$$

where θ is a positive parameter.

(a) Find the maximum likelihood estimator $\hat{\theta}$.

(b) Give an expression for the observed information matrix.

(c) Give an asymptotically exact confidence interval for θ based on the observed information matrix.

5.24 Candice conducts a life test in which n items are simultaneously placed on test at time 0. The test is concluded at time $c > 0$. Assuming that the lifetimes of the items are from an exponential population with mean θ, find the distribution of the *number* of failures that occur by time c.

5.25 Show that when a random sample is drawn an exponential(λ) population with Type II right censoring

$$\frac{2r\lambda}{\hat{\lambda}} \sim \chi^2(2r),$$

where $\chi^2(2r)$ is the chi-square distribution with $2r$ degrees of freedom.

5.26 Consider a Type II right censored sample of n items on test and r observed failures drawn from an exponential population with mean θ. Show that the maximum likelihood estimate $\hat{\theta}$ is unbiased.

5.27 Assume that a life test without replacement is conducted on n items from an exponential population with failure rate λ. The exact failure times are not known, but the test is terminated upon the rth ordered failure at time $t_{(r)}$. Find a point estimator for λ.

5.28 Consider a population of items with exponential(λ) lifetimes. A life test with replacement is terminated when r failures occur or at time c, whichever occurs first. This is a combination of Type I and Type II right censoring. Find the expected number of items that fail during the test as a function of λ.

5.29 For a life test of n items with exponential(λ) lifetimes (items are not replaced upon failure) which is continued until all items fail, show that

$$E\left[\hat{\lambda}\right] = \frac{n}{n-1}\lambda,$$

where λ is the population failure rate and $\hat{\lambda}$ is the maximum likelihood estimator for λ. Thus, an unbiasing constant for $\hat{\lambda}$ is $u_n = (n-1)/n$. Equivalently,

$$E\left[\frac{n-1}{n}\hat{\lambda}\right] = E\left[u_n\hat{\lambda}\right] = \lambda.$$

Find an unbiasing constant for the case of Type II right censoring.

5.30 Give a point and 95% interval estimator for the median lifetime of the 6–MP treatment group assuming that the data have been drawn from an exponential(λ) population.

5.31 Consider the following Type II right censored data set for the lifetime of a product ($n = 5$ and $r = 3$) drawn from an exponential population with failure rate λ:

$$3.6 \qquad 3.9 \qquad 8.5.$$

    (a) Find the maximum likelihood estimator for the mean of the population.

    (b) Find the maximum likelihood estimator for $S(5)$.

    (c) Find an exact two-sided 80% confidence interval for $E\left[T^3\right]$.

    (d) Find an exact one-sided 95% lower confidence interval for $S(5)$.

    (e) Find the p-value for the test $H_0 : \lambda = 0.04$ versus $H_1 : \lambda > 0.04$.

    (f) Find the value of the log likelihood function at the maximum likelihood estimate.

    (g) Find the value of the observed information matrix.

    (h) Assume the data values

$$3.8 \qquad 4.6 \qquad 6.0 \qquad 9.6$$

    constitute a complete data set for a different product. Find an exact two-sided 90% confidence interval for the ratio of the failure rates of the two products if both are assumed to come from exponential populations.

5.32 Sara observes a single observed lifetime T from an exponential(λ) population, where λ is a positive unknown rate parameter. Find an exact two-sided 95% confidence interval for λ.

5.33 Justin places a single item is placed on test ($n = 1$). The only information that is available is that the item failed between times a and b, where $a < b$. In other words, the single item's lifetime is *interval censored*. Assuming that the population time to failure is exponential(λ), what is the maximum likelihood estimator of λ?

5.34 Natalie conducts a life test with $n = 19$ items on test and random right censoring. Let $t_1, t_2, \ldots, t_{19}$ be the independent exponential(2) times to failure. Let $c_1, c_2, \ldots, c_{19}$ be the independent exponential(1) censoring times, which are independent of the times to failure. Use Monte Carlo simulation to estimate the actual coverage of the following approximate confidence interval procedures for the population failure rate λ at for $\alpha = 0.05$.

    (a) The confidence interval consisting of all λ satisfying

$$\frac{\hat{\lambda}\chi^2_{2r,1-\alpha/2}}{2r} < \lambda < \frac{\hat{\lambda}\chi^2_{2r,\alpha/2}}{2r}.$$

    (b) The confidence interval consisting of all λ satisfying

$$\hat{\lambda} - z_{\alpha/2}O(\hat{\lambda})^{-1/2} < \lambda < \hat{\lambda} + z_{\alpha/2}O(\hat{\lambda})^{-1/2}.$$

    (c) The confidence interval consisting of all λ satisfying

$$2[\ln L(\hat{\lambda}) - \ln L(\lambda)] < \chi^2_{1,\alpha}.$$

Replicate the experiment so as to estimate the actual coverages to three digits of accuracy.

5.35 Sixty-watt light bulb lifetimes are known to be exponentially distributed with unknown positive population mean θ from previous test results. The company that produces these light bulbs would like to estimate θ by testing n bulbs to failure at one facility and m bulbs to failure at a second facility. Let $X_1, X_2, \ldots, X_n$ be the independent lifetimes of the bulbs tested at the first facility; let $Y_1, Y_2, \ldots, Y_m$ be the independent lifetimes of the bulbs tested at the second facility. An unbiased estimate of θ is the convex combination

$$\hat{\theta} = p\hat{\theta}_X + (1-p)\hat{\theta}_Y,$$

where $0 < p < 1$, $\hat{\theta}_X = \bar{X}$ is the maximum likelihood estimator of θ for the data from the first facility, and $\hat{\theta}_Y = \bar{Y}$ is the maximum likelihood estimator of θ for the data from the second facility. Find the value of p that minimizes $V\left[\hat{\theta}\right]$.

5.36 Ash would like to test the hypothesis

$$H_0 : \lambda = 17$$

versus

$$H_1 : \lambda > 17$$

using a single value T from an exponential(λ) population, where λ is a positive unknown population failure rate. The null hypothesis is rejected if $T < 0.01$. Find the significance level α for the test.

5.37 Let T be an observation from an exponential population with positive unknown population mean θ. This observation is used to test

$$H_0 : \theta = 6$$

versus

$$H_1 : \theta = 2.$$

(a) Find the critical value for the test for a fixed significance level α.

(b) Find β for a fixed significance level α.

5.38 Paul collects a random sample $t_1, t_2, \ldots, t_n$ from an exponential population with positive unknown mean θ. Show that the sample mean, $\bar{t}$, and n times the first order statistic, $nt_{(1)}$, are both unbiased estimators of θ.

5.39 Jessica and Mary collect a random sample $t_1, t_2, \ldots, t_n$ of light bulb lifetimes drawn from an exponential(λ) population, where λ is a positive unknown failure rate. The bulbs are stamped with "1000 hour MTTF," indicating that the *mean time to failure* equals 1000 hours. They would like to determine whether there is statistical evidence in the sample that indicates the bulbs last *longer* than 1000 hours.

(a) State the appropriate H_0 and H_1.

(b) Jessica uses the test statistic $\bar{t}$ and Mary uses the test statistic $nt_{(1)}$ to test the hypothesis. Find the critical values for their test statistics when $\alpha = 0.05$ and $n = 10$.

(c) Draw the power curves associated with each of the test statistics from part (b) on the same set of axes using a computer. Again assume that $\alpha = 0.05$ and $n = 10$.

5.40   Camille observes a single lifetime T from an exponential population with a positive unknown population mean θ. She would like to test

$$H_0 : \theta = 1$$

versus

$$H_1 : \theta > 1$$

at $\alpha = 0.07$ using T as a test statistic.

  (a) Find the critical value c for this test.

  (b) Plot the power function for this test.

5.41   Ellen collects a random sample $t_1, t_2, \ldots, t_{10}$ of light bulb lifetimes from an exponential(λ) population, where λ is a positive unknown failure rate. Ellen is a reliability engineer. She is confident from previous test results that the time to failure for these light bulbs is exponentially distributed. She is interested in testing whether a manufacturer's claim that the population mean time to failure for the bulbs is 1000 hours. So she would like to test

$$H_0 : \lambda = 0.001$$

versus

$$H_1 : \lambda > 0.001.$$

She is in a hurry. She places ten bulbs on test and only observes the first bulb fail at $t_{(1)} = 14$ hours, and would like to draw a conclusion at 14 hours. Give the p-value for the test based on the value of this single order statistic.

5.42   Liz collects a random sample of lifetimes $t_1, t_2, \ldots, t_n$ from an exponential(λ) distribution, where λ is a positive unknown failure rate parameter. She conducts a significance test of

$$H_0 : \lambda = 1$$

versus

$$H_0 : \lambda \neq 1,$$

which achieves a p-value of $p = 0.07$ for a particular data set. If she then computes an exact two-sided 95% confidence interval for λ for this particular data set, will the confidence interval contain 1?

5.43   Karen fits the ball bearing data set to the Weibull distribution parameterized as

$$S(t) = e^{-(\lambda t)^\kappa} \qquad t \geq 0,$$

yielding maximum likelihood estimates $\hat{\lambda} = 0.0122$ and $\hat{\kappa} = 2.10$. Ute also wants to fit the same data set to the Weibull distribution, but she uses the parameterization

$$S(t) = e^{-\rho t^\beta} \qquad t \geq 0.$$

What will be the maximum likelihood estimates $\hat{\rho}$ and $\hat{\beta}$ that Ute obtains for the ball bearing data set?

5.44 Jay conducts a life test with $n = 5$ items on test which is terminated when $r = 3$ items have failed. Failed items are not replaced in this traditional Type II right-censored data set. Assuming that the time to failure of an item in the population has a Weibull(λ, κ) distribution with known, positive parameters λ and κ, what is the probability density function of the time to complete the life test?

5.45 Jennie collects a random sample $t_1, t_2, \ldots, t_7$ from a Rayleigh population with probability density function

$$f(t) = 2\theta^{-2} t e^{-(t/\theta)^2} \qquad t > 0,$$

where θ is a positive unknown parameter. She would like to test

$$H_0 : \theta = 10$$

versus

$$H_1 : \theta > 10$$

using the test statistic $t_{(1)} = \min\{t_1, t_2, \ldots, t_7\}$, which assumes the value $t_{(1)} = 6$. Find the p-value for her test.

5.46 Mildred collects a random sample $t_1, t_2, \ldots, t_n$ from a Rayleigh(λ) population with survivor function

$$S(t) = e^{-(\lambda t)^2} \qquad t > 0,$$

where λ is a positive unknown parameter.

   (a) Find the maximum likelihood estimator of λ.

   (b) Show that the log likelihood function is maximized at the maximum likelihood estimator $\hat{\lambda}$.

   (c) Given that the expected value of T is $E[T] = \sqrt{\pi}/(2\lambda)$, find the method of moments estimator of λ.

5.47 Find the elements of the score vector for the log logistic distribution for a randomly right-censored data set.

5.48 Bryan places n items on test and observes r failures. Assuming that the failure times of the items follow the log logistic distribution and censoring is random, set up an expression for the boundary of a 95% confidence region for the shape parameter κ and scale parameter λ of the log logistic distribution based on the likelihood ratio statistic. Assume that the survivor function for the log logistic distribution is

$$S(t) = \frac{1}{1 + (\lambda t)^\kappa} \qquad t \geq 0,$$

for $\lambda > 0$ and $\kappa > 0$. It is not necessary to solve for the maximum likelihood estimators.

5.49 Consider a proportional hazards model with $n = 3$ items on test and distinct failure times t_1, t_2, t_3. Compute the joint probability mass function values for the $3! = 6$ possible rank vectors, and show that they sum to 1.

5.50 Give the equations that must be solved in order to find the maximum likelihood estimators $\hat{\lambda}$, $\hat{\kappa}$, and $\hat{\boldsymbol{\beta}}$ for a proportional hazards model with log logistic baseline distribution and log linear link function. A random right-censoring scheme is used.

5.51   Joyce fits the Cox proportional hazards model with unknown baseline distribution given in
Examples 5.14, 5.15, and 5.16 to the $n = 3$ light bulb failure times. The purpose of the study
was to determine the effect of wattage on survival for 60-watt and 100-watt light bulbs.

(a) What is the value of the regression coefficient for wattage if it were coded as $z = 60$
and $z = 100$ rather than as a binary covariate?

(b) Write a short paragraph indicating whether or not these two approaches are fundamentally equivalent ways of coding the covariate. If they differ, is one method of coding
the covariate superior to the other for the purpose of the study?

5.52   Survival times (in weeks) for two groups of leukemia patients (AG positive and AG neg-
ative blood types), along with an additional covariate, white blood cell count are given in
Feigl, P. and Zelen, M., "Estimation of Exponential Survival Probabilities with Concomitant
Information," *Biometrics*, Vol. 21, No. 4, pp. 826–838, 1965, and are displayed below.

AG positive group		AG negative group	
Survival time	White blood count	Survival time	White blood count
65	2300	56	4400
156	750	65	3000
100	4300	17	4000
134	2600	7	1500
16	6000	16	9000
108	10500	22	5300
121	10000	3	10000
4	17000	4	19000
39	5400	2	27000
143	7000	3	28000
56	9400	8	31000
26	32000	4	26000
22	35000	3	21000
1	100000	30	79000
1	100000	4	100000
5	52000	43	100000
65	100000		

(a) Fit the Cox proportional hazards model to the survival times. Code the blood type as
the indicator variable z_1, using 1 for AG positive and 0 for AG negative. The second
covariate z_2 is the natural logarithm of the white blood cell counts minus the sample
mean of the natural logarithms of the white blood cell counts. Include the interaction
term $(z_1 - \bar{z}_1)z_2$ in the model. Use the Breslow method for handling tied survival times.

(b) Write a sentence interpreting the *sign* of $\hat{\beta}_1$, $\hat{\beta}_2$, and $\hat{\beta}_3$ in terms of risk to the patient.

(c) Give a 95% confidence interval for β_1.

(d) If covariates associated with p-values that are less than 0.10 are considered statistically
significant, what is the fitted hazard function for a leukemia patient with baseline hazard
function $h_0(t)$, white blood cell count 9000 who has AG positive blood type? *Hint*: The
sample mean of the natural logarithms of the white blood cell types is 9.52 and the mean
of the blood types coded as an indicator variable is $17/33 = 0.515$.

5.53 Consider the Cox proportional hazards model with a single ($q = 1$) binary covariate z_1, an exponential(λ) baseline distribution, and a log linear link function. The baseline distribution can be absorbed into the link function by creating an artificial covariate $z_0 = 1$ and setting $\lambda = e^{\beta_0 z_0}$.

(a) For a randomly right-censored data set, find the score vector.

(b) For a randomly right-censored data set, find closed-form expressions for the maximum likelihood estimators $\hat{\beta}_0$ and $\hat{\beta}_1$.

(c) For the $n = 3$ observations given in vector form below, calculate the maximum likelihood estimates $\hat{\beta}_0$ and $\hat{\beta}_1$.

$$x = \begin{bmatrix} 80 \\ 20 \\ 50 \end{bmatrix} \qquad \delta = \begin{bmatrix} 1 \\ 1 \\ 1 \end{bmatrix} \qquad Z = \begin{bmatrix} 1 \\ 1 \\ 0 \end{bmatrix}.$$

(d) What is the hazard function of the fitted model for the data from part (c)?

(e) Use the observed information matrix to give approximate two-sided 95% confidence intervals for β_0 and β_1 for the data from part (c).

(f) Give the p-values for testing the hypotheses

$$H_0 : \beta_i = 0$$
$$H_1 : \beta_i \neq 0$$

for $i = 0, 1$, for the data from part (c).

5.54 The wattage of the $n = 3$ light bulbs in Example 5.16 was coded as the covariate $z_1 = 0$ for a 60-watt bulb and $z_1 = 1$ for a 100-watt bulb. When the Cox proportional hazards model with an unspecified baseline hazard function was fit to the data set, the point estimate for the regression parameter was $\hat{\beta}_1 = -0.347$. Without doing the derivation from scratch, what is the point estimate for the regression parameter if the wattage (that is, 60 watts or 100 watts) of the bulb were used as the covariate.

5.55 Mark fits a Cox proportional hazards model with unknown baseline distribution to a data set of drill bit failure times (measured in number of items drilled) with $q = 2$, for which the covariates denote the turning speed (revolutions per minute, rpm) and the hardness of the material (Brinell hardness number, BHN) being drilled. The turning speeds range from 2400 to 4800 rpm and the hardness of the materials ranges from 250 to 440 BHN. Interactions are not considered and the variables are not centered. The fitted model has estimated regression vector $\hat{\beta} = (0.014, 0.45)'$, and the inverse of the observed information matrix is

$$O^{-1}(\hat{\beta}) = \begin{bmatrix} 0.000081 & 0.000016 \\ 0.000016 & 0.010000 \end{bmatrix}.$$

Write a paragraph interpreting these results.

Chapter 6

Topics in Survival Analysis

The previous two chapters have introduced some probabilistic models and statistical methods that arise in survival analysis. This chapter surveys some topics that would be a part of a full-semester course in survival analysis. The first section considers nonparametric methods that arise in survival analysis, with a focus on estimating the survivor function. The empirical survivor function is used in the case of a complete data set and the Kaplan–Meier product–limit estimator is used in the case of a right-censored data set. These methods require no parametric assumptions from the modeler. The log-rank test, which is a nonparametric hypothesis test used to compare two survivor functions, is also introduced. The second section introduces the competing risks model, which is appropriate when multiple risks compete for the lifetime of an item. The third section considers not just a single failure time, but items which undergo multiple failures, such as an automobile.

6.1   Nonparametric Methods

Nonparametric methods require no parametric assumptions (for example, exponential or Weibull lifetime models) concerning the lifetime of an item. The emphasis is to let the data speak for itself, rather than approximating the lifetime distribution by a parametric model. In many applications, the modeler does not have any clues revealing an appropriate parametric model, so a nonparametric approach is warranted. The first subsection considers the estimation of the survivor function for a complete data set of n items placed on test. The second subsection considers the estimation of the survivor function for a randomly right-censored data set of n items placed on test. Two different types of derivations both lead to the Kaplan–Meier product–limit estimator. The third subsection considers the problem of comparing the estimated survivor functions of two different types of items. In the reliability setting, this might be to compare the lifetimes of Product A versus Product B. In the biostatistical setting, this might be to compare the survival times of patients undergoing radiation and chemotherapy for a particular type of cancer.

6.1.1   Survivor Function Estimation for Complete Data Sets

Consider the nonparametric estimation of the survivor function from a complete data set of n lifetimes with no ties. The risk set $R(t)$ contains the indexes of all items at risk just prior to time t. Let $n(t) = |R(t)|$ be the cardinality of $R(t)$. In other words, $n(t)$ is the number of elements in $R(t)$. The simplest and most popular nonparametric estimate for the survivor function is

$$\hat{S}(t) = \frac{n(t)}{n} \qquad t \geq 0,$$

which is often referred to as the *empirical survivor function*. This step function takes a downward step of size $1/n$ at each observed lifetime. It is also the survivor function corresponding to a discrete distribution with n equally likely mass values. Ties are not difficult to adjust for because the formula for $\hat{S}(t)$ remains the same, but the function will take a downward step of d/n if there are d tied observations at a particular time value.

When there are no ties in the data set, one method for determining asymptotically exact confidence intervals for the survivor function is based on the normal approximation to the binomial distribution. Recall that a binomial random variable X models the number of successes in n independent Bernoulli trials, each with probability of success p. The expected value and population variance of the number of successes are $E[X] = np$ and $V[X] = np(1-p)$. The fraction of successes, X/n, on the other hand, has expected value $E[X/n] = p$ and population variance $V[X/n] = p(1-p)/n$.

Survival to a fixed time t can be considered a Bernoulli trial for each of the n items on test. An item either survives to time t or it does not. Thus, the number of items that survive to time t, which is $n(t)$, has the binomial distribution with parameters n and probability of success $S(t)$, where success is defined to be survival to time t. The empirical survivor function introduced earlier, $\hat{S}(t) = n(t)/n$, is the fraction of successes, which has expected value

$$E\left[\hat{S}(t)\right] = S(t)$$

and population variance

$$V\left[\hat{S}(t)\right] = \frac{S(t)\left(1 - S(t)\right)}{n}.$$

So $\hat{S}(t)$ is an unbiased and consistent estimator of $S(t)$ for all values of t. Furthermore, when the number of items on test n is large and $S(t)$ is not too close to 0 or 1, the binomial distribution assumes a shape that is closely approximated by a normal probability density function and thus can be used to find an interval estimate for $S(t)$. Notice that such an interval estimate is most accurate, in terms of coverage, around the median of the distribution because the normal approximation to the binomial distribution works best when the probability of success is about $1/2$, where the binomial distribution is symmetric. Replacing $S(t)$ by $\hat{S}(t)$ in the population variance formula, an asymptotically exact two-sided $100(1 - \alpha)\%$ confidence interval for the probability of survival to time t is

$$\hat{S}(t) - z_{\alpha/2}\sqrt{\frac{\hat{S}(t)\left(1 - \hat{S}(t)\right)}{n}} < S(t) < \hat{S}(t) + z_{\alpha/2}\sqrt{\frac{\hat{S}(t)\left(1 - \hat{S}(t)\right)}{n}}.$$

This confidence interval is also appropriate when there are tied observations, although it becomes more approximate as the number of ties increases. Confidence limits greater than 1 or less than 0 are typically truncated, as illustrated in the following example.

Example 6.1 For the ball bearing data set from Example 5.5 with $n = 23$ bearings placed on test until failure, find a nonparametric survivor function estimator and an approximate two-sided 95% confidence interval for the probability that a ball bearing will survive 50,000,000 cycles.

Recall that the ball bearing failure times in 10^6 revolutions are

17.88	28.92	33.00	41.52	42.12	45.60	48.48	51.84	51.96
54.12	55.56	67.80	68.64	68.64	68.88	84.12	93.12	98.64

105.12   105.84   127.92   128.04   173.40.

The nonparametric survivor function estimate $\hat{S}(t)$ is shown as the solid line in Figure 6.1. In this figure and others, the downward steps in $\hat{S}(t)$ have been connected by vertical lines. Many analysts find this useful when visually comparing a nonparametric estimator of $S(t)$ to a fitted parametric model. The empirical survivor function takes a downward step of size $1/23$ at each data value, with the exception of the tied value, 68.64, where it takes a downward step of $2/23$. By convention, the survivor function estimate cuts off after the largest observed failure time. Since the data is given in 10^6 revolutions, a point estimate for the survivor function at $t = 50$ is

$$\hat{S}(50) = \frac{16}{23} = 0.6957,$$

and an approximate two-sided 95% confidence interval for the survivor function at $t = 50$ is

$$\hat{S}(50) - 1.96\sqrt{\frac{\hat{S}(50)\left(1 - \hat{S}(50)\right)}{23}} < S(50) < \hat{S}(50) + 1.96\sqrt{\frac{\hat{S}(50)\left(1 - \hat{S}(50)\right)}{23}},$$

which reduces to

$$0.5076 < S(50) < 0.8837.$$

This process can be performed for all t values, yielding the approximate two-sided 95% confidence bands for $S(t)$ given by the dashed lines in Figure 6.1. The confidence bands are truncated at 0 and 1. Also, the lower confidence band appears to be absent prior to the first observed failure at $t_{(1)} = 17.88$. This is due to the fact that $\hat{S}(t)$ is 1 for t values between 0 and the first failure time, so the upper and lower confidence limits are both equal to 1.

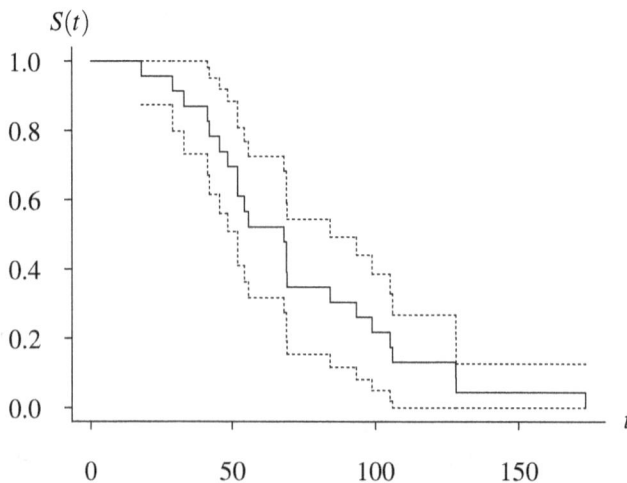

Figure 6.1: Nonparametric survivor function estimate for the ball bearing data set.

Even though the confidence interval for $S(t)$ from a complete data set of n lifetimes used in the previous example is intuitive and easy to compute, its performance in terms of its actual coverage is notoriously poor. One particular instance of its poor performance occurs at time $t = 10$, where the approximate two-sided 95% confidence interval has lower and upper bounds equal to 1. It is known as the *Wald* confidence interval, and its use is generally frowned upon because better alternatives exist. Four such alternatives are outlined (without derivation) in the next four paragraphs.

The approximate two-sided $100(1 - \alpha)\%$ Clopper–Pearson confidence interval for $S(t)$ has bounds that can be expressed as the fractiles of beta distributions:

$$B_{n(t), n-n(t)+1, 1-\alpha/2} < S(t) < B_{n(t)+1, n-n(t), \alpha/2},$$

for $n(t) = 0, 1, 2, \ldots, n$, where the first two values in the subscripts of B are the parameters of the beta distribution and the third value in the subscripts is a right-hand tail probability. The Clopper–Pearson confidence interval bounds can also be written as functions of fractiles of the F distribution.

The bounds on the Wilson–score approximate two-sided $100(1 - \alpha)\%$ confidence interval for $S(t)$ are

$$\frac{1}{1 + z_{\alpha/2}^2/n} \left[\hat{S}(t) + \frac{z_{\alpha/2}^2}{2n} \pm z_{\alpha/2} \sqrt{\frac{\hat{S}(t)\left(1 - \hat{S}(t)\right)}{n} + \frac{z_{\alpha/2}^2}{4n^2}} \right],$$

where $z_{\alpha/2}$ is the $1 - \alpha/2$ fractile of the standard normal distribution. The center of the Wilson–score confidence interval is

$$\frac{\hat{S}(t) + z_{\alpha/2}^2/(2n)}{1 + z_{\alpha/2}^2/n},$$

which is a weighted average of the point estimator $\hat{S}(t) = n(t)/n$ and $1/2$, with more weight on $\hat{S}(t)$ as n increases.

The Jeffreys approximate two-sided $100(1 - \alpha)\%$ interval estimate for $S(t)$ is a Bayesian credible interval that uses a Jeffreys non-informative prior distribution for $S(t)$. As was the case with the Clopper–Pearson confidence interval, the bounds of the Jeffreys interval for $S(t)$ are fractiles of beta random variables:

$$B_{n(t)+1/2, n-n(t)+1/2, 1-\alpha/2} < S(t) < B_{n(t)+1/2, n-n(t)+1/2, \alpha/2}$$

for $n(t) = 1, 2, \ldots, n-1$. When $n(t) = 0$, the lower bound is set to zero and the upper bound calculated using the formula above; when $n(t) = n$, the upper bound is set to one and the lower bound calculated using the formula above.

The bounds of the Agresti–Coull approximate two-sided $100(1 - \alpha)\%$ confidence interval for $S(t)$ are

$$\tilde{S}(t) \pm z_{\alpha/2} \sqrt{\frac{\tilde{S}(t)\left(1 - \tilde{S}(t)\right)}{\tilde{n}}},$$

where $\tilde{n} = n + z_{\alpha/2}^2$ and $\tilde{S}(t) = \left(n(t) + z_{\alpha/2}^2/2\right)/\tilde{n}$. In the special case of $\alpha = 0.05$, if one is willing to round $z_{\alpha/2} = 1.96$ to 2, this interval can be interpreted as "add two successes and add two failures and use the Wald confidence interval formula."

Example 6.2 Find the Clopper–Pearson, Wilson–score, Jeffreys, and Agresti–Coull approximate two-sided 95% confidence intervals for $S(50)$ for the ball bearing lifetimes from Example 5.5.

As in the previous example, the point estimator for $S(50)$ is $\hat{S}(50) = 16/23 = 0.6957$. Using the parameters $n = 23$, $n(50) = 16$, $\alpha = 0.05$, the approximate two-sided 95% Clopper–Pearson, Wilson–score, Jeffreys, and Agresti–Coull confidence intervals for $S(50)$ are given in Table 6.1. The R code to compute these confidence intervals is given below. All confidence intervals are calculated using the `binomTest` function from the `conf` package.

```
library(conf)
binomTest(23, 16, alpha = 0.05, intervalType = "Wald")
binomTest(23, 16, alpha = 0.05, intervalType = "Clopper-Pearson")
binomTest(23, 16, alpha = 0.05, intervalType = "Wilson-Score")
binomTest(23, 16, alpha = 0.05, intervalType = "Jeffreys")
binomTest(23, 16, alpha = 0.05, intervalType = "Agresti-Coull")
```

Method	95% confidence interval
Wald	$0.508 < S(50) < 0.884$
Clopper–Pearson	$0.471 < S(50) < 0.868$
Wilson–score	$0.491 < S(50) < 0.844$
Jeffreys	$0.493 < S(50) < 0.852$
Agresti–Coull	$0.489 < S(50) < 0.846$

Table 6.1: Approximate 95% confidence intervals for $S(50)$ for the ball bearing data.

The confidence interval bounds vary significantly between the techniques. The narrowest confidence interval is the Wilson–score and the widest confidence interval is the Clopper–Pearson. Some analysts prefer the Clopper–Pearson confidence interval because it is *conservative* in the sense that its actual coverage always exceeds the stated coverage (which is 95% in this example) for all values of $S(t)$. This implies that you will never claim more precision with your confidence interval than is implied by the stated coverage. The Clopper–Pearson 95% confidence intervals for $S(t)$ for all values of t are plotted as confidence bands in Figure 6.2. Unlike the Wald confidence bands depicted in Figure 6.1, these confidence intervals are not symmetric about the associated point estimators given by the solid lines, and this non-symmetry is particularly pronounced at the extremes.

There are dozens of confidence interval procedures for calculating an approximate confidence interval for $S(t)$. The intervals illustrated in the previous example were selected because of (*a*) their popularity with statisticians, (*b*) their availability in statistical software packages, and (*c*) their statistical properties, particularly their actual coverage. The four confidence interval procedures illustrated in the previous example all possess the following properties.

- For a fixed number of items on test n, the confidence intervals are complementary for any particular $n(t)$ and $n - n(t)$ values.

- The confidence intervals are asymptotically exact for $0 < S(t) < 1$.

- The confidence intervals do not degenerate to a confidence intervals of width zero for $n(t) = 0$ or $n(t) = n$ as was the case with the Wald confidence interval.

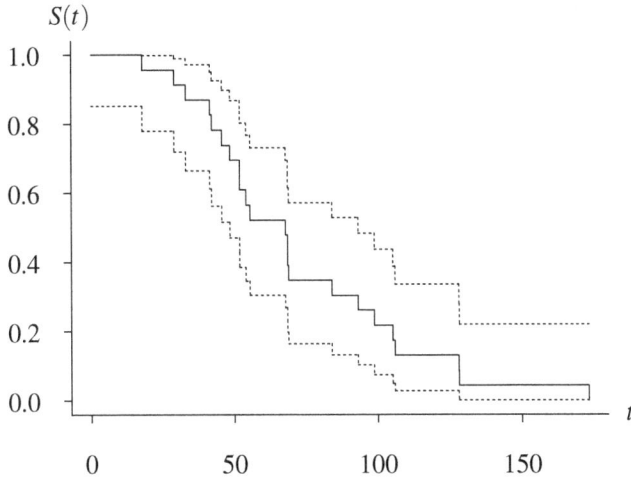

Figure 6.2: Clopper–Pearson confidence bands for the ball bearing data set.

This concludes the discussion concerning finding point and interval estimators for $S(t)$ from a complete data set of lifetimes. We now introduce techniques for estimating $S(t)$ from a right-censored data set.

6.1.2   Survivor Function Estimation for Right-Censored Data Sets

The general case in which there are both ties and right-censored data values is now considered. Some new notation must be established in order to derive the nonparametric estimator for $S(t)$. As before, assume that n items are on test. Let $y_1 < y_2 < \cdots < y_k$ denote the k distinct observed failure times, and let d_j denote the number of observed failures at y_j, for $j = 1, 2, \ldots, k$. Let $n_j = n(y_j)$ denote the number of items on test just before time y_j, for $j = 1, 2, \ldots, k$, and it is customary to include any values that are right censored at y_j in this count.

The search for a survivor function estimator begins by assuming that the data arose from a discrete distribution with mass values $y_1 < y_2 < \cdots < y_k$. For a discrete distribution, $h(y_j)$ is a conditional probability with interpretation $h(y_j) = P(T = y_j \mid T \geq y_j)$. The survivor function can be written in terms of the hazard function at the mass values as

$$S(t) = \prod_{j \mid y_j \leq t} \left[1 - h(y_j)\right] \qquad t \geq 0.$$

Thus, a reasonable estimator for $S(t)$ is $\prod_{j \mid y_j < t} \left[1 - \hat{h}(y_j)\right]$, which reduces the problem of estimating the survivor function to that of estimating the hazard function at each mass value. An appropriate element in the likelihood function at mass value y_j is

$$h(y_j)^{d_j} \left[1 - h(y_j)\right]^{n_j - d_j}$$

for $j = 1, 2, \ldots, k$. The above expression is correct because d_j is the number of failures at y_j, $h(y_j)$ is the conditional probability of failure at y_j, $n_j - d_j$ is the number of items on test not failing at y_j, and $1 - h(y_j)$ is the probability of failing after time y_j conditioned on survival to time y_j. Thus, the likelihood function for $h(y_1), h(y_2), \ldots, h(y_k)$ is

$$L\big(h(y_1), h(y_2), \ldots, h(y_k)\big) = \prod_{j=1}^{k} h(y_j)^{d_j} \left[1 - h(y_j)\right]^{n_j - d_j}$$

and the log likelihood function is

$$\ln L\big(h(y_1), h(y_2), \ldots, h(y_k)\big) = \sum_{j=1}^{k} \Big\{ d_j \ln h(y_j) + (n_j - d_j) \ln \left[1 - h(y_j)\right] \Big\}.$$

The ith element of the score vector is

$$\frac{\partial \ln L\big(h(y_1), h(y_2), \ldots, h(y_k)\big)}{\partial h(y_i)} = \frac{d_i}{h(y_i)} - \frac{n_i - d_i}{1 - h(y_i)}$$

for $i = 1, 2, \ldots, k$. Equating this element of the score vector to zero and solving for $h(y_i)$ yields the maximum likelihood estimate

$$\hat{h}(y_i) = \frac{d_i}{n_i},$$

for $i = 1, 2, \ldots, k$. This estimate for $\hat{h}(y_i)$ is sensible because d_i of the n_i items on test at time y_i fail, so the ratio of d_i to n_i is an appropriate estimate of the conditional probability of failure at time y_i. This derivation may strike a familiar chord because at each time y_i, estimating $h(y_i)$ with d_i divided by n_i is equivalent to estimating the probability of success (that is, failing at time y_i) for each of the n_i items on test. Thus, this derivation is equivalent to finding the maximum likelihood estimators for the probability of success for k binomial random variables.

Using this particular estimate for the hazard function at y_i, the survivor function estimate becomes

$$\hat{S}(t) = \prod_{j \mid y_j \le t} \left[1 - \hat{h}(y_j)\right] = \prod_{j \mid y_j \le t} \left[1 - \frac{d_j}{n_j}\right],$$

for $t \ge 0$, commonly known as the Kaplan–Meier or product–limit estimator. When the largest data value recorded corresponds to a failure, the product–limit estimator drops to zero; when the largest data value recorded corresponds to a right-censored observation, a common convention is to cut off the product–limit estimator at the current positive value of $\hat{S}(t)$. The original journal article by American mathematician Edward Kaplan and American statistician Paul Meier in 1958 that established the product–limit estimator is one of the most heavily cited papers in the statistics literature. The following example illustrates the process of calculating the product–limit estimate.

Example 6.3 Use the product–limit estimator to calculate a point estimate of the probability that a remission time in the treatment group in the 6–MP clinical trial described in Example 5.6 exceeds 14 weeks. In other words, estimate $S(14)$ using the Kaplan–Meier estimator.

The data set contains $n = 21$ patients on test, $r = 9$ observed failures (leukemia relapses), and $k = 7$ distinct observed failure times. The data values, in weeks, are

$$\begin{array}{ccccccccccc}
6 & 6 & 6 & 6^* & 7 & 9^* & 10 & 10^* & 11^* & 13 & 16 \\
17^* & 19^* & 20^* & 22 & 23 & 25^* & 32^* & 32^* & 34^* & 35^*.
\end{array}$$

Table 6.2 gives the values of y_j, d_j, n_j, and $1 - d_j/n_j$ for $j = 1, 2, \ldots, 7$. The product–limit survivor function estimate at $t = 14$ weeks is

$$\hat{S}(14) = \prod_{j|y_j \leq 14} \left[1 - \frac{d_j}{n_j}\right]$$

$$= \left[1 - \frac{3}{21}\right]\left[1 - \frac{1}{17}\right]\left[1 - \frac{1}{15}\right]\left[1 - \frac{1}{12}\right]$$

$$= \frac{176}{255}$$

$$= 0.69.$$

The product–limit survivor function estimate for all t values is plotted in Figure 6.3. Downward steps occur at the $k = 7$ observed failure times. Some software packages place a vertical hash mark on the Kaplan–Meier estimate to highlight censored values that occur between observed failure times; these occur at times 9, 11, 17, 19, 20, 25, 32, and 34 in Figure 6.3. The effect of censored observations in the survivor function estimate is a larger downward step at the next subsequent observed failure time. If there is a tie between an observed failure time and censoring time (as there is at time 6 in this example) the standard convention of including the censored value(s) in the risk set when computing the number of items at risk means that there will be a larger downward step in the survivor function estimate following the tied value. Since the last observed data value, 35*, corresponds to a right-censored observation, the survivor function estimate is truncated at time 35 and is assumed to be undefined for $t > 35$.

The R code to generate this plot uses the `survfit` function from the `survival` package. The failure and censoring times $x_1, x_2, \ldots, x_n$ are held in the vector named `time`. The indicator variables $\delta_1, \delta_2, \ldots, \delta_n$ are held in the vector named `status`. The `Surv` function creates a survival object, which is used in the left-hand side of the `formula` argument passed to `survfit`. The right-hand side of the `formula` argument to `survfit` contains just 1 to indicate that there are no covariates being considered when computing the product–limit estimator for just the remission times in the treatment group. The `summary` function reveals the calculations used in estimating the product–limit estimate

j	y_j	d_j	n_j	$1 - \frac{d_j}{n_j}$
1	6	3	21	$1 - \frac{3}{21}$
2	7	1	17	$1 - \frac{1}{17}$
3	10	1	15	$1 - \frac{1}{15}$
4	13	1	12	$1 - \frac{1}{12}$
5	16	1	11	$1 - \frac{1}{11}$
6	22	1	7	$1 - \frac{1}{7}$
7	23	1	6	$1 - \frac{1}{6}$

Table 6.2: Product–limit calculations for 6–MP treatment case.

and the `plot` function generates a graph of the product–limit estimate, which is given in Figure 6.3.

```
library(survival)
time   = c(6, 6, 6, 6, 7, 9, 10, 10, 11, 13, 16, 17, 19, 20, 22,
           23, 25, 32, 32, 34, 35)
status = c(1, 1, 1, 0, 1, 0, 1, 0, 0, 1, 1, 0, 0, 0, 1,
           1, 0, 0, 0, 0, 0)
kmest = survfit(Surv(time, status) ~ 1, conf.type = "none")
summary(kmest)
plot(kmest)
```

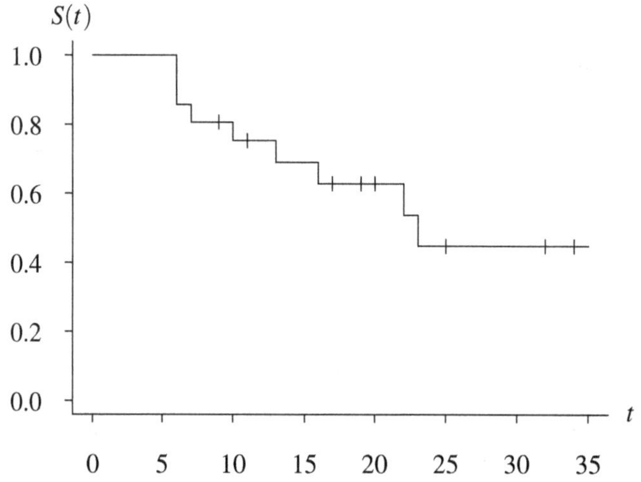

Figure 6.3: Product–limit survivor function estimate for the 6–MP treatment group.

There is a second and perhaps more intuitive way of deriving the product–limit estimator, often referred to as the "redistribute-to-the-right" algorithm. This technique begins by defining an initial probability mass function that apportions equal probability to each of the n data values. In subsequent passes through the data, this probability mass function estimate is modified as the probability is redistributed to the right, with special treatment given to right-censored observations. The algorithm is illustrated next on the 6–MP treatment group data set from Example 5.6.

Example 6.4 Implement the redistribute-to-the-right algorithm for calculating the Kaplan–Meier product–limit estimate of the survivor function for the remission time in the treatment group in the 6–MP clinical trial from Example 5.6.

For the $n = 21$ individuals in the treatment group for the 6–MP experiment, each failure or censoring time is initially assigned a mass value of $1/n$ as follows:

6	6	6	6*	7	9*	10	10*	11*	13	...
$\frac{1}{21}$	$\frac{1}{21}$	$\frac{1}{21}$	$\frac{1}{21}$	$\frac{1}{21}$	$\frac{1}{21}$	$\frac{1}{21}$	$\frac{1}{21}$	$\frac{1}{21}$	$\frac{1}{21}$	...

If there were no censored observations, the fractions would be the appropriate estimators for the probability mass function values. This probability mass function corresponds to the empirical survivor function described earlier in this section. Combining
the three tied observed failures at $t = 6$ yields

6	6*	7	9*	10	10*	11*	13	...
$\frac{1}{7}$	$\frac{1}{21}$	$\frac{1}{21}$	$\frac{1}{21}$	$\frac{1}{21}$	$\frac{1}{21}$	$\frac{1}{21}$	$\frac{1}{21}$	...

As indicated earlier, there are mass values in the product–limit estimator only at observed failure times. Since the random censoring model is assumed, the mass associated with the individual whose remission time is right censored at 6 weeks can be split
evenly among each of the 17 subsequent failure/censoring times:

6	6*	7	9*	10	10*	11*	13	...
$\frac{1}{7}$	0	$\frac{6}{119}$	$\frac{6}{119}$	$\frac{6}{119}$	$\frac{6}{119}$	$\frac{6}{119}$	$\frac{6}{119}$	...

because $\frac{1}{21} + \frac{1}{17} \cdot \frac{1}{21} = \frac{6}{119}$. The probability mass function estimates at $t = 6$ and $t = 7$
have now been determined. The mass value $\frac{6}{119}$ associated with the right censored
observation at time 9 can be allocated among the 15 subsequent failure/censoring times
as

6	6*	7	9*	10	10*	11*	13	...
$\frac{1}{7}$	0	$\frac{6}{119}$	0	$\frac{32}{595}$	$\frac{32}{595}$	$\frac{32}{595}$	$\frac{32}{595}$	...

because $\frac{6}{119} + \frac{1}{15} \cdot \frac{6}{119} = \frac{96}{1785} = \frac{32}{595}$. After allocating the mass at 10* to the subsequent
13 data values and the mass at 11* to the subsequent 12 data values, the estimator
becomes

6	6*	7	9*	10	10*	11*	13	...
$\frac{1}{7}$	0	$\frac{6}{119}$	0	$\frac{32}{595}$	0	0	$\frac{16}{255}$	...

When this process is continued through all the data values, the resulting probability
mass function defined on the observed failure times corresponds to the product–limit
estimator. To check this for one specific time value, the survivor function estimate at
time 14 is

$$\hat{S}(14) = 1 - \frac{1}{7} - \frac{6}{119} - \frac{32}{595} - \frac{16}{255} = \frac{176}{255} = 0.69,$$

which matches the result from the previous example.

Since we now have a point estimate for the survivor function, our attention turns to estimating its
population variance in order to construct confidence intervals and conduct hypothesis tests. To find
an estimate for the population variance of the product–limit estimate is significantly more difficult
than for the uncensored case. The Fisher and observed information matrices require the following
partial derivative of the score vector:

$$\frac{\partial^2 \ln L\big(h(y_1), h(y_2), \ldots, h(y_k)\big)}{\partial h(y_i)\partial h(y_j)} = -\frac{d_i}{h(y_i)^2} - \frac{n_i - d_i}{\big(1 - h(y_i)\big)^2}$$

when $i = j$ and 0 otherwise, for $i = 1, 2, \ldots, k$ and $j = 1, 2, \ldots, k$. Both the Fisher and observed
information matrices are diagonal. Replacing $h(y_i)$ by its maximum likelihood estimate, the diagonal
elements of the observed information matrix are

$$\left[-\frac{\partial^2 \ln L\big(h(y_1), h(y_2), \ldots, h(y_k)\big)}{\partial h(y_i)^2} \right]_{h(y_i) = d_i/n_i} = \frac{n_i^3}{d_i(n_i - d_i)}$$

for $i = 1, 2, \ldots, k$. Using some approximations, an estimate for the variance of the estimated survivor function is

$$\hat{V}\left[\hat{S}(t)\right] = \left[\hat{S}(t)\right]^2 \sum_{j \mid y_j \leq t} \frac{d_j}{n_j(n_j - d_j)},$$

commonly referred to as *Greenwood's formula*. The formula can be used to find an asymptotically exact two-sided confidence interval for $S(t)$ by using the normal critical values as in the uncensored case:

$$\hat{S}(t) - z_{\alpha/2}\sqrt{\hat{V}\left[\hat{S}(t)\right]} < S(t) < \hat{S}(t) + z_{\alpha/2}\sqrt{\hat{V}\left[\hat{S}(t)\right]}.$$

As was the case with the Wald confidence interval for $S(t)$ in the case of a complete data set, the confidence interval bounds should be truncated when they are greater than 1 or less than 0, as illustrated in the next example.

Example 6.5 Use Greenwood's formula to construct an approximate two-sided 95% confidence interval for the probability that a remission time in the treatment group in the 6–MP clinical trial described in Example 5.6 exceeds 14 weeks.

The point estimator for the probability of survival to time 14 from the previous two examples is $\hat{S}(14) = 176/255 = 0.69$. The estimated variance of the survivor function estimator at time 14 via Greenwood's formula is

$$\hat{V}\left[\hat{S}(14)\right] = \left[\hat{S}(14)\right]^2 \sum_{j \mid y_j \leq 14} \frac{d_j}{n_j(n_j - d_j)}$$

$$= \left(\frac{176}{255}\right)^2 \left[\frac{3}{21(21-3)} + \frac{1}{17(17-1)} + \frac{1}{15(15-1)} + \frac{1}{12(12-1)}\right]$$

$$= 0.011.$$

Thus, an estimate for the standard deviation of the survivor function estimate at $t = 14$ is $\sqrt{0.011} = 0.11$. An approximate two-sided 95% confidence interval for $S(14)$ is

$$\hat{S}(14) - z_{0.025}\sqrt{\hat{V}\left[\hat{S}(14)\right]} < S(14) < \hat{S}(14) + z_{0.025}\sqrt{\hat{V}\left[\hat{S}(14)\right]}$$

$$0.69 - 1.96\sqrt{0.011} < S(14) < 0.69 + 1.96\sqrt{0.011}$$

$$0.48 < S(14) < 0.90.$$

Using this confidence interval procedure for all values of t, Figure 6.4 shows the 95% confidence bands for the survivor function. These confidence intervals have also been cut off after $t = 35$ because the last observation corresponds to a right-censored individual. The bounds are particularly wide because there are only $r = 9$ observed failure times.

The R code to calculate this confidence interval for $S(14)$ and plot confidence bands around the product–limit estimate is given below. Setting the conf.type argument to "plain" in the call to survfit results in the calculations for the 95% confidence interval for $S(14)$ presented here. These are displayed in the two right-hand columns in the call to the summary function. Setting the mark.time argument to TRUE in the call to plot results in hash marks on the estimated survivor function.

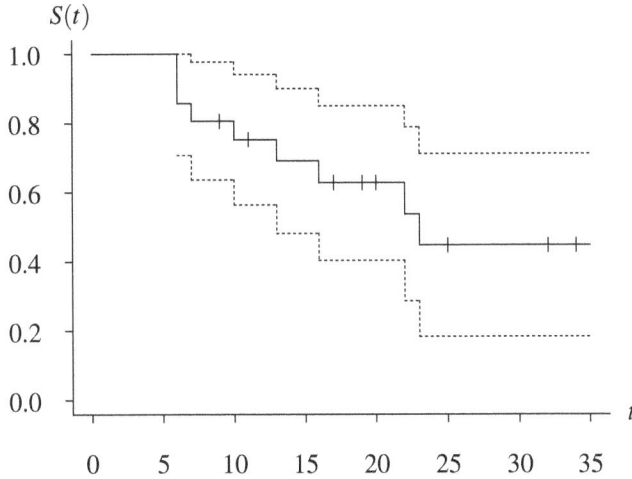

Figure 6.4: Confidence bands for the product–limit estimate for the 6–MP treatment group.

```
library(survival)
time   = c(6, 6, 6, 6, 7, 9, 10, 10, 11, 13, 16, 17, 19, 20, 22,
           23, 25, 32, 32, 34, 35)
status = c(1, 1, 1, 0, 1, 0, 1, 0, 0, 1, 1, 0, 0, 0, 1,
           1, 0, 0, 0, 0, 0)
kmest = survfit(Surv(time, status) ~ 1, conf.type = "plain")
summary(kmest)
plot(kmest, mark.time = TRUE)
```

6.1.3   Comparing Two Survivor Functions

This subsection introduces a nonparametric statistical test for determining whether samples of life-times from two populations arose from the same probability distribution. This test is nonparametric in the sense that it places no assumptions on the lifetime distribution of either population. The *log-rank test* (also known as the Mantel–Cox test, named after American biostatistician Nathan Mantel and British statistician David Cox or the Mantel–Haenszel test, named after American epidemiologist William Haenszel) is a nonparametric statistical test that can be used to test the equality of two survivor functions based on two randomly right-censored data sets collected from the two populations.

The null and alternative hypotheses for the log-rank test are

$$H_0 : S_1(t) = S_2(t)$$
$$H_1 : S_1(t) \neq S_2(t),$$

where $S_1(t)$ is the survivor function of the lifetimes of items from population 1 and $S_2(t)$ is the survivor function of the lifetimes of items from population 2. A randomly right-censored data set is collected from each population. The notation established below is similar to that used in the

Kaplan–Meier product–limit estimator. Let $y_1 < y_2 < \cdots < y_k$ be the observed failure times in the *combined* data set. Let

- n_{1j} be the number of items from data set 1 at risk just prior to time y_j,

- n_{2j} be the number of items from data set 2 at risk just prior to time y_j,

- $n_j = n_{1j} + n_{2j}$,

- d_{1j} be the number of items from data set 1 that fail at time y_j,

- d_{2j} be the number of items from data set 2 that fail at time y_j,

- $d_j = d_{1j} + d_{2j}$,

for $j = 1, 2, \ldots, k$.

Just before time y_j, there are n_j items in the combined sample that are at risk and subject to potential failure, for $j = 1, 2, \ldots, k$. Of the n_j items at risk just before time y_j, there are n_{1j} items from population 1 and n_{2j} items from population 2 that are at risk, for $j = 1, 2, \ldots, k$. Under H_0, each of the n_j items at risk has an identical conditional time to failure (conditioned on survival to time y_j), for $j = 1, 2, \ldots, k$. Under H_0, the random number of failures from population 1 at time y_j, d_{1j}, is equivalent to sampling d_j items without replacement from n_j items, n_{1j} of which are type 1 and $n_j - n_{1j}$ of which are type 2. Thus, d_{1j} has the *hypergeometric distribution* under H_0 with parameters n_j, n_{1j}, and d_j, for $j = 1, 2, \ldots, k$.

The population mean of the hypergeometric random variable d_{1j} under H_0 is

$$E[d_{1j}] = \frac{d_j n_{1j}}{n_j}$$

for $j = 1, 2, \ldots, k$. The population variance of d_{1j} under H_0 is

$$V[d_{1j}] = \frac{d_j (n_{1j}/n_j)(1 - n_{1j}/n_j)(n_j - d_j)}{n_j - 1}$$

for $j = 1, 2, \ldots, k$. So the random variables $d_{11}, d_{12}, \ldots, d_{1k}$ are marginally hypergeometric with population means and variances given above. Standardizing and summing, the log-rank test statistic

$$Z = \frac{\sum_{j=1}^{k} \left(d_{1j} - E[d_{1j}] \right)}{\sqrt{\sum_{j=1}^{k} V[d_{1j}]}}$$

is asymptotically standard normal in k under H_0. Large and small values of the test statistic Z correspond to departures from H_0.

Example 6.6 Perform a log-rank test to compare the survivor functions of the remission times in the treatment and control groups in the 6–MP clinical trial data from Example 5.6.

Recall from Example 5.6 that the remission times (in weeks) for the treatment group (population 1) are

$$
\begin{array}{ccccccccccc}
6 & 6 & 6 & 6^* & 7 & 9^* & 10 & 10^* & 11^* & 13 & 16 \\
17^* & 19^* & 20^* & 22 & 23 & 25^* & 32^* & 32^* & 34^* & 35^* &
\end{array}
$$

and the remission times for the control group (population 2) are

$$
\begin{array}{cccccccccc}
1 & 1 & 2 & 2 & 3 & 4 & 4 & 5 & 5 & 8 & 8 \\
8 & 8 & 11 & 11 & 12 & 12 & 15 & 17 & 22 & 23.
\end{array}
$$

The estimated survivor functions for the control and treatment groups are displayed in Figure 6.5. The use of 6–MP appears to be effective in prolonging remission times, but is the difference between the two survivor functions statistically significant? The log-rank test will answer this question.

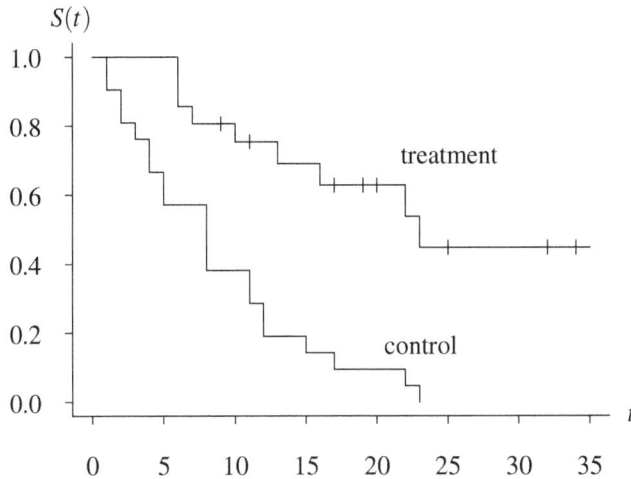

Figure 6.5: Estimated survivor functions for the 6–MP control and treatment and groups.

The $k = 17$ distinct observed failure times $y_1 < y_2 < \cdots < y_{17}$ in the combined sample are given in the second column of Table 6.3. The next three columns give the number of patients at risk just prior to time y_j in the combined data set (n_j), the number of patients at risk just prior to time y_j from population 1 (n_{1j}), and number of patients at risk just prior to time y_j from population 2 (n_{2j}). The final three columns give the number of overall observed remission times at time y_j in the combined data set (d_j), the number of observed remission times at time y_j from population 1 (d_{1j}), and the number of observed remission times at time y_j from population 2 (d_{2j}).

The null and alternative hypotheses for the test are

$$
H_0 : S_1(t) = S_2(t)
$$
$$
H_1 : S_1(t) \neq S_2(t),
$$

and the test statistic is

$$
Z = \frac{\sum_{j=1}^{17} \left(d_{1j} - E[d_{1j}] \right)}{\sqrt{\sum_{j=1}^{17} V[d_{1j}]}} = -4.1.
$$

This test statistic is negative because the observed d_{1j} values are smaller than their expected values. Fewer remissions occur in sample 1 (those patients treated with 6–MP)

j	y_j	n_j	n_{1j}	n_{2j}	d_j	d_{1j}	d_{2j}
1	1	42	21	21	2	0	2
2	2	40	21	19	2	0	2
3	3	38	21	17	1	0	1
4	4	37	21	16	2	0	2
5	5	35	21	14	2	0	2
6	6	33	21	12	3	3	0
7	7	29	17	12	1	1	0
8	8	28	16	12	4	0	4
9	10	23	15	8	1	1	0
10	11	21	13	8	2	0	2
11	12	18	12	6	2	0	2
12	13	16	12	4	1	1	0
13	15	15	11	4	1	0	1
14	16	14	11	3	1	1	0
15	17	13	10	3	1	0	1
16	22	9	7	2	2	1	1
17	23	7	6	1	2	1	1

Table 6.3: Data for calculating the log-rank test statistic.

than expected if the remission time distributions in the two populations were identical. Since the test statistic is 4.1 standard deviation units from its population mean under H_0, we expect a small p-value, and a rejection of the null hypothesis H_0. The p-value is

$$p = 2 \cdot P(Z < -4.1) = 0.00004,$$

so the conclusion is to reject H_0. There is statistical evidence that the survivor functions for the control and treatment groups differ. Figure 6.5 shows that the patients taking 6–MP have longer remission times.

Here are three final observations on the log-rank test. First, the test has been extended from testing the equality of two populations to testing the equality of several populations. Second, the Peto log-rank test statistic (named after British statistician Julian Peto) gives differing weights to the observed failure times. Third, there are several competitors to the log-rank test which should be considered when using this test.

6.2 Competing Risks

In *competing risks* models, several causes of failure compete for the lifetime of an item. These models are also useful for analyzing the relationships between the causes of failure. In addition, competing risks models are one way of combining several distributions to achieve a lifetime distribution with, for example, a bathtub-shaped hazard function.

In some situations, causes of failure can be grouped into k classes. An electrical engineer, for instance, might use failure by short and failure by open as a two-element competing risks model for the lifetime of a diode. Likewise, an actuary might use heart disease, cancer, accidents, and all other causes as a four-element competing risks model for human lifetimes. In *competing risks* analysis, an item is assumed to be subject to k competing risks (or causes) denoted by $C_1, C_2, \ldots, C_k$. Competing risks, often called *multiple decrements* by actuaries, can be viewed as a series system of components. Each risk can be thought of as a component in a series system in which system failure occurs when any component fails. Analyzing problems by competing risks might require the modeler to include an "all other risks" classification in order to study the effect of reduction or elimination of one risk. The origins of competing risks theory can be traced to a study by Daniel Bernoulli in the 1700s concerning the impact of eliminating smallpox on mortality for various age groups.

A second and equally appealing use of competing risks models is that they can be used to combine component distributions to form more complicated models. Although a distribution with a bathtub-shaped hazard function is often cited as an appropriate lifetime model, none of the five most popular lifetime distribution models (exponential, Weibull, gamma, log normal, and log logistic) can achieve this shape. Competing risks models are one way of combining several distributions to achieve a bathtub-shaped lifetime distribution. As shown in Figure 6.6, if a DFR Weibull distribution is used to model manufacturing defect failures and an IFR Weibull distribution is used to model wear-out failures, then a competing risks model with $k = 2$ risks yields a bathtub-shaped hazard function because the hazard functions are summed. We will formally develop this result later in this section.

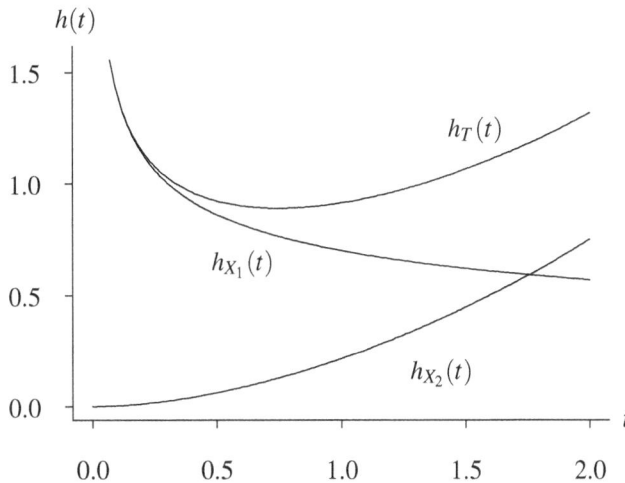

Figure 6.6: Hazard functions for a competing risks model.

6.2.1 Net Lifetimes

Competing risks theory is complicated by the existence of *net* and *crude* lifetimes. When working with *net lifetimes* or *net probabilities*, the causes $C_1, C_2, \ldots, C_k$ are viewed individually; that is, risk C_j, $j = 1, 2, \ldots, k$, is analyzed as if it is the only risk acting on the population. When working with *crude lifetimes* or *crude probabilities*, the lifetimes are considered in the presence of all other risks. The random variables associated with *net* lifetimes are defined next.

Definition 6.1 Let the random variable X_j, having probability density function $f_{X_j}(t)$, survivor function $S_{X_j}(t)$, hazard function $h_{X_j}(t)$, cumulative hazard function $H_{X_j}(t)$, and corresponding risk C_j, be the *net life* denoting the lifetime that occurs if only risk j is present, for $j = 1, 2, \ldots, k$.

Unless all risks except j are eliminated, X_j is not necessarily observed. In this sense, each net lifetime is a *potential* lifetime that is observed with certainty only if all the other $k - 1$ risks are eliminated. The observed lifetime of an item, T, is the minimum of $X_1, X_2, \ldots, X_k$. When the net lives are independent random variables, the hazard function for the observed time to failure is $h_T(t) = \sum_{j=1}^{k} h_{X_j}(t)$, because $S_T(t) = \prod_{j=1}^{k} S_{X_j}(t)$ for a series system of k independent components, $H_T(t) = -\ln S_T(t)$, and $h_T(t) = H'_T(t)$:

$$
\begin{aligned}
h_T(t) &= \frac{d}{dt} H_T(t) \\
&= \frac{d}{dt} \left[-\ln S_T(t) \right] \\
&= \frac{d}{dt} \left[-\ln \left(\prod_{j=1}^{k} S_{X_j}(t) \right) \right] \\
&= \frac{d}{dt} \left[\sum_{j=1}^{k} -\ln S_{X_j}(t) \right] \\
&= \frac{d}{dt} \left[\sum_{j=1}^{k} H_{X_j}(t) \right] \\
&= \sum_{j=1}^{k} \frac{d}{dt} H_{X_j}(t) \\
&= \sum_{j=1}^{k} h_{X_j}(t) \qquad t \geq 0.
\end{aligned}
$$

The *net probability* of failure in the time interval $[a, b)$ from risk j, denoted by $q_j(a, b)$, is the probability of failure in $[a, b)$ from risk j if risk j is the only risk present, conditioned on survival to time a. So

$$
\begin{aligned}
q_j(a, b) &= P(a \leq X_j < b \mid X_j \geq a) \\
&= 1 - P(X_j \geq b \mid X_j \geq a) \\
&= 1 - \frac{P(X_j \geq b)}{P(X_j \geq a)} \\
&= 1 - \frac{S_{X_j}(b)}{S_{X_j}(a)} \\
&= 1 - \frac{e^{-H_{X_j}(b)}}{e^{-H_{X_j}(a)}} \\
&= 1 - e^{-\left(H_{X_j}(b) - H_{X_j}(a) \right)} \\
&= 1 - e^{-\int_a^b h_{X_j}(t) \, dt}
\end{aligned}
$$

for $j = 1, 2, \ldots, k$.

6.2.2 Crude Lifetimes

Crude lifetimes are more difficult to work with than net lifetimes because they consider each of the causes of failure in the presence of all other causes of failure. Crude lifetimes are observed when lifetime data values are collected in a competing risks model in which all causes of failure are acting simultaneously in the population.

> **Definition 6.2** Let the random variable Y_j, having probability density function $f_{Y_j}(t)$, survivor function $S_{Y_j}(t)$, hazard function $h_{Y_j}(t)$, cumulative hazard function $H_{Y_j}(t)$, and corresponding risk C_j, be the *crude life* denoting the lifetime conditioned on risk j being the cause of failure in the presence of all other risks, for $j = 1, 2, \ldots, k$.

The *crude probability* of failure in the time interval $[a, b)$ from cause j, denoted by $Q_j(a, b)$, is the probability of failure in $[a, b)$ from risk j in the presence of all other risks, conditioned on survival of all risks to time a. A well-known result in competing risks theory gives this probability as

$$Q_j(a, b) = P(a \leq X_j < b, X_j < X_i \text{ for all } i \neq j \,|\, T \geq a)$$

$$= \int_a^b h_{X_j}(x) \, e^{-\int_a^x h_T(t)\,dt}\, dx$$

for $j = 1, 2, \ldots, k$. Rather than isolating individual risks, as in the case of net lifetimes, this quantity considers risk j as it works in the presence of the $k - 1$ other risks. The *probability of failure due to risk j* is defined by $\pi_j = P(X_j = T)$, for $j = 1, 2, \ldots, k$. Since failure will occur from one of the causes,

$$\sum_{j=1}^k \pi_j = 1.$$

A simple example to illustrate some of the concepts in competing risks is given next before the general theory is developed.

> **Example 6.7** Consider an item that is subject to $k = 2$ causes of failure. Let the random variables X_1 and X_2 be the net lives for causes C_1 and C_2. If the item under consideration is a cell phone, for instance, cause 1 might be dropping the cell phone and cause 2 might be all other causes (for example, battery or display failure). In this case, X_1 is the life of the cell phone if the only way it can fail is by being dropped. The second net life, X_2, is the lifetime of the cell phone if it is bolted to a desk and cannot be dropped. The first crude life, Y_1, is the failure time of a cell phone that failed due to being dropped in the presence of the second cause of failure. Likewise, Y_2 is the lifetime of a cell phone that failed by some mode other than being dropped, but was not bolted to a desk to avoid its being dropped. Let the observed lifetime, T, be the minimum of X_1 and X_2. Also, assume that X_1 and X_2 are independent and have exponential distributions with population means 1 and $1/2$, respectively. Thus,
>
> $$S_{X_1}(t) = e^{-t} \qquad\qquad f_{X_1}(t) = e^{-t} \qquad\qquad h_{X_1}(t) = 1$$
>
> and
>
> $$S_{X_2}(t) = e^{-2t} \qquad\qquad f_{X_2}(t) = 2e^{-2t} \qquad\qquad h_{X_2}(t) = 2$$
>
> for $t \geq 0$. The net probabilities of failure in the interval $[a,b)$ are
>
> $$q_1(a, b) = 1 - e^{-\int_a^b 1\,dt} = 1 - e^{-(b-a)}$$

and

$$q_2(a, b) = 1 - e^{-\int_a^b 2\, dt} = 1 - e^{-2(b-a)}$$

for $0 < a < b$. The crude probability of failure due to the first risk in the interval $[a, b)$, $Q_1(a, b)$, is the integral of the joint probability density function of X_1 and X_2 over the shaded area in Figure 6.7 (illustrated for $a = 0.5$ and $b = 1.2$), divided by the integral of the joint probability density function of X_1 and X_2 over the area to the northeast of the point (a, a). Thus,

$$
\begin{aligned}
Q_1(a, b) &= P(a \le X_1 < b \text{ and } X_1 < X_2 \,|\, T \ge a) \\
&= \frac{P(a \le X_1 < b \text{ and } X_1 < X_2)}{P(X_1 \ge a, X_2 \ge a)} \\
&= \frac{\int_a^b \int_{x_1}^\infty e^{-w_1} 2e^{-2w_2}\, dw_2\, dw_1}{\int_a^\infty \int_a^\infty e^{-w_1} 2e^{-2w_2}\, dw_2\, dw_1} \\
&= \frac{1}{3}\left[1 - e^{-3(b-a)}\right]
\end{aligned}
$$

for $0 < a < b$. Similarly,

$$Q_2(a, b) = \frac{2}{3}\left[1 - e^{-3(b-a)}\right]$$

for $0 < a < b$. The $Q_j(a, b)$ expressions have been determined by using their definitions. Alternatively, the formula given earlier,

$$Q_j(a, b) = \int_a^b h_{X_j}(x)\, e^{-\int_a^x h_T(t)\, dt}\, dx,$$

for $j = 1, 2$, can be used to determine these quantities. For this particular example,

$$Q_1(a, b) = \int_a^b e^{-\int_a^x 3\, dt}\, dx = \frac{1}{3}\left[1 - e^{-3(b-a)}\right]$$

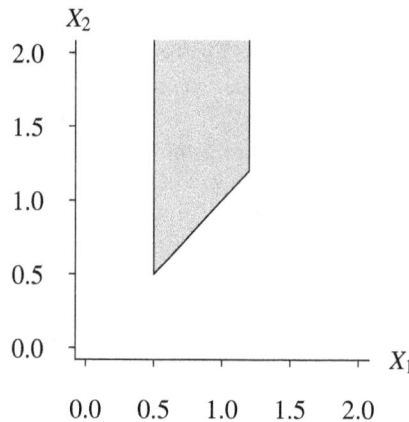

Figure 6.7: Numerator integration region for $Q_1(a, b)$ for $a = 0.5$ and $b = 1.2$.

and

$$Q_2(a, b) = \int_a^b 2e^{-\int_a^x 3\,dt}\,dx = \frac{2}{3}\left[1 - e^{-3(b-a)}\right]$$

for $0 < a < b$ because $h_T(t) = h_{X_1}(t) + h_{X_2}(t) = 1 + 2 = 3$ for $t \geq 0$. The probability of failure due to risk 1, π_1, can be found by integrating the joint density of the net lives $f(x_1, x_2)$ over the area $X_1 < X_2$ or, equivalently, using $a = 0$ and $b = \infty$ as arguments in $Q_1(a, b)$, yielding $\pi_1 = 1/3$. Similarly, $\pi_2 = 2/3$.

The focus now shifts to the determination of the distribution of the crude lives, Y_1 and Y_2. What is the survivor function for items that fail from one risk in the presence of the other risk? This survivor function is important because data collected in competing risks models often come in pairs: the cause of failure and the time of failure. The observed time of failure is typically a crude lifetime because it is observed in the presence of the other cause(s). The survivor function for the first crude lifetime, $S_{Y_1}(y_1)$, corresponds to a cell phone that fails by being dropped in the presence of risk C_2. If an analyst had a large data set of cell phone failure times for those cell phones that failed by being dropped, an empirical survivor function will converge to $S_{Y_1}(y_1)$ as the sample size increases. The survivor function for Y_1 is

$$\begin{aligned}
S_{Y_1}(y_1) &= P(T \geq y_1 \,|\, X_1 = T) \\
&= \frac{P(T \geq y_1, \, X_1 = T)}{\pi_1} \\
&= \frac{\displaystyle\int_{y_1}^{\infty} \int_{x_1}^{\infty} e^{-w_1} 2e^{-2w_2}\,dw_2\,dw_1}{1/3} \\
&= e^{-3y_1} \qquad\qquad y_1 \geq 0.
\end{aligned}$$

Similarly,

$$S_{Y_2}(y_2) = e^{-3y_2} \qquad\qquad y_2 \geq 0.$$

This surprising result, that both Y_1 and Y_2 have the same exponential distribution with population mean $1/3$, can be attributed to the definition of a crude lifetime. Since the two crude lifetimes are the minimum of two exponential random variables (the exponential net lifetimes), each will have an exponential distribution with a parameter being the sum of the rates. The crude lifetime Y_1, for example, consists of only those exponential(1) random variables that are smaller than another independent exponential(2) random variable. Likewise, the crude lifetime Y_2 consists of only those exponential(2) random variables that are smaller than another independent exponential(1) random variable. Theorem 4.4 provides the basis for the fact that the minimum of independent exponential random variables is also exponentially distributed.

As a result, there are two valid ways to generate a random lifetime T for use in Monte Carlo simulation. First, taking the net lifetime perspective, generate an exponential(1) random variate and an exponential(2) random variate and choose the minimum as T. Second, taking the crude lifetime perspective, generate an exponential(3) random variate T and indicate this is failure from risk 1 with probability $1/3$ and failure from risk 2 with probability $2/3$.

6.2.3   General Case

A general theory for competing risks is now developed based on the definitions for net and crude life-times given previously. Let $X_1, X_2, \ldots, X_k$ be the k continuous net lives and $T = \min\{X_1, X_2, \ldots, X_k\}$ be the observed failure time of the item. The X_j's are not necessarily independent as they were in Example 6.7. Letting the net lives have joint probability density function $f(x_1, x_2, \ldots, x_k)$, the joint survivor function is

$$S(x_1, x_2, \ldots, x_k) = P(X_1 \geq x_1, X_2 \geq x_2, \ldots, X_k \geq x_k)$$

$$= \int_{x_k}^{\infty} \cdots \int_{x_2}^{\infty} \int_{x_1}^{\infty} f(t_1, t_2, \ldots, t_k) \, dt_1 \, dt_2 \ldots dt_k$$

and the marginal net survival function is

$$S_{X_j}(x_j) = P(X_j \geq x_j) = S(0, \ldots, x_j, \ldots, 0)$$

for $j = 1, 2, \ldots, k$. The survivor function for the observed lifetime T is

$$S_T(t) = P(T \geq t) = S(t, t, \ldots, t).$$

The probability of failure from risk j can be determined from the joint survivor function because

$$-\frac{\partial}{\partial x_j} S(x_1, \ldots, x_j, \ldots, x_k) = \lim_{\Delta x \to 0} \frac{S(x_1, \ldots, x_j, \ldots, x_k) - S(x_1, \ldots, x_j + \Delta x, \ldots, x_k)}{\Delta x}$$

for $j = 1, 2, \ldots, k$ by the definition of the derivative. Thus,

$$\pi_j = \int_0^{\infty} -\left[\frac{\partial}{\partial x_j} S(x_1, \ldots, x_j, \ldots, x_k)\right]_{x_1 = x_2 = \cdots = x_k = x} dx$$

for $j = 1, 2, \ldots, k$. To derive a survivor function for the crude lifetimes, let the random variable J be the index of the cause of failure so that

$$P(T \geq t, J = j) = P(X_j \geq t, X_j < X_i \text{ for all } i \neq j)$$

$$= \int_t^{\infty} \left[\int_{x_j}^{\infty} \cdots \int_{x_j}^{\infty} \int_{x_j}^{\infty} f(x_1, x_2, \ldots, x_k) \prod_{i \neq j} dx_i\right] dx_j$$

for $j = 1, 2, \ldots, k$, where the survivor function for T is obtained by conditioning:

$$S_T(t) = P(T \geq t)$$

$$= \sum_{j=1}^{k} P(T \geq t \,|\, J = j) P(J = j)$$

$$= \sum_{j=1}^{k} P(T \geq t, J = j).$$

When $t = 0$, each term in the last summation is one of the π_j's because

$$\pi_j = P(J = j) = P(T \geq 0, J = j)$$

for $j = 1, 2, \ldots, k$. Thus, the distribution of the jth crude life, Y_j, is the distribution of T conditioned on $J = j$:

$$S_{Y_j}(y_j) = P(T \geq y_j \,|\, J = j) = \frac{P(T \geq y_j, J = j)}{P(J = j)}$$

for $j = 1, 2, \ldots, k$.

Example 6.8 The competing risks model from Example 6.7, which considered two independent, exponentially distributed risks, is used to illustrate the use of the formulas developed thus far. As before, let the net lives have marginal survivor functions

$$S_{X_1}(t) = e^{-t} \qquad\qquad S_{X_2}(t) = e^{-2t}$$

for $t \geq 0$. Since the risks are independent, the joint survivor function is

$$S(x_1, x_2) = S_{X_1}(x_1) \cdot S_{X_2}(x_2) = e^{-x_1 - 2x_2} \qquad\qquad x_1 \geq 0, x_2 \geq 0.$$

The probability of failure from the first risk is

$$
\begin{aligned}
\pi_1 &= \int_0^\infty -\left[\frac{\partial}{\partial x_1} S(x_1, x_2)\right]_{x_1 = x_2 = x} dx \\
&= \int_0^\infty -\left[-e^{-x_1 - 2x_2}\right]_{x_1 = x_2 = x} dx \\
&= \int_0^\infty e^{-3x} dx \\
&= \frac{1}{3}.
\end{aligned}
$$

Since $\pi_2 = 1 - \pi_1$,

$$\pi_2 = \frac{2}{3}.$$

The probability of survival to time t and risk 1 being the cause of failure is

$$
\begin{aligned}
P(T \geq t, J = 1) &= P(X_1 \geq t, X_1 < X_2) \\
&= \int_t^\infty \left[\int_{x_1}^\infty f(x_1, x_2) dx_2\right] dx_1 \\
&= \int_t^\infty \int_{x_1}^\infty 2e^{-x_1 - 2x_2} dx_2 dx_1 \\
&= \int_t^\infty e^{-3x_1} dx_1 \\
&= \frac{1}{3} e^{-3t} \qquad\qquad t \geq 0.
\end{aligned}
$$

Similarly,

$$P(T \geq t, J = 2) = \frac{2}{3} e^{-3t} \qquad\qquad t \geq 0.$$

Thus, the survival function for the first crude lifetime is

$$S_{Y_1}(y_1) = \frac{P(T \geq y_1, J = 1)}{P(J = 1)} = e^{-3y_1} \qquad\qquad y_1 \geq 0$$

and the survival function for the second crude lifetime is

$$S_{Y_2}(y_2) = \frac{P(T \geq y_2, J = 2)}{P(J = 2)} = e^{-3y_2} \qquad\qquad y_2 \geq 0.$$

These results are identical to those derived from first principles in the previous example. Both crude lifetimes have an exponential(3) distribution.

To this point, it has been shown how the distribution of the net lives $X_1, X_2, \ldots, X_k$ determines the distribution of the crude lives $Y_1, Y_2, \ldots, Y_k$. Net lives can be interpreted as *potential lifetimes*, while crude lives are the *observed lifetimes*. When lifetime data for the known cause of failure are collected, the observed values are Y_j's. An important question is whether the distribution of each Y_j contains enough information to determine the distribution of the X_j's. In general, the answer is no, but under the assumption of independence of the net lives, the answer is yes. The following discussion considers results under the assumption of independent net lives. This independence can often be attained by grouping the k risks so that dependencies occur within, but not between, risks.

Theorem 6.1 Let $X_1, X_2, \ldots, X_k$ be independent net lifetimes. Let $Y_1, Y_2, \ldots, Y_k$ be the associated crude lifetimes with known marginal probability density functions $f_{Y_1}(t), f_{Y_2}(t), \ldots, f_{Y_k}(t)$. The probability of failure from risk j, $\pi_j = P(J = j)$, is known. Then the hazard function for the net lifetime j is

$$h_{X_j}(t) = \frac{\pi_j f_{Y_j}(t)}{\sum_{i=1}^{k} \pi_i S_{Y_i}(t)} \qquad t \geq 0$$

for $j = 1, 2, \ldots, k$.

The proof of this result is given in a reference listed in the preface. This result is useful for determining the effect of removing one or more risks when the distributions of the crude lives are determined from a data set, as illustrated in the next example.

Example 6.9 Consider again the competing risks model from Example 6.8 in which the $k = 2$ risks were assumed to be independent. If a large number of failure times are collected, and the cause of failure is *identifiable* (that is, both the failure time and the index of the risk that caused failure are known), it might be possible to determine the distribution of the two crude lifetimes. If both are well fitted with an exponential distribution with failure rate $\lambda = 3$, and approximately one-third of the failures are from cause 1, then

$$\pi_1 = P(J = 1) = \frac{1}{3} \qquad\qquad \pi_2 = P(J = 2) = \frac{2}{3}$$

and

$$S_{Y_1}(t) = e^{-3t} \qquad\qquad S_{Y_2}(t) = e^{-3t}$$

for $t \geq 0$. Therefore, by Theorem 6.1 the hazard functions for the net lives are

$$h_{X_1}(t) = \frac{\frac{1}{3} \cdot 3e^{-3t}}{\frac{1}{3}e^{-3t} + \frac{2}{3}e^{-3t}} = 1 \qquad t \geq 0$$

and

$$h_{X_2}(t) = \frac{\frac{2}{3} \cdot 3e^{-3t}}{\frac{1}{3}e^{-3t} + \frac{2}{3}e^{-3t}} = 2 \qquad t \geq 0.$$

This result is consistent with the previous two examples.

The previous three examples have considered competing risks models with $k = 2$ risks and exponentially distributed net and crude lifetimes. This section concludes with an example of a competing risks model with $k = 3$ risks and non-exponential net lifetimes.

Example 6.10 An item is subject to $k = 3$ competing risks C_1, C_2, and C_3 with the three associated independent net lifetimes: $X_1 \sim \text{Weibull}(1, 2)$, $X_2 \sim \text{exponential}(1)$, and $X_3 \sim \text{Weibull}(1, 3)$.

(a) What is the population mean time to failure of the item?

(b) If one of the risks could be eliminated, the elimination of which risk results in the greatest increase in the population mean time to failure of the item?

(a) The hazard function associated with a Weibull(λ, κ) random variable is

$$h(t) = \kappa \lambda^{\kappa} t^{\kappa - 1} \qquad t \geq 0.$$

The hazard functions for the three net lifetimes are

$$h_{X_1}(t) = 2t \qquad h_{X_2}(t) = 1 \qquad h_{X_3}(t) = 3t^2$$

for $t \geq 0$. The hazard function for the time to failure of the item T is the sum of the hazard functions for the three net lifetimes:

$$h_T(t) = 3t^2 + 2t + 1 \qquad t \geq 0.$$

The associated cumulative hazard function is

$$\begin{aligned} H_T(t) &= \int_0^t h_T(\tau) \, d\tau \\ &= \int_0^t \left(3\tau^2 + 2\tau + 1 \right) d\tau \\ &= \left[\tau^3 + \tau^2 + \tau \right]_0^t \\ &= t^3 + t^2 + t \qquad t \geq 0. \end{aligned}$$

The associated survivor function is

$$S_T(t) = e^{-H_T(t)} = e^{-t^3 - t^2 - t} \qquad t \geq 0.$$

So the population mean time to failure of the item is

$$E[T] = \int_0^{\infty} S_T(t) \, dt = \int_0^{\infty} e^{-t^3 - t^2 - t} \, dt \cong 0.4630,$$

where the integral must be evaluated numerically.

(b) The same procedure given in part (a) can be used to assess the effect of removing risks. The results are given in Table 6.4. Removing risk 2 makes the greatest improvement on the population mean lifetime of the item.

Risk eliminated	Risk 1	Risk 2	Risk 3
$E[T]$	0.5689	0.6637	0.5456

Table 6.4: Mean lifetimes associated with eliminated risks.

To summarize this section, competing risks models are appropriate when there are k causes of failure and the occurrence of failure due to any risk causes the item to fail. These k risks can be thought of conceptually as a k-component series system. The probabilities of failure from the various causes are denoted by $\pi_1, \pi_2, \ldots, \pi_k$. The net lifetimes $X_1, X_2, \ldots, X_k$ occur if only one risk is evident at a time in the population. The crude lifetimes $Y_1, Y_2, \ldots, Y_k$ occur in the presence of all other risks. If the net lives are independent, once the distributions of $Y_1, Y_2, \ldots, Y_k$ and $\pi_1, \pi_2, \ldots, \pi_k$ are determined, the distribution of the net lives $X_1, X_2, \ldots, X_k$ can be determined.

6.3   Point Processes

So far, the focus has been on a single random variable T, generically referred to as a *lifetime*, and methods for estimating its probability distribution. In a reliability setting, T might be the lifetime of a light bulb. In a biostatistical setting, T might be the post-surgery remission time for a patient having a particular type of cancer. In an actuarial setting, T might be the time of death for a insured individual having a life insurance policy. In all of these examples, there is only a single random variable T that is of interest.

Occasions arise, however, when there are multiple events of interest. In a reliability setting, the sequence of events might be the repair times for an automobile. In a biostatistical setting, the sequence of events might be the times at which a cortisone injection is administered to a patient. In an actuarial setting, the sequence of events might be the times of insurance claims on an insured dwelling. In all of these examples, the probability mechanism governing the sequence of observations is of interest.

Point process models are often used to describe the probability mechanism governing a series of event times. The three elementary point process models considered in this section are Poisson processes, renewal processes, and nonhomogeneous Poisson processes.

Point process models can be applied to more than just failure times of repairable systems. Point processes have been used to describe arrival times to queues, earthquake times, hurricane landfall times, pothole positions on a highway, and other physical phenomena. They have also been used to describe the occurrence times of sociological events such as crimes, strikes, bankruptcies, and wars.

The examples in this section use reliability jargon, leaving it to the reader to extend the models to other disciplines. The reliability-centric term "failure" is used instead of the more generic term "event" for all of the point processes described in this section. The object of interest will continue to be referred to generically as an "item."

When the time to repair or replace an item is negligible, point processes are appropriate for modeling the probabilistic mechanism underlying the failure times. This would be the case for an automobile that works without failure for months, and then is in the shop for one hour for a repair in which no mileage is accrued while the maintenance is being performed. These models would not be appropriate, for example, for an aircraft that spends several months having its engine overhauled before being placed back into service if availability is of interest. The down time needs to be explicitly modeled in this case.

A small but important bit of terminology is used to differentiate between nonrepairable items, which were considered in the previous chapters, and repairable items, which are considered here. A nonrepairable item, such as a light bulb, has *one* failure, and the term *burn in* is used if its hazard function is decreasing and the term *wear out* is used if its hazard function is increasing. Figure 6.8 shows hazard functions for an item that undergoes burn in and another that wears out. The × on the time axis denotes a realization of one possible failure time. The lifetimes of nonrepairable items are described by the distribution of a *single* nonnegative random variable, usually denoted by T.

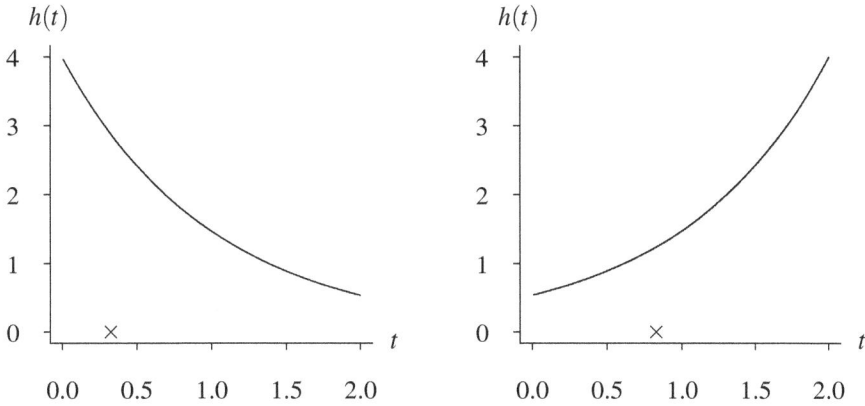

Figure 6.8: Hazard functions for nonrepairable items in the DFR and IFR classes.

In contrast, a repairable item, such as an automobile, typically fails at *several* points in time. In many situations, a nonhomogeneous Poisson process, which is governed by the intensity function $\lambda(t)$ that reflects the rate of occurrence of failures, might be the appropriate probabilistic mechanism for modeling the failure history of the item. The intensity function is analogous to the hazard function in the sense that higher levels of $\lambda(t)$ indicate an increased probability of failure. The term *improvement* is used if the intensity function is decreasing, and the term *deterioration* is used if the intensity function is increasing. Figure 6.9 shows intensity functions for an item that improves and another that deteriorates. Each $\times$ on the time axis denotes a failure time associated with a realization. The improving item has failures that tend to be less frequent as time passes; the deteriorating item has failures that tend to be more frequent as time passes. The failure times of repairable items are described by the probability mechanism underlying a *sequence* of random variables, often denoted by $T_1, T_2, \ldots$. These terms are summarized in Table 6.5.

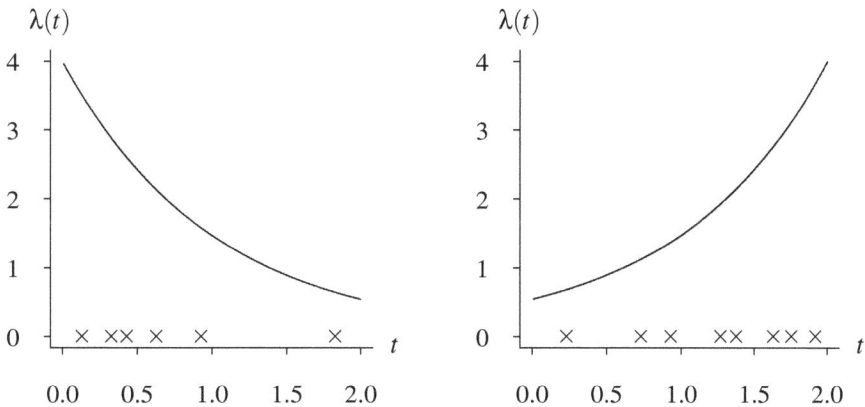

Figure 6.9: Intensity functions for a repairable improving and deteriorating items.

	Nonrepairable	Repairable
Item gets better as time passes	Burn in, $h'(t) \leq 0$	Improving, $\lambda'(t) \leq 0$
Item gets worse as time passes	Wear out, $h'(t) \geq 0$	Deteriorating, $\lambda'(t) \geq 0$

Table 6.5: Terminology for nonrepairable and repairable items in the reliability setting.

The notation that applies to all three point process models surveyed in this section is presented next.

In the point processes discussed in this section, failures occur at times $T_1, T_2, \ldots$, and the time to replace or repair an item is assumed to be negligible. The origin is defined to be $T_0 = 0$. The times between the failures are $X_1, X_2, \ldots$, so $T_k = X_1 + X_2 + \cdots + X_k$, for $k = 1, 2, \ldots$. The counting function $N(t)$ is the number of failures that occur in the time interval $(0, t]$. In other words,

$$N(t) = \max\{k \,|\, T_k \leq t\}$$

for $t > 0$. The nondecreasing, integer-valued stochastic process described by $\{N(t), t > 0\}$ is often called a *counting process* and satisfies the following two properties.

1. If $t_1 < t_2$, then $N(t_1) \leq N(t_2)$.

2. If $t_1 < t_2$, then $N(t_2) - N(t_1)$ is the number of failures in the time interval $(t_1, t_2]$.

Let $\Lambda(t) = E[N(t)]$ be the expected number of failures that occur in the interval $(0, t]$. The derivative of $\Lambda(t)$, which is $\lambda(t) = \Lambda'(t)$, is the rate of occurrence of failures. Figure 6.10 shows one realization of a point process, where $N(t)$ is shown as a step function and the $\times$s denote the failure times on the horizontal axis. The curve for the expected number of events by time t, $\Lambda(t)$,

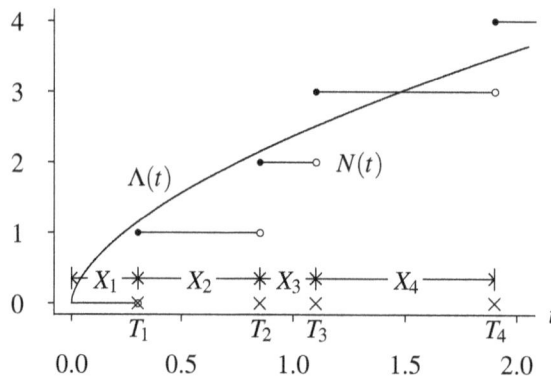

Figure 6.10: Point process realization.

is also on the same axis as $N(t)$. It should be kept in mind that $N(t)$ on this axis is a realization that will change from one item to another item, but $\Lambda(t)$ is the underlying population probabilistic mechanism describing the sequence of events and does not change from one item to another item.

The behavior of the interevent times $X_1, X_2, \ldots$ is always of interest in analyzing a repairable item. If the interevent times tend to increase with time, the item is improving; if the interevent times tend to decrease with time, the item is deteriorating. Other variables, such as a new untrained operator or repairman, must be considered when analyzing the failure times of a repairable item. These variables are ignored in the presentation of the point process models, but can result in erroneous conclusions if not considered along with the observed times between failures.

There are two properties that are important to discuss before introducing specific point processes. The first property is called *independent increments*. A point process has independent increments if the number of failures in mutually exclusive intervals are independent. As shown in the realization depicted in Figure 6.11, this property implies that the number of failures (the failure times are depicted by ×s) between times t_1 and t_2 are independent of the number of failures between times t_3 and t_4 because the intervals $(t_1, t_2]$ and $(t_3, t_4]$ are nonoverlapping. A second property is called *stationarity*. A point process is stationary if the distribution of the number of failures in any time interval depends only on the length of the time interval. Equivalently, failures are no more or less likely to occur at one time than another for an item. This is a rather restrictive assumption for an item because the item can neither deteriorate nor improve.

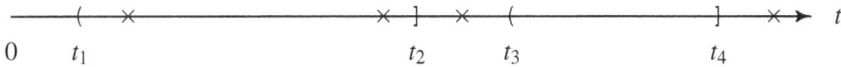

Figure 6.11: Independent increments illustration.

The three point process models, Poisson processes, renewal processes, and nonhomogeneous Poisson processes, are introduced in separate subsections.

6.3.1 Poisson Processes

The well-known Poisson process is a popular model due to its mathematical tractability, although it applies only to limited situations. These limited situations include replacement models with exponential standby items and repairable items with exponential times to failure and negligible repair times.

Definition 6.3 A counting process $N(t)$ is a Poisson process with rate parameter $\lambda > 0$ if

- $N(0) = 0$,
- the process has independent increments, and
- the number of failures in any interval of length t has the Poisson distribution with mean λt.

There are several implications of this definition of a Poisson process. First, by the last condition for a Poisson process, the distribution of the number of failures in the interval $(t_1, t_2]$ has the Poisson distribution with parameter $\lambda(t_2 - t_1)$. Therefore, the probability mass function of the number of failures in the interval $(t_1, t_2]$ is

$$P\big(N(t_2) - N(t_1) = x\big) = \frac{\left[\lambda(t_2 - t_1)\right]^x e^{-\lambda(t_2 - t_1)}}{x!} \qquad x = 0, 1, 2, \dots .$$

Second, the number of failures by time t, denoted by $N(t)$, has the Poisson distribution with population mean

$$\Lambda(t) = E[N(t)] = \lambda t \qquad t > 0,$$

where λ is often called the rate of occurrence of failures. The intensity function is therefore given by $\lambda(t) = \Lambda'(t) = \lambda$ for $t > 0$. Third, if $X_1, X_2, \dots$ are independent and identically distributed exponential random variables, then $N(t)$ corresponds to a Poisson process.

> **Example 6.11** Consider a socket model in which an infinite supply of light bulbs is used in a single-component standby system composed of a single socket. As each bulb fails, it is immediately replaced by a new bulb, and each bulb has an exponential(λ) time to failure. Find the probability that there are n or fewer failures by time t.
>
> Since the light bulb failure time distributions are each exponential, and the replacement time is negligible, a Poisson process is the appropriate model here. The probability that there are n or fewer failures by time t is therefore
>
> $$P\big(N(t) \leq n\big) = \sum_{k=0}^{n} \frac{(\lambda t)^k e^{-\lambda t}}{k!} \qquad n = 0, 1, 2, \dots$$
>
> for $t > 0$. When $n = 0$, this solution reduces to the survivor function for an exponential distribution (the nonrepairable case). It is easily recognized here that the time of the nth failure has the Erlang distribution with scale parameter λ and shape parameter n because $T_n = X_1 + X_2 + \dots + X_n$ and $X_1, X_2, \dots, X_n$ are independent and identically distributed exponential(λ) random variables.

This model is sometimes also called a *homogeneous Poisson process* because the failure rate λ does not change with time (that is, the model is stationary). The next two models are generalizations of homogeneous Poisson processes. In a renewal process, the assumption of exponentially distributed times between failures is relaxed; in a nonhomogeneous Poisson process, the stationarity assumption is relaxed.

6.3.2 Renewal Processes

A renewal process is a natural extension of a Poisson process in which the times between failure are assumed to have any lifetime distribution, rather than just the exponential distribution.

> **Definition 6.4** A point process is a *renewal process* if the times between failures $X_1, X_2, \dots$ are independent and identically distributed nonnegative random variables.

The term *renewal* is appropriate for these models because an item is assumed to be renewed to its original state after it fails. This is typically not the case for a repairable system consisting of many components, because only a few of the components are typically replaced upon failure. The remaining components that did not fail will only be as good as new if they have exponential lifetimes. Renewal processes are often used, for example, to determine the number of spare components to take on a mission or to determine the timing of a sequence of repairs.

One classification of renewal processes that is useful in the study of socket models concerns the coefficient of variation $\gamma = \sigma/\mu$ of the distribution of the times between failures. This classification divides renewal processes into underdispersed and overdispersed processes.

Definition 6.5 A renewal process is *underdispersed (overdispersed)* if the coefficient of variation of the distribution of the times between failures is less than (greater than) 1.

Figure 6.12 displays realizations of three different renewal process. The first process is under-dispersed because the coefficient of variation of the distribution of the time between failures is less than 1. An extreme case of an underdispersed process is one in which the coefficient of variation of the distribution of the time between failures is 0 (that is, a deterministic failure time for each item because $\sigma/\mu = 0$ implies that $\sigma = 0$), which would yield a deterministic renewal process. The un-derdispersed process is much more regular in its failure times; hence, it is easier to determine when it is appropriate to replace an item if failure is catastrophic or expensive. A design engineer's goal might be to reduce the variability of the lifetime of an item, which in turn decreases the coefficient of variation. Reduced variation with increased mean is desirable for most items. The second axis in Figure 6.12 corresponds to a realization of a renewal process that has a coefficient of variation of the distribution of the time between failures equal to 1. This case sits in between the underdispersed and overdispersed cases. There is more clumping of failures than in the underdispersed case. The third axis corresponds to a realization of an overdispersed distribution. There is extreme clumping of failures here, and many failures occur soon after an item is placed into service. Fortunately, the overdispersed case occurs less often in practice than the underdispersed case.

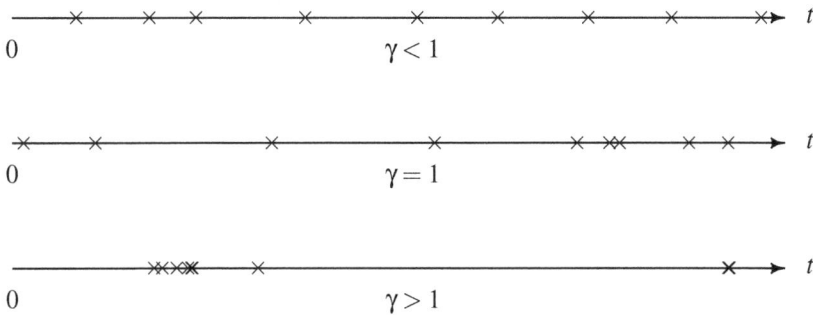

Figure 6.12: Classifying renewal processes based on the coefficient of variation.

Example 6.12 Consider a renewal process with Weibull(λ, κ) time between failures. Classify the renewal processes into the underdispersed and overdispersed cases.

The shape parameter κ partitions the renewal process into the various cases. When $\kappa > 1$, the process is underdispersed; when $\kappa < 1$, the process is overdispersed; when $\kappa = 1$, the process reduces to a Poisson process because the time between failures is exponential(λ). The sequence of failures depicted on the time axes in Figure 6.12 were generated by Monte Carlo simulation using Weibull(λ, κ) times between failures. The top axis corresponds to $\kappa = 5$ (an IFR time between failures distribution) and the pattern of failures reflects the relatively small standard deviation of the time between failures distribution. The middle axis corresponds to $\kappa = 1$ (the exponential special case of the Weibull distribution), so this is a realization of a Poisson process. This case is the dividing line between an underdispersed renewal process and an overdispersed renewal process. There is more clumping of failures than in the underdispersed case because the mode of the exponential distribution is zero. A replacement policy is ineffective in this case because of the memoryless property of the exponential distribution. The bottom axis corresponds to $\kappa = 1/2$ (a DFR time between failures distribution), and the clumping of failures becomes even more extreme. On all three axes, there are nine failures, but the pattern of failures differs significantly for the various values of κ.

Two measures of interest that often arise when using a renewal process are the distribution of T_n, the time of the nth failure, and the distribution of the number of failures by time t. In terms of the distribution of T_n, there are simple results for the expected value and population variance of T_n, but the tractability of the distribution of T_n depends on the tractability of the distribution of the times between failures. Since $T_n = X_1 + X_2 + \cdots + X_n$, and the X_i's are mutually independent and identically distributed, the expected value and population variance of T_n are

$$E[T_n] = nE[X] \qquad \text{and} \qquad V[T_n] = nV[X],$$

where $E[X]$ and $V[X]$ are the expected value and population variance of the time between failures. The survivor function for the time of the nth failure, $S_{T_n}(t) = P(T_n \geq t)$, can be found as a function of the distribution of the X_i's and is tractable only for simple time between failure distributions.

The distribution of the number of failures by time t can be calculated by finding the values of the mass function $P(N(t) = n)$ for all values of n. Since exactly n failures occurring by time t is equivalent to T_n being less than or equal to t and T_{n+1} being greater than t,

$$\begin{aligned} P(N(t) = n) &= P(T_n \leq t < T_{n+1}) \\ &= P(T_{n+1} \geq t) - P(T_n \geq t) \\ &= S_{T_{n+1}}(t) - S_{T_n}(t) \end{aligned}$$

for $n = 0, 1, 2, \ldots$ and $t > 0$, and continuous time between failures distribution. Although using the exponential distribution as the time to failure for each item reduces a renewal process to a Poisson process, it will be used in the next example because it is one of the few distributions for which these measures can easily be calculated.

Example 6.13 Consider again a socket model for which the time to failure of each light bulb inserted in the socket has an exponential distribution with failure rate λ. Find the expected value and the population variance of T_n, the survivor function of T_n, and the probability mass function of the number of failures by time t.

First, since each item has population mean time to failure $E[X] = 1/\lambda$, and population variance of the time to failure $V[X] = 1/\lambda^2$, the expected value and the population variance of the time of failure n are

$$E[T_n] = nE[X] = \frac{n}{\lambda} \qquad \text{and} \qquad V[T_n] = nV[X] = \frac{n}{\lambda^2}$$

for $n = 0, 1, 2, \ldots$. Since T_n is the sum of independent and identically distributed exponential random variables, it has the Erlang distribution with survivor function

$$S_{T_n}(t) = \sum_{k=0}^{n-1} \frac{(\lambda t)^k}{k!} e^{-\lambda t} \qquad t > 0$$

for $n = 1, 2, \ldots$. To find the probability mass function for the number of failures by time t,

$$P\big(N(t) = n\big) = S_{T_{n+1}}(t) - S_{T_n}(t)$$

$$= \sum_{k=0}^{n} \frac{(\lambda t)^k}{k!} e^{-\lambda t} - \sum_{k=0}^{n-1} \frac{(\lambda t)^k}{k!} e^{-\lambda t}$$

$$= \frac{(\lambda t)^n}{n!} e^{-\lambda t}$$

for $n = 0, 1, 2, \ldots$, and $t > 0$, which is recognized as the Poisson distribution with rate parameter λt. This simplest case of a renewal process corresponds to a Poisson process.

A more mathematically complicated situation occurs when the gamma distribution is used to model the time between failures.

Example 6.14 Consider a socket model with a single socket in which the lifetime of each light bulb to be placed in the socket has the gamma distribution with scale parameter $\lambda = 0.001$ and shape parameter $\kappa = 5.2$, where time is measured in hours. Find the probability that three light bulbs are sufficient to light the system for 8760 hours (one year).

Since the mean of the gamma distribution is κ/λ, each light bulb has mean time to failure $\mu = 5.2/0.001 = 5200$ hours. Thus, the expected time of failure number $n = 3$ is $E[T_3] = 3E[X] = 3(5200) = 15{,}600$ hours, or almost two years. This preliminary analysis indicates that the probability that three bulbs will be sufficient for one year of operation should be fairly high.

A result that can be used to determine the exact probability is that the sum of n independent and identically distributed gamma random variables also has a gamma distribution. This result is most easily derived by using the moment generating function approach to determine the distribution of the sum of independent random variables. Let the random variable X have a gamma distribution with parameters λ and κ. The moment generating function of X is

$$M_X(s) = E\left[e^{sX}\right]$$

$$= \int_0^\infty e^{sx} \frac{\lambda}{\Gamma(\kappa)} (\lambda x)^{\kappa-1} e^{-\lambda x} \, dx$$

$$= \frac{\lambda^\kappa}{\Gamma(\kappa)} \int_0^\infty x^{\kappa-1} e^{-x(\lambda - s)} \, dx$$

$$= \frac{\lambda^\kappa}{\Gamma(\kappa)} \int_0^\infty \left(\frac{u}{\lambda - s}\right)^{\kappa-1} e^{-u} \frac{1}{\lambda - s} \, du$$

$$= \left(\frac{\lambda}{\lambda - s}\right)^\kappa \frac{1}{\Gamma(\kappa)} \int_0^\infty u^{\kappa-1} e^{-u} \, du$$

$$= \left(\frac{\lambda}{\lambda - s}\right)^\kappa$$

for all $s < \lambda$. Since $X_1, X_2, \ldots, X_n$ are mutually independent and identically distributed gamma random variables, the moment generating function of $T_n = X_1 + X_2 + \cdots + X_n$ is the product of n of these moment generating functions:

$$M_{T_n}(s) = \prod_{i=1}^{n} M_{X_i}(s) = \left(\frac{\lambda}{\lambda - s}\right)^{n\kappa}$$

for all $s < \lambda$. Thus, if $X_1, X_2, \ldots, X_n$ are independent and identically distributed gamma random variables with parameters λ and κ, then the probability distribution of their sum has the gamma distribution with parameters λ and $n\kappa$. For the problem at hand, the time to the third failure, T_3, has a gamma distribution with scale parameter $\lambda = 0.001$ and shape parameter $n\kappa = (3)(5.2) = 15.6$. To find the probability that T_3 exceeds 8760,

$$P(T_3 \geq 8760) = S_{T_3}(8760) = 1 - I(15.6, 8.76) = 0.9771,$$

where I is the incomplete gamma function. The R statement below computes this probability using the pgamma function, which returns the cumulative distribution function of a random variable having the gamma distribution.

```
pgamma(15.6, 8.76)
```

This completes the brief introduction to renewal processes. The final subsection introduces nonhomogeneous Poisson processes.

6.3.3   Nonhomogeneous Poisson Processes

The third and final point process introduced here is the nonhomogeneous Poisson process. There are at least four reasons that a nonhomogeneous Poisson process should be considered for modeling the sequence of failures of a repairable item.

1. A homogeneous Poisson process is a special case of a nonhomogeneous Poisson process.

2. The probabilistic model for a nonhomogeneous Poisson process is mathematically tractable.

3. The statistical methods for a nonhomogeneous Poisson process are mathematically tractable.

4. Unlike a homogeneous Poisson process or a renewal process, a nonhomogeneous Poisson process is able to model the failure times of improving and deteriorating items.

One disadvantage with both Poisson processes and renewal processes is that they assume that the distribution of the time to failure for each item in a socket model with a single socket is identical. This means that it is not possible for the item to improve or deteriorate. A nonhomogeneous Poisson process is another generalization of the homogeneous Poisson process for which the stationarity assumption is relaxed. Instead of a constant rate of occurrence of failures λ, as in a homogeneous Poisson process, this rate varies over time according to $\lambda(t)$, which is often called the *intensity function*. The *cumulative intensity function* is defined by

$$\Lambda(t) = \int_0^t \lambda(\tau)d\tau$$

and is interpreted as the expected number of failures by time t. These two functions are generally used to describe the probabilistic mechanism governing the failure times of the item, as opposed to the five distribution representations used to describe the time to failure of nonrepairable items.

Definition 6.6 A counting process is a *nonhomogeneous Poisson process* with intensity function $\lambda(t) \geq 0$ defined on $t > 0$ if

- $N(0) = 0$,
- the process has independent increments, and
- the probability of exactly n events occurring in the interval $(t_1, t_2]$ is given by

$$P\big(N(t_2) - N(t_1) = n\big) = \frac{\left[\int_{t_1}^{t_2} \lambda(t)\, dt\right]^n e^{-\int_{t_1}^{t_2} \lambda(t)\, dt}}{n!}$$

for $n = 0, 1, 2, \ldots$.

Thus, if the intensity function is decreasing, the item is improving; if the intensity function is increasing, the item is deteriorating. For nonhomogeneous Poisson processes, the times between failures are neither independent nor identically distributed. The time to the first failure in a non-homogeneous Poisson process has the same distribution as the time to failure of a nonrepairable item with hazard function $h(t) = \lambda(t)$. Subsequent failures follow a conditional version of the intensity function that does not depend on previous values of $\lambda(t)$. The times between these subsequent failures do not necessarily follow any of the probability distributions (for example, the Weibull distribution) used in survival analysis.

Since the independent increments property has been retained from the definition of a homogeneous Poisson process, this model assumes that previous failure times do not affect the future failure times of the item. Although this may not be exactly true in practice, the nonhomogeneous Poisson process model is still valuable because it is mathematically tractable and allows for improving and deteriorating systems. In addition, parameter estimation for the nonhomogeneous Poisson process model is simple, which is another attractive feature.

Example 6.15 Consider a nonhomogeneous Poisson process with intensity function

$$\lambda(t) = \kappa \lambda^\kappa t^{\kappa-1} \qquad t > 0,$$

where λ and κ are positive parameters. This intensity function can be recognized as the same functional form as the hazard function for a Weibull random variable with scale parameter λ and shape parameter κ, and is often referred to as a *power law process*. For this intensity function, if $\kappa < 1$, the item is improving because the intensity function is decreasing, if $\kappa > 1$, the item is deteriorating because intensity function is increasing, and if $\kappa = 1$, it reduces to a homogeneous Poisson process with rate parameter λ. Find the probability that there will be exactly n failures by time t. Also, if failure n occurs at time t_n, find the conditional survivor function for the time to the next failure.

Using Definition 6.6, the number of failures by time t, $N(t)$, has probability mass function

$$P\big(N(t) = n\big) = \frac{\left[\int_0^t \lambda(\tau)\, d\tau\right]^n e^{-\int_0^t \lambda(\tau)\, d\tau}}{n!} \qquad n = 0, 1, 2, \ldots$$

for $t > 0$. Using the fact that $\Lambda(t) = \int_0^t \kappa\lambda^\kappa\tau^{\kappa-1}d\tau = (\lambda t)^\kappa$ for $t > 0$,

$$P\big(N(t) = n\big) = \frac{(\lambda t)^{\kappa n}e^{-(\lambda t)^\kappa}}{n!} \qquad n = 0, 1, 2, \ldots$$

for $t > 0$. Finding the survivor function for the time to the next failure involves conditioning. Using independent increments, the fact that $S(t) = e^{-H(t)}$, and conditioning, the conditional survivor function for the time to the next failure is

$$S_{T|T>t_n}(t) = e^{-(\Lambda(t)-\Lambda(t_n))} = e^{-((\lambda t)^\kappa-(\lambda t_n)^\kappa)} = e^{-\lambda^\kappa(t^\kappa-t_n^\kappa)} \qquad t > t_n.$$

It was stated earlier that point process models are used in applications outside of reliability. Figure 6.13 provides an example in which the event of interest is an arrival of a car to a drive-up window at a fast food restaurant rather than the usual failure time of a repairable item. The intensity function $\lambda(t)$ models the *arrival rate* to the drive-up window, which has peaks at breakfast, lunch, and dinner times. The highest peak is at lunch when the intensity function is about 6 cars per hour. Each $\times$ along the time axis denotes an arrival time of a car to the drive-up window, and the clusters during the three meal times are apparent in the realization. The height of the intensity function is proportional to the probability of an arrival rate in the next instant. As was the case before, the arrival time values will vary from one realization to the next for the fixed intensity function illustrated in Figure 6.13.

The renewal process and the nonhomogeneous Poisson process are the two most popular point process models for modeling the underlying probability mechanism associated with the failure times of a repairable item. The two models are at the extremes of the repair action associated with a repairable item with a negligible repair time. One can think of the failures and repairs in a renewal process as *perfect repairs*, in which the item is completely restored to a new item. One can think of the failures and repairs in a nonhomogeneous Poisson process as *minimal repairs*, in which the item continues along the same intensity function track that was in play prior to the failure. The terms perfect repair and minimal repair for an item (that is, a component or a system) are defined next.

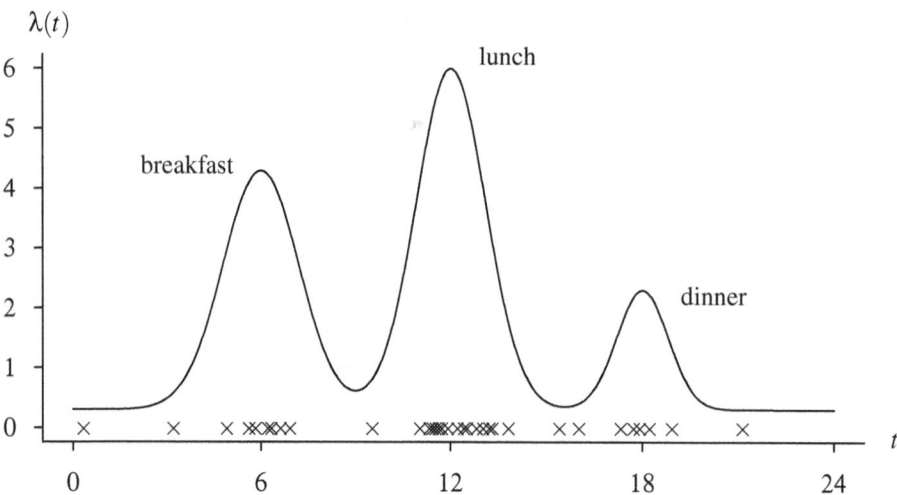

Figure 6.13: Intensity function for arrivals to a fast food restaurant drive-up window.

> **Definition 6.7** A *perfect repair* corresponds to a repair action that returns a failed item to a like new state in terms of its lifetime distribution.

There are two ways to think about a perfect repair. One way to perform a perfect repair is to discard the failed item and simply replace it with a new item. This is the case with *replacement* or *socket* model. A second way to perform a perfect repair is to perform the repair in a manner which makes the item as good as new with respect to its lifetime distribution. Regardless of which of these options occurs in practice, a renewal process is the appropriate probabilistic model to capture the time evolution of the repairable item. If all repairs on an item are perfect, then the times between failure are mutually independent and identically distributed random variables.

> **Definition 6.8** A *minimal repair* corresponds to a repair action that restores a failed item to the same condition as it was just prior to the failure in terms of its future risk of failure.

A nonhomogeneous Poisson process model can provide a reasonable underlying probability model for the failure sequence for a series system comprised of hundreds, or even thousands of components with roughly equal reliabilities. A component which fails and is replaced or repaired leaves the system in nearly the same condition as it was just prior to the failure. The failed component is such a small part of the overall system that using a minimal repair is appropriate.

Figure 6.14 shows the relationship between the three point process models that have been presented in this section. A renewal process is defined by the probability distribution of the time between events. These events are failures in reliability modeling. This probability distribution can be defined by any of the five lifetime distribution representations defined in Section 4.1. Figure 6.14 uses the hazard function to define the probability distribution of the time between failures. A renewal process collapses to a homogeneous Poisson process with positive rate λ when

$$h(t) = \lambda \qquad t > 0.$$

A nonhomogeneous Poisson process can be defined by the intensity function $\lambda(t)$ or the cumulative intensity function $\Lambda(t)$. Figure 6.14 uses the intensity function $\lambda(t)$ to define the probabilistic mechanism governing the failure times. A nonhomogeneous Poisson process collapses to a homogeneous Poisson process with positive rate λ when

$$\lambda(t) = \lambda \qquad t > 0.$$

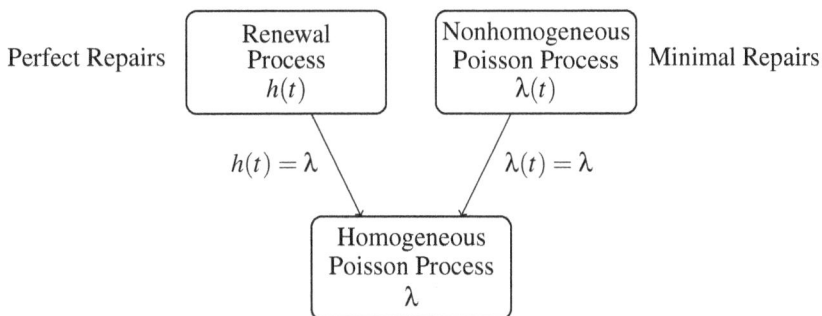

Figure 6.14: Relationships between point processes.

The hazard function was introduced in Section 4.1 as a rate of failure using limits. The conditional intensity function can be defined in a similar fashion. Let $\mathcal{H}_{t-}$ represent the history of an item from time zero until just prior to time t. The most informative way to think of $\mathcal{H}_{t-}$ is to consider it a record of the failure times associated with the counting process $N(t)$ for all time values from zero until just prior to t. Given this history, we can define the conditional intensity function as

$$\lambda(t \mid \mathcal{H}_{t-}) = \lim_{\Delta t \to 0} \frac{P(\text{failure in the interval } (t, t + \Delta t] \mid \mathcal{H}_{t-})}{\Delta t}.$$

The conditional intensity function for an item having independent and identically distributed times to failure in the IFR class and perfect repairs is illustrated in Figure 6.15. This corresponds to the risk profile associated with one realization of a renewal process. Each $\times$ on the time axis corresponds to a failure and repair. Each failure and repair restores the item to a like new state, so the conditional intensity function evolves after the failure like that of a new item.

The opposite extreme in terms of repair action is illustrated in Figure 6.16. This corresponds to the risk profile associated with one realization of a nonhomogeneous Poisson process. Each $\times$ on the time axis corresponds to a failure and a minimal repair which occurs in a negligible period of time. Each failure and minimal repair takes the item to the same condition as it was just prior to the failure in terms of its future risk. So each failure does not change the trajectory of $\lambda(t)$, which corresponds to a deteriorating item in this illustration.

There have been several schemes proposed for interpolating between renewal processes (to model perfect repairs) and nonhomogeneous Poisson processes (to model minimal repairs). One such scheme assigns a probability p to replacement of the entire item with a new item, which corresponds to a perfect repair, so that replacement or repair of just one of many components, which corresponds to a minimal repair, occurs with probability $1 - p$. The extreme case of $p = 0$ corresponds to a nonhomogeneous Poisson process; the extreme case of $p = 1$ corresponds to a renewal process.

One final topic, *superpositioning*, can be applied to any of the three point process models considered thus far. Poisson, renewal, and nonhomogeneous Poisson processes are useful for modeling

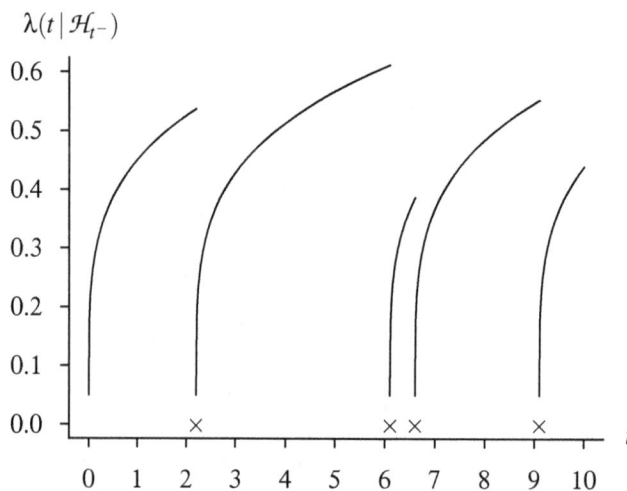

Figure 6.15: Conditional intensity function for perfect repairs.

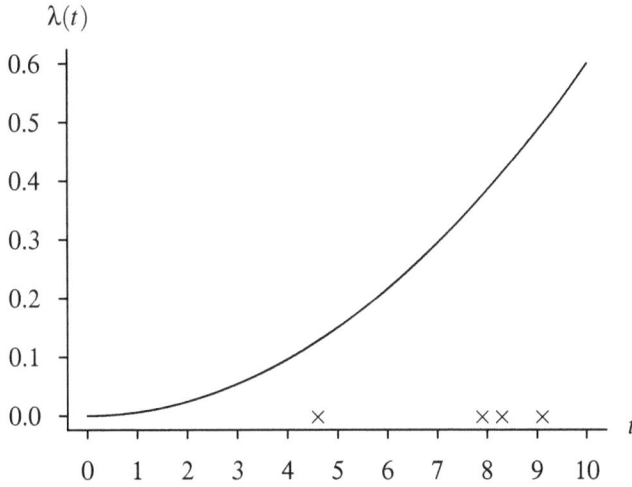

Figure 6.16: Conditional intensity function for minimal repairs.

the failure pattern of a *single* repairable item. In some situations it is important to model the failure pattern of several items simultaneously. Examples include a job shop with k machines, a military mission with k weapons, or an item failing from one of k causes of failure. Figure 6.17 shows a superposition of the failure times of $k = 3$ items. The bottom axis contains the superposition of the three point process realizations on the top three axes. The superposition of several point processes is the ordered sequence of all failures that occur in any of the individual point processes. An important result that applies to superpositions of nonhomogeneous Poisson processes is: if $\lambda_1(t), \lambda_2(t), \ldots, \lambda_k(t)$ are the intensity functions for k independent items, then the intensity function for the superposition is $\lambda(t) = \sum_{i=1}^{k} \lambda_i(t)$ for $t > 0$. This result is similar to the result concerning the hazard functions for net lives in competing risks.

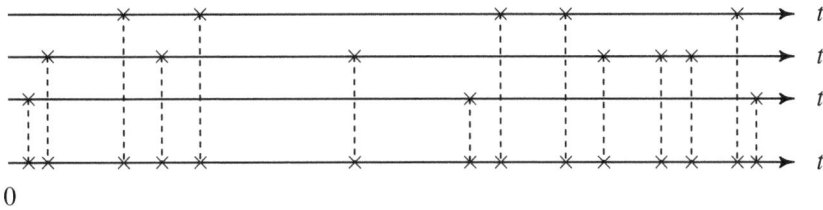

Figure 6.17: Superposition of three point processes.

This ends the presentation of the three point process models considered here: Poisson processes, renewal processes, and nonhomogeneous Poisson processes. The first two models are only capable of modeling socket models for which the time between failures has a common distribution, whereas nonhomogeneous Poisson processes are capable of modeling improving and deteriorating systems, which are more common in practice. All three of these models are appropriate when there is a negligible down time (that is, failure and return to service occur at essentially the same point in time).

6.4   Exercises

6.1   The failure times of $n = r = 20$ electric generators (in hours) placed on an accelerated life test are

$$
\begin{array}{ccccccccc}
7.5 & 121.5 & 279.8 & 592.1 & 711.5 & 848.2 & 1051.7 & 1425.5 & 1657.2 \\
1883.6 & 2311.1 & 2951.2 & 5296.6 & 5637.9 & 6054.3 & 6303.9 & 6853.7 \\
& & 7201.9 & 9068.5 & 10{,}609.7,
\end{array}
$$

as given on page 101 of Zacks, S., *Introduction to Reliability Analysis: Probabilistic Models and Statistical Methods*, Springer–Verlag, Inc., New York, 1992.

(a) Assuming that the time to failure of the population of generators at the accelerated conditions has the exponential distribution, find a point and 95% confidence interval estimate for the probability of survival to 6000 hours.

(b) Using nonparametric methods, find a point and 95% confidence interval estimate for the probability of survival to 6000 hours under the accelerated conditions.

6.2   Show that the product–limit estimate reduces to the survivor function estimate for a complete data set when the failure times are distinct.

6.3   Show that the diagonal elements of the observed information matrix associated with the product–limit estimate are

$$
\left[-\frac{\partial^2 \ln L\big(h(y_1), h(y_2), \ldots, h(y_k)\big)}{\partial h(y_i)^2} \right]_{h(y_i) = d_i / n_i} = \frac{n_i^3}{d_i(n_i - d_i)}
$$

for $i = 1, 2, \ldots, k$.

6.4   Find a point estimate and an approximate two-sided 95% confidence interval estimate for the probability that the remission times exceed 20 weeks for the control and treatment groups of the 6–MP data set from Example 5.6. Are there any conclusions that can be drawn from the two confidence intervals concerning the drug 6–MP's influence on remission times exceeding 20 weeks?

6.5   Thirteen aircraft components are placed on a life test that is discontinued after the tenth failure. The failure times, in hours, are

$$
\begin{array}{cccccccccc}
0.22 & 0.50 & 0.88 & 1.00 & 1.32 & 1.33 & 1.54 & 1.76 & 2.50 & 3.00,
\end{array}
$$

as given on page 43 of Crowder, M.J., Kimber, A.C., Smith, R.L., and Sweeting, T.J., *Statistical Analysis of Reliability Data*, Chapman and Hall, New York, 1991.

(a) Assuming that the time to failure of the components in the population has the exponential distribution, find a point and 95% confidence interval estimate for the probability of survival to 1.6 hours.

(b) Using nonparametric methods, find a point and 95% confidence interval estimate for the probability of survival to 1.6 hours.

6.6 The survival times (in weeks) of patients with acute myelogenous leukemia in the nonmaintained group are

$$5, 5, 8, 8, 12, 16^*, 23, 27, 30, 33, 43, 45,$$

where the $*$ superscript signifies a right-censored observation, as given on page 49 of Miller, R., *Survival Analysis*, John Wiley & Sons, Inc., New York, 1981. Give a nonparametric estimator of the expected survival time of a patient with acute myelogenous leukemia in the nonmaintained group.

6.7 The failure times (in minutes) of electrical insulating fluid subjected to constant voltages below are from Nelson, W. B., "Graphical Analysis of Accelerated Life Test Data with the Inverse Power Law Model," *IEEE Transactions on Reliability*, Vol. R–21, pp. 2–11, 1972.

Voltage (kV)	Failure time (minutes)							
28	68.85	426.07	110.29	108.29	1067.6			
30	17.05	22.66	21.02	175.88	139.07	144.12	20.46	43.40	194.90
	47.30	7.74						
32	0.40	82.85	9.88	89.29	215.10	2.75	0.79	15.93	3.91
	0.27	0.69	100.58	27.80	13.95	53.24		

(a) Plot a nonparametric survivor function estimate for each of the three voltage levels on a single set of axes.

(b) Fit the accelerated life model with an exponential baseline to the failure times. Report the maximum likelihood estimators and 95% confidence intervals (based on the observed information matrix) for all unknown parameters. Also, predict the mean time to failure at 26 kV voltage.

6.8 The lifetimes, in days, on $n = 10$ identical pieces of equipment are

$$2 \quad 72^* \quad 51 \quad 60^* \quad 33 \quad 27 \quad 14 \quad 24 \quad 4 \quad 21^*,$$

as given in Lawless, J.F., *Statistical Models and Methods for Lifetime Data*, 2nd ed., John Wiley & Sons, Inc., Hoboken, NJ, 2003. The asterix denotes a right-censored observation. Assume that a random right censoring scheme is appropriate.

(a) Find a nonparametric point estimate for $S(25)$.

(b) Find a nonparametric 95% confidence interval for $S(25)$.

(c) Use the relationship $H(t) = -\ln S(t)$ to determine a point estimate for $H(25)$ using your solution to part (a).

(d) One other most popular technique for estimating $H(t)$ is the Nelson–Aalen estimator

$$\hat{H}(t) = \sum_{j \mid y_j \leq t} \frac{d_j}{n_j},$$

where y_j, d_j, and n_j have the same meaning as in the Kaplan–Meier product–limit estimate. Give a point estimate for $H(25)$ using the Nelson–Aalen estimator.

(e) Give a point estimate for $S(25)$ using the Nelson–Aalen estimator.

6.9   Find all possible values of the Kaplan–Meier product–limit estimate for the survivor function for n distinct failure and censoring values for $n = 1, 2, \ldots, 10$ and display these values in a graphic. Research in this area has been conducted by Qin, Y., Sasinowska, H., Leemis, L., "The Probability Mass Function of the Kaplan–Meier Product–Limit Estimator," Forthcoming, *The American Statistician*, 2023.

6.10   One nonparametric technique for estimating the cumulative hazard function $H(t)$ is the Nelson–Aalen estimator:

$$\hat{H}(t) = \sum_{j \mid y_j \leq t} \frac{d_j}{n_j},$$

where y_j, d_j, and n_j have the same meaning as in the Kaplan–Meier product–limit estimate. The observed failure times (in hours) of $n = 4156$ integrated circuits placed on a test that was terminated at 1370 hours given on page 5 of Meeker, W.Q., Escobar, L.A., Pascual, F.G., *Statistical Methods for Reliability Data*, 2nd ed., John Wiley & Sons, Inc., New York, 2022 are given in the table below. The ordered observed failure times are arranged in a row-wise fashion.

0.10	0.10	0.15	0.60	0.80	0.80	1.20
2.50	3.00	4.00	4.00	6.00	10.00	10.00
12.50	20.00	20.00	43.00	43.0	48.00	48.00
54.00	74.00	84.00	94.00	168.00	263.00	593.00

Give a point estimate for $H(0.5)$ using the Nelson–Aalen estimator.

6.11   Calculate the test statistic and the p-value for the log-rank test associated with the leukemia remission times in the control and treatment groups for the clinical trial involving 6–MP.

6.12   The remission times (in weeks) for 40 leukemia patients, with 20 patients selected at random and assigned to treatment A:

$$1, 3, 3, 6, 7, 7, 10, 12, 14, 15, 18, 19, 22, 26, 28^*, 29, 34, 40, 48^*, 49^*$$

and the other 20 patients assigned to treatment B:

$$1, 1, 2, 2, 3, 4, 5, 8, 8, 9, 11, 12, 14, 16, 18, 21, 27^*, 31, 38^*, 44.$$

are given on page 346 of Lawless, J.F., *Statistical Models and Methods for Lifetime Data*, 2nd ed., John Wiley & Sons, Inc., Hoboken, N.J., 2003. Conduct the log-rank test to compare the survivor functions for the two populations and report the appropriate p-value.

6.13   Three independent risks act on a population. The net lives, X_j, are exponential(λ_j), for $j = 1, 2, 3$. Find

(a) $q_j(a, b)$,

(b) $Q_j(a, b)$,

(c) π_j,

for $j = 1, 2, 3$.

6.14 A component can fail from one of two identifiable causes. Ken collects a large number of failures of both types and determined that

$$\pi_1 = 0.73 \qquad \pi_2 = 0.27$$

and the crude lives have probability density functions

$$f_{Y_1}(t) = 1 \qquad 0 \le t \le 1$$

and

$$f_{Y_2}(t) = \theta t^{\theta-1} \qquad 0 \le t \le 1$$

for a positive parameter θ. Find the hazard functions of the net lives, X_1 and X_2, assuming that the risks are independent.

6.15 In a competing risks model, the distributions of the k crude lifetimes are exponential with identical parameter λ. In addition, the probabilities of failure from each of the risks (the π_j's) are known. Assuming that the net lives are independent, find

(a) the mean lifetime of the item,

(b) the mean lifetime of the item if risk j is eliminated, for $j = 1, 2, \ldots, k$.

6.16 Assume that the following are known in a competing risks model with two causes of failure:

$$\pi_1 = \frac{1}{4} \qquad \pi_2 = \frac{3}{4} \qquad S_{Y_1}(t) = e^{-\alpha t} \qquad S_{Y_2}(t) = e^{-(\lambda t)^\kappa}$$

for $t \ge 0$. Assuming that the risks are independent, find

(a) the hazard function for the first net lifetime,

(b) an expression for $E[T]$,

(c) an expression for the expected time to failure if risk 2 is removed.

The solutions to some parts of this problem might not be closed form.

6.17 If $X_1, X_2, \ldots, X_k$ are independent net lives and $X_j \sim$ exponential (λ_j) for $j = 1, 2, \ldots, k$, find $\pi_j = P(X_j = \min\{X_1, X_2, \ldots, X_k\})$ for $j = 1, 2, \ldots, k$.

6.18 Beth considers independent net lives X_1 and X_2, where $X_j \sim$ Weibull(λ_j, κ) for $j = 1, 2$. Find

$$\pi_1 = P(X_1 = \min\{X_1, X_2\}).$$

6.19 Rick uses a competing risks model with two independent risks C_1 and C_2. The net lifetime for risk 1 has a log logistic distribution with parameters λ_1 and κ_1. The net lifetime for risk 2 has a log logistic distribution with parameters λ_2 and κ_2. Write expressions for π_1 and π_2.

6.20 Let T be the lifetime of an item that is subject to three independent competing risks, for which the hazard functions for the crude lives are

$$h_{Y_1}(t) = \frac{a}{t+\alpha} \qquad t \ge 0,$$

$$h_{Y_2}(t) = bt \qquad t \ge 0,$$

and

$$h_{Y_3}(t) = \lambda \qquad t \ge 0$$

for positive parameters a, α, b, and λ. Given the values of π_1, π_2, and π_3, give an expression for the hazard function for the first net lifetime.

6.21 Consider a competing risks model with $k = 2$ independent risks. The first net lifetime, X_1, has a Weibull distribution with parameters λ_1 and κ_1. The second net lifetime, X_2, has a Weibull distribution with parameters λ_2 and κ_2. Find the expected remaining lifetime for an item that is operating at time t_0. Simplify your answer as much as possible.

6.22 Consider a competing risks model with $k = 2$ risks. Let X_1 be the net lifetime associated with risk 1 and let X_2 be the net lifetime associated with risk 2. The joint probability density function of X_1 and X_2 is

$$f(x_1, x_2) = 1 \qquad (x_1, x_2) \in A,$$

where A is the triangular region determined by connecting the points $(1, 2)$, $(3, 2)$, and $(3, 1)$ in the (x_1, x_2) plane with lines.

(a) Find π_1.
(b) Find the survivor function of $T = \min\{X_1, X_2\}$.

6.23 In a competing risks model with $k = 2$ independent risks, give an expression for $Q_2(a, b)$, where both net lives have Weibull distributions, that is, X_1 has a Weibull distribution with parameters λ_1 and κ_1 and X_2 has a Weibull distribution with parameters λ_2 and κ_2. Evaluate $Q_2(a, b)$ to four decimal places when $a = 1$, $b = 2$, $\lambda_1 = 3$, $\kappa_1 = 1/3$, $\lambda_2 = 4$, and $\kappa_2 = 1/4$.

6.24 Consider the lifetime T having the *bi-Weibull* distribution with survivor function

$$S(t) = e^{-(\lambda_1 t)^{\kappa_1} - (\lambda_2 t)^{\kappa_2}} \qquad t \geq 0,$$

where λ_1, λ_2, κ_1, and κ_2 are positive parameters. The distribution has a bathtub-shaped hazard function if $\min\{\kappa_1, \kappa_2\} < 1 < \max\{\kappa_1, \kappa_2\}$. Find the time value where $h'(t) = 0$ for a bi-Weibull distribution with a bathtub-shaped hazard function.

6.25 Bonnie models the lifetime of an automobile using two dependent competing risks. The first risk is from accidents and the second risk is from all other causes. The joint survivor function of the two net lives associated with the two risks on their support is

$$S(x_1, x_2) = (1 - x_1/2)(1 - x_2/2)\left(1 + x_1^2 x_2/8\right) \qquad 0 < x_1 < 2, 0 < x_2 < 2,$$

where x_1 and x_2 are the odometer readings measured in hundreds of thousands of miles.

(a) Find the marginal survivor function for X_1.
(b) Find the joint probability density function of the net lifetimes.
(c) Find π_2.
(d) Find $Q_2(0.5, 1.2)$.
(e) Find the expected lifetime of the automobile.
(f) Perform a Monte Carlo experiment to support your solution to part (e).

6.26 The formula for the hazard function of the ith net life in a competing risks model is

$$h_{X_i}(t) = \frac{1}{S(t, t, \ldots, t)}\left[\frac{-\partial S(x_1, x_2, \ldots, x_k)}{\partial x_i}\right]_{x_1 = x_2 = \cdots = x_k = t}$$

for $t \geq 0$ and $i = 1, 2, \ldots, k$. Find $h_{X_1}(t)$ for $k = 2$ risks and joint survivor function

$$S(x_1, x_2) = e^{-\lambda_1 x_1 - \lambda_2 x_2 - \lambda_3 x_1 x_2} \qquad x_1 \geq 0, x_2 \geq 0,$$

for $\lambda_1 > 0$, $\lambda_2 > 0$, and $\lambda_3 > 0$.

6.27 An item is subject to three competing risks with the associated three independent net life-times: $X_1 \sim$ Weibull$(1, 2)$, $X_2 \sim$ exponential(1), and $X_3 \sim$ Weibull$(1, 3)$. Write computer code in a language of your choice to provide the numerical methods necessary to compute

(a) the mean time to failure of the item,

(b) the mean time to failure of the item with each of the risks eliminated individually.

In addition, calculate the mean of each net lifetime and provide a plausible explanation of why the net lifetime with the largest mean corresponds to the risk that has the greatest impact on $E[T]$ when it is eliminated.

6.28 An item can fail from one of two competing risks. The first net lifetime is associated with accidents and is modeled by $X_1 \sim$ exponential(λ). The second net lifetime is associated with wear out and is modeled by the random variable X_2 with hazard function

$$h_{X_2}(t) = \beta t \qquad t \geq 0,$$

for $\beta > 0$. The net lifetimes are independent random variables.

(a) Give an expression for the mean time to failure of the item; that is, find $E[T]$, where $T = \min\{X_1, X_2\}$. This expression will not be in closed form.

(b) Give an expression for π_1. This expression will also not be in closed form.

(c) Use numerical methods to calculate the two quantities given in parts (a) and (b) to seven digits when $\lambda = 2$ and $\beta = 1$.

(d) Use Monte Carlo simulation to support your solutions in part (c).

6.29 A repairable item with negligible repair time fails according to a Poisson process with rate $\lambda = 0.001$ failures per hour. Find the probability of two or fewer failures between 3000 and 6000 hours.

6.30 A repairable item with negligible repair time fails according to a renewal process with inter-failure time having the gamma distribution with parameters λ and κ. Find the probability of n or fewer failures between times 0 and c.

6.31 Verify that the derivative of the renewal equation is satisfied when the items in a socket model have exponential lifetimes.

6.32 Consider a renewal process for which the times between failures have the Weibull distribution with scale parameter λ and shape parameter κ. Find the expected value and population variance of the time of failure n for $n = 1, 2, \ldots$.

6.33 A repairable item with negligible repair time fails according to a nonhomogeneous Poisson process with intensity function $\lambda(t) = 0.001 + 0.000001t$ failures per hour, for $t > 0$. Find the probability of two or fewer failures between 3000 and 6000 hours.

6.34 For a nonhomogeneous Poisson process with power law intensity function

$$\lambda(t) = \lambda\kappa(\lambda t)^{\kappa-1} \qquad t > 0,$$

find the probability mass function for the number of events between times a and b.

6.35   Consider an age replacement policy model for which items are replaced at failure or at time c, whichever comes first. Assuming that the time to failure has the Pareto distribution with parameters λ and κ, find the expected number of failures by time b. Assume that $\lambda \ll c \ll b$.

6.36   Two different maintenance procedures are to be compared for a repairable item: age replacement and block replacement. Assume that the item has a lifetime that is a mixture of three distributions: a Weibull distribution with $\lambda = 0.01$ and $\kappa = 0.6$ (with $p_1 = 0.05$), an exponential distribution with $\lambda = 0.002$ (with $p_2 = 0.45$), and a Weibull distribution with $\lambda = 0.001$ and $\kappa = 3.0$ (with $p_3 = 0.50$). Assume that time is measured in hours.

(a) Calculate the theoretical mean lifetime of the item.

(b) Use Monte Carlo simulation to compare the age replacement and block replacement maintenance strategies for the item with $c = 1000$ hours.

6.37   Georgie drives a car whose failure times are governed by a nonhomogeneous Poisson process with power law cumulative intensity function

$$\Lambda(t) = (\lambda t)^\kappa \qquad t > 0,$$

where t is measured in miles. If the car has 100,000 miles on the odometer, find the probability that Georgie can make a 1000-mile trip without a failure.

6.38   Bedrock Motors, Inc. is introducing their new "Tyrano-Taurus Rex" automobile, complete with a three-year warranty. Each Rex has a failure mechanism governed by a nonhomogeneous Poisson process with power law intensity function

$$\lambda(t) = \lambda^\kappa \kappa t^{\kappa-1} \qquad t > 0,$$

where t is time (in years), and λ and κ are positive parameters. If Fred, Wilma, Barney, and Betty each buy a Rex, find the expected number of failures under warranty that Bedrock Motors will experience for these four Rexes.

6.39   Cynthia is going camping. She takes along a flashlight which requires two batteries in order to operate. Cynthia's batteries each have an exponential time to failure with a mean of $1/\lambda$. If Cynthia takes five batteries with her (two batteries in the flashlight and three spare batteries), what is the distribution of time that she will be able to use her flashlight? You may assume that (a) her battery replacement time is negligible, (b) her flashlight bulb never fails, and (c) she has a battery tester that allows her to determine which of the two batteries in the flashlight has failed. Write a short paragraph on the reasonableness of the assumption of exponential battery lifetimes.

6.40   The event times $T_1, T_2, \ldots$ in a nonhomogeneous Poisson process with cumulative intensity function $\Lambda(t)$ can be simulated with

$$T_i = \Lambda^{-1}(E_i)$$

for $i = 1, 2, \ldots$, where $E_1, E_2, \ldots$ are the event times in a unit homogeneous Poisson process. Use this result to simulate the event times in a nonhomogeneous Poisson process with cumulative intensity function $\Lambda(t) = t^2$ for $0 < t < 2$. Provide convincing numerical evidence that you have correctly implemented the algorithm by conducting a Monte Carlo simulation experiment to estimate the number of events that have occurred by time $t = 1.5$.

6.41   Consider a nonhomogeneous Poisson process with intensity function $\lambda(t)$ and cumulative intensity function $\Lambda(t)$ defined on the time interval $0 < t < s$, where s is a prescribed, fixed, positive real number. Define the scaled intensity function as

$$\lambda^\star(t) = \frac{\lambda(t)}{\Lambda(s)} \qquad 0 < t < s.$$

(a) Show that $\int_0^s \lambda^\star(t)\,dt = 1$.

(b) Find the functional forms of $\lambda^\star(t)$ for the following intensity functions:
   - $\lambda(t) = \lambda$ for $0 < t < s$,
   - $\lambda(t) = \kappa \lambda^\kappa t^{\kappa-1}$ for $0 < t < s$,
   - $\lambda(t) = \frac{\lambda\kappa(\lambda t)^{\kappa-1}}{1+(\lambda t)^\kappa}$ for $0 < t < s$,

   for positive parameters λ and κ.

6.42   Barbara models the failure times of a digital camera by a nonhomogeneous Poisson process. Previous data has revealed that the intensity function for the times of warranty claims is well-modeled by the intensity function

$$\lambda(t) = 0.124t \qquad t > 0,$$

where time is measured in years. The camera company is considering offering three consecutive one-year term warranties: one upon purchase of the camera, a second after one year of use, and the third after two years of use. In order to be competitive, the camera company has decided to make no profit on their warranty policies. Give the three revenue-neutral premiums that a customer has to pay for these warranties. Make the following assumptions to make your calculations simpler.

- Each repair costs exactly $100.
- A repair is instantaneous.
- Ignore the effect of the time value of money.
- Each repair is a *minimal repair* in the sense that the repair to the camera does not reset the intensity function to $t = 0$ but rather the camera's age and intensity function are unaltered by the repair.

6.43   A truck requires a particular nonrepairable electrical component that has an exponential lifetime with a positive failure rate λ failures per hour. A site supports a large fleet of n trucks, each operating 24 hours a day, 7 days a week. A parts manager can make an order for spare parts once a week. Assuming that the lead time is 0 (that is, immediate delivery), what is the minimum number of parts that should be ordered up to each week to ensure that there is a probability of at least 0.9999 that a truck is not down for lack of this particular nonrepairable electrical component?

(a) Write a paragraph describing an algorithm that the parts manager should select an order quantity.

(b) Apply the algorithm from part (a) for $n = 20$ trucks and $\lambda = 0.001$ failures per hour.

(c) Support your solution to part (b) via Monte Carlo simulation.

6.44   Natasha models power failures in the state of Virginia during the month of March with a Poisson process with rate $\lambda = 7$ failures per month. Given that there have been a total of 4 failures during the first 10 days of March, what is the probability that Virginia will have more than 12 power outages during the entire month of March?

6.45   The number of annual failures of a particular brand of carburetor is $X_1 \sim \text{Poisson}(\lambda_1)$. The number of annual failures of a second brand of carburetor is $X_2 \sim \text{Poisson}(\lambda_2)$. Assuming that the number of annual failures of the two types of carburetors are independent and that there are n annual failures observed for both types of carburetors (that is, $X_1 + X_2 = n$), what is the probability distribution of the number of failures of the first carburetor during that particular year?

6.46   Which repairable system described below is the best candidate for being an improving system?

     (a) Automobile.

     (b) Wooden chair.

     (c) Blender.

     (d) Operating system.

     (e) Lawn mower.

6.47   Consider the three-component series system of repairable components with four cold-standby spares depicted below. All components are identical with failure rates 0.005 failure per hour. In this particular system, the failure detection and switching times are negligible. There is a repair facility with two repairmen. The repair rate is 0.01 repair per hour, and only one repairman can work on a failed component at a time. Find the expected time to system failure (that is, the expected time when there are fewer than three operating components) assuming all components are new at $t = 0$.

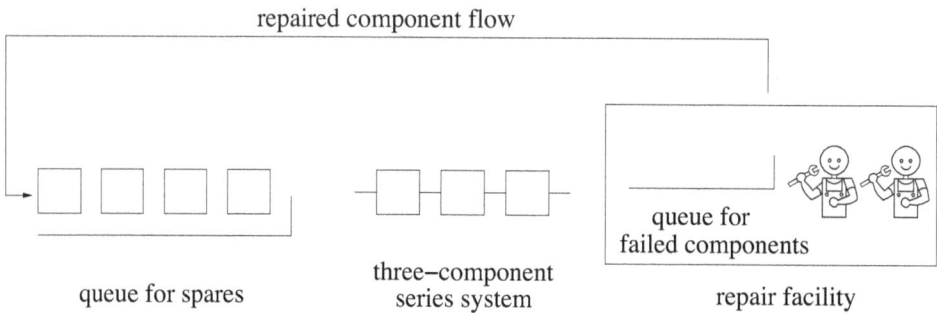

repaired component flow

queue for spares   three–component series system   queue for failed components   repair facility

Part III

TIME SERIES ANALYSIS

Chapter 7

Time Series Basics

This chapter defines a time series, describes where a time series falls with respect to other stochastic processes, introduces some basic properties of time series, and introduces some basic operations that can be applied to a time series. The presentation given here does not replace a full-semester class on time series analysis, but does provide an elementary introduction to the topic. Each section ends with a brief review of the computational tools available in R for time series analysis.

7.1 The Big Picture

A time series is a sequence of observed data values that are collected over time. The analysis of a time series is an important sub-discipline of statistics and data science that has applications in a variety of areas, including economics, business, science, and engineering.

Classical statistical methods rely on the assumption that the data values collected constitute a simple random sample, which implies that data values are realizations of mutually independent and identically distributed random variables. This is nearly universally not the case in time series analysis because nearby observations collected over time tend to be correlated. Special probability models and associated statistical methods have been developed to account for this correlation. When the focus is on the correlation between observations in the time series, analysis tools from the *time domain* are employed. When the focus is on the periodic behavior in the time series, analysis tools from the *frequency domain* are employed. We begin our exploration of time series with a subsection containing examples.

7.1.1 What is a Time Series?

The essence of a time series is best captured in a sequence of examples. The first example is the monthly number of kilowatt hours required for powering the utilities in my home from 2011 through 2018. The second example is the monthly number of international airline passengers between 1949 and 1960. The third example is a realization of what is known as *Gaussian white noise*. The fourth example is a realization of what is known as a *random walk*.

> **Example 7.1** The monthly number of kilowatt hours to power my home in Williamsburg, Virginia between 2011 and 2018 are given in Table 7.1. Scan the table carefully to see if you can determine any patterns in the time series. To provide more context, here is a little more information about the house that my family lived in between 2011

Year	Jan	Feb	Mar	Apr	May	Jun	Jul	Aug	Sep	Oct	Nov	Dec
2011	2731	1822	1189	1229	1260	2204	2518	1960	1032	788	1508	2279
2012	1667	1695	1220	872	1189	1164	1851	1789	1370	962	1716	1678
2013	2254	2362	1916	1293	1253	1635	1809	1562	1348	872	1290	2619
2014	3089	2217	2072	1270	1543	1642	1892	1688	1658	984	1609	2577
2015	2712	3363	1887	1494	1260	1127	1865	1741	1430	1588	1535	1626
2016	3004	2344	1969	1431	1456	2029	2294	2036	2173	1132	1834	1713
2017	2583	1810	1728	1145	1253	1696	1936	1875	956	1010	1751	1506
2018	3698	1767	1871	1270	966	1141	1463	1452	1484	1043	1378	1499

Table 7.1: Number of kilowatt hours to power a home.

and 2018. There were no additions made to the home during these years, nor were there any new windows or insulation installed. The two-zone heat pump system that cools the house in the summer and heats the house in the winter did not change during these years. The smallest value in the series occurred in October 2011, when 788 kilowatt hours were consumed. The largest value in the series occurred in January 2018, when a spectacular 3698 kilowatt hours were consumed. January 2018 was one of the coldest Januarys on record in Virginia, which caused the spike in kilowatt hours consumed. The pipes in my neighbor's house burst on one cold night in January. Viewed in table form, the data just looks like a mass of numbers. But plotting the data on a time axis reveals some of the patterns associated with the data values over time.

We use R to plot the observations over time. The first step is to get the time series observations into an R vector. This can be done for a small data set with the c function. The R statement below places the time series into the vector kwh in chronological order.

```
kwh = c(2731, 1822, 1189, ... , 1499)
```

Alternatively, if the data set is large and is contained in an external file, the scan function can be used to read the time series observations from the external file as

```
kwh = scan("kwh.d")
```

where kwh.d is a file in the current working directory that contains the energy consumption values in kilowatt hour values, one per line.

The next step is to use the ts function to convert the data values in the vector kwh to a time series object, which will allow us to use many of the R time series analysis operators included in the base language.

```
kwh.ts = ts(kwh, frequency = 12, start = c(2011, 1))
```

Setting the frequency argument in ts to 12 lets R know that the data is collected monthly. If the time series consisted of quarterly observations, for example, then frequency would be set to 4. Setting the start argument to c(2011, 1) lets R know that the time series starts in January of 2011. If the first observation in the time series was sampled in March of 2013, for example, then start would be set to c(2013, 3).

The next step is to plot the time series, which can be done with the plot.ts function.

```
plot.ts(kwh.ts)
```

The time series plot is shown in Figure 7.1. The horizontal axis is the time t, measured in years. The tick marks on the horizontal axis correspond to the January observation from each year. The vertical axis is the observed number of kilowatt hours consumed at time t, denoted by x_t. The points in the time series are connected with lines, but this is largely a matter of personal taste.

Plotting the time series in this fashion is a critical initial step in the analysis of an observed time series. Carefully examining this plot often informs the analyst of the type of probability model that might be appropriate. Unusually small and large observations should be carefully assessed to determine whether the x_t value was recorded properly. Shifts in the heights of the x_t values might correspond to events associated with the time series, such as installing a more efficient heat pump or adding a new room to the house for kwh.ts. The plot also allows the analyst to inspect the time series for any *trends* which correspond to gradual increases or decreases in the mean level of the process. In addition, the plot allows the analyst to identify any *seasonal* components, that is, fluctuations which are periodic in nature, that might be present in the time series. Some time series have a change in the variability of the observations that may also be identified in the plot.

There are some preliminary conclusions that can be drawn from the time series plot in Figure 7.1. First, there does appear to be a 12-month cyclical pattern, that is surely influenced by the annual outdoor temperature cycles. The peak energy consumption typically occurs in January. This is consistent with the fact that heat pumps have difficulty during the winter months because there is not much heat to pump, making them inefficient. Second, after accounting for the annual cycle, there does not seem to be any systematic increase or decrease to the amount of energy consumed over the 8-year period. This is consistent with the fact that no energy improvements were made to the home during the 8-year period. There is, of course, short-term change in the mean value of the time series due to the seasonal component of the time series, but there does not

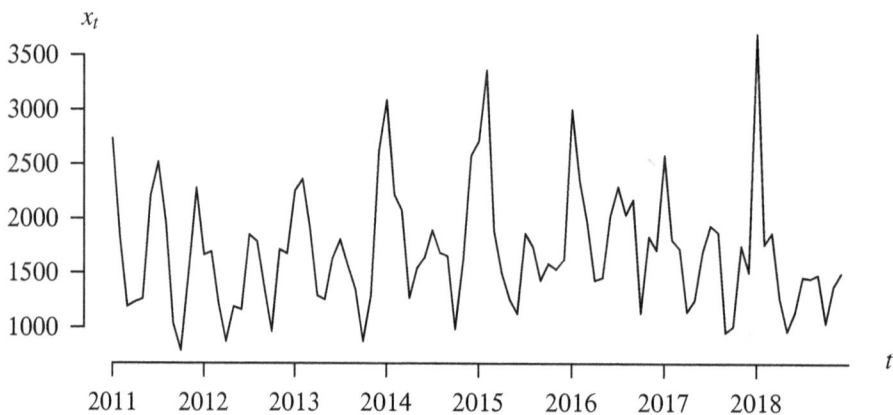

Figure 7.1: Single home energy consumption (in kilowatt hours) 2011–2018.

appear to be any long-term change in the mean value of the series. Third, there is significant random sampling variability associated with the time series. Although forecasts could be made from this data set for values of the time series into the future, they would be fairly imprecise predictions because of the random sampling variability. Some of the variability could be explained by the average outdoor monthly temperature during a particular month, with hot summer months and cold winter months requiring more energy to cool and heat the home. Some of the variability could also be explained by the fact that all months do not have the same number of days, which is easily accounted for. Time series analysts refer to the random sampling variability that remains after all of the *signal* has been accounted for as *noise*. The terms signal and noise are familiar terms in fields such as statistics, data science, astrophysics, and electrical engineering. The term *noise* is analogous to *error* from regression theory.

The home energy consumption example has shown that significant insight concerning a time series can be gleaned by an understanding of the context associated with the time series and the crucial step of making a simple plot of the data values over time. We will return to the home energy consumption time series later for further analysis. The next time series illustrates the increase in international airline travel between 1949 and 1960.

Example 7.2 The number of monthly international airline passengers, in thousands, between January 1949 and December 1960 resides in an R built-in data set named AirPassengers. All that is necessary to see the observations is to type the name of the data set. (Notice that all R commands are case sensitive.)

```
AirPassengers
```

The output is shown below.

```
     Jan Feb Mar Apr May Jun Jul Aug Sep Oct Nov Dec
1949 112 118 132 129 121 135 148 148 136 119 104 118
1950 115 126 141 135 125 149 170 170 158 133 114 140
1951 145 150 178 163 172 178 199 199 184 162 146 166
1952 171 180 193 181 183 218 230 242 209 191 172 194
1953 196 196 236 235 229 243 264 272 237 211 180 201
1954 204 188 235 227 234 264 302 293 259 229 203 229
1955 242 233 267 269 270 315 364 347 312 274 237 278
1956 284 277 317 313 318 374 413 405 355 306 271 306
1957 315 301 356 348 355 422 465 467 404 347 305 336
1958 340 318 362 348 363 435 491 505 404 359 310 337
1959 360 342 406 396 420 472 548 559 463 407 362 405
1960 417 391 419 461 472 535 622 606 508 461 390 432
```

Again, scan these data values and look for patterns. The lowest number of international air travelers was 104,000 in November 1949. The highest number of international air travelers was 622,000 in July 1960. Unlike the previous example, it is not necessary to convert AirPassengers from a vector to a time series object. Typing

```
str(AirPassengers)
```

shows that `AirPassengers` is already a time series object, as shown below, by using the `str` (structure) function in R.

```
Time-Series [1:144] from 1949 to 1961: 112 118 132 129 121 135 ...
```

This useful function tells you that there are $12 \times 12 = 144$ observations in the time series and lists the first few observations. By using the `help` function

```
help(AirPassengers)
```

additional information about the time series reveals that the observations are the number of international airline passengers (in thousands) per month during 1949–1960. Using the `plot.ts` function as in the previous example gives a graph of the time series over time, which is given in Figure 7.2.

Unlike the previous time series, this time series displays a trend. The number of international airline passengers is increasing over time. Although the number of passengers is increasing over time, it is not clear whether the increase is linear, quadratic, or exponential, and this would require further analysis to determine which functional form provides the best model for the increase. As was the case with the time series from the first example, there is also a 12-month cycle that is apparent in the data. The months of July and August–when school is typically not in session–tend to be the busiest months. The cyclic variation appears to be less sinusoidal in nature than the energy consumption time series because there is not a sinusoidal external time series (outdoor temperature) driving the international airline travel time series.

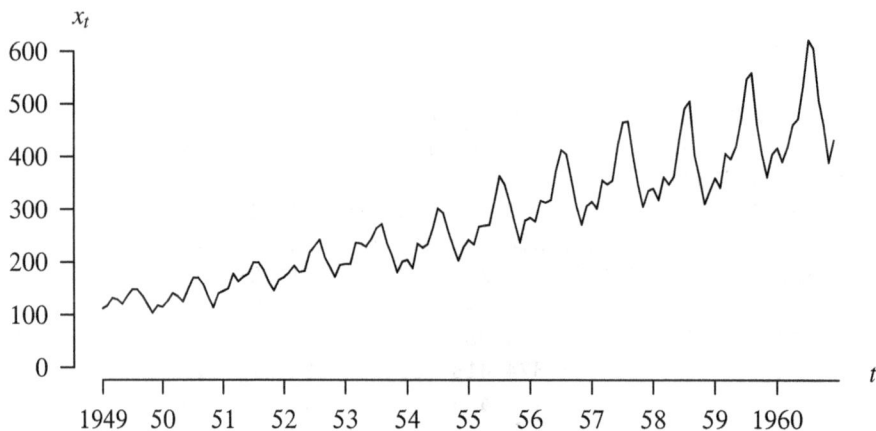

Figure 7.2: International airline passengers (in thousands) 1949–1960.

Before describing a third example of a time series, we define some notation. A common convention in probability theory is to use uppercase letters, such as X and Y, to denote random variables. When encountering a time series, it is often helpful to subscript the random variable denoting the time series observations with t, for time, as X_t. So a time series consisting of the time-ordered observations $X_1, X_2, \ldots, X_n$ can be referred to in the abstract as $\{X_t\}$, which is more compact. But when referring to a specific realization of a time series (either collected as data or generated by a computer via simulation) consisting of the time-ordered observations $x_1, x_2, \ldots, x_n$, we switch to the lowercase version $\{x_t\}$. This is why the vertical axes in Figures 7.1 and 7.2 were labeled x_t for the $n = 96$ home energy consumption observations from Example 7.1 and the $n = 144$ airline passenger counts from Example 7.2. The notation developed in this paragraph will be apparent in the formal definition of noise, which is defined next.

Instead of analyzing a realization of a time series as in the previous two examples, time series analysts often formulate and fit a *probability model* for a time series. This process is roughly the time series equivalent of fitting a univariate probability distribution, such as the normal distribution, to a data set. We will again refer to the time series as $\{X_t\}$ when constructing such a time series model, but when the index values for t are not obvious by context, we add the additional parameter T, which is the set of allowable values for t using $\{X_t, t \in T\}$. The set T will almost universally be either the set of all integers (when it is necessary to consider observations with negative t values) or the set of all nonnegative integers (when a time origin is necessary).

Most time series cannot be described by a deterministic function. In order to inject randomness into the time series model, it has become common practice to define *noise*, which consists of random shocks that will make a time series model stochastic rather than deterministic. Three varieties of noise used by time series analysts are defined next. In some application areas, noise might be referred to as *error* or *disturbance*.

Definition 7.1 The time series $\{Z_t\}$ that is a sequence of mutually independent random variables $Z_1, Z_2, \ldots, Z_n$, each with population mean 0 and finite population variance σ_Z^2, is known as *white noise*, and is denoted by

$$Z_t \sim WN\left(0, \sigma_Z^2\right).$$

The time series $\{Z_t\}$ that is a sequence of mutually independent and identically distributed random variables $Z_1, Z_2, \ldots, Z_n$, each with population mean 0 and finite population variance σ_Z^2, is known as *iid noise*, and is denoted by

$$Z_t \sim IID\left(0, \sigma_Z^2\right).$$

The time series $\{Z_t\}$ that is a sequence of mutually independent and identically normally distributed random variables $Z_1, Z_2, \ldots, Z_n$, each with population mean 0 and finite population variance σ_Z^2, is known as *Gaussian white noise*, and is denoted by

$$Z_t \sim GWN\left(0, \sigma_Z^2\right).$$

It is clear from Definition 7.1 that the three varieties of noise were defined from the more general case to the more specific case so that

$$GWN \subset IID \subset WN.$$

All three varieties share common population means and variances:

$$E\left[Z_t\right] = 0 \qquad \text{and} \qquad V\left[Z_t\right] = \sigma_Z^2.$$

These three probability models are, in some sense, the simplest possible time series models, although time series that are well-modeled by the three models are very rare in practice. Rather than approximating a real-world time series, they serve as building blocks for more realistic models. They are often used to describe the probability distribution of *error terms* in a probability model for a time series. In the next example, you will see a plot of a realization of Gaussian white noise.

Example 7.3 White noise, iid noise, and Gaussian white noise are just sequences of mutually independent observations centered around zero. A time series of $n = 100$ Gaussian white noise observations with population variance $\sigma_Z^2 = 1$, for example, can be generated and placed in the vector x with the R command

```
x = rnorm(100)
```

The choice of $n = 100$ was arbitrary. Defining x after an optional call to set.seed(8) to establish the random number stream, the values contained in the vector x can be converted to a time series with the ts function and then plotted as in the two previous examples with the plot.ts function. The resulting plot is given in Figure 7.3. The time series that consists of Gaussian white noise values has a minimum value of $x_{90} = -3.015$ and a maximum value of $x_{79} = 2.376$. Of the 100 Gaussian white noise values, 46 are positive and 54 are negative. Since the time series consists of mutually independent and identically distributed random variables, $x_1, x_2, \ldots, x_{100}$ are of no use in predicting the next value in the time series, x_{101}.

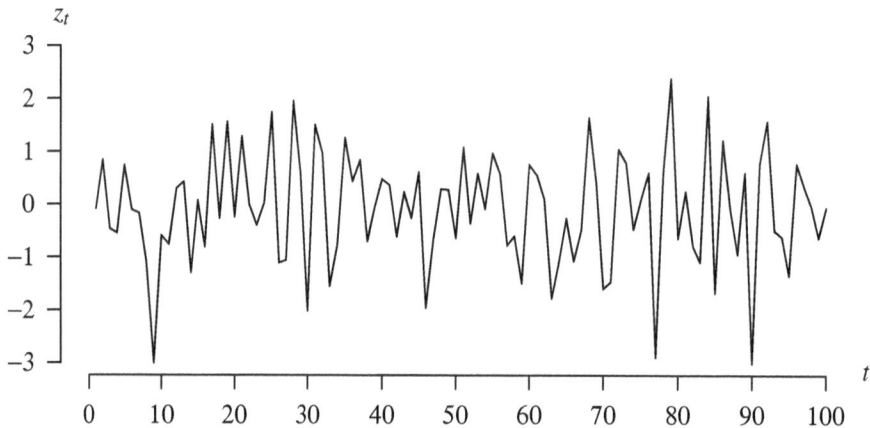

Figure 7.3: Time series plot of $n = 100$ Gaussian white noise observations.

Most time series that are encountered in practice do not behave in a fashion that approximates white noise, iid noise, or Gaussian white noise. It is more often the case that the value of the observation in the time series at time t will depend on the values of one or more of the previous observations in the time series. One time series model that exhibits this dependency is known as a *random walk*, which is defined and illustrated next.

Example 7.4 A time series that is a random walk $\{X_t\}$ is generated by the recursive equation

$$X_t = X_{t-1} + Z_t,$$

where $\{Z_t\}$ is Gaussian white noise. This is to say that the current value of the time series is the previous value of the time series plus a Gaussian white noise term. An algorithm for generating a random walk $\{X_t\}$ using Gaussian white noise is given in the pseudocode below. Indentation denotes nesting. In the second step, the initial observation, X_1, is arbitrarily set to 0. This step could instead have been $X_1 \leftarrow \mu$, where μ is the mean value of the time series. Alternatively, this start-up condition could instead have been $X_1 \leftarrow Z_1$, which starts the time series at a random position rather than 0. These two alternatives can be combined into $X_1 \leftarrow \mu + Z_1$.

$$
\begin{aligned}
&t \leftarrow 1 \\
&X_1 \leftarrow 0 \\
&\text{while } (t < n) \\
&\qquad t \leftarrow t + 1 \\
&\qquad \text{generate } Z_t \sim N\left(0, \sigma_Z^2\right) \\
&\qquad X_t \leftarrow X_{t-1} + Z_t
\end{aligned}
$$

This algorithm for generating a random walk can be implemented in R using the code given below. It is assumed that the population variance of the Gaussian white noise is $\sigma_Z^2 = 1$.

```
set.seed(8)
n    = 100
x    = numeric(n)
time = 1
x[1] = 0
while (time < n) {
  time = time + 1
  z = rnorm(1)
  x[time] = x[time - 1] + z
}
x = ts(x)
plot.ts(x)
```

The realization of the random walk stored in the vector **x** is plotted in Figure 7.4. The time series associated with the random walk takes on a decidedly different pattern than that of the Gaussian white noise from Figure 7.3. This realization of a random walk looks quite a bit like some graphs of economic data, such as the daily closing price of a stock or a stock market average. This makes sense because it might be the case that the probability model for the value of the closing price of the stock might be expressed as

$$[\text{today's closing price}] = [\text{yesterday's closing price}] + [\text{noise}]$$

which is equivalent to the random walk model

$$X_t = X_{t-1} + Z_t.$$

The random walk model does such a good job of approximating certain economic data that it can be difficult to distinguish a real set of economic data from a realization of a random walk generated by simulation.

As an illustration, Figure 7.5 contains graphs of three time series. One of these time series is the first $n = 100$ closing values of the Dow Jones Industrial Average during the

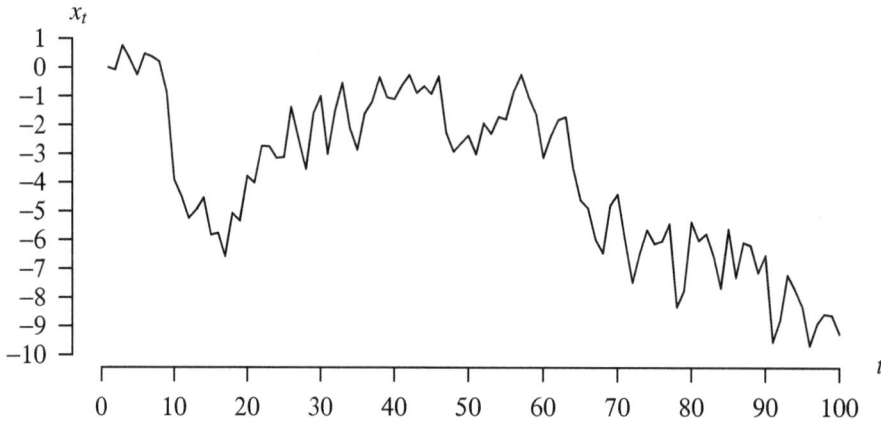

Figure 7.4: Time series plot of $n = 100$ observations from a random walk.

year 2000. The other two are random walks of length $n = 100$ with mutually independent standard normal noise terms generated by simulation. Spend some time looking
the three graphs over and try to determine which of the three is the real time series values and which are the random walks generated in R via simulation. In order to make
your task of identifying the Dow Jones Industrial Averages more difficult, the labels on
the vertical axes have been suppressed, and the scales have been set to stretch from the
smallest value to the largest value in the time series.

If you are having a difficult time identifying the stock market average values, it is because the random walk, which is a very simple time series model, is adequately approximating the time evolution of the stock market average values. The real data corresponding to the stock market average closes is in the top graph. The two lower graphs
in Figure 7.5 are random walks generated by Monte Carlo simulation.

In this section, we have encountered four examples of time series:

- a time series of $n = 96$ monthly home energy consumption observations,

- a time series of $n = 144$ monthly international airline passenger counts,

- a time series of $n = 100$ observations of Gaussian white noise, and

- two time series of $n = 100$ observations generated from a random walk, which were compared
  to a time series consisting of $n = 100$ closing values for the Dow Jones Industrial Average.

R has built-in data structures and functions that are useful in the analysis of a time series. Additional
tools beyond just the plotting of a time series will be introduced subsequently.

It is often the case that we want to formulate a hypothetical population probability model for a
time series from observed values of a time series, such as using the random walk model to model the
Dow Jones Industrial Average closing values. This notion of formulating a population probability
model is completely analogous to using the normal distribution to approximate the adult heights of
Swedish women, for example. Once a tentative model has been identified, any unknown parameters

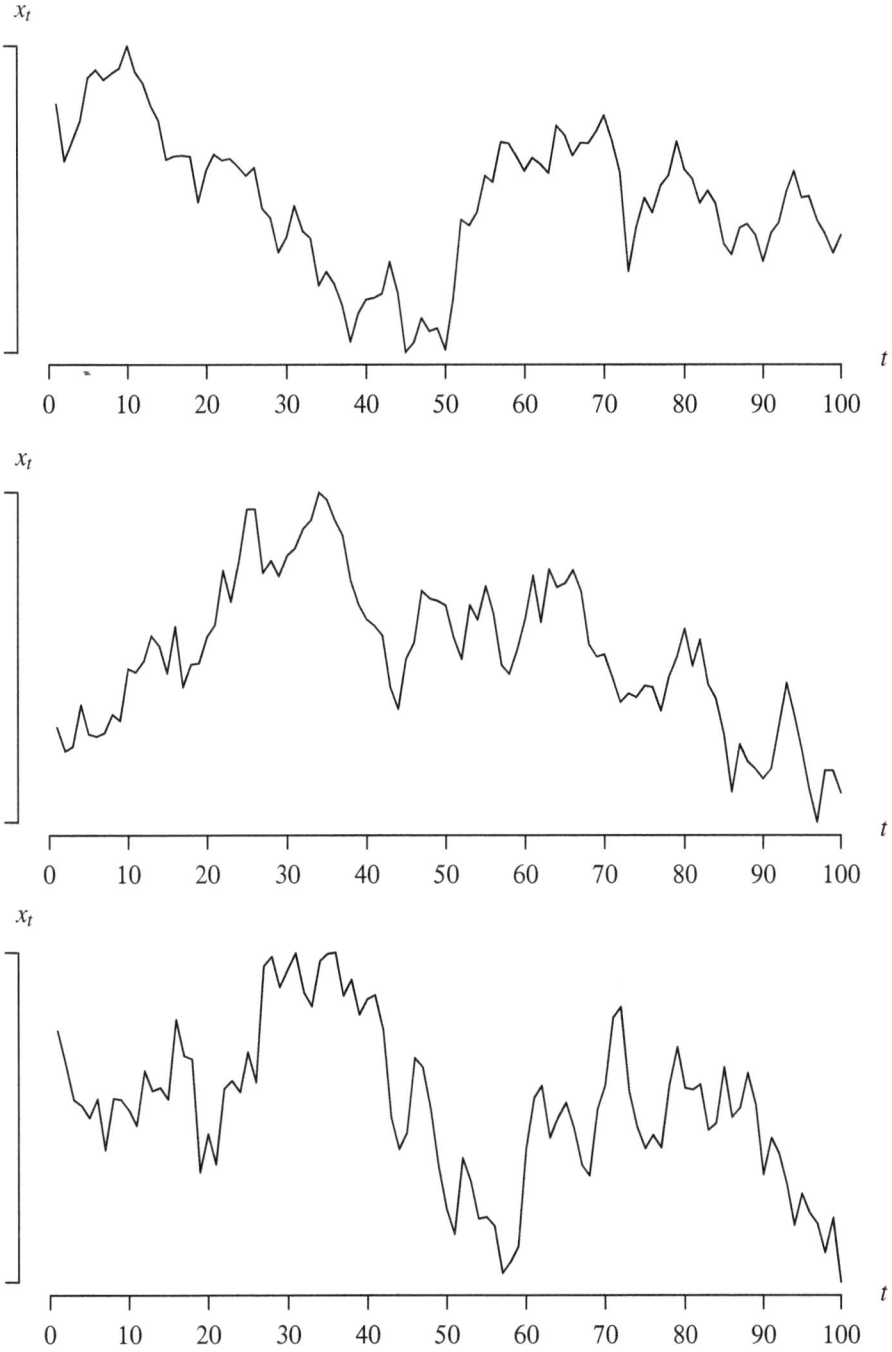

Figure 7.5: Which time series of length $n = 100$ consists of real data?

are estimated and goodness-of-fit tests are conducted. After a fitted probability model is accepted as a reasonable population probability model to describe a time series, a wide variety of statistical inference procedures can be employed. Broad categories of these statistical inference procedures are listed and described in the next subsection.

7.1.2   Why Analyze a Time Series?

There is not one universal purpose for conducting a time series analysis. Some of the more common purposes for conducting a time series analysis include description, explanation, prediction, simulation, control, signal estimation, and segmentation. This list is not comprehensive, but certainly covers the vast majority of the applications of time series analysis.

- **Description.** Time series analysis is often useful for *describing* the time evolution of the observations in a time series. Plotting the values in the time series, as we have done in the four examples, is a critical first step for observing trends, seasonal effects, extreme observations, etc. The time series plot also allows an analyst to easily identify outliers in a time series and decide whether these outliers were due to a coding error that occurred when inputting the time series or just random extreme observations. More sophisticated techniques that are helpful in describing a time series, such as the sample autocorrelation function to detect and quantify serial dependence in the values of a time series, or the periodogram to detect and quantify cyclic variation in the values of a time series, will be introduced subsequently.

- **Explanation.** It is often the case that one time series can be used to explain another time series. The home energy consumption time series from Example 7.1, for instance, might be partially explained by a time series of monthly average outdoor temperatures in Williamsburg, Virginia in 2011–2018. Another time series that might partially explain the home energy consumption values is the average number of hours of daylight in Williamsburg in a particular month.

- **Prediction.** Certain application areas, such as quantitative finance and seismology, engage in the analysis of a time series for the purpose of *forecasting*. The prediction of the next value of the time series, or perhaps the value of the time series h time units into the future, is often of interest. In quantitative finance, predicting the future value of a particular stock based on its history to date might be of interest. In seismology, predicting the time of the next earthquake might be of interest. Forecasted values are typically given by a point estimate and an associated confidence interval that measures the precision of the point estimate.

- **Simulation.** Simulating a time series in a discrete-event simulation might be the ultimate goal. A time series model that adequately mimics the real-world time series is critical in building a valid simulation model. As an example of such a simulation, financial planners often turn to simulation to estimate the probability that an individual or a married couple will have enough money to pay their expenses in retirement. This simulation requires, among other elements, a time series model that is capable of generating the annual inflation rate over the lifetimes of the individual or couple. The generation of simulated future annual inflation rates is based on building a time series model from previous annual inflation rates. Other elements, such as annual stock market returns or interest rates, would require a separate time series model. The values in these various time series are often correlated.

- **Control.** Time series analysis can be performed with the goal of controlling a particular variable. Examples include keeping ball bearing diameters between two prescribed thresholds, keeping delivery times below a prescribed threshold, and keeping unemployment in an economy between two thresholds. A branch of quality control known as *statistical process control* refers to the time series plots given earlier as *control charts*.

- **Signal estimation.** Certain application areas, such as astrophysics and electrical engineering, are particularly interested in separating signal from noise. The techniques using spectral analysis, which is presented subsequently, are particularly adept at detecting cyclic variation in a time series. Sometimes a very weak signal can be detected in a very noisy time series using these techniques.

- **Segmentation.** Economists often find it useful to classify a period of economic activity as a period of expansion or a period of contraction. They do so by breaking a time series into a sequence of segments. The challenge here is to identify the boundary points at which times the economy switches from expansion to contraction and then back again. Determining these points in time in which the changes in the time series model occur is one of the goals of segmentation.

A time series is just one instance of a process that evolves randomly over time known as a *stochastic process*. The next section classifies stochastic processes based on whether time passes in a discrete or continuous fashion, and whether the variable of interest at each time step is discrete or continuous.

7.1.3 Where Does Time Series Analysis Fall in the Modeling Matrix?

The purpose of this subsection is to step back and consider where time series analysis fits in the larger arena of stochastic processes. The common elements between the four time series we have encountered so far are that time is measured as an integer (representing months for the first two time series, the first 100 positive integers for the Gaussian white noise, and trading days for the stock market averages) and the values of the time series observations are measured on a continuous scale. So a *time series* is a sequence of observed data values, measured on a continuous scale, which are collected over time. In most instances, the observations are taken at equally-spaced points in time. The observations in the time series are denoted generically by

$$X_1, X_2, \ldots, X_n,$$

and can be referred to more compactly as just $\{X_t\}$. Table 7.2 shows the position that time series analysis resides in the 2×2 table in which the nature of time (discrete or continuous) defines the rows and the nature of the observed variable (discrete or continuous) defines the columns. Time series analysis occupies the discrete-time, continuous-state entry in the table. Popular stochastic

		state	
		discrete	continuous
time	discrete	Markov chains	a time series model
	continuous	continuous-time Markov chains	Brownian motion

Table 7.2: Four types of stochastic models in the modeling matrix.

process models, such as Markov chains which are often used to model discrete-time and discrete-state stochastic processes, occupy the other three positions in the table.

The defining characteristic of a time series model is that time is measured on a discrete scale and the observations are measured on a continuous scale. You have seen four examples of time series. Here are examples of applications of stochastic models in the other three boxes in Table 7.2.

- The classic example of a discrete-time, discrete-state stochastic process with two states is the weather on a particular day. If the two states are "rainy" and "sunny" (actually "not rainy" should be the second state so as to partition the state space so that a cloudy day with no rain is classified as "not rainy"), then a Markov chain is a potential model for the evolution of the weather from one day to the next.

- A continuous-time, discrete-state stochastic process that we have all encountered is that of a single-server queueing system. The state of the system is the number of customers in the system. If the number of customers in the system can either go up by one (via a customer arrival) or down by one (via a customer departure), then the state of the system is discrete. Furthermore, since customer arrivals and departures can occur at any instant, time is measured on a continuous scale.

- One well-known example of a continuous-time, continuous-state stochastic process is *Brownian motion*, which is named after Scottish botanist Robert Brown (1753–1858). He described the motion in 1827 while observing the pollen of the plant Clarkia pulchella immersed in water through a microscope. Physicist Albert Einstein (1879–1955) explained that the motion of the pollen was caused by individual water molecules in 1905. Brownian motion can be thought of as a random walk in which the time between subsequent observations approaches zero.

Figure 7.6 contains a 2×2 array of graphs that are analogous to the 2×2 array of stochastic process models in Table 7.2. The values plotted are one particular *realization*, also known as a *sample path*, of a stochastic process. A stochastic process can be thought of as a probability model which evolves over time. The dashed lines indicate that time or state is measured discretely. In this sense, the techniques for analyzing a time series represent 25% of the techniques for analyzing all of the stochastic processes.

7.1.4 Computing

We review some of the R functions that have been used in this section and also introduce some additional functions that are useful in time series analysis. All of these functions are available in the base distribution of R, so they are immediately available upon initiating an R session. These functions will typically be illustrated here for the built-in time series of monthly international air passenger counts from 1949 to 1960 named `AirPassengers` which was first encountered in Example 7.2, but they could be applied to any time series.

The time series function `ts` is used to convert data to the internal time series data structure in R. It takes arguments for the time series observations, the time of the first observation, the time of the last observation, the number of observations per unit of time, etc. As an illustration, the quarterly time series observations contained in the vector named `data` which begin in the second quarter of 1984 is converted to a time series named `x` via the `ts` function with the R command

```
x = ts(data, start = c(1984, 2), frequency = 4)
```

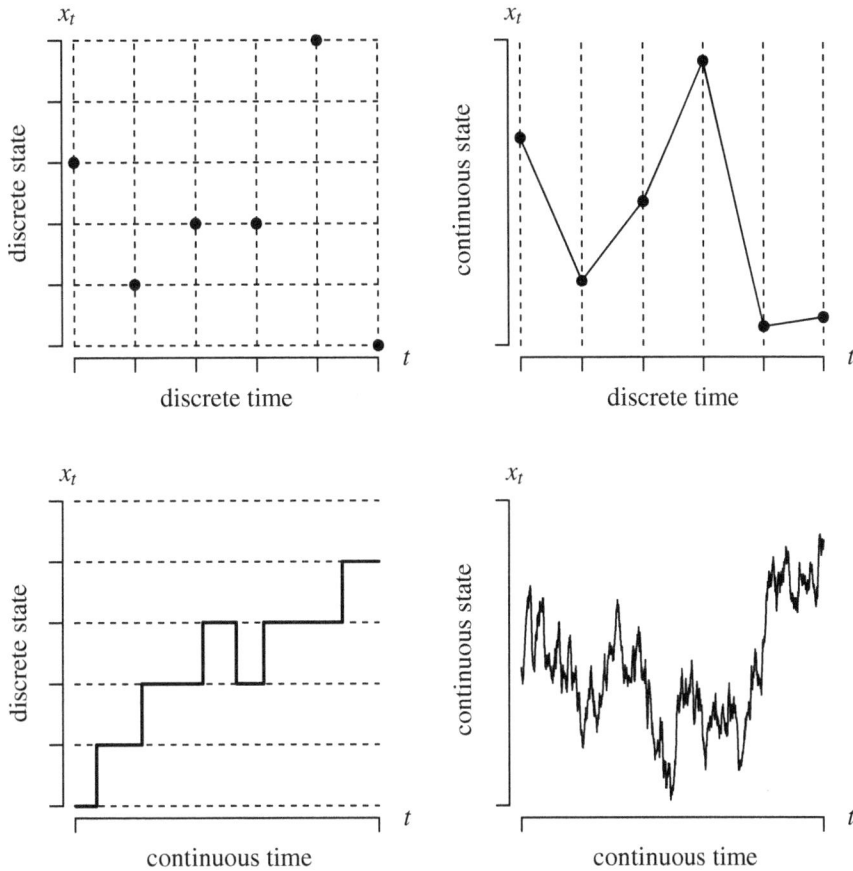

Figure 7.6: Graphs of realizations of four classes of stochastic processes.

Once a time series is in the R time series data structure, the `plot.ts` function can be used to provide a plot of the time series, for example,

`plot.ts(AirPassengers)`

The `ts.plot` command can be used to plot several time series on a single set of axes, for example,

`ts.plot(ldeaths, mdeaths, fdeaths)`

where the three time series that are built into R being plotted are monthly total deaths, male deaths, and female deaths from bronchitis, emphysema, and asthma in the United Kingdom from 1974 to 1979.

The next group of time series functions can be thought of as *utilities*, which are used to extract information about a time series. Illustrations of the application of these utility functions on the `AirPassengers` time series are given below.

`length(AirPassengers)`

```
start(AirPassengers)
end(AirPassengers)
frequency(AirPassengers)
deltat(AirPassengers)
time(AirPassengers)
cycle(AirPassengers)
```

The `length` function returns the length of a time series, which in this case is 144 observations. The `start` function returns the time associated with the first observation in the time series, which in this case is a vector comprised of the two elements 1949 and 1. The `end` function returns the time associated with the last observation in the time series. The `frequency` function shows the period length of a time series, which in this case is 12 because `AirPassengers` consists of monthly data. The `deltat` function returns the time increment associated with the time series, which in this case is $1/12$. The `time` function returns the time values for each observation as a time series, which in this case is $1949, 1949\frac{1}{12}, 1949\frac{2}{12}, \ldots, 1960\frac{11}{12}$. The `cycle` function returns integers indicating the position of each observation in a cycle, which in this case is 12 iterations of the first 12 integers.

7.2   Basic Properties of a Time Series

A time series has some unique properties that will require some special tools to aid in its modeling and analysis. Central to these properties is the notion of *stationarity*, which is the subject of one of the subsections that follow. Once stationarity is established for a time series, then the population autocorrelation function, and its statistical counterpart, the sample autocorrelation function, can be helpful in characterizing a time series. These autocorrelation functions give the correlation as a function of the distance between the values in the time series. We begin by defining the population autocovariance and autocorrelation.

7.2.1   Population Autocovariance and Autocorrelation

Traditional statistical methods rely on the assumption that observations are mutually independent and identically distributed. In most practical time series applications, this assumption is violated because adjacent or nearby values in a time series are correlated. Thus, the special analysis tools for time series known as the population autocovariance function and the population autocorrelation function are introduced in this subsection. Before motivating and defining these new notions, we briefly review the definitions of population covariance and correlation from probability theory in the next paragraph. The link to time series analysis will be made subsequently.

   The defining formula for the population covariance between the random variables X and Y is

$$\text{Cov}(X, Y) = E\left[(X - \mu_X)(Y - \mu_Y)\right],$$

where $\mu_X = E[X]$ is the expected value of X and $\mu_Y = E[Y]$ is the expected value of Y. The units on the population covariance are the units of X times the units of Y. When X and Y are independent random variables,

$$\text{Cov}(X, Y) = E\left[(X - \mu_X)(Y - \mu_Y)\right] = E\left[X - \mu_X\right]E\left[Y - \mu_Y\right] = (\mu_X - \mu_X)(\mu_Y - \mu_Y) = 0.$$

Zero covariance does not imply that X and Y are independent. The population correlation between the random variables X and Y is

$$\rho = \text{Corr}(X, Y) = \frac{\text{Cov}(X, Y)}{\sigma_X \sigma_Y} = \frac{E\left[(X - \mu_X)(Y - \mu_Y)\right]}{\sigma_X \sigma_Y} = E\left[\left(\frac{X - \mu_X}{\sigma_X}\right)\left(\frac{Y - \mu_Y}{\sigma_Y}\right)\right],$$

where σ_X is the population standard deviation of X and σ_Y is the population standard deviation of Y. The population correlation is a measure of the linear association between X and Y. The population correlation is unitless and satisfies $-1 \le \rho \le 1$, where the extremes indicate a perfect linear association.

A time series $\{X_t\}$ consists of a sequence of observations $X_1, X_2, \ldots, X_n$ indexed over time. Since we are working with time series *models* rather than time series data values, the time series values will typically be set in uppercase in this subsection. The observations are continuous random variables that have been drawn from some population probability distribution. This probability distribution can be described by a joint probability density function

$$f(x_1, x_2, \ldots, x_n)$$

or an associated joint cumulative distribution function

$$F(x_1, x_2, \ldots, x_n) = P(X_1 \le x_1, X_2 \le x_2, \ldots, X_n \le x_n).$$

The most well-known probability distribution for modeling n random variables $X_1, X_2, \ldots, X_n$ is the *multivariate normal distribution*, which is illustrated in the next example.

Example 7.5 Consider an $n \times 1$ vector of random variables $\boldsymbol{X} = (X_1, X_2, \ldots, X_n)'$, an associated $n \times 1$ vector of population means $\boldsymbol{\mu} = (\mu_1, \mu_2, \ldots, \mu_n)'$, and an associated $n \times n$ variance–covariance matrix $\boldsymbol{\Sigma}$. Matrix notation makes the expression of the joint probability density function of $X_1, X_2, \ldots, X_n$ much more compact than an entirely algebraic approach. The random vector $\boldsymbol{X}$ has the multivariate normal distribution if its joint probability density function has the form

$$f(x_1, x_2, \ldots, x_n) = \frac{1}{(2\pi)^{n/2}|\boldsymbol{\Sigma}|^{1/2}} \, e^{-\frac{1}{2}(\boldsymbol{x}-\boldsymbol{\mu})'\boldsymbol{\Sigma}^{-1}(\boldsymbol{x}-\boldsymbol{\mu})} \qquad (x_1, x_2, \ldots, x_n) \in \mathcal{A},$$

where

- $\boldsymbol{x} = (x_1, x_2, \ldots, x_n)'$,
- $\boldsymbol{\Sigma}^{-1}$ is the inverse of the variance–covariance matrix,
- $|\boldsymbol{\Sigma}|$ is the determinant of the variance–covariance matrix,
- the support is

$$\mathcal{A} = \{(x_1, x_2, \ldots, x_n) \,|\, -\infty < x_i < \infty, \text{ for } i = 1, 2, \ldots, n\},$$

- the parameter space is

$$\Omega = \{(\boldsymbol{\mu}, \boldsymbol{\Sigma}) \,|\, \boldsymbol{\mu} \in \mathcal{R}^n, \boldsymbol{\Sigma} \text{ is an } n \times n \text{ symmetric, positive semi-definite matrix}\}.$$

Although the multivariate normal distribution has some very appealing mathematical and statistical properties, it has one very significant drawback when it comes to being used as a time series model. That drawback concerns the number of parameters. There are n mean parameters $\mu_1, \mu_2, \ldots, \mu_n$ and $n(n+1)/2$ parameters in the symmetric variance–covariance matrix $\boldsymbol{\Sigma}$. If an analyst has collected just a single realization of a time series $x_1, x_2, \ldots, x_n$, then there are many more parameters to estimate than data values. So one of the goals for the rest of the section is to establish properties of a time series which allow us to formulate *parsimonious models* that adequately model a particular time series with as few parameters as possible. We begin the process of establishing these properties by defining the *population mean function* and the *population autocovariance function* associated with a time series $\{X_t\}$. As you will see, the focus is on the first two population moments associated with the time series.

Definition 7.2 A time series $\{X_t\}$ has a *population mean function*

$$\mu(t) = E\left[X_t\right]$$

provided that the expected values exist for all values of the index t.

So as long as the observations in the time series have expected values that are finite, the population mean function gives the expected observed value of the time series at time t. In other words, the mean values

$$\mu(1), \mu(2), \ldots, \mu(n)$$

are the expected values of $X_1, X_2, \ldots, X_n$. This defines what is essentially the first moment of the time series. The second moment of the time series is defined by the population autocovariance function.

Definition 7.3 A time series $\{X_t\}$ has a *population autocovariance function*

$$\gamma(s, t) = \mathrm{Cov}\left(X_s, X_t\right)$$

provided that the population covariances exist for all values of the indexes s and t.

Notice that the order associated with the arguments in the population autocovariance function is immaterial, so $\gamma(s, t) = \gamma(t, s)$. Notice also that when the two arguments in the population autocovariance function are identical, the expression reduces to the population variance, that is,

$$\gamma(s, s) = \mathrm{Cov}\left(X_s, X_s\right) = V[X_s].$$

The prefix "auto" means "self." This prefix is attached to covariance to signify that the population covariance is being taken between two members of the same time series. The value of $\gamma(s, t)$ is the population covariance between two snapshots of the same time series at times s and t.

We now consider a sequence of three examples in which we (a) define a time series model, (b) calculate the population mean function, and (c) calculate the population autocovariance function. The three examples, which will be in order of increasing complexity, are

- white noise,

- a three-point moving average, and

- a random walk.

We begin with a process that consists of just white noise.

Example 7.6 Recall from Definition 7.1 that the time series $\{Z_t\}$ that is a sequence of mutually independent random variables $Z_1, Z_2, \ldots, Z_n$, each with population mean 0 and population variance σ_Z^2, is known as white noise, and is denoted by

$$Z_t \sim WN\left(0, \sigma_Z^2\right).$$

Assume that our time series of interest $\{X_t\}$ is just this white noise; that is, $X_t = Z_t$. Find the population mean function and the population autocovariance function.

The population mean function is

$$\mu(t) = E[X_t] = E[Z_t] = 0$$

because each term in the white noise time series has expected value 0. The population autocovariance function is

$$\gamma(s, t) = \mathrm{Cov}(X_s, X_t) = \mathrm{Cov}(Z_s, Z_t) = \begin{cases} \sigma_Z^2 & t = s \\ 0 & t \neq s \end{cases}$$

because the observations in the time series are mutually independent random variables and

$$\gamma(s, s) = \mathrm{Cov}(X_s, X_s) = \mathrm{Cov}(Z_s, Z_s) = V[Z_s] = \sigma_Z^2$$

when $t = s$.

So the population mean and population autocovariance functions take on a particularly tractable form in the case of a time series that consists of white noise terms. We now consider calculating the population mean and population autocovariance functions for a three-point moving average of white noise.

Example 7.7 We again let the time series $\{Z_t\}$ denote mutually independent random variables $Z_1, Z_2, \ldots, Z_n$, each with population mean 0 and population variance σ_Z^2. This is again white noise, and is denoted by

$$Z_t \sim WN\left(0, \sigma_Z^2\right).$$

This time, however, our time series of interest $\{X_t\}$ is a three-point moving average of the white noise, that is,

$$X_t = \frac{Z_{t-1} + Z_t + Z_{t+1}}{3}$$

for $t = 2, 3, \ldots, n-1$. Find the population mean function and the population autocovariance function.

The population mean function is

$$\mu(t) = E[X_t] = E\left[\frac{Z_{t-1} + Z_t + Z_{t+1}}{3}\right] = \frac{1}{3}\left(E[Z_{t-1}] + E[Z_t] + E[Z_{t+1}]\right) = 0$$

because each term in the white noise time series has expected value 0. The population autocovariance function is more difficult than in the previous example because identical white noise terms are used in adjacent three-point moving averages. Assuming that all appropriate expected values exist, we rely on the formula

$$\mathrm{Cov}\left(\sum_{i=1}^n a_i X_i, \sum_{j=1}^m b_j Y_j\right) = \sum_{i=1}^n \sum_{j=1}^m a_i b_j \mathrm{Cov}(X_i, Y_j)$$

to help with the calculations. The population autocovariance function is

$$\gamma(s,t) = \text{Cov}\,(X_s, X_t)$$
$$= \text{Cov}\left(\frac{Z_{s-1} + Z_s + Z_{s+1}}{3}, \frac{Z_{t-1} + Z_t + Z_{t+1}}{3}\right)$$
$$= \frac{1}{9}\text{Cov}\,(Z_{s-1} + Z_s + Z_{s+1}, Z_{t-1} + Z_t + Z_{t+1}).$$

It is clear that $\gamma(s,t) = 0$ when $|t - s| > 2$ because there is no overlap in the white noise terms. The mutual independence of $Z_1, Z_2, \ldots, Z_n$ implies $\text{Cov}\,(Z_i, Z_j) = 0$ when $i \neq j$. So let's check the other cases individually using the formula concerning the population covariance between sums of random variables. First, the case of $t = s$:

$$\gamma(s,s) = \frac{1}{9}\text{Cov}\,(Z_{s-1} + Z_s + Z_{s+1}, Z_{s-1} + Z_s + Z_{s+1})$$
$$= \frac{1}{9}\left[\text{Cov}\,(Z_{s-1}, Z_{s-1}) + \text{Cov}\,(Z_s, Z_s) + \text{Cov}\,(Z_{s+1}, Z_{s+1})\right]$$
$$= \frac{1}{9}\left(V\,[Z_{s-1}] + V\,[Z_s] + V\,[Z_{s+1}]\right)$$
$$= \frac{1}{9}\left(\sigma_Z^2 + \sigma_Z^2 + \sigma_Z^2\right)$$
$$= \frac{1}{9} \cdot 3\sigma_Z^2$$
$$= \frac{\sigma_Z^2}{3}$$

based on the mutual independence of $Z_1, Z_2, \ldots, Z_n$. Next, consider the case of $t = s+1$:

$$\gamma(s,s+1) = \frac{1}{9}\text{Cov}\,(Z_{s-1} + Z_s + Z_{s+1}, Z_s + Z_{s+1} + Z_{s+2})$$
$$= \frac{1}{9}\left[\text{Cov}\,(Z_s, Z_s) + \text{Cov}\,(Z_{s+1}, Z_{s+1})\right]$$
$$= \frac{1}{9}\left(V\,[Z_s] + V\,[Z_{s+1}]\right)$$
$$= \frac{1}{9}\left(\sigma_Z^2 + \sigma_Z^2\right)$$
$$= \frac{1}{9} \cdot 2\sigma_Z^2$$
$$= \frac{2\sigma_Z^2}{9}.$$

Finally, consider the case of $t = s+2$:

$$\gamma(s,s+2) = \frac{1}{9}\text{Cov}\,(Z_{s-1} + Z_s + Z_{s+1}, Z_{s+1} + Z_{s+2} + Z_{s+3})$$
$$= \frac{1}{9}\text{Cov}\,(Z_{s+1}, Z_{s+1})$$
$$= \frac{1}{9}V\,[Z_{s+1}]$$
$$= \frac{\sigma_Z^2}{9}.$$

So to summarize, using the symmetry of the population autocovariance function in its arguments, the population autocovariance function is

$$\gamma(s,t) = \begin{cases} \sigma_Z^2/3 & |t-s| = 0 \\ 2\sigma_Z^2/9 & |t-s| = 1 \\ \sigma_Z^2/9 & |t-s| = 2 \\ 0 & |t-s| > 2. \end{cases}$$

The effect of using common terms from the time series Z_t consisting of white noise in constructing the three-point moving average time series X_t is apparent in the positive values in the population autocovariance function. There is positive population auto-covariance at lag 0 ($t = s$), slightly weaker positive population autocovariance at lag 1 ($|t-s| = 1$), still slightly weaker positive population autocovariance at lag 2 ($|t-s| = 2$), and zero population autocovariance at lags greater than 2. The decreasing magnitude of the population autocovariance function is due to the fewer common terms in the three-point moving average as the distance between values in $\{X_t\}$ increases. The diagram in Figure 7.7 conveys the intuition associated with the values in the population autocovariance function. The brackets show the mutually independent values of the white noise terms $Z_1, Z_2, \ldots, Z_n$ used in each term in the three-point moving average time series. Terms in the three-point moving average that are three time units apart, such as X_2 and X_5, have no white noise terms in common, and hence have population autocovariance zero.

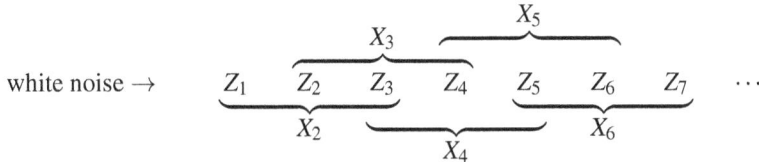

Figure 7.7: Relationship between white noise and three-point moving average.

The third and final example concerns the calculation of the population mean function and the population autocovariance function for a random walk.

Example 7.8 We now return to the random walk model first introduced in Example 7.4. The time series model for a random walk $\{X_t\}$ is the recursive equation

$$X_t = X_{t-1} + Z_t,$$

where $\{Z_t\}$ is white noise. A graph of a realization of a random walk was given in Figure 7.4. Find the population mean function and the population autocovariance function.

The first step is to write the model in a slightly different fashion. The random walk model can be written as a summation of the white noise terms:

$$X_t = \sum_{i=1}^{t} Z_i.$$

This formula can be verified by plugging it back into the original random walk model, yielding

$$\sum_{i=1}^{t} Z_i = \sum_{i=1}^{t-1} Z_i + Z_t.$$

This alternative formulation of the random walk model aids in the derivation of the population mean function and the population autocovariance function. Using the alternative formulation, the population mean function is

$$\mu(t) = E\left[X_t\right] = E\left[\sum_{i=1}^{t} Z_i\right] = \sum_{i=1}^{t} E\left[Z_i\right] = 0$$

because each term in the white noise time series has expected value 0. Again using the alternative formulation of the random walk model and the result from the previous example concerning the population covariance of sums of random variables, the population autocovariance function is

$$\gamma(s, t) = \text{Cov}\left(X_s, X_t\right)$$

$$= \text{Cov}\left(\sum_{i=1}^{s} Z_i, \sum_{j=1}^{t} Z_j\right)$$

$$= \sum_{i=1}^{s} \sum_{j=1}^{t} \text{Cov}\left(Z_i, Z_j\right)$$

$$= \sum_{i=1}^{\min\{s,t\}} V\left[Z_i\right]$$

$$= \min\left\{s, t\right\} \sigma_Z^2.$$

The population autocovariance function $\gamma(s, t)$ is the smaller of the arguments s and t multiplied by the population variance of the white noise.

The three examples have illustrated how to find the population mean function and the population autocovariance function for a time series model. Sometimes the population autocorrelation function is also of interest because population correlation is unitless and always lies between -1 and 1. The population autocorrelation function is defined next.

Definition 7.4 A time series $\{X_t\}$ has a *population autocorrelation function*

$$\rho(s, t) = \text{Corr}\left(X_s, X_t\right) = \frac{\text{Cov}\left(X_s, X_t\right)}{\sqrt{V\left[X_s\right] V\left[X_t\right]}} = \frac{\gamma(s, t)}{\sqrt{\gamma(s, s)\gamma(t, t)}}$$

provided that the population covariance exists for all indexes s and t.

We now revisit the previous three examples to compute the population autocorrelation function for the white noise, three-point moving average, and random walk models.

Example 7.9 The white noise model used

$$Z_t \sim WN\left(0, \sigma_Z^2\right)$$

and a time series $\{X_t\}$ which was just white noise, that is, $X_t = Z_t$. Find the population autocorrelation function.

The population autocovariance function for the white noise time series from Example 7.6 was

$$\gamma(s, t) = \begin{cases} \sigma_Z^2 & t = s \\ 0 & t \neq s. \end{cases}$$

Since $\gamma(s, s) = \sigma_Z^2$, the population autocorrelation function is

$$\rho(s, t) = \frac{\gamma(s, t)}{\sqrt{\gamma(s, s)\gamma(t, t)}} = \begin{cases} 1 & t = s \\ 0 & t \neq s. \end{cases}$$

There is perfect positive population correlation between each observation in the time series and itself because $\rho(s, s) = 1$. Furthermore, there is zero population correlation between distinct terms in the time series because $\rho(s, t) = 0$ for all $t \neq s$.

Although it lacks practical application in most real-world settings, the population autocorrelation function in the case of white noise is one of the most fundamental population autocorrelation functions possible. Since iid noise and Gaussian white noise are subsets of white noise, they also share this same population autocorrelation function. The next example considers the three-point moving average.

Example 7.10 Find the population autocorrelation function $\rho(s, t)$ for the three-point moving average time series model

$$X_t = \frac{Z_{t-1} + Z_t + Z_{t+1}}{3},$$

where

$$Z_t \sim WN\left(0, \sigma_Z^2\right).$$

The population autocovariance function for the three-point moving average time series from Example 7.7 was

$$\gamma(s, t) = \begin{cases} \sigma_Z^2/3 & |t - s| = 0 \\ 2\sigma_Z^2/9 & |t - s| = 1 \\ \sigma_Z^2/9 & |t - s| = 2 \\ 0 & |t - s| > 2. \end{cases}$$

Since $\gamma(s, s) = \sigma_Z^2/3$, the population autocorrelation function is

$$\rho(s, t) = \frac{\gamma(s, t)}{\sqrt{\gamma(s, s)\gamma(t, t)}} = \begin{cases} 1 & |t - s| = 0 \\ 2/3 & |t - s| = 1 \\ 1/3 & |t - s| = 2 \\ 0 & |t - s| > 2. \end{cases}$$

There is perfect positive population correlation between each observation in the time series and itself because $\rho(s, s) = 1$. The population autocorrelation function is positive and decreases linearly for lags 1 and 2 because of the common terms in the 3-point moving average, as illustrated previously in Figure 7.7. There is 0 population autocorrelation for lags of 3 or more because the moving averages contain no common white noise terms.

The third and final example concerns the calculation of the population autocorrelation function for a random walk model for a time series $\{X_t\}$.

Example 7.11 Find the population autocorrelation function $\rho(s, t)$ for the random walk time series model

$$X_t = X_{t-1} + Z_t,$$

where

$$Z_t \sim WN\left(0, \sigma_Z^2\right).$$

The population autocovariance function for the random walk time series from Example 7.8 was

$$\gamma(s, t) = \min\{s, t\}\sigma_Z^2.$$

Since $\gamma(s, s) = s\sigma_Z^2$, the population autocorrelation function is

$$\rho(s, t) = \frac{\gamma(s, t)}{\sqrt{\gamma(s, s)\gamma(t, t)}} = \frac{\min\{s, t\}\sigma_Z^2}{\sqrt{s\sigma_Z^2 \cdot t\sigma_Z^2}} = \frac{\min\{s, t\}}{\sqrt{st}}.$$

Since $\rho(s, s) = s/s = 1$, this can be written as

$$\rho(s, t) = \begin{cases} 1 & t = s \\ \min\{s, t\}/\sqrt{st} & t \neq s. \end{cases}$$

Once again, there is perfect positive population correlation between each observation in the time series and itself because $\rho(s, s) = 1$. This will be the case with any time series model.

This ends the introduction to three important functions that are associated with a time series model:

- the population mean function $\mu(t) = E[X_t]$,

- the population autocovariance function $\gamma(s, t) = \text{Cov}(X_s, X_t)$, and

- the population autocorrelation function $\rho(s, t) = \text{Corr}(X_s, X_t)$.

An important property of a time series, known as *stationarity*, will be defined and illustrated in the next subsection. A stationary time series is one in which there is no long-term change in the probability mechanism governing the time series. Knowing that a time series is stationary will have an important effect on $\mu(t), \gamma(s, t)$, and $\rho(s, t)$.

7.2.2 Stationarity

A time series $\{X_t\}$ is stationary if the underlying probability mechanism that governs the time series is independent of a shift in time. In other words, if you select two different time windows in which to view a number of observations from the time series, the probability distribution of the observations in those two time windows will be identical.

Definition 7.5 The time series $\{X_t\}$ is *strictly stationary* if

$$X_1, X_2, \ldots, X_n$$

and the shifted observations in the time series

$$X_{k+1}, X_{k+2}, \ldots, X_{k+n}$$

have the same joint probability distribution for all integers k and all positive integers n.

A strictly stationary time series is also known as a *strongly stationary* or *completely stationary* time series. The next two examples contain the type of thought experiment that is appropriate for determining whether a time series is strictly stationary.

> **Example 7.12** Strict stationarity implies that the probability mechanism that governs the time series does not change with a shift in time. Would the time series of monthly international airline passengers (in thousands) contained in the `AirPassengers` time series in R be likely to have been drawn from a strictly stationary time series model?
>
> Here is the thought associated with making such a judgment. Consider a specific instance of the time series from Definition 7.5 with $n = 3$ and $k = 18$ in order to develop a counterexample. So the question is whether
>
> $$X_1, X_2, X_3$$
>
> and the shifted observations in the time series
>
> $$X_{19}, X_{20}, X_{21}$$
>
> have the same trivariate probability distribution. In the case of the `AirPassengers` observed time series, these values correspond to January, February, and March of 1949 versus July, August, and September of 1950. The first three values in the time series are
>
> $$x_1 = 112, x_2 = 118, x_3 = 132,$$
>
> and the three time series observations shifted 18 months into the future are
>
> $$x_{19} = 170, x_{20} = 170, x_{21} = 158.$$
>
> From a cursory inspection, the three earlier values in the time series appear to be less than the three later values. In addition, Figure 7.2 showed a significant upward trend in the time series as time progresses. Furthermore, a careful inspection of the values in the `AirPassengers` time series from Figure 7.2 reveals that the annual peak travel occurs during the months of July and August. Based on this evidence, we conclude that the `AirPassengers` time series is *not* drawn from a strictly stationary time series. The hypothesis of an underlying stationary time series model can be rejected because of the trend and seasonal component that are clearly apparent in Figure 7.2. The underlying probability mechanism governing the time series appears to be changing over time.

The discussion above would indicate that very few time series which occur in practice would be strictly stationary. The previous example asks whether a *realization* appears to be drawn from a stationary time series model. The next example gives a simple time series model which is strictly stationary.

Example 7.13 Consider the time series from Example 7.3 which consists of Gaussian white noise with population variance $\sigma_Z^2 = 1$, that is,

$$X_t \sim GWN(0, 1).$$

Is this time series strictly stationary?

As in the last example, consider $n = 3$ and $k = 18$ from Definition 7.5. If the time series is strictly stationary, then

$$X_1, X_2, X_3$$

and the shifted observations in the time series

$$X_{19}, X_{20}, X_{21}$$

have the same trivariate probability distribution. In the case of Gaussian white noise with $\sigma_X^2 = \sigma_Z^2 = 1$, $(X_1, X_2, X_3)'$ has a trivariate normal distribution with 3×1 vector of population means $\boldsymbol{\mu} = (0, 0, 0)'$, and an associated 3×3 variance–covariance matrix which is the identity matrix. Using the formulation from Example 7.5, the joint probability density function of X_1, X_2, X_3 is

$$f(x_1, x_2, x_3) = \frac{1}{(2\pi)^{3/2}} e^{-\left(x_1^2 + x_2^2 + x_3^2\right)/2} \qquad -\infty < x_1 < \infty, \ -\infty < x_2 < \infty, \ -\infty < x_3 < \infty.$$

Because the values in the time series model are mutually independent and identically distributed, this is also the joint probability density function of X_{19}, X_{20}, X_{21}. So for this particular choice of n and k, the conditions of Definition 7.5 are satisfied. The probability mechanism governing X_1, X_2, X_3 is exactly the same as the probability mechanism governing X_{19}, X_{20}, X_{21}, so the realization of such a process in Figure 7.3 displays no trend, no seasonality, no change in variability, and no change in the marginal distributions of $X_1, X_2, \ldots, X_n$. But the choices of $n = 3$ and $k = 18$ were arbitrary. The joint probability distributions would be identical regardless of the choices for n and k, so we conclude that a time series consisting of Gaussian white noise is strictly stationary.

So the international airline passengers data set, just from observing the time series, is not strictly stationary. The Gaussian white noise process is strictly stationary. There are several implications of a strictly stationary time series, some of which are listed below.

- The initial n values of the time series $X_1, X_2, \ldots, X_n$ and their associated observations shifted k time units to the left or right $X_{k+1}, X_{k+2}, \ldots, X_{k+n}$ having the same joint probability distribution implies that each must have the same joint cumulative distribution function, that is,

$$P(X_1 \le x_1, X_2 \le x_2, \ldots, X_n \le x_n) = P(X_{k+1} \le x_1, X_{k+2} \le x_2, \ldots, X_{k+n} \le x_n)$$

for all values of $x_1, x_2, \ldots, x_n$.

- The marginal distribution of each value in the time series is identical. Symbolically,

$$P(X_s \le x) = P(X_t \le x)$$

for all integer time values s and t and all real-valued x.

- The population mean function $\mu(t) = E[X_t]$ is constant in time; that is, there is no trend.

- The population autocovariance function $\gamma(s,t) = \text{Cov}(X_s, X_t)$ is constant with respect to a shift; that is,

$$\gamma(s,t) = \gamma(s+k, t+k).$$

- Any time series consisting of mutually independent and identically distributed random variables must be strictly stationary.

Strict stationarity is a lot to ask of a time series, and is difficult to establish based on an observed realization of a time series. So time series analysts have defined a weaker version of strict stationarity which we will refer to here as just stationarity. Other terms used for this type of stationarity are

- weakly stationary,

- second-order stationary, and

- covariance stationary.

Whereas a strictly stationary time series required that the entire multivariate distribution remain the same on any time window, a stationary time series only places requirements on the first and second moments. The population mean function must be constant over time, and the population autocovariance function must depend only on the lag between the observations.

Definition 7.6 A time series $\{X_t\}$ is said to be *stationary* if the following two conditions are satisfied.

(a) The population mean function $\mu(t) = E[X_t]$ exists and is constant in t; that is, there is a real-valued constant c such that $E[X_t] = c$ for all values of t.

(b) The population autocovariance function $\gamma(s,t) = \text{Cov}(X_s, X_t)$ exists and depends only on $|t-s|$; that is, for integers s_1, t_1, s_2, and t_2,

$$\gamma(s_1, t_1) = \gamma(s_2, t_2)$$

if $|t_1 - s_1| = |t_2 - s_2|$.

The first condition implies that the time series has no trend because each observation in the time series has the same expected value. The second condition implies that the population covariance between two observations is a function of only the absolute difference between the two time indexes of the observations. This second condition implies that a stationary time series model only requires a single argument, which is known as the lag k, when defining the population autocovariance and autocorrelation function. We will use the same names for these functions, but only use a single argument when the time series is stationary.

Definition 7.7 For a stationary time series $\{X_t\}$, the population autocovariance function is

$$\gamma(k) = \text{Cov}(X_t, X_{t+k}),$$

and the population autocorrelation function is

$$\rho(k) = \text{Corr}(X_t, X_{t+k}) = \frac{\gamma(k)}{\gamma(0)}$$

for $k = 0, \pm 1, \pm 2, \ldots$.

The population autocorrelation function for a stationary time series provides an important reflection of the structure of a time series model. It gives the modeler a view of how observations in the time series are correlated based on their distance away from one another in the time series. The next example relates the population *correlation matrix* and the population autocorrelation function for a time series model.

Example 7.14 Consider the stationary time series $\{X_t\}$ with population correlation matrix

$$\begin{bmatrix} 1.0 & 0.4 & -0.2 & 0.1 \\ 0.4 & 1.0 & 0.4 & -0.2 \\ -0.2 & 0.4 & 1.0 & 0.4 \\ 0.1 & -0.2 & 0.4 & 1.0 \end{bmatrix}.$$

There is a population and a sample version of this matrix, but we will refer to this matrix as a population correlation matrix. This particular matrix is a special population correlation matrix because it corresponds to a stationary time series with equal-valued elements that are equal distance from the diagonal. The lag 2 population autocorrelations, for example, are all -0.2, and are two positions away from the diagonal elements in the population correlation matrix. The population autocorrelation function values for lags 4 and higher are all zero. Some notes on the population correlation matrix for a stationary time series model are given below.

- The population correlation matrix is symmetric and positive definite, with ones on the diagonal, and identical elements at a fixed number of entries from the diagonal.

- The population correlation between adjacent observations in the time series is given by the elements that are just off of the diagonal.

- This particular population correlation matrix has positive eigenvalues $\lambda_1 = 1.5$, $\lambda_2 = 1.3685$, $\lambda_3 = 1$, and $\lambda_4 = 0.1315$, which is consistent with the matrix being positive definite.

Convert this population correlation matrix to a population autocorrelation function.

Figure 7.8 shows the population correlation matrix rotated 45° clockwise. With this rotation, the identical elements are now aligned vertically. The vertical dashed lines show how the elements of the matrix are translated to a population autocorrelation function, which has nonzero spikes at lags $k = -3, -2, \ldots, 3$. The associated population autocorrelation function is symmetric. As expected, the lag k population autocorrelation satisfies $-1 \le \rho(k) \le 1$ for $k = 0, \pm 1, \pm 2, \ldots$. There is no information conveyed by including the population autocorrelation values for negative values of k. It is convention in time series analysis that the spike of height 1 associated with lag $k = 0$ is included in the graph of the population autocorrelation function. Furthermore, we always extend the vertical axis from -1 to 1 so that all population autocorrelation functions are viewed on an equal footing. Figure 7.9 shows the format that we will use for the plot of the population autocorrelation function for nonnegative lags k from this point forward.

You might have noticed that the word *population* precedes autocorrelation function. This convention is not universal, but we do so in order to distinguish the population autocorrelation function from its statistical counterpart, the *sample* autocorrelation function. An analogy in the realm of univariate probability distributions is the distinction between the population mean μ, which is a constant, and its statistical counterpart, the sample mean $\bar{X}$, which is a random variable. In the same

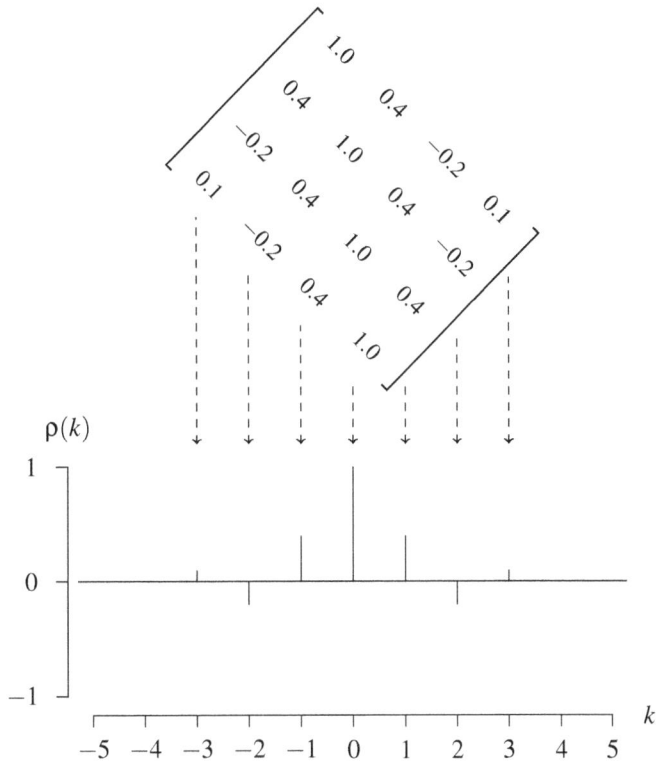

Figure 7.8: Mapping a correlation matrix to an autocorrelation function.

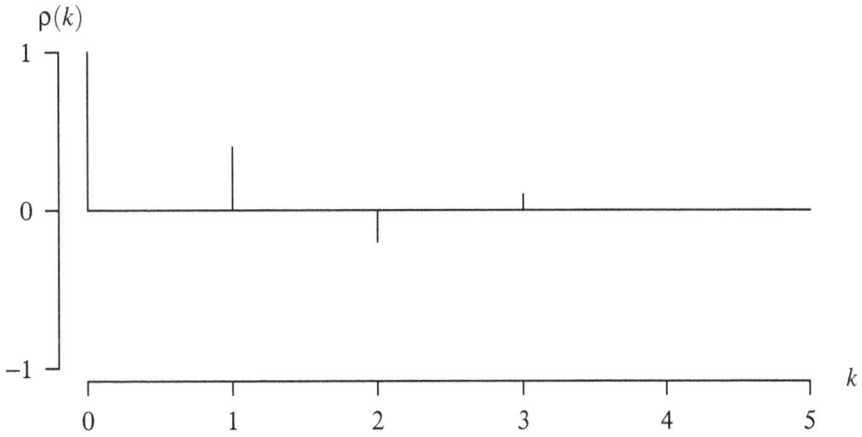

Figure 7.9: Population autocorrelation function for nonnegative lag values.

sense, the population autocorrelation function $\rho(k)$ is a sequence of population correlations, which are fixed constants that are indexed by the lag k. The sample autocorrelation function, which will be introduced in a subsequent section as r_k, is a sequence of sample correlations, which are random variables that are indexed by the lag k.

Several properties of the population autocorrelation function for *all* stationary time series models are given next.

Theorem 7.1 For a stationary time series $\{X_t\}$ with population mean μ and population autocorrelation function $\rho(k)$,

- $\rho(0) = 1$,

- $-1 \leq \rho(k) \leq 1$ for $k = 0, \pm 1, \pm 2, \ldots$,

- $\rho(k) = \rho(-k)$ for $k = 0, 1, 2, \ldots$,

- $\rho(k)$ is unitless, and

- $\rho(k)$ does not uniquely determine an underlying time series model.

Proof Consider a stationary time series $\{X_t\}$ with mean μ, population autocovariance function $\gamma(k)$, and population autocorrelation function $\rho(k)$.

- The lag zero population autocorrelation is $\rho(0) = 1$ because

$$\rho(0) = \text{Corr}(X_t, X_t) = \frac{\text{Cov}(X_t, X_t)}{\sigma_{X_t}\sigma_{X_t}} = \frac{V[X_t]}{\sigma_{X_t}^2} = \frac{\gamma(0)}{\gamma(0)} = 1.$$

- The lag k population autocorrelation must lie on the closed interval $[-1, 1]$. In other words, $-1 \leq \rho(k) \leq 1$, because $\rho(k)$ is defined as a population correlation. This can also be proved by first principles as follows. The inequality

$$V[c_1 X_t + c_2 X_{t+k}] \geq 0$$

holds for any real-valued constants c_1 and c_2 because all variances are nonnegative. This is equivalent to

$$c_1^2 V[X_t] + c_2^2 V[X_{t+k}] + 2c_1 c_2 \text{Cov}(X_t, X_{t+k}) \geq 0.$$

or

$$\left(c_1^2 + c_2^2\right)\sigma_{X_t}^2 + 2c_1 c_2 \gamma(k) \geq 0.$$

When $c_1 = c_2 = 1$, this inequality reduces to $\sigma_{X_t}^2 + \gamma(k) \geq 0$, which implies that $\rho(k) = \gamma(k)/\sigma_{X_t}^2 \geq -1$. Similarly, when $c_1 = 1$ and $c_2 = -1$, the inequality reduces to $\sigma_{X_t}^2 - \gamma(k) \geq 0$, which implies that $\rho(k) = \gamma(k)/\sigma_{X_t}^2 \leq 1$. Combining these two inequalities gives $-1 \leq \rho(k) \leq 1$.

- Since the time series $\{X_t\}$ is stationary,

$$\rho(k) = \text{Corr}(X_t, X_{t+k}) = \text{Corr}(X_{t-k}, X_t) = \rho(-k)$$

for $k = 0, 1, 2, \ldots$.

- The lag k population autocorrelation is unitless because the units of the numerator of

$$\rho(k) = \frac{\text{Cov}(X_t, X_{t+k})}{\sigma_{X_t} \sigma_{X_{t+k}}}$$

  are the square of the units of X_t, and the units of both σ_{X_t} and $\sigma_{X_{t+k}}$ are the units of X_t. Thus, the units cancel and $\rho(k)$ is unitless.

- This final property can be proved by counterexample. Consider two time series models. The first is $X_t \sim GWN(0, 1)$ (that is, Gaussian white noise with $\sigma_{X_t} = 1$). The second is $X_t \sim IID(0, 1)$, for example, iid noise with $\sigma_{X_t} = 1$ and error terms $U(-\sqrt{3}, \sqrt{3})$. These two time series models have identical population autocorrelation functions but are not identical time series models. Therefore, $\rho(k)$ does not uniquely determine an underlying time series model. $\square$

These properties of $\rho(k)$ have important implications in time series analysis. The first result from Theorem 7.1 indicates that there is perfect positive population autocorrelation between an observation and itself (that is, an observation at lag $k = 0$). The initial spike in the population autocorrelation function at $\rho(0) = 1$ is generally included in a graph of the population autocorrelation function, although it conveys no information. The second result indicates that all population autocorrelation functions must lie between -1 and 1. Subsequent plots of $\rho(k)$ will always stretch the vertical axis from -1 to 1 so that they can easily be compared with one another. The third result indicates that $\rho(k)$ is an even function in k, so although k can be any integer, it is common practice to only graph $\rho(k)$ for $k = 0, 1, 2, \dots$ because we know that the reflection about the $\rho(k)$ axis is identical. There is no need to graph the population autocorrelation function for negative lags because no additional information is conveyed. The fourth result explains why $\rho(k)$ tends to be more popular than $\gamma(k)$ because it is free of the units selected for X_t. The fifth result indicates that a time series model cannot be determined from its population autocorrelation function. Every stationary time series model has a population autocorrelation function, but knowing the autocorrelation function does not necessarily determine the underlying time series model.

We can now revisit the three examples from the previous subsection, namely white noise, a three-point moving average, and a random walk, to see if they are stationary time series models. In addition, we will make a plot of their population autocorrelation functions if they happen to be stationary.

Example 7.15 Consider the white noise time series model

$$Z_t \sim WN\left(0, \sigma_Z^2\right),$$

and the time series of interest is just $\{X_t\} = \{Z_t\}$. Determine whether this time series model is stationary, and plot the population autocorrelation function if it is stationary.

Recall from Example 7.6 that the population mean function for the white noise time series was

$$\mu(t) = 0$$

for all values of t, so the first condition of Definition 7.6 is satisfied. Recall also that the population autocovariance function was

$$\gamma(s, t) = \begin{cases} \sigma_Z^2 & t = s \\ 0 & t \neq s. \end{cases}$$

Since the value of $\gamma(s, t)$ depends only on $|t - s|$, the second condition of Definition 7.6 is satisfied, and we conclude that this time series model is stationary. Because the time series model is stationary, the population autocovariance function can be written in terms of the single argument k, the lag, as

$$\gamma(k) = \left\{ \begin{array}{ll} \sigma_Z^2 & k = 0 \\ 0 & k = 1, 2, \dots \end{array} \right. .$$

Since $\gamma(0) = \sigma_Z^2$, the population autocorrelation function written in terms of the lag k is

$$\rho(k) = \left\{ \begin{array}{ll} 1 & k = 0 \\ 0 & k = 1, 2, \dots \end{array} \right. .$$

It would be perfectly reasonable to consider the range of k to be $k = 0, \pm 1, \pm 2, \dots$, but Theorem 7.1 indicates that the population autocorrelation function for a stationary time series model is *always* an even function, so we will only report the nonnegative values of k. A graph of $\rho(k)$ for the white noise process is shown in Figure 7.10. A horizontal line has been drawn at $\rho(k) = 0$ for reference. There is a single spike of height 1 at lag $k = 0$ which indicates that each observation is perfectly positively correlated with itself. There are spikes of height 0 at $k = 1, 2, \dots$, which indicates that distinct observations in the time series are uncorrelated, as expected from the time series model consisting of white noise values.

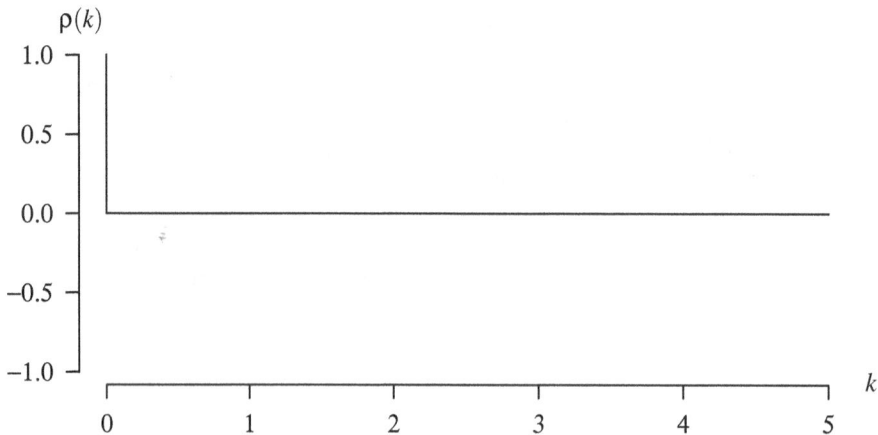

Figure 7.10: Population autocorrelation function for a white noise time series.

The population autocorrelation function for the white noise time series model is identical to that for iid noise and Gaussian white noise because those time series models are subsets of the white noise time series model. We now consider the three-point moving average model.

Example 7.16 Consider the white noise time series model

$$Z_t \sim WN\left(0, \sigma_Z^2\right),$$

and the time series of interest $\{X_t\}$ is the three-point moving average of the white noise; that is,

$$X_t = \frac{Z_{t-1} + Z_t + Z_{t+1}}{3}.$$

Determine whether this time series model is stationary, and plot the population autocorrelation function if it is stationary.

Recall from Example 7.7 that the population mean function for the white noise time series was

$$\mu(t) = 0$$

for all values of t, so the first condition of Definition 7.6 is satisfied. Recall also that the population autocovariance function was

$$\gamma(s, t) = \begin{cases} \sigma_Z^2/3 & |t-s| = 0 \\ 2\sigma_Z^2/9 & |t-s| = 1 \\ \sigma_Z^2/9 & |t-s| = 2 \\ 0 & |t-s| > 2. \end{cases}$$

Since the value of $\gamma(s, t)$ depends only on $|t - s|$, the second condition of Definition 7.6 is satisfied, and we conclude that this time series model is stationary. Because the time series model is stationary, the population autocovariance function can be written in terms of the single argument k, the lag, as

$$\gamma(k) = \begin{cases} \sigma_Z^2/3 & k = 0 \\ 2\sigma_Z^2/9 & k = 1 \\ \sigma_Z^2/9 & k = 2 \\ 0 & k = 3, 4, \ldots. \end{cases}$$

Since $\gamma(0) = \sigma_Z^2/3$, the population autocorrelation function written in terms of the lag k is

$$\rho(k) = \begin{cases} 1 & k = 0 \\ 2/3 & k = 1 \\ 1/3 & k = 2 \\ 0 & k = 3, 4, \ldots. \end{cases}$$

A graph of $\rho(k)$ for the three-point moving average model is shown in Figure 7.11. As with all population autocorrelation functions, there is a spike of height 1 at lag $k = 0$, which indicates that each observation is perfectly positively correlated with itself. The spikes at $k = 1$ and $k = 2$ reflect the effect of the nearby moving averages being functions of common white noise observations. Observations in $\{X_t\}$ that are three or more indexes apart are uncorrelated because they do not contain any common white noise terms. This corresponds to $\rho(k) = 0$ for $k = 3, 4, \ldots$.

    The previous example concerning a three-point moving average of white noise generalizes to an m-point moving average of white noise, where m is an odd, positive integer. The more general time series model is also stationary, and the population autocorrelation function also decreases linearly, and cuts off at lag m. The derivation of this result is given as an exercise at the end of this chapter. The third example considers a random walk time series model.

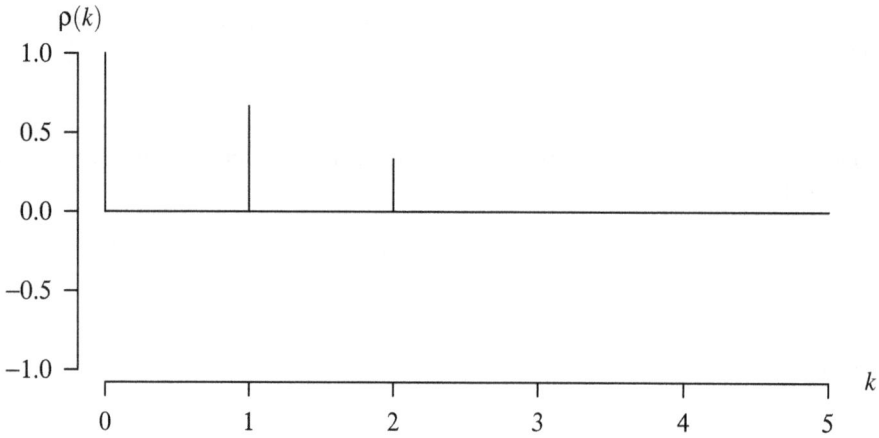

Figure 7.11: Population autocorrelation function for a three-point moving average time series.

Example 7.17 Consider the white noise time series model

$$Z_t \sim WN\left(0, \sigma_Z^2\right),$$

and the time series of interest $\{X_t\}$ is the random walk model; that is,

$$X_t = X_{t-1} + Z_t.$$

Determine whether this time series model is stationary, and plot the population autocorrelation function if it is stationary.

Recall from Example 7.8 that the population mean function for the white noise time series was

$$\mu(t) = 0$$

for all values of t, so the first condition of Definition 7.6 is satisfied. Recall also that the population autocovariance function was

$$\gamma(s, t) = \min\{s, t\}\sigma_Z^2.$$

Since the value of $\gamma(s, t)$ does *not* depend only on $|t - s|$, the second condition of Definition 7.6 is *not* satisfied, so we conclude that this time series model is nonstationary. Because the time series model is nonstationary, we are not able to write the population autocovariance function in terms of the single argument k. An example of the population autocorrelation function not being a function of the lag is

$$\gamma(1, 4) = \sigma_Z^2 \qquad \text{and} \qquad \gamma(2, 5) = 2\sigma_Z^2.$$

Equivalently,

$$\text{Cov}(X_1, X_4) = \sigma_Z^2 \qquad \text{and} \qquad \text{Cov}(X_2, X_5) = 2\sigma_Z^2.$$

Since observations that are three time indexes apart have different values of the population autocovariance function, $\gamma(s, t)$ does not depend only on $|t - s|$.

The statistical analogs of the population autocovariance and autocorrelation functions are the sample autocovariance and autocorrelation functions, which are calculated from an observed time series from a stationary model. These two functions are defined in the next subsection.

7.2.3   Sample Autocovariance and Autocorrelation

This section takes up the estimation of the population autocovariance function and the population autocorrelation function from a single realization of a time series denoted by the observations $x_1, x_2, \ldots, x_n$. In addition to the vital plot of a time series, a plot of the sample autocorrelation function, which is known as the *correlogram*, can yield additional insight concerning the underlying probability model governing the time series. The approach that we will take here is to review the sample versions of the covariance and correlation in terms of data pairs in the next paragraph, and then adapt these notions to their associated analogs in time series analysis.

This paragraph reviews the estimation of the population covariance and correlation from a data set of data pairs (X_i, Y_i), for $i = 1, 2, \ldots, n$. The population covariance is estimated by the *sample covariance*

$$\widehat{\text{Cov}}(X, Y) = \frac{1}{n} \sum_{i=1}^{n} (X_i - \bar{X})(Y_i - \bar{Y}),$$

where $\bar{X}$ and $\bar{Y}$ are the sample means of the associated sample values:

$$\bar{X} = \frac{1}{n} \sum_{i=1}^{n} X_i \qquad \text{and} \qquad \bar{Y} = \frac{1}{n} \sum_{i=1}^{n} Y_i.$$

This formula is the statistical analog to the formula

$$\text{Cov}(X, Y) = E[(X - \mu_X)(Y - \mu_Y)]$$

from probability theory. There are two formulas for estimating the population variance from a random sample—one with n in the denominator and one with $n-1$ in the denominator. Since n is required to be fairly large in time series analysis, the choice between the two is not critical. The formula with $n-1$ in the denominator is more prevalent in statistics because

$$E\left[\frac{1}{n-1} \sum_{i=1}^{n} (X_i - \bar{X})^2 \right] = \sigma_X^2$$

for mutually independent and identically distributed observations $X_1, X_2, \ldots, X_n$; that is, the sample variance S^2 is an unbiased estimator of the population variance σ_X^2. We use n in the denominator here because, in spite of being a biased estimator of the population variance in the non-times-series setting, it leads to certain terms dropping out of a subsequent formula. The population variances can be estimated by the maximum likelihood estimators

$$\hat{\sigma}_X^2 = \frac{1}{n} \sum_{i=1}^{n} (X_i - \bar{X})^2 \qquad \text{and} \qquad \hat{\sigma}_Y^2 = \frac{1}{n} \sum_{i=1}^{n} (Y_i - \bar{Y})^2.$$

An estimate for the population correlation ρ is given by the *sample correlation*

$$r = \hat{\rho} = \frac{\widehat{\text{Cov}}(X, Y)}{\hat{\sigma}_X \hat{\sigma}_Y} = \frac{\sum_{i=1}^{n} (X_i - \bar{X})(Y_i - \bar{Y})}{\sqrt{\left[\sum_{i=1}^{n} (X_i - \bar{X})^2 \right] \left[\sum_{i=1}^{n} (Y_i - \bar{Y})^2 \right]}}.$$

Notice that the $1/n$ terms in the numerator and the denominator cancel. Had a denominator of $n-1$, rather than n, been used in the formulas for $\widehat{\text{Cov}}(X,Y)$, $\hat{\sigma}_X^2$, and $\hat{\sigma}_Y^2$, the same cancellation would occur. Table 7.3 summarizes the results from Section 7.2.1 and this paragraph.

	covariance	correlation
population	$E\left[(X-\mu_X)(Y-\mu_Y)\right]$	$\dfrac{E\left[(X-\mu_X)(Y-\mu_Y)\right]}{\sigma_X\sigma_Y}$
sample	$\dfrac{1}{n}\sum_{i=1}^{n}(X_i-\bar{X})(Y_i-\bar{Y})$	$\dfrac{\sum_{i=1}^{n}(X_i-\bar{X})(Y_i-\bar{Y})}{\sqrt{\left[\sum_{i=1}^{n}(X_i-\bar{X})^2\right]\left[\sum_{i=1}^{n}(Y_i-\bar{Y})^2\right]}}$

Table 7.3: Population and sample covariance and correlation.

We now translate the concepts from the previous paragraph into the context of the analysis of a time series. Consider the estimation of $\gamma(k)$ and $\rho(k)$ from a realization of observations $x_1, x_2, \ldots, x_n$, which are assumed to be observed values from a stationary time series model. The *lag k sample autocovariance*, which estimates $\gamma(k)$, is

$$c_k = \frac{1}{n}\sum_{t=1}^{n-k}(x_t-\bar{x})(x_{t+k}-\bar{x}),$$

where $\bar{x}$ is the sample mean of the observations in the time series. This is not a universal choice for c_k. Since there are $n-k$ terms in the summation, some time series analysts prefer to divide by $n-k$ rather than n. Because of the two different options for the denominator, it is important to only calculate c_k for k values that are significantly smaller than n. Generally speaking, there should be at least 60 to 70 observations in a time series to use the techniques described here. Having a large value of n means that having n or $n-k$ in the denominator is not critical for small values of k. The units on c_k are the square of the units of the observations in the time series. Notice that when $k=0$, the lag 0 sample autocovariance reduces to

$$c_0 = \frac{1}{n}\sum_{t=1}^{n}(x_t-\bar{x})^2,$$

which is an estimate for $\gamma(0)=\sigma_X^2$. The *lag k sample autocorrelation*, which estimates $\rho(k)$, is

$$r_k = \frac{c_k}{c_0}$$

for integer values of k which are significantly smaller than n. As was the case with $\rho(k)$, the lag k sample autocorrelation is a unitless quantity. When $k=0$, $r_0=c_0/c_0=1$, as desired. The notation developed here to calculate $\gamma(k)$ and $\rho(k)$ for a stationary time series model and to estimate these functions with c_k and r_k for an observed time series $x_1, x_2, \ldots, x_n$ is summarized in Table 7.4.

	lag k autocovariance	lag k autocorrelation
population	$\gamma(k) = E\left[(X_t - \mu_X)(X_{t+k} - \mu_X)\right]$	$\rho(k) = \dfrac{\gamma(k)}{\gamma(0)}$
sample	$c_k = \dfrac{1}{n}\displaystyle\sum_{t=1}^{n-k}(x_t - \bar{x})(x_{t+k} - \bar{x})$	$r_k = \dfrac{c_k}{c_0}$

Table 7.4: Population and sample lag k autocovariance and autocorrelation.

Computing Sample Autocovariance and Autocorrelation

We now consider the estimation of the lag k sample autocovariance c_k and the lag k sample autocorrelation r_k in R. We write an R function named `autocovariance` below that has two arguments: the vector containing the time series x and the lag k. The first statement in the function uses the `length` function to determine the number of observations in the time series. The second statement uses the `mean` function to calculate the sample mean of the values in the time series. The third statement uses the formula

$$c_k = \frac{1}{n}\sum_{t=1}^{n-k}(x_t - \bar{x})(x_{t+k} - \bar{x})$$

to calculate the lag k sample autocovariance.

```
autocovariance = function(x, k) {
  n = length(x)
  xbar = mean(x)
  sum((x[1:(n - k)] - xbar) * (x[(k + 1):n] - xbar)) / n
}
```

We can now write an R function named `autocorrelation` below that has the same arguments as the `autocovariance` function. It uses the formula

$$r_k = \frac{c_k}{c_0}$$

to calculate the lag k sample autocorrelation.

```
autocorrelation = function(x, k) {
  autocovariance(x, k) / autocovariance(x, 0)
}
```

Time series analysts typically plot the sample autocorrelation function values for the first few lags. This plot is known as either the *sample autocorrelation function* or the *correlogram*. We illustrate the calculation and plotting of the correlogram for a simulated time series whose elements are Gaussian white noise, so $\sigma_X = \sigma_Z$. Recall that Gaussian white noise, denoted by

$$X_t \sim GWN\left(0, \sigma_Z^2\right),$$

consists of mutually independent $N\left(0, \sigma_Z^2\right)$ observations. Recall from Example 7.15 that the population autocovariance function is

$$\gamma(k) = \left\{ \begin{array}{ll} \sigma_Z^2 & k = 0 \\ 0 & k = 1, 2, \ldots \end{array} \right.$$

and the population autocorrelation function is

$$\rho(k) = \left\{ \begin{array}{ll} 1 & k = 0 \\ 0 & k = 1, 2, \ldots \end{array} \right. .$$

We expect the sample autocorrelation function to be similar to the population autocorrelation function except for random sampling variability. The R code below plots the correlogram for the first 20 lags for a time series that consists of $n = 100$ observations of a Gaussian white noise time series. We have assumed here that the population variance of the Gaussian white noise is equal to one (that is, $\sigma_Z = 1$). The first statement in the R code below uses the set.seed function to set the random number seed to 8. The second statement uses the rnorm function to generate a time series consisting of 100 mutually independent standard normal random variates. The vector correlogram is initialized to a vector of length 21. This will hold the lag 0 sample autocorrelation function value (which is always $r_0 = 1$) and the sample autocorrelation function values for lags 1 to 20. The autocorrelation function defined previously will compute the sample autocorrelation values. Finally, the plot function is used to plot the sample autocorrelation function. Using the type = "h" argument in the call to plot graphs the sample autocorrelation values as spikes. This is largely a matter of personal taste. Some time series analysts prefer to connect them with a line. We take the spike approach to emphasize that a non-integer value for the lag has no meaning in the context described here. The ylim = c(-1, 1) argument is included so that the entire potential range of the sample autocorrelation values $-1 \leq r_k \leq 1$ is included. The abline function is used to draw a horizontal line at $r_k = 0$, and two other dashed lines that will be described subsequently.

```
set.seed(8)
n = 100
x = rnorm(n)
correlogram = numeric(21)
for (i in 1:21) correlogram[i] = autocorrelation(x, i - 1)
plot(0:20, correlogram, type = "h", ylim = c(-1, 1))
abline(h = 0)
abline(h = c(-1, 1) * 2 / sqrt(n), lty = 2)
```

The plot of the time series and the correlogram for the first 20 lags are given in Figure 7.12. The time series plot displays the typical pattern for Gaussian white noise. The observations are mutually independent, so equally likely to be positive or negative. There are just a handful of observations more than 2 units away from the population mean function $\mu(t) = E[X_t] = 0$. The correlogram is exactly what we anticipated for a time series consisting of Gaussian white noise based on our population autocorrelation function $\rho(k)$, which was one at lag zero and zero at all other lags. We have $r_0 = 1$, as expected, and then small spikes associated with values of r_k at other lag values k that reflect the random sampling variability in the specific time series values that were generated by the rnorm function. The correlogram has a horizontal line drawn at correlation 0 to make it clearer which spikes are positive and which are negative. In addition, the correlogram would be identical if all of the points in the time series were translated to have arbitrary population mean μ rather than population mean 0. Correlograms are not influenced by a shift in the time series. Since drawing a

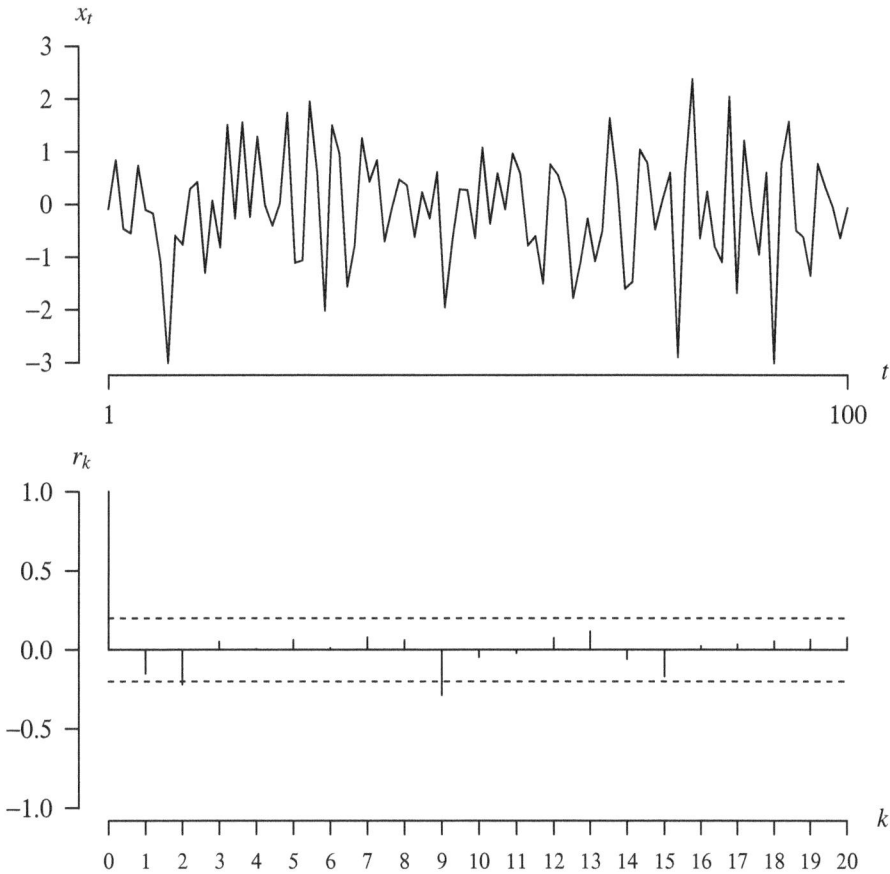

Figure 7.12: Time series plot and correlogram for $n = 100$ Gaussian white noise observations.

correlogram occurs so frequently in the analysis of a time series, R has automated the process with the `acf` function (acf is an abbreviation for autocorrelation function). After a call to `set.seed(8)`, all of the previous calculations can be performed with the single R statement

```
acf(rnorm(100), ylim = c(-1, 1), lag.max = 20)
```

When you make this call to the `acf` function, you will notice that two dashed lines are drawn just above and just below the horizontal line $r_k = 0$, as was the case in Figure 7.12. These two lines are 95% confidence limits that are helpful for determining whether the sample autocorrelation function values are statistically different from zero. Even for a time series consisting of Gaussian white noise, the probability distribution of r_k is complicated because the formula for r_k is complicated. However, when the time series $\{X_t\}$ consists of mutually independent observations, the population mean and variance of r_k are

$$E[r_k] \cong \frac{1}{n}$$

and

$$V[r_k] \cong \frac{1}{n}$$

for $k = 1, 2, \ldots$, and these values are asymptotically normally distributed. This means that an approximate two-sided 95% confidence interval for r_k when n is large and k is significantly less than n is

$$\frac{1}{n} - 1.96\frac{1}{\sqrt{n}} < r_k < \frac{1}{n} + 1.96\frac{1}{\sqrt{n}},$$

where 1.96 is the 0.975 fractile of the standard normal distribution. Since n is typically large in time series analysis, the $1/n$ term is often assumed to be small enough to ignore. Furthermore, if 1.96 is rounded to 2, then this approximate 95% confidence interval simplifies to

$$-\frac{2}{\sqrt{n}} < r_k < \frac{2}{\sqrt{n}},$$

The limits at $\pm 2/\sqrt{n}$ are plotted in Figure 7.12 as dashed lines at $\pm 2/\sqrt{100} = \pm 0.2$. We notice that the spikes in the correlogram in Figure 7.12 fall outside of the confidence interval limits for lag 2 (just barely) and lag 9. We should not be concerned about this occurring. Since these are approximate 95% confidence intervals, we would expect to have about 1 in 20 values fall outside of the limits even if we had mutually independent observations in the time series. Since it appears that there is little or no pattern to the spikes in Figure 7.12, we conclude that the two spikes that exceeded the confidence limits are just due to random sampling variability. Significant spikes at lower lags, for example, lag 1 and lag 2 should be scrutinized more carefully than other lone significant spikes, such as the one that we saw at lag 9. Furthermore, a significant spike at a lag associated with possible seasonal variation (for example, lag 12 for monthly data with a suspected annual variation) should also be scrutinized more carefully than other statistically significant spikes.

Correlogram Examples

Experience is critical in interpreting correlograms. Four examples are given here to illustrate the recommended thought process associated with the interpretation of a time series and its correlogram. The four examples are

- a time series with a linear trend illustrated by the population of Australia from 1971–1993,

- a time series of the first 100 Dow Jones Industrial Average closing values in the year 2000,

- a time series of chemical yields, and

- a seasonal time series illustrated by the home energy consumption values from 2011–2018.

For all four time series, we (a) plot the time series, (b) plot the associated correlogram, (c) interpret the statistically significant spikes in the correlogram, and (d) interpret the shape of the spikes in the correlogram.

> **Example 7.18** This example considers the calculation of the sample autocorrelation function for a time series with a linear trend. The time series consists of $n = 89$ quarterly observations, which are the quarterly number of Australian residents (in thousands) from the second quarter of 1971 through the second quarter of 1993. This time series is built into R and has the name `austres`. Plot the time series and associated correlogram, and interpret the significance of the spikes and shape formed by the values of r_k.
>
> We can view the observations in the time series by just typing the name of the data set.
>
> `austres`

The output from this command is given below.

```
          Qtr1      Qtr2      Qtr3      Qtr4
1971              13067.3  13130.5  13198.4
1972  13254.2  13303.7  13353.9  13409.3
1973  13459.2  13504.5  13552.6  13614.3
1974  13669.5  13722.6  13772.1  13832.0
1975  13862.6  13893.0  13926.8  13968.9
1976  14004.7  14033.1  14066.0  14110.1
1977  14155.6  14192.2  14231.7  14281.5
1978  14330.3  14359.3  14396.6  14430.8
1979  14478.4  14515.7  14554.9  14602.5
1980  14646.4  14695.4  14746.6  14807.4
1981  14874.4  14923.3  14988.7  15054.1
1982  15121.7  15184.2  15239.3  15288.9
1983  15346.2  15393.5  15439.0  15483.5
1984  15531.5  15579.4  15628.5  15677.3
1985  15736.7  15788.3  15839.7  15900.6
1986  15961.5  16018.3  16076.9  16139.0
1987  16203.0  16263.3  16327.9  16398.9
1988  16478.3  16538.2  16621.6  16697.0
1989  16777.2  16833.1  16891.6  16956.8
1990  17026.3  17085.4  17106.9  17169.4
1991  17239.4  17292.0  17354.2  17414.2
1992  17447.3  17482.6  17526.0  17568.7
1993  17627.1  17661.5
```

The plot of the time series and the plot of the sample autocorrelation function are graphed one above the another using the R `plot.ts` and `acf` functions. The `par` function called with the argument `mfrow = c(2, 1)` creates a template for a 2×1 matrix of graphs.

```
par(mfrow = c(2, 1))
plot.ts(austres, type = "p", cex = 0.4)
abline(h = mean(austres))
acf(austres)
```

The default on the `plot.ts` function is to connect the time series values with lines. That default is modified here by setting the `type` argument to `"p"` in order to just plot points instead. The `cex` (character expand) parameter controls the size of the points. A horizontal line has been added to the time series plot using the `abline` function at the sample mean value of the time series, $\bar{x} = 15,273$, which will aid in the interpretation of the values of r_k. The plots are displayed in Figure 7.13. The time series is plotted as just points because of the linear growth in the population. The first 46 of the $n = 89$ observations are below the sample mean, and the remainder are above the sample mean. Consider now the calculation of c_1, the lag 1 sample autocovariance. The formula for c_1 is

$$c_1 = \frac{1}{n} \sum_{t=1}^{n-1} (x_t - \bar{x})(x_{t+1} - \bar{x}).$$

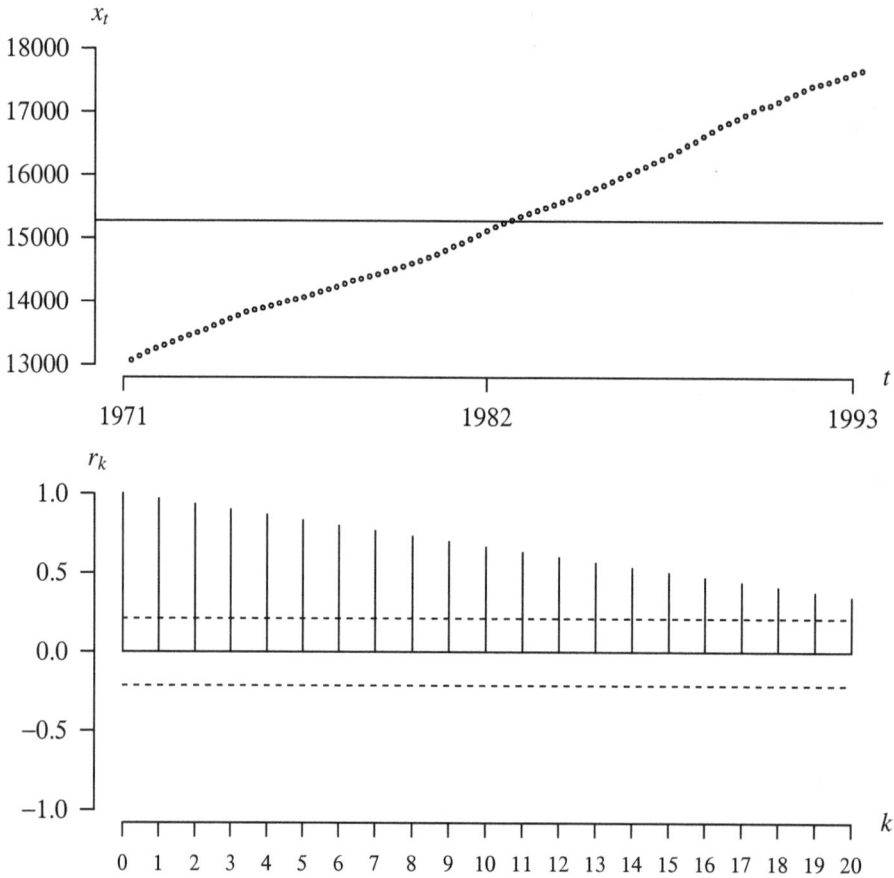

Figure 7.13: Time series plot and correlogram for $n = 89$ quarterly population observations.

Consider adjacent observations in the time series (that is, observations that are just one lag apart). The first two observations, x_1 and x_2, are both less than $\bar{x}$, so the product $(x_1 - \bar{x})(x_2 - \bar{x})$ makes a positive contribution to c_1. Likewise, x_2 and x_3 make a positive contribution to c_1. Likewise, x_3 and x_4 make a positive contribution to c_1. In fact, all of the adjacent observations make a positive contribution to c_1 except for x_{46} and x_{47}, which are on opposite sides of $\bar{x}$, so this pair makes a negative contribution. It is for this reason that c_1 will be positive for this particular time series, and the associated correlation r_1 will be positive and statistically significant. The terms in c_1 and c_0 are very similar for this time series, so r_1 will be close to 1.

Now consider observations in the time series that are two lags apart. There are now $n-2$ terms in the summation for c_2. The first two observations, x_1 and x_3, are both less than $\bar{x}$, so the product $(x_1 - \bar{x})(x_3 - \bar{x})$ makes a positive contribution to c_2. Likewise, x_2 and x_4 make a positive contribution to c_2. Likewise, x_3 and x_5 make a positive contribution to c_2. In fact, all of the observations make a positive contribution to c_1 except for two pairs, x_{45} and x_{47} and x_{46} and x_{48}, which are on opposite sides of $\bar{x}$. These pairs make a negative contribution. So there will be a significant positive value for r_2, but it will be slightly smaller in magnitude than r_1. This pattern continues for $r_3, r_4, \ldots, r_{32}$ as the r_k values are a decreasing value of k. Then at lag 33, which is beyond the lags displayed in Figure 7.13, there is an approximately equal number of positive and negative terms in the summation to calculate c_{33}. This results in r_{33} being the first negative value in the correlogram. So r_{33} and the sample autocorrelation function values that follow it are negative. So in conclusion, a time series with a linear trend (either increasing or decreasing) has a correlogram in which the initial spikes of r_k are linearly decreasing in k. The correlogram for a time series with a linear or nonlinear trend will not have a traditional interpretation that will be seen in the other examples because the trend overwhelms the values in the correlogram. It is often the case that the trend is first removed, and then the correlogram of the detrended series is analyzed. It is a good exercise to use the `acf` function with a bigger `lag.max` argument than the default to see what the autocorrelation function does for larger values of k.

The next example considers a time series that has statistically significant positive autocorrelation values at small lags.

Example 7.19 Consider again the time series of the first $n = 100$ Dow Jones Industrial Averages during 2000 that was first detailed in Example 7.4. Plot the time series and associated correlogram, and interpret the significance of the spikes and shape formed by the values of r_k.

The time series (with a horizontal line drawn at the mean value $\bar{x} = 10{,}766$) and the correlogram are shown in Figure 7.14. Consider the lag 1 sample autocorrelation. Considering the adjacent observations in the time series plot, the vast majority lie on the same side of $\bar{x}$. This results in a statistically significant positive lag 1 sample autocorrelation r_1. Similar thinking should convince you that there will also be a statistically significant positive lag 2 sample autocorrelation r_2. This time series exhibits runs of significant length above and below the population mean, so it has a dozen statistically significant initial spikes on the correlogram. So a time series that is well-modeled by a random walk (as shown in Example 7.4) has a correlogram with statistically significant early positive sample autocorrelation values.

The next example considers a stationary time series in which adjacent observations tend to be on opposite sides of the sample mean.

Example 7.20 Consider the time series consisting of $n = 70$ consecutive yields from a batch chemical process from page 32 of Box and Jenkins (1976) given in Table 7.5 (read row-wise). Plot the time series and associated correlogram. Interpret the statistical significance of the spikes and the shape formed by the values of r_k.

The time series plot of the yields, along with a horizontal line at $\bar{x} = 51.1$, is given in the top graph in Figure 7.15. The bottom graph contains the associated correlogram. The time series plot indicates that a large yield is followed by a small yield in a majority of the observations, so we expect a negative lag 1 sample autocorrelation function value. The lag 1 sample autocorrelation function value is $r_1 = -0.390$. Since observations that are two apart tend to be on the same side of the sample mean, the lag 2 sample autocorrelation function value is $r_2 = 0.304$. So a time series which tends to jump above and below the sample mean results in a correlogram whose r_k values alternate

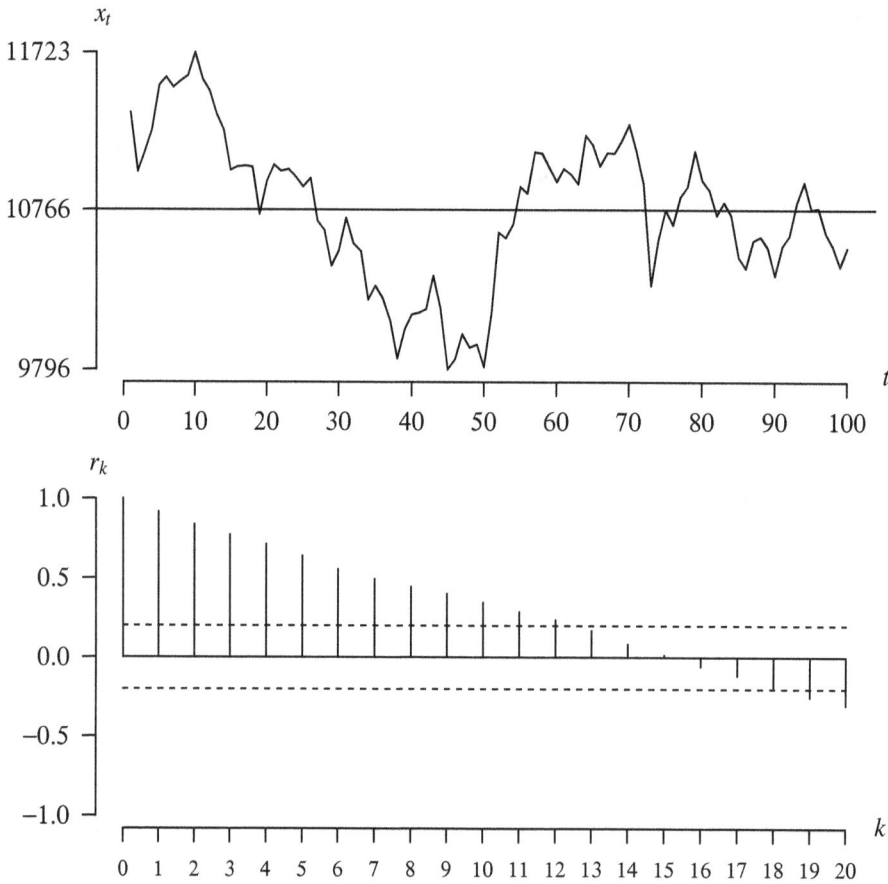

Figure 7.14: Time series plot and correlogram for $n = 100$ DJIA closing prices.

in sign. These are the only two values of the correlogram that show a statistically significant difference from zero because they lie outside of the 95% confidence limits. The dashed horizontal lines corresponding to 95% confidence bounds that determine

47	64	23	71	38	64	55	41	59	48	71	35	57	40
58	44	80	55	37	74	51	57	50	60	45	57	50	45
25	59	50	71	56	74	50	58	45	54	36	54	48	55
45	57	50	62	44	64	43	52	38	59	55	41	53	49
34	35	54	45	68	38	50	60	39	59	40	57	54	23

Table 7.5: A time series of $n = 70$ consecutive yields from a chemical process.

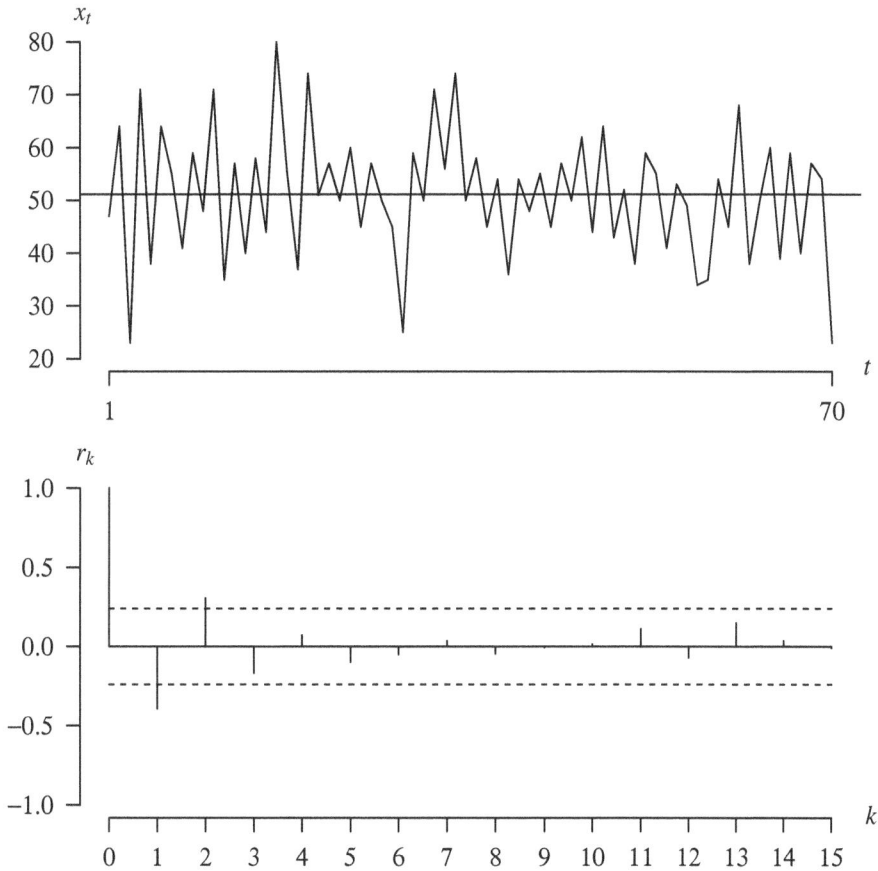

Figure 7.15: Time series plot and correlogram of $n = 70$ yields from a chemical process.

statistical significance are drawn on the correlogram at heights

$$\pm \frac{2}{\sqrt{70}} \cong \pm 0.239.$$

These two sample autocorrelation function values might be due to overcorrection by the personnel running the batch chemical process.

   The final example illustrates the impact of a time series with a seasonal component on the shape of the correlogram.

   Example 7.21 Consider again the home energy consumption time series from Example 7.1 consisting of $n = 84$ monthly observations (measured in kilowatt hours) collected between 2011 and 2018. Plot the time series and associated correlogram. Interpret the statistical significance of the spikes and the shape formed by the values of r_k.

The time series and correlogram are shown in Figure 7.16, with a horizontal line drawn at the mean value $\bar{x} = 1703$ kilowatt hours. The time series displays a seasonal pattern, with higher energy consumption during the winter months and the summer months. The winter months tend to draw more energy than the summer months. The correlogram for a time series with a seasonal component is also cyclic, with a frequency that matches the frequency in the time series. Since the summer and winter months draw more energy from the heat pump, the cycle on the correlogram repeats itself with a wavelength of 6. The fact that the magnitude of r_{12} is larger than the magnitude of r_6 is due to the fact that the winter months draw more energy than the summer months. For this particular time series, the shape of the correlogram does not provide much information beyond confirming that this is a time series with a seasonal component. The more valuable information is typically gleaned by first detrending the time series (that is, removing the seasonal component) and inspecting the correlogram of the detrended time series. The statistically significant sample autocorrelation function value at lag 3, which is $r_3 = -0.397$, indicates that energy consumption in months that differ by 3 (for example,

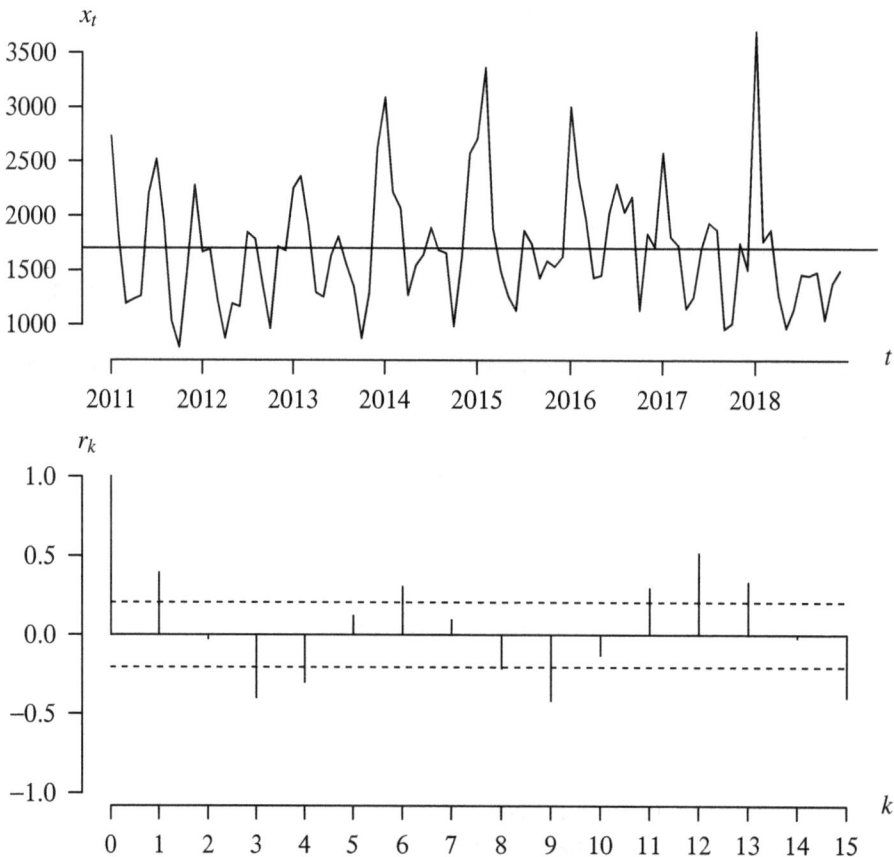

Figure 7.16: Time series plot and correlogram for $n = 84$ home energy consumption values.

February and May), tend to have energy consumptions that lie on the opposite side of the sample mean.

To summarize the section thus far, the population mean function $\mu(t) = E[X_t]$ is the expected value of the time series at time t, and indicates whether a trend and/or cyclic variation is present. The population autocovariance and population autocorrelation functions reflect the linear relationship between two values, X_s and X_t, in the time series. A time series is stationary if the population mean function is constant in t and the population autocovariance function $\gamma(s,t) = \text{Cov}(X_s, X_t)$ depends only on $|t - s|$. For a stationary time series, the population autocovariance function and the population autocorrelation function can be written as a function of the lag $k = |t - s|$ as $\gamma(k)$ and $\rho(k)$. The sample autocorrelation function estimates the population autocorrelation function.

Occasions arise in time series analysis in which we are also interested in the correlation between X_t and X_{t+k} with the linear dependency on the values between these two values, namely $X_{t+1}, X_{t+2}, \ldots, X_{t+k-1}$, removed. This leads to what is known as the partial autocorrelation function, which is presented next.

7.2.4   Population Partial Autocorrelation

One key characteristic of nearly all time series is that nearby observations tend to be correlated. This makes the notion of autocorrelation crucial in time series analysis because it captures the correlation between observations in a stationary time series that are separated in time by a prescribed number of lags. The population autocorrelation function was introduced in Section 7.2.1; its statistical counterpart, the sample autocorrelation function, was introduced in Section 7.2.3.

It is often difficult to distinguish between population autocorrelation functions for two different competing, tentative stationary time series models in practice because

- the population autocorrelation functions for the two models are nearly identical, and/or

- there is significant sampling variability in the sample autocorrelation functions making it difficult to determine which of the two models provides a better fitted model.

As will be seen in subsequently, the sample autocorrelation function is particularly helpful for determining the number of terms to include in a popular time series model known as a *moving average model*. However, the sample autocorrelation function is much less helpful for determining the number of terms to include in another popular time series model known as an *autoregressive model*. A second type of autocorrelation function, the *partial autocorrelation function*, is an ancillary diagnostic tool to help determine the number of terms to include in an autoregressive model. As was the case with the autocorrelation function, there is a population and a sample version of the partial autocorrelation function. The population partial autocorrelation function is introduced in this subsection.

The notion behind partial autocorrelation is intuitive. Before describing the interpretation of partial autocorrelation in the context of time series analysis, we present a scenario involving just partial correlation in a more general setting. Let's say you are interested in the correlation between a full-time employee's age, X, and their annual income, Y. Intuition suggests that this correlation is positive because annual income tends to rise with age. But there are other factors that influence income, such as the employee's education level achieved, the number of years on the job, specific industry of employment, etc. To simplify, let's consider just one of these factors, say, the employee's years of education achieved, Z. The population partial correlation is the population correlation between age X and annual income Y with the linear relationship associated with the number of years of education Z removed. We are effectively controlling for the influence of education as we measure the correlation between X and Y. We regress X on Z to obtain $\tilde{X}$. We regress Y on Z to obtain $\tilde{Y}$. This regression is in the sense of probability rather than its statistical counterpart (which typically uses least squares for parameter estimation). Finally, we calculate the population partial correlation $\mathrm{Corr}\left(X - \tilde{X}, Y - \tilde{Y}\right)$, which is the population correlation between X and Y with the linear influence of Z removed. Like the ordinary population correlation, the population partial correlation falls in the closed interval $[-1, 1]$. The extreme values on this interval correspond to perfect negative population correlation and perfect positive population correlation, respectively. To summarize, the partial correlation measures the degree of linear association between two variables, with the linear association of one or more other variables removed.

Returning to the time series context, the partial autocorrelation reflects the relationship between observations at a particular lag in a time series with the linear relationship associated with intervening observations removed. The variables whose influence is being removed are the observations between the two values of the time series of interest. Stated in another fashion, the partial autocorrelation at lag k is the population correlation between two observations in the time series that are k time units apart after the removal of the linear influence of the time series observations at lags $1, 2, \ldots, k-1$. As was the case with autocorrelation, we want to find the population and sample versions of the partial autocorrelation. We will then have an inventory of possible population partial autocorrelation shapes that we can match to sample partial autocorrelation functions, which will help determine the number of terms to include in a time series model. The main role of the sample partial autocorrelation function is to determine the number of terms to include in an autoregressive model.

We now develop some general notation concerning partial autocorrelation. Although many authors use ϕ_{kk} to denote the population lag k partial autocorrelation, we will instead use $\rho^*(k)$ to emphasize that this quantity is still a correlation and to use the symbol ϕ exclusively for the coefficients in an autoregressive time series model. The superscript $*$ is used to distinguish the population partial autocorrelation function from the population autocorrelation function $\rho(k)$. Likewise, we will use r_k^* to denote the sample lag k partial autocorrelation in the next subsection. The superscript $*$ is used here to distinguish the sample partial autocorrelation function from the sample autocorrelation function r_k.

The next example illustrates the calculation of a population partial autocorrelation for a stationary time series model.

Example 7.22 Consider the stationary time series model

$$X_t = 0.8 X_{t-1} + Z_t,$$

where $\{Z_t\} \sim WN\left(0, \sigma_Z^2\right)$. The current value in the time series is 0.8 times the previous value in the time series, plus a random shock of white noise Z_t. This time series model is similar to the random walk time series model that was introduced in Example 7.4 and analyzed in Examples 7.8, 7.11, and 7.17. The random walk time series model was determined to be nonstationary. The one small difference between this time series model and the random walk is the 0.8 coefficient associated with the X_{t-1} term. This small alteration makes this time series model stationary. What is the population lag 2 partial autocorrelation?

The population lag 2 partial autocorrelation is the population correlation between X_t and X_{t-2} with the linear effect of the intervening observation X_{t-1} removed. This is the population correlation between $X_t - 0.8 X_{t-1}$ and $X_{t-2} - 0.8 X_{t-1}$, which can be written as

$$\rho^*(2) = \text{Corr}\,(X_t - 0.8X_{t-1}, X_{t-2} - 0.8X_{t-1})$$

$$= \frac{\text{Cov}\,(X_t - 0.8X_{t-1}, X_{t-2} - 0.8X_{t-1})}{\sqrt{V\,[X_t - 0.8X_{t-1}]\,V\,[X_{t-2} - 0.8X_{t-1}]}}$$

$$= \frac{\text{Cov}\,(Z_t, X_{t-2} - 0.8X_{t-1})}{\sqrt{V\,[Z_t]\,V\,[X_{t-2} - 0.8X_{t-1}]}}$$

$$= \frac{0}{\sqrt{V\,[Z_t]\,V\,[X_{t-2} - 0.8X_{t-1}]}}$$

$$= 0$$

because the population covariance between the white noise term at time t, which is Z_t, and the linear combination of the two previous values of the time series, which is $X_{t-2} - 0.8X_{t-1}$, is zero.

We now pivot from the calculation of population partial autocorrelation for a specific time series model to the calculation of the population partial autocorrelation for a general stationary time series model. Let's begin with the calculation of the lag 1 population partial autocorrelation $\rho^*(1)$ for a stationary time series model. By definition, this is the population correlation between X_t and X_{t-1} after removing the linear effect of any observations between X_t and X_{t-1}. But there aren't any observations between X_t and X_{t-1}, so the lag 1 population partial autocorrelation is simply the lag 1 population autocorrelation: $\rho^*(1) = \rho(1)$. The population partial autocorrelation for higher lags uses the best linear estimate of each of the two values of interest as a function of the intervening values. Minimizing the associated mean square error, the population partial autocorrelation can be determined by solving a set of linear equations. Using Cramer's rule to solve these equations, the population lag 2 partial autocorrelation function value is given by the ratio of determinants

$$\rho^*(2) = \frac{\begin{vmatrix} 1 & \rho(1) \\ \rho(1) & \rho(2) \end{vmatrix}}{\begin{vmatrix} 1 & \rho(1) \\ \rho(1) & 1 \end{vmatrix}}.$$

Notice that the denominator is the determinant of the correlation matrix of any two adjacent observations. The numerator is the determinant of this same matrix with the last column replaced by the first two population autocorrelation values. This pattern continues for the population lag 3 partial autocorrelation function value, which is

$$\rho^*(3) = \frac{\begin{vmatrix} 1 & \rho(1) & \rho(1) \\ \rho(1) & 1 & \rho(2) \\ \rho(2) & \rho(1) & \rho(3) \end{vmatrix}}{\begin{vmatrix} 1 & \rho(1) & \rho(2) \\ \rho(1) & 1 & \rho(1) \\ \rho(2) & \rho(1) & 1 \end{vmatrix}}.$$

Again, the denominator is the determinant of the correlation matrix of any three sequential observations. The numerator is the determinant of this same matrix with the last column replaced by the first three population autocorrelation values (where the lag number matches the row number). This pattern continues for higher lag values, which leads to the following definition.

Definition 7.8 For a stationary time series model, the lag 0 population partial autocorrelation is 1, the lag 1 population partial autocorrelation is $\rho(1)$, and the lag k population partial autocorrelation is

$$\rho^*(k) = \frac{\begin{vmatrix} 1 & \rho(1) & \rho(2) & \cdots & \rho(1) \\ \rho(1) & 1 & \rho(1) & \cdots & \rho(2) \\ \rho(2) & \rho(1) & 1 & \cdots & \rho(3) \\ \vdots & \vdots & \vdots & \ddots & \vdots \\ \rho(k-1) & \rho(k-2) & \rho(k-3) & \cdots & \rho(k) \end{vmatrix}}{\begin{vmatrix} 1 & \rho(1) & \rho(2) & \cdots & \rho(k-1) \\ \rho(1) & 1 & \rho(1) & \cdots & \rho(k-2) \\ \rho(2) & \rho(1) & 1 & \cdots & \rho(k-3) \\ \vdots & \vdots & \vdots & \ddots & \vdots \\ \rho(k-1) & \rho(k-2) & \rho(k-3) & \cdots & 1 \end{vmatrix}},$$

for $k = 2, 3, \ldots$.

The next example illustrates the calculation of the population partial autocorrelation function for a stationary time series model.

Example 7.23 Consider the time series model for $\{X_t\}$ described by

$$X_t = Z_t - Z_{t-1} + \frac{1}{2}Z_{t-2},$$

where $\{Z_t\} \sim WN\left(0, \sigma_Z^2\right)$. The current value of the time series is a linear combination of the current and two previous shock values. Find the population autocorrelation function and the population partial autocorrelation function for the first eight lags.

The population mean function is

$$\mu(t) = E[X_t] = E\left[Z_t - Z_{t-1} + \frac{1}{2}Z_{t-2}\right] = E[Z_t] - E[Z_{t-1}] + \frac{1}{2}E[Z_{t-2}] = 0.$$

The population autocovariance function is

$$\begin{aligned}
\gamma(s,t) &= \text{Cov}(X_s, X_t) \\
&= E[(X_s - E[X_s])(X_t - E[X_t])] \\
&= E[X_s X_t] \\
&= E\left[\left(Z_s - Z_{s-1} + \frac{1}{2}Z_{s-2}\right)\left(Z_t - Z_{t-1} + \frac{1}{2}Z_{t-2}\right)\right] \\
&= E[Z_s Z_t] - E[Z_s Z_{t-1}] + \frac{1}{2}E[Z_s Z_{t-2}] - E[Z_{s-1}Z_t] + E[Z_{s-1}Z_{t-1}] - \\
&\quad \frac{1}{2}E[Z_{s-1}Z_{t-2}] + \frac{1}{2}E[Z_{s-2}Z_t] - \frac{1}{2}E[Z_{s-2}Z_{t-1}] + \frac{1}{4}E[Z_{s-2}Z_{t-2}].
\end{aligned}$$

When $t = s$,

$$\gamma(t,t) = E\left[Z_t^2\right] + E\left[Z_{t-1}^2\right] + \frac{1}{4}E\left[Z_{t-2}^2\right] = V[Z_t] + V[Z_{t-1}] + \frac{1}{4}V[Z_{t-2}] = \frac{9}{4}\sigma_Z^2$$

because of the mutual independence of the white noise terms and the fact that the expected value of each white noise term is zero. When $|t - s| = 1$, for example, when $t = s - 1$,

$$\gamma(s, t) = -E\left[Z_{s-1}Z_t\right] - \frac{1}{2}E\left[Z_{s-2}Z_{t-1}\right] = -\frac{3}{2}\sigma_Z^2.$$

When $|t - s| = 2$, for example, when $t = s - 2$,

$$\gamma(s, t) = \frac{1}{2}E\left[Z_{s-2}Z_t\right] = \frac{1}{2}\sigma_Z^2.$$

When $|t - s| = 3, 4, \ldots$, the population autocovariance is $\gamma(s, t) = 0$ because each expected value in the expansion of $\gamma(s, t)$ contains independent random variables. Since the population mean function is constant in time and the population autocovariance is a function of the lag $k = |t - s|$ (as required by Definition 7.6), we have established that the time series model is stationary with population autocovariance function

$$\gamma(k) = \begin{cases} 9\sigma_Z^2/4 & k = 0 \\ -3\sigma_Z^2/2 & k = 1 \\ \sigma_Z^2/2 & k = 2 \\ 0 & k = 3, 4, \ldots \,. \end{cases}$$

Since $\rho(k) = \gamma(k)/\gamma(0)$, the population autocorrelation function is

$$\rho(k) = \begin{cases} 1 & k = 0 \\ -2/3 & k = 1 \\ 2/9 & k = 2 \\ 0 & k = 3, 4, \ldots, \end{cases}$$

Notice that the population autocorrelation function is independent of the population variance of the white noise σ_Z^2.

We now turn to calculation the population partial autocorrelation function. The lag 0 population partial autocorrelation is $\rho^*(0) = 1$. From Definition 7.8, the lag 1 population partial autocorrelation is $\rho^*(1) = \rho(1) = -2/3$. The lag 2 population partial autocorrelation is the ratio of the determinants of two 2×2 matrices:

$$\rho^*(2) = \frac{\begin{vmatrix} 1 & \rho(1) \\ \rho(1) & \rho(2) \end{vmatrix}}{\begin{vmatrix} 1 & \rho(1) \\ \rho(1) & 1 \end{vmatrix}} = \frac{\begin{vmatrix} 1 & -2/3 \\ -2/3 & 2/9 \end{vmatrix}}{\begin{vmatrix} 1 & -2/3 \\ -2/3 & 1 \end{vmatrix}} = \frac{-2/9}{5/9} = -\frac{2}{5} = -0.4.$$

The lag 3 population partial autocorrelation is the ratio of the determinants of two 3×3 matrices:

$$\rho^*(3) = \frac{\begin{vmatrix} 1 & \rho(1) & \rho(1) \\ \rho(1) & 1 & \rho(2) \\ \rho(2) & \rho(1) & \rho(3) \end{vmatrix}}{\begin{vmatrix} 1 & \rho(1) & \rho(2) \\ \rho(1) & 1 & \rho(1) \\ \rho(2) & \rho(1) & 1 \end{vmatrix}} = \frac{\begin{vmatrix} 1 & -2/3 & -2/3 \\ -2/3 & 1 & 2/9 \\ 2/9 & -2/3 & 0 \end{vmatrix}}{\begin{vmatrix} 1 & -2/3 & 2/9 \\ -2/3 & 1 & -2/3 \\ 2/9 & -2/3 & 1 \end{vmatrix}} = \frac{-8/243}{7/27} = -\frac{8}{63} \cong -0.1270.$$

Computing determinants by hand gets more tedious as the size of the matrices grows.
The R code below automates this process, using the det function to calculate the determinants. After executing the code, the vector rhostar contains the first eight population partial autocorrelation values. Examine the subscripts carefully because there is a
lag zero autocorrelation but R begins its subscripts at 1.

```
rho = c(1, -2 / 3, 2 / 9, 0, 0, 0, 0, 0, 0)
rhostar = rho
for (k in 2:(length(rho) - 1)) {
  a = matrix(1, k, k)
  for (i in 1:(k - 1)) a[abs(row(a) - col(a)) == i] = rho[i + 1]
  denominator = det(a)
  a[ , k] = rho[2:(k + 1)]
  numerator = det(a)
  rhostar[k + 1] = numerator / denominator
}
rhostar
```

The calculations in this example have been automated in the ARMAacf function in R.
The ar and ma parameters will be described in a subsequent chapter, but notice that
the elements in the ma vector are the coefficients of Z_{t-1} and Z_{t-2} in the original time
series model. The first R command below calculates the values of $\rho(1), \rho(2), \ldots, \rho(8)$,
and the second R command calculates the values of $\rho^*(1), \rho^*(2), \ldots, \rho^*(8)$ because the
pacf argument in the call to ARMAacf is set to TRUE.

```
ARMAacf(ar = 0, ma = c(-1, 1 / 2), lag.max = 8)
ARMAacf(ar = 0, ma = c(-1, 1 / 2), lag.max = 8, pacf = TRUE)
```

The resulting population autocorrelation and partial autocorrelation functions are plotted in Figure 7.17. The population autocorrelation function cuts off at lag 2 and the
population partial autocorrelation function has correlations that appear to behave in a

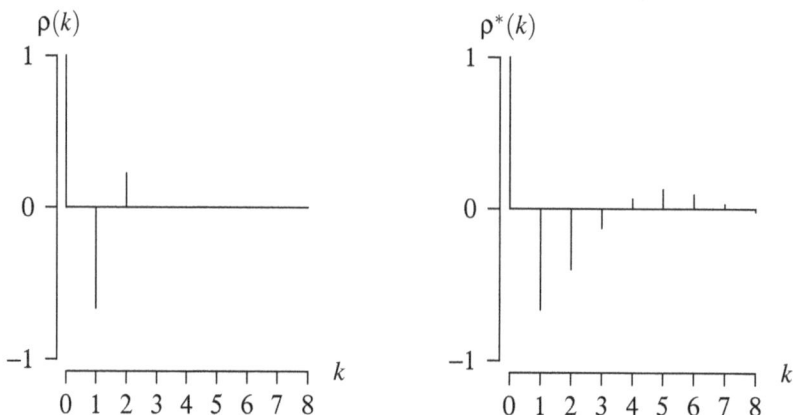

Figure 7.17: Population autocorrelation and partial autocorrelation functions.

damped sinusoidal fashion. Time series analysts refer to this type of population partial autocorrelation function as one that "tails off."

7.2.5   Sample Partial Autocorrelation

We now transition from considering the *population* partial autocorrelation function to considering its statistical counterpart, the *sample* partial autocorrelation function. Calculating the sample partial autocorrelation function is just a matter of replacing the population values with the sample values in the determinants given in Definition 7.8.

Definition 7.9   For a stationary time series model, the lag 0 sample partial autocorrelation is 1, the lag 1 sample partial autocorrelation is r_1, and the lag k sample partial autocorrelation is

$$r^*(k) = \frac{\begin{vmatrix} 1 & r_1 & r_2 & \cdots & r_1 \\ r_1 & 1 & r_1 & \cdots & r_2 \\ r_2 & r_1 & 1 & \cdots & r_3 \\ \vdots & \vdots & \vdots & \ddots & \vdots \\ r_{k-1} & r_{k-2} & r_{k-3} & \cdots & r_k \end{vmatrix}}{\begin{vmatrix} 1 & r_1 & r_2 & \cdots & r_{k-1} \\ r_1 & 1 & r_1 & \cdots & r_{k-2} \\ r_2 & r_1 & 1 & \cdots & r_{k-3} \\ \vdots & \vdots & \vdots & \ddots & \vdots \\ r_{k-1} & r_{k-2} & r_{k-3} & \cdots & 1 \end{vmatrix}},$$

for $k = 2, 3, \ldots$.

Given an observed time series, the R code from the previous example that used the `det` function to calculate the determinants could be used to perform these calculations. The sample partial autocorrelation function can be calculated much more efficiently, however, by using the built-in `pacf` function in R, as illustrated in the next example. The lag in which the sample partial autocorrelation function values become statistically indistinguishable from zero can be determined by using the approximate result that for a time series of white noise values, $r_k^* \sim N(0, 1/n)$, for $k = 1, 2, \ldots$. For this reason, the `pacf` function in R plots dashed lines at the approximate 95% bands at $\pm 1.96/\sqrt{n}$. These dashed lines are useful to a time series analyst in determining which sample partial autocorrelation values differ significantly from zero.

Example 7.24 Plot the time series, sample autocorrelation function, and sample partial autocorrelation function for the $n = 70$ chemical yield values from Example 7.20, repeated in Table 7.6 for convenience. The values in the time series should be read row-wise.

It would be reasonable to simply use the code from the previous example to compute the sample partial autocorrelation function, but we instead illustrate the use of R's built-in `pacf` function here. In addition, the `layout` function can be used to stretch the time series plot from left-to-right on the graphic, but yet compress the plots of the sample autocorrelation function and the sample partial autocorrelation function. The elements in the matrix given as the first argument to `layout` indicate the plot number being displayed. Stretching the time series plot is important in order to visually detect

47	64	23	71	38	64	55	41	59	48	71	35	57	40
58	44	80	55	37	74	51	57	50	60	45	57	50	45
25	59	50	71	56	74	50	58	45	54	36	54	48	55
45	57	50	62	44	64	43	52	38	59	55	41	53	49
34	35	54	45	68	38	50	60	39	59	40	57	54	23

Table 7.6: A time series of $n = 70$ consecutive yields from a chemical process.

patterns in the time series. Nothing is lost by horizontally compressing the sample autocorrelation function and the sample partial autocorrelation function as long as the spike values r_k and r_k^* are distinct on the plots.

```
chemical = scan("chemical.d")
layout(matrix(c(1, 1, 2, 3), 2, 2, byrow = TRUE))
plot.ts(chemical)
acf(chemical)
pacf(chemical)
```

If you prefer confidence limits other than the default 95% limits, both `acf` and `pacf` accept a `ci` argument that accepts any argument between 0 and 1, but defaults to 0.95. The resulting plots are displayed in Figure 7.18. There are two statistically significant spikes in the sample autocorrelation function and one statistically significant spike in the sample partial autocorrelation function.

There will be more examples of computing the partial autocorrelation function and its interpretation subsequently.

7.2.6 Computing

The R `plot.ts` function generates a plot of a realization of a time series, which is an important initial step in formulating an appropriate stochastic model for the time series. Many time series analysts prefer to also see the sample autocorrelation and partial autocorrelation functions along with the plot of the time series. The `layout` function can be used to stretch the plot of the time series horizontally, while horizontally compressing the plots of the sample autocorrelation and partial autocorrelation functions. The `acf` function computes the sample autocorrelation function and has arguments that control the number of lags to display, whether to suppress the plot, etc. The `pacf` function computes the sample partial autocorrelation function and has similar arguments. Notice that the `acf` function includes the lag 0 sample autocorrelation, which is always 1, but the `pacf` function does not include the lag 0 sample partial autocorrelation. The statements below apply these functions to the built-in R `AirPassengers` time series.

```
layout(matrix(c(1, 1, 2, 3), 2, 2, byrow = TRUE))
plot.ts(AirPassengers)
acf(AirPassengers)
pacf(AirPassengers)
```

The vertical axes on all three plots are autoscaled. Use the `ylim = c(-1, 1)` argument in the `acf` and `pacf` functions in order to stretch the vertical axis from -1 to 1. Finally, the `ARMAacf` function can be used to compute the population autocorrelation and partial autocorrelation function values for a prescribed time series model.

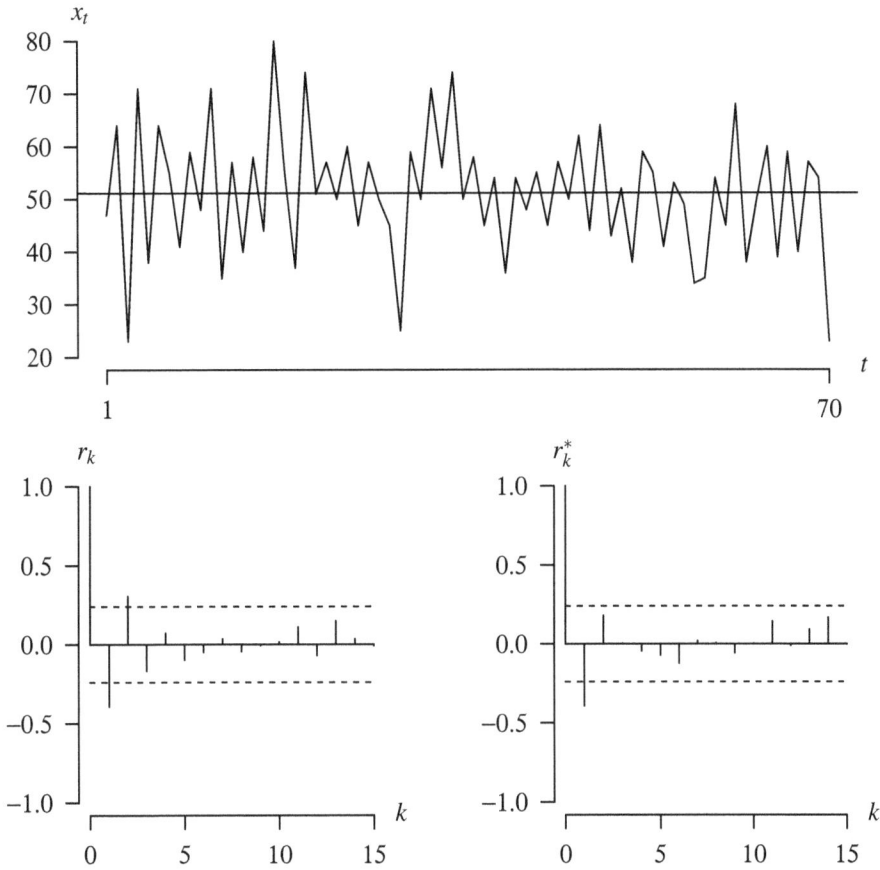

Figure 7.18: Time series plot, r_k, and r_k^* for $n = 70$ yields from a chemical process.

7.3   Operations on a Time Series

This section considers operations that can be performed on a time series. The first subsection introduces *filters* that can be applied to a time series. We have already encountered an example of a filter when we considered a three-point moving average of a time series consisting of Gaussian white noise. The second subsection introduces *decomposition*, which is the process of breaking an observed time series into its component parts. The `AirPassengers` time series that is built into R, for example, can be decomposed into an trend, a seasonal component, and any remaining noise in the process once the trend and seasonal components have been removed. The third subsection concerns R functions that can be helpful in implementing these operations.

7.3.1   Filtering

This section takes up the important topic of *filtering*, which can be thought of as the process of converting one observed time series $\{x_t\}$ to another time series $\{y_t\}$. The mathematical operations required to convert one time series to another can be abstractly depicted as

$$\{x_t\} \xrightarrow{\quad \text{filter} \quad} \{y_t\}$$

It is often the case that several of these filters can be applied sequentially to a particular observed time series. Filter 1, for example, converts $\{x_t\}$ to $\{y_t\}$, and then Filter 2 converts $\{y_t\}$ to $\{z_t\}$.

$$\{x_t\} \xrightarrow{\quad \text{filter 1} \quad} \{y_t\} \xrightarrow{\quad \text{filter 2} \quad} \{z_t\}$$

The resultant time series $\{z_t\}$ is not associated with the white noise terms Z_t from Definition 7.1. The purpose of such a series of filters applied to a time series might be to remove the trend with the first filter, and then to remove some seasonal variation with the second filter. If the resulting time series $\{z_t\}$ looks like random noise values, then the two filters applied in series have successfully removed the trend and the seasonal variation.

Three different general classes of filters will be considered: transformations, detrending, and linear filters. These classes of filters allow for the manipulation of a times series for a particular purpose, such as smoothing or variance stabilization.

Transformations

One simple filter that can be applied to a time series is to apply a transformation to each element of the time series. Two transformations that are commonly applied to a time series are the logarithmic and square root transformations, which are

$$y_t = \ln x_t$$

and

$$y_t = \sqrt{x_t}.$$

Some common purposes of applying such a transformation are to

- stabilize the variance of the time series (for example, when larger values of x_t tend to have greater variability than smaller values of x_t),

- make the trend and seasonal components of a time series appear to be additive, rather than multiplicative, in nature, and

- make the values in the filtered time series appear to be similar to white noise, iid noise, or Gaussian white noise (see Definition 7.1). The advantage to having values of the fitted time series be approximately mutually independent and normally distributed is to enable the use of easier statistical inference procedures concerning, for example, forecasted values.

The transformation of the values in a time series given in the next example is a variance-stabilizing transformation which makes significant improvement to a time series in terms of its visualization and interpretation.

Example 7.25 The Dow Jones Industrial Average (DJIA), also known as the Dow 30, was devised by Charles Dow and was initiated on May 26, 1896. The average bears Dow's name and that of statistician and business associate Edward Jones. The DJIA is the average stock price of 30 U.S.-based, publicly traded companies, adjusted for stock splits and the swapping of companies in and out of the average so that it adequately reflects the composition of the domestic stock market. These adjustments are made by altering the average's denominator for historical continuity, which is now much less than 30.

The evolution of the DJIA is not a true reflection of the yield of the 30 stocks because two important factors are not incorporated into the average. First, the average does not factor in dividends that are paid by some of the 30 companies. Second, the average does not factor in inflation, which erodes the true return that a stock investment provides. If dividends were factored into the calculation, the DJIA would be much higher than it is presently; if inflation were factored into the calculation, the DJIA would be much lower than it is presently.

This example considers the time series plot of the average annual DJIA closing values during the 20th century. This plot is generated with the R code given below.

```
x = 1901:2000
y = ts(scan("dow.d"))
plot.ts(x, y)
```

The file dow.d contains the 100 annual average closing values. The resulting graph is shown in Figure 7.19.

The DJIA had a sample mean closing value of 69.52 during 1901 and a sample mean closing value of 10731.15 during 2000. The linear scale that is used in Figure 7.19 obscures most of the variability of the DJIA during the first half of the century. The graph can be made more meaningful by using a logarithmic scale on the vertical axis. This is accomplished by calling the plot.ts function as plot.ts(x, y, log = "y"), resulting in the graph shown in Figure 7.20. This is equivalent to plotting the filtered time series

$$y_t = \log_{10} x_t$$

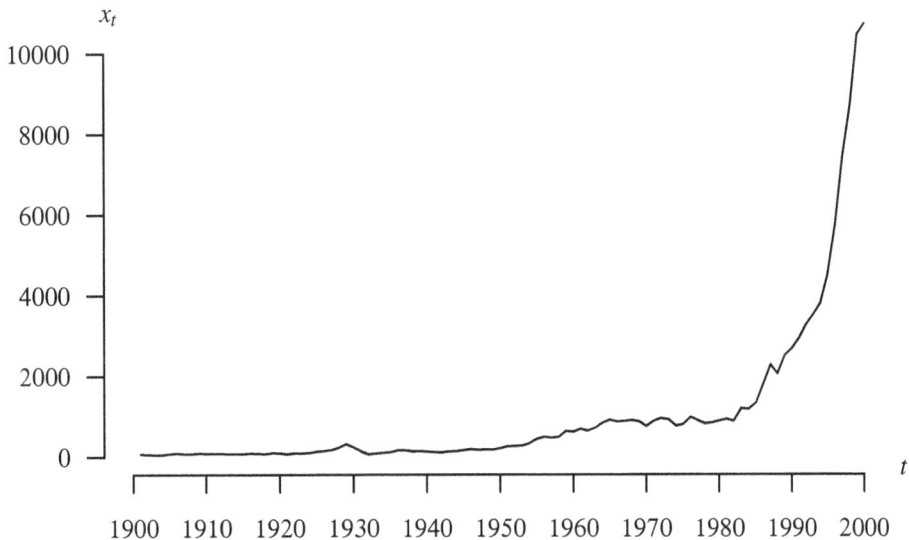

Figure 7.19: Dow Jones Industrial Average (1901–2000).

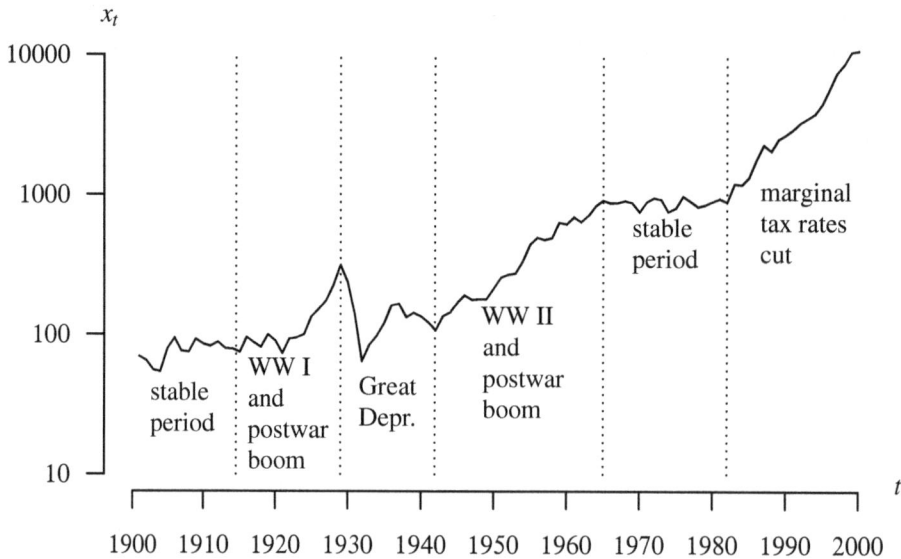

Figure 7.20: Dow Jones Industrial Average (1901–2000) with logarithmic scale.

on a linear vertical scale. An equal percent change now covers the same vertical distance with the logarithmic vertical scale. Labels have been added to help highlight events that might have influenced the DJIA.

The stock market crash in October of 1929 is much more pronounced in Figure 7.20. The DJIA had peaked with a close of 381.20 on September 3, 1929. The market bottomed out on July 8, 1932 when it closed at 41.20, which corresponds to a loss of almost 90%. Each of the two World Wars fought during the twentieth century was followed by a sustained bull market in the DJIA. The top marginal income tax rate was lowered from 70% to 28% and the federal budget was brought into balance in the 1980s and 1990s, resulting in a prolonged growth in the DJIA.

Detrending Filters

When a consistent trend in a time series is apparent, as was the case with the international airline passengers time series from Example 7.2, an analyst typically would like to estimate the trend. Once the trend has been estimated, the residual time series remaining after detrending can often be fitted to a time series model. There are two popular types of filters that can be used to detrend a time series: curve fitting and differencing. These will be considered individually. Time series analysts often use the term *secular trend* to describe a long-term, non-periodic trend, but we will refer to it as just a trend here.

One way to detrend a time series is to fit a curve that approximates the mean value of the time series. As a simple example, consider a time series that appears to have a linear trend. In this case a simple linear regression statistical model

$$X_t = \beta_0 + \beta_1 t + \varepsilon_t$$

can be fitted to the time series in order to estimate the slope β_0 and intercept β_1 of the regression line. The time t plays the role of the predictor in the regression model; the time series observations X_t play the role of the response in the regression model. It is also possible to have a non-linear trend in a time series. A quadratic trend in time, for example, could be modeled via

$$X_t = \beta_0 + \beta_1 t + \beta_2 t^2 + \varepsilon_t.$$

Note that this model is linear in the β parameters. The statistical models used to detrend a time series are not limited to just polynomials in time. It is also possible to have an exponential model such as

$$X_t = \beta_0 e^{\beta_1 t} + \varepsilon_t.$$

This model is not linear in the β parameters. The potential statistical models are endless. A working knowledge of regression modeling is helpful in constructing an appropriate model for formulating, estimating, and assessing a model for the trend in a time series.

> **Example 7.26** This example considers the simplest case of detrending a data set, which is a linear trend. The data set consists of $n = 89$ quarterly observations which are the quarterly number of Australian residents (in thousands) from March 1971 to March 1993 which was first encountered in Example 7.18. This time series is built into R and has the name austres. Use simple linear regression to estimate the trend in the data set and calculate the detrended time series.
>
> The raw data values and the time series plot are given in Example 7.18. The time series plot is repeated in Figure 7.21 for convenience, plotting individual points but not connecting them with lines. It is clear that the population of Australia is increasing in a linear fashion over this time period.
>
> The next step is to fit a simple linear regression model to the time series model
>
> $$X_t = \beta_0 + \beta_1 t + \varepsilon_t.$$
>
> This can be accomplished in R using the lm (for *linear model*) function.

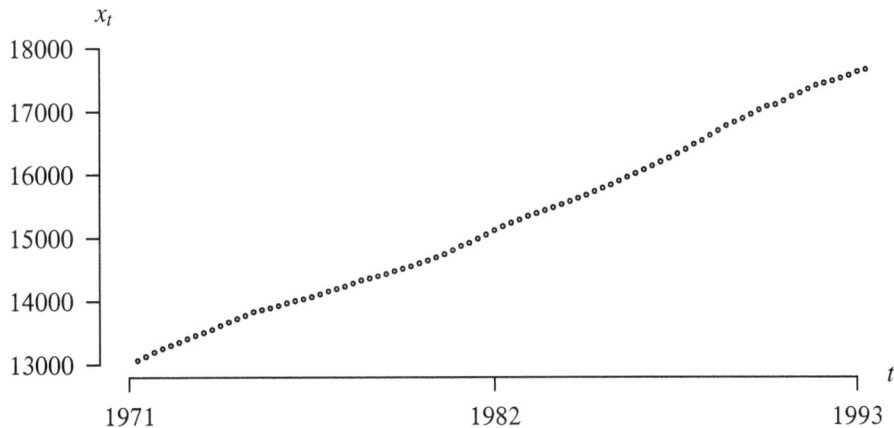

Figure 7.21: Quarterly population of Australia (in thousands) 1971–1993.

```
plot.ts(austres, type = "p", cex = 0.4)
fit = lm(austres ~ seq(1971.25, 1993.25, by = 0.25))
abline(fit$coefficients)
coef(fit)
```

The fitted slope and intercept of the least square regression line are

$$\hat{\beta}_0 = -399{,}861 \qquad \text{and} \qquad \hat{\beta}_1 = 209.426.$$

The interpretation of the estimated intercept is that in the year 0 the population of Australia was negative 400 million. (This is a good illustration that the model should not be extrapolated significantly outside of the range of the time values in the time series.) The interpretation of the estimated slope is that the population of Australia increases by an estimated 209,426 each year over the time window 1971–1993. The plot that includes the regression line plotted via the `abline` function is given in Figure 7.22. The regression line reveals some very slight nonlinear trends in the time series that were not immediately apparent in the original time series plot in Figure 7.21.

The final step in the analysis is to examine the residual time series after detrending. Viewing that residual series as a filter, the new time series after detrending is

$$y_t = x_t - \left(\hat{\beta}_0 + \hat{\beta}_1 t\right).$$

The time series $\{y_t\}$ can be calculated and plotted with the additional R statements

```
austres.detrend = austres - fit$fitted
plot.ts(austres.detrend)
```

This residual series is plotted in Figure 7.23. The residual series is connected by lines. In addition, a horizontal dashed line is added at $y_t = 0$. Clearly, the residual series does not consist of mutually independent noise terms. Its shape might have been influenced by Australian immigration policies or the Australian economy between 1971 and 1993.

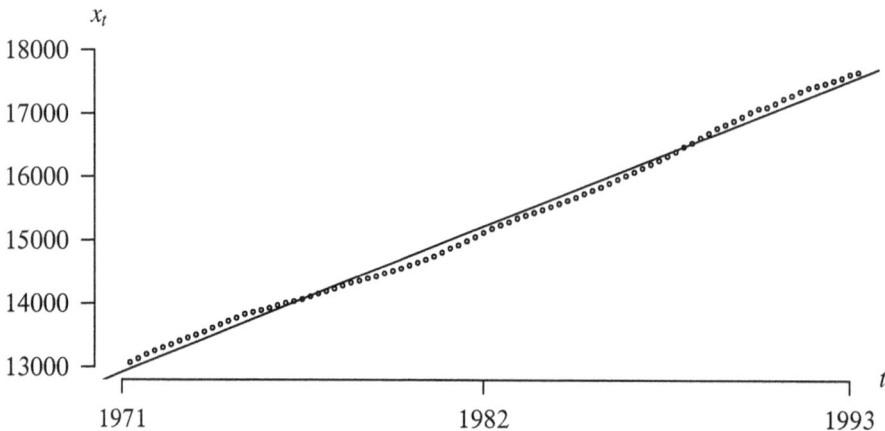

Figure 7.22: Quarterly population of Australia (in thousands) 1971–1993 with regression line.

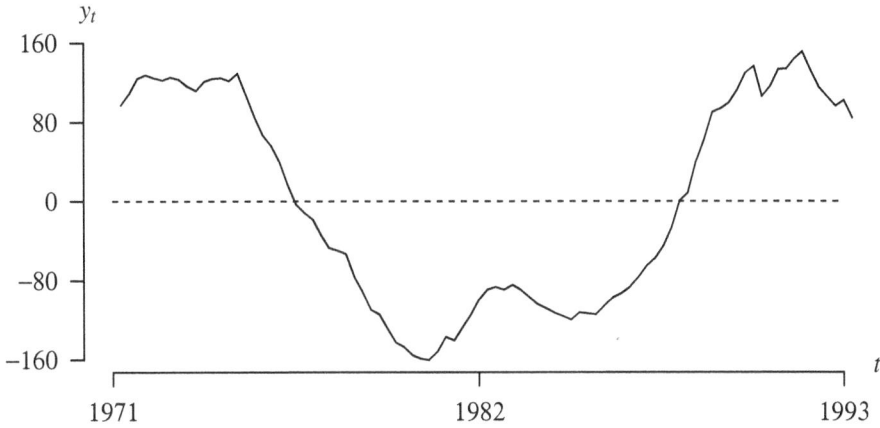

Figure 7.23: Residual time series after detrending for the quarterly population of Australia.

A second way to detrend a time series is to use differencing. The difference operator ∇, is defined as a filter by

$$y_t = \nabla x_t = x_t - x_{t-1} = (1 - B)x_t,$$

where B is the backshift operator defined by $Bx_t = x_{t-1}$. Differencing a time series is the discrete analog of taking a derivative of a continuous function. So a time series that exhibits a linear trend, for example, will pass through the differencing filter ∇ and result in a time series without a trend. Notice that there will be one fewer observation in the new time series after applying this filter. If the original time series observations are $x_1, x_2, \ldots, x_n$, then the differenced series will be $y_2, y_3, \ldots, y_n$. Likewise, a time series with a quadratic trend can be detrended by applying the differencing operator ∇ twice to the original time series:

$$y_t = \nabla^2 x_t = \nabla\left(\nabla x_t\right) = \nabla\left(x_t - x_{t-1}\right) = \left(x_t - x_{t-1}\right) - \left(x_{t-1} - x_{t-2}\right) = x_t - 2x_{t-1} + x_{t-2}.$$

This detrending filter can be written with the backshift operator as $y_t = (1 - 2B + B^2)x_t$. There will be two fewer observations in the new time series after applying this filter. If the original time series observations are $x_1, x_2, \ldots, x_n$, then the twice differenced series will be $y_3, y_4, \ldots, y_n$.

A time series that exhibits a seasonal component can have a seasonal differencing filter applied. Monthly observations from a time series with an annual seasonal component, for example, can have the seasonal differencing filter

$$y_t = \nabla_{12} x_t = x_t - x_{t-12} = \left(1 - B^{12}\right) x_t$$

applied to eliminate the seasonal effects. There will be 12 fewer observations in the new time series after applying this filter. If the original time series observations are $x_1, x_2, \ldots, x_n$, then the time series observations that result from applying the filter associated with the ∇_{12} operator are $y_{13}, y_{14}, \ldots, y_n$.

We now illustrate the application of a differencing filter. The next example applies a single difference filter to the Australian population time series.

Example 7.27 Consider again the quarterly population of Australia between 1971 and 1993 from Example 7.26 given in the R built-in data set `austres`. Apply the single difference filter $y_t = \nabla x_t = x_t - x_{t-1}$ and make a time series plot of the differenced time series.

The filter

$$y_t = \nabla x_t = x_t - x_{t-1}$$

is appropriate for detrending because the time series $\{x_t\}$ is approximately linear, as seen in Figure 7.22. The `diff` function in R differences the time series. So the differenced time series $\{y_t\}$ can be plotted with the single R statement

```
plot.ts(diff(austres))
```

The plot of the differenced time series is given in Figure 7.24. A dashed horizontal line has height equal to the slope of the line (with respect to quarters) connecting the first and last points in the time series [that is, a horizontal line at $(x_n - x_1)/(n - 1) = 52.2$] can be added with the `abline` function with an `h` (for horizontal) and `lty = 2` (for a dashed line) arguments.

```
n = length(austres)
abline(h = (austres[n] - austres[1]) / (n - 1), lty = 2)
```

The original time series increases by $4 \cdot 52.2 = 208.8$, or $208,800$ residents annually. This is roughly equal to the slope of the regression line $\hat{\beta}_1 = 209.4$, or an increase of $209,400$ residents annually from Example 7.26. The filtered time series $\{y_t\}$ does not appear to exhibit any long-term trend, which was the original purpose of using the differencing filter.

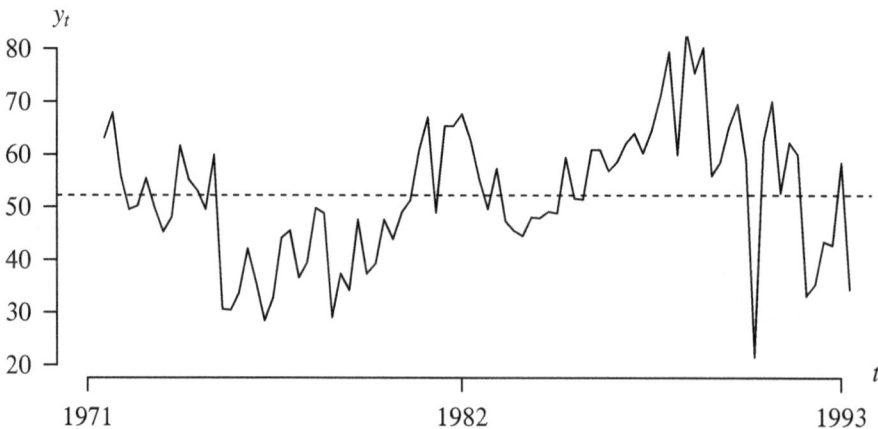

Figure 7.24: Filtered time series after differencing the quarterly population of Australia.

So far, two classes of filters have been introduced. The first class is known as a transformation; the second class is known as a detrending filter. Two types of detrending filters were introduced: curve fitting and differencing. We now turn to a third class of filter which is known as a linear filter. Differencing is a special case of a linear filter.

Linear Filters

Linear filters provide a wide array of options for converting the original time series $\{x_t\}$ to another time series $\{y_t\}$. The general form of a linear filter is

$$y_t = \sum_{s=t_0}^{t_1} c_s x_{t+s},$$

where the coefficients c_s are real-valued constants. It is often the case that $t_0 < 0$ and $t_1 > 0$, which means that the time series $\{y_t\}$ is a linear combination of the chronological *current*, *previous*, and *future* values of $\{x_t\}$ in time.

One purpose of a linear filter is *smoothing* the original time series by using what is known as a *moving average*. When the coefficients sum to one, written symbolically as

$$\sum_{s=t_0}^{t_1} c_s = 1,$$

this linear filter is a moving average. One elementary example of a *symmetric moving average* is when $t_0 = -t_1$ with identical weights

$$c_s = \frac{1}{2t_1 + 1}.$$

In this case, the smoothed time series $\{y_t\}$ is the arithmetic mean of

- the current value of the time series $\{x_t\}$,

- the t_1 previous values of the time series $\{x_t\}$,

- the t_1 future values of the time series $\{x_t\}$,

for a total of $2t_1 + 1$ values averaged. The symmetric moving average $\{y_t\}$ will have $2t_1$ fewer observations than the original time series $\{x_t\}$ because the average cannot be computed for the first and last t_1 observations in $\{x_t\}$. The smoothed values of the first 100 Dow Jones Industrial Average closing values during the year 2000, introduced in Example 7.4, will be illustrated next.

Example 7.28 Consider the time series $\{x_t\}$ consisting of the first $n = 100$ closing values of the Dow Jones Industrial Average during the year 2000 that appeared in the top graph of Figure 7.5. Graph the original time series $\{x_t\}$ and the linear filter which is a symmetric moving average of five adjacent values (that is, $t_1 = 2$)

$$y_t = \frac{x_{t-2} + x_{t-1} + x_t + x_{t+1} + x_{t+2}}{5}.$$

Notice that the coefficients in this linear filter are all $1/5$. This symmetric moving average is sometimes known as a *five-point moving average*.

The first step is to write an R function to compute the five-point moving average. The R function `movingAverage5` given below calculates the five-point moving average.

```
movingAverage5 = function(x) {
  n = length(x)
  (x[1:(n - 4)] + x[2:(n - 3)] + x[3:(n - 2)] + x[4:(n - 1)] + x[5:n]) / 5
}
```

The original time series of Dow Jones Industrial Averages and the five-point moving average are plotted in Figure 7.25. The original time series $\{x_t\}$ is plotted as a solid black line connecting the points. The linear filter $\{y_t\}$ smooths the original time series and is given by the thicker gray curve, which is a piecewise linear function that connects the five-point moving average points, given as dots within the gray curve. The original time series consists of $n = 100$ points. The five-point moving average loses two points at the beginning and two points at the end, resulting in just the points $y_3, y_4, \ldots, y_{98}$. The moving average successfully smooths the original time series. By averaging the current value, two previous values, and two future values, the significant variations in the original time series are damped, and the trend of the Dow Jones Industrial during the year 2000 is more apparent with the five-point moving average.

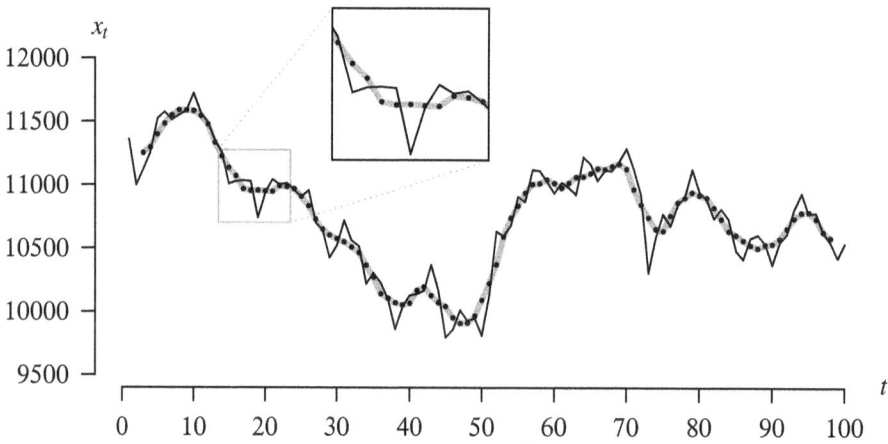

Figure 7.25: The first $n = 100$ DJIA closes in 2000 and a five-point moving average.

We now illustrate the case in which filters are applied to a time series in a serial fashion as illustrated below.

$$\{x_t\} \xrightarrow{\text{filter 1}} \{y_t\} \xrightarrow{\text{filter 2}} \{z_t\}$$

Let the values in the original time series be

$$x_1, x_2, \ldots, x_n.$$

Consider the linear filter which is a three-point moving average (in which $t_1 = 1$)

$$y_t = \frac{x_{t-1} + x_t + x_{t+1}}{3}.$$

This linear filter results in the time series $\{y_t\}$ consisting of the observations

$$\frac{x_1 + x_2 + x_3}{3}, \frac{x_2 + x_3 + x_4}{3}, \ldots, \frac{x_{n-2} + x_{n-1} + x_n}{3}.$$

Now consider applying this same linear filter again, but this time to $\{y_t\}$:

$$z_t = \frac{y_{t-1} + y_t + y_{t+1}}{3}.$$

The second linear filter results in the time series $\{z_t\}$ consisting of the observations

$$\frac{x_1 + 2x_2 + 3x_3 + 2x_4 + x_5}{9}, \frac{x_2 + 2x_3 + 3x_4 + 2x_5 + x_6}{9}, \ldots, \frac{x_{n-4} + 2x_{n-3} + 3x_{n-2} + 2x_{n-1} + x_n}{9}$$

when written in terms of the original time series $\{x_t\}$. Notice that the serial application of the two linear filters is the same as the application of a single linear filter with the coefficients

$$\frac{1}{9}, \frac{2}{9}, \frac{3}{9}, \frac{2}{9}, \frac{1}{9}.$$

The R `convolve` function calculates the coefficients of the linear filter associated with two linear filters applied to a time series. In the example described here, the coefficients can be determined with the R statements

```
a = c(1 / 3, 1 / 3, 1 / 3)
b = c(1 / 3, 1 / 3, 1 / 3)
convolve(a, b, type = "open")
```

The application of two linear filters in sequence is illustrated in the next example.

Example 7.29 Consider again the time series $\{x_t\}$ consisting of the first $n = 100$ closing values of the Dow Jones Industrial Average during the year 2000. Graph the original time series $\{x_t\}$ and two serial applications of the five-point moving average

$$y_t = \frac{x_{t-2} + x_{t-1} + x_t + x_{t+1} + x_{t+2}}{5}.$$

As in the case of the three-point moving average, we can use the R `convolve` function to calculate the coefficients in the convolution of the two linear filters.

```
a = rep(1 / 5, 5)
convolve(a, a, type = "open")
```

The `convolve` function results in the coefficients

$$\frac{1}{25}, \frac{2}{25}, \frac{3}{25}, \frac{4}{25}, \frac{5}{25}, \frac{4}{25}, \frac{3}{25}, \frac{2}{25}, \frac{1}{25}$$

associated with the two linear filters applied in series. These coefficients sum to one as expected. The convolution of the two moving average filters remains a moving average. The application of the two linear filters in series has two potential benefits over the five-point moving average alone:

- the serial application of the two linear filters provides more smoothing than in the previous example because more observations are used in the moving average, and the coefficients are all smaller than in the five-point moving average, and

- the serial application of the two linear filters provides a mechanism in which more distant observations get less weight than nearby observations.

The result of applying this moving average to the Dow Jones Industrial Average closes is shown in Figure 7.26. As expected, this filter provides more smoothing than the five-point moving average from the previous example.

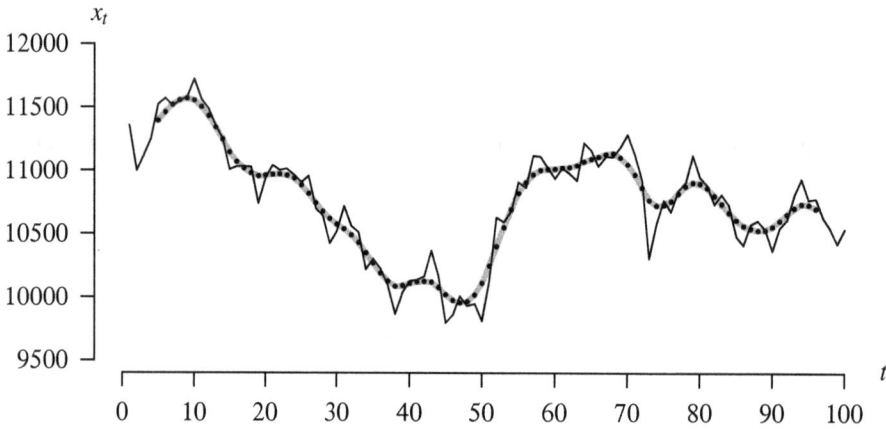

Figure 7.26: The first $n = 100$ DJIA closes in 2000 and a nine-point moving average.

Some experimentation is often necessary to achieve a moving average that provides the appropriate amount of smoothing. The amount of smoothing desired is problem specific. Another type of symmetric moving average that places the most weight on the current value x_t and decreasing weight to more distant observations is to use the terms in the expansion of

$$\left(\frac{1}{2} + \frac{1}{2}\right)^{2t_1}$$

as coefficients in a weighted moving average. When $t_1 = 2$, for example, the coefficients are

$$\frac{1}{16}, \frac{4}{16}, \frac{6}{16}, \frac{4}{16}, \frac{1}{16}.$$

The numerators can be recognized as one row in Pascal's triangle. The weights must sum to 1 because $1/2 + 1/2 = 1$.

For a time series without a trend that contains a seasonal component, a special linear filter can be applied. Consider a time series of monthly observations with seasonal variation. A common way to eliminate the seasonal component is the linear filter

$$y_t = \frac{\frac{1}{2}x_{t-6} + x_{t-5} + x_{t-4} + \cdots + x_{t+4} + x_{t+5} + \frac{1}{2}x_{t+6}}{12}.$$

This linear filter has 13 coefficients

$$\frac{1}{24}, \frac{1}{12}, \frac{1}{12}, \ldots, \frac{1}{12}, \frac{1}{12}, \frac{1}{24},$$

which places a weight of $1/12$ on the current observation x_t and each observation within five months of x_t and splits the weight between the two months that are six months before and six months after the current observation. Notice that the filtered time series $\{y_t\}$ will have 12 fewer observations than the original time series $\{x_t\}$ because the moving average loses six points at the beginning of the time series and six points at the end of the time series. This seasonal weighted average will be illustrated in the next example for the monthly home energy consumption time series.

Example 7.30 Consider the time series $\{x_t\}$ of $n = 96$ monthly home energy consumption observations, in kilowatt hours, from Example 7.1. Compute the seasonal moving average

$$y_t = \frac{\frac{1}{2}x_{t-6} + x_{t-5} + x_{t-4} + \cdots + x_{t+5} + x_{t+6} + \frac{1}{2}x_{t+6}}{12}$$

and plot the original time series and the seasonal moving average on the same set of axes.

The original time series and the seasonal moving average are graphed in Figure 7.27. The seasonal moving average effectively removes the seasonal component, revealing a slight upward trend in the first half of the time series and a slight downward trend toward the end of the time series. Since the time series was collected over an eight-year period, there are a total of $8 \cdot 12 - 12 = 84$ observations in the seasonal moving average. An observation from each of the 12 months plays a role in every value calculated in the seasonal moving average $\{y_t\}$.

The R code below calculates the seasonal moving average of the energy consumption values, which are stored in the file named kwh.d. The coefficients of the seasonal moving average which control the weights allocated to each value in the time series are stored in the w vector. The seasonal moving average values are stored in the y vector.

```
x = scan("kwh.d")
w = c(1 / 24, rep(1 / 12, 11), 1 / 24)
n = length(x)
y = numeric(n - 12)
for (i in 1:(n - 12)) y[i] = sum(w * x[i:(i + 12)])
print(y)
```

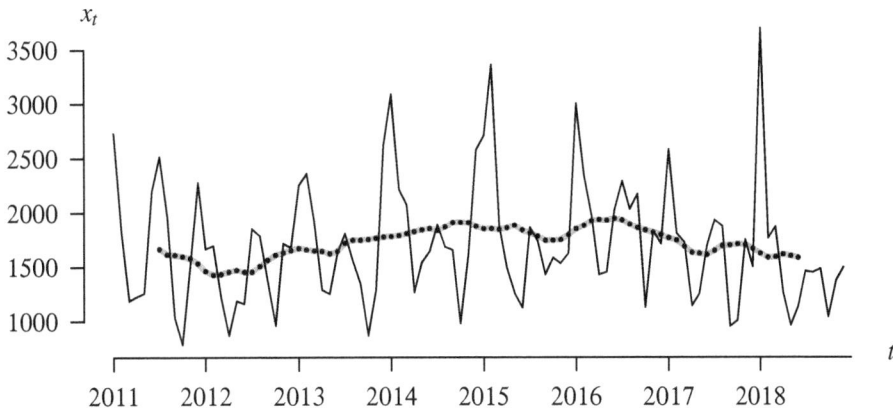

Figure 7.27: Home energy consumption time series and seasonal moving average.

All of the linear filters we have encountered so far have been *symmetric moving averages*. Each of them has reached as far into the past as they have into the future. One weakness of these filters is that in many settings we often do not have any future observations. So in many practical problems the linear filter

$$y_t = \sum_{s=t_0}^{t_1} c_s x_{t+s}$$

is rewritten to avoid any observations in the future as

$$y_t = \sum_{s=t_0}^{0} c_s x_{t+s}$$

for some negative integer index t_0. The most well-known of these filters is known as the *exponentially-weighted moving average*, often abbreviated EWMA, which can be written as

$$y_t = \sum_{s=-\infty}^{0} c_s x_{t+s},$$

where the weights c_s are given by

$$c_s = \alpha(1-\alpha)^{-s}$$

for $0 < \alpha < 1$ and $s = -\infty, \ldots, -2, -1, 0$. The weights in the exponentially-weighted moving average must sum to one because they form a geometric series:

$$\sum_{s=-\infty}^{0} c_s = \sum_{s=-\infty}^{0} \alpha(1-\alpha)^{-s} = \alpha \sum_{s=0}^{\infty} (1-\alpha)^s = \frac{\alpha}{1-(1-\alpha)} = 1.$$

The exponentially-weighted moving average gives weight α to the current observation, and then geometrically declining weights to previous observations. So the parameter α can be thought of as a dial or tuning parameter which controls the amount of smoothing. Large values of α mean very little smoothing; small values of α mean significant smoothing. While it is daunting to think about values of a time series $\{x_t\}$ running back in time to $-\infty$, it is possible to avoid the infinite summation. The exponentially-weighted moving average filter can be rewritten as

$$\begin{aligned} y_t &= \alpha x_t + \alpha(1-\alpha)x_{t-1} + \alpha(1-\alpha)^2 x_{t-2} + \cdots \\ &= \alpha x_t + (1-\alpha)[\alpha x_{t-1} + \alpha(1-\alpha)x_{t-2} + \cdots] \\ &= \alpha x_t + (1-\alpha)y_{t-1}; \end{aligned}$$

that is, the exponentially-weighted moving average is α times the current value in the time series x_t plus $1-\alpha$ times the previous value in the moving average. Arbitrarily setting $y_1 = x_1$ to initiate this recursive relationship, the initial terms in the moving average are

$$\begin{aligned} y_2 &= \alpha x_2 + (1-\alpha)y_1 \\ y_3 &= \alpha x_3 + (1-\alpha)y_2 \end{aligned}$$

$$\vdots \qquad \vdots$$

Notice that the extreme case of $\alpha = 1$ is possible in this recursive equation, and this corresponds to a moving average that is identical to the original time series: $y_1 = x_1$, $y_2 = x_2$, $y_3 = x_3$, etc. This extreme case corresponds to no smoothing at all.

The next example applies the exponentially-weighted moving average to the first $n = 100$ Dow Jones Industrial Average closing observations during the year 2000.

Example 7.31 Consider yet again the time series $\{x_t\}$ consisting of the first $n = 100$ closing values of the Dow Jones Industrial Average during the year 2000. Graph the original time series $\{x_t\}$ and the exponentially-weighted moving average with $\alpha = 0.2$ on the same set of axes.

The original time series $\{x_t\}$ and the exponentially-weighted moving average $\{y_t\}$ are plotted in Figure 7.28. The exponentially-weighted moving average values are generated by

$$y_t = \alpha x_t + (1 - \alpha)y_{t-1} = 0.2x_t + 0.8y_{t-1}$$

for $t = 2, 3, \ldots, n$. The first point in the exponentially-weighted moving average, y_2, for example, is a convex combination of x_1 and x_2 with coefficients 0.8 and 0.2. Unlike the symmetric moving averages shown in the previous examples, Figure 7.28 shows that the exponentially-weighted moving average smooths the original time series, but also lags the original time series because it is not a symmetric moving average. Adjusting α can make this exponentially-weighted moving average respond more quickly or more slowly than that shown in Figure 7.28.

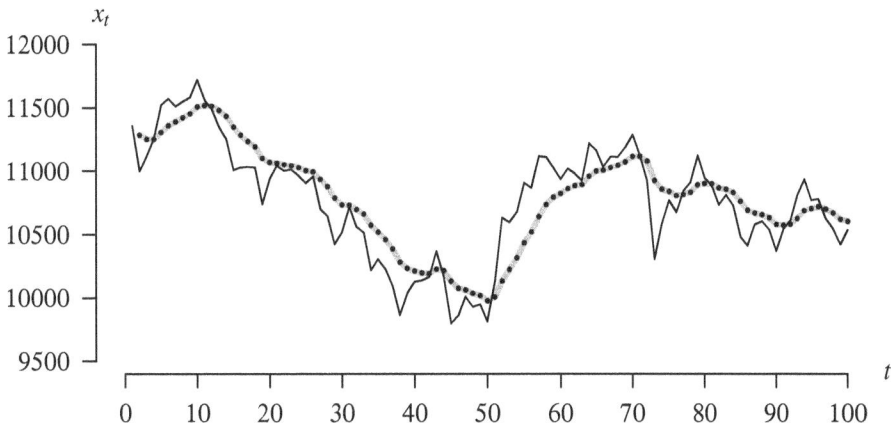

Figure 7.28: The first $n = 100$ DJIA closes in 2000 and an exponentially-weighted moving average.

This concludes the discussion of filters that can be applied to a time series. The three classes of filters that were presented in this section are

- transformations,

- detrending filters, and

- linear filters.

Filters can be applied for several different purposes, including: (a) to stabilize the variance of a time series whose variance increases with time, (b) to stabilize the variance of a time series in which larger values are more variable than smaller values, (c) to make error terms look approximately normally distributed, (d) to express a time series with a seasonal component as an additive model, (e) to detrend a time series containing a trend, (f) to estimate and eliminate a seasonal component in a time series without a trend, and (g) to smooth a time series. Many of these reasons for using a filter on a time series will be discussed subsequently.

The next section uses two filters in series in order to decompose a time series into trend and seasonal components. Decomposition will be applied to the international air travel time series and the home energy consumption time series from the first two examples in this chapter.

7.3.2 Decomposition

The time series plots on the four examples from Section 7.1.1 have revealed certain types of patterns that appear in many time series. The partial list below contains the most common types of variation in time series. One common approach to decomposing a time series into these various types of variation is to remove the detected types of variation one-by-one until only noise (that is, random variation) remains.

- **Trend.** The time series consisting of the number of international airline passengers between 1949 and 1961 from Example 7.2 had a clear and obvious long-term systemic increase in its mean value as air travel became more popular over that time period. Detecting trends and including them in a time series model is an important part of the analysis of a time series that will be illustrated in the next example.

- **Seasonal variation.** Both the home energy consumption time series from Example 7.1 and the international airline passenger time series from Example 7.2 exhibit seasonal variation. The period associated with the seasonal variation in both cases was one year. A single time series is capable of having multiple seasonal variation cycles. Outdoor temperature, for example, has both an annual cycle (warmer during the summer and cooler during the winter) and a daily cycle (warmer during the day and cooler during the night). The frequency of the daily cycle is 365 times greater than the frequency of the annual cycle. (To be more careful, the frequency is actually 365.2422 times greater.)

- **Other cyclical variation.** There are other types of cyclical variation that have an unknown period that might be included in a mathematical model that describes the time series. Economists, for example, often refer to *business cycles* that might influence the values in a time series. Business cycles typically have a varying and unknown period generally ranging from a few years to decades.

- **Remaining variation.** Once the trend, seasonal variation, and other cyclical variation have been removed from the original time series, a time series with no trend, seasonal, or cyclic variation is obtained. Once this new time series is obtained, it is common practice to plot these values to see how closely they approximate noise terms. The final step in constructing a time series model for the original process is often fitting a time series model to the residual time series, which reflects the noise terms in the time series model.

We now consider mathematical models for decomposing a time series into these constituent parts. Just as a probability model like the normal or exponential distribution is used to approximate the probability distribution from which a data set is drawn in classical statistics, we want to develop a probability model for the time series $\{X_t\}$. This probability model will be more complicated than the random walk model because we would like it to include both trends and seasonal variation. An *additive model* to describe $\{X_t\}$ is

$$X_t = m_t + s_t + \varepsilon_t,$$

where the m_t term models the trend, the s_t term models the seasonal variation with fixed period, and the ε_t term models the noise. In an ideal probabilistic modeling sequence, once the trend and seasonality terms have been estimated and removed from the time series, only random variation remains. We have ignored other cyclic variation (for example, business cycles) in this particular mathematical model. Notice that the trend term m_t and the seasonal variation term s_t are set in lowercase because these are assumed to be deterministic functions of t in the model. The stochastic element of the time series model comes from the ε_t term. A *multiplicative model* to describe $\{X_t\}$ is

$$X_t = m_t \cdot s_t \cdot \varepsilon_t.$$

The multiplicative model is often appropriate in the case in which the variance of the time series increases with time. This is because taking the natural logarithm of both sides of this model yields

$$\ln X_t = \ln m_t + \ln s_t + \ln \varepsilon_t,$$

which is an additive model for the time series $\{\ln X_t\}$.

Decomposing a time series into its constituent parts can be preformed in R with the `decompose` function. The additive model is the default. The next example applies the `decompose` function to the time series consisting of the international airline passengers.

Example 7.32 Consider again the time series of the number of international airline passengers (in thousands) between 1949 and 1961 from Example 7.2. Decompose this time series into its trend, seasonal, and random components using the R `decompose` function.

Since the plot of the time series in Figure 7.2 shows that the variance of the time series is increasing with time, we elect to use the multiplicative model

$$X_t = m_t \cdot s_t \cdot \varepsilon_t.$$

The call to the `decompose` function applied to the `AirPassengers` built-in data set is

```
fit = decompose(AirPassengers, type = "multiplicative")
```

The fitted model that extracts the trend and seasonal components, and calculates the residual time series once those two components are extracted, is held in a list named `fit`. The `type` argument in the call to the `decompose` function is used to invoke the multiplicative model. A plot of the original time series and its decomposition into its component parts is obtained by the additional R statement

```
plot(fit)
```

The associated plot is given in Figure 7.29, which gives four time series stacked one above the another. The first time series, labeled *observed*, is the original time series $\{x_t\}$, the number of monthly international airline passengers (in thousands). The lowercase variable name x_t is used here because these are observed values of the time series. The second time series, labeled *trend*, is the estimate of the trend $\{m_t\}$ returned by the `decompose` function. The third time series, labeled *seasonal*, is the estimate of the seasonal component of the multiplicative model $\{s_t\}$ returned by the `decompose` function. The seasonal component is identical from one year to the next with period 12 extracted

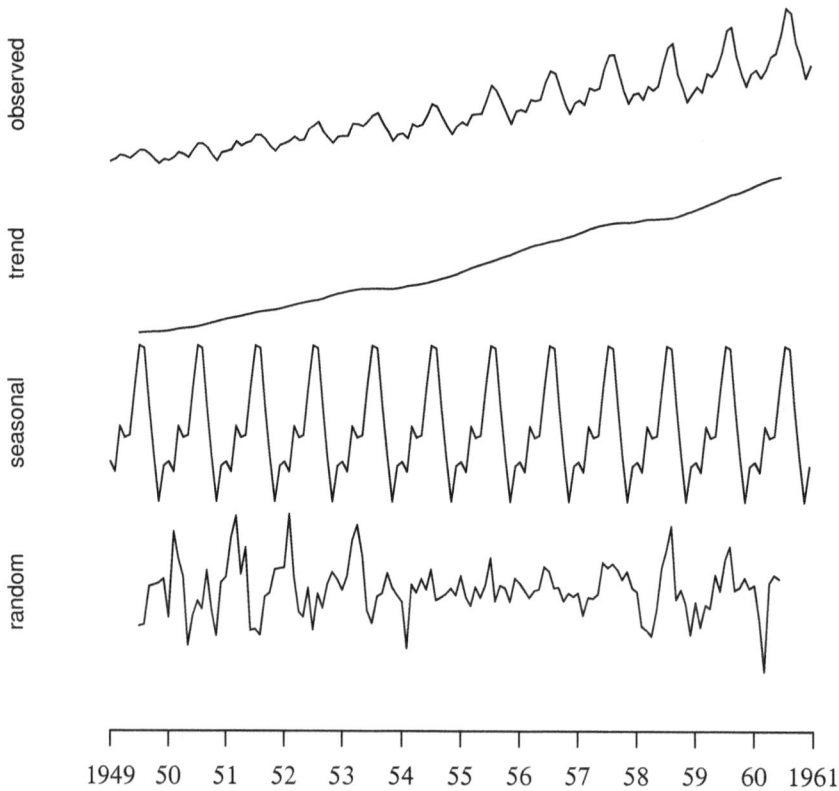

Figure 7.29: Decomposition of the international airline passengers time series.

from the AirPassengers time series. Finally, the fourth time series, labeled *random*, is the remaining time series $\{e_t\}$ once the trend and seasonal components of the model have been removed.

A call to the str (structure) function

```
str(fit)
```

reveals that the object named fit is a list that consists of six components. One of these components is named seasonal, which contains the seasonal components of the time series. This means that fit$seasonal is a time series, which is identical over every year in the time series. In order to investigate the seasonal component, Figure 7.30 contains one cycle of the seasonal component of the decomposed model. The additional R statement which can be used to plot the first cycle of the seasonal component is given below. The subscripts 1:12 extract the values in the first annual cycle.

```
plot.ts(fit$seasonal[1:12])
```

The resulting graph displayed in Figure 7.30 shows that the largest number of international airline passengers during the years from 1949 through 1960 occurs in the months of July and August when school is typically not in session. It also shows a local maximum in March which might correspond to families traveling over spring break week. The global minimum occurs in the month of November.

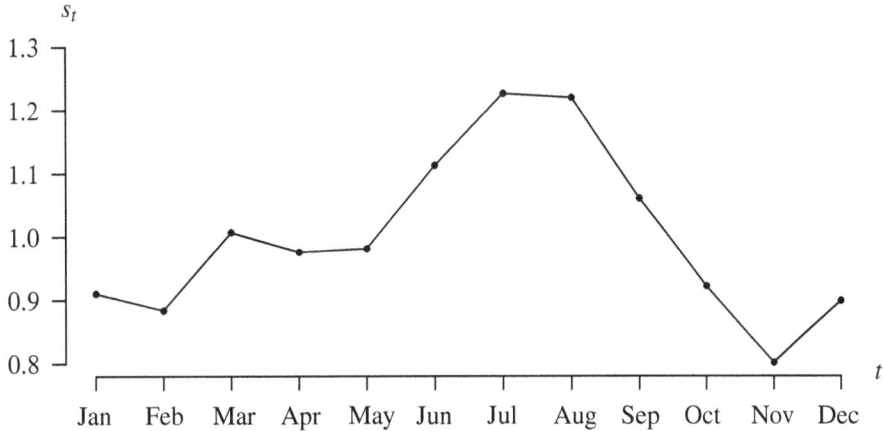

Figure 7.30: Seasonal component of the international airline passengers time series.

The international airline passengers data set had its own distinct signature for its seasonal component. The next example plots the analogous seasonal component for the home energy consumption data.

Example 7.33 Consider the home energy consumption data from Example 7.1. Using similar code to the previous example, we can plot the seasonal component of the time series with the R statements below.

```
kwh    = scan("kwh.d")
kwh.ts = ts(kwh, frequency = 12, start = c(2011, 1))
fit    = decompose(kwh.ts)
plot.ts(fit$seasonal[1:12])
```

The first statement reads the monthly energy consumption observations into a vector named kwh. The second statement uses the R ts function to convert the observations into a time series. The third statement uses the R decompose function to decompose the time series into its constituent trend, seasonal, and remaining variation components using an additive model (the default). The last statement uses the R plot.ts function to plot the first 12 elements of the seasonal component of the decomposed time series. The resulting graph shown in Figure 7.31 reveals a distinctly different seasonal pattern than the associated graph for the international airline passengers data set. The peak energy consumption is clearly in January. This peak is consistent with the intuition for the time series because (a) the outdoor temperature in the winter in Williamsburg is further from a comfortable indoor temperature than the outdoor in the summer in Williamsburg, and

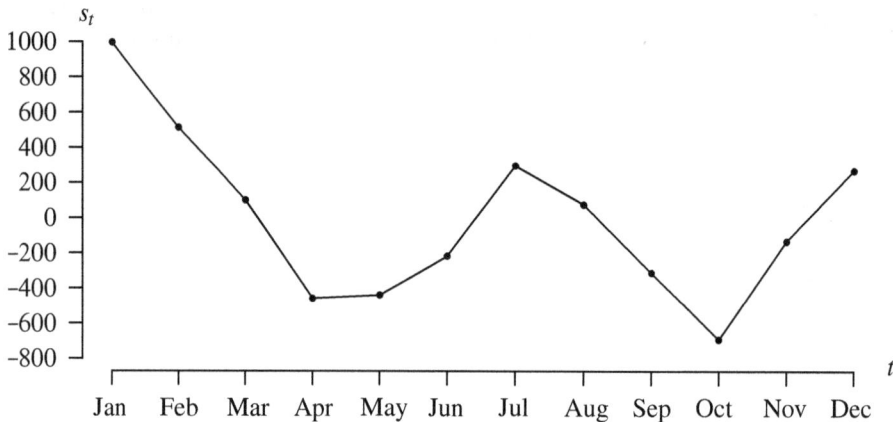

Figure 7.31: Seasonal component of the home energy consumption time series.

(b) the heat pumps are more energy efficient in the summer than they are in the winter because there is more heat available to capture. Not surprisingly, the two peaks in the time series occur in the winter and summer when demand on the heat pumps is the greatest. Since an additive model was selected, the units on the vertical axis are in kilowatt hours. The energy consumption bump associated with the number of kilowatt hours consumed in January, for example, is about 1000 kilowatt hours above the average monthly energy consumption over the entire eight-year period.

Recall that the additive model is

$$X_t = m_t + s_t + \varepsilon_t$$

and the decompose function has provided a fitted time series for the trend component m_t, the seasonal component s_t, and the error component ε_t. This indicates that the original time series can be reconstructed with the R statement

```
fit$trend + fit$seasonal + fit$random
```

which yields the original time series fit$x, with the exception of some NA values at the extreme values, which will be investigated next.

It is important to not treat a function like decompose as a just a black box that decomposes a time series without knowing the internal workings of the function. It is crucial to know exactly what is going on inside of decompose for (a) proper interpretation of the output of the function, and (b) the ability to modify the function. In that light, we now show the intermediate steps that occurred in decompose using the object names in decompose that resulted in the values plotted in Figure 7.31.

```
kwh      = scan("kwh.d")
kwh.ts   = ts(kwh, frequency = 12, start = c(2011, 1))
l        = length(kwh.ts)
```

```
f        = frequency(kwh.ts)
trend    = filter(kwh.ts, c(0.5, rep(1, f - 1), 0.5) / f)
season   = kwh.ts - trend
periods  = l / f
index    = seq(1, 1, by = f) - 1
figure   = numeric(f)
for (i in 1:f) figure[i] = mean(season[index + i], na.rm = TRUE)
figure   = figure - mean(figure)
seasonal = ts(rep(figure, periods + 1)[seq(l)], start = start(kwh.ts),
              frequency = f)
```

Try typing these statements into R and viewing the resulting objects. The bullet points below give a line-by-line explanation of the algorithm associated with this R code.

- The first statement reads the home energy consumption time series into the vector kwh.

- The second statement converts the vector named kwh to a time series named kwh.ts with monthly values beginning in January of 2011 via the ts function.

- The third statement calculates the length of the time series as $l = 96$.

- The fourth statement extracts the frequency of the time series as $f = 12$.

- The fifth statement uses the filter function to apply a 13-point moving average to the original time series, which results in a time series named trend. The extreme values in this moving average are identical months separated by one year, each getting weight $1/24$, and the interior 11 months each get weight $1/12$. Notice that the first six and last six values of the resulting time series trend are NA, as expected. The 13-point moving average is first reported in July of 2011, using the 13 values from the original time series from January 2011 to January 2012. Each value in trend is effectively an annual average of energy consumption, so this is a fairly flat time series for the energy consumption time series data because there does not appear to be any significant trend. The values in trend are plotted in Figure 7.27.

- The sixth statement creates a time series named season which is the difference between the original time series kwh.ts and the time series trend. In time series analysis, this step is known as *detrending*. This new time series season isolates the empirical seasonal component, which will change from one year to the next. The remaining R statements are designed to average these seasonal components.

- The seventh statement calculates the number of periods in the original time series: $l/f = 96/12 = 8$.

- The eighth statement creates a vector named index that will be used in the calculation of the seasonal component averages. For the energy consumption data, the eight elements of index are $0, 12, 24, \ldots, 84$.

- The ninth statement uses the numeric function to initialize the 12-element vector named figure, which will contain the seasonal component averages.

- The tenth statement contains a for loop which uses the mean function to calculate the seasonal component averages for each of the 12 months.

- The eleventh statement centers these seasonal component averages around zero.

- The twelfth statement uses the `ts` function to create a time series named `seasonal` which contains 8 periods of the seasonal component averages with a frequency of $f = 12$ and a start time of January of 2011. These are the values that are plotted in Figure 7.31.

View the `decompose` function by simply typing

```
decompose
```

In addition to the R statements described above, you will see (a) error trapping at the top of the function, (b) several conditional statements to account for the additive and the multiplicative models, and (c) code that will adjust to a time series consisting of incomplete periods.

The figures that we have seen so far have plotted a time series $\{x_t\}$ or a filter applied to create another time series $\{y_t\}$. The critical initial step of plotting the time series and making a careful examination of the plot should never be skipped.

7.3.3 Computing

Regression can be used to fit a model to a time series using the built-in `lm` (linear model) function in R. This is illustrated for the `AirPassengers` time series below.

```
fit = lm(AirPassengers ~ time(AirPassengers))
plot.ts(AirPassengers)
abline(fit$coefficients)
fitted(fit)
```

The first statement sets the object `fit` to a list that contains the results of a simple linear regression of `time(AirPassengers)` as the independent variable (the predictor) and `AirPassengers` as the dependent variable (the response). The second statement plots the time series in the usual fashion using the `plot.ts` function. The third statement appends the plot with the regression line using the `abline` function. Finally, the fitted values can be extracted by the call to `fitted(fit)` as shown above or by using `fit$fitted.values`.

A second way to remove a trend from a time series $x_1, x_2, \ldots, x_n$ is differencing. The difference operator ∇ defined by

$$\nabla x_t = x_t - x_{t-1} = (1 - B)x_t$$

(where B is the backshift operator defined by $Bx_t = x_{t-1}$) can be used to remove a linear trend. The R function `diff` can be used to difference a time series. The following statement creates a time series that contains the differences between adjacent values in the `AirPassengers` time series.

```
diff(AirPassengers)
```

There will be one fewer observation in the differenced time series than in the original time series. A quadratic trend in a time series can be detrended by applying the differencing operator ∇ twice to the original time series:

$$\nabla^2 x_t = \nabla(\nabla x_t) = \nabla(x_t - x_{t-1}) = (x_t - x_{t-1}) - (x_{t-1} - x_{t-2}) = x_t - 2x_{t-1} + x_{t-2}.$$

This detrending filter can be written with the backshift operator as $(1 - 2B + B^2)x_t$. Second-order differences, denoted by $\nabla^2 x_t$, can be calculated by applying the `diff` function twice:

```
diff(diff(AirPassengers))
```

or by using the `differences` argument in the `diff` function:

```
diff(AirPassengers, differences = 2)
```

Monthly observations from a time series with an annual seasonal component, for example, can have the seasonal differencing filter

$$\nabla_{12}x_t = x_t - x_{t-12} = \left(1 - B^{12}\right)x_t$$

applied to eliminate the seasonal effects. The `lag` argument is added to the `diff` function in order to do this type of seasonal differencing.

```
diff(AirPassengers, lag = 12)
```

There will be 12 fewer observations in the new time series after applying this filter.

The backshift operator B, which is useful in writing differencing operations compactly, is defined by $Bx_t = x_{t-1}$, or more generally as $B^m x_t = x_{t-m}$. A single application of the B operator can be achieved by a call to the `lag` function.

```
lag(AirPassengers)
```

The first observation in the resulting time series is December of 1948. The time series has simply been shifted back in time by one month. To apply the B operator twice, denoted by B^2, the second argument should be set to 2.

```
lag(AirPassengers, 2)
```

The first observation in the resulting time series is November of 1948. In order to shift a time series forward in time, a negative value for the second argument is used.

```
lag(AirPassengers, -3)
```

The first observation in the resulting time series is April of 1949.

The intersection of several time series can be achieved with the `ts.intersect` function, which is illustrated below.

```
ts.intersect(lag(AirPassengers), AirPassengers, lag(AirPassengers, -1))
```

The elements of the resulting intersection of the three time series will only be defined on common time values. A related function named `ts.union`, will take the union of the constituent time series, appending an `NA` to positions in any of the constituent time series without observations.

Occasions arise in which only a subset of a time series is of interest. The `window` function shown below extracts the portion of the `AirPassengers` time series between January of 1951 and June of 1957.

```
window(AirPassengers, start = c(1951, 1), end = c(1957, 6))
```

A linear filter can also be applied to a time series using the `filter` function. The example below calculates a 12-point moving average of the `AirPassengers` time series.

```
filter(AirPassengers, filter = rep(1 / 12, 12), sides = 1)
```

A time series can be decomposed into trend, seasonal, and random components using the `decompose` function.

```
decompose(AirPassengers)
```

These components can be viewed by embedding this command into the plot function.

```
plot(decompose(AirPassengers))
```

This provides only a graph of the four components. The `decompose` function provides a rather elementary way of decomposing time series. A more sophisticated approach using the loess (locally estimated scatterplot smoothing) method is the `stl` (which is an abbreviation for seasonal decomposition of a time series by loess) function.

```
stl(AirPassengers, s.window = "periodic")
```

7.4 Exercises

7.1 White noise (*WN*), iid noise (*IID*), and Gaussian white noise (*GWN*) were introduced in Definition 7.1. The three classes are related by

$$GWN \subset IID \subset WN.$$

Indicate the strongest class of noise associated with the following three time series.

    (a) $X_1, X_2, \ldots, X_n$ are mutually independent and identically distributed $N(0, 1)$ random variables.

    (b) $X_1, X_2, \ldots, X_n$ are mutually independent random variables with $X_t \sim N(0, 1)$ when t is even and $X_t \sim U\left(-\sqrt{3}, \sqrt{3}\right)$ when t is odd.

    (c) $X_1, X_2, \ldots, X_n$ are mutually independent and identically distributed $U(-2, 2)$ random variables.

7.2 Let $X_1, X_2, \ldots, X_n$ be n observations from the random walk model described in Example 7.4. Find $V\left[\bar{X}\right]$, where $\bar{X}$ is the sample mean.

7.3 The realization of the random walk in Example 7.4 was generated by using a `while` loop in R. Write R code to generate the same time series values without using a loop.

7.4 The time series of the number of monthly accidental deaths in the United States from 1973 to 1978 is given in `USAccDeaths` in R. Make a plot of the time series values and comment on any features you can glean from the plot.

7.5 Classify each of the following stochastic processes by time (discrete or continuous) and state (discrete or continuous).

    (a) The number of eastbound cars stopped at a particular stoplight over time.

    (b) The location of a taxi cab (classified as city, airport, or suburbs) at the end of each passenger's ride.

    (c) The temperature of a puppy measured at 20-minute intervals.

    (d) A person's internal body temperature over time.

    (e) The number of goldfish on inventory at a local pet shop.

7.6 Consider the random variables $X_1, X_2, \ldots, X_n$ and $Y_1, Y_2, \ldots, Y_m$ with associated finite population means, population variances, and population covariances. Show that

$$\text{Cov}\left(\sum_{i=1}^{n} a_i X_i, \sum_{j=1}^{m} b_j Y_j\right) = \sum_{i=1}^{n} \sum_{j=1}^{m} a_i b_j \text{Cov}\left(X_i, Y_j\right)$$

for real-valued constants $a_1, a_2, \ldots, a_n$ and $b_1, b_2, \ldots, b_m$.

7.7 The time series $\{X_t\}$ consists of the number of spots on the up face of sequential rolls of a fair die.

    (a) Is this time series strictly stationary?

    (b) What is the population mean function?

    (c) What is the population autocovariance function?

    (d) What is the population autocorrelation function?

7.8 Consider the (tiny) time series X_1, X_2, X_3, whose values are the cumulative number of spots showing in three rolls of a fair die. More specifically, if R_1, R_2, and R_3 denote the outcomes of the three rolls, then $X_1 = R_1$, $X_2 = R_1 + R_2$, and $X_3 = R_1 + R_2 + R_3$.

    (a) What is the population mean function?

    (b) Is this time series strictly stationary?

    (c) Is this time series stationary?

    (d) What is the population variance–covariance matrix of X_1, X_2, X_3?

    (e) Perform a Monte Carlo simulation which provides convincing numerical evidence that the population variance–covariance matrix from part (d) is correct.

7.9 Argue whether the time series of monthly home energy consumption observations from Example 7.1 is a stationary time series.

7.10 Consider the time series $\{X_t\}$ consisting of white noise terms defined by

$$X_t \sim \begin{cases} N(0, 1) & t \text{ odd} \\ U\left(-\sqrt{3}, \sqrt{3}\right) & t \text{ even.} \end{cases}$$

    (a) Is this time series strictly stationary?

    (b) Is this time series stationary?

7.11 The energy consumption time series introduced in Example 7.1 was given in number of kilowatt hours per month. The varying number of days per month was not taken into account in the analysis performed on this data set in this chapter. Adjust the time series so that the varying month length has been taken into account and answer the following.

    (a) The original time series had the maximum monthly energy consumption in January of 2018. Which month has the maximum monthly energy consumption in the adjusted time series?

    (b) Make a plot of the time series using the units average daily number of kilowatt hours.

    (c) Use the R `acf` function to calculate the sample autocorrelation at lag 3. Interpret the sign of the sample lag 3 autocorrelation.

7.12   The R built-in data set `nhtemp` contains the average annual temperatures in New Hampshire from 1912 to 1971. Plot the time series and correlogram.

7.13   Consider a time series of $n = 100$ observations which are Gaussian white noise with $\sigma_Z = 1$. Use Monte Carlo simulation to estimate the probability that the lag 3 sample autocorrelation falls between $-z_{\alpha/2}/\sqrt{100}$ and $z_{\alpha/2}/\sqrt{100}$, where $\alpha = 0.05$. Report your estimate to 3-digit accuracy.

7.14   The sojourn time of a customer in a single-server queue is defined as the waiting time plus the service time. Consider 100 consecutive sojourn times for customers in a single-server queue with exponential times between arrivals with population mean 1 and exponential service times with population mean 0.9. (This is a special case of what is known in queueing theory as an M/M/1 queue. The first M is for Markov, and indicates that the times between arrivals is exponentially distributed. The second M is also for Markov, and indicates that the service times are exponentially distributed. The 1 indicates that there is a single server.) Also assume that the first 1000 customer sojourn times have been discarded so that the system has "warmed up." A realization of these 100 consecutive sojourn times can be generated in R and placed into the vector `x` with the statements

```
install.packages("simEd")
library(simEd)
x = ssq(maxArrivals = 1100, saveSojournTimes = TRUE,
        showProgress = FALSE, seed = 12345)$sojournTimes[1001:1100]
```

     (a) Write a paragraph that outlines whether or not the stationarity assumption is appropriate in this setting.

     (b) Before running a simulation, predict whether the sample lag 3 autocorrelation will be positive, zero, or negative.

     (c) Make three runs of this simulation with three different seeds and plot the three sample autocorrelation functions on the same set of axes (plot them as points connected by lines rather than spikes) for the first 20 lags.

7.15   Consider a stationary time series $\{X_t\}$. For $k = 1, 2, \ldots$, the lag k population partial autocorrelation $\rho^*(k)$ can be written as the last component of the vector defined by

$$\Gamma_k^{-1}\gamma_k,$$

where Γ_k is the $k \times k$ variance–covariance matrix of any k elements of the time series and $\gamma_k = \left(\gamma(1), \gamma(2), \ldots, \gamma(k)\right)'$. Show that this way of calculating the lag k population partial autocorrelation is equivalent to that given in Definition 7.8 for $k = 1$ and $k = 2$.

7.16   Consider a three-point moving average.

     (a) What are the coefficients associated with three of these moving averages applied in series.

     (b) Check your solution using the R `convolve` function.

     (c) Apply this series of three filters to the `AirPassengers` data set and plot the smoothed series.

7.17 The logarithm and square root transformations are commonly used on a time series whose variability increases with time. Propose a transformation for a time series whose variability decreases with time.

7.18 Find the population autocorrelation function for an m-point moving average of a white noise time series, where m is an odd, positive integer.

7.19 The R decompose function can be used to decompose a time series. Another R function named stl is a more sophisticated function for decomposing a time series. Apply this function to the built-in R time series JohnsonJohnson, which contains the quarterly earnings, in dollars, for one share of stock in Johnson & Johnson from 1960 to 1980. Plot the trend and seasonal components of the decomposed time series. Which quarter tends to have the highest quarterly earnings? Considering the seasonal part of the decomposed model, what is the difference between the best and worst quarter's earnings to the nearest penny?

Chapter 8

Time Series Modeling

This chapter presents several popular probability models for describing a time series, along with the associated statistical methods. Analogous to using the univariate normal distribution to model a quantitative variable which has a bell-shaped probability distribution, no time series model will provide a perfect fit to the data. The goal is to identify a probability model which provides a reasonable approximation to the time series, fit the model to an observed time series, and then use the fitted model for statistical inference, which is often forecasting.

8.1   Probability Models

A suite of probability models for time series known as *linear models* are introduced in this section. The unifying characteristic of these models is that they express the current value of the time series as a linear function of (*a*) the current noise term, (*b*) previous noise terms, and (*c*) previous values of the time series. We begin by taking a birds-eye view of these linear time series models by introducing *general linear models* (often abbreviated glm) and some of their properties. This is followed by a section that introduces a suite of time series models that are special cases of general linear models that are known as *ARMA* (autoregressive moving average) models. ARMA models are *parsimonious* in the sense that they are able to specify a wide variety of underlying probability models that govern a stationary time series with only a few parameters. With both general linear models and ARMA models, you will see a great deal of symmetry and some mathematics that works out beautifully on the road to developing time series models that can be implemented in real-world applications.

8.1.1   General Linear Models

General linear models provide an important way of thinking about how to define a time series model in a simple and general manner. Working with general linear models also provides some practice with using the backshift operator B, which was defined in Section 7.3.1. We also consider the causal and invertible form of general linear models. The causal form is important for establishing stationarity. The invertible form is important for ensuring a one-to-one relationship between parameter values and the associated population autocorrelation function.

The concepts of white noise from Definition 7.1 and linear filters from Section 7.3.1 are tied together in this section to define general linear models. White noise is a time series of mutually independent random variables denoted by $\{Z_t\}$. Each element in the white noise time series has common population mean 0 and common population variance σ_Z^2. Time series analysts often refer to the Z_t values as *random shocks* whose purpose is to inject randomness into a time series model. Without these shocks, the time series model would be purely deterministic. Linear filters are a general way of expressing one time series as a linear combination of the values in another time series. White noise and linear filters are the key concepts in the definition of general linear models. As you will see in the next paragraph, there are two distinctly different ways of defining general linear models.

More specifically, one way to describe a general linear model is to define the current value in the time series X_t as the current white noise term Z_t plus a linear combination of the previous white noise terms:

$$X_t = Z_t + \psi_1 Z_{t-1} + \psi_2 Z_{t-2} + \cdots,$$

where the coefficients $\psi_1, \psi_2, \ldots$ in the infinite series are real-valued constants. This time series model is stationary when appropriate restrictions are placed on the $\psi_1, \psi_2, \ldots$ values. Since this description of a general linear model is valid at time t, it is also valid at other time values, for example,

$$X_{t-1} = Z_{t-1} + \psi_1 Z_{t-2} + \psi_2 Z_{t-3} + \cdots,$$

or

$$X_{t-2} = Z_{t-2} + \psi_1 Z_{t-3} + \psi_2 Z_{t-4} + \cdots.$$

Solving these equations for the current white noise value and sequentially substituting into the first formulation of the general linear model, you can see that there is a second way to formulate a general linear model:

$$X_t = Z_t + \pi_1 X_{t-1} + \pi_2 X_{t-2} + \cdots,$$

where the coefficients $\pi_1, \pi_2, \ldots$ are real-valued constants and appropriate restrictions are placed on the $\pi_1, \pi_2, \ldots$ values in order to achieve stationarity. In this second formulation of a general linear model, the current value of the time series is a linear combination of the previous values of the time series plus the current white noise term. This formulation is analogous to that of a multiple linear regression model with an infinite number of predictor variables.

A reasonable question to ask at this point is why there is no coefficient associated with Z_t in both formulations of the general linear model. Although some authors associate a coefficient ψ_0 with Z_t, we avoid this practice and simply assume that $\psi_0 = 1$. Including a ψ_0 parameter is redundant because a nonzero constant multiplied by a white noise term is still a white noise term. The population variance of the white noise σ_Z^2 is essentially absorbed into the ψ_0 parameter. Also, some authors use a $-$ rather than a $+$ between terms on the right-hand side of the second formulation of the general linear model.

The two formulations for the general linear model involve a random variable on the left-hand side of the model and random variables on the right-hand side of the model. In some settings, this might be viewed as a transformation of random variables, but this is not the correct interpretation of the model in the time series setting. The general linear model formulations define a hypothesized relationship between the random variable on the left-hand side of the model and the random variables on the right-hand side of the model. In the first formulation of the general linear model, the current value of the time series X_t is hypothesized to be a linear combination of the current and previous noise values. In the second formulation of the general linear model, the current value of the time series X_t is hypothesized to be a linear combination of the previous values in the time series plus a noise term. This probability model is hypothesized to govern the process over time. The goal in constructing a time series model is to write a formula for a model which adequately captures the probabilistic relationship that governs the time series. Estimation of the model parameters will be followed by assessments to see if the model holds in an empirical sense.

The coefficients in the two formulations of a general linear model are related. To make these two formulations of the general linear model more concrete, we will now look at a specific instance.

Example 8.1 Consider the special case of the first formulation of the general linear model

$$X_t = Z_t + \psi_1 Z_{t-1}.$$

This model only has one coefficient ψ_1. The subsequent coefficients are $\psi_j = 0$ for $j = 2, 3, \ldots$. Find the equivalent form of the general linear model using the second formulation.

Recall from Section 7.3.1 that the backshift operator B shifts a time series value back one unit in time, for example,

$$BX_t = X_{t-1}.$$

When the backshift operator includes a superscript, the superscript accounts for multiple backshifts, for example,

$$B^4 Z_t = Z_{t-4}.$$

The special case of the general linear model considered here can be converted from its original form,

$$X_t = Z_t + \psi_1 Z_{t-1},$$

to a form using the backshift operator,

$$X_t = Z_t + \psi_1 B Z_t$$

or

$$X_t = (1 + \psi_1 B) Z_t.$$

Although it might seem like an unusual operation involving B, both sides of this equation can be divided by $1 + \psi_1 B$, which gives

$$\frac{X_t}{1 + \psi_1 B} = Z_t.$$

For ψ_1 values on the interval $-1 < \psi_1 < 1$, this can be expanded as a geometric series with common ratio $-\psi_1 B$:

$$\left(1 - \psi_1 B + \psi_1^2 B^2 - \cdots \right) X_t = Z_t$$

or

$$X_t - \psi_1 X_{t-1} + \psi_1^2 X_{t-2} - \cdots = Z_t$$

or

$$X_t = Z_t + \psi_1 X_{t-1} - \psi_1^2 X_{t-2} + \cdots.$$

This is the second formulation of the general linear model with coefficients $\pi_j = (-1)^{j-1} \psi_1^j$ for $j = 1, 2, \ldots$ and $-1 < \psi < 1$.

A sleight of hand has occurred in the previous example with respect to the use of the backshift operator B, first as an operator and then as a variable. This paragraph concerns that dual use. When B is used as an operator, it has a domain or input, for instance, X_t, and a range or output, for instance, $BX_t = X_{t-1}$. In this case, the effect of the operator B on a time series value is to go back in the time series one unit of time. The input to B is the value of the time series at time t, and the output from B is the value of the time series at time $t-1$. The full domain of the operator B is the entire sequence of time series values. Why is it acceptable to take an operator like the backshift operator B and use it as a variable? It can be demonstrated that the backshift operator B functions like a linear map in the sense of allowing the standard multiplication and addition operations in its domain. In addition to the standard operations such addition, multiplication, and inversion, we may thus treat polynomials in B as polynomials in real variables.

For the particular case of the general linear model considered in the previous example, there was a relationship between the coefficients in the two formulations of the general linear model. We now consider whether there is a relationship between the coefficients $\psi_1, \psi_2, \ldots$ and $\pi_1, \pi_2, \ldots$ in the general setting. We continue with our use of the backshift operator B. The first formulation of the general linear model is

$$X_t = Z_t + \psi_1 Z_{t-1} + \psi_2 Z_{t-2} + \cdots,$$

which can be rewritten using the backshift operator as

$$X_t = Z_t + \psi_1 B Z_t + \psi_2 B^2 Z_t + \cdots$$

or

$$X_t = \left(1 + \psi_1 B + \psi_2 B^2 + \cdots\right) Z_t.$$

The polynomial in B in this formulation of the model is denoted by $\psi(B)$, so the first formulation of the general linear model can be written compactly as

$$X_t = \psi(B) Z_t,$$

where $\psi(B) = 1 + \psi_1 B + \psi_2 B^2 + \cdots$.

Now consider the second formulation of the general linear model:

$$X_t = Z_t + \pi_1 X_{t-1} + \pi_2 X_{t-2} + \cdots.$$

Separating the time series terms on the left-hand side of the equation and the white noise term on the right-hand side of the equation results in

$$X_t - \pi_1 X_{t-1} - \pi_2 X_{t-2} - \cdots = Z_t,$$

which can be rewritten using the backshift operator as

$$X_t - \pi_1 B X_t - \pi_2 B^2 X_t - \cdots = Z_t$$

or

$$\left(1 - \pi_1 B - \pi_2 B^2 - \cdots\right) X_t = Z_t.$$

The polynomial in B in this formulation of the model is denoted by $\pi(B)$, so the second formulation of the general linear model can be written compactly as

$$\pi(B) X_t = Z_t,$$

where $\pi(B) = 1 - \pi_1 B - \pi_2 B^2 - \cdots$.

Definition 8.1 gives the two formulations of the general linear model expressed in purely algebraic form and in terms of polynomials in the backshift operator.

Definition 8.1 A time series $\{X_t\}$ can be expressed as a *general linear model* as

$$X_t = Z_t + \psi_1 Z_{t-1} + \psi_2 Z_{t-2} + \cdots,$$

where $\psi_1, \psi_2, \ldots$ are real-valued constants and $Z_t \sim WN\left(0, \sigma_Z^2\right)$, or, equivalently, as

$$X_t = \left(1 + \psi_1 B + \psi_2 B^2 + \cdots\right) Z_t = \psi(B) Z_t.$$

Alternatively, the general linear model for a time series can be written as

$$X_t = Z_t + \pi_1 X_{t-1} + \pi_2 X_{t-2} + \cdots$$

for certain values of the real-valued constants $\pi_1, \pi_2, \ldots$, or, equivalently, as

$$\left(1 - \pi_1 B - \pi_2 B^2 - \cdots\right) X_t = \pi(B) X_t = Z_t.$$

In the previous example, we were able to perform algebraic steps to determine the relationship between the coefficients in the first formulation of the general linear model (that is, $\psi_1, \psi_2, \ldots$) and the coefficients in the second formulation (that is, $\pi_1, \pi_2, \ldots$). This can also be done in the more general setting. The equations that define the two formulations of the general linear model in Definition 8.1 written in terms of the backshift operator are

$$X_t = \psi(B) Z_t \qquad \text{and} \qquad \pi(B) X_t = Z_t.$$

Applying the $\psi(B)$ polynomial to both sides of the second equation gives

$$\psi(B)\pi(B) X_t = \psi(B) Z_t$$

or

$$\psi(B)\pi(B) X_t = X_t$$

or

$$\psi(B)\pi(B) = 1$$

for nonzero X_t. Since the product of the polynomials $\psi(B)$ and $\pi(B)$ is one, they are inverses. For suitable values of the coefficients, this allows us to calculate the coefficients $\psi_1, \psi_2, \ldots$ from the coefficients $\pi_1, \pi_2, \ldots$ and vice versa. The inverse relationship between $\psi(B)$ and $\pi(B)$ will now be confirmed for the polynomials identified in the previous example.

Example 8.2 Verify that $\psi(B)\pi(B) = 1$ for the time series model for $\{X_t\}$ from the previous example:

$$X_t = Z_t + \psi_1 Z_{t-1},$$

where $-1 < \psi_1 < 1$ and $\{Z_t\}$ is a time series of white noise.

From Example 8.1, the polynomials in the backshift operator are

$$\psi(B) = 1 + \psi_1 B$$

and

$$\pi(B) = 1 - \psi_1 B + \psi_1^2 B^2 - \cdots.$$

The product of $\psi(B)$ and $\pi(B)$ is

$$
\begin{aligned}
\psi(B)\pi(B) &= (1 + \psi_1 B)\left(1 - \psi_1 B + \psi_1^2 B^2 - \cdots\right) \\
&= \left(1 - \psi_1 B + \psi_1^2 B^2 - \cdots\right) + \left(\psi_1 B - \psi_1^2 B^2 + \psi_1^3 B^3 - \cdots\right) \\
&= 1
\end{aligned}
$$

as expected.

The previous discussion constitutes a proof of the following theorem concerning writing the two forms of the general linear model in terms of polynomials in the backshift operator and the relationship between the two polynomials $\psi(B)$ and $\pi(B)$.

Theorem 8.1 The two formulations of the general linear model from Definition 8.1 associated with the two polynomials $\psi(B)$ and $\pi(B)$ are equivalent time series models and are related by

$$
\psi(B)\pi(B) = 1
$$

for certain values of the coefficients.

We will toggle between the purely algebraic formulations of the general linear model and the associated formulations using the backshift operator B based on which is more convenient and effective for the mathematics in a particular setting. Definition 8.1 gives two different ways of writing a general linear model, but is vague concerning any constraints placed on the coefficients. Some constraints on the coefficients that give the general linear model certain important characteristics are outlined next. Stationarity will play a central role in these constraints. The stationarity property implies that the time series is stable over time; this stability allows us to predict how the time series will behave in the future.

Causality and Invertibility

The general linear model is formulated in two different fashions in Definition 8.1. But we have not yet defined any general constraints on the coefficients in the two different formulations of the general linear model. We begin the consideration of appropriate constraints on the coefficients with some calculations on the first formulation of the general linear model.

The first formulation of the general linear model from Definition 8.1 using the purely algebraic form is

$$
X_t = Z_t + \psi_1 Z_{t-1} + \psi_2 Z_{t-2} + \cdots .
$$

We would like to determine constraints on the coefficients $\psi_1, \psi_2, \ldots$ that will result in a stationary model and also find expressions for quantities associated with the stationary version of this model, such as $E[X_t]$, $V[X_t]$, $\gamma(k)$, and $\rho(k)$. Taking the expected value of both sides of the defining formula results in

$$
\begin{aligned}
E[X_t] &= E[Z_t + \psi_1 Z_{t-1} + \psi_2 Z_{t-2} + \cdots] \\
&= E[Z_t] + E[\psi_1 Z_{t-1}] + E[\psi_2 Z_{t-2}] + \cdots \\
&= E[Z_t] + \psi_1 E[Z_{t-1}] + \psi_2 E[Z_{t-2}] + \cdots \\
&= 0
\end{aligned}
$$

because each of the white noise terms has expected value 0. This is a promising first step toward achieving stationarity. So far, no constraints are needed on the coefficients $\psi_1, \psi_2, \ldots$. That will change when we compute the population variance of X_t. Taking the population variance of both sides of the defining formula results in

$$\begin{aligned} V[X_t] &= V[Z_t + \psi_1 Z_{t-1} + \psi_2 Z_{t-2} + \cdots] \\ &= V[Z_t] + V[\psi_1 Z_{t-1}] + V[\psi_2 Z_{t-2}] + \cdots \\ &= V[Z_t] + \psi_1^2 V[Z_{t-1}] + \psi_2^2 V[Z_{t-2}] + \cdots \\ &= \left(1 + \psi_1^2 + \psi_2^2 + \cdots\right)\sigma_Z^2 \end{aligned}$$

because the white noise terms are mutually independent random variables with common finite population variance σ_Z^2 (see Definition 7.1). Not all values of $\psi_1, \psi_2, \ldots$ will result in a finite population variance of X_t. Setting $\psi_1 = \psi_2 = \cdots = 1$, for example, results in an infinite population variance of X_t. In order to get a finite population variance, the ψ values must decrease in magnitude rapidly enough so that

$$\psi_1^2 + \psi_2^2 + \cdots < \infty.$$

One way to achieve this condition is to have finite values for the first q coefficients $\psi_1, \psi_2, \ldots, \psi_q$ then zeros thereafter. Any general linear model of the first formulation with coefficients that "cut off" in this fashion will satisfy the constraint. Another way of considering this constraint is to write this model using the backshift operator. Using Definition 8.1, the first formulation of the general linear model is

$$X_t = \psi(B)Z_t = \left(1 + \psi_1 B + \psi_2 B^2 + \cdots\right)Z_t.$$

The polynomial in the backshift operator

$$\psi(B) = 1 + \psi_1 B + \psi_2 B^2 + \cdots$$

will be considered for B values that can assume complex values. So B can have the form $B = a + bi$. The constraint on the coefficients $\psi_1, \psi_2, \ldots$ is equivalent to $\psi(B)$ converging for all B values falling on or inside the unit circle. In other words, $|B| \leq 1$.

The population autocovariance function for the general linear model stated in the form

$$X_t = Z_t + \psi_1 Z_{t-1} + \psi_2 Z_{t-2} + \cdots$$

with coefficients $\psi_1, \psi_2, \ldots$ satisfying the constraint can be calculated by using the definition of the population covariance:

$$\begin{aligned} \gamma(k) &= \text{Cov}(X_t, X_{t+k}) \\ &= \text{Cov}(Z_t + \psi_1 Z_{t-1} + \psi_2 Z_{t-2} + \cdots, Z_{t+k} + \psi_1 Z_{t+k-1} + \psi_2 Z_{t+k-2} + \cdots) \\ &= \text{Cov}(Z_t, \psi_k Z_{t+k-k}) + \text{Cov}(\psi_1 Z_{t-1}, \psi_{k+1} Z_{t+k-(k+1)}) + \cdots \\ &= \psi_k \sigma_Z^2 + \psi_1 \psi_{k+1} \sigma_Z^2 + \psi_2 \psi_{k+2} \sigma_Z^2 + \cdots \\ &= (\psi_k + \psi_1 \psi_{k+1} + \psi_2 \psi_{k+2} + \cdots)\sigma_Z^2 \end{aligned}$$

for $k = 1, 2, \ldots$ because of the mutual independence of the terms in the white noise time series. As expected from the previous derivation, the autocovariance at lag 0 is the population variance of X_t:

$$\gamma(0) = V[X_t] = \left(1 + \psi_1^2 + \psi_2^2 + \cdots\right)\sigma_Z^2,$$

where ψ_0, the coefficient of Z_t, equals 1. The associated autocorrelation function is

$$\rho(k) = \frac{\gamma(k)}{\gamma(0)} = \frac{\sigma_Z^2 (\psi_k + \psi_1 \psi_{k+1} + \psi_2 \psi_{k+2} + \cdots)}{\sigma_Z^2 (1 + \psi_1^2 + \psi_2^2 + \cdots)} = \frac{\psi_k + \psi_1 \psi_{k+1} + \psi_2 \psi_{k+2} + \cdots}{1 + \psi_1^2 + \psi_2^2 + \cdots}$$

for $k = 1, 2, \ldots$. Notice that $\rho(0) = 1$ as expected.

The derivation so far has been general, so when specific values of the coefficients $\psi_1, \psi_2, \ldots$ are specified, we now have formulas to determine the population autocovariance function and the population autocorrelation function. Computing these two functions will be illustrated in the next example.

Example 8.3 Consider a time series model $\{X_t\}$ described by

$$X_t = Z_t - \frac{3}{2} Z_{t-1} + \frac{3}{4} Z_{t-2},$$

where $\{Z_t\} \sim WN\left(0, \sigma_Z^2\right)$. Determine whether this time series is stationary and calculate the population autocovariance function and autocorrelation function.

This time series model is a special case of the first formulation of the general linear model from Definition 8.1 which expresses X_t as a linear combination of the white noise terms with coefficients $\psi_1 = -3/2$, $\psi_2 = 3/4$ and $\psi_j = 0$ for $j = 3, 4, \ldots$. The time series is stationary because

$$\psi_1^2 + \psi_2^2 + \cdots = \left(-\frac{3}{2}\right)^2 + \left(\frac{3}{4}\right)^2 = \frac{45}{16} < \infty.$$

The population autocovariance function is

$$\gamma(k) = (\psi_k + \psi_1 \psi_{k+1} + \psi_2 \psi_{k+2} + \cdots) \sigma_Z^2$$
$$= \begin{cases} \left(1 + (-3/2)^2 + (3/4)^2\right) \sigma_Z^2 & k = 0 \\ \left(-3/2 + (-3/2)(3/4)\right) \sigma_Z^2 & k = 1 \\ (3/4) \sigma_Z^2 & k = 2 \\ 0 & k = 3, 4, \ldots \end{cases}$$
$$= \begin{cases} 61\sigma_Z^2/16 & k = 0 \\ -21\sigma_Z^2/8 & k = 1 \\ 3\sigma_Z^2/4 & k = 2 \\ 0 & k = 3, 4, \ldots, \end{cases}$$

where $\psi_0 = 1$ is the coefficient of Z_t. The associated population autocorrelation function $\rho(k) = \gamma(k)/\gamma(0)$ is

$$\rho(k) = \begin{cases} 1 & k = 0 \\ -42/61 & k = 1 \\ 12/61 & k = 2 \\ 0 & k = 3, 4, \ldots, \end{cases}$$

which is graphed in Figure 8.1. The population autocorrelation function "cuts off" after spikes at lags 1 and 2.

The constraint that has been placed on the values of $\psi_1, \psi_2, \ldots$ can be formalized in this definition of the *causal* representation of the general linear model.

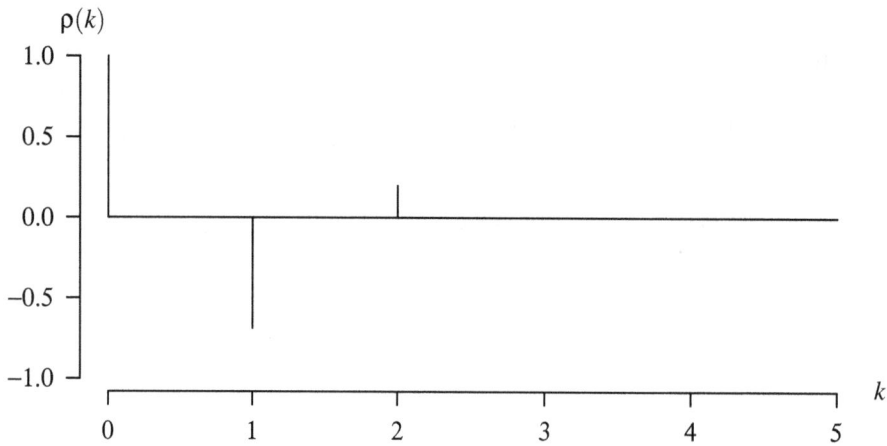

Figure 8.1: Population autocorrelation function for $X_t = Z_t - \frac{3}{2}Z_{t-1} + \frac{3}{4}Z_{t-2}$.

Definition 8.2 A time series $\{X_t\}$ is *causal* if it can be written as

$$X_t = Z_t + \psi_1 Z_{t-1} + \psi_2 Z_{t-2} + \cdots,$$

where $\psi_1, \psi_2, \ldots$ are real-valued coefficients that satisfy

$$\psi_1^2 + \psi_2^2 + \cdots < \infty.$$

A time series model that can be written in the causal form is stationary.

The next example illustrates how to convert a general linear model into the causal form in order to establish stationarity.

Example 8.4 Consider the special case of the general linear model

$$\left(1 - \frac{2}{5}B\right) X_t = Z_t.$$

Convert this time series model to the causal representation.

The causal form from Definition 8.2 is

$$X_t = Z_t + \psi_1 Z_{t-1} + \psi_2 Z_{t-2} + \cdots.$$

So for the specific case given here,

$$\left(1 - \frac{2}{5}B\right)(Z_t + \psi_1 Z_{t-1} + \psi_2 Z_{t-2} + \cdots) = Z_t.$$

Expanding the left-hand side of this equation gives

$$Z_t + \left(\psi_1 - \frac{2}{5}\right)Z_{t-1} + \left(\psi_2 - \frac{2}{5}\psi_1\right)Z_{t-2} + \left(\psi_3 - \frac{2}{5}\psi_2\right)Z_{t-3} + \cdots = Z_t.$$

Equating the coefficients on the left-hand side and right-hand side of this equation as illustrated in Table 8.1 allows us to solve for $\psi_1, \psi_2, \ldots$. So the causal form of the time series model is

$$X_t = Z_t + \left(\frac{2}{5}\right) Z_{t-1} + \left(\frac{2}{5}\right)^2 Z_{t-2} + \left(\frac{2}{5}\right)^3 Z_{t-3} + \cdots,$$

which has coefficients $\psi_j = (2/5)^j$, for $j = 1, 2, \ldots$. Notice that

$$\psi_1^2 + \psi_2^2 + \psi_3^2 + \cdots = \left(\frac{2}{5}\right)^2 + \left(\frac{2}{5}\right)^4 + \left(\frac{2}{5}\right)^6 + \cdots = \frac{4/25}{1 - 4/25} = \frac{4}{21} < \infty,$$

so the time series is causal because Definition 8.2 is satisfied. Since the time series is causal, this implies that it is also stationary.

term	equation	solution
Z_{t-1}	$\psi_1 - \dfrac{2}{5} = 0$	$\psi_1 = \dfrac{2}{5}$
Z_{t-2}	$\psi_2 - \dfrac{2}{5}\psi_1 = 0$	$\psi_2 = \left(\dfrac{2}{5}\right)^2$
Z_{t-3}	$\psi_3 - \dfrac{2}{5}\psi_2 = 0$	$\psi_3 = \left(\dfrac{2}{5}\right)^3$
$\vdots$	$\vdots$	$\vdots$

Table 8.1: Matching coefficients.

When the second formulation of the general linear model that uses the coefficients $\pi_1, \pi_2, \ldots$ is used, there is an analogous property known as *invertibility* which is defined next. In this case the coefficients $\pi_1, \pi_2, \ldots$ need to decrease in magnitude rapidly enough so that

$$\pi_1^2 + \pi_2^2 + \cdots < \infty.$$

Loosely speaking, a time series model is invertible if there is a one-to-one correspondence between the coefficients $\pi_1, \pi_2, \ldots$ and the associated population autocorrelation function.

Definition 8.3 A time series $\{X_t\}$ is *invertible* if it can be written as

$$X_t = Z_t + \pi_1 X_{t-1} + \pi_2 X_{t-2} + \cdots,$$

where $\pi_1, \pi_2, \ldots$ are real-valued coefficients that satisfy

$$\pi_1^2 + \pi_2^2 + \cdots < \infty.$$

An invertible time series model has a one-to-one correspondence between the coefficients and the autocorrelation function.

So causality and invertibility are dual properties. Causality indicates that a time series model can be written in the first formulation of the general linear model from Definition 8.1 with coefficients that result in a stationarity model. Invertibility indicates that a time series model can be written in the second formulation of the general linear model from Definition 8.1 with coefficients that ensure a one-to-one correspondence between the coefficients and the population autocorrelation function.

There are three unsettling aspects to the general linear model. First, it only considers *linear* relationships between the X's and the Z's. Situations might arise in which a quadratic term, for example, might be appropriate. Second, the general linear model has an infinite number of parameters: the coefficients $\psi_1, \psi_2, \ldots$ for the first formulation and the coefficients $\pi_1, \pi_2, \ldots$ for the second formulation. ARMA (autoregressive moving average) models, which are special cases of general linear models that are introduced in the next section, limit the number of parameters in the model. The third shortcoming concerns the population mean. Taking the expected value of both sides of the first formulation of the general linear model

$$X_t = Z_t + \psi_1 Z_{t-1} + \psi_2 Z_{t-2} + \cdots,$$

for example, gives $E[X_t] = 0$. But the vast majority of real-world time series are not centered around zero. These problems associated with an infinite number of parameters and nonzero mean value will be overcome by the ARMA models introduced in the next section.

8.1.2 An Introduction to ARMA Models

The autoregressive moving average time series model, universally known as the ARMA model, provides two twists on the general linear model. First, the ARMA model limits the number of terms, and therefore limits the number of parameters. Second, the ARMA model includes both types of terms in the two formulations of the general linear model given in Definition 8.1.

There are several reasons for the popularity of the ARMA time series model. First, the population autocorrelation function $\rho(k)$ for an ARMA model can take on a wide variety of shapes, which makes it an appropriate time series model in a wide variety of applications. Second, the ARMA model is parsimonious in the sense that it typically requires only a small number of parameters to achieve an adequate representation of the probability model governing a time series. The notion of parsimony appears in all branches of statistics in which there is interest in finding an approximate probability model using the smallest number of parameters. Third, the ARMA model has been around for several decades, which means that dozens of software packages have been developed over the years for model identification, parameter estimation, forecasting, etc. Although the emphasis here will be on the R language, there are many other software packages that support time series modeling.

The general linear model from Definition 8.1 used the parameters $\psi_1, \psi_2, \ldots$ for the first formulation and $\pi_1, \pi_2, \ldots$ for the second formulation. Of course both of these formulations have the additional parameter σ_Z^2, which is the population variance of the white noise. Tradition dictates that in the conversion from the first formulation of the general linear model to the ARMA model, the Greek letter ψ used for coefficients in the general linear model is replaced by θ, and there are q of these coefficients: $\theta_1, \theta_2, \ldots, \theta_q$. Likewise, in the conversion from the second formulation of the general linear model to the ARMA model, the Greek letter π used for the coefficients in the general model is replaced by ϕ, and there are p of these coefficients: $\phi_1, \phi_2, \ldots, \phi_p$.

So two key parameters in specifying an ARMA model are p and q, which are both nonnegative integers. The parameter p is the number of coefficient parameters in the autoregressive portion of the model; the parameter q is the number of coefficient parameters in the moving average portion of the model. The format for specifying the orders p and q of an ARMA model with p autoregressive terms and q moving average terms is ARMA(p, q).

Definition 8.4 The ARMA(p, q) time series model is

$$X_t = \overbrace{\phi_1 X_{t-1} + \phi_2 X_{t-2} + \cdots + \phi_p X_{t-p}}^{\text{autoregressive portion}} + \underbrace{Z_t + \theta_1 Z_{t-1} + \theta_2 Z_{t-2} + \cdots + \theta_q Z_{t-q}}_{\text{moving average portion}},$$

where $\{X_t\}$ is the time series of interest, $\{Z_t\}$ is a time series of white noise, $\phi_1, \phi_2, \ldots, \phi_p$ are real-valued parameters associated with the AR portion of the model, and $\theta_1, \theta_2, \ldots, \theta_q$ are real-valued parameters associated with the MA portion of the model.

The autoregressive portion of this time series model is aptly named because the current value of the time series X_t is regressed on the p previous values of itself. White noise is injected into the model through $\{Z_t\}$ because it is the widest class of the three noise processes from Definition 7.1 which gives the probabilistic properties that are derived in this chapter.

If an ARMA model only involves, for example, the autoregressive portion of the model with two terms (that is, no moving average terms because $\theta_1 = \theta_2 = \cdots = \theta_q = 0$), then this ARMA(2, 0) model is specified as an AR(2) model. Likewise, if an ARMA model only involves, for example, the moving average portion of the model with four terms (that is, no autoregressive terms because $\phi_1 = \phi_2 = \cdots = \phi_p = 0$), then this ARMA(0, 4) model is specified as an MA(4) model. An ARMA(0, 0) model is just a time series of white noise, which was analyzed in Examples 7.9 and 7.15.

The ARMA(p, q) time series model from Definition 8.4 can also be written in terms of the backshift operator B. Taking the original form of the ARMA(p, q) model

$$X_t = \phi_1 X_{t-1} + \phi_2 X_{t-2} + \cdots + \phi_p X_{t-p} + Z_t + \theta_1 Z_{t-1} + \theta_2 Z_{t-2} + \cdots + \theta_q Z_{t-q},$$

and separating the autoregressive terms on the left-hand side of the equation and the moving average terms on the right-hand side of the equation results in

$$X_t - \phi_1 X_{t-1} - \phi_2 X_{t-2} - \cdots - \phi_p X_{t-p} = Z_t + \theta_1 Z_{t-1} + \theta_2 Z_{t-2} + \cdots + \theta_q Z_{t-q}.$$

This can be written in terms of the backshift operator as

$$X_t - \phi_1 B X_t - \phi_2 B^2 X_t - \cdots - \phi_p B^p X_t = Z_t + \theta_1 B Z_t + \theta_2 B^2 Z_t + \cdots + \theta_q B^q Z_t$$

or

$$\left(1 - \phi_1 B - \phi_2 B^2 - \cdots - \phi_p B^p\right) X_t = \left(1 + \theta_1 B + \theta_2 B^2 + \cdots + \theta_q B^q\right) Z_t$$

or more compactly as

$$\phi(B) X_t = \theta(B) Z_t,$$

where the polynomials in B are

$$\phi(B) = 1 - \phi_1 B - \phi_2 B^2 - \cdots - \phi_p B^p$$

and

$$\theta(B) = 1 + \theta_1 B + \theta_2 B^2 + \cdots + \theta_q B^q,$$

and these are often referred to as the *characteristic polynomials*. This algebra constitutes a proof of the alternative representation of the ARMA(p, q) time series model using polynomials.

Theorem 8.2 The ARMA(p, q) time series model can be written using the backshift operator B as

$$\phi(B)X_t = \theta(B)Z_t,$$

where the characteristic polynomials in B are

$$\phi(B) = 1 - \phi_1 B - \phi_2 B^2 - \cdots - \phi_p B^p$$

and

$$\theta(B) = 1 + \theta_1 B + \theta_2 B^2 + \cdots + \theta_q B^q.$$

Being able to convert between the purely algebraic formulation of an ARMA(p, q) model and the backshift operator formulation is an important skill in time series analysis. The next three examples illustrate how to perform these conversions.

Example 8.5 For the ARMA time series model

$$X_t = 5X_{t-1} - 2X_{t-2} + Z_t - 4Z_{t-1} + 2Z_{t-2} - Z_{t-3},$$

(a) identify the time series model, and

(b) write the time series model in terms of the backshift operator B.

(a) Since there are two terms in the autoregressive portion of the time series model with coefficients

$$\phi_1 = 5 \qquad \text{and} \qquad \phi_2 = -2$$

and three terms in the moving average portion of the time series model with coefficients

$$\theta_1 = -4, \qquad \theta_2 = 2, \qquad \text{and} \qquad \theta_3 = -1,$$

this is an ARMA(2, 3) model.

(b) The time series model

$$X_t = 5X_{t-1} - 2X_{t-2} + Z_t - 4Z_{t-1} + 2Z_{t-2} - Z_{t-3}$$

can be separated into autoregressive and moving average portions as

$$X_t - 5X_{t-1} + 2X_{t-2} = Z_t - 4Z_{t-1} + 2Z_{t-2} - Z_{t-3}.$$

This can be written in terms of B as

$$X_t - 5BX_t + 2B^2 X_t = Z_t - 4BZ_t + 2B^2 Z_t - B^3 Z_t$$

or

$$\left(1 - 5B + 2B^2\right) X_t = \left(1 - 4B + 2B^2 - B^3\right) Z_t.$$

So the polynomials in B that define the coefficients for the ARMA(2, 3) time series model written in the form $\phi(B)X_t = \theta(B)Z_t$ are

$$\phi(B) = 1 - 5B + 2B^2 \qquad \text{and} \qquad \theta(B) = 1 - 4B + 2B^2 - B^3.$$

The previous example converted an ARMA time series model from a purely algebraic formulation to a formulation that uses the backshift operator. The next example goes in the other direction.

Example 8.6 For the ARMA time series model

$$\phi(B)X_t = \theta(B)Z_t,$$

where $\phi(B) = 1 - 0.3B$ and $\theta(B) = 1$,

(a) identify the time series model, and

(b) write the time series model in purely algebraic form.

(a) Since $\phi(B)$ is a first degree polynomial, $p = 1$. Since $\theta(B)$ is a zero degree polynomial, $q = 0$. So this is an ARMA$(1, 0)$ model, which is more commonly referred to as an AR(1) model.

(b) The time series model is

$$(1 - 0.3B)X_t = 1 \cdot Z_t$$

or

$$X_t - 0.3BX_t = Z_t$$

or

$$X_t - 0.3X_{t-1} = Z_t.$$

Isolating X_t on the left-hand side of the equation, the purely algebraic formulation of this AR(1) model with $\phi_1 = 0.3$ is

$$X_t = 0.3X_{t-1} + Z_t.$$

The third and final example of converting between the purely algebraic formulation and backshift formulation of the ARMA(p, q) model would certainly be classified as a trick question. The example emphasizes the importance of looking for common factors between the $\phi(B)$ and $\theta(B)$ polynomials.

Example 8.7 For the ARMA time series model

$$X_t = -3X_{t-1} + X_{t-2} + 3X_{t-3} + Z_t - 3Z_{t-1} - 4Z_{t-2},$$

(a) identify the time series model, and

(b) write the time series model using the backshift operator.

(a) Since there are three terms in the autoregressive portion of the model and two terms in the moving average portion of the model, one might be temped to conclude that this is an ARMA$(3, 2)$ model with autoregressive coefficients

$$\phi_1 = -3, \qquad \phi_2 = 1, \qquad \text{and} \qquad \phi_3 = 3,$$

and moving average coefficients

$$\theta_1 = -3 \qquad \text{and} \qquad \theta_2 = -4.$$

But that conclusion is wrong. It is actually an ARMA$(2, 1)$ model because $\phi(B)$ and $\theta(B)$ have a common factor, as will be seen in part (b). Writing the time series model using the backshift operator B makes it easier to recognize this common factor.

(b) The time series model

$$X_t = -3X_{t-1} + X_{t-2} + 3X_{t-3} + Z_t - 3Z_{t-1} - 4Z_{t-2}$$

can be separated into autoregressive and moving average portions as

$$X_t + 3X_{t-1} - X_{t-2} - 3X_{t-3} = Z_t - 3Z_{t-1} - 4Z_{t-2}$$

or

$$X_t + 3BX_t - B^2 X_t - 3B^3 X_t = Z_t - 3BZ_t - 4B^2 Z_t$$

or

$$\left(1 + 3B - B^2 - 3B^3\right) X_t = \left(1 - 3B - 4B^2\right) Z_t$$

or

$$\phi(B)X_t = \theta(B)Z_t,$$

where

$$\phi(B) = 1 + 3B - B^2 - 3B^3 \qquad \text{and} \qquad \theta(B) = 1 - 3B - 4B^2.$$

The model still looks like an ARMA(3, 2) model. But factoring $\phi(B)$ and $\theta(B)$ reveals that both polynomials contain a common factor:

$$\phi(B) = 1 + 3B - B^2 - 3B^3 = (1 + B)\left(1 + 2B - 3B^2\right)$$

and

$$\theta(B) = 1 - 3B - 4B^2 = (1 + B)(1 - 4B).$$

The common factor $(1 + B)$ in the two polynomials cancels, which means that the ARMA model can be reduced to

$$\phi(B)X_t = \theta(B)Z_t,$$

where

$$\phi(B) = 1 + 2B - 3B^2 \qquad \text{and} \qquad \theta(B) = 1 - 4B,$$

which is an ARMA(2, 1) model. Written in purely algebraic form, this ARMA(2, 1) model is

$$X_t + 2X_{t-1} - 3X_{t-2} = Z_t - 4Z_{t-1},$$

or

$$X_t = -2X_{t-1} + 3X_{t-2} + Z_t - 4Z_{t-1},$$

so the autoregressive coefficients are $\phi_1 = -2$ and $\phi_2 = 3$, and the moving average coefficient is $\theta_1 = -4$.

Based on this example involving a common factor in the $\phi(B)$ and $\theta(B)$ polynomials, we will henceforth assume that the modeler has removed any redundant factors in an ARMA(p, q) time series model. So any ARMA(p, q) model you see going forward will in this sense be presented in lowest terms with no common factors between $\phi(B)$ and $\theta(B)$.

Since an AR(p) model has a finite number of coefficients $\phi_1, \phi_2, \ldots, \phi_p$ in the autoregressive portion of the model, they always satisfy

$$\phi_1^2 + \phi_2^2 + \cdots + \phi_p^2 < \infty,$$

so AR(p) models are always invertible per Definition 8.3. Likewise, since an MA(q) model has a finite number of coefficients $\theta_1, \theta_2, \ldots, \theta_q$ in the moving average portion of the model, they always satisfy

$$\theta_1^2 + \theta_2^2 + \cdots + \theta_q^2 < \infty,$$

so MA(q) models are always stationary per Definition 8.2. In an advanced class in time series, you will prove that an AR(p) model is stationary when all of the p complex roots of the polynomial $\phi(B) = 0$ lie outside of the unit circle in the complex plane. Likewise, an MA(q) model is invertible when all of the q complex roots of the polynomial $\theta(B) = 0$ lie outside of the unit circle in the complex plane. An ARMA(p, q) model is stationary when all of the p complex roots of $\phi(B) = 0$ lie outside of the unit circle in the complex plane. An ARMA(p, q) model is invertible when all of the q complex roots of $\theta(B) = 0$ lie outside of the unit circle in the complex plane. These results are summarized below.

Theorem 8.3 The AR(p) model $\phi(B)X_t = Z_t$ is

- always invertible, and

- stationary when the p roots of $\phi(B) = 0$ lie outside the unit circle in the complex plane.

The MA(q) model $X_t = \theta(B)Z_t$ is

- always stationary, and

- invertible when the q roots of $\theta(B) = 0$ lie outside the unit circle in the complex plane.

The ARMA(p, q) model $\phi(B)X_t = \theta(B)Z_t$ is

- stationary when the p roots of $\phi(B) = 0$ lie outside the unit circle in the complex plane, and

- invertible when the q roots of $\theta(B) = 0$ lie outside the unit circle in the complex plane.

We now revisit the first numeric example of a time series model that we encountered earlier in this chapter to check and see if it is both stationary and invertible.

Example 8.8 Consider the time series model for $\{X_t\}$ that first appeared in Example 8.3 described by

$$X_t = Z_t - \frac{3}{2}Z_{t-1} + \frac{3}{4}Z_{t-2},$$

where $\{Z_t\} \sim WN\left(0, \sigma_Z^2\right)$. Identify this time series model and determine whether it is stationary and invertible.

Since the current and two previous white noise values included in this time series model, this is an MA(2) model. By Theorem 8.3, all MA(2) models are stationary. To see whether this model is invertible, we want to calculate the roots of $\theta(B) = 0$ and see if they lie outside of the unit circle in the complex plane. The purely algebraic form of the time series model

$$X_t = Z_t - \frac{3}{2}Z_{t-1} + \frac{3}{4}Z_{t-2},$$

can be written in terms of the backshift operator as

$$X_t = Z_t - \frac{3}{2}BZ_t + \frac{3}{4}B^2Z_t$$

or

$$X_t = \left(1 - \frac{3}{2}B + \frac{3}{4}B^2\right)Z_t,$$

so $\theta(B) = 1 - \frac{3}{2}B + \frac{3}{4}B^2$. To find the values of B that solve $\theta(B) = 0$ requires solving

$$\frac{3}{4}B^2 - \frac{3}{2}B + 1 = 0,$$

which is equivalent to the quadratic equation

$$3B^2 - 6B + 4 = 0.$$

Using the quadratic formula, the roots of this quadratic equation are

$$B = \frac{6 \pm \sqrt{36 - 48}}{6}$$

or

$$B = 1 \pm \frac{\sqrt{3}}{3}i.$$

Since $\theta(B)$ is a second-order polynomial, the complex roots are necessarily complex conjugates. We now need to determine whether these two roots lie outside of the unit circle in the complex plane. There are two ways to proceed. The first is to simply plot these two roots in the complex plane and see if they fall outside of the unit circle. Figure 8.2 shows that the two roots do indeed fall outside of the unit circle. The second way to determine whether the roots fall outside the unit circle is to take the sum of squares of the real and imaginary parts of the roots and see if they exceed 1. In this case,

$$(1)^2 + \left(\frac{\sqrt{3}}{3}\right)^2 = 1 + \frac{1}{3} = \frac{4}{3} > 1.$$

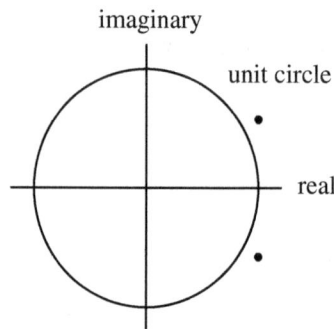

Figure 8.2: Unit circle in the complex plane and the roots of $\theta(B) = 0$.

Both techniques draw the same conclusion: the two roots of $\theta(B) = 0$ fall outside of the unit circle in the complex plane, which means that the time series model is invertible. In conclusion, this MA(2) time series model is both stationary and invertible.

We will get some further practice with these calculations involving the polynomials $\phi(B)$ and $\theta(B)$ when we investigate special cases of the ARMA(p, q) model in more detail in the sections that follow.

Shifted ARMA models

We now address a major shortcoming of the ARMA(p, q) model that–fortunately–is easily overcome. For a stationary ARMA(p, q) model as it has been defined in Definition 8.4, the expected value of X_t is $E[X_t] = 0$. But most real-world stationary time series are not centered about 0; they are typically centered about some nonzero constant value. The reason that we have waited this long to bring up the topic of a time series centered around a value other than zero is that when we shift the time series, there will be no change in the population autocovariance and autocorrelation functions because population covariance and correlation are unaffected by shifting the time series. The mathematics involved with determining these important functions is much cleaner if you assume that the time series model is centered about zero. There are two ways to tweak the ARMA(p, q) model to allow for it to be centered about some constant value. These two alterations are presented next.

The first way to introduce a nonzero central value for an ARMA(p, q) time series model is to subtract μ from all of the values in the time series. In other words, transform the usual ARMA(p, q) time series model

$$X_t = \phi_1 X_{t-1} + \phi_2 X_{t-2} + \cdots + \phi_p X_{t-p} + Z_t + \theta_1 Z_{t-1} + \theta_2 Z_{t-2} + \cdots + \theta_q Z_{t-q}$$

to the shifted ARMA(p, q) time series model

$$X_t - \mu = \phi_1 (X_{t-1} - \mu) + \phi_2 (X_{t-2} - \mu) + \cdots + \phi_p (X_{t-p} - \mu) + Z_t + \theta_1 Z_{t-1} + \theta_2 Z_{t-2} + \cdots + \theta_q Z_{t-q}.$$

This can be written compactly in terms of the backshift operator B as

$$\phi(B) (X_t - \mu) = \theta(B) Z_t,$$

where $\phi(B)$ is the usual polynomial of degree p in B associated with the autoregressive portion of the model:

$$\phi(B) = 1 - \phi_1 B - \phi_2 B^2 - \cdots - \phi_p B^p,$$

and $\theta(B)$ is the usual polynomial of degree q in B associated with the moving average portion of the model:

$$\theta(B) = 1 + \theta_1 B + \theta_2 B^2 + \cdots + \theta_q B^q.$$

In this particular formulation of a shifted ARMA(p, q) model, the population mean of the process is $E[X_t] = \mu$ when the model is stationary. This can be established by taking the expected value of both sides of the shifted ARMA(p, q) time series model.

A second way to formulate a shifted ARMA(p, q) time series model with a nonzero population mean is to simply add a constant, denoted by $\tilde{\mu}$, to the right-hand side of the model:

$$X_t = \tilde{\mu} + \phi_1 X_{t-1} + \phi_2 X_{t-2} + \cdots + \phi_p X_{t-p} + Z_t + \theta_1 Z_{t-1} + \theta_2 Z_{t-2} + \cdots + \theta_q Z_{t-q}$$

This can be written in terms of the backshift operator as

$$\phi(B) X_t = \tilde{\mu} + \theta(B) Z_t.$$

The reason that a tilde has been placed above μ in this formulation is that $\tilde{\mu}$ is *not* the population mean of the time series model. The two ways of formulating a shifted ARMA(p, q) time series model in these two fashions are summarized as follows.

Definition 8.5 A shifted ARMA(p, q) time series model with a nonzero population mean μ can be written in purely algebraic form as

$$X_t - \mu = \phi_1 (X_{t-1} - \mu) + \phi_2 (X_{t-2} - \mu) + \cdots + \phi_p (X_{t-p} - \mu) + Z_t + \theta_1 Z_{t-1} + \theta_2 Z_{t-2} + \cdots + \theta_q Z_{t-q},$$

or equivalently using the backshift operator B as

$$\phi(B) (X_t - \mu) = \theta(B) Z_t.$$

A second way to formulate a shifted ARMA(p, q) time series model with a nonzero population mean can be written in purely algebraic form as

$$X_t = \tilde{\mu} + \phi_1 X_{t-1} + \phi_2 X_{t-2} + \cdots + \phi_p X_{t-p} + Z_t + \theta_1 Z_{t-1} + \theta_2 Z_{t-2} + \cdots + \theta_q Z_{t-q},$$

or equivalently using the backshift operator B as

$$\phi(B) X_t = \tilde{\mu} + \theta(B) Z_t,$$

where $\phi(B)$ and $\theta(B)$ are the usual polynomials in the backshift operator B given in Theorem 8.2.

The example that follows illustrates how to convert a shifted time series model from one of these forms to the other.

Example 8.9 The shifted ARMA(1, 1) model defined by

$$X_t = 8 + 0.6 X_{t-1} + Z_t - 0.1 Z_{t-1}$$

is written in the second form from Definition 8.5 with $\tilde{\mu} = 8$. Convert it to the first form.

Moving all autoregressive terms and the constant term to the left-hand side of the equation results in
$$X_t - 0.6 X_{t-1} - 8 = Z_t - 0.1 Z_{t-1}.$$

Using the backshift operator, this can be written as

$$(1 - 0.6B) X_t - 8 = (1 - 0.1B) Z_t.$$

We would like to fold the constant 8 into position on the left-hand side of the equation to match the first formulation from Definition 8.5. We multiply and divide 8 by $(1 - 0.6B)$, keeping in mind that the backshift operator applied to a constant is just the constant:

$$(1 - 0.6B) X_t - 8 \cdot \frac{1 - 0.6B}{1 - 0.6B} = (1 - 0.1B) Z_t$$

or

$$(1 - 0.6B) X_t - (1 - 0.6B) \cdot \frac{8}{0.4} = (1 - 0.1B) Z_t$$

or

$$(1 - 0.6B) (X_t - 20) = (1 - 0.1B) Z_t.$$

So this shifted ARMA$(1, 1)$ time series model is now written in the first formulation from Definition 8.5, which is $\phi(B)(X_t - \mu) = \theta(B)$. The expected value of X_t is $\mu = E[X_t] = 20$. One way to check that we have done all of the algebra correctly is to use $\mu = 20$ as an argument in the first formulation of the model from Definition 8.5 and perform the algebra to see whether it is equivalent to the second formulation.

The previous example can be generalized from the shifted ARMA$(1, 1)$ model to the shifted ARMA(p, q) model. The following theorem gives the relationship between μ and $\tilde{\mu}$ for the two formulations of the shifted ARMA(p, q) models in Definition 8.5.

Theorem 8.4 The parameters $\mu = E[X_t]$ and $\tilde{\mu}$ for the two shifted ARMA(p, q) models from Definition 8.5 are related by

$$\mu = \frac{\tilde{\mu}}{1 - \phi_1 - \phi_2 - \cdots - \phi_p}$$

when the coefficients $\phi_1, \phi_2, \ldots, \phi_p$ correspond to a stationary model.

Proof The second shifted ARMA(p, q) model from Definition 8.5 is

$$X_t = \tilde{\mu} + \phi_1 X_{t-1} + \phi_2 X_{t-2} + \cdots + \phi_p X_{t-p} + Z_t + \theta_1 Z_{t-1} + \theta_2 Z_{t-2} + \cdots + \theta_q Z_{t-q}.$$

Taking the expected value of both sides of this equation yields

$$E[X_t] = \tilde{\mu} + \phi_1 E[X_{t-1}] + \phi_2 E[X_{t-2}] + \cdots + \phi_p E[X_{t-p}] + 0$$

because all of the white noise terms have expected value zero. Since the time series is assumed to be stationary, $E[X_t] = E[X_{t-1}] = E[X_{t-2}] = \cdots = E[X_{t-p}]$, and this equation becomes

$$E[X_t] = \tilde{\mu} + \phi_1 E[X_t] + \phi_2 E[X_t] + \cdots + \phi_p E[X_t].$$

Solving for $\mu = E[X_t]$ gives

$$\mu = \frac{\tilde{\mu}}{1 - \phi_1 - \phi_2 - \cdots - \phi_p}. \qquad \square$$

In the previous example, the value of $\mu = E[X_t]$ could have been calculated by appealing to Theorem 8.4 with $\tilde{\mu} = 8$ and $\phi_1 = 0.6$, which gives

$$\mu = E[X_t] = \frac{8}{1 - 0.6} = 20.$$

This provides an illustration of how Theorem 8.4 provides a mechanism for converting between the two forms of the shifted ARMA(p, q) models given in Definition 8.5.

This section has provided an introduction to linear models. The first subsection surveyed the two formulations of the general linear model and introduced the causality and invertibility properties. The second subsection introduced a special case of the general linear model known as the ARMA (autoregressive moving average) model. These time series models are inherently probabilistic in nature. The next section introduces some of the associated statistical topics in time series analysis: parameter estimation, forecasting, model assessment, and model selection.

8.2 Statistical Methods

The previous section introduced two linear probability models for time series: the general linear model and the ARMA model. These models contain parameters which can be used to tune the model to a particular application. This chapter introduces the statistical methods that are used to estimate these parameters and assess whether the model with its fitted parameters provides an adequate representation of the probabilistic mechanism governing the time series. As you read the rest of this book, you should be continually asking yourself whether the new material is associated with a probability model or presents a statistical method. The statistical methods are presented here in a somewhat generic manner; the specific implementations on a time series of observations occurs subsequently. The first subsection in this section introduces three methods for estimating the parameters in an ARMA model: the method of moments, least squares, and maximum likelihood. This is followed by a subsection that considers the important topic of forecasting future observations in a time series. Subsections on model assessment and model selection complete the section.

8.2.1 Parameter Estimation

The emphasis now shifts from a time series model, which is developed using probability theory, to statistical questions associated with a realization of a time series. The observed values of this realization are denoted by $X_1, X_2, \ldots, X_n$ when considered abstractly; when specific values are considered, they are denoted by $x_1, x_2, \ldots, x_n$.

Before considering parameter estimation, we consider the topic of *model identification*. Since p and q are nonnegative integers, there are an infinite number of ARMA(p, q) models from which to choose. Which model is appropriate for a particular application? Most statistical software packages that perform the analysis of a time series have functions that estimate parameters and forecast future values of the time series. So those two aspects of time series analysis are largely automated. The part of the process that requires some insight from the modeler is the specification of an appropriate time series model for a particular application. By what criteria do we decide whether an MA(1), AR(2), or ARMA(2, 1) is a tentative or a final time series model? The two steps associated with model identification for an ARMA(p, q) model are given next.

1. **Inspect the time series plot**. The process of identifying a time series model always begins with a careful inspection of a plot of the time series. Take a few minutes to look for cyclic variation, trends, step changes, outliers, and nonconstant variance in the plot of the time series. Visually assess the time series for any serial correlation. The human eye can spot subtleties that an algorithm might miss. Only you can perform this step. We assume for now that no trends, step changes, outliers, cyclic variation, or nonconstant variance in the time series have been identified, so a stationary model for the time series is sought. Modeling cyclic variation, trends, and nonconstant variance will be taken up subsequently.

2. **Inspect the plots of r_k and r_k^***. Inspecting plots of the sample autocorrelation function and the sample partial autocorrelation function is an attempt to conduct a visual pattern match between the sample autocorrelation patterns with a known inventory of population autocorrelation patterns for the various ARMA(p, q) models. The minimum length of a time series in order to make meaningful visual comparisons between the sample and population autocorrelation functions is about $n = 60$ or $n = 70$ observations. As will be seen in subsequent chapters, the shape of the sample autocorrelation function and the sample partial autocorrelation function can provide clues as to an appropriate time series model. In some cases, the values of p and q in the ARMA(p, q) model become immediately apparent upon viewing these

three plots. In other cases, the situation is murky, and there might be two or three potential ARMA(p, q) models that seem to be plausible. Since we have assumed that the time series is stationary in the previous paragraph, there is no need to transform or difference the data based on these plots in the current setting. The p and q values for the ARMA time series model identified from this step will be known as the *tentative model*. Once a tentative model has been identified, the next step is to estimate the parameters, which accounts for the remainder of this section.

We would like to estimate the parameters of a stationary and invertible tentative ARMA(p, q) model. It is assumed that the number of autoregressive terms p and the number of moving average terms q have been established for a tentative ARMA(p, q) time series model based on an inspection of the sample autocorrelation and sample partial autocorrelation functions. There are a total of $p + q + 1$ unknown parameters in a standard ARMA(p, q) model from Definition 8.4: the autoregressive coefficients $\phi_1, \phi_2, \ldots, \phi_p$, the moving average coefficients $\theta_1, \theta_2, \ldots, \theta_q$, and the population variance of the white noise σ_Z^2. The shifted ARMA(p, q) model from Definition 8.5 has the additional parameter μ.

Consistent with conventional notation in statistics, hats on unknown parameters denote their point estimators. The point estimator of the unknown parameter ϕ_1, for example, is $\hat{\phi}_1$. The point estimators developed here are random variables that take on one particular value for an observed time series $x_1, x_2, \ldots, x_n$. Point estimators are typically paired with a $100(1 - \alpha)\%$ confidence interval that gives a sense of the precision of the point estimator. A confidence interval for the unknown parameter ϕ_1, for example, is typically expressed in the form $L < \phi_1 < U$, where L is the random lower bound of the confidence interval and U is the random upper bound of the confidence interval.

In most practical problems involving a time series model, a shifted ARMA(p, q) model is used because very few time series are centered around zero. Since the ARMA(p, q) time series model is generally assumed to be stationary and invertible, it is common practice in time series analysis to estimate the population mean parameter μ with the sample mean $\bar{X}$. This is justified by the fact that $E[X_t] = \mu$ for a stationary and invertible shifted ARMA(p, q) model. This is consistent with the method of moments approach. Once μ has been estimated, the new time series which is shifted by $\hat{\mu} = \bar{X}$ is

$$x_1 - \bar{x}, x_2 - \bar{x}, \ldots, x_n - \bar{x}.$$

This time series can be fitted to a standard ARMA(p, q) model from Definition 8.4. This new time series has a sample mean value of zero because

$$\frac{1}{n} \sum_{i=1}^{n} (x_i - \bar{x}) = \frac{1}{n} \sum_{i=1}^{n} x_i - \frac{1}{n} \sum_{i=1}^{n} \bar{x} = \bar{x} - \frac{1}{n} \cdot n \cdot \bar{x} = 0.$$

So for now we dispatch with the parameter μ and assume that it will typically be estimated by $\bar{x}$ for a stationary and invertible ARMA(p, q) model by centering the time series as described above. Both the original time series and the centered time series will be denoted by as $\{X_t\}$ or $\{x_t\}$ in order to avoid introducing a new letter (Y_t or y_t) into the notation. The parameter estimation techniques that follow will be applied to a standard ARMA(p, q) model centered around zero, which assumes that μ has been estimated in the shifted model. This will make the notation somewhat more compact. The population variance of $\bar{X}$ for mutually independent and identically distributed observations $X_1, X_2, \ldots, X_n$ is the well-known formula

$$V[\bar{X}] = \frac{\sigma_X^2}{n}.$$

But for a stationary ARMA(p, q) time series model with population autocovariance function $\gamma(k)$ and population autocorrelation function $\rho(k)$, the population variance of the sample mean is

$$
\begin{aligned}
V\left[\bar{X}\right] &= V\left[\frac{1}{n}\left(X_1 + X_2 + \cdots + X_n\right)\right] \\
&= \frac{1}{n^2} V\left[X_1 + X_2 + \cdots + X_n\right] \\
&= \frac{1}{n^2} \sum_{i=1}^{n} \sum_{j=1}^{n} \mathrm{Cov}\left(X_i, X_j\right) \\
&= \frac{1}{n^2}\left[\sum_{i=1}^{n} V\left[X_i\right] + 2\sum_{i=1}^{n-1} \sum_{j=i+1}^{n} \mathrm{Cov}\left(X_i, X_j\right)\right] \\
&= \frac{1}{n^2}\left[n\gamma(0) + 2\sum_{k=1}^{n-1}(n-k)\gamma(k)\right] \\
&= \frac{\sigma_X^2}{n}\left[1 + 2\sum_{k=1}^{n-1}\left(1 - \frac{k}{n}\right)\rho(k)\right].
\end{aligned}
$$

Notice that this formula collapses to $V\left[\bar{X}\right] = \sigma_X^2/n$ when $\rho(1) = \rho(2) = \cdots = \rho(n-1) = 0$ as expected. This formula should be kept in mind whenever statistical inferences, such as confidence intervals or hypothesis tests, are made concerning the population mean from a realization of a time series. The sample mean is a meaningful summary statistic for a time series only when appropriate transformations have been applied to the time series in order to reduce it to a stationary time series.

Three techniques for the estimation of parameters in a time series model will be introduced here: the method of moments, least squares, and maximum likelihood estimation. There are three reasons why just one parameter estimation technique is not adequate. First, an AR(3) model, for example, might be well fitted with one estimation technique, but an MA(2) model, on the other hand, might be more compatible with another estimation technique. Second, it is often the case that one technique will provide initial estimates for a numerical method associated with a second technique. Third, some of the estimation techniques provide estimators which have degraded statistical properties near the boundaries of the stationarity or invertibility regions. The three techniques will be discussed generally below, and then will be illustrated with examples subsequently using real time series data.

Method of Moments

The essence of the method of moments technique is to equate low-order population and sample moments and solve for all unknown parameters. This method was developed by English mathematician and biostatistician Karl Pearson. This approach often seems arresting to those encountering it for the first time because population moments are constants and sample moments are random variables. Equating constants and random variables is simply a device that is used to get a perfect match between low-order population and sample moments.

In a non-time-series context with data values $X_1, X_2, \ldots, X_n$ and m unknown population parameters, the m equations

$$E\left[X_t\right] = \frac{1}{n}\sum_{t=1}^{n} X_t$$

$$E\left[X_t^2\right] = \frac{1}{n}\sum_{t=1}^{n} X_t^2$$

$$\vdots$$

$$E\left[X_t^m\right] = \frac{1}{n}\sum_{t=1}^{n} X_t^m$$

can be solved to arrive at the m method of moments estimators of the unknown parameters. In some settings this can be done analytically, but in other settings numerical methods are required.

Returning to a time-series context, the stationarity assumption (see Definition 7.6) places requirements on only the first two population moments $E\left[X_t\right]$ and $E\left[X_t^2\right]$. Stationarity places no requirements on the third and higher order moments. But stationarity does imply that the autocorrelation between two observations depends only on the lag, and this can be exploited to generate the necessary number of equations to employ the method of moments technique. Consider a stationary and invertible ARMA(p, q) model, for example, that has four unknown parameters. Solving the set of four equations in the four unknown parameters

$$E\left[X_t\right] = \frac{1}{n}\sum_{t=1}^{n} X_t$$

$$E\left[X_t^2\right] = \frac{1}{n}\sum_{t=1}^{n} X_t^2$$

$$\rho(1) = r_1$$

$$\rho(2) = r_2$$

yields the method of moments estimators for the four unknown parameters. The usual approach to fitting a time series model to a realization of a time series by the method of moments technique is to use the first two of these equations, and then equate population and sample autocorrelations at enough low-order lags in order to account for all unknown parameters. In this way the population and the sample autocorrelations will match at lower-order lags.

Least Squares Estimation

The least squares estimation technique is used nearly universally in regression analysis. This method developed by German mathematician Carl Friedrich Gauss. The essence of the least squares technique is to find the values of the unknown parameters that minimize the sum of squares of the error terms in a model. In the time series setting, we want to find the values of the parameters that minimize

$$S = \sum_{t=1}^{n} Z_t^2.$$

The use of least squares for ARMA(p, q) models requires two steps. First, solve the target model for Z_t, and then substitute that expression into the equation above. At this point, S is written in terms of the unknown parameters. Second, take the partial derivatives of S with respect to all unknown parameters and solve for the unknown parameters. The set of equations to solve is often referred to as the *orthonormal equations*. The solution to these equations yields the least squares estimates of the unknown parameters. In some cases these equations can be solved analytically; in other cases numerical methods are required.

Maximum Likelihood Estimation

Maximum likelihood estimation is the most prevalent technique for estimating unknown parameters from a data set in the field of statistics, particularly outside of regression. The method was popularized by English statistician Sir Ronald Fisher. The essence of the maximum likelihood estimation technique, whether applied in time series analysis or otherwise, is to select the parameters in a hypothesized model that are the most likely ones to have resulted in the observed data values. The *maximum likelihood estimators* of the unknown parameters are found by maximizing the *likelihood function*, which is the joint probability density function of the data values evaluated at their observed values. The likelihood function is a function of the unknown parameters in the model with the data values fixed at their observed values. We begin by using maximum likelihood estimation on an ARMA(0, 0) model in order to establish some of the issues associated with the use of the maximum likelihood estimation technique to estimate the parameters in a time series model.

Example 8.10 Let $x_1, x_2, \ldots, x_n$ be a realization of observations from an ARMA(0, 0) time series model that is simply white noise:

$$X_t = Z_t,$$

where $Z_t \sim WN\left(0, \sigma_Z^2\right)$. Find the maximum likelihood estimator of σ_Z^2, determine whether the maximum likelihood estimator is unbiased and consistent, and derive an exact two-sided $100(1 - \alpha)\%$ confidence interval for σ_Z^2.

The ARMA(0, 0) time series model has just a single unknown parameter σ_Z^2, the population variance of the white noise, that needs to be estimated. The likelihood function is the joint probability density function of the observations:

$$L\left(\sigma_Z^2\right) = f(x_1, x_2, \ldots, x_n).$$

The $x_1, x_2, \ldots, x_n$ arguments on L and the σ_Z^2 argument on f are suppressed for brevity. We are lucky with the ARMA(0, 0) model because we can exploit the fact that the observations in the time series are mutually independent, which means that the joint probability density function of the observed values $x_1, x_2, \ldots, x_n$ is the product of the marginal probability density functions:

$$L\left(\sigma_Z^2\right) = f(x_1, x_2, \ldots, x_n) = f(x_1)f(x_2)\ldots f(x_n),$$

where $f(x)$ is the probability density function of a single observation in the time series, which is just white noise. We won't be so lucky for general ARMA(p, q) models. The assumption of white noise is vague in the sense that we do not know the functional form of $f(x)$. We only know that it is a probability distribution with population mean 0 and population variance σ_Z^2. In order to apply the maximum likelihood estimation technique, we must make an additional assumption about the distribution of $X_1, X_2, \ldots, X_n$. So at this point we make the additional assumption that the white noise terms are in fact Gaussian white noise terms:

$$f(x_i) = \frac{1}{\sqrt{2\pi\sigma_Z^2}}\, e^{-x_i^2/(2\sigma_Z^2)} \qquad -\infty < x_i < \infty,$$

for $i = 1, 2, \ldots, n$, which is the probability density function of a $N\left(0, \sigma_Z^2\right)$ random variable. The assumption of normally-distributed error terms in order to use the maximum likelihood estimation technique is nearly universal in time series analysis. The associated likelihood function is

$$L\left(\sigma_Z^2\right) = \prod_{i=1}^{n} f(x_i) = \left(2\pi\sigma_Z^2\right)^{-n/2} e^{-\sum_{i=1}^{n} x_i^2/(2\sigma_Z^2)}.$$

The maximum likelihood estimator of σ_Z^2 is the value of σ_Z^2 that maximizes the likelihood function:

$$\hat{\sigma}_Z^2 = \underset{\Omega}{\operatorname{argmax}} L\left(\sigma_Z^2\right),$$

where Ω is the parameter space $\Omega = \left\{\sigma_Z^2 \mid \sigma_Z^2 > 0\right\}$. It is often the case that the mathematics associated with maximizing the natural logarithm of the likelihood function is easier than the mathematics of maximizing the likelihood function. Both functions are maximized at the same value because the natural logarithm is a monotonic transformation. The log likelihood function is

$$\ln L\left(\sigma_Z^2\right) = -\frac{n}{2}\ln\left(2\pi\sigma_Z^2\right) - \frac{1}{2\sigma_Z^2}\sum_{i=1}^{n} x_i^2.$$

The derivative of the log likelihood function with respect to the unknown parameter σ_Z^2 is

$$\frac{\partial \ln L\left(\sigma_Z^2\right)}{\partial \sigma_Z^2} = -\frac{n}{2\sigma_Z^2} + \frac{1}{2\sigma_Z^4}\sum_{i=1}^{n} x_i^2.$$

Equating this derivative to zero and solving for σ_Z^2 gives the maximum likelihood estimator

$$\hat{\sigma}_Z^2 = \frac{1}{n}\sum_{i=1}^{n} x_i^2.$$

The maximum likelihood estimator is an unbiased estimator of σ_Z^2 because

$$E\left[\hat{\sigma}_Z^2\right] = E\left[\frac{1}{n}\sum_{i=1}^{n} X_i^2\right] = \frac{1}{n}E\left[\sum_{i=1}^{n} X_i^2\right] = \frac{1}{n}\sum_{i=1}^{n} E\left[X_i^2\right] = \frac{1}{n}\sum_{i=1}^{n} V\left[X_i\right] = \frac{1}{n}\cdot n \cdot \sigma_Z^2 = \sigma_Z^2$$

based on the shortcut formula for the population variance and the fact that $E[X_i] = 0$. This means that although the maximum likelihood estimator might miss the true parameter value σ_Z^2 on the low side or on the high side, it is pointing at the correct target because its expected value (long-run average) is the true parameter value.

By standardizing the X_i values, we find that a function of the maximum likelihood estimator has the chi-square distribution because it can be written as the sum of squares of mutually independent standard normal random variables:

$$\frac{n\hat{\sigma}_Z^2}{\sigma_Z^2} = \sum_{i=1}^{n}\left(\frac{X_i - 0}{\sigma_Z}\right)^2 = \sum_{i=1}^{n}\left(\frac{X_i}{\sigma_Z}\right)^2 \sim \chi^2(n).$$

The population variance of the maximum likelihood estimator is

$$V\left[\hat{\sigma}_Z^2\right] = \frac{\sigma_Z^4}{n^2}\cdot V\left[\frac{n\hat{\sigma}_Z^2}{\sigma_Z^2}\right] = \frac{\sigma_Z^4}{n^2}\cdot 2n = \frac{2\sigma_Z^4}{n}$$

because the population variance of a chi-square random variable with n degrees of freedom is $2n$. The maximum likelihood estimator is a consistent estimator of σ_Z^2 because it is unbiased and $\lim_{n\to\infty} V\left[\hat{\sigma}_Z^2\right] = 0$. The maximum likelihood estimator $\hat{\sigma}_Z^2$ will approach the true parameter value σ_Z^2 in the limit as n increases. In other words, for any positive constant ε,

$$\lim_{n\to\infty} P\left(\left|\hat{\sigma}_Z^2 - \sigma_Z^2\right| < \varepsilon\right) = 1.$$

The unbiased and consistent point estimator $\hat{\sigma}_Z^2$ does not convey any sense of the precision of the point estimator, however. That information is best conveyed in this setting by a confidence interval. An appropriate pivotal quantity is

$$\frac{n\hat{\sigma}_Z^2}{\sigma_Z^2} \sim \chi^2(n),$$

which implies that

$$\chi^2_{n,\,1-\alpha/2} < \frac{n\hat{\sigma}_Z^2}{\sigma_Z^2} < \chi^2_{n,\,\alpha/2}$$

with probability $1 - \alpha$. The second subscript on the quantile of the chi-square distribution is a right-hand tail probability. Performing the algebra required to isolate σ_Z^2 in the center of the inequality results in the exact two-sided $100(1-\alpha)\%$ confidence interval

$$\frac{n\hat{\sigma}_Z^2}{\chi^2_{n,\,\alpha/2}} < \sigma_Z^2 < \frac{n\hat{\sigma}_Z^2}{\chi^2_{n,\,1-\alpha/2}}.$$

Common values for α are 0.1, 0.05, and 0.01, which are known as 90%, 95%, and 99% confidence intervals, respectively. The proper interpretation of a confidence interval like this one is critical. An *incorrect* interpretation of this exact confidence interval for, say, $\alpha = 0.05$, is:

"The probability that this confidence interval contains σ_Z^2 is 0.95"

because once the data has been collected and the interval is calculated, it either contains the unknown parameter σ_Z^2 or it does not. A probability statement like this one does not make sense because there are no random variables after the data values are collected. The correct interpretation of this exact confidence interval for σ_Z^2 with nominal coverage 0.95 is as follows.

"The confidence interval I have calculated might contain σ_Z^2 or it might not. However, if (a) all of the assumptions that I have made concerning the ARMA(0, 0) time series model with Gaussian white noise are correct, (b) many realizations of the time series of size n are collected, and (c) the same procedure was used for calculating a confidence interval for each of the realizations, then 0.95 is the expected fraction of these confidence intervals that will contain the true parameter σ_Z^2."

Obviously, one would not want to repeat this tedious explanation every time a confidence interval is calculated. So statisticians shorten this by simply saying:

"I am 95% confident that my confidence interval contains the unknown parameter σ_Z^2."

The brevity and avoidance of the use of "probability" in this statement aids the proper interpretation of the confidence interval.

Finally, we consider an application area in which the ARMA(0, 0) might be appropriate. The ARMA(0, 0) model has industrial applications in quality control. When formulating a model for a continuous measurement associated with a product (such as a ball bearing diameter or the pre-cooked weight of a quarter-pound hamburger) that is produced repeatedly over time, management prefers a stationary time series model with mutually independent consecutive observations. In this particular setting, a shifted ARMA(0, 0) is appropriate and justified. This model is used in practice to help detect when the continuous measurement trends away from the mean value in a shifted ARMA(0, 0) time series model in what is known in quality control as a *control chart*.

Applying the maximum likelihood estimation technique to the ARMA(0, 0) time series model was ideal in that the point estimator for σ_Z^2 could be expressed in closed form and an exact two-sided confidence interval for σ_Z^2 could be derived to give an indication of the precision of the point estimator. There are three key take-aways from the ARMA(0, 0) example involving maximum likelihood estimation.

- We needed to narrow the assumption of white noise error terms to Gaussian white noise error terms in order to implement the maximum likelihood estimation technique.

- We were fortunate that the likelihood function could be factored into the product of the marginal probability density functions because of the mutual independence of the observations. This will not be the case with the ARMA(p, q) model with $p > 0$ and/or $q > 0$.

- We were fortunate in the sense that we could establish an exact two-sided $100(1 - \alpha)\%$ confidence interval for σ_Z^2 based on a pivotal quantity. For ARMA(p, q) models with $p > 0$ and/or $q > 0$ we will generally have only approximate confidence intervals which are based on asymptotic results.

We now address the third take-away concerning confidence intervals for parameters in ARMA models that go beyond the ARMA(0, 0) model illustrated in the previous example. The mathematics associated with deriving the exact distribution of some pivotal quantity becomes too difficult once autocorrelation is injected into a model, so we use asymptotic results concerning the parameter estimates in order to arrive at approximate confidence intervals. To frame the conversation concerning these asymptotic results, some notation must be established. Let

$$\beta = (\beta_1, \beta_2, \ldots, \beta_r)'$$

be a vector that denotes the r unknown parameters in a time series model. In the case of a shifted ARMA(p, q) model, for example, the elements of β are the $p+q+2$ unknown parameters $\phi_1, \phi_2, \ldots, \phi_p, \theta_1, \theta_2, \ldots, \theta_q, \mu$, and σ_Z^2. Let $x_1, x_2, \ldots, x_n$ denote a realization of the time series observations. The likelihood function is

$$L(\beta) = f(x_1, x_2, \ldots, x_n)$$

and the associated log likelihood function is

$$\ln L(\beta) = \ln f(x_1, x_2, \ldots, x_n).$$

The jth element of the score vector is

$$\frac{\partial \ln L(\beta)}{\partial \beta_j}$$

for $j = 1, 2, \ldots, r$. Equating the elements of the score vector to zero and solving for the unknown parameters yields the maximum likelihood estimators $\hat{\beta}_1, \hat{\beta}_2, \ldots, \hat{\beta}_r$. The (j, k) element of the Fisher information matrix $I(\beta)$ is

$$E\left[-\frac{\partial^2 \ln L(\beta)}{\partial \beta_j \partial \beta_k}\right]$$

for $j = 1, 2, \ldots, r$ and $k = 1, 2, \ldots, r$, when the expected values exist. The Fisher information matrix is estimated by the observed information matrix $O(\hat{\beta})$, whose (j, k) element is

$$\left[-\frac{\partial^2 \ln L(\beta)}{\partial \beta_j \partial \beta_k} \right]_{\beta = \hat{\beta}}$$

for $j = 1, 2, \ldots, r$ and $k = 1, 2, \ldots, r$. The inverse of the observed information matrix is the asymptotic variance–covariance matrix of the parameter estimates. If one is willing to ignore the off-diagonal elements of this matrix, the square roots of the diagonal elements are estimates of the standard errors of the point estimators. The asymptotic normality of maximum likelihood estimators allows one to construct approximate confidence intervals for the unknown parameters.

We were able to obtain an exact two-sided confidence interval for σ_Z^2 for the ARMA(0, 0) model in the previous example; the next example goes through the appropriate steps for the model had we not been so lucky. We return to the analysis of the standard ARMA(0, 0) time series model because it is the only ARMA(p, q) model with a single unknown parameter and associated tractable mathematics.

> **Example 8.11** Find an asymptotically exact two-sided $100(1 - \alpha)\%$ confidence interval for σ_Z^2 for an ARMA(0, 0) model based on the asymptotic normality of the maximum likelihood estimator $\hat{\sigma}_Z^2$. Estimate the actual coverage of this confidence interval for $n = 100$, $\sigma_Z^2 = 1$, and $\alpha = 0.05$. What is the impact of n on the actual coverage?

Although we know that there is an exact confidence interval for σ_Z^2 from the previous example, we pretend that we are unaware of such an interval and try to find an asymptotically exact interval based on the inverse of the observed information matrix. This is done to illustrate the mechanics of constructing the asymptotically exact confidence interval. From Example 8.10, the maximum likelihood estimator of σ_Z^2 is

$$\hat{\sigma}_Z^2 = \frac{1}{n} \sum_{i=1}^{n} x_i^2.$$

Once again treating σ_Z^2 as a unit, the second partial derivative of the log likelihood function with respect to σ_Z^2 is

$$\frac{\partial^2 \ln L\left(\sigma_Z^2\right)}{\partial \left(\sigma_Z^2\right)^2} = \frac{n}{2\sigma_Z^4} - \frac{1}{\sigma_Z^6} \sum_{i=1}^{n} x_i^2.$$

The single entry in the 1×1 Fisher information matrix is the expected value of the negative of this partial derivative:

$$I\left(\sigma_Z^2\right) = E\left[-\frac{\partial^2 \ln L\left(\sigma_Z^2\right)}{\partial \left(\sigma_Z^2\right)^2} \right] = -\frac{n}{2\sigma_Z^4} + \frac{1}{\sigma_Z^6} \sum_{i=1}^{n} V\left[X_i\right] = \frac{n}{2\sigma_Z^4}.$$

Since σ_Z^2 is an unknown parameter, the Fisher information matrix cannot be determined from the observations from a time series. The 1×1 observed information matrix provides an estimate of the Fisher information matrix from the data values:

$$O\left(\hat{\sigma}_Z^2\right) = \left[-\frac{\partial^2 \ln L\left(\sigma_Z^2\right)}{\partial \left(\sigma_Z^2\right)^2} \right]_{\sigma_Z^2 = \hat{\sigma}_Z^2} = -\frac{n}{2\hat{\sigma}_Z^4} + \frac{1}{\hat{\sigma}_Z^6} \sum_{i=1}^{n} x_i^2 = \frac{n^3}{2\left(\sum_{i=1}^{n} x_i^2\right)^2}.$$

The inverse of this 1×1 matrix is just the reciprocal of the single entry:

$$O^{-1}\left(\hat{\sigma}_Z^2\right) = \frac{2\left(\sum_{i=1}^n x_i^2\right)^2}{n^3}.$$

For large values of n, this quantity converges to the variance of $\hat{\sigma}_Z^2$. So since

$$\hat{\sigma}_Z^2 \xrightarrow{D} N\left(\sigma_Z^2, \frac{2\left(\sum_{i=1}^n x_i^2\right)^2}{n^3}\right),$$

an asymptotically exact $100(1-\alpha)\%$ confidence interval for σ_Z^2 is

$$\hat{\sigma}_Z^2 - z_{\alpha/2}\sqrt{\frac{2\left(\sum_{i=1}^n x_i^2\right)^2}{n^3}} < \sigma_Z^2 < \hat{\sigma}_Z^2 + z_{\alpha/2}\sqrt{\frac{2\left(\sum_{i=1}^n x_i^2\right)^2}{n^3}}.$$

We know that the actual coverage of this two-sided confidence interval converges to the exact coverage as $n \to \infty$. But how does the confidence interval perform for finite values of n? This can only be assessed by a Monte Carlo simulation experiment.

The Monte Carlo simulation given by the R code below simulates four million time series of length $n = 100$ generated from an ARMA(0, 0) model with Gaussian white noise having variability $\sigma_Z^2 = 1$ and estimates the actual coverage of the approximate 95% confidence interval by printing the fraction of the simulated confidence intervals that contain the arbitrarily-assigned true parameter value $\sigma_Z^2 = 1$.

```
nrep  = 4000000
count = 0
n     = 100
alpha = 0.05
crit  = qnorm(1 - alpha / 2)
for (i in 1:nrep) {
  x   = rnorm(n)
  ssq = sum(x ^ 2)
  mle = ssq / n
  std = sqrt(2 * ssq ^ 2 / n ^ 3)
  lo  = mle - crit * std
  hi  = mle + crit * std
  if (lo < 1 && hi > 1) count = count + 1
}
print(count / nrep)
```

After a call to set.seed(3) to establish the random number stream, five runs of this simulation yield the following estimated confidence interval coverages:

0.9402	0.9399	0.9400	0.9401	0.9401.

Although the stated (or nominal) coverage for this confidence interval is 0.95, the Monte Carlo simulation reveals that the actual coverage is 0.940.

The final question concerns the impact of n on the actual coverage. The Monte Carlo simulation experiment given above is executed for several other values of n. The actual coverage values are shown in Figure 8.3. These values confirm what we suspect about an asymptotic confidence interval: the actual coverage asymptotically approaches the stated coverage (indicated by the dashed horizontal line in Figure 8.3). This behavior is typical of asymptotic confidence intervals.

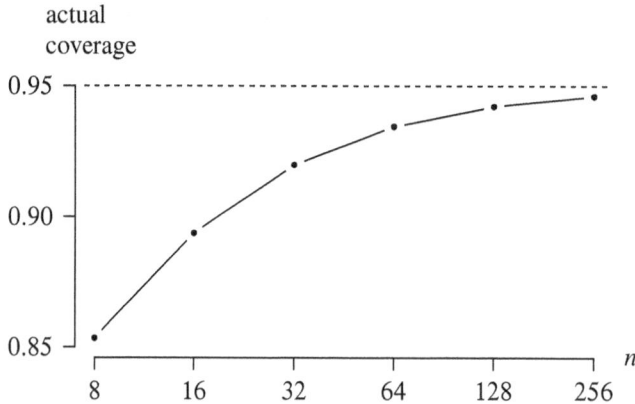

Figure 8.3: Asymptotic 95% confidence interval actual coverage for $n = 8, 16, 32, \ldots, 256$.

This ends the discussion of the important topic of parameter estimation. The time series model that emerges from this step is known as a *fitted tentative model*. Three techniques for parameter estimation have been introduced: the method of moments, least squares, and maximum likelihood estimation. In time series analysis, exact confidence intervals for the unknown parameters are typically mathematically intractable, so we must settle for asymptotically exact confidence intervals.

The next section introduces another important statistical topic that arises frequently in time series analysis: the prediction of future values in a time series based on a realization of n observations of a time series, which is typically known as *forecasting*.

8.2.2   Forecasting

The purpose of forecasting is to predict one or more future values of a time series based on observed values of a time series $x_1, x_2, \ldots, x_n$. Forecasting future values of a time series often plays a critical role in policy decisions. The closing price of the Dow Jones Industrial Average tomorrow, the number of oysters in the Chesapeake Bay next year, the high temperature in Tuscaloosa on Saturday, and a company's profit next quarter are examples of applications of forecasting.

The term "forecasting" is synonymous with "prediction" and the two terms will be used interchangeably. Forecasting is a slightly more popular term in the time series literature. Both terms can be interpreted as "telling before."

Forecasting involves extrapolation of the time series model outside of the time frame associated with the observed values $x_1, x_2, \ldots, x_n$, typically into the future. The notion of backcasting, which is predicting values in the past, will not be considered here. Care must be taken to ensure that the fitted probability model still applies in the time range in which the extrapolation occurs. If future observations are governed by the same probability model as previous observations, then a forecasted value is meaningful. Furthermore, if an ARMA(p, q) model is used, it is subject to errors in identification (for example, the wrong values of p and q or perhaps an ARMA model is used when a non-ARMA model is appropriate) and estimation (for example, due to random sampling variability or choosing an inferior parameter estimation procedure).

There are several choices for forecasting notation. We assume that the values of a time series $\{X_t\}$ are given by the observed values $x_1, x_2, \ldots, x_n$. We would like to predict the value of the time series h (for "horizon") time units into the future, given that we know the values of $x_1, x_2, \ldots, x_n$ and our forecast is being made at time n. The notation that we will use for this future value of the time series will be the random variable X_{n+h}. Its associated predicted value will be denoted by $\hat{X}_{n+h}$. This predicted value is defined as the conditional expected value of the future value given the values of the n observed values:

$$\hat{X}_{n+h} = E\left[X_{n+h} \mid X_1 = x_1, X_2 = x_2, \ldots, X_n = x_n\right].$$

We will use the alternative notation $\hat{X}_n(h)$ for the forecast whenever there might be some ambiguity associated with the origin of the forecast. The default assumption for forecasting in this book is that we are making a forecast based on n observed values, and the forecast is being made at time origin n for h time units into the future. The forecasted value at time $n + h$ can be thought of as the average of all future possibilities given the history up to time n. But why use the conditional expectation? Might a quantile of the probability distribution of X_{n+h}, for example, the population median, provide a better forecast? The rationale behind using the conditional expectation is that it minimizes the mean square error of the predicted value, which is defined as

$$E\left[\left(X_{n+h} - \hat{X}_{n+h}\right)^2\right],$$

among all linear functions of the observed values $x_1, x_2, \ldots, x_n$. For this reason, the forecasted value given by the conditional expectation is often known as the *best linear predictor* of X_{n+h} in the sense of minimizing the mean square error of the predicted value.

Figure 8.4 illustrates the case of a (tiny) time series of just $n = 4$ observations: x_1, x_2, x_3, x_4. (Recall that $n = 60$ or $n = 70$ is the minimum value of n in practice. This example with a tiny value of n is for illustrative purposes only.) The observed values of the time series are indicated by points which are connected by lines. Each of the three forecasted values, $\hat{X}_5, \hat{X}_6, \hat{X}_7$, is indicated by a ∘.

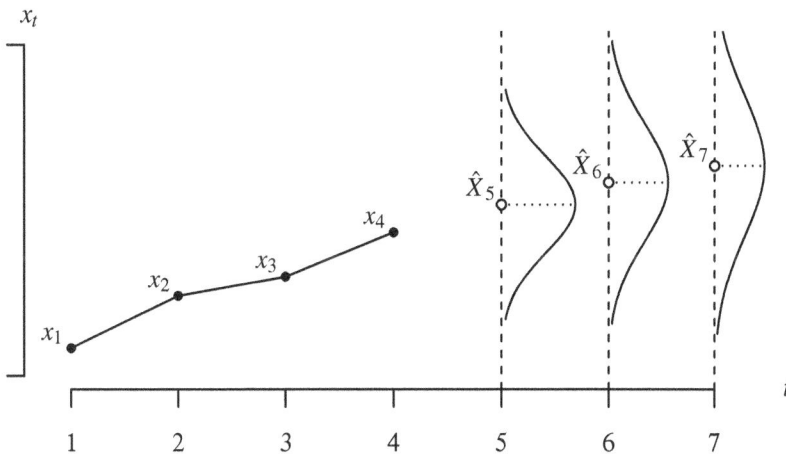

Figure 8.4: Forecasting three future values from $n = 4$ observations.

The three forecasts, associated with $h = 1$, $h = 2$, and $h = 3$, are made at time $t = n = 4$. In addition, there are three probability density functions, each rotated clockwise 90°, which indicate the probability distributions of the random future observations X_5, X_6, X_7. There are three key observations associated with this figure.

- The time series values x_1, x_2, x_3, x_4 increase over time, and the associated forecasted values $\hat{X}_5, \hat{X}_6, \hat{X}_7$ continue this trend.

- The population variance of the probability distributions of X_5, X_6, X_7 increases as the forecasting time horizon increases. This is consistent with weather prediction, for example, in that the weather prediction three days from now is less precise than the weather prediction tomorrow.

- The random sampling variability that is apparent in the four observed values x_1, x_2, x_3, x_4 is not apparent in the forecasted values $\hat{X}_5, \hat{X}_6, \hat{X}_7$. Observed time series values typically exhibit random sampling variability; forecasted values tend to be smooth.

Our goal in this subsection is to discuss forecasting generally and to introduce techniques for determining point estimates and interval estimates for future values in a time series. The example that follows assumes that a valid ARMA model has been specified and the parameters in a time series model are known, rather than estimated from a realization of the time series. For a long realization (large n) or significant amounts of previous history associated with a particular time series, this assumption might not pose any problem. In order to derive a prediction interval for X_{n+h}, the white noise terms are assumed to be Gaussian white noise for mathematical tractability. The reason for this assumption will be apparent in the following example.

Example 8.12 Consider the shifted stationary AR(1) time series model

$$X_t - \mu = \phi \left(X_{t-1} - \mu \right) + Z_t,$$

where $\{Z_t\}$ is Gaussian white noise and $-1 < \phi < 1$, μ, and $\sigma_Z^2 > 0$ are fixed, known parameters. Let $x_1, x_2, \ldots, x_n$ be one realization of the time series.

(a) Find a point estimate and an exact two-sided $100(1 - \alpha)\%$ prediction interval for X_{n+1}.

(b) Find a point estimate and an exact two-sided $100(1 - \alpha)\%$ prediction interval for X_{n+2}.

Notice that ϕ is a constant here and should not be confused with the polynomial $\phi(B)$. This is an unusual case because the three parameters ϕ, μ, and σ_Z^2 are known. In addition, it is assumed that the AR(1) model is a perfect stochastic model to govern the time series. Neither of these assumptions are typically satisfied perfectly in practice.

(a) Writing the AR(1) time series model with X_{n+1} on the left-hand side:

$$X_{n+1} - \mu = \phi \left(x_n - \mu \right) + Z_{n+1}$$

or

$$X_{n+1} = \mu + \phi \left(x_n - \mu \right) + Z_{n+1}.$$

Notice that X_{n+1} and Z_{n+1} are random future values which are set in uppercase, but x_n has already been observed, so it is set in lowercase. Taking the conditional expected value of both sides of this equation yields the one-step-ahead forecast

$$E \left[X_{n+1} \mid X_1 = x_1, X_2 = x_2, \ldots, X_n = x_n \right] = \mu + \phi \left(x_n - \mu \right)$$

because the expected value of a constant is a constant and the future Gaussian white noise term has conditional expected value 0. Taking the conditional population variance of both sides of the equation yields

$$V[X_{n+1} | X_1 = x_1, X_2 = x_2, \ldots, X_n = x_n] = \sigma_Z^2$$

because μ, ϕ, and x_n are all constants and the population variance is unaffected by a shift. So the point estimate of X_{n+1} is

$$\hat{X}_{n+1} = \mu + \phi(x_n - \mu).$$

Since X_{n+1} is a constant, $\mu + \phi(x_n - \mu)$, plus a normal random variable, Z_{n+1}, it too is normally distributed with conditional mean $\hat{X}_{n+1}$ and conditional population variance σ_Z^2. So an exact two-sided $100(1-\alpha)\%$ prediction interval for X_{n+1} is

$$\hat{X}_{n+1} - z_{\alpha/2} \sigma_Z < X_{n+1} < \hat{X}_{n+1} + z_{\alpha/2} \sigma_Z,$$

where $z_{\alpha/2}$ is the $1 - \alpha/2$ quantile of the standard normal distribution.

(b) Writing the AR(1) time series model with X_{n+2} on the left-hand side:

$$X_{n+2} - \mu = \phi(X_{n+1} - \mu) + Z_{n+2}$$

or

$$X_{n+2} = \mu + \phi(X_{n+1} - \mu) + Z_{n+2}.$$

All of the X and Z variables are random future values, so they are set in uppercase. Taking the conditional expected value of both sides of this equation yields the two-step-ahead forecast

$$
\begin{aligned}
E[X_{n+2} | X_1 &= x_1, X_2 = x_2, \ldots, X_n = x_n] \\
&= \mu + \phi\big(E[X_{n+1} | X_1 = x_1, X_2 = x_2, \ldots, X_n = x_n] - \mu\big) \\
&= \mu + \phi\big(\phi(x_n - \mu)\big) \\
&= \mu + \phi^2(x_n - \mu)
\end{aligned}
$$

because the conditional expected value of Z_{n+2} is zero. Taking the conditional population variance of both sides of the equation yields

$$
\begin{aligned}
V[X_{n+2} | X_1 &= x_1, X_2 = x_2, \ldots, X_n = x_n] \\
&= \phi^2 V[X_{n+1} | X_1 = x_1, X_2 = x_2, \ldots, X_n = x_n] + V[Z_{n+2} | X_1 = x_1, X_2 = x_2, \ldots, X_n = x_n] \\
&= (\phi^2 + 1)\sigma_Z^2.
\end{aligned}
$$

So the point estimate of X_{n+2} is

$$\hat{X}_{n+2} = \mu + \phi^2(x_n - \mu).$$

Since X_{n+2} is written as a constant, μ, plus the linear combination of two normally distributed random variables, $\phi(X_{n+1} - \mu)$ and Z_{n+2}, which is itself normally distributed, an exact two-sided $100(1-\alpha)\%$ prediction interval for X_{n+2} is

$$\hat{X}_{n+2} - z_{\alpha/2}\sqrt{\phi^2 + 1}\,\sigma_Z < X_{n+2} < \hat{X}_{n+2} + z_{\alpha/2}\sqrt{\phi^2 + 1}\,\sigma_Z.$$

Notice that for $\phi \neq 0$, the prediction interval for X_{n+2} is wider than the prediction interval for X_{n+1} for the same time series values and the same α value. This is consistent with intuition because we are less certain as we forecast further out into the future. This is the typical case in practice. On the other hand, the two prediction intervals have identical width when $\phi = 0$ because the AR(1) time series model reduces to Gaussian white noise in this case, and each future observation will have the same precision because of the mutual independence of the X_t values in this case.

This case was ideal in the sense that all three of the parameters, ϕ, μ, and σ_Z^2, are fixed and known. When these parameters are replaced by their point estimates, $\hat{\phi}$, $\hat{\mu}$, and $\hat{\sigma}_Z^2$, the prediction intervals become approximate rather than exact.

The previous example has illustrated the process for determining forecasted values and associated prediction intervals for an AR(1) time series model with known parameters. Consider generalizing this process for the h-step-ahead forecast. In order to obtain a point estimate for the forecast, take the conditional expected value of both sides of the model with X_{n+h} isolated on the left-hand side, which effectively results in: (a) present and past values of X_t are replaced by their observed values; (b) future values of Z_t are replaced by their conditional expected values, which are zero; and (c) future values of X_t are replaced by their conditional expected values. After simplification, this results in the forecast value $\hat{X}_{n+h}$.

As is typically the case in statistics, a point estimate is usually accompanied by an interval estimate which gives an indication of the precision of the point estimate. In a time series setting, a *prediction interval* for X_{n+h} has the generic form

$$\hat{X}_{n+h} \pm z_{\alpha/2} \sqrt{V\left[X_{n+h} \mid X_1 = x_1, X_2 = x_2, \ldots, X_n = x_n\right]}.$$

This formula assumes that the random future value at time $n + h$, denoted by X_{n+h}, is normally distributed. This is usually achieved by assuming that the white noise terms consist of Gaussian white noise. Unlike confidence intervals, prediction intervals typically do not have widths that shrink to zero as the sample size n increases.

This ends the important topic of forecasting. Many more examples of forecasting will appear in subsequent sections in this chapter when special cases of ARMA(p, q) models are introduced. We now turn to another important statistical topic, which is model assessment.

8.2.3 Model Assessment

It is often the case that we have little or no information concerning the underlying physical mechanism governing a time series, so we must resort to an entirely data-driven approach to developing a time series model that adequately approximates the underlying probability mechanism. The usual approach to building a times series model consists of iterating through the following steps until a suitable model is formulated. The model building process is—by design—both iterative and interactive, making R an ideal platform for carrying out the process.

1. Identify a tentative time series model.

2. Estimate the unknown parameters of the tentative time series model.

3. Assess the adequacy of the fitted time series model.

The third step is considered in this section. As an instance of this approach, let's say we decide (based on inspecting plots of the time series, the sample autocorrelation function, and the sample partial autocorrelation function) that a shifted AR(2) time series model is a strong candidate for modeling a particular time series. After the parameters μ, ϕ_1, ϕ_2, and σ_Z^2 are estimated, we hope that the fitted model adequately models the underlying probability mechanism for the time series. If this is the case, then the *signal* associated with the time series has been captured, and all that should remain is *noise*. So how do we test whether or not the fitted model provides an adequate representation of the time series? One common approach taken in time series modeling is to assess whether the random shocks $\{Z_t\}$ are mutually independent and identically distributed random variables with population mean zero and common population variance σ_Z^2. But these Z_t values are not observed by the modeler, so instead we inspect the *residuals*, which are estimates of the Z_t values. In time series analysis, this important step is known as *diagnostic checking* or *residual analysis*. (This step is analogous to the similar step in regression analysis.) This process is the rough equivalent of *goodness-of-fit testing* from classical statistical theory. A residual value is defined as

$$[\text{residual}] = [\text{observed value}] - [\text{predicted value}].$$

The predicted value is the one-step-ahead forecast from the time $t - 1$. Using the notation from the forecasting section, the residual at time t can be written as

$$\hat{Z}_t = X_t - \hat{X}_t.$$

This is one instance in which a more precise notation for a forecasted value would be helpful; this is more clearly written as

$$\hat{Z}_t = X_t - \hat{X}_{t-1}(1).$$

The hat is added to Z_t in order to indicate that the parameters in the fitted model have been estimated from the observed time series. Only in a simulated time series with known parameters do we observe Z_t. The residuals are ordered in time, so they can be viewed as a time series in their own right. If the hypothesized and fitted model are adequate, then the time series plot of the residuals will approximate a time series of white noise. The question here is how closely the residuals resemble white noise terms.

The behavior of the residuals is an indicator of whether the time series has been adequately modeled. If the model has been specified correctly and the parameter estimates are near their associated population values, then the residuals should appear to be white noise values, with common population mean zero and common population standard deviation. If this is not the case, then the search for an adequate time series model should continue.

A plot of the residuals over time is a crucial initial step in assessing whether they resemble white noise terms. Carefully examine the plot for any signs of trend, seasonality, or serial correlation. An example of a plot of Gaussian white noise was given in Figure 7.3. This step is just as important in residual analysis as was the inspection of the plot of the original time series. In addition, a plot of the sample autocorrelation function and the sample partial autocorrelation function of the residuals can be helpful in assessing whether the residuals closely approximate white noise. But rather than just a subjective visual inspection, we also want to confirm our intuition with a formal statistical test. The next four paragraphs briefly survey four statistical tests to assess the following null and alternative hypotheses:

H_0 : the residuals are mutually independent and identically distributed random variables

versus

H_1 : the residuals are not mutually independent and identically distributed random variables.

If there is no apparent visual trend, seasonality, or serial correlation in the residuals, then any one of the four hypothesis tests that follow can be conducted to confirm that the residuals do not exhibit any of these characteristics.

Count the number of significant spikes in the sample autocorrelation function. This test begins with a plot of the sample autocorrelation function of the residuals. If the residuals are well approximated by white noise terms, then the time series model can be judged to be adequate. The sample autocorrelation function values for white noise terms are approximately mutually independent and identically distributed $N(0, 1/n)$ random variables. So if the residuals closely approximate white noise, then any sample autocorrelation function value will fall between $-1.96/\sqrt{n}$ and $1.96/\sqrt{n}$ with approximate probability 0.95. We would like to conduct a hypothesis test in which the null hypothesis is that the sample autocorrelation function values of the residuals are independent $N(0, 1/n)$ random variables. A large number of sample autocorrelation values falling outside of the limits (which serves as the test statistic here) will result in rejecting the null hypothesis. So if each sample autocorrelation function value can be thought of as a toss of a biased coin in the case of the residuals being approximately white noise, then for, say, the first $m = 40$ such values, we expect $40 \cdot 0.05 = 2$ to fall outside of the limits $\pm 1.96/\sqrt{n}$. (Of course, the lag 0 sample autocorrelation $r_0 = 1$ is not included in the count.) In order to achieve an approximate level of significance $\alpha = 0.05$, if four or fewer of the 40 sample autocorrelation function values associated with the residuals fall outside of $\pm 1.96/\sqrt{n}$, we fail to reject H_0. The time series model is deemed to be adequate. But if five or more of the 40 sample autocorrelation function values associated with the residuals fall outside of $\pm 1.96/\sqrt{n}$, this is evidence *against* the hypothesized model and we reject H_0. The time series model is deemed to be inadequate. The p-value associated with four or fewer of the 40 sample autocorrelation function values associated with the residuals falling outside of the limits $\pm 1.96/\sqrt{40}$ can be calculated with the R statement

```
1 - pbinom(4, 40, 0.05)
```

This statement returns

```
[1] 0.04802826
```

So the exact level of significance for this test is $\alpha = 0.048$, which is quite close to the desired level of significance of 0.05. Rather than using trial and error with the pbinom function to determine the number of lags to use as the critical value, the qbinom function can be used to determine the cutoff.

```
qbinom(0.95, 40, 0.05)
```

This statement returns

```
[1] 4
```

A similar analysis can be applied to lag counts other than the $m = 40$ sample autocorrelation function values illustrated above. This analysis assumes that the sample autocorrelation function values of the residuals are independent and identically distributed normal random variables. One weakness of this approach is that it simply counts the number of sample autocorrelation function values falling outside the 95% confidence interval limits and ignores (a) how far outside of the limits the values fall or (b) how close to the limits they fall when they lie within the limits. This weakness prompts us to seek a statistical test that captures all of the sample autocorrelation function values associated with the residuals and includes their magnitudes.

Box–Pierce test. Let r_k be the lag k sample autocorrelation function value associated with the *residuals* of the fitted time series. As before, we only consider the first m such sample autocorrelation function values $r_1, r_2, \ldots, r_m$. It is approximately true that for mutually independent and identically distributed residuals,

$$r_k \sim N(0, 1/n).$$

By the transformation technique, this implies that

$$\sqrt{n}\, r_k \sim N(0, 1).$$

Squaring this random variable gives

$$n r_k^2 \sim \chi^2(1).$$

Assuming that the sample autocorrelation function values are uncorrelated, the sum of the first m of these random variables is

$$n \sum_{k=1}^{m} r_k^2 \sim \chi^2(m).$$

In the case in which r unknown model parameters have been estimated, the degrees of freedom are reduced by r:

$$n \sum_{k=1}^{m} r_k^2 \sim \chi^2(m-r).$$

This is the test statistic for the Box–Pierce test for serial correlation. Large values of this test statistic lead to rejecting H_0 and indicate a poor fit. The null hypothesis is rejected at level of significance α when this test statistic is greater than $\chi^2_{m-r,\alpha}$, where the first subscript is the number of degrees of freedom and the second subscript is the right-hand tail probability associated with this quantile of the chi-square distribution. There have been several approximations that occurred in formulating this statistical test. First, the r_k values are only approximately normally distributed. Second, the r_k values have variances which are less than $1/n$ for small lag values k. To compound this approximation, these smaller initial variances are dependent on the model under consideration. Third, the r_k values exhibit some serial correlation even when the residuals are mutually independent and identically distributed. These three weaknesses prompted a modification of the Box–Pierce test which provides a test statistic whose distribution more closely approximates the $\chi^2(m-r)$ distribution.

Ljung–Box test. The Box–Pierce test statistic was modified by Ljung and Box as

$$n(n+2) \sum_{k=1}^{m} \frac{r_k^2}{n-k},$$

which is approximately $\chi^2(m-r)$, where r is the number of parameters estimated in the model. Comparing the Box–Pierce and Ljung–Box test statistics, since

$$\frac{n+2}{n-k} > 1$$

for $k = 1, 2, \ldots, m$, the Ljung–Box test statistic always exceeds the Box–Pierce test statistic. The Box–Pierce test is more likely to accept a time series model with a poor fit than the Ljung–Box test for the same set of residuals. The Ljung–Box test should be used over the Box–Pierce because the probability distribution of its test statistic is closer to a $\chi^2(m-r)$ random variable under H_0.

Turning point test. As opposed to focusing on the sample autocorrelation function associated with the residuals, the turning point test considers the number of turning points in the time series of residuals. A turning point in a time series is defined to be a value associated with a local minimum or a local maximum. A local minimum occurs when $\hat{Z}_{t-1} > \hat{Z}_t$ and $\hat{Z}_t < \hat{Z}_{t+1}$. A local maximum occurs when $\hat{Z}_{t-1} < \hat{Z}_t$ and $\hat{Z}_t > \hat{Z}_{t+1}$. The random number of turning points in a time series of length n comprised of strictly continuous observations is denoted by T. The strictly continuous assumption is in place to avoid ties in adjacent values. A turning point cannot occur at the first or last value of the time series. Keep in mind that there might be fewer residuals than original observations. The n that is used here is the number of residuals. As given in an exercise at the end of this chapter, if the residuals are mutually independent and identically distributed continuous random variables, then

$$E[T] = \frac{2(n-2)}{3} \qquad \text{and} \qquad V[T] = \frac{16n - 29}{90}.$$

Furthermore, even though T is a discrete random variable, it is well approximated by the normal distribution with population mean $E[T]$ and population variance $V[T]$ for a time series of mutually independent and identically distributed observations and large n. Thus, an appropriate test statistic for testing H_0 is

$$\frac{T - 2(n-2)/3}{\sqrt{(16n - 29)/90}},$$

which is approximately standard normal for large values of n. The null hypothesis is rejected in favor of the alternative hypothesis whenever the test statistic is less than $-z_{\alpha/2}$ (which indicates fewer turning points than expected, which is an indicator of *positive* serial correlation among the residuals) or the test statistic is greater than $z_{\alpha/2}$ (which indicates more turning points than expected, which is an indicator of *negative* serial correlation among the residuals).

This completes the brief introduction to four statistical tests concerning the mutual independence of the residuals. There are several other such tests, some of which are introduced in the exercises at the end of the chapter, but these four are representative of how such tests work. Three questions are given below concerning issues associated with the analysis of the residuals.

1. What if two time series models are deemed adequate by these statistical tests?

   Instances frequently arise in which two or more candidate time series models fail to be rejected by the statistical tests on residuals that were just surveyed. In these cases, the modeler has four guiding principles. First, there might be physical considerations that might favor one model over another. An engineer, for example, might provide some engineering design insight concerning why one time series model would be favored over another. Second, the model with the best value of one of the *model-selection statistics* outlined in the next section, might be the appropriate choice. Third, if the modeler is torn between two time series models, selecting the model with the fewer parameters follows the parsimony principle. We would like a time series model that adequately captures the probabilistic aspects of the time series with the minimum number of parameters. Fourth, the purpose of the model, for example, description, explanation, prediction, or simulation, might drive the final choice of the model.

2. If a time series model is deemed inadequate, can the analysis of the residuals guide the modeler toward a more suitable model?

   In some cases, the analysis of the residuals can indeed guide the modeler toward a more suitable time series model. Here is one instance. Let's say that a shifted AR(1) model is being considered as a potential time series model:

$$X_t - \mu = \phi(X_{t-1} - \mu) + Z_t.$$

   Isolating the white noise term, this model can be written as

$$X_t - \mu - \phi(X_{t-1} - \mu) = Z_t.$$

The parameters μ, ϕ, and σ_Z^2 are estimated from the observed time series, and the associated residuals are calculated and plotted. Rather than appearing as white noise, let's say that the residuals appear to look like observations from an MA(1) time series model

$$Z_t = W_t + \theta W_{t-1},$$

where $\{W_t\}$ is a time series of white noise. Combining the two previous equations, this would lead us in the direction of considering the model

$$X_t - \mu - \phi(X_{t-1} - \mu) = W_t + \theta W_{t-1},$$

which can be recognized as a shifted ARMA(1, 1) time series model. Thus, the ARMA(1, 1) composite model has been constructed from the two simpler models. We would then revisit parameter estimation procedures for the parameters μ, ϕ, θ, and σ_Z^2, and perform model adequacy tests on the associated residual values on the fitted ARMA(1, 1) model.

3. If a time series model is deemed adequate, should the noise terms be modeled as white noise or Gaussian white noise?

   The four statistical tests for autocorrelation do not assess the *normality* of the residuals. Drawing a histogram of the residuals is an important first step in terms of determining whether the residuals are normally distributed. If the histogram appears to be bell-shaped, then the Gaussian white noise aspect of the model is justified. Some time series analysts prefer to view a histogram of the standardized residuals, and the vast majority of these values should lie between -3 and 3. A QQ (quantile–quantile) plot is also useful for visually assessing normality, which can be graphed with the R function qqnorm. A QQ plot which is linear is an indication of normality. The behavior at the extremes of a QQ plot is typically more variable than at the center, so some analysts prefer to focus on the behavior between, say, the first and third quartiles. Assessing the normality of a histogram or the linearity of a QQ plot is subjective. Objective statistical tests for the normality of the residuals include the Shapiro–Wilk, Anderson–Darling, Cramer–von Mises, and Kolmogorov–Smirnov tests.

Analyzing the residuals is not the only way to assess the adequacy of a time series model. Another technique is known as *overfitting*. ARMA models with a single additional term are fitted to the original time series. This approach is analogous to *forward selection* in the stepwise approach to multiple regression. We will refer to the time series model under consideration as the *tentative* model and the overfitted models as *enhanced* models. For example, if an MA(1) model is being contemplated as a tentative time series model, then

- adding an additional moving average term yields the enhanced MA(2) model, and

- adding an autoregressive term yields the enhanced ARMA(1, 1) model.

The parameters for these two enhanced models should be fit to the original time series in the usual fashion. If both of the following two criteria are met, then the tentative time series model should be accepted as the final model.

- The parameter estimates in the enhanced models are close to the parameter estimates in the tentative model.

- The additional parameter in the enhanced models does not differ significantly from zero.

So in the example given above, the parameters in the tentative MA(1) model, θ_1 and σ_Z^2, should be estimated from the original time series. Then the parameters in the enhanced MA(2) model, θ_1, θ_2, and σ_Z^2, should be estimated from the original time series. If a confidence interval for θ_2 contains zero (or you fail to reject the null hypothesis $H_0 : \theta_2 = 0$ versus the alternative hypothesis $H_1 : \theta_2 \neq 0$), and the other parameter estimates do not vary significantly between the two models, then the modeler concludes that the extra parameter in the MA(2) model is not necessary. The same type of thinking applies to the enhanced ARMA(1, 1) model. So in addition to a careful examination of the residuals, it is also helpful to overfit the model in the autoregressive and moving average directions to assess whether the additional term significantly improves the fit.

The model assessment techniques described in this subsection will be applied to actual time series later in this chapter.

8.2.4 Model Selection

Model-selection statistics are helpful when there are two or more tentative fitted ARMA(p, q) models for a time series which have been deemed adequate by the model assessment techniques outlined in the previous subsection. One naive approach to model selection is to just add additional terms to an ARMA(p, q) model and check the resulting sum of the squared residuals. This approach violates the parsimony principle because it is typically the case that adding parameters to a model results in a lower sum of squared residuals. Just blindly adding terms to minimize the sum of squares is likely to produce time series models with superfluous terms that contain no real explanatory value, which can potentially cause problems in the application of the model.

We seek some statistical measure that strikes a balance between simplicity and capturing the essence of the probabilistic mechanism governing the time series model. Some statistical measure which reflects the benefit of an additional parameter, but extracts a penalty for adding parameters would be helpful to strike this balance.

In the case in which the analyst is presented with multiple plausible tentative fitted models, a model-selection statistic such as Akaike's Information Criterion might prove helpful in determining the best model. This statistic strikes a harmony between a simple model (which might not capture certain probabilistic aspects of the mechanism governing the time series) and a more complex model (which might contain unnecessary terms). This is the notion of a *parsimonious* model which uses as few parameters as possible to achieve adequate explanatory power. Akaike's Information Criterion (AIC), named after Japanese statistician Hirotugu Akaike (1927–2009), extracts a penalty for each additional parameter that is added to the model. The AIC is

$$\text{AIC} = -2 \ln \left(L(\cdot) \right) + 2r,$$

where r is the number of unknown parameters that are estimated and L is the likelihood function evaluated at the maximum likelihood estimators for the r unknown parameters. Since $L(\cdot)$ is maximized at the maximum likelihood estimators, the first part of the AIC statistic, namely $-2 \ln \left(L(\cdot) \right)$, is minimized at the maximum likelihood estimator values because of the negative sign. The $2r$ term can be thought of as a penalty term for adding additional parameters to the model. Each additional parameter added to the model will probably decrease the first term in the AIC involving the log likelihood function, but will also increase the penalty term because r has been increased. The model with the lowest value of AIC is deemed by this model-selection statistic to be the most appropriate parsimonious time series model.

There are two variants of the AIC that provide improved ability to correctly identify a time series model.

- The AIC estimates the expected value of the Kullback–Leibler divergence of the estimated model from the true model, and there is a slight bias in the AIC which is significant for small values of n. The *corrected* Akaike Information Criterion, usually denoted by AICC, replaces the $2r$ penalty term with $2rn/(n-r-1)$, resulting in

$$\text{AICC} = -2\ln\left(L(\cdot)\right) + \frac{2rn}{n-r-1}.$$

Since $n/(n-r-1) > 1$, the AICC always exceeds the AIC for the same time series, meaning that the penalty for adding parameters is increased. The AICC will be more stingy than the AIC when it comes to adding parameters. The AICC model-selection statistic compensates for the AIC's tendency to overfit models.

- Another variant to the AIC is the Bayesian Information Criterion (BIC) which replaces the penalty term $2r$ with $r\ln n$, resulting in

$$\text{BIC} = -2\ln\left(L(\cdot)\right) + r\ln n.$$

As shown in Figure 8.5 for a time series of length $n = 50$ and $r = 0, 1, 2, \ldots, 5$ unknown parameters, the BIC places an even higher penalty on additional terms in the time series model than the AIC and the AICC, which will result, on average, with time series models with fewer terms. Since the use of maximum likelihood estimation is required for calculating AIC, AICC, and BIC because all three are a function of the likelihood function L, the white noise terms are assumed to be normally distributed (that is, Gaussian white noise). A visual check of this assumption can be made by looking at a histogram of the residuals or a QQ plot of the residuals.

The time series analyst should consult with people who are familiar with the time series in order to glean whether there might be some aspects of the data set that might suggest one particular model or another. The analyst should also not necessarily assume that one of the models suggested in this

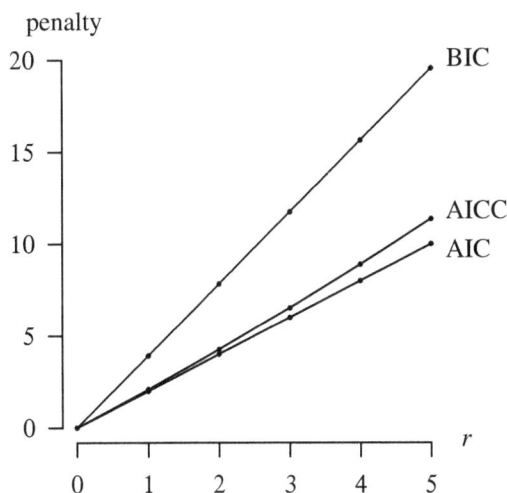

Figure 8.5: Penalty terms for model-selection statistics AIC, AICC, and BIC.

chapter might be appropriate for every setting. There are seldom uniquely correct values for p and q but rather these model-selection statistics are helpful in comparing two fitted tentative models.

In principle, the general linear model and its associated statistical methods are all that is necessary to fit and assess an ARMA(p, q) model. Since each specific ARMA(p, q) model has its own idiosyncrasies, the first few special cases of the autoregressive and moving average models will be examined in the next chapter.

8.3 Exercises

8.1 Show that the general linear model

$$X_t = Z_t + \psi_1 Z_{t-1} + \psi_2 Z_{t-2} + \cdots$$

can be written in the form

$$X_t = Z_t + \pi_1 X_{t-1} + \pi_2 X_{t-2} + \cdots .$$

8.2 For the ARMA time series model

$$X_t = 4X_{t-1} - 3X_{t-2} - 2X_{t-3} + Z_t - 5Z_{t-1} + 6Z_{t-2}.$$

(a) identify the time series model, and

(b) write the time series model in terms of the backshift operator B.

8.3 For the ARMA time series model

$$\phi(B)X_t = \theta(B)Z_t,$$

where $\phi(B) = 1$ and $\theta(B) = 1 - 0.6B + 0.1B^2$,

(a) identify the time series model, and

(b) write the time series model in purely algebraic form.

8.4 For the ARMA time series model

$$X_t = 2X_{t-1} - X_{t-2} + Z_t - Z_{t-2},$$

(a) identify the time series model, and

(b) write the time series model using the backshift operator.

8.5 Consider the special case of the general linear model

$$X_t = \frac{1}{2}X_{t-1} + Z_t - \frac{1}{3}Z_{t-1}.$$

(a) Write this model in its causal representation.

(b) Write this model in its invertible representation.

8.6 Show that $E[X_t] = \mu$ for the stationary shifted ARMA(p, q) model

$$X_t - \mu = \phi_1(X_{t-1} - \mu) + \phi_2(X_{t-2} - \mu) + \cdots + \phi_p(X_{t-p} - \mu) + Z_t + \theta_1 Z_{t-1} + \theta_2 Z_{t-2} + \cdots + \theta_q Z_{t-q}.$$

8.7 Find $E[X_t]$ for the shifted ARMA(2, 1) model

$$X_t = 7 + 0.4X_{t-1} - 0.1X_{t-2} + Z_t + 0.3Z_{t-1}.$$

8.8 Let $X_1, X_2, \ldots, X_n$ be observations from an ARMA(0, 0) time series model with Gaussian white noise. The maximum likelihood estimator of the population variance of the Gaussian white noise derived in Example 8.10 is

$$\hat{\sigma}_Z^2 = \frac{1}{n} \sum_{i=1}^{n} X_i^2.$$

An asymptotically exact confidence interval for σ_Z^2 derived in Example 8.11 is

$$\hat{\sigma}_Z^2 - z_{\alpha/2} \sqrt{\frac{2\left(\sum_{i=1}^{n} X_i^2\right)^2}{n^3}} < \sigma_Z^2 < \hat{\sigma}_Z^2 + z_{\alpha/2} \sqrt{\frac{2\left(\sum_{i=1}^{n} X_i^2\right)^2}{n^3}}.$$

Calculate and plot the actual coverage of a 95% confidence interval for σ_Z^2 as a function of n for $n = 8, 9, \ldots, 256$. Use analytic methods rather than Monte Carlo simulation.

8.9 Let $X_1, X_2, \ldots, X_n$ be observations from an ARMA(0, 0) time series model with Gaussian white noise. Find the probability density function of the maximum likelihood estimator of the population variance of the Gaussian white noise

$$\hat{\sigma}_Z^2 = \frac{1}{n} \sum_{i=1}^{n} X_i^2.$$

8.10 Let $X_1, X_2, \ldots, X_n$ be observations from an ARMA(0, 0) time series model with Gaussian white noise. As shown in Example 8.10, the maximum likelihood estimator of the population variance of the Gaussian white noise is

$$\hat{\sigma}_Z^2 = \frac{1}{n} \sum_{i=1}^{n} X_i^2$$

and a pivotal quantity for developing an exact two-sided $100(1 - \alpha)\%$ confidence interval for σ_Z^2 is

$$\frac{n\hat{\sigma}_Z^2}{\sigma_Z^2} \sim \chi^2(n).$$

Find an exact two-sided $100(1 - \alpha)\%$ confidence interval for σ_Z^2.

8.11 Let $X_1, X_2, \ldots, X_n$ be observations from an ARMA(0, 0) time series model with Gaussian white noise having finite positive population variance σ_Z^2. The maximum likelihood estimator of the population variance of the Gaussian white noise is

$$\hat{\sigma}_Z^2 = \frac{1}{n} \sum_{i=1}^{n} X_i^2.$$

Conduct a Monte Carlo simulation experiment that provides convincing numerical evidence that

$$\frac{n\hat{\sigma}_Z^2}{\chi^2_{n,\alpha/2}} < \sigma_Z^2 < \frac{n\hat{\sigma}_Z^2}{\chi^2_{n,1-\alpha/2}}$$

is an *exact* $100(1 - \alpha)\%$ confidence interval for σ_Z^2 for one particular set of n, α, and σ_Z^2 of your choice.

8.12 Let X_1 and X_2 be jointly distributed random variables. The population mean and variance of X_1 are μ_{X_1} and $\sigma_{X_1}^2$. The population mean and variance of X_2 are μ_{X_2} and $\sigma_{X_2}^2$. The population correlation between X_1 and X_2 is $\rho = \text{Corr}(X_1, X_2)$. The value of X_2 is to be predicted as a linear function of X_1 with $mX_1 + b$. Find the values of m and b which minimize the mean square error of the prediction. In other words, find m and b which minimize

$$ E\left[(X_2 - mX_1 - b)^2 \right]. $$

8.13 Consider an ARMA(0, 0) model with $U(-1, 1)$ white noise terms. Find an exact two-sided 95% prediction interval for X_{n+h}.

8.14 Suppose an ARMA(2, 1) time series model is a strong candidate for modeling a particular time series. A long time series is available for analysis, so n is large. The ARMA(2, 1) model is fitted and residuals are calculated. If the sample autocorrelation function associated with the residuals is calculated for the first 100 lags, how many values need to fall outside of $\pm 1.96/\sqrt{n}$ in order to reject the null hypothesis H_0, which corresponds to a good fit at a significance level that is less than $\alpha = 0.05$?

8.15 Compare the expected p-values for the Box–Pierce and Ljung–Box tests for serial independence of a time series consisting of $n = 100$ mutually independent and identically distributed standard normal random variables. Consider only the first $k = 40$ lag values.

8.16 Let $\hat{Z}_1, \hat{Z}_2, \ldots, \hat{Z}_n$ be residual values associated with a fitted time series model. The Durbin–Watson test statistic defined by

$$ D = \sum_{t=2}^{n} \left(\hat{Z}_t - \hat{Z}_{t-1} \right)^2 \Big/ \sum_{t=1}^{n} \hat{Z}_t^2 $$

is useful for testing the serial independence of the residuals.

   (a) Conduct a Monte Carlo simulation experiment to estimate the expected value of D when $\hat{Z}_1, \hat{Z}_2, \ldots, \hat{Z}_{1000}$ are $n = 1000$ mutually independent and identically distributed standard normal random variables.

   (b) Give an explanation for the result that you obtained in part (a).

8.17 The *turning point test* for serial dependence counts the number of turning points (the number of local minima and maxima) T in a time series of length n comprised of strictly continuous observations. A turning point cannot occur at the first or last value of the time series.

   (a) Show that $E[T] = 2(n-2)/3$ when the observations in the time series are mutually independent and identically distributed.

   (b) Show that $V[T] = (16n - 29)/90$ when the observations in the time series are mutually independent and identically distributed.

   (c) Perform a Monte Carlo simulation that supports the values of $E[T]$ and $V[T]$ for a time series of length $n = 101$.

   (d) Argue why T is approximately normally distributed with population mean $E[T]$ and population variance $V[T]$ for a time series of mutually independent and identically distributed observations and large n. Suggest an appropriate test statistic for testing the null hypothesis that there is no serial correlation in the time series.

8.18 Let X_1, X_2, X_3, X_4 be a time series of mutually independent and identically distributed continuous random variables. Let T be the number of turning points. Find the probability mass function of T.

8.19 The nonparametric *difference–sign test* for serial dependence counts the number of values in a time series of strictly continuous observations $X_1, X_2, \ldots, X_n$ in which $X_i > X_{i-1}$, for $i = 2, 3, \ldots, n$. Denote this count by T.

  (a) Show that $E[T] = (n-1)/2$ when the observations in the time series are mutually independent and identically distributed.

  (b) Show that $V[T] = (n+1)/12$ when the observations in the time series are mutually independent and identically distributed.

  (c) Perform a Monte Carlo simulation that supports the values of $E[T]$ and $V[T]$ for a time series of length $n = 101$.

  (d) Argue why T is approximately normally distributed with population mean $E[T]$ and population variance $V[T]$ for a time series of mutually independent and identically distributed observations and large n. Suggest an appropriate test statistic for testing the null hypothesis that there is no serial correlation in the time series.

8.20 Suppose an AR(1) model is being considered as a tentative time series model based on a realization of the time series. A single autoregressive parameter and a single moving average parameter is added to the tentative model, resulting in an ARMA(2, 1) enhanced model. Describe any problems that might arise by comparing the AR(1) time series model and the ARMA(2, 1) time series model.

Chapter 9

Topics in Time Series Analysis

This chapter presents several topics in time series analysis. These include several of the popular time series models which are special cases of the ARMA(p, q), including the software required for fitting these models. The first section surveys the probability models and statistical methods associated with autoregressive models, more specifically the AR(1), AR(2), and AR(p) models. The second section surveys the probability models and statistical methods associated with autoregressive models, more specifically the MA(1), MA(2), and MA(q) models. It is important to know the properties of these special cases of the ARMA(p, q) model in order to successfully fit such a model to a realization of a time series. This will allow us to build an inventory of population autocorrelation and partial autocorrelation functions for these models that can be matched to their statistical counterparts for building a time series model. Time series analysts tend to use the smallest possible p and q values that adequately describe a time series. For this reason, separate subsections are devoted to the AR(1), AR(2), MA(1), and MA(2) time series models.

9.1   Autoregressive Models

Autoregressive models for a time series $\{X_t\}$ will be considered in this section. An autoregressive model of order p is a special case of an ARMA(p, q) model with no moving average terms (that is, $q = 0$), specified as

$$X_t = \phi_1 X_{t-1} + \phi_2 X_{t-2} + \cdots + \phi_p X_{t-p} + Z_t,$$

where $\phi_1, \phi_2, \ldots, \phi_p$ are real-valued parameters and $\{Z_t\}$ is a time series of white noise with population mean zero and population variance σ_Z^2. The formulation of the AR(p) time series model looks quite similar to that of a multiple linear regression model with p independent variables. These independent variables are also known as predictors, regressors, or covariates in regression analysis. That is the genesis of the term *autoregressive* to describe this model. The prefix *auto* means *self*, indicating that this model has the current value of the time series $\{X_t\}$ written as a linear function of the p previous versions of itself plus a white noise term Z_t. The white noise term is critical to the model because without it, there would be no randomness in the model.

   Rather than diving right into an AR(p) model, we first introduce the AR(1) and AR(2) models in separate sections because the mathematics are somewhat easier than the general case and some important geometry and intuition can be developed in these restricted models. In addition, an AR(1) or AR(2) model is often an adequate time series model in a particular setting. We always want a model with the fewest possible number of parameters that adequately approximates the underlying time series probability model. In the sections that follow, we will

- define the time series model for $\{X_t\}$,

- determine the values of the parameters associated with a stationary model,

- derive the population autocorrelation and partial autocorrelation functions,

- develop algorithms for simulating observations from the time series,

- inspect simulated realizations to establish patterns,

- estimate the parameters from a time series realization $\{x_t\}$,

- assess the adequacy of the time series model, and

- forecast future values of the time series using both point and interval estimates.

The purpose of deriving the population autocorrelation and partial autocorrelation functions is to build an inventory of shapes and patterns for these functions that can be used to identify tentative time series models from their sample counterparts by making a visual comparison between population and sample versions. This inventory of shapes and patterns plays an analogous role to knowing the shapes of various probability density functions (for example, the bell-shaped normal probability density function or the rectangular-shaped uniform distribution) in the analysis of univariate data in which the shape of the histogram is visually compared to the inventory of probability density function shapes.

In each section that follows, a single example of a time series will be carried through the various statistical procedures given in the list above. Stationarity plays a critical role in time series analysis because we are not able to forecast future values of the time series without knowing that the probability model is stable over time. This is why the visual assessment of a plot of the time series is always a critical first step in the analysis of a time series.

9.1.1   The AR(1) Model

The autoregressive model of order 1 is defined next. It has a closed-form expression for the population autocorrelation function and is frequently used in applications.

Definition 9.1 A *first-order autoregressive time series model*, denoted by AR(1), for the time series $\{X_t\}$ is defined by

$$X_t = \phi X_{t-1} + Z_t,$$

where ϕ is a real-valued parameter and $\{Z_t\}$ is a time series of white noise:

$$Z_t \sim WN\left(0, \sigma_Z^2\right).$$

No subscript is necessary on the ϕ parameter because there is only one ϕ parameter in the AR(1) model. So there are two parameters that define an AR(1) model: the coefficient ϕ and the population variance of the white noise σ_Z^2.

The current value in the time series, X_t, is given by the parameter ϕ multiplied by the previous observed value in the time series, ϕX_{t-1}, plus the current white noise term Z_t. This model has the form of a simple linear regression model forced through the origin in which X_t is being predicted by the previous value of the time series X_{t-1}. The parameter ϕ plays the role of the slope of the regression line. Thinking about an AR(1) model as a simple linear regression model suggests a statistical graphic that can be helpful in determining whether it is an appropriate model for a particular time series. A plot of x_t on the vertical axis against x_{t-1} on the horizontal axis should be approximately linear if the AR(1) model is appropriate. The slope of the regression line on this plot corresponds to ϕ, and the magnitude of the variability of the points about the regression line is determined by the population variance of the white noise σ_Z^2.

Some authors prefer to parameterize the AR(1) model as

$$X_t = \phi_1 X_{t-1} + \phi_0 Z_t,$$

where ϕ_0 and ϕ_1 are real-valued parameters. We avoid this parameterization because the ϕ_0 parameter is redundant in the sense that the population variance of the white noise σ_Z^2 is absorbed into the ϕ_0 parameter. Also, some authors use a $-$ rather than a $+$ between the terms on the right-hand side of the model.

To illustrate the thinking behind the AR(1) model in a specific context, let X_t represent the closing price of a particular stock on day t. The AR(1) model indicates that today's closing price, denoted by X_t, equals ϕ multiplied by yesterday's closing price (ϕX_{t-1}), plus today's random white noise term Z_t.

Stationarity

One initial important question concerning the AR(1) model is whether or not the model is stationary. Consider a thought experiment that determines whether an AR(1) model is stationary for specific values of ϕ. For one particular instance, consider $\phi = 0$. In this case the AR(1) time series model reduces to

$$X_t = Z_t,$$

which is a time series model consisting solely of white noise. We know from Example 7.15 that a time series model of white noise terms is stationary. Now consider another instance, $\phi = 1$. In this case the AR(1) time series model reduces to

$$X_t = X_{t-1} + Z_t,$$

which indicates that each value in the time series is the previous value plus the current white noise. In this case the population variance of the process is increasing with time because the number of white noise terms accumulate over time (see Example 7.8), so the AR(1) model with $\phi = 1$ violates one of the stationarity conditions given in Definition 7.6. The AR(1) model with $\phi = 1$ can be recognized as a random walk model from Example 7.4, and it was determined to be nonstationary in Example 7.17. So we have established that the AR(1) time series model is stationary for $\phi = 0$ and nonstationary for $\phi = 1$. We now try to determine general restrictions on ϕ associated with a stationary AR(1) time series model. We take four different approaches to establishing the values of the coefficient ϕ that lead to a stationary model. The four approaches provide a review of several concepts defined previously.

Approach 1: Causality. In the derivations concerning the AR(1) time series model that follow, it will be beneficial to write the time series value X_t as a linear combination of the current and previous white noise values. This will allow us to use the definition of causality in Definition 8.2 to determine the values of ϕ associated with a stationary AR(1) model. To begin, recall that the AR(1) model given by

$$X_t = \phi X_{t-1} + Z_t$$

can be shifted in time and is equally valid for other t values, for example,

$$X_{t-1} = \phi X_{t-2} + Z_{t-1}$$
$$X_{t-2} = \phi X_{t-3} + Z_{t-2}$$
$$\vdots$$

Using successive substitutions into the AR(1) model results in

$$
\begin{aligned}
X_t &= \phi X_{t-1} + Z_t \\
&= \phi \left(\phi X_{t-2} + Z_{t-1} \right) + Z_t \\
&= \phi^2 X_{t-2} + \phi Z_{t-1} + Z_t \\
&= \phi^2 \left(\phi X_{t-3} + Z_{t-2} \right) + \phi Z_{t-1} + Z_t \\
&= \phi^3 X_{t-3} + \phi^2 Z_{t-2} + \phi Z_{t-1} + Z_t \\
&\qquad\qquad \vdots \\
&= Z_t + \phi Z_{t-1} + \phi^2 Z_{t-2} + \phi^3 Z_{t-3} + \cdots .
\end{aligned}
$$

This can be recognized as an MA(∞) time series model. Representing an AR(1) model as an MA(∞) model is known as *duality*. We now determine the constraints on the parameter ϕ which are required for stationarity. This is the form that is required for causality from Definition 8.2. The coefficients $\psi_1, \psi_2, \ldots$ for the AR(1) model from Definition 8.2 are

$$\psi_1 = \phi, \ \psi_2 = \phi^2, \ \psi_3 = \phi^3, \ldots,$$

or in general, $\psi_j = \phi^j$, for $j = 1, 2, \ldots$. Definition 8.2 requires that

$$\sum_{j=1}^{\infty} \psi_j^2 = \sum_{j=1}^{\infty} \phi^{2j} = \phi^2 + \phi^4 + \phi^6 + \cdots < \infty$$

for the time series model to be written in causal form. This summation is a geometric series that converges when $|\phi| < 1$, or equivalently, when $-1 < \phi < 1$, so these are the values of ϕ for which the AR(1) model is causal, which also implies that the model is stationary. Expressing the AR(1) model as an MA(∞) model will also be helpful in the subsequent derivation of the population auto-covariance and autocorrelation functions.

Approach 2: Backshift operator. Although the purely algebraic derivation of the causal form of the AR(1) time series model using standard algebra techniques from Approach 1 works fine for establishing stationarity, there is an alternative approach which is slightly more elegant that exploits the backshift operator B. The AR(1) model

$$X_t = \phi X_{t-1} + Z_t$$

can be rewritten as

$$X_t - \phi X_{t-1} = Z_t,$$

which can be expressed using the backshift operator as

$$(1 - \phi B)X_t = Z_t.$$

The first-order polynomial $\phi(B) = 1 - \phi B$ generalizes to a polynomial in B of order p for an AR(p) model. Dividing both sides of this equation by $1 - \phi B$ gives

$$X_t = \frac{Z_t}{1 - \phi B}.$$

For values of ϕ satisfying $-1 < \phi < 1$, this can be recognized as a geometric series in B:

$$X_t = \left(1 + \phi B + \phi^2 B^2 + \phi^3 B^3 + \cdots\right) Z_t.$$

Executing the B operator converts this to the form

$$X_t = Z_t + \phi Z_{t-1} + \phi^2 Z_{t-2} + \phi^3 Z_{t-3} + \cdots,$$

which is the same form that we encountered using the successive substitutions in the causality approach.

   Approach 3: Unit roots analysis. Theorem 8.3 indicates that all AR(1) models are invertible and they are stationary when the root of

$$\phi(B) = 1 - \phi B = 0$$

lies outside of the unit circle in the complex plane. The solution to this equation is

$$B = \frac{1}{\phi}.$$

This root falls on the real axis in the complex plane and lies outside of the unit circle when

$$-1 < \phi < 1,$$

which is consistent with Approaches 1 and 2.

   Approach 4: Definition of stationarity. We can also return to first principles to establish the values of ϕ associated with a stationary AR(1) model. This approach also results in the derivation of the population autocorrelation function. Recall from Definition 7.6 that a time series model is stationary if (*a*) the expected value of X_t is constant for all t, and (*b*) the population covariance between X_s and X_t depends only on the lag $|t - s|$. Using the causal formula for the AR(1) time series model expressed as an MA(∞) time series model from Approach 1, the expected value of X_t is

$$
\begin{aligned}
E\left[X_t\right] &= E\left[Z_t + \phi Z_{t-1} + \phi^2 Z_{t-2} + \phi^3 Z_{t-3} + \cdots\right] \\
&= E\left[Z_t\right] + \phi E\left[\phi Z_{t-1}\right] + \phi^2 E\left[Z_{t-2}\right] + \phi^3 E\left[Z_{t-3}\right] + \cdots \\
&= 0
\end{aligned}
$$

for all values of the parameters ϕ and σ_Z^2, and all values of t. Again using the causal formula for the AR(1) time series model expressed as an MA(∞) time series model,

$$
\begin{aligned}
\gamma(s,t) &= \text{Cov}\left(X_s, X_t\right) \\
&= \text{Cov}\left(Z_s + \phi Z_{s-1} + \phi^2 Z_{s-2} + \cdots, Z_t + \phi Z_{t-1} + \phi^2 Z_{t-2} + \cdots\right) \\
&= \text{Cov}\left(Z_s, \phi^{|t-s|} Z_s\right) + \text{Cov}\left(\phi Z_{s-1}, \phi^{|t-s|+1} Z_{s-1}\right) + \text{Cov}\left(\phi^2 Z_{s-2}, \phi^{|t-s|+2} Z_{s-2}\right) + \cdots \\
&= \phi^{|t-s|} \sigma_Z^2 + \phi^{|t-s|+2} \sigma_Z^2 + \phi^{|t-s|+4} \sigma_Z^2 + \cdots \\
&= \left(\phi^{|t-s|} + \phi^{|t-s|+2} + \phi^{|t-s|+4} + \cdots\right) \sigma_Z^2 \\
&= \phi^{|t-s|} \left(\frac{1}{1 - \phi^2}\right) \sigma_Z^2 \qquad\qquad |t-s| = 0, 1, 2, \ldots
\end{aligned}
$$

for $-1 < \phi < 1$. Since $E[X_t] = 0$ for all values of t and the population autocovariance function depends only on the lag $|t - s|$, we conclude that the AR(1) process is stationary when $-1 < \phi < 1$. So the population autocovariance function can be expressed in terms of the lag k as

$$\gamma(k) = \left(\frac{\phi^k}{1 - \phi^2}\right) \sigma_Z^2 \qquad k = 0, 1, 2, \dots .$$

Dividing by the population autocovariance function by

$$\gamma(0) = \left(\frac{1}{1 - \phi^2}\right) \sigma_Z^2$$

gives the population autocorrelation function

$$\rho(k) = \frac{\gamma(k)}{\gamma(0)} = \phi^k \qquad k = 0, 1, 2, \dots .$$

Based on the four approaches, we now know beyond a shadow of doubt that an AR(1) model is stationary for values of the parameter ϕ satisfying $-1 < \phi < 1$. This derivation constitutes a proof of the following result, which will be stated for just the nonnegative lags. Many authors list the lags as $k = \pm 1, \pm 2, \dots$, but we appeal to Theorem 7.1 to cover the negative lags and only report the nonnegative lags in all of the population autocorrelation functions given in this chapter.

Theorem 9.1 The AR(1) time series model for $\{X_t\}$ is stationary for values of the parameter ϕ satisfying $-1 < \phi < 1$ and σ_Z^2 satisfying $\sigma_Z^2 > 0$ with population autocorrelation function

$$\rho(k) = \phi^k \qquad k = 0, 1, 2, \dots .$$

The derivation of $\rho(k) = \phi^k$ for $k = 0, 1, 2, \dots$ provides still further evidence of the restriction that $-1 < \phi < 1$. If ϕ were equal to a value outside of this range, say $\phi = 2$, this would result in population correlation values outside of the range $-1 \le \rho(k) \le 1$.

For all admissible values of ϕ on the interval $-1 < \phi < 1$, we see from the formula $\rho(k) = \phi^k$ for $k = 0, 1, 2, \dots$ that there will be a geometric decline in the *magnitude* of the values in the population autocorrelation function as the lag k increases. There are two distinct cases for ϕ, however, which will result in population autocorrelation functions with distinctly different shapes. The first case is $0 < \phi < 1$, which gives positive population autocorrelation values at all lags. This is associated with a time series that lingers on one side of the mean. How long it lingers depends on the magnitude of ϕ. Larger values of ϕ indicate that nearby observations will tend to be more likely to be on the same side of the mean, and therefore the time series will tend to linger longer on one side of the mean. The second case is $-1 < \phi < 0$, which gives population autocorrelation function values which alternate in sign and is associated with a time series that is likely to jump from one side of the mean to the other for adjacent observations. These two cases are illustrated in Figure 9.1 for the first 8 lags of the population autocorrelation function for $\phi = 0.8$ and $\phi = -0.8$.

Population Partial Autocorrelation Function

We now determine the population partial autocorrelation function for an AR(1) model. By Definition 7.4, the population lag 0 partial autocorrelation value is $\rho^*(0) = 1$. The population lag 1

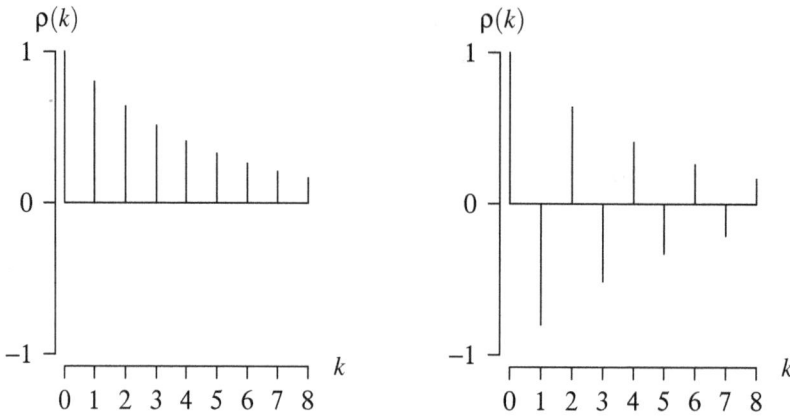

Figure 9.1: AR(1) population autocorrelation functions for $\phi = 0.8$ (left) and $\phi = -0.8$ (right).

partial autocorrelation value is $\rho^*(1) = \rho(1) = \phi$. The population lag 2 partial autocorrelation is

$$
\rho^*(2) = \frac{\begin{vmatrix} 1 & \rho(1) \\ \rho(1) & \rho(2) \end{vmatrix}}{\begin{vmatrix} 1 & \rho(1) \\ \rho(1) & 1 \end{vmatrix}} = \frac{\begin{vmatrix} 1 & \phi \\ \phi & \phi^2 \end{vmatrix}}{\begin{vmatrix} 1 & \phi \\ \phi & 1 \end{vmatrix}} = 0.
$$

This is consistent with the result from Example 7.22 from first principles. Notice that the second column of the matrix in the numerator is a multiple of the first column of the matrix in the numerator. This is why the determinant of the numerator is zero. The population lag 3 partial autocorrelation is

$$
\rho^*(3) = \frac{\begin{vmatrix} 1 & \rho(1) & \rho(1) \\ \rho(1) & 1 & \rho(2) \\ \rho(2) & \rho(1) & \rho(3) \end{vmatrix}}{\begin{vmatrix} 1 & \rho(1) & \rho(2) \\ \rho(1) & 1 & \rho(1) \\ \rho(2) & \rho(1) & 1 \end{vmatrix}} = \frac{\begin{vmatrix} 1 & \phi & \phi \\ \phi & 1 & \phi^2 \\ \phi^2 & \phi & \phi^3 \end{vmatrix}}{\begin{vmatrix} 1 & \phi & \phi^2 \\ \phi & 1 & \phi \\ \phi^2 & \phi & 1 \end{vmatrix}} = 0.
$$

Again, the determinant in the numerator is zero because the third column is a multiple of the first column. This pattern continues for the lag k population partial autocorrelation function, which has a first column of the numerator matrix $\left[1, \phi, \phi^2, \ldots, \phi^{k-1}\right]'$ and last column $\left[\phi, \phi^2, \phi^3, \ldots, \phi^k\right]'$. Since the last column of the numerator matrix is a multiple of the first column of the numerator matrix, the determinant of the numerator matrix is zero. This constitutes a proof of the following result.

Theorem 9.2 The population partial autocorrelation function for a stationary AR(1) process is

$$
\rho^*(k) = \begin{cases} 1 & k = 0 \\ \phi & k = 1 \\ 0 & k = 2, 3, \ldots \, . \end{cases}
$$

    Figure 9.2 shows the first 8 lags of the population partial autocorrelation function for $\phi = 0.8$ and $\phi = -0.8$. These are the same parameter settings as in Figure 9.1. Unlike the population autocorrela-

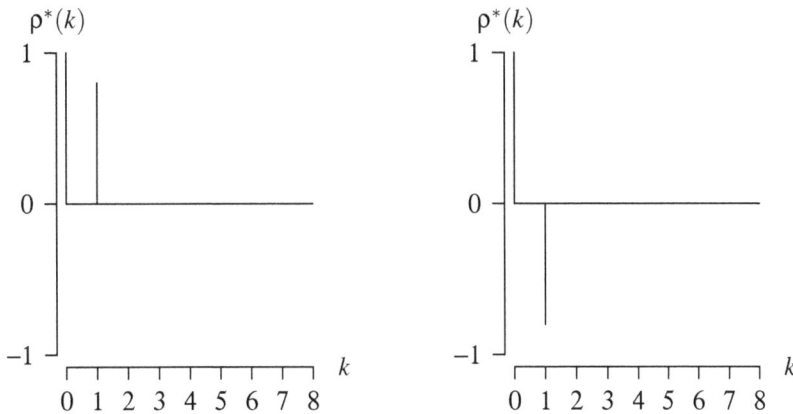

Figure 9.2: Population partial autocorrelation functions for $\phi = 0.8$ (left) and $\phi = -0.8$ (right).

tion function which tails off in magnitude for increasing lags, the population partial autocorrelation cuts off after lag 1. When plotting the corresponding sample analogs, it is typically easier to visually assess a function cutting off rather than tailing off, particularly if there is significant random sampling variability in the observed time series.

The Shifted AR(1) Model

The population mean function for the AR(1) model is $E[X_t] = 0$. This model is not of much use in practice because most real-world time series are not centered around zero. Adding a third parameter μ overcomes this shortcoming. Since population variance and covariance are unaffected by a shift, the associated population autocorrelation and partial autocorrelation functions remain the same as those given in Theorems 9.1 and 9.2. Likewise, the condition for stationarity is unchanged.

Theorem 9.3 A *shifted first-order autoregressive model* for the time series $\{X_t\}$ is defined by

$$X_t - \mu = \phi(X_{t-1} - \mu) + Z_t,$$

where ϕ, μ, and $\sigma_Z^2 > 0$ are real-valued parameters and $\{Z_t\}$ is a time series of white noise. This model is stationary when $-1 < \phi < 1$. The expected value of X_t is $E[X_t] = \mu$. The population autocorrelation function is

$$\rho(k) = \phi^k \qquad k = 0, 1, 2, \ldots$$

and the population partial autocorrelation function is

$$\rho^*(k) = \begin{cases} 1 & k = 0 \\ \phi & k = 1 \\ 0 & k = 2, 3, \ldots \end{cases}$$

Simulation

An AR(1) time series can be simulated by appealing to the defining formula for the AR(1) model.

Iteratively applying the defining formula for an AR(1) model

$$X_t = \phi X_{t-1} + Z_t$$

results in the simulated values $X_1, X_2, \ldots, X_n$. The difficult aspect of this algorithm is how to generate the first value X_1 because there is no X_0 available. For simplicity, assume that the white noise terms are Gaussian white noise. Since the expected value of X_t is $E[X_t] = 0$, the population variance of X_t is

$$V[X_t] = \gamma(0) = \left(\frac{1}{1-\phi^2}\right)\sigma_Z^2,$$

and linear combinations of mutually independent normally distributed random variables are normal, then the first simulated observation

$$X_1 \sim N\left(0, \left(\frac{1}{1-\phi^2}\right)\sigma_Z^2\right).$$

The algorithm given as pseudocode below generates an initial time series observation X_1 as indicated above, and then uses an additional $n-1$ Gaussian white noise terms $Z_2, Z_3, \ldots, Z_n$ to generate the remaining time series values $X_2, X_3, \ldots, X_n$ using the AR(1) defining formula from Definition 9.1. Indentation denotes nesting in the algorithm.

$t \leftarrow 1$
generate $X_t \sim N\left(0, \left(\frac{1}{1-\phi^2}\right)\sigma_Z^2\right)$
while $(t < n)$
    $t \leftarrow t+1$
    generate $Z_t \sim N\left(0, \sigma_Z^2\right)$
    $X_t \leftarrow \phi X_{t-1} + Z_t$

The three-parameter shifted AR(1) time series model which includes a population mean parameter μ can be simulated by simply adding μ to each time series observation generated by this algorithm. The next example implements this algorithm in R.

Example 9.1 Generate a realization of $n = 100$ observations from an AR(1) time series model with $\phi = 0.8$ and Gaussian white noise error terms with $\sigma_Z^2 = 9$.

Since $\phi = 0.8$ lies in the interval $-1 < \phi < 1$, this is a stationary AR(1) time series model via Theorem 9.1. The first (optional) statement in the R code below uses the `set.seed` function to establish the random number seed. The second statement sets the AR(1) coefficient to $\phi = 0.8$. The third statement sets the standard deviation of the Gaussian white noise to $\sigma_Z = 3$. The fourth statement sets the number of simulated values to $n = 100$. The fifth statement defines the vector x of length $n = 100$ to hold the simulated time series values. The sixth statement generates the first simulated time series observation X_1 with a call the `rnorm` function. Finally, the `for` loop iterates through the defining formula for the AR(1) model generating the remaining observations $X_2, X_3, \ldots, X_{100}$.

```
set.seed(3)
phi  = 0.8
sigz = 3
n    = 100
x    = numeric(n)
x[1] = rnorm(1, 0, sigz / sqrt(1 - phi ^ 2))
for (t in 2:n) x[t] = phi * x[t - 1] + rnorm(1, 0, sigz)
```

Using the plot.ts function to make a plot of the time series contained in x, the acf function to plot the associated correlogram, the pacf function to plot the associated sample partial autocorrelation function, and the layout function to arrange the graphs as in Example 7.24, the resulting trio of graphs are displayed in Figure 9.3. The points that have been added to the time series plot can be helpful in identifying patterns. Consistent with an AR(1) model with $\phi = 0.8$ the time series plot shows that the observations tend to have runs of observations that linger above and below the population mean of 0, which is indicated by a horizontal line. The associated sample autocorrelation function tails off as expected from Figure 9.1. The associated sample partial autocorrelation function has a statistically significant spike at lag 1 with $r_1^* = 0.8187$, and then cuts off after lag 1 as expected from Figure 9.2. The spike at lag 1 on both autocorrelation graphs is approximately $\phi = 0.8$, as expected. The 95% confidence intervals indicated by the dashed lines show that the values of the sample partial autocorrelation function do not significantly differ from zero at lags beyond lag 1.

We recommend running the simulation code from the previous example several dozen times in a

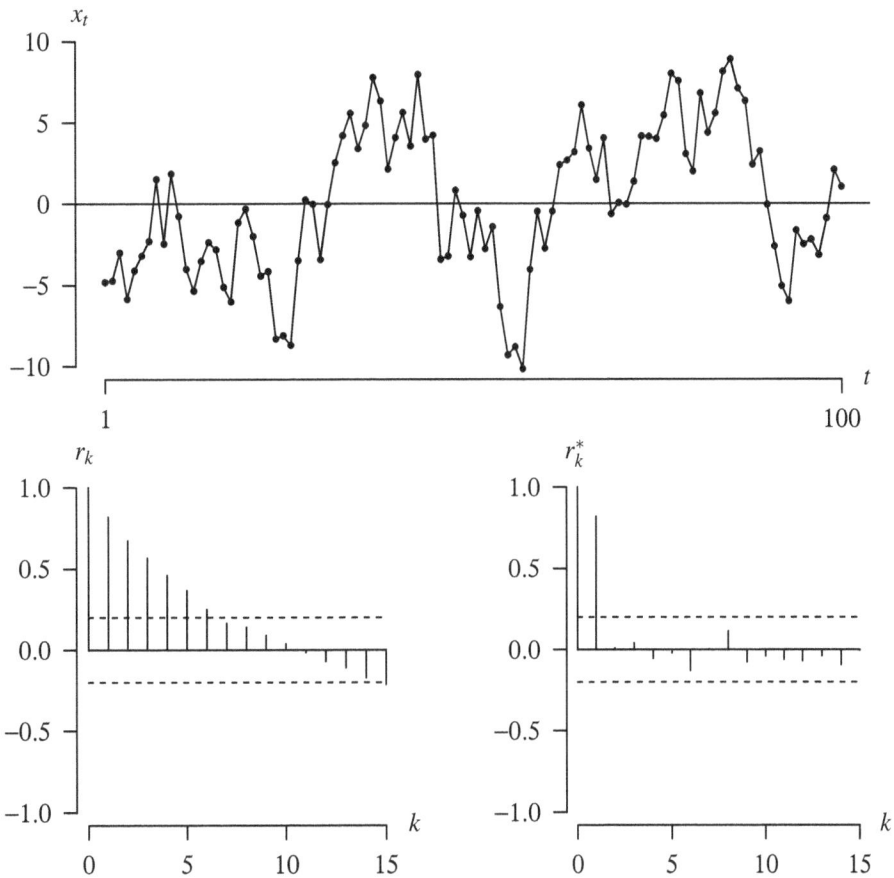

Figure 9.3: Time series plot, r_k, and r_k^* for $n = 100$ simulated values from an AR(1) model.

loop and viewing the associated plots of x_t, r_k, and r_k^* in search of patterns. A call to the R function Sys.sleep between the displays of the trio of plots can be used to include an artificial time delay to allow you to inspect the plots. This will allow you to see how various realizations of a simulated AR(1) time series model vary from one realization to the next. So when you then view a *single* realization of a real-life time series, you will have a better sense of how far these plots might deviate from their expected patterns.

There is a second way to simulate observations from an AR(1) time series. This second technique starts the time series at an initial arbitrary value, and then allows the time series to "warm up" or "burn in" for several time periods before producing the first observation X_1. A reasonable initial arbitrary value for the standard AR(1) model is 0; a reasonable initial arbitrary value for the shifted AR(1) model is μ. This is the approach taken by the built-in R function named arima.sim (for autoregressive moving average simulation), which simulates a realization of a time series. Using the arima.sim function saves a few keystrokes over the approach taken in the previous example, as illustrated next.

> **Example 9.2** Generate a realization of $n = 100$ observations from a shifted AR(1) time series model with $\phi = -0.8$, Gaussian white noise error terms with $\sigma_Z^2 = 9$, and mean value $\mu = 10$.
>
> Since there is now a nonzero population mean value, the shifted AR(1) model is
>
> $$X_t - \mu = \phi(X_{t-1} - \mu) + Z_t,$$
>
> where $\mu = 10$, $\phi = -0.8$, and $\sigma_Z = 3$. Since $\phi = -0.8$ lies in the interval $-1 < \phi < 1$, this is a stationary AR(1) time series model. The model argument in the arima.sim function is a list containing the value of ϕ. Although the default probability distribution for the white noise is normal (that is, Gaussian white noise) with population variance σ_Z^2, the function allows for other distributions. The second argument to arima.sim is n, the number of time series observations to be generated. The sd argument defines the standard deviation of the white noise. The n.start argument gives the number of observations in the warm-up period, which we specify here as 50. The R code to generate $n = 100$ values from the shifted AR(1) model is given below.

```
set.seed(10)
x = 10 + arima.sim(model = list(ar = -0.8), n = 100, sd = 3, n.start = 50)
```

> Figure 9.4 shows the three plots associated with the simulated values using the plot.ts, acf, and pacf functions. The time series plot shows a radically different pattern than the time series in the previous example in two manners. First, this simulated time series is centered around $\mu = 10$ (indicated by a horizontal line) rather than $\mu = 0$. Second, adjacent observations in the time series tend to jump from one side of the population mean to other side of the population mean, which is consistent with the population autocorrelation function from the right-hand plot in Figure 9.1. Consistent with the time series plot, the values in the sample autocorrelation function alternate in sign and decrease in magnitude. The sample partial autocorrelation function has a statistically significant spike at lag 1 of $r_1^* = -0.8330$, and nonsignificant spikes thereafter. This is consistent with the right-hand plot in Figure 9.2. Type

```
getAnywhere(arima.sim)
```

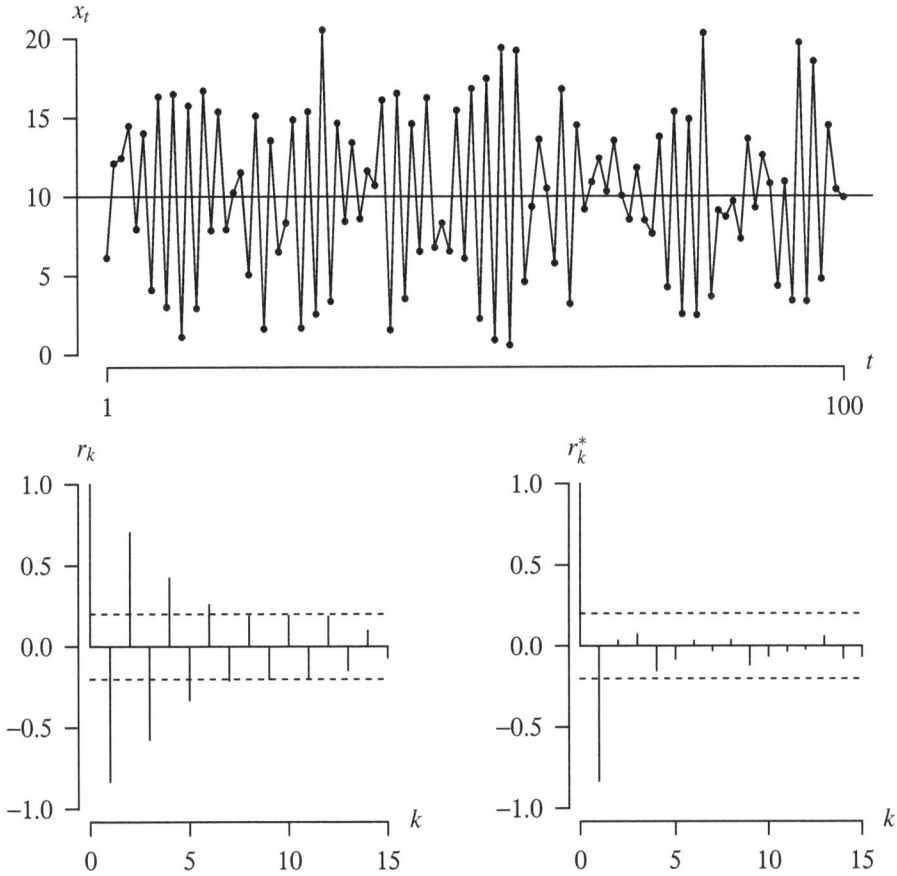

Figure 9.4: Time series plot, r_k, and r_k^* for $n = 100$ simulated values from an AR(1) model.

in order to view the R source code associated with the `arima.sim` function. This function can simulate a realization of any ARMA(p, q) time series model, and we will use it for simulations subsequently.

The remaining topics associated with the AR(1) time series model are statistical in nature: parameter estimation, model assessment, model selection, and forecasting. A sample time series that will be revisited throughout these topics will be introduced next.

Example 9.3 The temperature in degrees Celsius of a beaver (Castor canadensis) in Wisconsin was taken every ten minutes by telemetry on November 3–4, 1990. The resulting time series of $n = 100$ observations is contained in the built-in data frame named `beaver2` in R. The data frame includes columns that contain the temperatures recorded in degrees Celsius (`beaver2$temp`) and an indicator variable (`beaver2$activ`) that reports whether or not the beaver was active outside of its lodge at the associated observation time. The data frame can be viewed by typing `beaver2`. More information about the data set can be viewed by typing `help(beaver2)`. The R statement

```
plot.ts(beaver2$temp)
```

generates the time series plot of the temperature readings given in Figure 9.5. A vertical dashed line has been added between x_{38} and x_{39} to signify when the beaver transitioned from an inactive state to an active state. It is clear that a stationary time series model is *not* appropriate for the entire time series because the population mean appears to increase significantly between the inactive and active periods. So we limit our modeling effort to just the temperatures that were recorded while the beaver was in the active state. The $n = 62$ beaver temperatures during the active period, ordered row-wise, are given in Table 9.1.

37.98	38.02	38.00	38.24	38.10	38.24	38.11	38.02	38.11
38.01	37.91	37.96	38.03	38.17	38.19	38.18	38.15	38.04
37.96	37.84	37.83	37.84	37.74	37.76	37.76	37.64	37.63
38.06	38.19	38.35	38.25	37.86	37.95	37.95	37.76	37.60
37.89	37.86	37.71	37.78	37.82	37.76	37.81	37.84	38.01
38.10	38.15	37.92	37.64	37.70	37.46	37.41	37.46	37.56
37.55	37.75	37.76	37.73	37.77	38.01	38.04	38.07	

Table 9.1: Beaver temperatures at ten-minute intervals in the active state.

The question posed in this example is whether an AR(1) model is appropriate time series model for the 62 temperatures taken during the active period.

The time series of temperatures of the beaver in the active state, the sample autocorrelation function, and the sample partial autocorrelation function can be graphed with the R statements

```
x = beaver2$temp[beaver2$activ == 1]
layout(matrix(c(1, 1, 2, 3), 2, 2, byrow = TRUE))
```

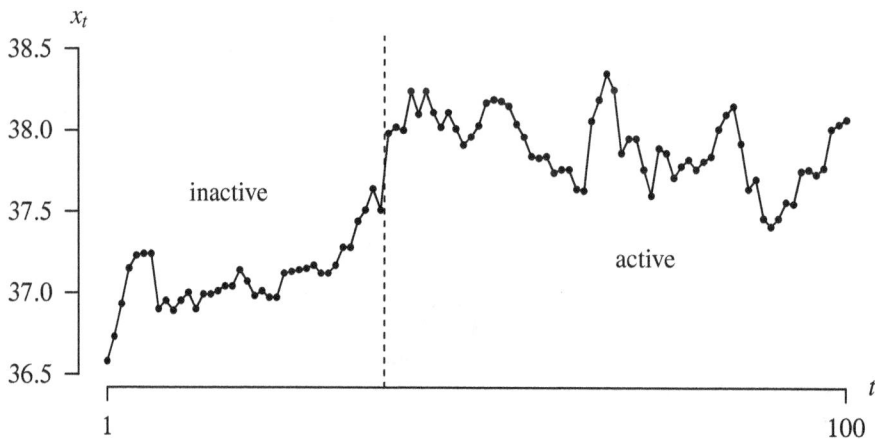

Figure 9.5: Wisconsin beaver temperatures recorded at 10-minute intervals.

```
plot.ts(x)
abline(h = mean(x))
acf(x)
pacf(x)
```

The trio of graphs is displayed in Figure 9.6. A horizontal line has been added to the time series plot at $\bar{x} = 37.9$. A visual assessment of the $n = 62$ observations from the time series indicates that the mean value does not appear to be systematically increasing or decreasing over the time period. From the documentation of the time series, one can see that the active period for the beaver began at 3:50 PM on November 3, 1990 and ended at 2:00 AM on November 4, 1990. The ambient temperature might have an effect on the beaver's temperature, but this will not be pursued further. For now, we will assume that there is no systematic, secular trend in the mean value. The variance of the observations in the time series also seems to be stable over time. Based on this cursory analysis, it appears plausible that the beaver's temperature during the active period could have been drawn from a stationarity time series model. Now we turn to the

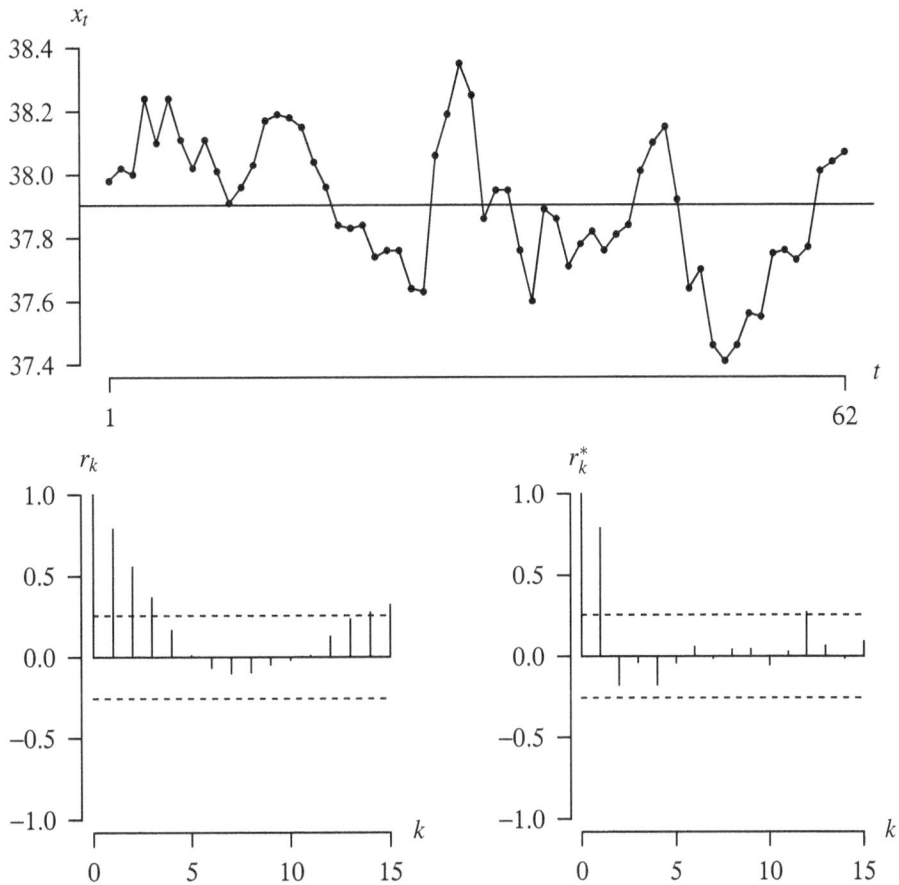

Figure 9.6: Time series plot, r_k, and r_k^* for $n = 62$ temperatures of an active beaver.

interpretation of the sample autocorrelation function and the sample partial autocorrelation function. The sample autocorrelation function has three initial positive statistically significant spikes which decrease in magnitude. The correlogram tails off. This is consistent with an AR(1) model. We discount the barely significant spikes at lags $k = 14$ and $k = 15$, although we could pursue other studies concerning beaver temperatures over time to see if there might be a cyclic variation component present. Furthermore, the partial autocorrelation function has a positive statistically significant spike at lag 1, then cuts off after lag 1. This is also consistent with an AR(1) model. The fact that $r_1 = r_1^* = 0.79$ is positive is consistent with the time series plot of the beaver's temperature, which tends to linger above and below the sample mean value $\bar{x} = 37.9$ for significant periods of time. So far, the evidence points to the AR(1) time series model being a reasonable model for the beaver's temperature during its active period.

The AR(1) model gives us a secondary manner to visually assess whether or not it is an appropriate time series model for the beaver temperatures. Since the shifted AR(1) model is

$$X_t - \mu = \phi(X_{t-1} - \mu) + Z_t,$$

the aforementioned interpretation of this time series model as a simple linear regression model means that a plot of $x_{t-1} - \bar{x}$ versus $x_t - \bar{x}$ (or x_{t-1} versus x_t) should be approximately linear. The plot displayed in Figure 9.7 contains the $n - 1 = 62 - 1 = 61$ pairs (x_{t-1}, x_t) of adjacent points on a set of axes, which is generated by the R statements

```
x = beaver2$temp[beaver2$activ == 1]
n = length(x)
plot(x[2:n], x[1:(n - 1)])
```

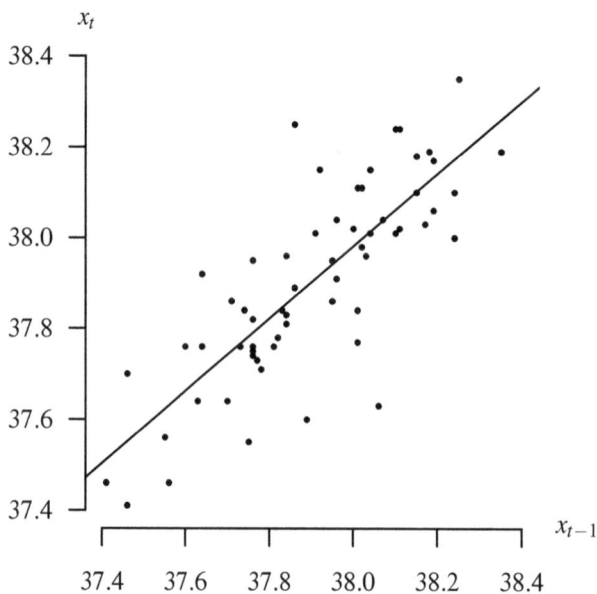

Figure 9.7: Scatterplot of adjacent pairs of temperatures of an active beaver.

```
abline(lm(x[2:n] ~ x[1:(n - 1)]))
```

The paucity of points in the upper-left and lower-right portions of the scatterplot and the approximately linear relationship between the adjacent observations lends further evidence that the AR(1) model might be an appropriate time series model. The `lm` function fits a linear model to the data pairs and the `abline` function plots the associated regression line. The additional R statement

```
summary(lm(x[2:n] ~ x[1:(n - 1)]))
```

indicates that the slope of the regression line differs statistically from 0 with p-value $2 \cdot 10^{-14}$, indicating a strong linear relationship between adjacent observations.

In conclusion, a preliminary graphical analysis of the $n = 62$ temperatures of the beaver in the active state indicates that an AR(1) time series model should be on the short list of potential time series models. The next step is to estimate the parameters in the model.

Parameter Estimation

There are two parameters, ϕ and σ_Z^2, to estimate in the standard AR(1) model

$$X_t = \phi X_{t-1} + Z_t.$$

There are three parameters, μ, ϕ, and σ_Z^2, to estimate in the shifted AR(1) model

$$X_t - \mu = \phi(X_{t-1} - \mu) + Z_t.$$

The three parameter estimation techniques outlined in Section 8.2.1 are applied to the shifted AR(1) time series model next.

Approach 1: Method of moments. In the case of the shifted AR(1) model, we match the population and sample (*a*) first-order moments, (*b*) second-order moments, and (*c*) lag 1 autocorrelation. Placing the population moments on the left-hand side of the equation and the associated sample moments on the right-hand side of the equation results in three equations in three unknowns:

$$E[X_t] = \frac{1}{n}\sum_{t=1}^{n} X_t$$

$$E[X_t^2] = \frac{1}{n}\sum_{t=1}^{n} X_t^2$$

$$\rho(1) = r_1$$

or

$$\mu = \bar{X}$$

$$V[X_t] + E[X_t]^2 = \gamma(0) + \mu^2 = \left(\frac{1}{1-\phi^2}\right)\sigma_Z^2 + \mu^2 = \frac{1}{n}\sum_{t=1}^{n} X_t^2$$

$$\phi = r_1.$$

These equations can be solved in closed form for the three unknown parameters μ, ϕ, and σ_Z^2 yielding the method of moments estimators

$$\hat{\mu} = \bar{X}, \qquad \hat{\phi} = r_1, \qquad \hat{\sigma}_Z^2 = \left(1 - \hat{\phi}^2\right)\left(\frac{1}{n}\sum_{t=1}^{n} X_t^2 - \hat{\mu}^2\right) = \left(1 - r_1^2\right)\left(\frac{1}{n}\sum_{t=1}^{n} X_t^2 - \bar{X}^2\right).$$

This constitutes a proof of the following result.

Theorem 9.4 The method of moments estimators of the parameters in a shifted AR(1) model are

$$\hat{\mu} = \bar{X} \qquad \hat{\phi} = r_1 \qquad \hat{\sigma}_Z^2 = \left(1 - r_1^2\right)\left(\frac{1}{n}\sum_{t=1}^{n} X_t^2 - \bar{X}^2\right).$$

These point estimators are random variables and have been written as a function of the random time series values $X_1, X_2, \ldots, X_n$. For observed time series values $x_1, x_2, \ldots, x_n$, the lowercase versions of the formulas will be used.

Example 9.4 For the time series of $n = 62$ temperature observations of the beaver in the active state, find the method of moments estimators of μ, ϕ, and σ_Z^2 for the AR(1) model.

The R code below calculates and prints the point estimates of the μ, ϕ, and σ_Z^2 parameters.

```
x       = beaver2$temp[beaver2$activ == 1]
muhat   = mean(x)
phihat  = acf(x, plot = FALSE)$acf[2]
sig2hat = (1 - phihat ^ 2) * (mean(x ^ 2) - muhat ^ 2)
print(c(muhat, phihat, sig2hat))
```

The point estimates for the unknown parameters computed by this code are

$$\hat{\mu} = 37.90 \qquad \hat{\phi} = 0.7894 \qquad \hat{\sigma}_Z^2 = 0.01734.$$

These point estimates are reported to four digits because the data values were given to four-digit accuracy. The positive value for $\hat{\phi}$ is consistent with the fact that the beaver's temperature lingers above and below the sample mean in the time series plot in Figure 9.6. The estimated standard deviation of the white noise error terms,

$$\hat{\sigma}_Z = \sqrt{0.01734} \cong 0.1317,$$

reflects the dispersion of the observations in Figure 9.7 about the regression line.

Approach 2: Least squares. Consider the shifted stationary AR(1) model

$$X_t - \mu = \phi(X_{t-1} - \mu) + Z_t.$$

For least squares estimation, we first establish the sum of squares S as a function of the parameters μ and ϕ and use calculus to find the least squares estimators of μ and ϕ. This will result in a slight difference between the usual pattern of using the sample mean $\bar{x}$ to estimate the population mean μ. Once these least squares estimators have been determined, the population variance of the white noise σ_Z^2 will be estimated.

The sum of squared errors is

$$S = \sum_{t=2}^{n} Z_t^2 = \sum_{t=2}^{n} [X_t - \mu - \phi(X_{t-1} - \mu)]^2.$$

The partial derivatives of S with respect to μ and ϕ are

$$\frac{\partial S}{\partial \mu} = \sum_{t=2}^{n} 2\left[X_t - \mu - \phi\left(X_{t-1} - \mu\right)\right]\left(-1 + \phi\right)$$

and

$$\frac{\partial S}{\partial \phi} = \sum_{t=2}^{n} -2\left[X_t - \mu - \phi\left(X_{t-1} - \mu\right)\right]\left(X_{t-1} - \mu\right).$$

Equating the first of the partial derivatives to zero yields

$$\sum_{t=2}^{n}\left[X_t - \mu - \phi\left(X_{t-1} - \mu\right)\right] = 0$$

or

$$\sum_{t=2}^{n} X_t - \phi\sum_{t=2}^{n} X_{t-1} - (n-1)\mu(1-\phi) = 0$$

or

$$\bar{X}_2 - \phi\bar{X}_1 - \mu(1-\phi) = 0$$

or

$$\hat{\mu} = \frac{\bar{X}_2 - \hat{\phi}\bar{X}_1}{1 - \hat{\phi}},$$

where

$$\bar{X}_1 = \frac{1}{n-1}\sum_{t=1}^{n-1} X_t \qquad \text{and} \qquad \bar{X}_2 = \frac{1}{n-1}\sum_{t=2}^{n} X_t.$$

Equating the second of the partial derivatives to zero yields

$$\sum_{t=2}^{n}\left(X_t - \mu\right)\left(X_{t-1} - \mu\right) - \phi\sum_{t=2}^{n}\left(X_{t-1} - \mu\right)^2 = 0$$

or

$$\hat{\phi} = \frac{\sum_{t=2}^{n}\left(X_t - \hat{\mu}\right)\left(X_{t-1} - \hat{\mu}\right)}{\sum_{t=2}^{n}\left(X_{t-1} - \hat{\mu}\right)^2}.$$

So the ordinary least squares estimators for μ and ϕ can be determined by numerically solving the simultaneous equations

$$\hat{\mu} = \frac{\bar{X}_2 - \hat{\phi}\bar{X}_1}{1 - \hat{\phi}} \qquad \text{and} \qquad \hat{\phi} = \frac{\sum_{t=2}^{n}\left(X_t - \hat{\mu}\right)\left(X_{t-1} - \hat{\mu}\right)}{\sum_{t=2}^{n}\left(X_{t-1} - \hat{\mu}\right)^2}$$

for $\hat{\mu}$ and $\hat{\phi}$.

The last parameter to estimate is σ_Z^2. Since

$$\gamma(0) = \left(\frac{1}{1 - \phi^2}\right)\sigma_Z^2$$

for an AR(1) time series model, the population variance of the white noise can be expressed as

$$\sigma_Z^2 = \left(1 - \phi^2\right)\gamma(0).$$

Replacing ϕ by the estimator r_1 because $\rho(1) = \phi$, and replacing $\gamma(0) = V[X_t]$ by the estimator $c_0 = \frac{1}{n} \sum_{t=1}^{n} (X_t - \bar{X})^2$ gives the point estimator

$$\hat{\sigma}_Z^2 = \left(1 - r_1^2\right) c_0,$$

which matches the method of moments estimator from Theorem 9.4. This derivation constitutes a proof of the following result.

Theorem 9.5 The least squares estimators of the parameters in a shifted AR(1) model are the $\hat{\mu}$ and $\hat{\phi}$ values that satisfy

$$\hat{\mu} = \frac{\bar{X}_2 - \hat{\phi} \bar{X}_1}{1 - \hat{\phi}} \qquad \text{and} \qquad \hat{\phi} = \frac{\sum_{t=2}^{n} (X_t - \hat{\mu})(X_{t-1} - \hat{\mu})}{\sum_{t=2}^{n} (X_{t-1} - \hat{\mu})^2},$$

where $\bar{X}_1 = \frac{1}{n-1} \sum_{t=1}^{n-1} X_t$ and $\bar{X}_2 = \frac{1}{n-1} \sum_{t=2}^{n} X_t$. The least squares estimator of σ_Z^2 is

$$\hat{\sigma}_Z^2 = \left(1 - r_1^2\right) c_0.$$

We now apply these techniques to the beaver temperature data set from Example 9.3.

Example 9.5 Find the least squares estimators of μ, ϕ, and σ_Z^2 for the time series of $n = 62$ beaver temperatures from Example 9.3.

The code below contains a function s which calculates the sum of squares, and then uses the R optim function to minimize the sum of squares using the method of moments estimates as initial estimates. The optim function minimizes the objective function by default.

```
x = beaver2$temp[beaver2$activ == 1]
n = length(x)
s = function(parameters) {
  mu  = parameters[1]
  phi = parameters[2]
  sum((x[2:n] - mu - phi * (x[1:(n - 1)] - mu)) ^ 2)
}
optim(c(37.90, 0.7894), s)
r1 = acf(x, plot = FALSE)$acf[2]
sig2hat = (1 - r1 ^ 2) * mean((x - mean(x)) ^ 2)
```

The point estimates for the unknown parameters computed by this code are

$$\hat{\mu} = 37.91 \qquad \hat{\phi} = 0.7972 \qquad \hat{\sigma}_Z^2 = 0.01762.$$

Figure 9.8 shows the sum of squares for fixed $\hat{\mu} = 37.91$ as a function of ϕ. The sum of squares is minimized at $\hat{\phi} = 0.7972$. The least squares parameter estimates are very close to the method of moments parameter estimators. We now consider why the two estimators are so close to one another.

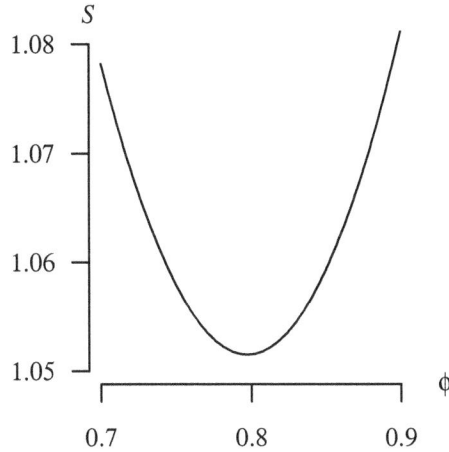

Figure 9.8: Sum of squares as a function of ϕ.

Since

$$\bar{X}_1 = \frac{1}{n-1} \sum_{t=1}^{n-1} X_t \qquad \text{and} \qquad \bar{X}_2 = \frac{1}{n-1} \sum_{t=2}^{n} X_t$$

contain the $n-2$ common values $X_2, X_3, \ldots, X_{n-1}$, one approximation that can be applied to the least squares estimates is to assume that $\bar{X}_1 \cong \bar{X}_2 \cong \bar{X}$ for large values of n, which allows for closed-form approximate least squares estimators:

$$\hat{\mu} = \bar{X} \qquad \text{and} \qquad \hat{\phi} = \frac{\sum_{t=2}^{n} (X_t - \bar{X})(X_{t-1} - \bar{X})}{\sum_{t=2}^{n} (X_{t-1} - \bar{X})^2}.$$

As a secondary additional approximation, the denominator of $\hat{\phi}$ with the first approximation in place,

$$\sum_{t=2}^{n} (X_{t-1} - \bar{X})^2,$$

is approximately equal to

$$\sum_{t=1}^{n} (X_t - \bar{X})^2$$

for large values of n. With this additional assumption, the least squares estimate for ϕ reduces to the approximate least squares estimate

$$\hat{\phi} = \frac{c_1}{c_0} = r_1,$$

which is the method of moments estimator of ϕ because $\rho(1) = \phi$ for an AR(1) model. With both approximations in place, the least squares estimators exactly match the method of moments estimators. This is why the estimates from the two techniques are so close.

Approach 3: Maximum likelihood estimation. The likelihood function is the joint probability density function of the observed values in the time series $x_1, x_2, \ldots, x_n$ in a shifted AR(1) model is

$$L\left(\mu, \phi, \sigma_Z^2\right) = f(x_1, x_2, \ldots, x_n),$$

where the $x_1, x_2, \ldots, x_n$ arguments on L and the μ, ϕ, and σ_Z^2 arguments on f have been dropped for brevity. It is not possible to simply multiply the marginal probability density functions because the values in the AR(1) time series model are correlated. In order to use maximum likelihood estimation, we make the additional assumption that the white noise terms $Z_1, Z_2, \ldots, Z_n$ are in fact Gaussian white noise terms:

$$f_{Z_t}(z_t) = \frac{1}{\sqrt{2\pi\sigma_Z^2}} \, e^{-z_t^2/(2\sigma_Z^2)} \qquad -\infty < z_t < \infty$$

for $t = 1, 2, \ldots, n$, which is the probability density function of a $N\left(0, \sigma_Z^2\right)$ random variable. Ignoring Z_1 temporarily, the joint probability density function of the mutually independent white noise random variables $Z_2, Z_3, \ldots, Z_n$ is

$$f_{Z_2, Z_3, \ldots, Z_n}(z_2, z_3, \ldots, z_n) = \left(2\pi\sigma_Z^2\right)^{-(n-1)/2} e^{-\sum_{t=2}^{n} z_t^2/(2\sigma_Z^2)}$$

for $(z_2, z_3, \ldots, z_n) \in \mathcal{R}^{n-1}$. The shifted AR(1) model

$$X_t - \mu = \phi(X_{t-1} - \mu) + Z_t$$

applies for all values of t, so

$$X_2 - \mu = \phi(X_1 - \mu) + Z_2$$
$$X_3 - \mu = \phi(X_2 - \mu) + Z_3$$
$$\vdots$$
$$X_n - \mu = \phi(X_{n-1} - \mu) + Z_n.$$

Solving these equations for $X_2, X_3, \ldots, X_n$, consider the transformation of the $Z_2, Z_3, \ldots, Z_n$ values

$$X_2 = \mu + \phi(X_1 - \mu) + Z_2$$
$$X_3 = \mu + \phi(X_2 - \mu) + Z_3$$
$$\vdots$$
$$X_n = \mu + \phi(X_{n-1} - \mu) + Z_n$$

conditioned on $X_1 = x_1$, which is a one-to-one transformation from $\mathcal{R}^{n-1}$ to $\mathcal{R}^{n-1}$ with inverse transformation

$$Z_2 = X_2 - \mu - \phi(X_1 - \mu)$$
$$Z_3 = X_3 - \mu - \phi(X_2 - \mu)$$
$$\vdots$$
$$Z_n = X_n - \mu - \phi(X_{n-1} - \mu)$$

and Jacobian

$$J = \begin{vmatrix} 1 & -\phi & 0 & \cdots & 0 & 0 \\ 0 & 1 & -\phi & \cdots & 0 & 0 \\ 0 & 0 & 1 & \cdots & 0 & 0 \\ \vdots & \vdots & \vdots & \ddots & \vdots & \vdots \\ 0 & 0 & 0 & \cdots & 1 & -\phi \\ 0 & 0 & 0 & \cdots & 0 & 1 \end{vmatrix} = 1.$$

By the transformation technique, the joint probability density function of $X_2, X_3, \ldots, X_n$ conditioned on $X_1 = x_1$ is

$$f_{X_2, X_3, \ldots, X_n \mid X_1}(x_2, x_3, \ldots, x_n \mid X_1 = x_1) = \left(2\pi\sigma_Z^2\right)^{-(n-1)/2} e^{-\sum_{t=2}^{n}[x_t - \mu - \phi(x_{t-1} - \mu)]^2 / (2\sigma_Z^2)}$$

for $(x_2, x_3, \ldots, x_n) \in \mathcal{R}^{n-1}$ and $x_1 \in \mathcal{R}$. The final step in the derivation of the likelihood function involves determining the marginal distribution of X_1. Since

$$X_1 \sim N\left(\mu, \frac{\sigma_Z^2}{1 - \phi^2}\right),$$

the probability density function of X_1 is

$$f_{X_1}(x_1) = \sqrt{\frac{1 - \phi^2}{2\pi\sigma_Z^2}} \; e^{-(1 - \phi^2)(x_1 - \mu)^2 / (2\sigma_Z^2)} \qquad -\infty < x_1 < \infty.$$

The joint probability density function of $X_1, X_2, \ldots, X_n$ is the product of the conditional probability density function and the marginal probability density function:

$$f_{X_1, X_2, \ldots, X_n}(x_1, x_2, \ldots, x_n) = f_{X_2, X_3, \ldots, X_n \mid X_1}(x_2, x_3, \ldots, x_n \mid X_1 = x_1) f_{X_1}(x_1)$$

for $(x_1, x_2, \ldots, x_n) \in \mathcal{R}^n$. So the likelihood function is

$$L\left(\mu, \phi, \sigma_Z^2\right) = \left(2\pi\sigma_Z^2\right)^{-n/2} \sqrt{1 - \phi^2} \; e^{-S(\mu, \phi) / (2\sigma_Z^2)},$$

where the *unconditional sum of squares* is

$$S(\mu, \phi) = \left(1 - \phi^2\right)(x_1 - \mu)^2 + \sum_{t=2}^{n}\left[(x_t - \mu) - \phi(x_{t-1} - \mu)\right]^2.$$

The associated log likelihood function is

$$\ln L\left(\mu, \phi, \sigma_Z^2\right) = -\frac{n}{2}\ln\left(2\pi\sigma_Z^2\right) + \frac{1}{2}\ln\left(1 - \phi^2\right) - \frac{S(\mu, \phi)}{2\sigma_Z^2}.$$

The maximum likelihood estimators $\hat{\mu}$, $\hat{\phi}$, and $\hat{\sigma}_Z^2$ satisfy

$$\frac{\partial \ln L\left(\mu, \phi, \sigma_Z^2\right)}{\partial \mu} = \frac{\left(1 - \phi^2\right)(x_1 - \mu) + (1 - \phi)\sum_{t=2}^{n}\left[(x_t - \mu) - \phi(x_{t-1} - \mu)\right]}{\sigma_Z^2} = 0$$

$$\frac{\partial \ln L\left(\mu, \phi, \sigma_Z^2\right)}{\partial \phi} = -\frac{\phi}{1 - \phi^2} + \frac{\phi(x_1 - \mu)^2 + \sum_{t=2}^{n}\left[(x_t - \mu) - \phi(x_{t-1} - \mu)\right](x_{t-1} - \mu)}{\sigma_Z^2} = 0$$

$$\frac{\partial \ln L\left(\mu, \phi, \sigma_Z^2\right)}{\partial \sigma_Z^2} = -\frac{n}{2\sigma_Z^2} + \frac{S(\mu, \phi)}{2\sigma_Z^4} = 0.$$

Although the third equation satisfies

$$\hat{\sigma}_Z^2 = \frac{S(\hat{\mu}, \hat{\phi})}{n},$$

numerical methods are required to solve the equations.

Theorem 9.6 The maximum likelihood estimators of the parameters in a shifted AR(1) model with Gaussian white noise are the $\hat{\mu}$, $\hat{\phi}$, and $\hat{\sigma}_Z^2$ values that satisfy

$$\frac{\partial \ln L(\mu, \phi, \sigma_Z^2)}{\partial \mu} = \frac{(1-\phi^2)(x_1 - \mu) + (1-\phi)\sum_{t=2}^{n}[(x_t - \mu) - \phi(x_{t-1} - \mu)]}{\sigma_Z^2} = 0$$

$$\frac{\partial \ln L(\mu, \phi, \sigma_Z^2)}{\partial \phi} = -\frac{\phi}{1-\phi^2} + \frac{\phi(x_1 - \mu)^2 + \sum_{t=2}^{n}[(x_t - \mu) - \phi(x_{t-1} - \mu)](x_{t-1} - \mu)}{\sigma_Z^2} = 0$$

$$\frac{\partial \ln L(\mu, \phi, \sigma_Z^2)}{\partial \sigma_Z^2} = -\frac{n}{2\sigma_Z^2} + \frac{S(\mu, \phi)}{2\sigma_Z^4} = 0.$$

Maximum likelihood estimation will be illustrated in the next example.

Example 9.6 Find the maximum likelihood estimators of μ, ϕ, and σ_Z^2 for the time series of $n = 62$ beaver temperatures from Example 9.3.

The R code below again uses the `optim` function to maximize the likelihood function. Since the default for `optim` is to minimize, the likelihood function is negated within the L function. The method of moments estimators are used as initial estimates. The `optim` function uses the Nelder–Mead method by default to maximize the likelihood function.

```
x = beaver2$temp[beaver2$activ == 1]
n = length(x)
L = function(parameters) {
  mu   = parameters[1]
  phi  = parameters[2]
  sig2 = parameters[3]
  -(2 * pi * sig2) ^ (- n / 2) * (1 - phi ^ 2) ^ (1 / 2) *
    exp(-((1 - phi ^ 2) * (x[1] - mu) ^ 2 +
    sum(((x[2:n] - mu) - phi * (x[1:(n - 1)] - mu)) ^ 2) / (2 * sig2)))
}
optim(c(37.90, 0.7894, 0.01734), L)
```

The point estimates for the unknown parameters computed by this code are

$$\hat{\mu} = 37.91 \qquad \hat{\phi} = 0.7850 \qquad \hat{\sigma}_Z^2 = 0.01697.$$

Table 9.2 summarizes the point estimators that have been calculated in the previous three examples for the $n = 62$ beaver temperatures. The point estimators associated with the three methods are quite close for this particular time series.

Method	$\hat{\mu}$	$\hat{\phi}$	$\hat{\sigma}_Z^2$
Method of moments	37.90	0.7894	0.01734
Ordinary least squares	37.91	0.7972	0.01762
Maximum likelihood estimation	37.91	0.7850	0.01697

Table 9.2: Point estimators for the AR(1) parameters for the $n = 62$ beaver temperatures.

The R function ar fits autoregressive models. The parameter estimates from the three previous examples could have been calculated with the following four R statements.

```
x = beaver2$temp[beaver2$activ == 1]
ar(x, order.max = 1, aic = FALSE, method = "yule-walker")
ar(x, order.max = 1, aic = FALSE, method = "ols")
ar(x, order.max = 1, aic = FALSE, method = "mle")
```

Table 9.3 contains the point estimates returned by the ar function. The tiny differences between some of the entries in Tables 9.2 and 9.3 might be due to slightly different approximations and/or roundoff in the optimization routines.

Method	$\hat{\mu}$	$\hat{\phi}$	$\hat{\sigma}_Z^2$
Method of moments (Yule–Walker)	37.90	0.7894	0.01792
Ordinary least squares	37.90	0.7972	0.01724
Maximum likelihood estimation	37.90	0.7865	0.01699

Table 9.3: Point estimators for the $n = 62$ beaver temperatures via the ar function.

We have now derived and illustrated the three point estimation techniques, the method of moments, least squares, and maximum likelihood estimation, for the parameters in an AR(1) model from a realization of a time series consisting of n observations. Which of these techniques provides the best point estimators? This is not an easy question to answer because there are a large number of factors, such as the sample size n, the values of the parameters in the model, and the fact that there are three parameters to estimate. There will not necessarily be one universal answer to the question. We do a focused evaluation on the point estimator for ϕ because it typically differs for the three methods of point estimation. The mean square error associated with the point estimate for ϕ is

$$E\left[(\hat{\phi} - \phi)^2\right].$$

The following R code conducts a Monte Carlo simulation experiment which estimates the mean square error of the three point estimators for ϕ for 40,000 replications. We selected the time series model with $\mu = 38$, $\phi = 0.8$, $\sigma_Z = 0.13$, and $n = 62$, which are parameters that are near the estimated parameters in the last three examples involving the time series of beaver temperatures.

```
nrep = 40000
mse.mom = 0
mse.ols = 0
mse.mle = 0
for (i in 1:nrep) {
  x = 38 + arima.sim(model = list(ar = 0.8, sd = 0.13), n = 62, n.start = 10)
  mse.mom = mse.mom + (ar(x, order.max = 1, method = "yw")$ar[1]  - 0.8) ^ 2
  mse.ols = mse.ols + (ar(x, order.max = 1, method = "ols")$ar[1] - 0.8) ^ 2
  mse.mle = mse.mle + (ar(x, order.max = 1, method = "mle")$ar[1] - 0.8) ^ 2
}
print(c(mse.mom, mse.ols, mse.mle) / nrep)
```

After a call to set.seed(4) to establish the random number stream, three runs of this simulation yielded the following estimated mean squared error values:

Method of moments : 0.0135   Least squares : 0.0117   Maximum likelihood : 0.0113.

Furthermore, confidence intervals for the three methods do not overlap. Since small values of the mean square error are preferred, we conclude that the maximum likelihood estimator is the preferred estimator for these parameter settings, followed by the least squares estimator, followed by the method of moments estimator in a distant third place.

The focus on estimation thus far has been on point estimation techniques. We also want to report some indication of the *precision* associated with these point estimators. In the previous example, the sampling distributions of $\hat{\mu}$, $\hat{\phi}$, and $\hat{\sigma}_Z^2$ in the AR(1) model are too complicated to derive analytically. As an illustration of a confidence interval for one of the parameters, we use the asymptotic normality of the maximum likelihood estimator of ϕ in the result:

$$\hat{\phi} \xrightarrow{D} N\left(\phi, \frac{1-\phi^2}{n}\right).$$

This result leads to an asymptotically exact two-sided $100(1-\alpha)\%$ confidence interval for ϕ.

Theorem 9.7 For a stationary AR(1) time series model, an asymptotically exact two-sided $100(1-\alpha)\%$ confidence interval for ϕ is given by

$$\hat{\phi} - z_{\alpha/2}\sqrt{\frac{1-\hat{\phi}^2}{n}} < \phi < \hat{\phi} + z_{\alpha/2}\sqrt{\frac{1-\hat{\phi}^2}{n}},$$

where $\hat{\phi}$ is the maximum likelihood estimator of ϕ and $z_{\alpha/2}$ is the $1-\alpha/2$ fractile of the standard normal distribution.

This asymptotically exact confidence interval will now be illustrated with the time series of active beaver temperatures from the three previous examples.

Example 9.7 Find an approximate 95% confidence interval for ϕ for the AR(1) model associated with the $n = 62$ beaver temperature time series values from Example 9.3.

Recall from Table 9.3 that the maximum likelihood estimator returned by the `ar` function is $\hat{\phi} = 0.7865$. The following R code calculates a 95% confidence interval for ϕ.

```
x     = beaver2$temp[beaver2$activ == 1]
n     = length(x)
mle   = ar(x, order.max = 1, aic = FALSE, method = "mle")$ar
alpha = 0.05
crit  = qnorm(1 - alpha / 2)
lo    = mle - crit * sqrt((1 - mle ^ 2) / n)
hi    = mle + crit * sqrt((1 - mle ^ 2) / n)
print(c(lo, hi))
```

This code returns the approximate 95% confidence interval

$$0.6328 < \phi < 0.9402$$

which does not contain $\phi = 0$, giving further evidence that the AR(1) model is justified. An AR(1) model with $\phi = 0$ reduces to just white noise, and the beaver temperature time series is clearly not comprised of mutually independent observations.

Model Assessment

Now that techniques for point and interval estimates for the parameters in the AR(1) model have been established, we are interested in assessing the adequacy of the AR(1) time series model. This will involve an analysis of the residuals. Recall from Section 8.2.3 that the residuals are defined by

$$[\text{residual}] = [\text{observed value}] - [\text{predicted value}]$$

or

$$\hat{Z}_t = X_t - \hat{X}_t.$$

Since $\hat{X}_t$ is the one-step-ahead forecast from the time origin $t - 1$, this is more clearly written as

$$\hat{Z}_t = X_t - \hat{X}_{t-1}(1).$$

Therefore, for the time series $x_1, x_2, \ldots, x_n$ and the fitted AR(1) model with parameter estimates $\hat{\mu}$ and $\hat{\phi}$, the residual at time t is

$$\hat{Z}_t = x_t - \left[\hat{\mu} + \hat{\phi} \left(x_{t-1} - \hat{\mu} \right) \right]$$

for $t = 2, 3, \ldots, n$ via Example 8.12. The next example shows the steps associated with assessing the adequacy of the AR(1) model for the active beaver temperature time series.

> **Example 9.8** Fit the AR(1) model to the active beaver temperatures from Example 9.3 using the sample mean to estimate μ and the maximum likelihood estimators for ϕ and σ^2. Assess the fitted AR(1) model by the following five methods.
>
> (a) Calculate and plot the residuals, their sample autocorrelation function, and their sample partial autocorrelation function.
>
> (b) Conduct a test of independence on the residuals using the number of sample autocorrelation function values for the first $m = 40$ lags which fall outside of $\pm 1.96/\sqrt{n}$.
>
> (c) Conduct the Box–Pierce and Ljung–Box tests for independence of the residuals.
>
> (d) Conduct the turning point test for independence of the residuals.
>
> (e) Plot a histogram and a QQ plot of the standardized residuals in order to assess the normality of the residuals.
>
> (a) The following R commands calculate the $n - 1 = 61$ residuals and plot them as a time series, along with the associated sample autocorrelation function and sample partial autocorrelation function.
>
> ```
> x       = beaver2$temp[beaver2$activ == 1]
> n       = length(x)
> m       = 40
> muhat   = mean(x)
> phihat  = ar(x, order.max = 1, aic = FALSE, method = "mle")$ar
> zhat    = x[2:n] - (muhat + phihat * (x[1:(n - 1)] - muhat))
> layout(matrix(c(1, 1, 2, 3), 2, 2, byrow = TRUE))
> plot.ts(zhat)
> acf(zhat, lag.max = m)
> pacf(zhat, lag.max = m)
> ```

The results are displayed in Figure 9.9. The residuals do not appear to have any cyclic variation, trend, or serial correlation.

(b) There is just one sample autocorrelation function value that falls outside of the limits $\pm 1.96/\sqrt{n}$ (at lag 15) in the plot in Figure 9.9 of the first 40 sample autocorrelation function values associated with the residuals. Since we expect $40 \cdot 0.05 = 2$ values to fall outside of these limits in the case of a good fit, we fail to reject H_0 in this case. The adequacy of the fit of the AR(1) model is not rejected by this test.

(c) The additional R code below calculates the Box–Pierce and Ljung–Box test statistics and the associated p-values.

```
n          = length(zhat)
r          = acf(zhat, lag.max = m, plot = FALSE)$acf[2:(m + 1)]
boxpierce = n * sum(r ^ 2)
1 - pchisq(boxpierce, m - 2)
ljungbox  = n * (n + 2) * sum(r ^ 2 / seq(n - 1, n - m))
1 - pchisq(ljungbox, m - 2)
```

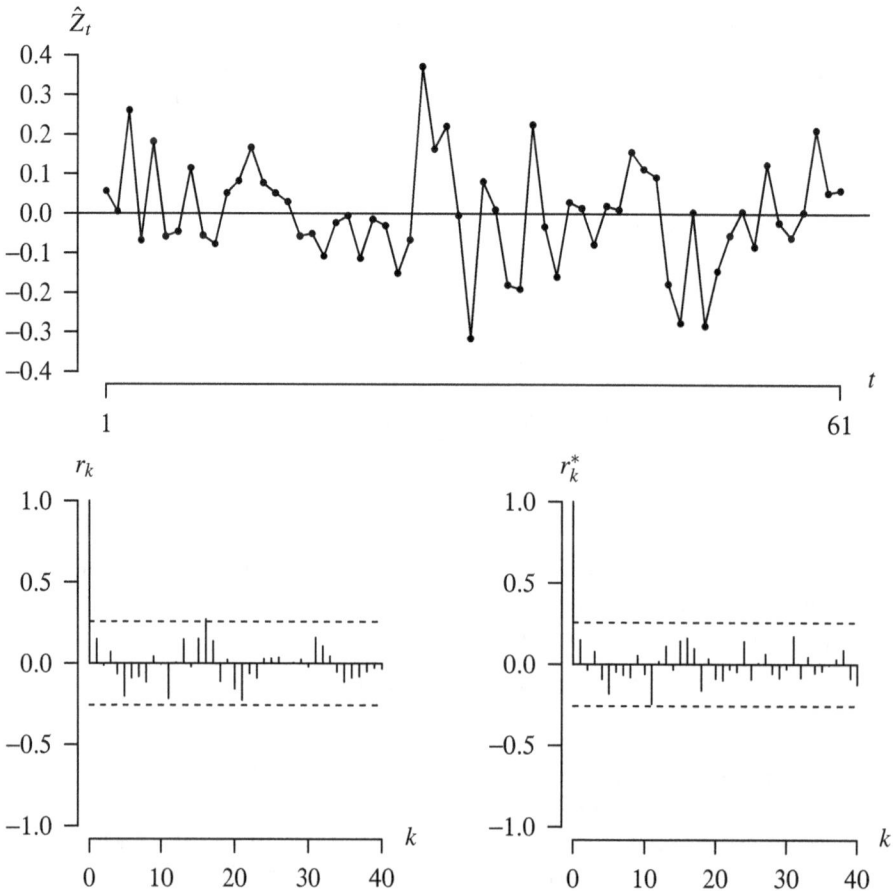

Figure 9.9: Time series plot, r_k, and r_k^* for $n - 1 = 61$ residuals from AR(1) fitted model.

The Box–Pierce test statistic is 27.9 and the associated p-value is $p = 0.89$. The Ljung–Box test statistic is 41.8 and the associated p-value is $p = 0.31$. We fail to reject H_0 in both tests based on the chi-square critical value with $40 - 2 = 38$ degrees of freedom. Some keystrokes can be saved by using the built-in Box.test function in R as shown below.

```
Box.test(zhat, lag = 40, type = "Box-Pierce", fitdf = 2)
Box.test(zhat, lag = 40, type = "Ljung-Box", fitdf = 2)
```

The Box.test function delivers identical test statistics and p-values. The adequacy of the fit of the AR(1) model is not rejected by these tests.

(d) The following additional R statements calculate the test statistic and the p-value for the turning point test applied to the time series consisting of the $n - 1 = 61$ residual values for the AR(1) fit to the beaver temperatures in the active state.

```
n = n - 1
m = (2 / 3) * (n - 2)
v = (16 * n - 29) / 90
T = 0
for (i in 2:(n - 1)) {
  if ((zhat[i - 1] < zhat[i] && zhat[i] > zhat[i + 1]) ||
      (zhat[i - 1] > zhat[i] && zhat[i] < zhat[i + 1])) T = T + 1
}
s = (T - m) / sqrt(v)
2 * (1 - pnorm(abs(s)))
```

The tail probability is doubled because the alternative hypothesis is two-tailed for the turning point test. The test statistic s is -0.83 and the p-value is $p = 0.41$. We again fail to reject the null hypothesis in this case. The adequacy of the fit of the AR(1) model is not rejected by this test.

(e) The residuals are standardized by dividing by their sample standard deviation. The following additional R statements plot a histogram of the standardized residuals using the hist function and a QQ plot to assess normality using the qqnorm function.

```
par(mfrow = c(1, 2))
hist(zhat / sd(zhat))
qqnorm(zhat / sd(zhat))
```

The plots are shown in Figure 9.10. The histogram shows that all standardized residuals fall between -3 and 3 and exhibit a bell-shaped probability distribution. The horizontal axis on the histogram is the standardized residual and the vertical axis is the frequency. The QQ plot is approximately linear, indicating a reasonable approximation to normality based on the $n - 1 = 61$ residuals plotted. The horizontal axis on the QQ plot is the standardized theoretical quantile and the vertical axis is the associated normal data quantile. Although a formal statistical goodness-of-fit test should be conducted, it appears that the assumption of Gaussian white noise is appropriate for the AR(1) time series model based on these two plots.

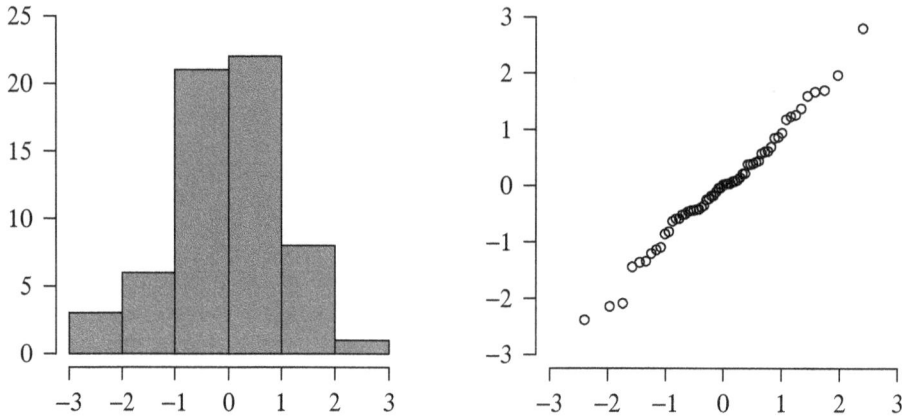

Figure 9.10: Histogram (left) and QQ plot (right) of the fitted AR(1) standardized residuals.

We have seen a number of indicators that the AR(1) time series model is an adequate model for the active beaver temperatures. But how do we know that there is not a better model with more terms lurking below the surface that might provide a better fit? The next subsection considers the process of model selection.

Model Selection

One way of eliminating the possibility of a better time series model is to overfit the tentative AR(1) time series model with ARMA(p, q) models of higher order. We have not yet surveyed the techniques for estimating the parameters in these models with additional terms, so for now we will let the `arima` function in R estimate their parameters and compare them via their AIC (Akaike's Information Criterion) statistics. The AIC statistic was introduced in Section 8.2.4.

> **Example 9.9** For the $n = 62$ temperatures of an active beaver given in Example 9.3, find the ARMA(p, q) model that minimizes the AIC.
>
> The R code below creates a 4×4 matrix a that will be populated with the AIC statistics for the ARMA(p, q) time series models, for $p = 0, 1, 2, 3$ and $q = 0, 1, 2, 3$ using nested `for` loops. The `arima` function is used to fit the models via maximum likelihood estimation, whose AIC values are placed in the matrix a.
>
> ```
> a = matrix(0, 4, 4)
> x = beaver2$temp[beaver2$activ == 1]
> for (p in 0:3)
>   for (q in 0:3)
>     a[p + 1, q + 1] = arima(x, order = c(p, 0, q), method = "ML")$aic
> ```
>
> The results of this code are given in Table 9.4. The smallest AIC value (barely!) is set in boldface type and corresponds to the AR(1) model. This provides further evidence that the AR(1) model is adequate, and a more complex model is probably not warranted. Although the AR(2) and ARMA(1, 1) models have nearly identical AIC values, the additional parameter did not overcome the penalty inflicted by AIC for its inclusion in the time series model.

	$q=0$	$q=1$	$q=2$	$q=3$
$p=0$	-10.9	-51.1	-60.9	-64.4
$p=1$	$\mathbf{-69.7}$	-69.6	-67.6	-67.0
$p=2$	-69.6	-67.6	-65.6	-66.4
$p=3$	-67.7	-65.7	-65.1	-64.7

Table 9.4: AIC statistics for ARMA(p, q) models for the $n = 62$ beaver temperatures.

The $ extractor with the `aic` argument was used to extract the AIC statistics from the results of the call to `arima`. If the `coef` and `sigma2` components are extracted from the list returned by the call to `arima`, our final model is the AR(1) model with maximum likelihood estimates for the parameters given by

$$\hat{\mu} = 37.9, \qquad \hat{\phi} = 0.787, \qquad \hat{\sigma}_Z^2 = 0.017.$$

The final model is therefore

$$X_t - 37.9 = 0.787\,(X_{t-1} - 37.9) + Z_t,$$

where Z_t is a time series of Gaussian white noise values with $\sigma_Z^2 = 0.017$, as established by the histogram and QQ plot in Example 9.8.

In some applications, just describing the time series model for the beaver temperatures in the active state with the fitted AR(1) model is adequate. In other applications, simulating the values in the fitted AR(1) model is the goal. But in many application areas, particularly economics, there is often an interest in forecasting future values of a time series from a realization. In our setting, we might be interested in this particular beaver's future temperature based on the $n = 62$ temperature values collected. The next subsection considers forecasting for the AR(1) model.

Forecasting

We now pivot to the development of a procedure to forecast future values of a time series that is governed by an AR(1) model. To review the notation for forecasting, the observed time series values are $x_1, x_2, \ldots, x_n$. The forecast is being made at time $t = n$. The random future value of the time series that is h time units in the future is denoted by X_{n+h}. The associated forecasted value is denoted by $\hat{X}_{n+h}$, and is the conditional expected value

$$\hat{X}_{n+h} = E\left[X_{n+h} \,|\, X_1 = x_1, X_2 = x_2, \ldots, X_n = x_n\right].$$

We would like to calculate this forecasted value and an associated prediction interval for the AR(1) model. As in Section 8.2.2, we assume that all parameters are known in the derivations that follow.

Recall from Example 8.12 that the forecasted value for one time unit into the future for a shifted AR(1) model is

$$\hat{X}_{n+1} = \mu + \phi\,(x_n - \mu).$$

We would like to generalize this so as to find the forecasted value h time units into the future. In other words, we want to find $\hat{X}_{n+h}$. The shifted AR(1) model is

$$X_t - \mu = \phi\,(X_{t-1} - \mu) + Z_t.$$

Replacing t by $n+h$, which is the time value of interest, gives

$$X_{n+h} - \mu = \phi(X_{n+h-1} - \mu) + Z_{n+h}.$$

Taking the conditional expected value of each side of this equation results in

$$\hat{X}_{n+h} = \mu + \phi(\hat{X}_{n+h-1} - \mu).$$

Iterating on this equation for time values that are sequentially one time unit closer to the present time $t = n$ yields

$$
\begin{aligned}
\hat{X}_{n+h} &= \mu + \phi(\hat{X}_{n+h-1} - \mu) \\
&= \mu + \phi[\mu + \phi(\hat{X}_{n+h-2} - \mu) - \mu] \\
&= \mu + \phi^2(\hat{X}_{n+h-2} - \mu) \\
&\vdots \\
&= \mu + \phi^{h-1}(\hat{X}_{n+1} - \mu) \\
&= \mu + \phi^{h-1}[\mu + \phi(x_n - \mu) - \mu] \\
&= \mu + \phi^h(x_n - \mu).
\end{aligned}
$$

Notice that the forecasted value is a function of x_n, but not a function of $x_1, x_2, \ldots, x_{n-1}$. This is a sensible forecast in the sense that for a long time horizon h into the future and a stationary shifted AR(1) model with $-1 < \phi < 1$,

$$\lim_{h \to \infty} \hat{X}_{n+h} = \mu.$$

If you were asked to forecast your temperature one year from now, you would probably say $98.6°$ Fahrenheit (or whatever your average temperature might be), regardless of whether you are healthy or have a fever right now. Long-term forecasts for stationary time series models always tend to the population mean.

As is typically the case in statistics, we would like to pair our point estimator $\hat{X}_{n+h}$ with an interval estimator, which is a prediction interval in this setting. The prediction interval gives us an indication of the precision of the forecast. In order to derive an exact two-sided $100(1-\alpha)\%$ prediction interval for X_{n+h}, it is helpful to write the shifted AR(1) model as a shifted MA(∞) model. Using successive substitutions, each one time unit prior to the previous substitution,

$$
\begin{aligned}
X_t - \mu &= \phi(X_{t-1} - \mu) + Z_t \\
&= \phi[\phi(X_{t-2} - \mu) + Z_{t-1}] + Z_t \\
&= \phi^2(X_{t-2} - \mu) + Z_t + \phi Z_{t-1} \\
&= \phi^2[(\phi X_{t-3} - \mu) + Z_{t-2}] + Z_t + \phi Z_{t-1} \\
&= \phi^3(X_{t-3} - \mu) + Z_t + \phi Z_{t-1} + \phi^2 Z_{t-2} \\
&\vdots
\end{aligned}
$$

For $-1 < \phi < 1$ corresponding to a stationary shifted AR(1) model, the limiting expression for X_t is

$$X_t = \mu + Z_t + \phi Z_{t-1} + \phi^2 Z_{t-2} + \cdots,$$

which is a shifted MA(∞) model. Replacing t with $n+h$ results in

$$X_{n+h} = \mu + Z_{n+h} + \phi Z_{n+h-1} + \phi^2 Z_{n+h-2} + \cdots,$$

Taking the conditional variance of both sides of this equation yields

$$
\begin{aligned}
V\left[X_{n+h} \mid X_1 = x_1, X_2 = x_2, \ldots, X_n = x_n\right] & \\
&= V\left[\mu + Z_{n+h} + \phi Z_{n+h-1} + \phi^2 Z_{n+h-2} + \cdots \mid X_1 = x_1, X_2 = x_2, \ldots, X_n = x_n\right] \\
&= \sigma_Z^2 + \phi^2 \sigma_Z^2 + \phi^4 \sigma_Z^2 + \cdots + \phi^{2h-2} \sigma_Z^2 \\
&= \left(1 + \phi^2 + \phi^4 + \cdots + \phi^{2h-2}\right) \sigma_Z^2 \\
&= \frac{1 - \phi^{2h}}{1 - \phi^2} \, \sigma_Z^2
\end{aligned}
$$

because the error terms at time n and prior are observed and can therefore be treated as constants. Assuming Gaussian white noise terms, an *exact* two-sided $100(1-\alpha)\%$ prediction interval for X_{n+h} is

$$\hat{X}_{n+h} - z_{\alpha/2} \sqrt{\frac{1 - \phi^{2h}}{1 - \phi^2}} \, \sigma_Z < X_{n+h} < \hat{X}_{n+h} + z_{\alpha/2} \sqrt{\frac{1 - \phi^{2h}}{1 - \phi^2}} \, \sigma_Z.$$

In most practical problems, the parameters in this prediction interval will be estimated from data, which results in the following *approximate* two-sided $100(1-\alpha)\%$ prediction interval.

Theorem 9.8 For a stationary shifted AR(1) time series model, the forecasted value of X_{n+h} is

$$\hat{X}_{n+h} = \hat{\mu} + \hat{\phi}^h \left(x_n - \hat{\mu}\right).$$

and an approximate two-sided $100(1-\alpha)\%$ prediction interval for X_{n+h} is

$$\hat{X}_{n+h} - z_{\alpha/2} \sqrt{\frac{1 - \hat{\phi}^{2h}}{1 - \hat{\phi}^2}} \, \hat{\sigma}_Z < X_{n+h} < \hat{X}_{n+h} + z_{\alpha/2} \sqrt{\frac{1 - \hat{\phi}^{2h}}{1 - \hat{\phi}^2}} \, \hat{\sigma}_Z.$$

Example 9.10 For the beaver temperature time series values $x_1, x_2, \ldots, x_{62}$ from Example 9.3, forecast the next six values in the time series and give approximate 95% prediction intervals for the forecasted values, assuming that the time series arises from a shifted AR(1) model fitted by maximum likelihood estimation.

The R code below calculates the forecasted values and associated approximate 95% prediction interval limits. The forecasts are stored in the vector `pred`, the lower and upper prediction limits are stored in the vectors `lo` and `hi`, respectively.

```
x       = beaver2$temp[beaver2$activ == 1]
model   = ar(x, order.max = 1, aic = FALSE, method = "mle")
phihat  = model$ar
muhat   = model$x.mean
sighat  = sqrt(model$var.pred)
n       = length(x)
hmax    = 6
```

```
alpha   = 0.05
crit    = qnorm(1 - alpha / 2)
pred    = numeric(hmax)
lo      = numeric(hmax)
hi      = numeric(hmax)
for (h in 1:hmax) {
  pred[h] = muhat + phihat ^ h * (x[n] - muhat)
  stderr  = sqrt((1 - phihat ^ (2 * h)) / (1 - phihat ^ 2)) * sighat
  lo[h]   = pred[h] - crit * stderr
  hi[h]   = pred[h] + crit * stderr
}
cat(lo, "\n", pred, "\n", hi, "\n")
```

Some keystrokes can be saved by using the R built-in generic `predict` function to compute the forecasts and the associated standard errors.

```
x       = beaver2$temp[beaver2$activ == 1]
model = ar(x, order.max = 1, aic = FALSE, method = "mle")
predict(model, n.ahead = 6)
```

The two code segments produce identical results, which are summarized in Table 9.5. Notice that the forecasts trend monotonically toward $\bar{x} = 37.90$ and the standard errors increase as the time horizon h increases. The increasing standard error is consistent with having less precision in the forecast as the time horizon h increases.

Time	$t = 63$	$t = 64$	$t = 65$	$t = 66$	$t = 67$	$t = 68$
Forecast	38.04	38.01	37.99	37.97	37.96	37.95
Standard error	0.130	0.166	0.184	0.195	0.201	0.205
Lower prediction bound	37.78	37.69	37.63	37.59	37.57	37.55
Upper prediction bound	38.29	38.34	38.35	38.36	38.36	38.35

Table 9.5: Forecasts and 95% prediction intervals for the beaver temperatures.

Figure 9.11 shows (a) the original time series $x_1, x_2, \ldots, x_{62}$ as points (•) connected by lines, (b) the first 12 forecasted temperatures $\hat{X}_{63}, \hat{X}_{64}, \ldots, \hat{X}_{74}$ as open circles (○), and (c) the 95% prediction intervals as a shaded region. There are three key observations concerning this figure.

- Even though the last five observations in the time series $x_{58}, x_{59}, \ldots, x_{62}$ show an increasing trend, the forecasts, which are a function only of $x_n = x_{62} = 38.07$, monotonically approach $\hat{\mu} = \bar{x} = 37.90$.

- The widths of the prediction intervals increase as the time horizon h increases. These widths do not increase indefinitely, but rather approach a limit as $h \to \infty$.

- The random sampling variability which is evident in the observed time series values $x_1, x_2, \ldots, x_{62}$ is not apparent in the forecasted values $\hat{X}_{63}, \hat{X}_{64}, \ldots, \hat{X}_{74}$. Observed time series values tend to exhibit the typical random sampling variability; forecasted values for a shifted AR(1) model with $0 < \phi < 1$ tend to be smooth.

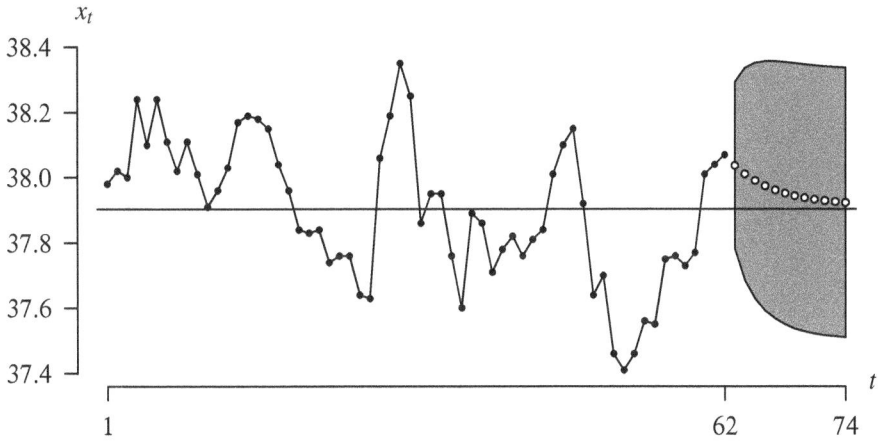

Figure 9.11: Wisconsin beaver forecasted temperatures and 95% prediction intervals.

This subsection has introduced the AR(1) time series model. The key results for an AR(1) model are listed below.

- The standard AR(1) model can be written algebraically and with the backshift operator B as

$$X_t = \phi X_{t-1} + Z_t \qquad \text{and} \qquad (1 - \phi B)X_t = Z_t,$$

where $Z_t \sim WN\left(0, \sigma_Z^2\right)$ and $\sigma_Z^2 > 0$.

- The shifted AR(1) model can be written algebraically and with the backshift operator B as

$$X_t - \mu = \phi\left(X_{t-1} - \mu\right) + Z_t \qquad \text{and} \qquad (1 - \phi B)\left(X_t - \mu\right) = Z_t.$$

- The AR(1) model is always invertible; the AR(1) model is stationary for $-1 < \phi < 1$.

- The stationary shifted AR(1) model can be written as an MA(∞) model for $-1 < \phi < 1$ as

$$X_t = \mu + Z_t + \phi Z_{t-1} + \phi^2 Z_{t-2} + \cdots .$$

- The AR(1) population autocorrelation function is $\rho(k) = \phi^k$ for $-1 < \phi < 1$ and $k = 1, 2, \ldots$.

- The AR(1) population partial autocorrelation function at lag one is $\rho^*(1) = \phi$ for $-1 < \phi < 1$ and $\rho^*(k) = 0$ for $k = 2, 3, \ldots$.

- The three parameters in the shifted AR(1) model, μ, ϕ, and σ_Z^2, can be estimated from a realization of a time series $x_1, x_2, \ldots, x_n$ by the method of moments, least squares, and maximum likelihood. The point estimators for μ, ϕ, and σ_Z^2 are denoted by $\hat{\mu}$, $\hat{\phi}$, and $\hat{\sigma}_Z^2$, and are typically paired with asymptotically exact two-sided $100(1 - \alpha)\%$ confidence intervals.

- The forecast value $\hat{X}_{n+h} = \hat{\mu} + \hat{\phi}^h\left(x_n - \hat{\mu}\right)$ for an AR(1) model approaches $\hat{\mu} = \bar{x}$ as $h \to \infty$. The associated prediction intervals have widths that increase as h increases and approach a limit as the time horizon $h \to \infty$.

If the time series of interest is the daily high temperatures in July in Tuscaloosa, then an AR(1) model would be appropriate if tomorrow's daily high temperature (X_t) can be modeled as a linear function of

- today's high temperature (X_{t-1}), and

- a random shock (Z_t).

But what if weather had more of a memory than just one day? What if tomorrow's daily high temperature (X_t) is better modeled as a linear function of

- today's high temperature (X_{t-1}),

- yesterday's high temperature (X_{t-2}), and

- a random shock (Z_t).

This is an example of the thinking that lies behind the AR(2) model, which is introduced in the next section.

9.1.2 The AR(2) Model

The second-order autoregressive model, denoted by AR(2), can be used for modeling a stationary time series in instances in which the current value of the time series is a linear combination of the two previous values plus a random shock. The mathematics associated with the AR(2) model is somewhat more difficult than that associated with the AR(1) model.

Definition 9.2 A *second-order autoregressive time series model*, denoted by AR(2), for the time series $\{X_t\}$ is defined by

$$X_t = \phi_1 X_{t-1} + \phi_2 X_{t-2} + Z_t,$$

where ϕ_1 and ϕ_2 are real-valued parameters and $\{Z_t\}$ is a time series of white noise:

$$Z_t \sim WN\left(0, \sigma_Z^2\right).$$

There are three parameters that define an AR(2) model: the real-valued coefficients ϕ_1 and ϕ_2, and the population variance of the white noise σ_Z^2. The AR(2) model can be written more compactly in terms of the backshift operator B as

$$\phi(B)X_t = Z_t,$$

where $\phi(B)$ is the second-order polynomial

$$\phi(B) = 1 - \phi_1 B - \phi_2 B^2.$$

The AR(2) model has the form of a multiple linear regression model with two independent variables and no intercept term. The current value X_t is modeled as a linear combination of the two previous values of the time series, X_{t-1} and X_{t-2}, plus a white noise term. The parameters ϕ_1 and ϕ_2 control the inclination of the regression plane in three-dimensional space. The parameter σ_Z^2 reflects the magnitude of the dispersion of the time series values from the regression plane.

To illustrate the thinking behind the AR(2) model in a specific context, let X_t represent the annual return of a particular stock market index in year t. The AR(2) model indicates that the annual return in year t equals ϕ_1 multiplied by the previous year's annual return ($\phi_1 X_{t-1}$), plus ϕ_2 multiplied by the annual return two years prior ($\phi_2 X_{t-2}$), plus the year t random white noise term Z_t.

Stationarity

Theorem 8.3 indicates that all AR(2) models are invertible, but are stationary when the roots of

$$\phi(B) = 1 - \phi_1 B - \phi_2 B^2$$

lie outside of the unit circle in the complex plane. Let B_1 and B_2 denote these two roots. Using the quadratic equation, the two roots are

$$B_1 = \frac{\phi_1 - \sqrt{\phi_1^2 + 4\phi_2}}{-2\phi_2} \qquad \text{and} \qquad B_2 = \frac{\phi_1 + \sqrt{\phi_1^2 + 4\phi_2}}{-2\phi_2}.$$

Since $\phi(B_1) = \phi(B_2) = 0$, the quadratic function $\phi(B)$ can also be written in factored form as

$$\phi(B) = \left(1 - B_1^{-1}B\right)\left(1 - B_2^{-1}B\right).$$

Equating the two versions of $\phi(B)$ above and matching coefficients results in

$$\phi_1 = B_1^{-1} + B_2^{-1} \qquad \text{and} \qquad \phi_2 = -(B_1 B_2)^{-1}.$$

These two equations define the mapping from the complex plane, which contains the roots B_1 and B_2, to the plane that contains the AR(2) parameters ϕ_1 and ϕ_2. To find the stationary region, we must find the mapping of the part of the complex plane outside of the unit circle to the (ϕ_1, ϕ_2) plane. The mapping yields a triangular-shaped stationary region, as specified in the following result.

Theorem 9.9 The AR(2) time series model is stationary when ϕ_1 and ϕ_2 satisfy

$$\phi_1 + \phi_2 < 1, \qquad \phi_2 - \phi_1 < 1, \qquad \text{and} \qquad \phi_2 > -1.$$

Proof The three cases considered below are based on whether the discriminant in the roots of $\phi(B) = 0$ is zero (two identical real roots), positive (two distinct real roots), or negative (two complex roots).

Case 1: Two identical real roots ($\phi_1^2 + 4\phi_2 = 0$). This is a concave-down parabola through the origin in the (ϕ_1, ϕ_2) plane. The single real root is

$$B_1 = B_2 = -\frac{\phi_1}{2\phi_2} = \frac{\phi_1}{\phi_1^2/2} = \frac{2}{\phi_1}.$$

Since this single real root must lie outside of the unit circle for a stationary model,

$$\left|\frac{2}{\phi_1}\right| > 1 \qquad \Rightarrow \qquad -2 < \phi_1 < 2.$$

This portion of the parabola is the leftmost graph in Figure 9.12.

Case 2: Two distinct real roots ($\phi_1^2 + 4\phi_2 > 0$). This is the region *above* the parabola $\phi_1^2 + 4\phi_2 = 0$ in the (ϕ_1, ϕ_2) plane. Since B_1 and B_2 are real-valued, the conditions for stationarity $|B_1| > 1$ and $|B_2| > 1$ that correspond to having both B_1 and B_2 falling outside of the unit circle are equivalent to

$$-1 < \frac{1}{B_1} < 1 \qquad \text{and} \qquad -1 < \frac{1}{B_2} < 1.$$

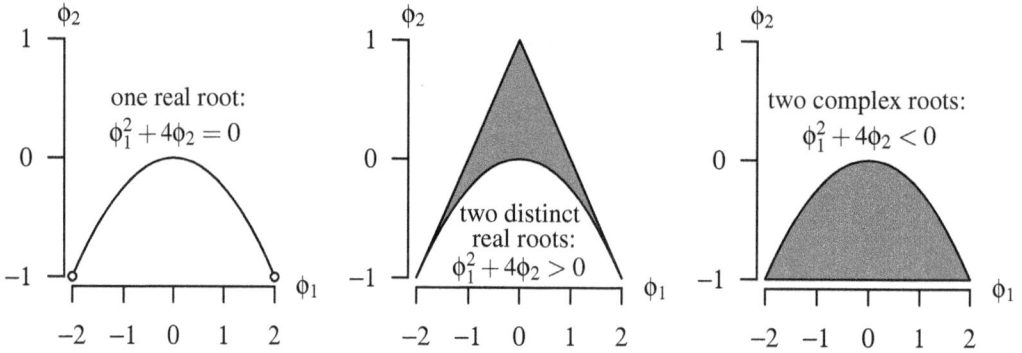

Figure 9.12: Partitioning the stationary region for the AR(2) time series model.

The reciprocals of the roots of $\phi(B) = 0$ are

$$\frac{1}{B_1} = \left[\frac{-2\phi_2}{\phi_1 - \sqrt{\phi_1^2 + 4\phi_2}} \right] \cdot \left[\frac{\phi_1 + \sqrt{\phi_1^2 + 4\phi_2}}{\phi_1 + \sqrt{\phi_1^2 + 4\phi_2}} \right] = \frac{\phi_1 + \sqrt{\phi_1^2 + 4\phi_2}}{2}$$

and

$$\frac{1}{B_2} = \left[\frac{-2\phi_2}{\phi_1 + \sqrt{\phi_1^2 + 4\phi_2}} \right] \cdot \left[\frac{\phi_1 - \sqrt{\phi_1^2 + 4\phi_2}}{\phi_1 - \sqrt{\phi_1^2 + 4\phi_2}} \right] = \frac{\phi_1 - \sqrt{\phi_1^2 + 4\phi_2}}{2}.$$

Since B_1 and B_2 are real and distinct, stationarity is achieved when

$$-1 < \frac{1}{B_2} < \frac{1}{B_1} < 1$$

or

$$-1 < \frac{\phi_1 - \sqrt{\phi_1^2 + 4\phi_2}}{2} < \frac{\phi_1 + \sqrt{\phi_1^2 + 4\phi_2}}{2} < 1.$$

Writing the leftmost inequality as $\sqrt{\phi_1^2 + 4\phi_2} < \phi_1 + 2$, squaring both sides of this inequality, and simplifying gives

$$\phi_2 - \phi_1 < 1.$$

Applying a similar approach to the rightmost inequality gives

$$\phi_1 + \phi_2 < 1.$$

So for the AR(2) model in the case of $\phi(B)$ having two distinct real roots, the model is stationary when the three inequalities

$$\phi_1^2 + 4\phi_2 > 0, \qquad \phi_2 - \phi_1 < 1, \qquad \text{and} \qquad \phi_1 + \phi_2 < 1$$

are satisfied. This region is shaded in the middle graph in Figure 9.12.

Case 3: Two complex conjugate roots ($\phi_1^2 + 4\phi_2 < 0$). This is the region *below* the parabola $\phi_1^2 + 4\phi_2 = 0$ in the (ϕ_1, ϕ_2) plane. For the model to be stationary, B_1 and B_2 must lie outside of the unit circle in the complex plane. Since the roots are complex conjugates, $|B_1| = |B_2|$, which is calculated as

$$|B_1| = \sqrt{\left(\frac{\phi_1}{-2\phi_2}\right)^2 + \left(\frac{\sqrt{-\phi_1^2 - 4\phi_2}}{-2\phi_2}\right)^2} = \sqrt{\frac{-4\phi_2}{4\phi_2^2}} = \sqrt{-\frac{1}{\phi_2}},$$

which is greater than 1 for $-1 < \phi_2 < 0$. (The imaginary part of the discriminant was negated to avoid taking the square root of a negative number because the discriminant is negative in Case 3.) The region associated with the inequalities

$$\phi_1^2 + 4\phi_2 < 0 \qquad \text{and} \qquad \phi_2 > -1$$

is the shaded region in the rightmost graph in Figure 9.12.

The union of these three regions is the interior of the triangular region described by the three inequalities in Theorem 9.9, which proves the result. $\qquad \square$

Population Autocorrelation Function

Now that the stationary region for an AR(2) time series model has been established, we turn to the derivation of the population autocorrelation function. Assuming that the parameters ϕ_1 and ϕ_2 fall in the stationary region, the AR(2) model

$$X_t = \phi_1 X_{t-1} + \phi_2 X_{t-2} + Z_t$$

can be multiplied by X_{t-k} to give

$$X_t X_{t-k} = \phi_1 X_{t-1} X_{t-k} + \phi_2 X_{t-2} X_{t-k} + Z_t X_{t-k}.$$

Taking the expected value of both sides of this equation results in the recursive equation

$$\gamma(k) = \phi_1 \gamma(k-1) + \phi_2 \gamma(k-2)$$

for $k = 1, 2, \ldots$ because Z_t has expected value zero and is independent of X_{t-k}. Dividing both sides of this equation by $\gamma(0) = V[X_t]$ gives the recursive equation

$$\rho(k) = \phi_1 \rho(k-1) + \phi_2 \rho(k-2)$$

for $k = 1, 2, \ldots$. These linear equations, whether written in terms of $\gamma(k)$ or $\rho(k)$, are known in time series analysis as the *Yule–Walker equations* after British statisticians George Udny Yule and Sir Gilbert Walker. Once the first two values of $\gamma(k)$ or $\rho(k)$ are known, these recursive equations can be used to calculate subsequent values. The next two paragraphs focus on determining the first two values of $\gamma(k)$ and $\rho(k)$, respectively.

For a stationary AR(2) time series model, we derive expressions for $\gamma(0)$ and $\gamma(1)$. The AR(2) model is

$$X_t = \phi_1 X_{t-1} + \phi_2 X_{t-2} + Z_t.$$

Squaring both sides of this equation and taking the expected value of both sides gives

$$\gamma(0) = \phi_1^2 \gamma(0) + \phi_2^2 \gamma(0) + \sigma_Z^2 + 2\phi_1 \phi_2 \gamma(1).$$

Using the symmetry of the population autocovariance function, the Yule–Walker equation with $k = 1$ is

$$\gamma(1) = \phi_1\gamma(0) + \phi_2\gamma(1) \qquad \Rightarrow \qquad \gamma(1) = \frac{\phi_1\gamma(0)}{1 - \phi_2}.$$

Replacing this expression for $\gamma(1)$ in the previous equation gives

$$\gamma(0) = \phi_1^2\gamma(0) + \phi_2^2\gamma(0) + \sigma_Z^2 + 2\phi_1\phi_2\frac{\phi_1\gamma(0)}{1 - \phi_2}.$$

Moving all terms involving $\gamma(0)$ to the left-hand side of this equation gives

$$\gamma(0)\left[1 - \phi_1^2 - \phi_2^2 - 2\phi_1^2\phi_2\frac{1}{1 - \phi_2}\right] = \sigma_Z^2.$$

Solving this equation for $\gamma(0)$,

$$\begin{aligned}
\gamma(0) &= \frac{(1 - \phi_2)\sigma_Z^2}{(1 - \phi_2) - (1 - \phi_2)\phi_1^2 - (1 - \phi_2)\phi_2^2 - 2\phi_1^2\phi_2} \\
&= \frac{(1 - \phi_2)\sigma_Z^2}{1 - \phi_2 - \phi_1^2 + \phi_1^2\phi_2 - \phi_2^2 + \phi_2^3 - 2\phi_1^2\phi_2} \\
&= \frac{(1 - \phi_2)\sigma_Z^2}{(1 + \phi_2)(1 + \phi_1 - \phi_2)(1 - \phi_1 - \phi_2)}.
\end{aligned}$$

An expression for $\gamma(1)$ is

$$\gamma(1) = \frac{\phi_1\gamma(0)}{1 - \phi_2} = \frac{\phi_1\sigma_Z^2}{(1 + \phi_2)(1 + \phi_1 - \phi_2)(1 - \phi_1 - \phi_2)}.$$

These two values can be used as arguments in the Yule–Walker equations to obtain subsequent values for $\gamma(k)$.

We now turn to the problem of finding $\rho(1)$ and $\rho(2)$. The first two Yule–Walker equations in terms of $\rho(k)$ are

$$\begin{aligned}
\rho(1) &= \phi_1\rho(0) + \phi_2\rho(-1) \\
\rho(2) &= \phi_1\rho(1) + \phi_2\rho(0).
\end{aligned}$$

Since $\rho(0) = 1$ and $\rho(-k) = \rho(k)$ via Theorem 7.1, these equations reduce to

$$\begin{aligned}
\rho(1) &= \phi_1 + \phi_2\rho(1) \\
\rho(2) &= \phi_1\rho(1) + \phi_2,
\end{aligned}$$

which are easily solved for $\rho(1)$ and $\rho(2)$:

$$\rho(1) = \frac{\phi_1}{1 - \phi_2} \qquad \text{and} \qquad \rho(2) = \frac{\phi_1^2}{1 - \phi_2} + \phi_2.$$

A general formula for $\rho(k)$ exists, but it can involve complex numbers and is unwieldy. An exercise concerning its calculation is given at the end of the chapter. These results are summarized in the following theorem.

Theorem 9.10 The population autocovariance function for a stationary AR(2) time series model is calculated by

$$\gamma(k) = \phi_1 \gamma(k-1) + \phi_2 \gamma(k-2)$$

for $k = 1, 2, \ldots$, where

$$\gamma(0) = \frac{(1-\phi_2)\sigma_Z^2}{(1+\phi_2)(1+\phi_1-\phi_2)(1-\phi_1-\phi_2)} \quad \text{and} \quad \gamma(1) = \frac{\phi_1\sigma_Z^2}{(1+\phi_2)(1+\phi_1-\phi_2)(1-\phi_1-\phi_2)}.$$

The population autocorrelation function for a stationary AR(2) time series model is calculated by

$$\rho(k) = \phi_1 \rho(k-1) + \phi_2 \rho(k-2)$$

for $k = 1, 2, \ldots$, where

$$\rho(1) = \frac{\phi_1}{1-\phi_2} \quad \text{and} \quad \rho(2) = \frac{\phi_1^2}{1-\phi_2} + \phi_2.$$

We now focus in on the values of $\rho(1)$ and $\rho(2)$. We can solve for ϕ_1 and ϕ_2 in terms of $\rho(1)$ and $\rho(2)$ as

$$\phi_1 = \frac{\rho(1)\big(1-\rho(2)\big)}{1-\rho(1)^2} \quad \text{and} \quad \phi_2 = \frac{\rho(2)-\rho(1)^2}{1-\rho(1)^2}.$$

These equations can be helpful in the three settings described below.

1. These equations bear some practical use in that the first two *sample* autocorrelation function values, r_1 and r_2, can be calculated from a time series and used as approximations for $\rho(1)$ and $\rho(2)$, yielding estimates for ϕ_1 and ϕ_2. These can in turn be used as initial estimates for finding point estimates for ϕ_1 and ϕ_2 by, for example, least squares or maximum likelihood estimation, should numerical methods be required.

2. Level surfaces (that is, contours) in the triangular-shaped stationary region from Theorem 9.9 can be determined by fixing values of $\rho(1)$ and $\rho(2)$. As an illustration, consider $\rho(1) = 0$. In this case, $\phi_1 = 0$ and $\phi_2 = \rho(2)$, which is a line segment in the stationary region. Continuing in this fashion for several fixed values of $\rho(1)$ [with varying values of $\rho(2)$] and then for several fixed values of $\rho(2)$ [with varying values of $\rho(1)$] results in the graph of the stationary region with the level surfaces included shown in Figure 9.13. The level surfaces associated with fixed values of $\rho(1)$ are lines; the level surfaces associated with fixed values of $\rho(2)$ are curves.

3. Since $\rho(1)$ and $\rho(2)$ are population correlations, the obvious constraints on their values for an AR(2) time series model are $-1 < \rho(1) < 1$ and $-1 < \rho(2) < 1$. Additionally, since $\phi_2 > -1$ in order to fall into the triangular-shaped stationary region defined in Theorem 9.9 for the AR(2) time series model,

$$\phi_2 > -1 \quad \Rightarrow \quad \frac{\rho(2)-\rho(1)^2}{1-\rho(1)^2} > -1 \quad \Rightarrow \quad \rho(2) > 2\rho(1)^2 - 1.$$

The boundary of this third constraint is a parabola in the $\big(\rho(1), \rho(2)\big)$ plane. The shaded region in Figure 9.14 shows the $\rho(1)$ and $\rho(2)$ values that are associated with stationary AR(2) time series models. Unlike the AR(1) population autocorrelation function, it is possible to

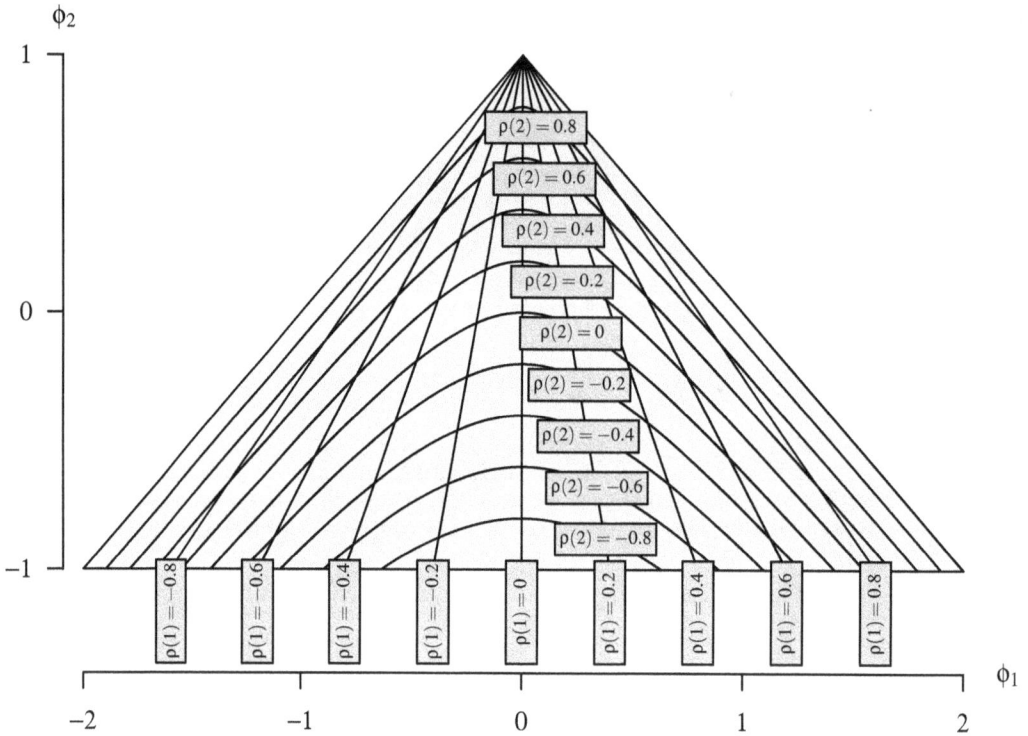

Figure 9.13: Stationary region for an AR(2) time series model with level surfaces.

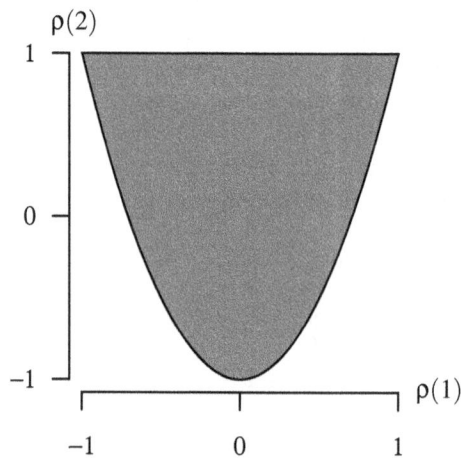

Figure 9.14: Values of $\rho(1)$ and $\rho(2)$ associated with a stationary AR(2) model.

achieve a stationary model with $|\rho(2)| > |\rho(1)|$. The AR(2) population autocorrelation function values are not necessarily monotonically decreasing in magnitude as they were in the AR(1) model.

Population Partial Autocorrelation Function

We now determine the population partial autocorrelation function for an AR(2) model. Using Definition 7.4, the population lag 0 partial autocorrelation is $\rho^*(0) = 1$. The population lag 1 partial autocorrelation is $\rho^*(1) = \rho(1) = \phi_1/(1 - \phi_2)$. After evaluating the determinants and simplifying, the population lag 2 partial autocorrelation is

$$\rho^*(2) = \frac{\begin{vmatrix} 1 & \rho(1) \\ \rho(1) & \rho(2) \end{vmatrix}}{\begin{vmatrix} 1 & \rho(1) \\ \rho(1) & 1 \end{vmatrix}} = \frac{\begin{vmatrix} 1 & \phi_1/(1-\phi_2) \\ \phi_1/(1-\phi_2) & (\phi_1^2 - \phi_2^2 + \phi_2)/(1-\phi_2) \end{vmatrix}}{\begin{vmatrix} 1 & \phi_1/(1-\phi_2) \\ \phi_1/(1-\phi_2) & 1 \end{vmatrix}} = \phi_2.$$

Appealing to the Yule–Walker equations from Theorem 9.10 to define the third column of the determinant of the numerator, the population lag 3 partial autocorrelation is

$$\rho^*(3) = \frac{\begin{vmatrix} 1 & \rho(1) & \rho(1) \\ \rho(1) & 1 & \rho(2) \\ \rho(2) & \rho(1) & \rho(3) \end{vmatrix}}{\begin{vmatrix} 1 & \rho(1) & \rho(2) \\ \rho(1) & 1 & \rho(1) \\ \rho(2) & \rho(1) & 1 \end{vmatrix}} = \frac{\begin{vmatrix} 1 & \rho(1) & \phi_1 + \phi_2\rho(1) \\ \rho(1) & 1 & \phi_1\rho(1) + \phi_2 \\ \rho(2) & \rho(1) & \phi_1\rho(2) + \phi_2\rho(1) \end{vmatrix}}{\begin{vmatrix} 1 & \rho(1) & \rho(2) \\ \rho(1) & 1 & \rho(1) \\ \rho(2) & \rho(1) & 1 \end{vmatrix}} = 0.$$

The determinant in the numerator is zero because the third column is a linear combination of the first two columns. This pattern continues for the higher lags. When computing $\rho^*(k)$ for $k = 3, 4, \ldots$, the first, second, and last columns of the matrix in the numerator are

$$\begin{bmatrix} 1 \\ \rho(1) \\ \rho(2) \\ \vdots \\ \rho(k-1) \end{bmatrix}, \begin{bmatrix} \rho(1) \\ 1 \\ \rho(1) \\ \vdots \\ \rho(k-2) \end{bmatrix}, \text{ and } \begin{bmatrix} \phi_1 + \phi_2\rho(1) \\ \phi_1\rho(1) + \phi_2 \\ \phi_1\rho(2) + \phi_2\rho(1) \\ \vdots \\ \phi_1\rho(k-1) + \phi_2\rho(k-2) \end{bmatrix}.$$

The last column of the matrix in the numerator is a linear combination of the first two columns. The matrix in the numerator of the calculation of $\rho^*(k)$ is singular, which means that its determinant is zero. This constitutes a proof of the following result.

Theorem 9.11 The population partial autocorrelation function for a stationary AR(2) time series model is

$$\rho^*(k) = \begin{cases} 1 & k = 0 \\ \phi_1/(1-\phi_2) & k = 1 \\ \phi_2 & k = 2 \\ 0 & k = 3, 4, \ldots . \end{cases}$$

The population partial autocorrelation function for the AR(2) model cuts off after lag 2. A graph of the sample partial autocorrelation function (that is, a graph of r_k^* for the first few values of k), should also cut off after lag 2 if the AR(2) model is appropriate. This sample partial auto-correlation function shape is easier to recognize than the associated sample autocorrelation function shape because cutting off is easier to recognize than tailing off in the presence of random sampling variability.

A careful inspection of Theorem 9.11 reveals that the signs of ϕ_1 and ϕ_2 match the signs of $\rho^*(1)$ and $\rho^*(2)$, respectively:

$$\text{sgn}(\phi_1) = \text{sgn}\big(\rho^*(1)\big) \qquad \text{and} \qquad \text{sgn}(\phi_2) = \text{sgn}\big(\rho^*(2)\big)$$

for ϕ_1 and ϕ_2 falling in the triangular-shaped stationary region. Figure 9.15 shows the stationary region from Theorem 9.9, along with plots of the representative population autocorrelation function and population partial autocorrelation function. There are four points, one in each quadrant, that are plotted. The population autocorrelation function and the population partial autocorrelation function associated with those four points are plotted in each of the quadrants. As expected, the signs of the values of ϕ_1 and $\rho^*(1)$ match and the signs of the values of ϕ_2 and $\rho^*(2)$ match. The *quadrant* in the stationary region determines the signs of $\rho^*(1)$ and $\rho^*(2)$, as illustrated by the four representative population partial autocorrelation functions graphed in Figure 9.15. As you can see by inspecting the shapes of $\rho(k)$ and $\rho^*(k)$ from Figure 9.15, the addition of the parameter ϕ_2 in the transition from the AR(1) model to the AR(2) model results in significant additional modeling capability. The following observations can be gleaned from Figure 9.15.

- As expected, all population partial autocorrelation functions cut off after lag two.

- When $\phi(B)$ has real roots, the population autocorrelation function consists of mixtures of damped exponentials.

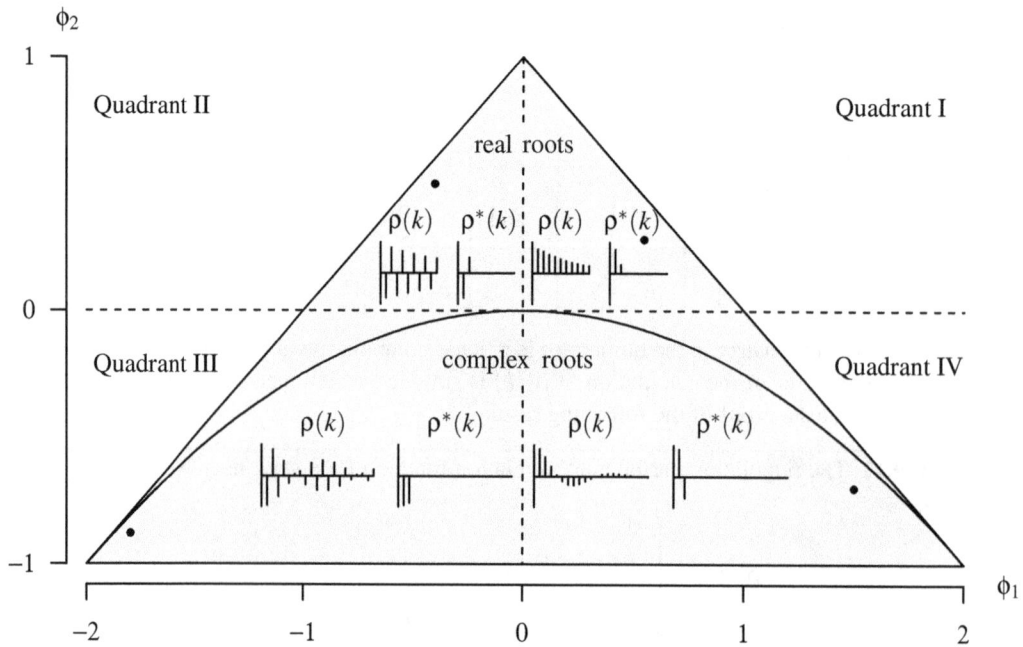

Figure 9.15: Stationary AR(2) time series model signature $\rho(k)$ and $\rho^*(k)$ shapes.

- When $\phi(B)$ has complex roots, the population autocorrelation function has a damped sinusoidal shape.

The population autocorrelations on the tiny inset plots of $\rho(k)$ and $\rho^*(k)$ in Figure 9.15 can be calculated using the recursive relationships from Theorem 9.10 [for $\rho(k)$] and Theorem 9.11 [for $\rho^*(k)$]. They can also be calculated using the R ARMAacf function. Consider the two inset plots in the fourth quadrant of the graph in Figure 9.15, for example, that correspond to $\phi_1 = 1.5$ and $\phi_2 = -0.7$. The graph of the first 20 lags of $\rho(k)$ can be plotted with the R command

```
plot(ARMAacf(ar = c(1.5, -0.7), ma = 0, lag.max = 20), type = "h")
```

Likewise, the graph of the first 20 lags of $\rho^*(k)$ can be plotted with

```
plot(ARMAacf(ar = c(1.5, -0.7), ma = 0, lag.max = 20, pacf = TRUE), type = "h")
```

The ar argument defines the ϕ_1 and ϕ_2 parameters of the AR(2) model, the ma argument is set to zero to indicate that there are no moving average terms in the AR(2) model, the lag.max argument is set to 20 to return the first 20 autocorrelations, and the type argument in the call to plot is set to "h" in order to graph the autocorrelations as spikes rather than points.

As was the case with the AR(1) time series model, the AR(2) time series can be written as an MA(∞) time series model. This alternative representation can be useful for deriving certain quantities associated with the AR(2) model, in particular, standard errors of forecasted values. The first form of a general linear model, which is equivalent to an MA(∞) model, is

$$X_t = Z_t + \theta_1 Z_{t-1} + \theta_2 Z_{t-2} + \cdots .$$

Our goal is to determine the values of $\theta_1, \theta_2, \ldots$ that correspond to fixed parameters ϕ_1 and ϕ_2. Since the MA(∞) model is valid at time t, it is also valid at times $t-1$ and $t-2$:

$$X_{t-1} = Z_{t-1} + \theta_1 Z_{t-2} + \theta_2 Z_{t-3} + \cdots$$

and

$$X_{t-2} = Z_{t-2} + \theta_1 Z_{t-3} + \theta_2 Z_{t-4} + \cdots .$$

So the AR(2) time series model

$$X_t = \phi_1 X_{t-1} + \phi_2 X_{t-2} + Z_t$$

as established in Definition 9.2, can be rewritten as

$$Z_t + \theta_1 Z_{t-1} + \theta_2 Z_{t-2} + \cdots = \phi_1 \left(Z_{t-1} + \theta_1 Z_{t-2} + \theta_2 Z_{t-3} + \cdots \right) + \phi_2 \left(Z_{t-2} + \theta_1 Z_{t-3} + \theta_2 Z_{t-4} + \cdots \right) + Z_t.$$

Equating the coefficients of Z_{t-1} gives

$$\theta_1 = \phi_1.$$

Equating the coefficients of Z_{t-2} gives

$$\theta_2 = \phi_1 \theta_1 + \phi_2 = \phi_1^2 + \phi_2.$$

Equating the coefficients of Z_{t-k} gives the recursive equation

$$\theta_k = \phi_1 \theta_{k-1} + \phi_2 \theta_{k-2}$$

for $k = 3, 4, \ldots .$

Theorem 9.12 A stationary AR(2) model with parameters ϕ_1 and ϕ_2 can be written as an MA(∞) model

$$X_t = Z_t + \theta_1 Z_{t-1} + \theta_2 Z_{t-2} + \cdots,$$

where $\theta_1 = \phi_1$, $\theta_2 = \phi_1^2 + \phi_2$, and

$$\theta_k = \phi_1 \theta_{k-1} + \phi_2 \theta_{k-2}$$

for $k = 3, 4, \ldots$.

An exercise at the end of the chapter highlights other methods for calculating the coefficients θ_1, $\theta_2, \ldots$ in the MA(∞) model which is equivalent to the stationary AR(2) model.

Example 9.11 Calculate the first six coefficients of an MA(∞) model associated with the stationary AR(2) model with $\phi_1 = 1$ and $\phi_2 = -1/2$.

The AR(2) model is stationary because the point $(\phi_1, \phi_2) = (1, -1/2)$ falls in the triangular-shaped stationary region defined by the inequalities in Theorem 9.9. Using Theorem 9.12, the first six coefficients of the MA(∞) model are

$$\theta_1 = 1, \qquad \theta_2 = \frac{1}{2}, \qquad \theta_3 = 0, \qquad \theta_4 = -\frac{1}{4}, \qquad \theta_5 = -\frac{1}{4}, \qquad \theta_6 = -\frac{1}{8}.$$

These coefficients can also be calculated in R with the `ARMAtoMA` function. The statement

```
ARMAtoMA(ar = c(1, -1 / 2), ma = 0, lag.max = 6)
```

returns the same coefficients calculated above. The `ar` argument defines the ϕ_1 and ϕ_2 parameters of the AR(2) model, the `ma` argument is set to zero to indicate that there are no moving average terms in the AR(2) model, and the `lag.max` argument is set to 6 in order to calculate the first six coefficients in the MA(∞) model and return these values in a vector in R.

The Shifted AR(2) Model

For a stationary AR(2) model expressed as an MA(∞) model, it is clear that $E[X_t] = 0$. This model is not of much use in practice because most real-world time series are not centered around zero. Adding a shift parameter μ overcomes this shortcoming. Since population variance and covariance are unaffected by a shift, the associated population autocorrelation and partial autocorrelation functions remain the same as those given in Theorems 9.10 and 9.11.

Theorem 9.13 A *shifted second-order autoregressive model* for the time series $\{X_t\}$ is defined by

$$X_t - \mu = \phi_1 (X_{t-1} - \mu) + \phi_2 (X_{t-2} - \mu) + Z_t,$$

where ϕ_1, ϕ_2, μ, and $\sigma_Z^2 > 0$ are real-valued parameters and $\{Z_t\}$ is a time series of white noise. This model is stationary when ϕ_1 and ϕ_2 satisfy the three inequalities given in Theorem 9.9. The expected value of X_t is $E[X_t] = \mu$. The population autocorrelation function can be calculated using the recursive equations in Theorem 9.10. The population partial autocorrelation function is given in Theorem 9.11.

The shifted AR(2) model can be written in terms of the backshift operator B as

$$\phi(B)(X_t - \mu) = Z_t,$$

where $\phi(B) = 1 - \phi_1 B - \phi_2 B^2$. The practical problem of fitting a shifted AR(2) model to an observed time series of n values $x_1, x_2, \ldots, x_n$ will be illustrated later in this subsection.

Simulation

An AR(2) time series can be simulated by appealing to the defining formula for the AR(2) model. Iteratively applying the defining formula for a standard AR(2) model

$$X_t = \phi_1 X_{t-1} + \phi_2 X_{t-2} + Z_t$$

from Definition 9.2 results in the simulated values $X_1, X_2, \ldots, X_n$. The primary difficult aspect of devising a simulation algorithm is generating the first two values, X_1 and X_2. For simplicity, assume that the white noise terms are Gaussian white noise terms. There are two approaches to overcome this initialization problem. The first approach generates X_1 and X_2 from a bivariate normal distribution with population mean vector $\mathbf{0} = (0, 0)'$ and variance–covariance matrix

$$\Sigma = \begin{bmatrix} \gamma(0) & \gamma(1) \\ \gamma(1) & \gamma(0) \end{bmatrix} = \frac{\sigma_Z^2}{(1+\phi_2)(1+\phi_1-\phi_2)(1-\phi_1-\phi_2)} \begin{bmatrix} 1-\phi_2 & \phi_1 \\ \phi_1 & 1-\phi_2 \end{bmatrix}$$

via Theorem 9.10. Notice that in the special case of $\phi_1 = \phi_2 = 0$ this matrix reduces to the variance–covariance matrix for Gaussian white noise, which is $I\sigma_Z^2$. The algorithm given as pseudocode below generates initial time series observations X_1 and X_2 as indicated above, and then uses an additional $n - 2$ Gaussian white noise terms $Z_3, Z_4, \ldots, Z_n$ to generate the remaining time series values $X_3, X_4, \ldots, X_n$ using the AR(2) defining formula from Definition 9.2. Indentation denotes nesting in the algorithm.

```
generate (X₁, X₂) ~ BVN(0, Σ)
t ← 2
while (t < n)
    t ← t + 1
    generate Zₜ ~ N(0, σ²_Z)
    Xₜ ← φ₁Xₜ₋₁ + φ₂Xₜ₋₂ + Zₜ
```

The four-parameter shifted AR(2) time series model which includes a population mean parameter μ can be simulated by simply adding μ to each time series observation generated by this algorithm. The next example implements this algorithm in R.

Example 9.12 Generate a realization of $n = 100$ observations from an AR(2) time series model with $\phi_1 = 1.5$, $\phi_2 = -0.7$, and Gaussian white noise error terms with $\sigma_Z^2 = 16$.

Since $(\phi_1, \phi_2) = (1.5, -0.7)$ lies in the triangular-shaped stationary region defined in Theorem 9.9, the simulated values will be generated from a stationary time series model. The population autocorrelation function $\rho(k)$ and the population partial autocorrelation function $\rho^*(k)$ are displayed in the fourth quadrant of Figure 9.15, and we expect similar shaped functions r_k and r_k^* from our simulated values. The optional first statement in the R code below uses the set.seed function to establish the random number seed. The second and third statements set the AR(2) coefficients to $\phi_1 = 1.5$ and

$\phi_2 = -0.7$. The fourth statement sets the standard deviation of the Gaussian white noise to $\sigma_Z = 4$. The fifth statement places the variance–covariance matrix of X_1 and X_2 in the 2×2 matrix sigma. The sixth statement sets the number of simulated values to $n = 100$. The seventh statement defines the vector x of length $n = 100$ to hold the simulated time series values. The eighth statement uses the mvrnorm function from the MASS package to generate the first two simulated time series observations X_1 and X_2 from the appropriate bivariate normal distribution. Finally, the for loop iterates through the defining formula for the AR(2) model generating the remaining observations $X_3, X_4, \ldots, X_{100}$.

```
set.seed(3)
phi1    = 1.5
phi2    = -0.7
sigz    = 4
sigma   = matrix(c(1 - phi2, phi1, phi1, 1 - phi2), 2, 2) * sigz /
          ((1 + phi2) * (1 + phi1 - phi2) * (1 - phi1 - phi2))
n       = 100
x       = numeric(n)
x[1:2] = MASS::mvrnorm(1, mu = c(0, 0), Sigma = sigma)
for (t in 3:n) x[t] = phi1 * x[t - 1] + phi2 * x[t - 2] + rnorm(1, 0, sigz)
```

Using the plot.ts function to make a plot of the time series contained in x, the acf function to plot the associated correlogram, the pacf function to plot the associated sample partial autocorrelation function, and the layout function to arrange the graphs as in Example 7.24, the resulting trio of graphs are displayed in Figure 9.16. The sample partial autocorrelation function has statistically significant spikes at lags 1 and 2 with $r_1^* = 0.8036$ and $r_2^* = -0.6229$, and then cuts off after lag 2 as expected from the population counterparts in Figure 9.15. The approximate 95% confidence intervals indicated by the dashed lines show that the values of the sample partial autocorrelation function do not significantly differ from zero at lags beyond lag 2. The sample autocorrelation function displays a damped sinusoidal shape as expected. The time series plot shows that observations tend to linger on one side of the population mean (indicated by a horizontal line), which is consistent with the two initial statistically significant positive spikes in the sample autocorrelation function. However, the time that the observations linger on one side of the mean is inhibited by the statistically significant negative spikes at lags 4, 5, 6, and 7 in the sample autocorrelation function. There is thus some tug exerted by the time series model to linger on one side of the mean for only a limited amount of time.

We recommend running the simulation code from the previous example several dozen times in a loop and viewing the associated plots of x_t, r_k, and r_k^* in search of patterns. This will allow you to see how various realizations of a simulated AR(2) time series model vary from one realization to the next. So when you then view a *single* realization of a real-life time series, you will have a sense of how far these plots might deviate from their expected patterns.

There is a second way to overcome the initialization problem in simulating observations from an AR(2) time series. This second technique starts the time series with two initial arbitrary values, and then allows the time series to "warm up" or "burn in" for several time periods before producing the first observation X_1. Reasonable initial arbitrary values for the standard AR(2) model are 0; reasonable initial arbitrary values for the shifted AR(2) model are μ. This is the approach taken by the built-in R function named arima.sim, which simulates a realization of a time series. Using the

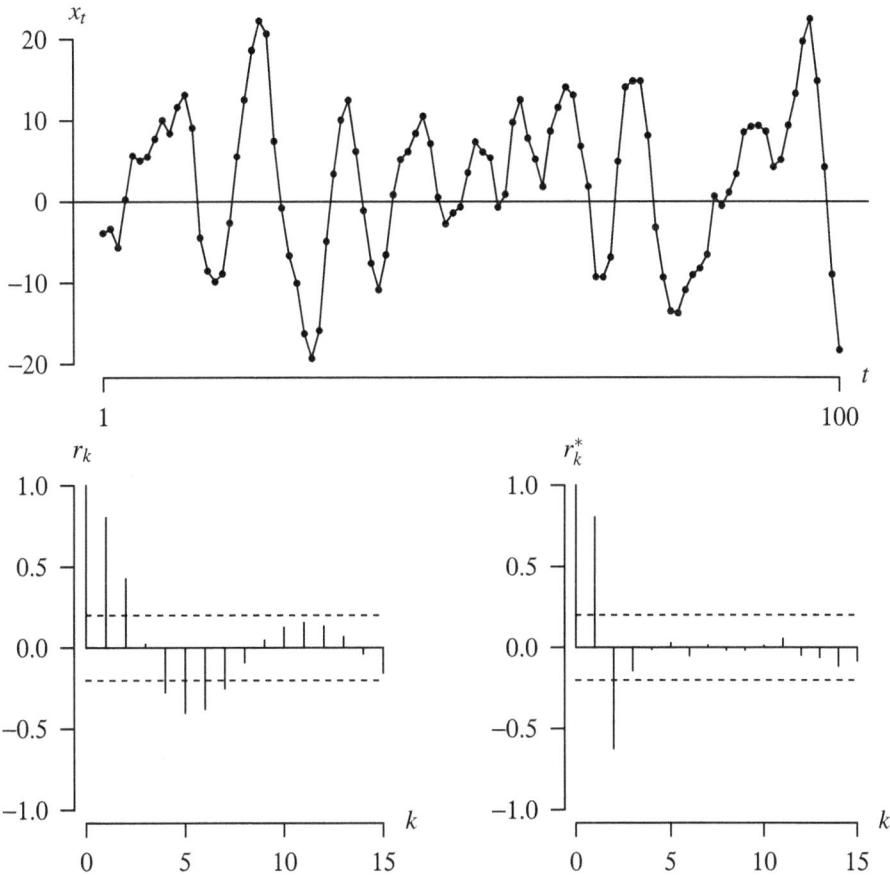

Figure 9.16: Time series plot, r_k, and r_k^* for $n = 100$ simulated values from an AR(2) model.

arima.sim function saves a few keystrokes over the approach taken in the previous example, as illustrated next.

Example 9.13 Generate a realization of $n = 100$ observations from a shifted AR(2) time series model with coefficients $\phi_1 = -1.8$ and $\phi_2 = -0.88$, population mean value $\mu = 10$, and Gaussian white noise error terms with $\sigma_Z^2 = 16$.

Since there is now a nonzero population mean value, the shifted AR(2) model is

$$X_t - \mu = \phi_1 \left(X_{t-1} - \mu \right) + \phi_2 \left(X_{t-2} - \mu \right) + Z_t,$$

where $\mu = 10$, $\phi_1 = -1.8$, $\phi_2 = -0.88$, and $\sigma_Z = 4$. Since $(\phi_1, \phi_2) = (-1.8, -0.88)$ lies in the triangular-shaped stationary region defined in Theorem 9.9, this is a stationary AR(2) time series model. The $\rho(k)$ and $\rho^*(k)$ values for this model are plotted in the third quadrant of Figure 9.15. The model argument in the arima.sim function is a list containing the values of the coefficients ϕ_1 and ϕ_2. The second argument to arima.sim is n, the number of time series observations to be generated. The sd argument defines the standard deviation of the white noise σ_Z. The n.start argument gives the number of observations in the burn-in period, which we specify here as 50. The R code to generate $n = 100$ values from the shifted AR(2) model is given below.

```
set.seed(9)
x = 10 + arima.sim(model = list(ar = c(-1.8, -0.88)),
                   n = 100, sd = 4, n.start = 50)
```

Figure 9.17 shows the three plots associated with the simulated values in the vector x using the plot.ts, acf, and pacf functions. The time series plot shows a radically different pattern than the time series in the previous example in several aspects. First, this simulated time series is centered around $\mu = 10$ (indicated by a horizontal line) rather than $\mu = 0$. Second, the time series jumps from one side of the population mean to the other from one observation to the next. This is consistent with the highly statistically significant negative lag 1 sample autocorrelation $r_1 = -0.9546$. The signs of the initial sample autocorrelation function values alternate, their magnitudes decrease, and sample

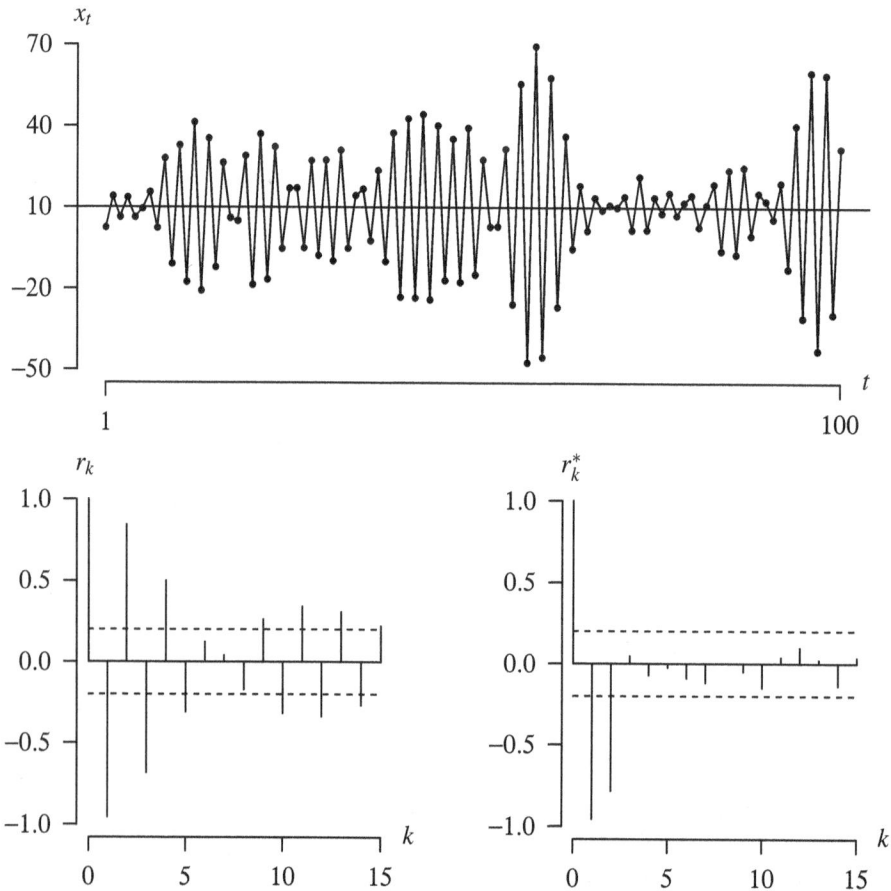

Figure 9.17: Time series plot, r_k, and r_k^* for $n = 100$ simulated values from an AR(2) model.

autocorrelations at subsequent lags follow a damped sinusoidal pattern. In addition, the lag 1 sample autocorrelation is so close to -1 that adjacent observations often tend to be about the same distance away from μ. Third, the two statistically significant spikes in the partial autocorrelation function, $r_1^* = -0.9546$ and $r_2^* = -0.7834$, have the same signs as ϕ_1 and ϕ_2. As expected, the partial autocorrelation function cuts off after lag 2.

The remaining topics associated with the AR(2) time series model are statistical in nature: parameter estimation, model assessment, model selection, and forecasting. A sample time series that will be revisited throughout these topics is introduced next.

Example 9.14 The five Great Lakes that lie along the U.S.–Canada border are Huron, Ontario, Michigan, Erie, and Superior. Their names are easily remembered with the acronym HOMES. The built-in R time series `LakeHuron` consists of $n = 98$ monthly mean levels (in feet) of the lake level of Lake Huron taken at the Harbor Beach, Michigan water level gauge every July from 1875–1972. The measurements are essentially the number of feet above sea level of Lake Huron over time. Plot the time series, sample autocorrelation function, and sample partial autocorrelation function, and suggest a tentative time series model.

For simplicity, we define time $t = 1$ to be the year 1875 and $t = 98$ to be the year 1972. The time series of levels, the sample autocorrelation function, and the sample partial autocorrelation function can be graphed with the R statements

```
x = LakeHuron
layout(matrix(c(1, 1, 2, 3), 2, 2, byrow = TRUE))
plot.ts(x)
acf(x)
pacf(x)
```

The trio of graphs is displayed in Figure 9.18. A horizontal line has been added to the time series plot at $\bar{x} = 579$ feet. A visual assessment of the $n = 98$ observations from the time series reveals that the population mean level might be systematically decreasing relative to sea level over the time period. The tied observations in the years 1925–1926 ($t = 51$ and $t = 52$), the local minimum in year 1934 ($t = 60$), and the global minimum in year 1964 ($t = 90$), along with nearby observations, provide a downward tug on the mean value of the time series as time advances. Alternatively, the initial 15 or so levels might be drawn from a non-representative population. The population variance of the observations in the time series seems to be stable over time.

Now we turn to the sample autocorrelation function and sample partial autocorrelation function. The sample autocorrelation function appears to be tailing out. The initial positive spikes in the sample autocorrelation function are consistent with nearby observations in the time series lingering above and below the sample mean. The sample partial autocorrelation function has a statistically significant positive spike at lag 1, and a marginally significant negative spike at lag 2. The sample partial autocorrelation function does not have any statistically significant spikes after lag 2. These two graphs indicate that a shifted AR(2) model might be a reasonable tentative time series model.

The question concerning the possible downward trend of the level of Lake Huron in the time series remains a stumbling block to an enthusiastic recommendation of the shifted

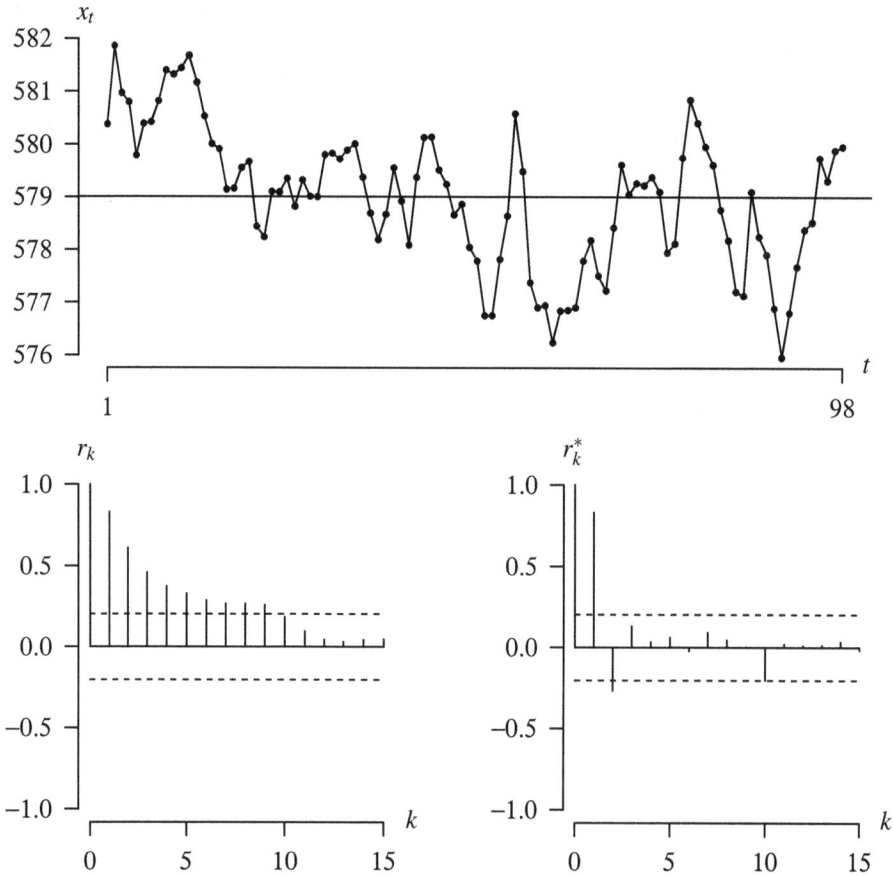

Figure 9.18: Time series plot, r_k, and r_k^* for $n = 98$ levels of Lake Huron (1875–1972).

AR(2) model. Perhaps a nonstationary model is appropriate. The list below provides four ways to proceed.

- Eliminate the first 15 or so observations from the time series if we can find an *assignable cause* that made the initial observations higher than the others. Here are some examples of potential assignable causes. Was the measuring equipment, location, procedure, or personnel changed at some point in the time series? There were various bridge, power, and flow control projects conducted during the early part of this time series. Some projects increased flow; others decreased flow. Could these projects account for the increased early observations in the time series? The primary driver of the year-to-year variability in water levels is the regional climate, which is influenced by global oceanic and atmospheric patterns. Lake Huron thermodynamics can be influenced by above-average lake evaporation rates. Might any of these factors account for the increased early observations in the time series? Did the episodic dredging of Lake Huron's outlet, the St. Clair River, result in a lowering of the level of Lake Huron? In the case of the beaver temperature time series from Example 9.3, it was easy to find an assignable cause, because the beaver's temperature was clearly lower when in the lodge than when outside of the lodge. Such an assignable cause might be more difficult to identify in the case of the time series of Lake Huron levels.

• If an assignable cause cannot be found, fit a simple linear regression model to the original time series and consider the time series consisting of the original time series values minus the fitted values to be a stationary time series. Figure 9.19 shows the fitted regression line using the model

$$Y = \beta_0 + \beta_1 X + \varepsilon,$$

where X is time (which is measured without error), Y is the random and continuous lake level, β_1 is the slope of the regression line, β_0 is the intercept of the regression line, and ε is an error term. The hypothesis test

$$H_0 : \beta_1 = 0$$

versus

$$H_1 : \beta_1 \neq 0$$

results in a tiny p-value ($p = 4 \cdot 10^{-8}$), which confirms our visual assessment. There does indeed appear to be a decrease in the level of the water in Lake Huron over time. The estimated regression coefficients and p-values can be calculated in R with the statement

```
summary(lm(LakeHuron ~ seq(1, 98)))
```

The estimated slope $\hat{\beta}_1 = -0.024$ indicates that the level of Lake Huron is decreasing by an average of about a quarter of an inch annually over this time horizon. Extrapolation of the simple linear regression model outside of the years 1875–1972 is probably not warranted in this setting. The usual regression assumption of independence is clearly violated in this setting because the observations in the time series are autocorrelated.

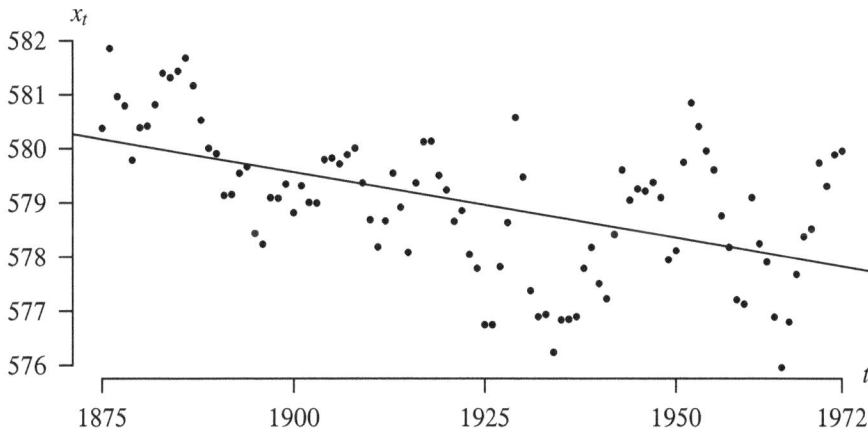

Figure 9.19: Lake Huron levels (1875–1972) with regression line.

- Again assuming that an assignable cause cannot be identified in order to remove the initial observations, the time series can be differenced (via $x_{i+1} - x_i$ for $i = 1, 2, \ldots, n-1$) in order to remove the possible linear trend. The $n - 1 = 97$ observations in this differenced series can then be evaluated as a stationary time series. This approach is analogous to the simple linear regression approach.

- Leave the time series alone, and assume that the early observations being larger than the rest of the observations is due to random sampling variability.

This might seem like a lot of fuss to establish a stationary time series model, but this crucial early detective work is common when trying to formulate a tentative time series model. We take the fourth approach from the list above for now and fit a tentative shifted AR(2) model to the $n = 98$ original observations in the time series. Fitting this tentative model will illustrate all of the steps involved with fitting, evaluating, and applying an AR(2) model: estimating the model parameters by the three techniques (method of moments, least squares, and maximum likelihood estimation), constructing confidence intervals for these point estimators, assessing the validity of the fitted model, performing model selection procedures for the AR(2) model, and forecasting the level of Lake Huron into the future. Later in the chapter, we will re-analyze this time series using a nonstationary time series model.

In conclusion, a preliminary graphical analysis of the $n = 98$ Lake Huron levels suggests a tentative AR(2) time series model should be on the short list. It is worthwhile investigating the possibility of an assignable cause which artificially elevates the initial 15 observations. There is significant concern about nonstationarity, which will be addressed later in the chapter. The next step is to estimate the parameters in the model.

Parameter Estimation

There are four parameters, μ, ϕ_1, ϕ_2, and σ_Z^2, to estimate in the shifted AR(2) model

$$X_t - \mu = \phi_1 \left(X_{t-1} - \mu \right) + \phi_2 \left(X_{t-2} - \mu \right) + Z_t.$$

The three parameter estimation techniques outlined in Section 8.2.1, method of moments, least squares, and maximum likelihood estimation, are applied to the shifted AR(2) time series model next.

Approach 1: Method of moments. In the case of estimating the four parameters in the shifted AR(2) model by the method of moments, we match the population and sample (*a*) first-order moments, (*b*) second-order moments, (*c*) lag 1 autocorrelation, and (*d*) lag 2 autocorrelation. Placing the population moments on the left-hand side of the equation and the associated sample moments on the right-hand side of the equation results in four equations in four unknowns:

$$E\left[X_t \right] = \frac{1}{n} \sum_{t=1}^{n} X_t$$

$$E\left[X_t^2 \right] = \frac{1}{n} \sum_{t=1}^{n} X_t^2$$

$$\rho(1) = r_1$$

$$\rho(2) = r_2.$$

The expected value of X_t is μ, the expected value of X_t^2 can be found by using the shortcut formula for the population variance and by using the value of $\gamma(0) = V[X_t]$ from Theorem 9.10, and the values of $\rho(1)$ and $\rho(2)$ are also obtained from Theorem 9.10. So the four equations become

$$\mu = \frac{1}{n}\sum_{t=1}^{n} X_t$$

$$\frac{(1-\phi_2)\sigma_Z^2}{(1+\phi_2)(1+\phi_1-\phi_2)(1-\phi_1-\phi_2)} + \mu^2 = \frac{1}{n}\sum_{t=1}^{n} X_t^2$$

$$\frac{\phi_1}{1-\phi_2} = r_1$$

$$\frac{\phi_1^2}{1-\phi_2} + \phi_2 = r_2.$$

Solving these equations for the four unknown parameters μ, ϕ_1, ϕ_2 and σ_Z^2 yields closed-form solutions for the method of moments estimators

$$\hat{\mu} = \bar{X}$$

$$\hat{\phi}_1 = \frac{r_1(1-r_2)}{1-r_1^2}$$

$$\hat{\phi}_2 = \frac{r_2 - r_1^2}{1-r_1^2}$$

$$\hat{\sigma}_Z^2 = \left[\frac{1}{n}\sum_{t=1}^{n} X_t^2 - \hat{\mu}^2\right]\frac{(1+\hat{\phi}_2)(1+\hat{\phi}_1-\hat{\phi}_2)(1-\hat{\phi}_1-\hat{\phi}_2)}{1-\hat{\phi}_2}.$$

This constitutes a proof of the following result.

Theorem 9.14 The method of moments estimators of the parameters in a shifted AR(2) model are

$$\hat{\mu} = \bar{X}, \qquad \hat{\phi}_1 = \frac{r_1(1-r_2)}{1-r_1^2}, \qquad \hat{\phi}_2 = \frac{r_2 - r_1^2}{1-r_1^2},$$

$$\hat{\sigma}_Z^2 = \left[\frac{1}{n}\sum_{t=1}^{n} X_t^2 - \hat{\mu}^2\right]\frac{(1+\hat{\phi}_2)(1+\hat{\phi}_1-\hat{\phi}_2)(1-\hat{\phi}_1-\hat{\phi}_2)}{1-\hat{\phi}_2}.$$

These estimators are random variables and have been written as a function of the random time series values $X_1, X_2, \ldots, X_n$. For observed time series values $x_1, x_2, \ldots, x_n$, the lowercase versions of the formulas will be used. These estimators are often known as the *Yule–Walker estimators* because their derivation involved the Yule–Walker equations from Theorem 9.10.

Example 9.15 For the time series of $n = 98$ Lake Huron levels from Example 9.14, find the method of moments estimators of μ, ϕ_1, ϕ_2, and σ_Z^2 for the AR(2) model.

The R code below calculates and prints the point estimates of the μ, ϕ_1, ϕ_2, and σ_Z^2 parameters using the method of moments estimators given in Theorem 9.14.

```
x        = LakeHuron
```

```
r1       = acf(x, plot = FALSE)$acf[2]
r2       = acf(x, plot = FALSE)$acf[3]
muhat    = mean(x)
phi1hat = r1 * (1 - r2) / (1 - r1 ^ 2)
phi2hat = (r2 - r1 ^ 2) / (1 - r1 ^ 2)
sig2hat = (mean(x ^ 2) - muhat ^ 2) * (1 + phi2hat) *
             (1 + phi1hat - phi2hat) * (1 - phi1hat - phi2hat) /
             (1 - phi2hat)
print(c(muhat, phi1hat, phi2hat, sig2hat))
```

The point estimates for the unknown parameters computed by this code are

$$\hat{\mu} = 579.00 \qquad \hat{\phi}_1 = 1.0538 \qquad \hat{\phi}_2 = -0.26675 \qquad \hat{\sigma}_Z^2 = 0.49199.$$

These point estimates are reported to five digits because the data values were given to five-digit accuracy. The positive value for $\hat{\phi}_1$ and the negative value for $\hat{\phi}_2$ are consistent with the sample partial autocorrelation function in Figure 9.18. Figure 9.20 is analogous to Figure 9.13 but contains just two of the level surfaces associated with the method of moments match on the population and sample autocorrelations at lags 1 and 2: the line associated with $\rho(1) = r_1 = 0.83191$ and the curve associated with $\rho(2) = r_2 = 0.60994$. These two level surfaces intersect at the point $(\hat{\phi}_1, \hat{\phi}_2) = (1.0538, -0.26675)$.

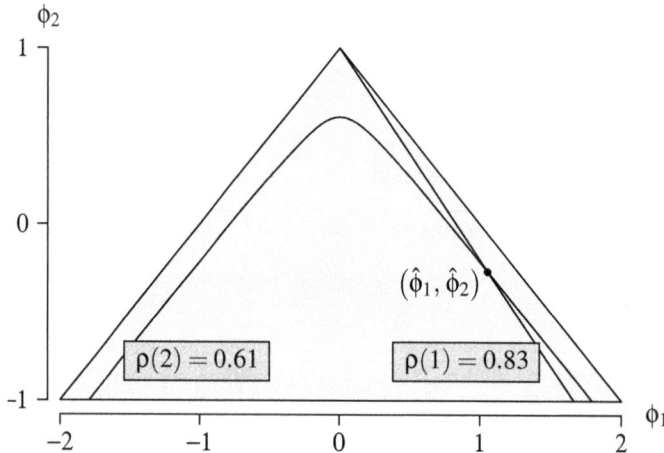

Figure 9.20: Level surfaces for the Lake Huron time series in the stationary region.

Approach 2: Least squares. Consider the shifted stationary AR(2) model

$$X_t - \mu = \phi_1 (X_{t-1} - \mu) + \phi_2 (X_{t-2} - \mu) + Z_t.$$

For least squares estimation, we first establish the sum of squares S as a function of the parameters μ, ϕ_1, and ϕ_2. This time, however, we forgo the calculus and leave the optimization to the R optim function in order to find the least squares estimators of μ, ϕ_1, and ϕ_2. Once these least squares estimators have been determined, the population variance of the white noise σ_Z^2 will be estimated.

The sum of squared errors is

$$S = \sum_{t=3}^{n} Z_t^2 = \sum_{t=3}^{n} [X_t - \mu - \phi_1 (X_{t-1} - \mu) - \phi_2 (X_{t-2} - \mu)]^2.$$

If this derivation were being done by hand, we would now calculate the partial derivatives of S with respect to the unknown parameters μ, ϕ_1, and ϕ_2, equate them to zero and solve. As was the case with the AR(1) model, there is no closed-form solution, so numerical methods are required to calculate the parameter estimates. In the example that follows, we will use the `optim` function in R to determine the least squares parameter estimates that minimize S.

The last parameter to estimate is σ_Z^2. Since

$$\gamma(0) = \frac{(1 - \phi_2)\sigma_Z^2}{(1 + \phi_2)(1 + \phi_1 - \phi_2)(1 - \phi_1 - \phi_2)}$$

from Theorem 9.10 for an AR(2) time series model, the population variance of the white noise can be expressed as

$$\sigma_Z^2 = \frac{(1 + \phi_2)(1 + \phi_1 - \phi_2)(1 - \phi_1 - \phi_2)\gamma(0)}{(1 - \phi_2)}.$$

Replacing ϕ_1 and ϕ_2 by their least squares estimators $\hat{\phi}_1$ and $\hat{\phi}_2$, respectively, and replacing the lag 0 autocovariance $\gamma(0) = V[X_t]$ by its estimator $c_0 = \frac{1}{n} \sum_{t=1}^{n} (X_t - \bar{X})^2$ gives the estimator

$$\hat{\sigma}_Z^2 = \frac{\left(1 + \hat{\phi}_2\right)\left(1 + \hat{\phi}_1 - \hat{\phi}_2\right)\left(1 - \hat{\phi}_1 - \hat{\phi}_2\right) c_0}{(1 - \hat{\phi}_2)}.$$

This derivation constitutes a proof of the following result.

Theorem 9.15 The least squares estimators of the parameters in a shifted AR(2) model are the $\hat{\mu}$, $\hat{\phi}_1$, and $\hat{\phi}_2$ values that minimize

$$S = \sum_{t=3}^{n} Z_t^2 = \sum_{t=3}^{n} [X_t - \mu - \phi_1 (X_{t-1} - \mu) - \phi_2 (X_{t-2} - \mu)]^2$$

and the population variance of the white noise is estimated by

$$\hat{\sigma}_Z^2 = \frac{\left(1 + \hat{\phi}_2\right)\left(1 + \hat{\phi}_1 - \hat{\phi}_2\right)\left(1 - \hat{\phi}_1 - \hat{\phi}_2\right) c_0}{(1 - \hat{\phi}_2)}.$$

We now use numerical methods to find the least squares estimates for the unknown parameters in the AR(2) time series model for the Lake Huron time series from Example 9.14.

Example 9.16 Find the least squares estimators of μ, ϕ_1, ϕ_2, and σ_Z^2 for the AR(2) time series model associated with the $n = 98$ lake level observations from Example 9.14.

The R code that follows contains a function s which calculates the sum of squares, and then uses the R `optim` function to minimize the sum of squares using the method of moments estimates as initial estimates. The `optim` function minimizes the objective function by default.

```
x = LakeHuron
n = length(x)
s = function(parameters) {
  mu   = parameters[1]
  phi1 = parameters[2]
  phi2 = parameters[3]
  sum((x[3:n] - mu - phi1 * (x[2:(n - 1)] - mu)
                   - phi2 * (x[1:(n - 2)] - mu)) ^ 2)
}
fit = optim(c(579, 1.0538, -0.26675), s)$par
muhat   = fit[1]
phi1hat = fit[2]
phi2hat = fit[3]
sig2hat = (1 + phi2hat) * (1 + phi1hat - phi2hat) *
          (1 - phi1hat - phi2hat) * mean((x - mean(x)) ^ 2) /
          (1 - phi2hat)
```

The point estimates for the unknown parameters computed by this code are

$$\hat{\mu} = 578.89 \qquad \hat{\phi}_1 = 1.0217 \qquad \hat{\phi}_2 = -0.23760 \qquad \hat{\sigma}_Z^2 = 0.51680,$$

which corresponds to a sum of squares $S = 43.58$. The optimal sum of squares can be extracted with the additional R command s(fit). These least squares point estimates of the unknown parameters in the AR(2) time series model are close to the associated method of moments point estimates. The left-hand graph in Figure 9.21 shows the sum of squares as a function of ϕ_1 for fixed values of the parameters $\hat{\mu} = 578.89$ and $\hat{\phi}_2 = -0.23760$. The sum of squares is minimized at $\hat{\phi}_1 = 1.0217$. The right-hand graph in Figure 9.21 shows the sum of squares as a function of ϕ_2 for fixed values of the parameters $\hat{\mu} = 578.89$ and $\hat{\phi}_1 = 1.0217$. The sum of squares is minimized at $\hat{\phi}_2 = -0.23760$.

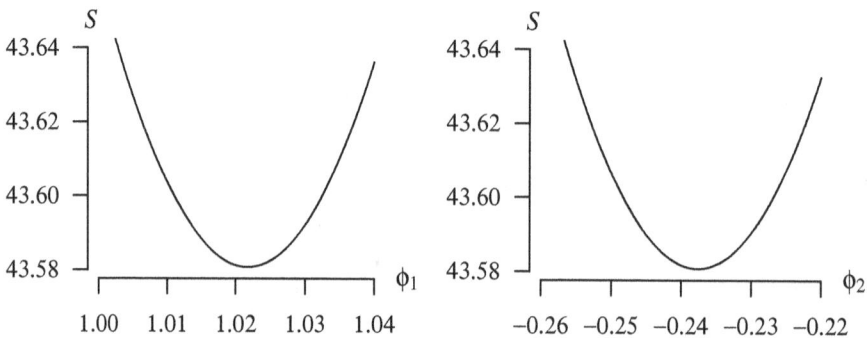

Figure 9.21: Sum of squares as a function of ϕ_1 and ϕ_2 for an AR(2) model.

Approach 3: Maximum likelihood estimation. The procedure for determining the maximum likelihood estimators for the unknown parameters in an AR(2) time series model follows along the same lines as in the AR(1) time series model from the previous section. Once again, to use maximum likelihood estimation, we must assume that the random shocks from the white noise are Gaussian white noise, with associated probability density function

$$f_{Z_t}(z_t) = \frac{1}{\sqrt{2\pi\sigma_Z^2}} e^{-z_t^2/(2\sigma_Z^2)} \qquad -\infty < z_t < \infty,$$

for $t = 1, 2, \ldots, n$. Determining the likelihood function, which is the joint probability density function of the observed values in the time series $X_1, X_2, \ldots, X_n$, involves finding

$$L(\mu, \phi_1, \phi_2, \sigma_Z^2) = f(x_1, x_2, \ldots, x_n),$$

where the $x_1, x_2, \ldots, x_n$ arguments on L and the μ, ϕ_1, ϕ_2, and σ_Z^2 arguments on f have been dropped for brevity. As before, it is not possible to simply multiply the marginal probability density functions because the values in the AR(2) time series model are correlated. As in the case of an AR(1) model, we use the transformation technique to find the conditional joint probability density function of $X_3, X_4, \ldots, X_n$ conditioned on $X_1 = x_1$ and $X_2 = x_2$, which is denoted by

$$f_{X_3, X_4, \ldots, X_n | X_1, X_2}(x_3, x_4, \ldots, x_n | X_1 = x_1, X_2 = x_2)$$

for $(x_3, x_4, \ldots, x_n) \in \mathcal{R}^{n-2}$. This conditional joint probability density function is multiplied by the marginal joint probability density function of X_1 and X_2 (which has the bivariate normal distribution) resulting in a joint probability density function of $X_1, X_2, \ldots, X_n$:

$$f_{X_1, X_2, \ldots, X_n}(x_1, x_2, \ldots, x_n) = f_{X_3, X_4, \ldots, X_n | X_1, X_2}(x_3, x_4, \ldots, x_n | X_1 = x_1, X_2 = x_2) f_{X_1, X_2}(x_1, x_2)$$

for $(x_1, x_2, \ldots, x_n) \in \mathcal{R}^n$. This function serves as the likelihood function, which should be maximized with respect to the unknown parameters μ, ϕ_1, ϕ_2, and σ_Z^2. One can easily imagine how complicated this expression is, based on the values of $\gamma(0)$ and $\gamma(1)$ from Theorem 9.10. So we forgo the tedious mathematics and leave the calculations to the `ar` function in R when determining the maximum likelihood estimates for the parameters in fitting the Lake Huron time series to the shifted AR(2) time series model.

Example 9.17 Find the maximum likelihood estimators of μ, ϕ_1, ϕ_2, and σ_Z^2 for the time series of $n = 98$ observations of the level of Lake Huron from Example 9.14 for a shifted AR(2) time series model

$$X_t - \mu = \phi_1(X_{t-1} - \mu) + \phi_2(X_{t-2} - \mu) + Z_t,$$

where ϕ_1, ϕ_2, μ, and $\sigma_Z^2 > 0$ are real-valued parameters and $\{Z_t\}$ is a time series of Gaussian white noise.

The point estimates for the unknown parameters in the shifted AR(2) time series model are computed by the single R command

```
ar(LakeHuron, order.max = 2, aic = FALSE, method = "mle")
```

Unlike the procedure described above, the `ar` function first subtracts the sample mean $\bar{x} = 579$ from each observation and then proceeds to fit the remaining parameters to the standard AR(2) time series model. This function call returns the maximum likelihood estimates for the parameters as

$$\hat{\mu} = 579.00 \qquad \hat{\phi}_1 = 1.0437 \qquad \hat{\phi}_2 = -0.2496 \qquad \hat{\sigma}_Z^2 = 0.4788.$$

These parameter estimates are near the associated method of moments and least squares estimates from the previous two examples. The fitted shifted AR(2) model by maximum likelihood estimation is

$$X_t - 579.00 = 1.0437\,(X_{t-1} - 579.00) - 0.2496\,(X_{t-2} - 579.00) + Z_t,$$

where $Z_t \sim N(0, 0.4788)$.

Table 9.6 summarizes the point estimators for the AR(2) model for the Lake Huron time series calculated by the R commands

```
ar(LakeHuron, order.max = 2, aic = FALSE, method = "yule-walker")
ar(LakeHuron, order.max = 2, aic = FALSE, method = "ols")
ar(LakeHuron, order.max = 2, aic = FALSE, method = "mle")
```

The point estimators associated with the three methods are quite close for this particular time series. The R function ar fits autoregressive models. There are tiny differences between some of the entries in Table 9.6 and those from Examples 9.15 and 9.16 which might be due to slightly different approximations and/or roundoff in the optimization routines.

Method	$\hat{\mu}$	$\hat{\phi}_1$	$\hat{\phi}_2$	$\hat{\sigma}_Z^2$
Method of moments (Yule–Walker)	579.0	1.0538	−0.2668	0.5075
Ordinary least squares	579.0	1.0217	−0.2376	0.4540
Maximum likelihood estimation	579.0	1.0437	−0.2496	0.4788

Table 9.6: AR(2) point estimators for the $n = 98$ Lake Huron levels via the ar function.

The focus on estimation thus far has been on point estimation techniques. We also want to report some indication of the *precision* associated with these point estimators. The sampling distributions of $\hat{\mu}$, $\hat{\phi}_1$, $\hat{\phi}_2$, and $\hat{\sigma}_Z^2$ in the AR(2) model are too complicated to derive analytically. As an illustration of how to construct an approximate confidence interval for ϕ_1 or ϕ_2, we use the asymptotic normality of the maximum likelihood estimator of ϕ_1 and ϕ_2 in the following result. The asymptotic variance–covariance matrix associated with the parameters ϕ_1 and ϕ_2 is

$$\frac{1}{n}\begin{bmatrix} 1 - \phi_2^2 & -\phi_1(1+\phi_2) \\ -\phi_1(1+\phi_2) & 1 - \phi_2^2 \end{bmatrix}.$$

Using just the diagonal elements of this matrix results in the following asymptotically exact two-sided $100(1-\alpha)\%$ confidence interval for ϕ_1 and ϕ_2.

Theorem 9.16 For a stationary AR(2) time series model, an asymptotically exact two-sided $100(1-\alpha)\%$ confidence interval for ϕ_i is given by

$$\hat{\phi}_i - z_{\alpha/2}\sqrt{\frac{1 - \hat{\phi}_2^2}{n}} < \phi_i < \hat{\phi}_i + z_{\alpha/2}\sqrt{\frac{1 - \hat{\phi}_2^2}{n}}$$

for $i = 1, 2$, where $\hat{\phi}_i$ is the maximum likelihood estimator of ϕ_i and $z_{\alpha/2}$ is the $1 - \alpha/2$ fractile of the standard normal distribution.

These asymptotically exact confidence intervals for ϕ_1 and ϕ_2 will now be illustrated for the lake levels from the Lake Huron time series from the previous four examples.

Example 9.18 Find an approximate 95% confidence interval for ϕ_1 for the AR(2) time series model associated with the $n = 98$ Lake Huron time series values from Example 9.14 and assess its actual coverage.

Recall from Table 9.6 that the maximum likelihood estimators of ϕ_1 and ϕ_2 returned by the ar function are $\hat\phi_1 = 1.0437$ and $\hat\phi_2 = -0.2496$. We seek an asymptotically exact two-sided 95% confidence interval for ϕ_1, which is given by

$$1.0437 - 1.96\sqrt{\frac{1 - (-0.2496)^2}{98}} < \phi_1 < 1.0437 + 1.96\sqrt{\frac{1 - (-0.2496)^2}{98}}$$

or

$$0.8519 < \phi_1 < 1.2354.$$

This confidence interval does not contain $\phi_1 = 0$, which leads us to conclude that ϕ_1 is a statistically significant parameter in the AR(2) model. A similar procedure could be used to find a confidence interval for ϕ_2.

To assess the actual coverage of this 95% confidence interval for ϕ_1 requires a Monte Carlo simulation experiment. The code below uses population parameters that are near the parameter estimates for the Lake Huron time series.

```
n      = 98
mu     = 579
phi1   = 1
phi2   = -1 / 4
sigz   = sqrt(1 / 2)
crit   = qnorm(0.975)
nrep   = 40000
count = 0
for (i in 1:nrep) {
   x        = mu + arima.sim(model = list(ar = c(phi1, phi2), sd = sigz),
                       n = n, n.start = 50)
   fit      = arima(x, order = c(2, 0, 0), method = "ML")
   phi1hat = fit$coef[1]
   phi2hat = fit$coef[2]
   std      = sqrt((1 - phi2hat ^ 2) / n)
   lo       = phi1hat - crit * std
   hi       = phi1hat + crit * std
   if (lo < phi1 && hi > phi1) count = count + 1
}
print(count / nrep)
```

After a call to set.seed(3) to establish the random number stream, five runs of this simulation yield:

0.9376	0.9390	0.9366	0.9385	0.9395.

The conclusion that can be drawn from these simulations is that the actual coverage of the approximate 95% confidence interval is about 93.8%. When this code is executed for larger values of n, the anticipated asymptotic results are achieved, as displayed in Figure 9.22. Keep in mind that these actual coverages are not for an AR(2) model in general, but rather an AR(2) model with these particular parameter settings.

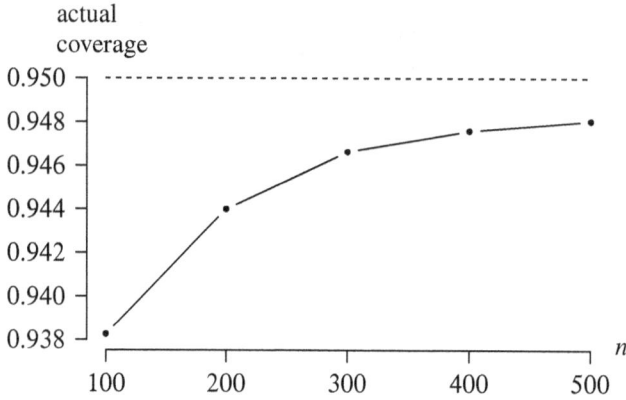

Figure 9.22: Asymptotic 95% confidence interval for ϕ_1 actual coverage.

Model Assessment

Now that techniques for point and interval estimates for the parameters in the AR(2) model have been established, we are interested in assessing the adequacy of the AR(2) time series model. This will involve an analysis of the residuals. Recall from Section 8.2.3 that the residuals are defined by

$$[\text{residual}] = [\text{observed value}] - [\text{predicted value}]$$

or

$$\hat{Z}_t = X_t - \hat{X}_t.$$

Since $\hat{X}_t$ is the one-step-ahead forecast from the time origin $t-1$, this is more clearly written as

$$\hat{Z}_t = X_t - \hat{X}_{t-1}(1).$$

From Theorem 9.13, the shifted AR(2) model is

$$X_t - \mu = \phi_1 (X_{t-1} - \mu) + \phi_2 (X_{t-2} - \mu) + Z_t$$

or

$$X_t = \mu + \phi_1 (X_{t-1} - \mu) + \phi_2 (X_{t-2} - \mu) + Z_t.$$

Taking the conditional expected value of both sides of this equation gives

$$E[X_t | X_1 = x_1, X_2 = x_2, \ldots, X_{t-1} = x_{t-1}] = \mu + \phi_1 (x_{t-1} - \mu) + \phi_2 (x_{t-2} - \mu).$$

Replacing the parameters by their point estimators, the one-step-ahead forecast from the time origin $t-1$ is

$$\hat{X}_{t-1}(1) = \hat{\mu} + \hat{\phi}_1 (x_{t-1} - \hat{\mu}) + \hat{\phi}_2 (x_{t-2} - \hat{\mu}).$$

Therefore, for the time series $x_1, x_2, \ldots, x_n$ and the fitted AR(2) model with parameter estimates $\hat{\mu}$, $\hat{\phi}_1$, and $\hat{\phi}_2$, the residual at time t is

$$\hat{Z}_t = x_t - \left[\hat{\mu} + \hat{\phi}_1 (x_{t-1} - \hat{\mu}) + \hat{\phi}_2 (x_{t-2} - \hat{\mu}) \right]$$

for $t = 3, 4, \ldots, n$. The next example shows the steps associated with assessing the adequacy of the AR(2) model for the Lake Huron lake level time series.

Example 9.19 Fit the AR(2) model to the Lake Huron levels from Example 9.14 using the sample mean to estimate μ and the maximum likelihood estimators for ϕ_1, ϕ_2, and σ_Z^2.

(a) Calculate and plot the residuals, their sample autocorrelation function, and their sample partial autocorrelation function.

(b) Conduct a test of independence on the residuals using the number of sample autocorrelation function values for the first $m = 40$ lags which fall outside of $\pm 1.96/\sqrt{n}$.

(c) Conduct the Box–Pierce and Ljung–Box tests for independence of the residuals.

(d) Conduct the turning point test for independence of the residuals.

(e) Plot a histogram and a QQ plot of the standardized residuals in order to assess the normality of the residuals.

(a) The following R commands calculate the $n - 2 = 96$ residuals and plot them as a time series, along with the associated sample autocorrelation function and sample partial autocorrelation function.

```
x        = LakeHuron
n        = length(x)
m        = 40
muhat    = mean(x)
fit      = ar(x, order.max = 2, aic = FALSE, method = "mle")
phi1hat  = fit$ar[1]
phi2hat  = fit$ar[2]
zhat     = x[3:n] - (muhat + phi1hat * (x[2:(n - 1)] - muhat) +
                            phi2hat * (x[1:(n - 2)] - muhat))
layout(matrix(c(1, 1, 2, 3), 2, 2, byrow = TRUE))
plot.ts(zhat)
acf(zhat, lag.max = m)
pacf(zhat, lag.max = m)
```

The results are displayed in Figure 9.23. The residuals do not appear to have any cyclic variation, trend, or serial correlation.

(b) There are no sample autocorrelation function values that fall outside of the limits $\pm 1.96/\sqrt{n}$ in the plot in Figure 9.23 of the first 40 sample autocorrelation function values associated with the residuals. Since we expect $40 \cdot 0.05 = 2$ values to fall outside of these limits in the case of a good fit, we fail to reject H_0 in this case. The fit of the AR(2) model is not rejected by this test.

(c) The additional R code below calculates the Box–Pierce test statistic and the Ljung–Box test statistic and the associated p-values using the built-in Box.test function.

```
Box.test(zhat, lag = 40, type = "Box-Pierce", fitdf = 3)
Box.test(zhat, lag = 40, type = "Ljung-Box", fitdf = 3)
```

The Box–Pierce test statistic is 18.7 and the associated p-value is $p = 0.995$. The Ljung–Box test statistic is 24.9 and the associated p-value is $p = 0.935$. We fail to reject H_0 in both tests based on the chi-square critical value with $40 - 3 = 37$ degrees of freedom. The fit of the AR(2) model is not rejected by these tests.

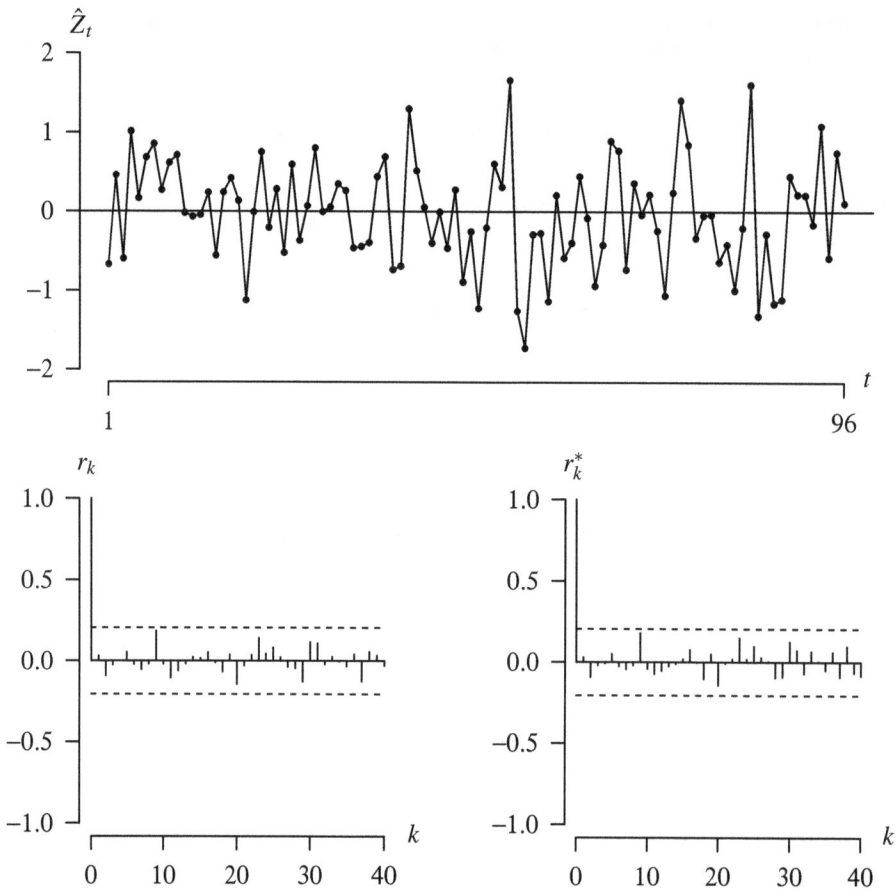

Figure 9.23: Time series plot, r_k, and r_k^* for $n-2 = 96$ residuals from AR(2) fitted model.

(d) The following additional R code calculates the test statistic and the p-value for the turning point test applied to the time series consisting of the $n-2 = 96$ residual values for the AR(2) fit to the Lake Huron time series.

```
n = n - 2
m = (2 / 3) * (n - 2)
v = (16 * n - 29) / 90
T = 0
for (i in 2:(n - 1)) {
  if ((zhat[i - 1] < zhat[i] && zhat[i] > zhat[i + 1]) ||
      (zhat[i - 1] > zhat[i] && zhat[i] < zhat[i + 1])) T = T + 1
}
s = (T - m) / sqrt(v)
2 * (1 - pnorm(abs(s)))
```

The tail probability is doubled because the alternative hypothesis is two-tailed for the turning point test. The test statistic s is 0.0815 and the p-value is $p = 0.94$. The turning point test found that there were $T = 63$ turning points in the time series of the residuals, and that is about the number that we expect to have if the residuals from the fitted AR(2) model were mutually independent random variables. We again fail to reject the null hypothesis in this case. The fit of the AR(2) model is not rejected by this test.

(e) The residuals are standardized by dividing by their sample standard deviation. The following additional R statements plot a histogram of the standardized residuals using the `hist` function and a QQ plot to assess normality using the `qqnorm` function.

```
hist(zhat / sd(zhat))
qqnorm(zhat / sd(zhat))
```

The plots are shown in Figure 9.24. The histogram shows that all standardized residuals fall between -3 and 3 and exhibit a roughly bell-shaped probability distribution. The horizontal axis on the histogram is the standardized residual and the vertical axis is the frequency. The QQ plot is approximately linear, indicating a reasonable approximation to normality based on the $n - 2 = 96$ residuals plotted. The horizontal axis on the QQ plot is the standardized theoretical quantile and the vertical axis is the associated normal data quantile. Although a formal statistical goodness-of-fit test (such as the Shapiro–Wilk or the Kolmogorov–Smirnov test) should be conducted, it appears that the assumption of Gaussian white noise is appropriate for the AR(2) time series model based on these two plots.

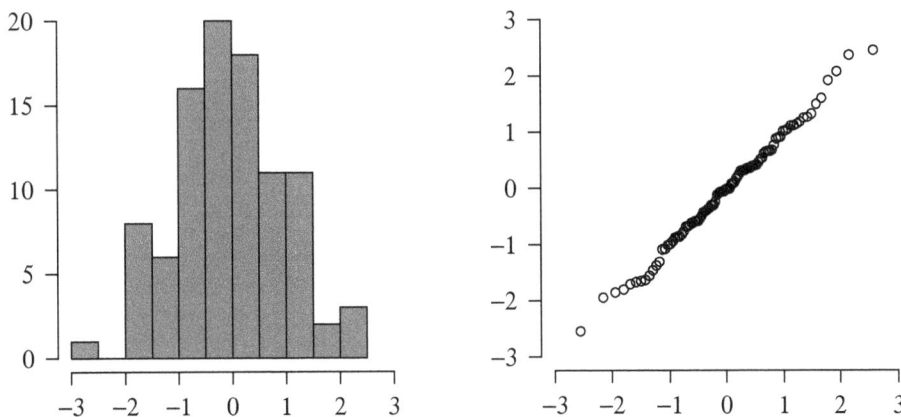

Figure 9.24: Histogram (left) and QQ plot (right) of the fitted AR(2) standardized residuals.

Model Selection

We have seen a number of indicators that the AR(2) time series model seems to be an adequate model for the Lake Huron lake level time series, with the exception of a linear trend apparent by viewing the time series in Figure 9.18. The model has not been rejected by any of the model adequacy tests. We now overfit the tentative AR(2) time series model with ARMA(p, q) models of higher order. We have not yet surveyed the techniques for estimating the parameters in these models with additional terms, so for now we will let the `arima` function in R estimate their parameters and compare them via their AIC (Akaike's Information Criterion) statistics. The AIC statistic was introduced in Section 8.2.4

Example 9.20 For the $n = 98$ levels of Lake Huron from Example 9.14, determine the ARMA(p, q) model that minimizes the AIC.

The R code below creates a 4×4 matrix a which will be populated with the AIC statistics for the ARMA(p, q) time series models, for $p = 0, 1, 2, 3$ and $q = 0, 1, 2, 3$ using nested for loops. The arima function is used to fit the models via maximum likelihood estimation, and the AIC values are placed in the matrix a.

```
a = matrix(0, 4, 4)
x = LakeHuron
for (p in 0:3)
  for (q in 0:3)
    a[p + 1, q + 1] = arima(x, order = c(p, 0, q), method = "ML")$aic
```

The results of this code are given in Table 9.7. The two smallest AIC values are set in boldface type; they correspond to the AR(2) and ARMA(1, 1) models. These two models seem to be close competitors for providing a probabilistic model for the time series.

	$q = 0$	$q = 1$	$q = 2$	$q = 3$
$p = 0$	335	255	231	222
$p = 1$	219	**214**	216	218
$p = 2$	**215**	216	218	220
$p = 3$	216	218	220	220

Table 9.7: AIC statistics for ARMA(p, q) models for the $n = 98$ lake water levels.

The $ extractor with the aic argument was used to extract the AIC statistics from the list returned by the call to arima. If the coef and sigma2 components are extracted from the list returned by the call to arima, our final model is the AR(2) model with maximum likelihood estimates for the parameters given by

$$\hat{\mu} = 579.05 \qquad \hat{\phi}_1 = 1.0436 \qquad \hat{\phi}_2 = -0.24949 \qquad \hat{\sigma}_Z^2 = 0.47882,$$

which corresponds to the fitted AR(2) model

$$X_t - 579.05 = 1.0436 (X_{t-1} - 579.05) - 0.24949 (X_{t-2} - 579.05) + Z_t,$$

where Z_t is a time series of Gaussian white noise values with $\sigma_Z^2 = 0.47822$, as established by the histogram and QQ plot in Example 9.19.

The analysis here suggests that this tentative fitted shifted AR(2) time series model should be compared with (a) a shifted ARMA(1, 1) model because of the lower value for its AIC for an identical number of parameters, and (b) a time series model based on removing the possible downward trend in the time series by using regression or differencing as described in Example 9.14.

Forecasting

We now consider forecasting future values of a time series that is governed by a shifted AR(2) time series model. In the case of the Lake Huron time series, this corresponds to the one-step-ahead forecast for 1973, the two-steps-ahead forecast for 1974, the three-steps-ahead forecast for 1975, etc. To review forecasting notation, the observed time series values are $x_1, x_2, \ldots, x_n$. The forecast is being made at time $t = n$. The random future value of the time series that is h time units in the future is denoted by X_{n+h}. The associated forecasted value is denoted by $\hat{X}_{n+h}$, and is the conditional expected value

$$\hat{X}_{n+h} = E[X_{n+h} | X_1 = x_1, X_2 = x_2, \ldots, X_n = x_n].$$

We would like to find this forecasted value and an associated prediction interval for a shifted AR(2) model. As in Section 8.2.2, we assume that all parameters are known in the derivations that follow. We also assume that the parameters ϕ_1 and ϕ_2 correspond to a stationary shifted AR(2) time series model.

The shifted AR(2) model is

$$X_t - \mu = \phi_1 (X_{t-1} - \mu) + \phi_2 (X_{t-2} - \mu) + Z_t.$$

Replacing t by $n+1$ and solving for X_{n+1}, this becomes

$$X_{n+1} = \mu + \phi_1 (X_n - \mu) + \phi_2 (X_{n-1} - \mu) + Z_{n+1}.$$

Taking the conditional expected value of each side of this equation results in the one-step-ahead forecast

$$\hat{X}_{n+1} = \mu + \phi_1 (x_n - \mu) + \phi_2 (x_{n-1} - \mu)$$

because x_{n-1} and x_n have already been observed in the time series $x_1, x_2, \ldots, x_n$. The forecasted value at time $n+1$ is a function of the last two values in the time series. Applying this same process to the predicted value at time $n+2$ results in the time series model

$$X_{n+2} = \mu + \phi_1 (X_{n+1} - \mu) + \phi_2 (X_n - \mu) + Z_{n+2}.$$

This time, the value of X_{n+1} has not been observed, so we replace it by its forecasted value when taking the conditional expected value of both sides of the equation

$$\hat{X}_{n+2} = \mu + \phi_1 (\hat{X}_{n+1} - \mu) + \phi_2 (x_n - \mu),$$

because x_n has already been observed. Continuing in this fashion, a recursive formula for the forecasted value of X_{n+h} is

$$\hat{X}_{n+h} = \mu + \phi_1 (\hat{X}_{n+h-1} - \mu) + \phi_2 (\hat{X}_{n+h-2} - \mu).$$

Although we would prefer an explicit formula, the recursive formula is easy to implement for an observed time series $x_1, x_2, \ldots, x_n$. As in the case of the AR(1) model, long-term forecasts for a stationary AR(2) time series model tend to μ as the time horizon $h \to \infty$.

We would like to pair our point estimator $\hat{X}_{n+h}$ with an interval estimator, which is a prediction interval in this setting. The prediction interval gives us an indication of the precision of the forecast. In order to derive an exact two-sided $100(1 - \alpha)\%$ prediction interval for X_{n+h}, it is helpful to write the shifted AR(2) model as a shifted MA(∞) model. The coefficients $\theta_1, \theta_2, \ldots$ of a stationary shifted AR(2) model written as an MA(∞) model

$$X_t = \mu + Z_t + \theta_1 Z_{t-1} + \theta_2 Z_{t-2} + \cdots$$

are given in terms of ϕ_1 and ϕ_2 in Theorem 9.12. Consider this model at time $t = n+1$. Since the error terms $Z_n, Z_{n-1}, Z_{n-2}, \ldots$ are unknown but fixed because they are associated with the observed time series $x_1, x_2, \ldots, x_n$, the conditional population variance of X_{n+1} is

$$V[X_{n+1}] = V[Z_{n+1}] = \sigma_Z^2$$

because the population variance of μ is zero and Z_{n+1} is the only random term in the model. The error terms at time n and prior are observed even though unknown and can therefore be treated as constants. Likewise, considering the MA(∞) model at time $t = n+2$, the conditional population variance of X_{n+2} is

$$V[X_{n+2}] = V[Z_{n+2} + \theta_1 Z_{n+1}] = \left(1 + \theta_1^2\right)\sigma_Z^2.$$

Similarly, the conditional population variance of X_{n+3} is

$$V[X_{n+3}] = V[Z_{n+3} + \theta_1 Z_{n+2} + \theta_2 Z_{n+1}] = \left(1 + \theta_1^2 + \theta_2^2\right)\sigma_Z^2.$$

Continuing in this fashion, the conditional population variance of X_{n+h} is

$$V[X_{n+h}] = \left(1 + \theta_1^2 + \theta_2^2 + \cdots + \theta_{h-1}^2\right)\sigma_Z^2.$$

If we assume that the white noise terms in the MA(∞) representation of the AR(2) time series model are Gaussian white noise terms, then X_{n+h} is also normally distributed because a linear combination of mutually independent normal random variables is also normally distributed. So an exact two-sided $100(1-\alpha)\%$ prediction interval for X_{n+h} is

$$\hat{X}_{n+h} - z_{\alpha/2}\sqrt{1 + \theta_1^2 + \theta_2^2 + \cdots + \theta_{h-1}^2}\,\sigma_Z < X_{n+h} < \hat{X}_{n+h} + z_{\alpha/2}\sqrt{1 + \theta_1^2 + \theta_2^2 + \cdots + \theta_{h-1}^2}\,\sigma_Z.$$

In most practical problems, the parameters in this prediction interval will be estimated from data, which results in the following *approximate* two-sided $100(1-\alpha)\%$ prediction interval.

Theorem 9.17 For a stationary shifted AR(2) time series model, a forecasted value of X_{n+h} can be found by the recursive equation

$$\hat{X}_{n+h} = \hat{\mu} + \hat{\phi}_1\left(\hat{X}_{n+h-1} - \hat{\mu}\right) + \hat{\phi}_2\left(\hat{X}_{n+h-2} - \hat{\mu}\right),$$

where $\hat{X}_{n+1} = \hat{\mu} + \hat{\phi}_1(x_n - \hat{\mu}) + \hat{\phi}_2(x_{n-1} - \hat{\mu})$. An approximate two-sided $100(1-\alpha)\%$ prediction interval for X_{n+h} is

$$\hat{X}_{n+h} - z_{\alpha/2}\sqrt{1 + \hat{\theta}_1^2 + \hat{\theta}_2^2 + \cdots + \hat{\theta}_{h-1}^2}\,\hat{\sigma}_Z < X_{n+h} < \hat{X}_{n+h} + z_{\alpha/2}\sqrt{1 + \hat{\theta}_1^2 + \hat{\theta}_2^2 + \cdots + \hat{\theta}_{h-1}^2}\,\hat{\sigma}_Z,$$

where $\hat{\theta}_1, \hat{\theta}_2, \ldots$ are the estimated coefficients in the MA(∞) model associated with the estimated AR(2) model.

Example 9.21 For the time series of Lake Huron levels $x_1, x_2, \ldots, x_{98}$ from Example 9.14, forecast the next five values (for years 1973–1977) in the time series and give approximate 95% prediction intervals for the forecasted values assuming that the time series arises from a shifted AR(2) model with parameters estimated by maximum likelihood.

The R code below uses the `ar` function to estimate the parameters in the shifted AR(2) model via maximum likelihood estimation. The `predict` function implements Theorem 9.17 to calculate the forecasted values and associated standard errors. These standard errors can be used to calculate approximate 95% prediction interval limits.

```
model = ar(LakeHuron, order.max = 2, aic = FALSE, method = "mle")
predict(model, n.ahead = 5)
```

The results are summarized in Table 9.8. Notice that the forecasts trend monotonically toward $\bar{x} = 579$ and the standard errors increase as the time horizon h increases. The increasing standard error is consistent with having less precision in the forecast as the time horizon h increases.

Time	$t = 99$	$t = 100$	$t = 101$	$t = 102$	$t = 103$
Year	1973	1974	1975	1976	1977
Forecast	579.79	579.59	579.43	579.31	579.23
Standard error	0.692	1.000	1.157	1.233	1.269
Lower prediction bound	578.43	577.63	577.16	576.89	576.74
Upper prediction bound	581.15	581.55	581.70	581.73	581.71

Table 9.8: Forecasts and 95% prediction intervals for the Lake Huron time series.

Figure 9.25 shows (a) the original time series $x_1, x_2, \ldots, x_{98}$ as points (•) connected by lines, (b) the first 10 forecasted lake levels $\hat{X}_{99}, \hat{X}_{100}, \ldots, \hat{X}_{108}$ as open circles (○), (c) the 95% prediction intervals as a shaded region, and (d) the next 10 actual average lake level values in July for the years 1973–1982 taken from the NOAA Great Lakes Experimental Research Laboratory website,

580.98, 581.04, 580.49, 580.52, 578.57, 578.96, 579.94, 579.77, 579.44, 578.97,

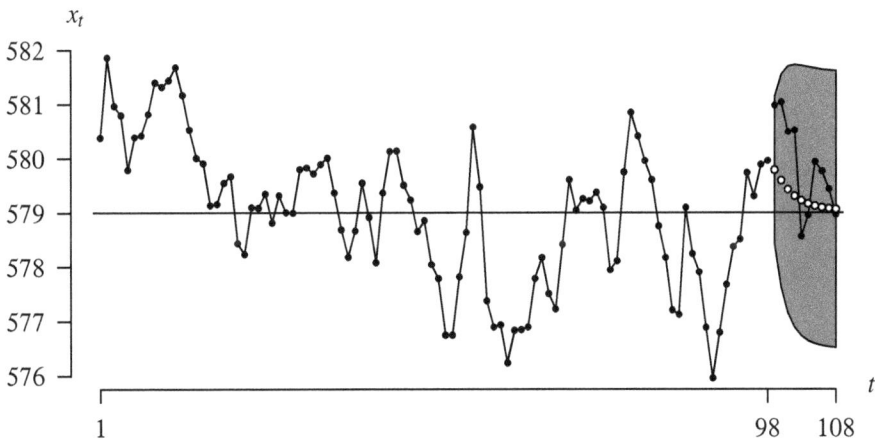

Figure 9.25: Lake Huron forecasts and 95% prediction intervals.

as points (•) connected by lines. There are four key observations concerning Figure 9.25.

- Even though the last three observations in the Lake Huron water level time series, $x_{96} = 579.31$, $x_{97} = 579.89$, and $x_{98} = 579.96$, show an increasing trend, the forecasts, which are a function only of $x_{n-1} = x_{97}$ and $x_n = x_{98}$, monotonically approach $\hat{\mu} = \bar{x} = 579$. The reason that the forecasts approach $\hat{\mu} = \bar{x} = 579$ in a damped exponential fashion is that the maximum likelihood estimators $\hat{\phi}_1$ and $\hat{\phi}_2$ satisfy $\hat{\phi}_1^2 + 4\hat{\phi}_2 > 0$, which indicates that the characteristic equation has two real roots which fall outside of the unit circle in the complex plane (see the proof of Theorem 9.9). Had the two roots been complex conjugates, the forecasted values would likewise approach $\hat{\mu} = \bar{x} = 579$, but in a damped sinusoidal fashion.

- The widths of the prediction intervals increase as the time horizon h increases. These widths do not increase indefinitely, but rather approach a limit as $h \to \infty$.

- The random sampling variability which is evident in the observed time series values $x_1, x_2, \ldots, x_{98}$ is not apparent in the forecasted values $\hat{X}_{99}, \hat{X}_{100}, \ldots, \hat{X}_{108}$. Observed time series values tend to exhibit the typical random sampling variability; forecasted values for a stationary shifted AR(2) time series model tend to be smooth.

- The first actual value in the forecast region, $x_{99} = 580.98$ for the year 1973, nearly falls outside of the associated 95% prediction interval. Even if the AR(2) model is a good fit for this time series, there is still a probability of approximately 0.05 that a future observation will fall outside of the associated 95% prediction interval. One value out of ten falling outside of the prediction intervals would not be shocking to see, assuming that a reasonable time series model has been formulated.

This section has introduced the AR(2) time series model. The important results for an AR(2) model are listed below.

- The standard AR(2) model can be written algebraically and with the backshift operator B as

$$X_t = \phi_1 X_{t-1} + \phi_2 X_{t-2} + Z_t \qquad \text{and} \qquad \phi(B)X_t = Z_t,$$

where $\phi(B) = 1 - \phi_1 B - \phi_2 B^2$ is the characteristic polynomial and $Z_t \sim WN\left(0, \sigma_Z^2\right)$ (Definition 9.2).

- The shifted AR(2) model can be written algebraically and with the backshift operator B as (Theorem 9.13)

$$X_t - \mu = \phi_1 (X_{t-1} - \mu) + \phi_2 (X_{t-2} - \mu) + Z_t \qquad \text{and} \qquad \phi(B)(X_t - \mu) = Z_t.$$

- The AR(2) model is always invertible; the AR(2) model is stationary when ϕ_1 and ϕ_2 fall in a triangular-shaped region in the (ϕ_1, ϕ_2) plane defined by the constraints $\phi_1 + \phi_2 < 1$, $\phi_2 - \phi_1 < 1$, and $\phi_2 > -1$ (Theorem 9.9).

- The AR(2) population autocorrelation function is a mixture of damped exponential functions, when $\phi(B)$ has real roots, or a damped sinusoidal function, when $\phi(B)$ has complex roots (Theorem 9.10).

- The AR(2) population partial autocorrelation function cuts off after lag 2 (Theorem 9.11), making its shape easier to recognize than the population autocorrelation function for the statistical counterparts associated with a realization of a time series.

- The stationary shifted AR(2) model can be written as a shifted MA(∞) model (Theorem 9.12).

- The four parameters in the shifted AR(2) model, μ, ϕ_1, ϕ_2 and σ_Z^2, can be estimated from a realization of a time series $x_1, x_2, \ldots, x_n$ by the method of moments (Theorem 9.14), least squares (Theorem 9.15), and maximum likelihood using at least $n = 60$ or $n = 70$ observations. The point estimators for μ, ϕ_1, ϕ_2, and σ_Z^2 are denoted by $\hat{\mu}$, $\hat{\phi}_1$, $\hat{\phi}_2$, and $\hat{\sigma}_Z^2$, and are typically paired with asymptotically exact two-sided $100(1 - \alpha)\%$ confidence intervals (Theorem 9.16).

- The forecasted value $\hat{X}_{n+h}$ in an AR(2) model is a function of x_{n-1} and x_n and can be calculated by a recursive formula. It approaches $\hat{\mu} = \bar{x}$ as the time horizon $h \to \infty$. The associated prediction intervals have widths that increase as h increases and approach a limit as the time horizon $h \to \infty$ (Theorem 9.17).

The AR(1) time series model expresses the current value in the time series X_t as a constant times the previous value in the time series plus a random shock. The AR(2) time series model expresses the current value in the time series X_t as a linear combination of the previous two values in the time series plus a random shock. There is conceptually no difficulty extending this thinking to the AR(p) time series model in which the current value in the time series X_t is expressed as a linear combination of the previous p values in the time series plus a random shock. The AR(p) time series model is the subject of the next section.

9.1.3 The AR(p) Model

The order p autoregressive model, denoted by AR(p), is a straightforward generalization of the AR(2) model. The use of matrices in the derivations will be novel, along with the inability to easily visualize the stationary region as a function of the parameters. The AR(p) model is appropriate in instances in which the current value of the time series is a linear combination of the p previous values in the time series plus a random shock.

Definition 9.3 An *order p autoregressive time series model*, denoted by AR(p), for the time series $\{X_t\}$ is defined by

$$X_t = \phi_1 X_{t-1} + \phi_2 X_{t-2} + \cdots + \phi_p X_{t-p} + Z_t,$$

where $\phi_1, \phi_2, \ldots, \phi_p$ are real-valued parameters and $\{Z_t\}$ is a time series of white noise:

$$Z_t \sim WN\left(0, \sigma_Z^2\right).$$

The $p + 1$ parameters that define an AR(p) model are the real-valued coefficients $\phi_1, \phi_2, \ldots, \phi_p$, and the population variance of the white noise σ_Z^2. The final coefficient, ϕ_p, must be nonzero. The AR(p) model can be written more compactly in terms of the backshift operator B as

$$\phi(B)X_t = Z_t,$$

where $\phi(B)$ is the order p characteristic polynomial

$$\phi(B) = 1 - \phi_1 B - \phi_2 B^2 - \cdots - \phi_p B^p.$$

The AR(p) model has the form of a multiple linear regression model with p independent variables and no intercept term. The current value X_t is being modeled as a linear combination of the p previous values of the time series, $X_{t-1}, X_{t-2}, \ldots, X_{t-p}$, plus a white noise term Z_t that provides a random shock to the model. The parameters $\phi_1, \phi_2, \ldots, \phi_p$ control the inclination of the regression line ($p = 1$), plane ($p = 2$), or hyperplane ($p > 2$). The σ_Z^2 parameter reflects the magnitude of the dispersion of the time series values about the regression plane.

Stationarity

Theorem 8.3 indicates that all AR(p) models are invertible, but are stationary when all of the roots of $\phi(B)$ lie outside of the unit circle in the complex plane. Let $B_1, B_2, \ldots, B_p$ denote the p solutions of $\phi(B) = 0$. For a stationary model, all of these roots will be real-valued or complex conjugate pairs that lie outside of the unit circle in the complex plane. Since $\phi(B_1) = \phi(B_2) = \cdots = \phi(B_p) = 0$, the order p characteristic polynomial $\phi(B)$ can also be written in factored form as

$$\phi(B) = \left(1 - B_1^{-1}B\right)\left(1 - B_2^{-1}B\right)\ldots\left(1 - B_p^{-1}B\right).$$

Unfortunately, except for the cases of $p = 1$ and $p = 2$, the region in the space of $(\phi_1, \phi_2, \ldots, \phi_p)$ corresponding to a stationary model cannot be expressed in a simple mathematical form. The following example illustrates how to determine whether an AR(4) model is stationary. This AR(4) model will be used in the next five examples.

Example 9.22 Determine whether the AR(4) model with characteristic polynomial

$$\phi(B) = 1 - \frac{21}{20}B - \frac{1}{20}B^2 + \frac{23}{40}B^3 - \frac{3}{10}B^4$$

is stationary.

The AR(4) model is stationary if all of the roots of $\phi(B)$ lie outside the unit circle in the complex plane. The characteristic polynomial can be factored as

$$\phi(B) = -\frac{1}{40}(4B-5)(3B+4)\left(B^2 - 2B + 2\right).$$

Using the quadratic formula, the solutions of $\phi(B) = 0$ are

$$B_1 = \frac{5}{4} \qquad B_2 = -\frac{4}{3} \qquad B_3 = 1+i \qquad B_4 = 1-i.$$

The first two roots are real-valued, and the other two roots are complex-valued conjugates. The four roots are plotted in Figure 9.26. Since all four roots lie outside of the unit circle in the complex plane, this AR(4) model is stationary.

Duality

As was the case with the AR(1) and AR(2) time series models, a stationary AR(p) time series model can be written as an MA(∞) time series model. This alternative representation can be useful for estimating standard errors of forecasted values. One way to frame the problem of writing an AR(p) time series model as an MA(∞) time series model is to write the compact form of the AR(p) model as

$$\phi(B)X_t = Z_t$$

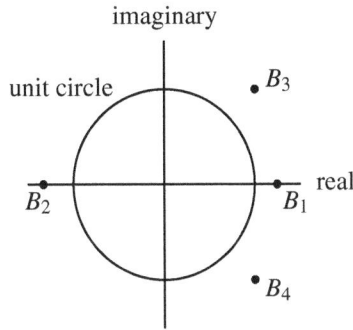

Figure 9.26: Unit circle in the complex plane and the solutions of $\phi(B) = 0$.

and divide both sides by $\phi(B)$, which results in

$$X_t = \frac{Z_t}{\phi(B)}.$$

Therefore, the conversion from the AR(p) form of the model to the MA(∞) form involves finding the coefficients $\theta_1, \theta_2, \ldots$ such that

$$X_t = \frac{Z_t}{\phi(B)} = \left(1 + \theta_1 B + \theta_2 B^2 + \cdots\right) Z_t.$$

The coefficients $\theta_1, \theta_2, \ldots$ essentially correspond to finding the inverse of the $\phi(B)$ characteristic polynomial. Taking the expected value of both sides of this equation leads to the important result: $E[X_t] = 0$ for all values of t. As was the case of the AR(2) time series model, the coefficients for the MA(∞) time series model are found by equating coefficients. This process will be illustrated in the next example for the AR(4) model. Generalization to the AR(p) model is straightforward.

Example 9.23 Calculate the first six coefficients of the MA(∞) model associated with the stationary AR(4) model from Example 9.22 with characteristic polynomial

$$\phi(B) = 1 - \frac{21}{20}B - \frac{1}{20}B^2 + \frac{23}{40}B^3 - \frac{3}{10}B^4.$$

The MA(∞) model has the form

$$X_t = Z_t + \theta_1 Z_{t-1} + \theta_2 Z_{t-2} + \theta_3 Z_{t-3} + \theta_4 Z_{t-4} + \cdots.$$

The AR(4) model

$$X_t = \phi_1 X_{t-1} + \phi_2 X_{t-2} + \phi_3 X_{t-3} + \phi_4 X_{t-4} + Z_t$$

can be written in terms of θ_1, θ_2, ... as

$$Z_t + \theta_1 Z_{t-1} + \theta_2 Z_{t-2} + \theta_3 Z_{t-3} + \theta_4 Z_{t-4} + \cdots =$$
$$\phi_1 \left(Z_{t-1} + \theta_1 Z_{t-2} + \theta_2 Z_{t-3} + \theta_3 Z_{t-4} + \cdots \right) +$$
$$\phi_2 \left(Z_{t-2} + \theta_1 Z_{t-3} + \theta_2 Z_{t-4} + \cdots \right) +$$
$$\phi_3 \left(Z_{t-3} + \theta_1 Z_{t-4} + \cdots \right) +$$
$$\phi_4 \left(Z_{t-4} + \cdots \right) + Z_t.$$

Equating the coefficients of Z_{t-1} gives

$$\theta_1 = \phi_1.$$

Equating the coefficients of Z_{t-2} gives

$$\theta_2 = \phi_1 \theta_1 + \phi_2 = \phi_1^2 + \phi_2.$$

Equating the coefficients of Z_{t-3} and simplifying gives

$$\theta_3 = \phi_1^3 + 2\phi_1\phi_2 + \phi_3.$$

Equating the coefficients of Z_{t-4} and simplifying gives

$$\theta_4 = \phi_1^4 + 3\phi_1^2\phi_2 + \phi_2^2 + 2\phi_1\phi_3 + \phi_4.$$

Equating the coefficients of Z_{t-k} gives the recursive equation

$$\theta_k = \phi_1 \theta_{k-1} + \phi_2 \theta_{k-2} + \phi_3 \theta_{k-3} + \phi_4 \theta_{k-4}$$

for $k = 5, 6, \ldots$. The coefficients of the AR(4) model of interest are

$$\phi_1 = \frac{21}{20} \qquad \phi_2 = \frac{1}{20} \qquad \phi_3 = -\frac{23}{40} \qquad \phi_4 = \frac{3}{10}.$$

Using the equations derived here, the first six coefficients of the associated MA(∞) model as exact fractions are

$$\theta_1 = \frac{21}{20}, \; \theta_2 = \frac{461}{400}, \; \theta_3 = \frac{5501}{8000}, \; \theta_4 = \frac{76141}{160000}, \; \theta_5 = \frac{596381}{3200000}, \; \theta_6 = \frac{10870221}{64000000}.$$

The `ARMAtoMA` function in R can also compute these coefficients as follows.

```
ARMAtoMA(ar = c(21 / 20, 1 / 20, -23 / 40, 3 / 10), ma = 0, lag.max = 6)
```

This R command returns the decimal approximations of the exact fractions:

$$\theta_1 = 1.05, \; \theta_2 = 1.1525, \; \theta_3 = 0.6876, \; \theta_4 = 0.4759, \; \theta_5 = 0.1864, \; \theta_6 = 0.1698.$$

Population Autocorrelation Function

We now pivot to the derivation of the population autocovariance and autocorrelation functions. Assuming that the parameters $\phi_1, \phi_2, \ldots, \phi_p$ are associated with a stationary model, the AR(p) model

$$X_t = \phi_1 X_{t-1} + \phi_2 X_{t-2} + \cdots + \phi_p X_{t-p} + Z_t$$

can be multiplied by X_{t-k} to give

$$X_t X_{t-k} = \phi_1 X_{t-1} X_{t-k} + \phi_2 X_{t-2} X_{t-k} + \cdots + \phi_p X_{t-p} X_{t-k} + Z_t X_{t-k}.$$

Taking the expected value of both sides of this equation for $k = 0$ results in

$$\gamma(0) = \phi_1 \gamma(1) + \phi_2 \gamma(2) + \cdots + \phi_p \gamma(p) + \sigma_Z^2$$

and the recursive equation

$$\gamma(k) = \phi_1 \gamma(k-1) + \phi_2 \gamma(k-2) + \cdots + \phi_p \gamma(k-p)$$

for $k = 1, 2, \ldots$ because Z_t has expected value zero and is independent of X_{t-k}. For $k = 1, 2, \ldots, p$, the recursive equation can be written as the system of linear equations

$$
\begin{aligned}
\gamma(1) &= \phi_1 \gamma(0) + \phi_2 \gamma(1) + \phi_3 \gamma(2) + \cdots + \phi_p \gamma(p-1) \\
\gamma(2) &= \phi_1 \gamma(1) + \phi_2 \gamma(0) + \phi_3 \gamma(1) + \cdots + \phi_p \gamma(p-2) \\
\gamma(3) &= \phi_1 \gamma(2) + \phi_2 \gamma(1) + \phi_3 \gamma(0) + \cdots + \phi_p \gamma(p-3) \\
\vdots \ &= \ \vdots \\
\gamma(p) &= \phi_1 \gamma(p-1) + \phi_2 \gamma(p-2) + \phi_3 \gamma(p-3) + \cdots + \phi_p \gamma(0),
\end{aligned}
$$

which relies on the symmetry of the population autocovariance function: $\gamma(-k) = \gamma(k)$. This linear system of p equations in $p+1$ unknowns can be written in matrix form as

$$\gamma = \Gamma \phi,$$

where

$$
\gamma = \begin{bmatrix} \gamma(1) \\ \gamma(2) \\ \gamma(3) \\ \vdots \\ \gamma(p) \end{bmatrix}, \qquad
\Gamma = \begin{bmatrix}
\gamma(0) & \gamma(1) & \gamma(2) & \cdots & \gamma(p-1) \\
\gamma(1) & \gamma(0) & \gamma(1) & \cdots & \gamma(p-2) \\
\gamma(2) & \gamma(1) & \gamma(0) & \cdots & \gamma(p-3) \\
\vdots & \vdots & \vdots & \ddots & \vdots \\
\gamma(p-1) & \gamma(p-2) & \gamma(p-3) & \cdots & \gamma(0)
\end{bmatrix}, \qquad
\phi = \begin{bmatrix} \phi_1 \\ \phi_2 \\ \phi_3 \\ \vdots \\ \phi_p \end{bmatrix}.
$$

Given the values of the parameters $\phi_1, \phi_2, \ldots, \phi_p$, and σ_Z^2, this set of linear equations and

$$\gamma(0) = \phi_1 \gamma(1) + \phi_2 \gamma(2) + \cdots + \phi_p \gamma(p) + \sigma_Z^2,$$

one can compute the first $p+1$ population autocovariances $\gamma(0), \gamma(1), \ldots, \gamma(p)$ by solving these linear equations. The recursion relationship can be used to compute subsequent autocovariances.

Example 9.24 Calculate the initial values of the population autocovariance function $\gamma(0), \gamma(1), \ldots, \gamma(6)$ associated with the stationary AR(4) model with characteristic polynomial

$$\phi(B) = 1 - \frac{21}{20} B - \frac{1}{20} B^2 + \frac{23}{40} B^3 - \frac{3}{10} B^4$$

and white noise variance $\sigma_Z^2 = 1$.

The coefficients of the AR(4) model of interest are

$$\phi_1 = \frac{21}{20} \qquad \phi_2 = \frac{1}{20} \qquad \phi_3 = -\frac{23}{40} \qquad \phi_4 = \frac{3}{10}.$$

To find the initial population autocovariances, solve the 5×5 set of linear equations

$$\gamma(0) = \phi_1\gamma(1) + \phi_2\gamma(2) + \phi_3\gamma(3) + \phi_4\gamma(4) + \sigma_Z^2$$
$$\gamma(1) = \phi_1\gamma(0) + \phi_2\gamma(1) + \phi_3\gamma(2) + \phi_4\gamma(3)$$
$$\gamma(2) = \phi_1\gamma(1) + \phi_2\gamma(0) + \phi_3\gamma(1) + \phi_4\gamma(2)$$
$$\gamma(3) = \phi_1\gamma(2) + \phi_2\gamma(1) + \phi_3\gamma(0) + \phi_4\gamma(1)$$
$$\gamma(4) = \phi_1\gamma(3) + \phi_2\gamma(2) + \phi_3\gamma(1) + \phi_4\gamma(0)$$

for $\gamma(0), \gamma(1), \gamma(2), \gamma(3), \gamma(4)$. The recursive equation

$$\gamma(k) = \phi_1\gamma(k-1) + \phi_2\gamma(k-2) + \phi_3\gamma(k-3) + \phi_4\gamma(k-4)$$

can be used to calculate $\gamma(k)$ values for $k = 5, 6, \ldots$. The initial population autocovariance values are

$$\gamma(0) = \frac{3520}{819} \cong 4.298, \ \gamma(1) = \frac{2960}{819} \cong 3.614, \ \gamma(2) = \frac{2260}{819} \cong 2.759, \ \gamma(3) = \frac{1385}{819} \cong 1.691,$$

$$\gamma(4) = \frac{3685}{3276} \cong 1.125, \ \gamma(5) = \frac{10001}{13104} \cong 0.763, \ \gamma(6) = \frac{186881}{262080} \cong 0.713.$$

These population autocovariances can be used to calculate the associated population autocorrelations by dividing each of them by $\gamma(0)$.

Dividing both sides of the recursive equation for calculating population autocovariance by $\gamma(0) = V[X_t]$ gives the recursive equation

$$\rho(k) = \phi_1\rho(k-1) + \phi_2\rho(k-2) + \cdots + \phi_p\rho(k-p)$$

for $k = 1, 2, \ldots$. Exploiting the symmetry of the $\rho(k)$ function, the first p of these equations are

$$\rho(1) = \phi_1\rho(0) + \phi_2\rho(1) + \phi_3\rho(2) + \cdots + \phi_p\rho(p-1)$$
$$\rho(2) = \phi_1\rho(1) + \phi_2\rho(0) + \phi_3\rho(1) + \cdots + \phi_p\rho(p-2)$$
$$\rho(3) = \phi_1\rho(2) + \phi_2\rho(1) + \phi_3\rho(0) + \cdots + \phi_p\rho(p-3)$$
$$\vdots \quad = \quad \vdots$$
$$\rho(p) = \phi_1\rho(p-1) + \phi_2\rho(p-2) + \phi_3\rho(p-3) + \cdots + \phi_p\rho(0).$$

Since $\rho(0) = 1$, this linear system of p equations in the p unknowns $\rho(1), \rho(2), \ldots, \rho(p)$ can be written in matrix form as

$$\rho = P\phi,$$

where

$$\rho = \begin{bmatrix} \rho(1) \\ \rho(2) \\ \rho(3) \\ \vdots \\ \rho(p) \end{bmatrix}, \quad P = \begin{bmatrix} 1 & \rho(1) & \rho(2) & \cdots & \rho(p-1) \\ \rho(1) & 1 & \rho(1) & \cdots & \rho(p-2) \\ \rho(2) & \rho(1) & 1 & \cdots & \rho(p-3) \\ \vdots & \vdots & \vdots & \ddots & \vdots \\ \rho(p-1) & \rho(p-2) & \rho(p-3) & \cdots & 1 \end{bmatrix}, \quad \phi = \begin{bmatrix} \phi_1 \\ \phi_2 \\ \phi_3 \\ \vdots \\ \phi_p \end{bmatrix}.$$

Given the values of the parameters $\phi_1, \phi_2, \ldots, \phi_p$, these linear equations can be solved for the initial p population autocorrelation function values $\rho(1), \rho(2), \ldots, \rho(p)$, and the recursive function can be used to calculate subsequent values of the population autocorrelation values.

As was the case with the AR(2) time series model, (a) the real roots of $\phi(B)$ correspond to contributions to the population autocorrelation function which are mixtures of damped exponential terms, and (b) the complex conjugate roots of $\phi(B)$ correspond to contributions to the population autocorrelation function which are damped sinusoidal terms.

These equations bear some practical use in that the first p *sample* autocorrelation function values, $r_1, r_2, \ldots, r_p$, can be calculated from an observed time series and used as approximations for $\rho(1), \rho(2), \ldots, \rho(p)$, yielding estimators for $\phi_1, \phi_2, \ldots, \phi_p$. These estimates are known as the *Yule–Walker estimators*. These can in turn be used as initial estimates for finding point estimates for $\phi_1, \phi_2, \ldots, \phi_p$ by, for example, least squares or maximum likelihood estimation, should numerical methods be required.

The results concerning the calculation of the population autocovariance function $\gamma(k)$ and the population autocorrelation function $\rho(k)$ are summarized below.

Theorem 9.18 The population autocovariance function for a stationary AR(p) time series model is calculated by

$$\gamma(k) = \phi_1\gamma(k-1) + \phi_2\gamma(k-2) + \cdots + \phi_p\gamma(k-p)$$

for $k = 1, 2, \ldots$. The first p of these equations can be written in matrix form as

$$\gamma = \Gamma\phi.$$

The population variance of X_t is

$$V[X_t] = \gamma(0) = \phi_1\gamma(1) + \phi_2\gamma(2) + \cdots + \phi_p\gamma(p) + \sigma_Z^2.$$

The population autocorrelation function for a stationary AR(p) time series model is calculated by

$$\rho(k) = \phi_1\rho(k-1) + \phi_2\rho(k-2) + \cdots + \phi_p\rho(k-p)$$

for $k = 1, 2, \ldots$. The first p of these equations can be written in matrix form as

$$\rho = P\phi.$$

The system of linear equations in Theorem 9.18, whether written in terms of $\gamma(k)$ or $\rho(k)$ as $\gamma = \Gamma\phi$ or $\rho = P\phi$, is known in time series analysis as the *Yule–Walker equations*.

Population Partial Autocorrelation Function

We now determine the population partial autocorrelation function for an AR(p) model. Using Definition 7.4, the initial population partial autocorrelation values are

$$\rho^*(0) = 1, \quad \rho^*(1) = \rho(1), \quad \rho^*(2) = \frac{\begin{vmatrix} 1 & \rho(1) \\ \rho(1) & \rho(2) \end{vmatrix}}{\begin{vmatrix} 1 & \rho(1) \\ \rho(1) & 1 \end{vmatrix}}, \quad \rho^*(3) = \frac{\begin{vmatrix} 1 & \rho(1) & \rho(1) \\ \rho(1) & 1 & \rho(2) \\ \rho(2) & \rho(1) & \rho(3) \end{vmatrix}}{\begin{vmatrix} 1 & \rho(1) & \rho(2) \\ \rho(1) & 1 & \rho(1) \\ \rho(2) & \rho(1) & 1 \end{vmatrix}},$$

etc. One distinctive characteristic of the $AR(p)$ population partial autocorrelation function is that it cuts off after lag p. To see why this is the case, consider the first p columns of the matrix in the numerator of $\rho^*(k)$ for $k > p$:

$$\begin{bmatrix} 1 \\ \rho(1) \\ \rho(2) \\ \vdots \\ \rho(k-1) \end{bmatrix}, \begin{bmatrix} \rho(1) \\ 1 \\ \rho(1) \\ \vdots \\ \rho(k-2) \end{bmatrix}, \dots, \begin{bmatrix} \rho(p-1) \\ \rho(p-2) \\ \rho(p-3) \\ \vdots \\ \rho(k-p) \end{bmatrix}.$$

Using Theorem 9.18, the last column of the matrix in the numerator of $\rho^*(k)$ is

$$\begin{bmatrix} \phi_1 + \phi_2\rho(1) + \phi_3\rho(2) + \cdots + \phi_p\rho(p-1) \\ \phi_1\rho(1) + \phi_2 + \phi_3\rho(1) + \cdots + \phi_p\rho(p-2) \\ \phi_1\rho(2) + \phi_2\rho(1) + \phi_3 + \cdots + \phi_p\rho(p-3) \\ \vdots \\ \phi_1\rho(k-1) + \phi_2\rho(k-2) + \phi_3\rho(k-3) + \cdots + \phi_p\rho(k-p) \end{bmatrix}.$$

The last column of the matrix in the numerator of $\rho^*(k)$ is a linear combination of the first p columns with coefficients $\phi_1, \phi_2, \dots, \phi_p$. Thus, the matrix in the numerator of the calculation of $\rho^*(k)$ is singular, which means that its determinant is zero. So $\rho^*(k) = 0$ for $k = p+1, p+2, \dots$ for an $AR(p)$ time series model. This constitutes a proof of the following result.

Theorem 9.19 The population partial autocorrelation function for a stationary $AR(p)$ time series model cuts off after lag p.

A graph of the *sample* partial autocorrelation function r_k^* for the first few values of k, should also cut off after lag p if the $AR(p)$ model is appropriate. This sample partial autocorrelation function shape is easier to recognize than the associated sample autocorrelation function shape because cutting off is typically easier to recognize than tailing off in the presence of random sampling variability.

There is a second interpretation of the partial autocorrelation function that ties it more closely to determining the order of the autoregressive portion of the model. The partial autocorrelation at lag k is the value of the final coefficient ϕ_k in an autoregressive model of order k. This coefficient measures the excess correlation at lag k which is not accounted for by an autoregressive model of order $k-1$. It is for this reason that many authors use the notation ϕ_{kk} for the population lag k partial autocorrelation.

The population autocorrelation function and the population partial autocorrelation functions can be calculated using the formulas given here, but can also be calculated using the R ARMAacf function, as illustrated in the next example.

> **Example 9.25** Calculate and plot the values of the population autocorrelation function and the population partial autocorrelation function associated with the $AR(4)$ model with characteristic polynomial
>
> $$\phi(B) = 1 - \frac{21}{20}B - \frac{1}{20}B^2 + \frac{23}{40}B^3 - \frac{3}{10}B^4.$$
>
> The coefficients of the $AR(4)$ model of interest are
>
> $$\phi_1 = \frac{21}{20} \qquad \phi_2 = \frac{1}{20} \qquad \phi_3 = -\frac{23}{40} \qquad \phi_4 = \frac{3}{10}.$$

The matrix equation $\rho = P\phi$ from Theorem 9.18 could be solved for the initial values of $\rho(k)$. Alternatively, the values of $\gamma(k)$ calculated in the previous example could be divided by $\gamma(0)$ to arrive at the population autocorrelation function values. The first two such autocorrelation function values, for example, are

$$\rho(1) = \frac{\gamma(1)}{\gamma(0)} = \frac{2960/819}{3520/819} = \frac{2960}{3520} \cong 0.8409,$$

$$\rho(2) = \frac{\gamma(2)}{\gamma(0)} = \frac{2260/819}{3520/819} = \frac{2260}{3520} \cong 0.6420,$$

The R function `ARMAacf` can also be used to calculate the population autocorrelation function values for the first 15 lags, as illustrated below. The `ar` argument is a vector containing the coefficients ϕ_1, ϕ_2, ϕ_3, and ϕ_4, and the `ma` argument is set to zero because there are no moving average terms.

```
ARMAacf(ar = c(21 / 20, 1 / 20, -23 / 40, 3 / 10), ma = 0, 15)
```

The population partial autocorrelation function values can be computed by taking the ratios of the determinants from Definition 7.4. Alternatively, the `pacf` argument to the `ARMAacf` function can be set to `TRUE` to compute the values of $\rho^*(k)$ for the first 15 lags.

```
ARMAacf(ar = c(21 / 20, 1 / 20, -23 / 40, 3 / 10), ma = 0, 15, pacf = TRUE)
```

Table 9.9 contains the numeric values of the first seven values of $\rho(k)$ and $\rho^*(k)$. Figure 9.27 contains a plot of $\rho(k)$ and $\rho^*(k)$ for the first 15 lags. The population autocorrelation function includes the effects of mixtures of damped exponential terms (associated with the two real roots $B_1 = 5/4$ and $B_2 = -4/3$ of $\phi(B) = 0$ computed in Example 9.22) and damped sinusoidal terms (associated with the two complex roots $B_3 = 1 + i$ and $B_4 = 1 - i$ of $\phi(B) = 0$ computed in Example 9.22). As expected, the population partial autocorrelation function cuts off after lag 4.

k	1	2	3	4	5	6	7
$\rho(k)$	0.8409	0.6420	0.3935	0.2617	0.1776	0.1659	0.1506
$\rho^*(k)$	0.8409	-0.2222	-0.2857	0.3000	0	0	0

Table 9.9: The first seven values of $\rho(k)$ and $\rho^*(k)$ for an AR(4) time series model.

The Shifted AR(p) Model

The standard AR(p) model from Definition 9.3 is not of much practical use because most real-world time series are not centered around zero. Adding a shift parameter μ overcomes this shortcoming. Since population variance and covariance are unaffected by a shift, the associated population autocorrelation and partial autocorrelation functions remain the same as those given in Theorems 9.18 and 9.19.

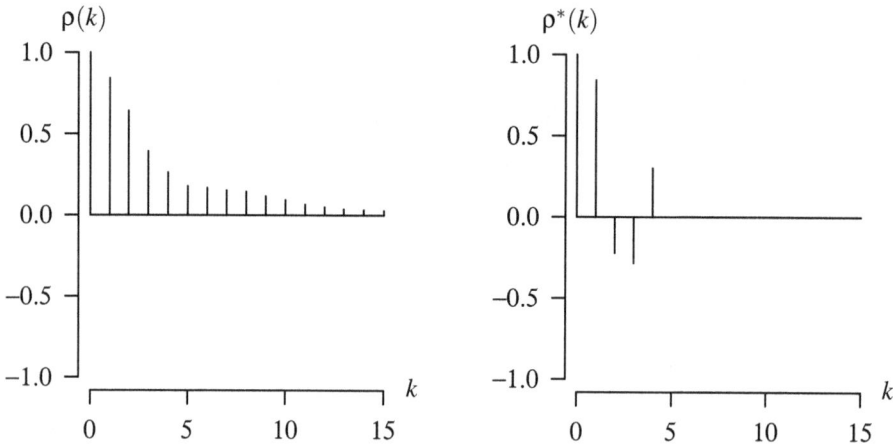

Figure 9.27: The first 15 values of $\rho(k)$ and $\rho^*(k)$ for an AR(4) time series model.

Theorem 9.20 A *shifted order p autoregressive model* for the time series $\{X_t\}$ is defined by

$$X_t - \mu = \phi_1\,(X_{t-1} - \mu) + \phi_2\,(X_{t-2} - \mu) + \cdots + \phi_p\,(X_{t-p} - \mu) + Z_t,$$

where $\phi_1, \phi_2, \ldots, \phi_p$, μ, and $\sigma_Z^2 > 0$ are real-valued parameters, and $\{Z_t\}$ is a time series of white noise. This model is stationary when all of the roots of the characteristic equation $\phi(B) = 0$ fall outside of the unit circle in the complex plane. The expected value of X_t is $E[X_t] = \mu$. The population autocorrelation function can be calculated using the recursive equations in Theorem 9.18. The population partial autocorrelation function can be calculated using the defining formulas in Definition 7.4.

The shifted AR(p) model can be written in terms of the backshift operator B as

$$\phi(B)(X_t - \mu) = Z_t,$$

where $\phi(B) = 1 - \phi_1 B - \phi_2 B^2 - \cdots - \phi_p B^p$. The practical problem of fitting a shifted AR(p) model to an observed time series of n values $x_1, x_2, \ldots, x_n$ will be illustrated later in this subsection.

Simulation

An AR(p) time series can be simulated by appealing to the defining formula for the AR(p) model. Iteratively applying the defining formula for a standard AR(p) model

$$X_t = \phi_1 X_{t-1} + \phi_2 X_{t-2} + \cdots + \phi_p X_{t-p} + Z_t$$

from Definition 9.3 results in the simulated values $X_1, X_2, \ldots, X_n$. The difficult aspect of devising a simulation algorithm is generating the first p simulated values, $X_1, X_2, \ldots, X_p$. For simplicity, assume that the white noise terms are Gaussian white noise terms. There are two approaches to overcome this initialization problem. The first approach generates $X_1, X_2, \ldots, X_p$ from a multivariate normal distribution with population mean p-vector $\mathbf{0} = (0, 0, \ldots, 0)'$ and $p \times p$ variance–covariance matrix

$$
\Gamma = \begin{bmatrix}
\gamma(0) & \gamma(1) & \gamma(2) & \cdots & \gamma(p-1) \\
\gamma(1) & \gamma(0) & \gamma(1) & \cdots & \gamma(p-2) \\
\gamma(2) & \gamma(1) & \gamma(0) & \cdots & \gamma(p-3) \\
\vdots & \vdots & \vdots & \ddots & \vdots \\
\gamma(p-1) & \gamma(p-2) & \gamma(p-3) & \cdots & \gamma(0)
\end{bmatrix},
$$

which was defined in Theorem 9.18. The algorithm given below generates initial time series observations $X_1, X_2, \ldots, X_p$ as indicated above, and then uses an additional $n - p$ Gaussian white noise terms $Z_{p+1}, Z_{p+2}, \ldots, Z_n$ to generate the remaining time series values $X_{p+1}, X_{p+2}, \ldots, X_n$ using the AR(p) defining formula from Definition 9.3. Indentation denotes nesting in the algorithm.

> generate $(X_1, X_2, \ldots, X_p) \sim N(\mathbf{0}, \Gamma)$
> $t \leftarrow p$
> while $(t < n)$
> > $t \leftarrow t + 1$
> > generate $Z_t \sim N\left(0, \sigma_Z^2\right)$
> > $X_t \leftarrow \phi_1 X_{t-1} + \phi_2 X_{t-2} + \cdots + \phi_p X_{t-p} + Z_t$

The $(p+2)$-parameter shifted AR(p) time series model which includes a population mean parameter μ can be simulated by simply adding μ to each time series observation generated by this algorithm. The next example implements this algorithm in R.

Example 9.26 Generate a realization of $n = 100$ observations from the stationary AR(4) time series model with

$$
\phi_1 = \frac{21}{20} \qquad \phi_2 = \frac{1}{20} \qquad \phi_3 = -\frac{23}{40} \qquad \phi_4 = \frac{3}{10}
$$

and Gaussian white noise error terms with $\sigma_Z^2 = 1$.

This model is stationary (see Example 9.22). The population autocorrelation function $\rho(k)$ and the population partial autocorrelation function $\rho^*(k)$ are displayed in Figure 9.27; we expect similar shaped functions r_k and r_k^* from our simulated values. The first statement in the R code below uses the `set.seed` function to establish the random number seed. The second statement sets $p = 4$, corresponding to an AR(4) model. The third statement sets the vector `phi` to the AR(4) coefficients $\phi_1 = 21/20$, $\phi_2 = 1/20$, $\phi_3 = -23/40$, and $\phi_4 = 3/10$. The fourth statement places the initial population autocovariance values from Example 9.24, namely $\gamma(0) = 3520/819$, $\gamma(1) = 2960/819$, $\gamma(2) = 2260/819$, and $\gamma(3) = 1385/819$, into the vector `gam`. The subsequent nested `for` loops place these population autocovariance values in the 4×4 variance–covariance matrix `GAMMA`. The next statement sets the standard deviation of the Gaussian white noise to $\sigma_Z = 1$. The next statement sets the number of simulated values to $n = 100$. The next statement defines the vector `x` of length $n = 100$ to hold the simulated time series values. The next statement uses the `mvrnorm` function from the `MASS` package to generate the first four simulated time series observations X_1, X_2, X_3, X_4 from the appropriate multivariate normal distribution. Finally, the `for` loop iterates through the defining formula for the AR(4) model generating the remaining observations $X_5, X_6, \ldots, X_{100}$.

```
set.seed(9)
p       = 4
```

```
phi     = c(21 / 20, 1 / 20, -23 / 40, 3 / 10)
gam     = c(3520 / 819, 2960 / 819, 2260 / 819, 1385 / 819)
GAMMA   = matrix(0, p, p)
for (i in 1:p) for (j in 1:p) GAMMA[i, j] = gam[abs(i - j) + 1]
sigz    = 1
n       = 100
x       = numeric(n)
x[1:p] = MASS::mvrnorm(1, mu = rep(0, p), Sigma = GAMMA)
for (t in (p + 1):n) x[t] = sum(phi * x[(t - 1):(t - p)]) +
                       rnorm(1, 0, sigz)
```

Using the plot.ts function to make a plot of the time series contained in x, the acf function to plot the associated correlogram, the pacf function to plot the associated sample partial autocorrelation function, and the layout function to arrange the graphs as in Example 7.24, the resulting trio of graphs are displayed in Figure 9.28. The sample partial autocorrelation function has four statisti-

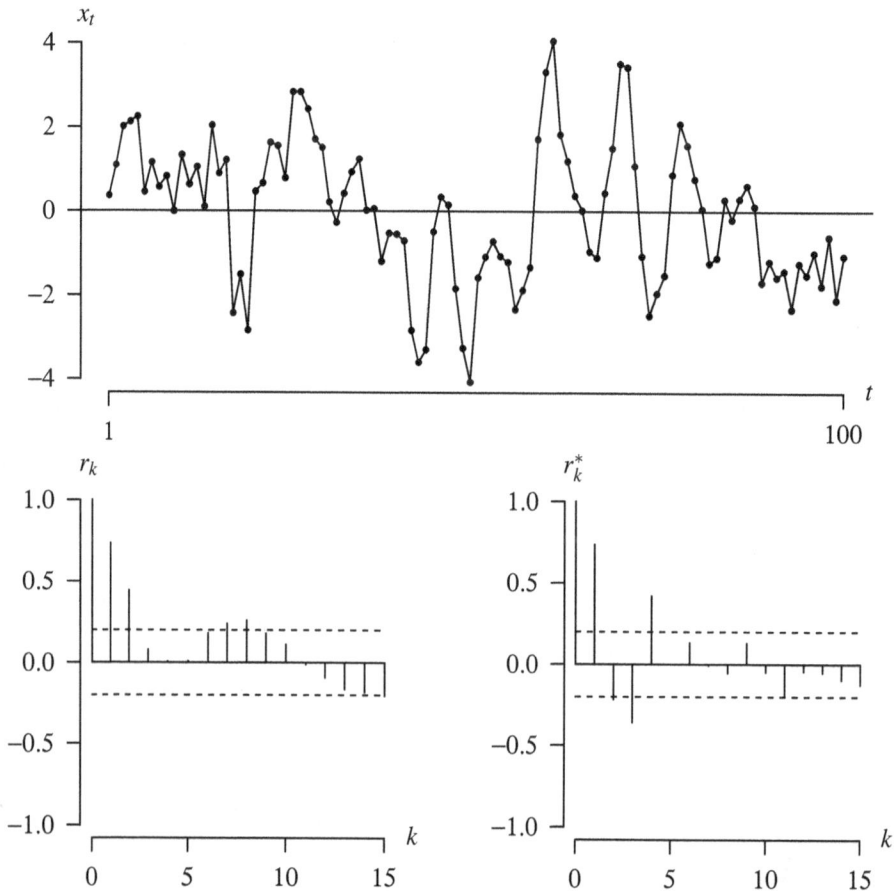

Figure 9.28: Time series plot, r_k, and r_k^* for $n = 100$ simulated values from an AR(4) model.

cally significant spikes at lags 1, 2, 3, and 4 which is consistent with an AR(4) model. The spikes cut off after lag 4 as expected from the population counterparts in Figure 9.27. The approximate 95% confidence intervals indicated by the dashed lines show that the values of the sample partial autocorrelation function do not significantly differ from zero at lags beyond lag 4. The sample autocorrelation function displays a mixture of damped exponential terms damped sinusoidal terms as expected, with statistically significant autocorrelations at the first two lags: $r_1^* = 0.7351$ and $r_2^* = 0.4417$. The time series plot shows that observations tend to linger on one side of the population mean (indicated by a horizontal line at $\mu = 0$), which is consistent with the two initial statistically significant positive spikes in the sample autocorrelation function.

We recommend running the simulation code from the previous example several dozen times in a loop and viewing the associated plots of x_t, r_k, and r_k^* in search of patterns. This will allow you to see how various realizations of this simulated AR(4) time series model vary from one realization to the next. So when you then view a *single* realization of a real-life time series, you will have a sense of how far these plots might deviate from their expected patterns.

There is a second way to overcome the initialization problem in simulating observations from an AR(p) time series. This second technique starts the time series with p initial arbitrary values, and then allows the time series to "warm up" or "burn in" for several time periods before producing the first observation X_1. Reasonable p initial arbitrary values for the standard AR(p) model are 0; reasonable p initial arbitrary values for the shifted AR(p) model are μ. This approach can be implemented in R with the `filter` function with `"recursive"` as the `method` argument. The code below generates $n = 100$ values in the AR(4) time series model from the previous example using a warm-up period of 50 observations.

```
phi = c(21 / 20, 1 / 20, -23 / 40, 3 / 10)
z   = rnorm(150)
x   = filter(z, filter = phi, method = "recursive")
x   = x[51:150]
```

This is also the approach taken by the built-in R function named `arima.sim`, which simulates a realization of a time series. Using the `arima.sim` function means that $n = 100$ observations from the AR(4) time series model from the previous example can be simulated using a single command, using a warm-up period of 50 observations.

```
x = arima.sim(model = list(ar = c(21 / 20, 1 / 20, -23 / 40, 3 / 10)),
              n = 100, sd = 1, n.start = 50)
```

The remaining topics associated with the AR(p) time series model are statistical in nature: parameter estimation, model assessment, model selection, and forecasting. We begin with parameter estimation.

Parameter Estimation

The $p + 2$ parameters to estimate in a shifted AR(p) time series model are $\phi_1, \phi_2, \ldots, \phi_p, \mu, \sigma_Z^2$. There are three techniques for estimating these parameters considered here: method of moments, least squares, and maximum likelihood estimation. These techniques were introduced in Section 8.2.1. These three techniques are outlined in the following paragraphs.

Approach 1: Method of moments. In the case of estimating the $p + 2$ parameters in the shifted AR(p) time series model by the method of moments, we match the population and sample first-order moments, second-order moments, lag 1 autocorrelation, lag 2 autocorrelation, ..., lag p autocorrelation. Placing the population moments on the left-hand side of the equation and the associated sample moments on the right-hand side of the equation results in $(p+2)$ equations in $(p+2)$ unknowns:

$$E[X_t] = \frac{1}{n}\sum_{t=1}^{n} X_t$$

$$E[X_t^2] = \frac{1}{n}\sum_{t=1}^{n} X_t^2$$

$$\rho(1) = r_1$$

$$\rho(2) = r_2$$

$$\vdots \;\; = \;\; \vdots$$

$$\rho(p) = r_p.$$

Since $E[X_t] = \mu$ for a stationary shifted AR(p) time series model, the first equation gives the method of moments estimator $\hat{\mu} = \bar{X}$. Recall from Theorem 9.18 that the relationship between $\phi_1, \phi_2, \ldots, \phi_p$ and $\rho(1), \rho(2), \ldots, \rho(p)$ is given by the matrix equation

$$\rho = P\phi,$$

where

$$\rho = \begin{bmatrix} \rho(1) \\ \rho(2) \\ \rho(3) \\ \vdots \\ \rho(p) \end{bmatrix}, \qquad P = \begin{bmatrix} 1 & \rho(1) & \rho(2) & \cdots & \rho(p-1) \\ \rho(1) & 1 & \rho(1) & \cdots & \rho(p-2) \\ \rho(2) & \rho(1) & 1 & \cdots & \rho(p-3) \\ \vdots & \vdots & \vdots & \ddots & \vdots \\ \rho(p-1) & \rho(p-2) & \rho(p-3) & \cdots & 1 \end{bmatrix}, \qquad \phi = \begin{bmatrix} \phi_1 \\ \phi_2 \\ \phi_3 \\ \vdots \\ \phi_p \end{bmatrix}.$$

Satisfying the method of moments criteria, the lag k population autocorrelation $\rho(k)$ can be replaced with its statistical analog r_k, for $k = 1, 2, \ldots, p$. The resulting matrix equation is

$$r = R\phi,$$

where

$$r = \begin{bmatrix} r_1 \\ r_2 \\ r_3 \\ \vdots \\ r_p \end{bmatrix}, \qquad R = \begin{bmatrix} 1 & r_1 & r_2 & \cdots & r_{p-1} \\ r_1 & 1 & r_1 & \cdots & r_{p-2} \\ r_2 & r_1 & 1 & \cdots & r_{p-3} \\ \vdots & \vdots & \vdots & \ddots & \vdots \\ r_{p-1} & r_{p-2} & r_{p-3} & \cdots & 1 \end{bmatrix}, \qquad \phi = \begin{bmatrix} \phi_1 \\ \phi_2 \\ \phi_3 \\ \vdots \\ \phi_p \end{bmatrix}.$$

This matrix equation can be solved for the method of moments estimators as

$$\hat{\phi} = R^{-1}r.$$

These are known as the *Yule–Walker estimators* because of their relationship to the Yule–Walker equations. Finally, the remaining parameter to estimate is the population variance of the white noise σ_Z^2. From Theorem 9.18,

$$\sigma_Z^2 = \gamma(0) - \phi_1\gamma(1) - \phi_2\gamma(2) - \cdots - \phi_p\gamma(p).$$

Multiplying and dividing the right-hand side of this equation by $\gamma(0)$ gives

$$\sigma_Z^2 = \gamma(0)\left[1 - \phi_1\rho(1) - \phi_2\rho(2) - \cdots - \phi_p\rho(p)\right].$$

Replacing these elements by their method of moments estimators gives

$$\hat{\sigma}_Z^2 = c_0\left[1 - \hat{\phi}_1 r_1 - \hat{\phi}_2 r_2 - \cdots - \hat{\phi}_p r_p\right],$$

which can be expressed in matrix form as

$$\hat{\sigma}_Z^2 = c_0\left(1 - r'\hat{\phi}\right).$$

Since the formula for these estimators does not require any iterative methods, the method of moments estimators are often used as initial parameter estimates for the least squares estimators and the maximum likelihood estimators, which do require iterative methods. These point estimators for the parameters in a shifted AR(p) model are summarized below.

Theorem 9.21 The method of moments estimators of the parameters in a shifted AR(p) model are

$$\hat{\mu} = \bar{X} \qquad \hat{\phi} = R^{-1}r \qquad \hat{\sigma}_Z^2 = c_0\left(1 - r'\hat{\phi}\right).$$

Example 9.27 We now revisit the modeling of the built-in R time series LakeHuron from Example 9.14 consisting of $n = 98$ monthly mean levels (in feet) of the lake level of Lake Huron from 1875–1972. An AR(2) time series model was fit to this time series using the method of moments in Example 9.15. The fitted AR(2) model was deemed to be a reasonable fit via the goodness-of-fit tests in Example 9.19. Calculate the method of moments parameter estimates for the overfitted AR(3) model.

Since estimating the parameters involves just a matrix inverse and a matrix multiplication, these estimators are easily computed in an R function. The user-written YuleWalker function given below has the time series observations in the vector x and the order of the AR(p) time series model p as arguments. It uses the built-in acf function to compute $r_1, r_2, \ldots, r_p$ and the solve function to compute the inverse of the R matrix. The R code below calculates and prints the point estimates of the parameters μ, ϕ_1, ϕ_2, ϕ_3, and σ_Z^2 parameters for the AR(3) time series model using the method of moments estimators given in Theorem 9.21.

```
YuleWalker = function(x, p) {
  muhat   = mean(x)
  r       = acf(x, plot = FALSE, lag.max = p)$acf
  R       = matrix(1, p, p)
  for (i in 1:p) for (j in 1:p) R[i, j] = r[abs(i - j) + 1]
  r       = r[2:(p + 1)]
  phihat  = solve(R) %*% r
  sig2hat = mean((x - muhat) ^ 2) * (1 - sum(r * phihat))
  c(muhat, phihat, sig2hat)
}
YuleWalker(LakeHuron, 3)
```

The method of moments point estimates for the unknown parameters computed by this code are

$$\hat{\mu} = 579.00, \quad \hat{\phi}_1 = 1.0887, \quad \hat{\phi}_2 = -0.40454, \quad \hat{\phi}_3 = 0.13075, \quad \hat{\sigma}_Z^2 = 0.48358.$$

Alternatively, some keystrokes can be saved by using the built-in ar function to estimate the parameters in the AR(3) time series model, as shown below. The results are identical except the estimate of the population variance of the white noise differs slightly because of differing assumptions made within the ar function.

```
fit = ar(LakeHuron, order.max = 3, aic = FALSE, method = "yw")
fit$x.mean
fit$ar
fit$var.pred
```

Approach 2: Least squares. Consider the shifted stationary AR(p) model

$$X_t - \mu = \phi_1 (X_{t-1} - \mu) + \phi_2 (X_{t-2} - \mu) + \cdots + \phi_p (X_{t-p} - \mu) + Z_t.$$

For least squares estimation, we first establish the sum of squares S as a function of the parameters $\mu, \phi_1, \phi_2, \ldots, \phi_p$. We leave the optimization to the R ar function in order to calculate the least squares estimators of $\mu, \phi_1, \phi_2, \ldots, \phi_p$. Once these least squares estimators have been determined, the population variance of the white noise σ_Z^2 will be estimated.

The sum of squared errors is

$$S = \sum_{t=p+1}^{n} Z_t^2 = \sum_{t=p+1}^{n} [X_t - \mu - \phi_1 (X_{t-1} - \mu) - \phi_2 (X_{t-2} - \mu) - \cdots - \phi_p (X_{t-p} - \mu)]^2.$$

If this derivation were being done by hand, we would now calculate the partial derivatives of S with respect to the unknown parameters $\mu, \phi_1, \phi_2, \ldots, \phi_p$, equate them to zero and solve. As was the case with the AR(1) and AR(2) models, there is no closed-form solution, so numerical methods are required to calculate the parameter estimates. In the example that follows, we will use the ar function in R to determine the least squares parameter estimates that minimize S.

The last parameter to estimate is the population variance of the white noise σ_Z^2. The same estimator as the method of moments will be used:

$$\hat{\sigma}_Z^2 = c_0 \left(1 - r'\hat{\phi}\right).$$

Least squares estimation for a shifted AR(p) time series model is summarized below.

Theorem 9.22 The least squares estimators of the parameters in a shifted AR(p) time series model are the $\hat{\mu}, \hat{\phi}_1, \hat{\phi}_2, \ldots, \hat{\phi}_p$ values that minimize

$$S = \sum_{t=p+1}^{n} Z_t^2 = \sum_{t=p+1}^{n} [X_t - \mu - \phi_1 (X_{t-1} - \mu) - \phi_2 (X_{t-2} - \mu) - \cdots - \phi_p (X_{t-p} - \mu)]^2$$

and the population variance of the white noise is estimated by

$$\hat{\sigma}_Z^2 = c_0 \left(1 - r'\hat{\phi}\right).$$

We now use numerical methods to find the least squares estimates for the unknown parameters in the AR(p) time series model for the Lake Huron time series from Example 9.14.

Example 9.28 Find the least squares estimates of μ, ϕ_1, ϕ_2, $\ldots$, ϕ_p, and σ_Z^2 from the AR(p) time series model for the time series of $n = 98$ Lake Huron lake level observations from Example 9.14, for $p = 1, 2, 3, 4$. Plot the sum of squares associated with the least squares estimates as a function of p.

The R code below uses a `for` loop to iterate over the various values of p. It uses the `ar` function with the `method` argument set to `"ols"` (for ordinary least squares) to calculate the least squares estimates of the unknown parameters. It uses a nested `for` loop to calculate the sum of squares at the values of the point estimates.

```
x = LakeHuron
n = length(x)
for (p in 1:4) {
   fit     = ar(x, order.max = p, aic = FALSE, method = "ols")
   muhat   = fit$x.mean
   phihat  = fit$ar
   sig2hat = fit$var.pred
   S       = 0
   for (t in (p + 1):n) {
     S = S + (x[t] - muhat - sum(phihat * (x[(t - 1):(t - p)] - muhat))) ^ 2
   }
   print(c(muhat, phihat, sig2hat, S))
}
```

The point estimates for the unknown parameters and the sums of squares at the point estimates that are computed by this code are given in Table 9.10. Notice that the least squares point estimators for the AR(3) model are close to the method of moments point estimators for the AR(3) model calculated in Example 9.27. The graph in Figure 9.29 shows the sum of squares as a function of the order of the autoregressive model p. The sum of squares shows a "law of diminishing returns" as p increases. There is a large decrease in the sum of squares on the transition from $p = 1$ term to $p = 2$ terms. Beyond $p = 2$, however, the decreases are substantially smaller. This pattern is consistent with the $q = 0$ column from Table 9.7 in Example 9.20, which indicated that the AIC statistic was minimized for $p = 2$, which corresponds to a shifted AR(2) time series model.

p	$\hat{\mu}$	$\hat{\phi}_1$	$\hat{\phi}_2$	$\hat{\phi}_3$	$\hat{\phi}_4$	$\hat{\sigma}_Z^2$	S
1	579.00	0.8364				0.5090	49.38
2	579.00	1.0217	−0.2376			0.4540	43.64
3	579.00	1.0719	−0.3653	0.1088		0.4488	42.66
4	579.00	1.0738	−0.3739	0.0569	0.0625	0.4475	42.12

Table 9.10: Least squares parameter estimates and sums of squares for AR(p) models.

Approach 3: Maximum likelihood estimation. The procedure for determining the maximum likelihood estimators for the unknown parameters in a shifted AR(p) time series model follows along the same lines as in the AR(1) and AR(2) time series models from the previous subsections. Once again, to use maximum likelihood estimation, we must assume that the random shocks from the

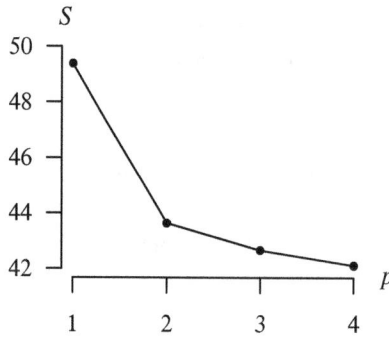

Figure 9.29: Sum of squares as a function of p for AR(p) models.

white noise are Gaussian white noise, with associated probability density function

$$f_{Z_t}(z_t) = \frac{1}{\sqrt{2\pi\sigma_Z^2}} \, e^{-z_t^2/(2\sigma_Z^2)} \qquad -\infty < z_t < \infty,$$

for $t = 1, 2, \ldots, n$. Determining the likelihood function, which is the joint probability density function of the observed values in the time series $X_1, X_2, \ldots, X_n$, involves finding

$$L\big(\mu, \phi_1, \phi_2, \ldots, \phi_p, \sigma_Z^2\big) = f(x_1, x_2, \ldots, x_n),$$

where the $x_1, x_2, \ldots, x_n$ arguments on L and the μ, ϕ_1, ϕ_2, $\ldots$, ϕ_p, and σ_Z^2 arguments on f have been dropped for brevity and $n > p$. As before, it is not possible to simply multiply the marginal probability density functions because the values in the AR(p) time series model are correlated. As in the case of the AR(1) and AR(2) models, we use the transformation technique to find the conditional joint probability density function of $X_{p+1}, X_{p+2}, \ldots, X_n$ conditioned on $X_1 = x_1, X_2 = x_2,$ $\ldots, X_p = x_p$, which is denoted by

$$f_{X_{p+1}, X_{p+2}, \ldots, X_n \mid X_1, X_2, \ldots, X_p}\big(x_{p+1}, x_{p+2}, \ldots, x_n \mid X_1 = x_1, X_2 = x_2, \ldots, X_p = x_p\big)$$

for $(x_{p+1}, x_{p+2}, \ldots, x_n) \in \mathcal{R}^{n-p}$. This conditional joint probability density function is multiplied by the marginal joint probability density function of $X_1, X_2, \ldots, X_p$ (which has the p-dimensional multivariate normal distribution) resulting in a joint probability density function of $X_1, X_2, \ldots, X_n$:

$$f_{X_1, X_2, \ldots, X_n}(x_1, x_2, \ldots, x_n) =$$
$$\qquad f_{X_{p+1}, X_{p+2}, \ldots, X_n \mid X_1, X_2, \ldots, X_p}\big(x_{p+1}, x_{p+2}, \ldots, x_n \mid X_1 = x_1, X_2 = x_2, \ldots, X_p = x_p\big) \times$$
$$\qquad f_{X_1, X_2, \ldots, X_p}(x_1, x_2, \ldots, x_p)$$

for $(x_1, x_2, \ldots, x_n) \in \mathcal{R}^n$. This function serves as the likelihood function, which should be maximized with respect to the unknown parameters μ, ϕ_1, ϕ_2, $\ldots$, ϕ_p, and σ_Z^2. We leave the maximization to the `ar` and `arima` functions in R when determining the maximum likelihood estimates for the parameters for a particular time series to be fitted to the shifted AR(p) time series model.

   In addition to point estimators for the parameters, we are also interested in confidence intervals that capture the precision of the point estimators. The population variance of the vector of parameter estimators $\hat{\phi} = (\hat{\phi}_1, \hat{\phi}_2, \ldots, \hat{\phi}_p)'$ is given by the variance–covariance matrix

$$V\big[\hat{\phi}\big] = \frac{1}{n}\big(1 - \rho'\phi\big)P^{-1}.$$

Since the maximum likelihood estimators for $\phi_1, \phi_2, \ldots, \phi_p$ are asymptotically unbiased and normally distributed under certain regularity conditions,

$$\hat{\phi} \xrightarrow{D} N\left(\phi, \frac{1}{n}\left(1 - \rho'\phi\right)P^{-1}\right).$$

For $p = 1$, this reduces to

$$\hat{\phi}_1 \xrightarrow{D} N\left(\phi_1, \frac{1 - \phi_1^2}{n}\right).$$

For $p = 2$, this reduces to

$$\begin{bmatrix} \hat{\phi}_1 \\ \hat{\phi}_2 \end{bmatrix} \xrightarrow{D} N\left(\begin{bmatrix} \phi_1 \\ \phi_2 \end{bmatrix}, \frac{1}{n}\begin{bmatrix} 1 - \phi_2^2 & -\phi_1(1 + \phi_2) \\ -\phi_1(1 + \phi_2) & 1 - \phi_2^2 \end{bmatrix}\right).$$

These asymptotic results for $p = 1$ and $p = 2$ were used in the confidence intervals given in Theorems 9.7 and 9.16. When the quantities in this expression are replaced by their statistical counterparts, the estimated variance–covariance matrix of the vector $\hat{\phi}$ is

$$\hat{V}\left[\hat{\phi}\right] = \frac{1}{n}\left(1 - r'\hat{\phi}\right)R^{-1}.$$

Using the diagonal elements of this matrix and the asymptotic normality of maximum likelihood estimators, an asymptotically exact $100(1 - \alpha)\%$ confidence interval for ϕ_i is easily constructed.

Theorem 9.23 For a stationary AR(p) time series model, an asymptotically exact two-sided $100(1 - \alpha)\%$ confidence interval for ϕ_i is given by

$$\hat{\phi}_i - z_{\alpha/2}\sqrt{\left[\frac{1}{n}\left(1 - r'\hat{\phi}\right)R^{-1}\right]_{i,i}} < \phi_i < \hat{\phi}_i + z_{\alpha/2}\sqrt{\left[\frac{1}{n}\left(1 - r'\hat{\phi}\right)R^{-1}\right]_{i,i}}$$

for $i = 1, 2, \ldots, p$, where $\hat{\phi}_i$ is the maximum likelihood estimator of ϕ_i and $z_{\alpha/2}$ is the $1 - \alpha/2$ fractile of the standard normal distribution.

The maximum likelihood estimates and associated confidence intervals will be illustrated for an economic time series in the next example.

Example 9.29 Table 9.11 contains the annual lynx (Lynx canadensis) pelt sales, read row-wise, at the Hudson's Bay Company in Canada from 1857 to 1911. Suggest a time series model for the annual pelt sales.

23362	31642	33757	23226	15178	7272	4448	4926	5437	16498
35971	76556	68392	37447	45686	7942	5123	7106	11250	18774
30508	42834	27345	17834	15386	9443	7599	8061	27187	51511
74050	78773	33899	18886	11520	8352	8660	12902	20331	36853
56407	39437	26761	15185	4473	5781	9117	19267	36116	58850
61478	36300	9704	3410	3774					

Table 9.11: Annual lynx pelt sales at the Hudson's Bay Company, 1857–1911.

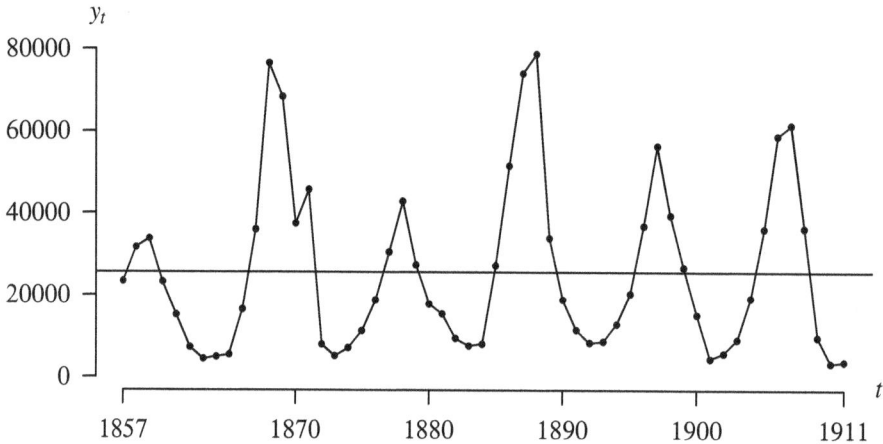

Figure 9.30: Time series plot for $n = 55$ annual lynx pelt sales (1857–1911).

The time series is plotted in Figure 9.30. The annual sales figures vary widely, from a minimum of 3410 pelts sold in 1910 to a maximum of 78,773 pelts sold in 1888. A horizontal line is drawn at the average sales over this time horizon at 25,600 pelts. The time series appears to have a periodic component that seems to cycle about every ten years or so, although nothing about the sales of lynx pelts would seem to account for an approximately decade-long periodicity. There are local maximums in the time series associated with the years 1859, 1868, 1878, 1888, 1897, and 1907. Is consumer behavior driving this periodicity? Are the prices of the pelts driving this periodic behavior? Is the availability of the pelts driving this periodic behavior?

Some further analysis of the time series indicates that ecology can answer some of the questions. Lynxes depend on the snowshoe rabbit (Lepus americanus) for food, and lynxes starve when the rabbits near extinction periodically. This is an example of one time series depending on another time series. We ignore the dependence on the snoeshoe rabbit in our analysis because multivariate time series analysis is a topic for a more advanced time series course. We consider an AR(p) model here and consider a time series model with a periodic component subsequently.

Based on Figure 9.30, is a stationary time series model appropriate? There does not appear to be any trend in the time series, but the population variance does not appear to be stable. The first local maximum (in 1859) and the third local maximum (in 1878) are not as pronounced as the others. One remedy to this nonconstant population variance is to transform the time series. Taking the logarithm of the time series values, $x_t = \ln y_t$, reduces the impact of the nonconstant variance. Figure 9.31 contains a time series plot of the logarithm of the sales figures, along with the associated sample autocorrelation and partial autocorrelation functions. For simplicity, we define time $t = 1$ to be the year 1857 and $t = 55$ to be the year 1911. The transformation has proven to be effective. As expected, the first and third peaks are still the smallest local maximums of the group, but they are less pronounced than those in the raw data. The original time series values are denoted by y_t, and the transformed time series values are denoted by $x_t = \ln y_t$. The sample autocorrelation function appears to have a damped sinusoidal shape, and the

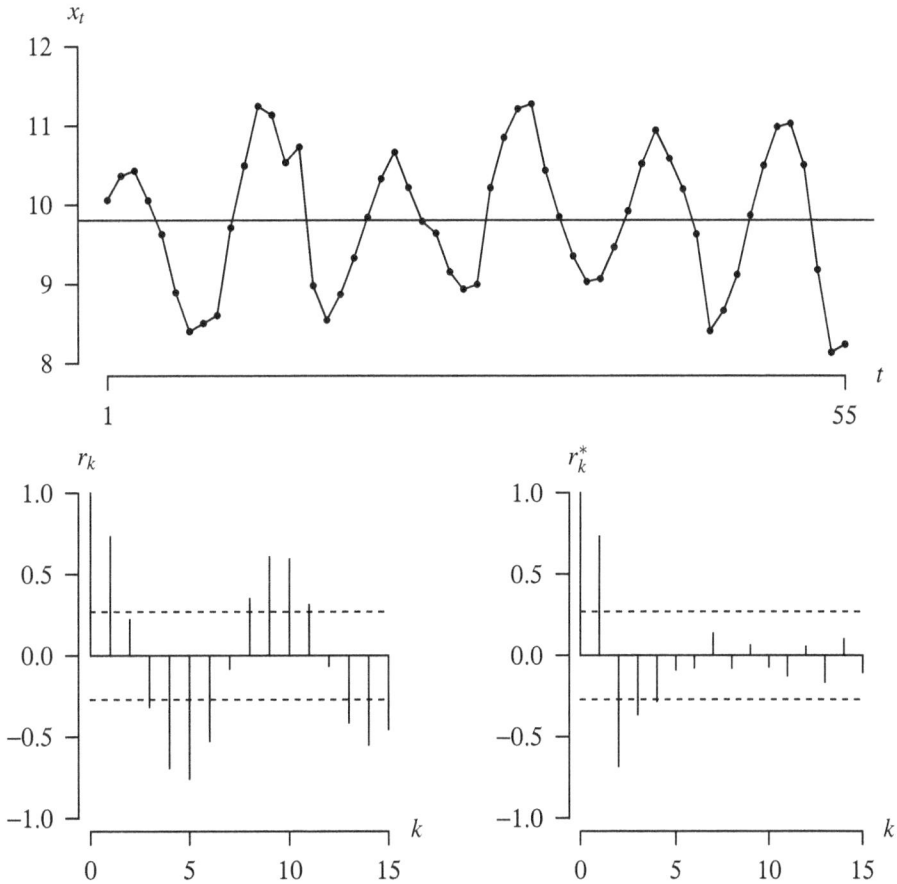

Figure 9.31: Time series plot, r_k, and r_k^* for $n = 55$ log annual lynx sales (1857–1911).

sample partial autocorrelation function cuts off after lag 4, although the spike at lag 4 is only marginally significant. The fact that the sample partial autocorrelation function cuts off leads us to consider an autoregressive time series model. Based on these graphs, we will attempt to fit tentative AR(3) and AR(4) models to the transformed time series $x_1, x_2, \ldots, x_{55}$. The R code to produce the graphs in Figure 9.31 is given below.

```
y = c(23362, 31642, 33757, 23226, 15178, 7272, 4448, 4926, 5437,
      16498, 35971, 76556, 68392, 37447, 45686, 7942, 5123, 7106,
      11250, 18774, 30508, 42834, 27345, 17834, 15386, 9443, 7599,
      8061, 27187, 51511, 74050, 78773, 33899, 18886, 11520, 8352,
      8660, 12902, 20331, 36853, 56407, 39437, 26761, 15185, 4473,
      5781, 9117, 19267, 36116, 58850, 61478, 36300, 9704, 3410, 3774)
x = log(y)
layout(matrix(c(1, 1, 2, 3), 2, 2, byrow = TRUE))
plot.ts(x)
abline(h = mean(x))
```

```
acf(x)
pacf(x)
```

We use the R arima function to estimate the parameters of the AR(p) time series models for the transformed time series using maximum likelihood estimation and to compute their associated standard errors. The additional R code below fits the AR(3) model to the transformed time series. Setting the method argument to "ML" indicates that the arima function should use maximum likelihood estimation. Various aspects of the fitted model are extracted using the $ extractor.

```
fit = arima(x, order = c(3, 0, 0), include.mean = TRUE, method = "ML")
fit$coef
fit$var.coef
fit$sigma2
sqrt(fit$var.coef[1, 1])
sqrt(fit$var.coef[2, 2])
sqrt(fit$var.coef[3, 3])
sqrt(fit$var.coef[4, 4])
```

The resulting fitted AR(3) model is

$$X_t - \hat{\mu} = \hat{\phi}_1 (X_{t-1} - \hat{\mu}) + \hat{\phi}_2 (X_{t-2} - \hat{\mu}) + \hat{\phi}_3 (X_{t-3} - \hat{\mu}) + Z_t$$

or

$$X_t - 9.809 = 0.957 (X_{t-1} - 9.809) - 0.126 (X_{t-2} - 9.809) - 0.470 (X_{t-3} - 9.809) + Z_t,$$
$$\quad (0.074) \quad (0.117) \qquad\qquad (0.176) \qquad\qquad\qquad (0.120)$$

where Z_t is white noise with estimated population variance $\hat{\sigma}_Z^2 = 0.119$. The numbers in parentheses just below the parameter estimates are the estimated standard errors of the associated parameter estimates. The associated approximate 95% confidence intervals are

$$0.728 < \phi_1 < 1.186,$$
$$-0.471 < \phi_2 < 0.219,$$
$$-0.705 < \phi_3 < -0.235,$$
$$9.663 < \mu < 9.955.$$

The fact that the confidence interval for ϕ_2 contains zero should not deter us from considering the AR(3) model because the confidence interval for ϕ_3 has bounds which do not include zero.

When this same procedure is applied to the fitting of an AR(4) model, the fitted model

$$X_t - \hat{\mu} = \hat{\phi}_1 (X_{t-1} - \hat{\mu}) + \hat{\phi}_2 (X_{t-2} - \hat{\mu}) + \hat{\phi}_3 (X_{t-3} - \hat{\mu}) + \hat{\phi}_4 (X_{t-4} - \hat{\mu}) + Z_t$$

is

$$X_t - 9.807 = 0.774 (X_{t-1} - 9.807) - 0.151 (X_{t-2} - 9.807) - 0.120 (X_{t-3} - 9.807) - 0.378 (X_{t-4} - 9.807) + Z_t,$$
$$\quad (0.051) \quad (0.125) \qquad\qquad (0.165) \qquad\qquad (0.163) \qquad\qquad (0.127)$$

where Z_t is white noise with estimated population variance $\hat{\sigma}_Z^2 = 0.102$. The associated approximate 95% confidence intervals are

$$0.529 < \phi_1 < 1.018,$$
$$-0.474 < \phi_2 < 0.172,$$
$$-0.439 < \phi_3 < 0.200,$$
$$-0.626 < \phi_4 < -0.130,$$
$$9.708 < \mu < 9.907.$$

Again, the confidence intervals for ϕ_2 and ϕ_3 containing zero should not deter us from considering the AR(4) model because the confidence interval for ϕ_4 has bounds which do not include zero.

So should the AR(3) or AR(4) model be considered the preferred stationary model? A check of the solutions of $\hat{\phi}(B) = 0$ indicates that all of the solutions lie outside of the unit circle in the complex plane for both the AR(3) and AR(4) models. Another way to select between the two models is to calculate the AIC statistic for these models. The additional R statement

```
for (p in 0:5) print(arima(x, order = c(p, 0, 0), method = "ML")$aic)
```

calculates the AIC statistic associated with the fitted AR(p) models, for $p = 0, 1, \ldots, 5$. The results are shown in Table 9.12.

$p = 0$	$p = 1$	$p = 2$	$p = 3$	$p = 4$	$p = 5$
145.3	101.5	63.4	52.3	46.2	48.2

Table 9.12: AIC values for AR(p) models for the transformed annual lynx pelt sales.

The AIC statistic is minimized for $p = 4$, indicating that the AR(4) model is selected over the AR(3) model. So to summarize, the tentative AR(p) time series model based on (a) the time series plot, (b) the sample autocorrelation function, (c) the sample partial autocorrelation function, and (d) the AIC statistic, is the fitted AR(4) model

$$\ln Y_t - 9.807 = 0.774 \left(\ln Y_{t-1} - 9.807 \right) - 0.151 \left(\ln Y_{t-2} - 9.807 \right) -$$

$$0.120 \left(\ln Y_{t-3} - 9.807 \right) - 0.378 \left(\ln Y_{t-4} - 9.807 \right) + Z_t,$$

where Y_t corresponds to the original time series consisting of annual lynx pelt sales at the Hudson's Bay Company, and Z_t is white noise with estimated variance $\hat{\sigma}_Z^2 = 0.102$.

Model Assessment

Now that techniques for point and interval estimates for the parameters in the AR(p) model have been established, we are interested in assessing the adequacy of the fitted AR(p) time series model. This will involve an analysis of the residuals. Recall from Section 8.2.3 that the residuals are defined by

$$[\text{residual}] = [\text{observed}] - [\text{predicted}]$$

or

$$\hat{Z}_t = X_t - \hat{X}_t.$$

Since $\hat{X}_t$ is the one-step-ahead forecast from the time origin $t-1$, this is more clearly written as

$$\hat{Z}_t = X_t - \hat{X}_{t-1}(1).$$

From Theorem 9.20, the shifted AR(p) model is

$$X_t - \mu = \phi_1 (X_{t-1} - \mu) + \phi_2 (X_{t-2} - \mu) + \cdots + \phi_p (X_{t-p} - \mu) + Z_t$$

or

$$X_t = \mu + \phi_1 (X_{t-1} - \mu) + \phi_2 (X_{t-2} - \mu) + \cdots + \phi_p (X_{t-p} - \mu) + Z_t.$$

Taking the conditional expected value of both sides of this equation gives

$$E[X_t \mid X_1 = x_1, X_2 = x_2, \ldots, X_{t-1} = x_{t-1}] = \mu + \phi_1 (x_{t-1} - \mu) + \phi_2 (x_{t-2} - \mu) + \cdots + \phi_p (x_{t-p} - \mu).$$

Replacing the parameters by their point estimators, the one-step-ahead forecast from the time origin $t-1$ is

$$\hat{X}_{t-1}(1) = \hat{\mu} + \hat{\phi}_1 (x_{t-1} - \hat{\mu}) + \hat{\phi}_2 (x_{t-2} - \hat{\mu}) + \cdots + \hat{\phi}_p (x_{t-p} - \hat{\mu}).$$

Therefore, for the time series $x_1, x_2, \ldots, x_n$ and the fitted AR(p) model with parameter estimates $\hat{\mu}$, $\hat{\phi}_1, \hat{\phi}_2, \ldots, \hat{\phi}_p$, the residual at time t is

$$\hat{Z}_t = x_t - \left[\hat{\mu} + \hat{\phi}_1 (x_{t-1} - \hat{\mu}) + \hat{\phi}_2 (x_{t-2} - \hat{\mu}) + \cdots + \hat{\phi}_p (x_{t-p} - \hat{\mu}) \right]$$

for $t = p+1, p+2, \ldots, n$. The next example shows the steps associated with assessing the adequacy of the AR(4) model for the time series of annual lynx pelt sales.

> **Example 9.30** Fit the AR(4) time series model to the transformed annual lynx sales from Example 9.29 via maximum likelihood estimation.
>
> (a) Calculate and plot the residuals, their sample autocorrelation function, and their sample partial autocorrelation function.
>
> (b) Conduct a test of independence on the residuals using the number of sample autocorrelation function values for the first $m = 40$ lags which fall outside of $\pm 1.96/\sqrt{n}$.
>
> (c) Conduct the Box–Pierce and Ljung–Box tests for independence of the residuals.
>
> (d) Conduct the turning point test for independence of the residuals.
>
> (e) Plot a histogram and a QQ plot of the standardized residuals in order to assess the normality of the residuals.
>
>
> (a) The following R commands calculate the $n - 4 = 51$ residuals of the transformed time series and plot them as a time series, along with the associated sample autocorrelation function and sample partial autocorrelation function.

```
y = c(23362, 31642, 33757, 23226, 15178, 7272, 4448, 4926, 5437,
      16498, 35971, 76556, 68392, 37447, 45686, 7942, 5123, 7106,
      11250, 18774, 30508, 42834, 27345, 17834, 15386, 9443, 7599,
      8061, 27187, 51511, 74050, 78773, 33899, 18886, 11520, 8352,
      8660, 12902, 20331, 36853, 56407, 39437, 26761, 15185, 4473,
      5781, 9117, 19267, 36116, 58850, 61478, 36300, 9704, 3410, 3774)
x = log(y)
n = length(x)
p = 4
m = 40
fit = arima(x, order = c(p, 0, 0), include.mean = TRUE, method = "ML")
phi1hat = fit$coef[1]
phi2hat = fit$coef[2]
phi3hat = fit$coef[3]
phi4hat = fit$coef[4]
muhat   = fit$coef[5]
zhat    = x[(p + 1):n] - (muhat + phi1hat * (x[4:(n - 1)] - muhat) +
                                  phi2hat * (x[3:(n - 2)] - muhat) +
                                  phi3hat * (x[2:(n - 3)] - muhat) +
                                  phi4hat * (x[1:(n - 4)] - muhat))
layout(matrix(c(1, 1, 2, 3), 2, 2, byrow = TRUE))
plot.ts(zhat)
acf(zhat, lag.max = m)
pacf(zhat, lag.max = m)
```

The results are displayed in Figure 9.32. The residuals do not appear to have any cyclic variation, trend, or serial correlation.

(b) There are no sample autocorrelation function values that fall outside of the limits $\pm 1.96/\sqrt{n}$ in the plot in Figure 9.32 of the first 40 sample autocorrelation function values associated with the residuals. Since we expect $40 \cdot 0.05 = 2$ values to fall outside of these limits in the case of a good fit, we fail to reject H_0 in this case. The fit of the AR(4) model is not rejected by this test.

(c) The additional R statements below calculate the Box–Pierce test statistic and the Ljung–Box test statistic and the associated p-values using the built-in `Box.test` function.

```
Box.test(zhat, lag = 40, type = "Box-Pierce", fitdf = 5)
Box.test(zhat, lag = 40, type = "Ljung-Box", fitdf = 5)
```

The Box–Pierce test statistic is 19.6 and the associated p-value is $p = 0.984$. The Ljung–Box test statistic is 36.1 and the associated p-value is $p = 0.418$. We fail to reject H_0 in both tests based on the chi-square critical value with $40 - 5 = 35$ degrees of freedom. Since both p-values exceed 0.05, the fit of the AR(4) model is not rejected by these tests.

(d) The following additional R statements calculate the test statistic and the p-value for the turning point test applied to the time series consisting of the $n - 4 = 51$ residual values for the AR(4) fit to the transformed annual lynx pelt sales time series.

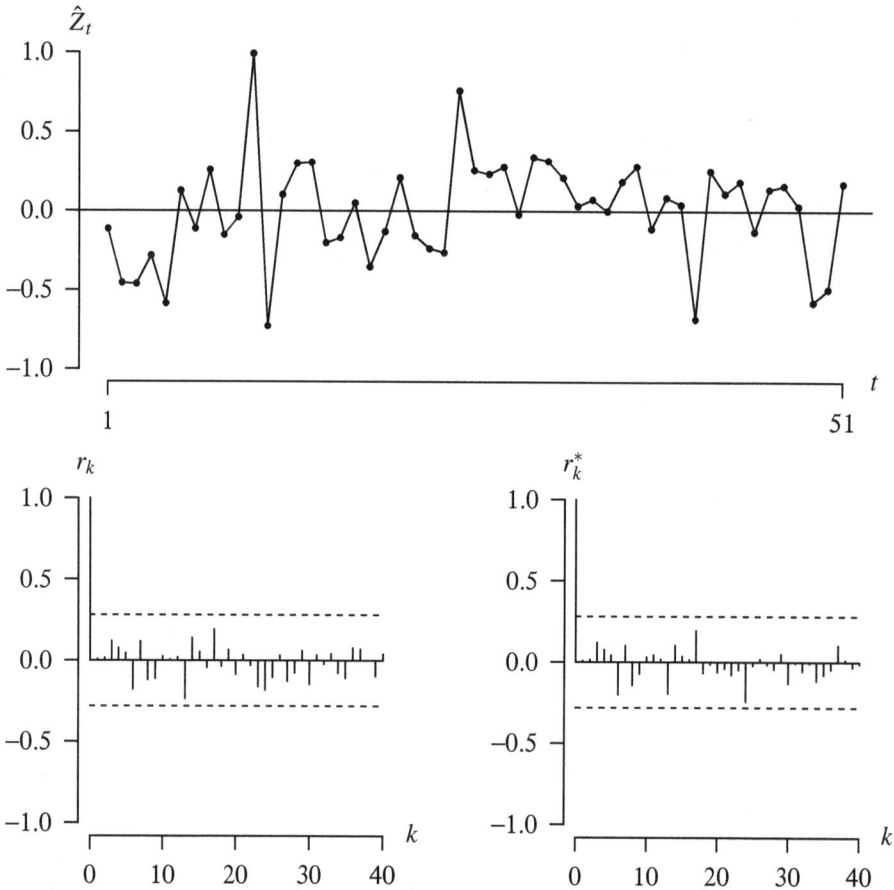

Figure 9.32: Time series plot, r_k, and r_k^* for $n - 4 = 51$ residuals from AR(4) fitted model.

```
n = n - 4
m = (2 / 3) * (n - 2)
v = (16 * n - 29) / 90
T = 0
for (i in 2:(n - 1)) {
  if ((zhat[i - 1] < zhat[i] && zhat[i] > zhat[i + 1]) ||
      (zhat[i - 1] > zhat[i] && zhat[i] < zhat[i + 1])) T = T + 1
}
s = (T - m) / sqrt(v)
2 * (1 - pnorm(abs(s)))
```

The tail probability is doubled because the alternative hypothesis is two-tailed for the turning point test. The test statistic is 0.1127 and the p-value is $p = 0.91$. The turning point test detected 33 turning points in the time series of the 51 residuals, and that is about the number that we expect to have if the residuals from the fitted AR(4) time series model of the transformed annual lynx pelt sales were mutually independent random variables. We again fail to reject the null hypothesis in this case. The fit of the AR(4) model is not rejected by this test.

(e) The residuals are standardized by dividing by their sample standard deviation. The following additional R statements plot a histogram of the standardized residuals using the `hist` function and a QQ plot to assess normality using the `qqnorm` function.

```
hist(zhat / sd(zhat))
qqnorm(zhat / sd(zhat))
```

The plots are shown in Figure 9.33. The histogram shows that all standardized residuals fall between -3 and 3, but deviate significantly from a bell-shaped probability distribution, particularly in the right-hand tail. The horizontal axis on the histogram is the standardized residual and the vertical axis is the frequency. The QQ plot shows considerable nonlinearity, indicating a possible departure from normality based on the $n - 4 = 51$ residuals plotted. The horizontal axis on the QQ plot is the standardized theoretical quantile and the vertical axis is the associated normal data quantile. These plots indicate that a formal statistical goodness-of-fit test for normality should be conducted in order to assess whether Gaussian white noise is appropriate for the residuals of the fitted AR(4) time series model based on these two plots.

In summary, the model adequacy tests applied to the residuals on the AR(4) time series model of the transformed observations of the annual lynx pelt sales have revealed that the mutual independence of the residuals cannot be rejected by four statistical tests. The histogram and QQ plot of the residuals appear to not support the assumption of normally distributed residuals. We conclude that the AR(4) time series model is an adequate model for the transformed annual lynx pelt sales at the Hudson's Bay Company time series, with the exception of non-Gaussian error terms apparent in Figure 9.33. Another way to visually assess the adequacy of the time series model is to inspect time series plots of simulations of the fitted model, which is left as an exercise.

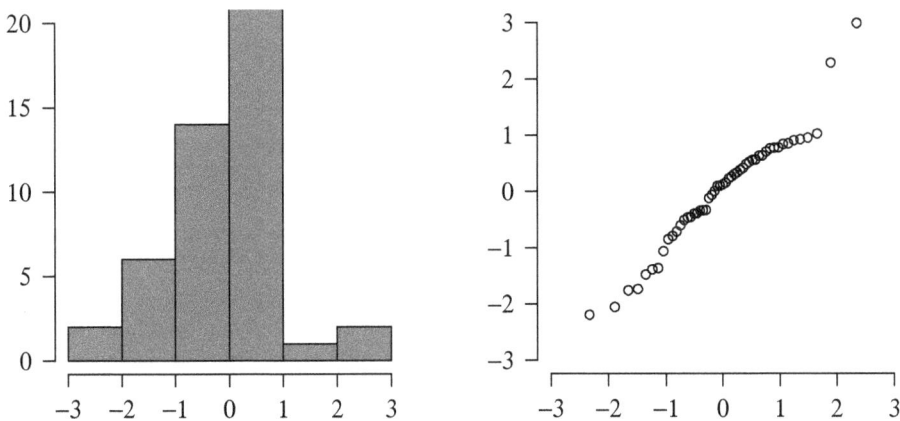

Figure 9.33: Histogram (left) and QQ plot (right) of the fitted AR(4) standardized residuals.

Forecasting

We now consider the question of forecasting future values of a time series that is governed by a shifted AR(p) time series model. In the case of the annual lynx pelt sales time series, this corresponds to the one-step-ahead forecast for 1912, the two-steps-ahead forecast for 1913, the three-steps-ahead forecast for 1914, etc. To review forecasting notation, the observed time series values are $x_1, x_2, \ldots, x_n$. The forecast is being made at time $t = n$. The random future value of the time series that is h time units in the future is denoted by X_{n+h}. The associated forecasted value is denoted by $\hat{X}_{n+h}$, and is the conditional expected value

$$\hat{X}_{n+h} = E\left[X_{n+h} \mid X_1 = x_1, X_2 = x_2, \ldots, X_n = x_n\right].$$

We would like to find this forecasted value and an associated prediction interval for a shifted AR(p) model. As in Section 8.2.2, we assume that all parameters are known in the derivations that follow. We also assume that the parameters $\phi_1, \phi_2, \ldots, \phi_p$ correspond to a stationary shifted AR(p) time series model and $p < n$.

The shifted AR(p) model is

$$X_t - \mu = \phi_1\left(X_{t-1} - \mu\right) + \phi_2\left(X_{t-2} - \mu\right) + \cdots + \phi_p\left(X_{t-p} - \mu\right) + Z_t.$$

Replacing t by $n+1$ and solving for X_{n+1}, this becomes

$$X_{n+1} = \mu + \phi_1\left(X_n - \mu\right) + \phi_2\left(X_{n-1} - \mu\right) + \cdots + \phi_p\left(X_{n-p+1} - \mu\right) + Z_{n+1}.$$

Taking the conditional expected value of each side of this equation results in the one-step-ahead forecast

$$\hat{X}_{n+1} = \mu + \phi_1\left(x_n - \mu\right) + \phi_2\left(x_{n-1} - \mu\right) + \cdots + \phi_p\left(x_{n-p+1} - \mu\right)$$

because the final p observations $x_{n-p+1}, x_{n-p+2}, \ldots, x_n$ in the time series $x_1, x_2, \ldots, x_n$ have already been observed. The forecasted value at time $n+1$ is a function of the final p values in the time series. Applying this same process to the predicted value at time $n+2$ results in the time series model

$$X_{n+2} = \mu + \phi_1\left(X_{n+1} - \mu\right) + \phi_2\left(X_n - \mu\right) + \cdots + \phi_p\left(X_{n-p+2} - \mu\right) + Z_{n+2}.$$

This time, the value of X_{n+1} has not been observed, so we replace it by its forecasted value when taking the conditional expected value of both sides of the equation

$$\hat{X}_{n+2} = \mu + \phi_1\left(\hat{X}_{n+1} - \mu\right) + \phi_2\left(x_n - \mu\right) + \cdots + \phi_p\left(x_{n-p+2} - \mu\right),$$

because $x_{n-p+2}, x_{n-p+3}, \ldots, x_n$ have already been observed. Continuing in this fashion, a recursive formula for the forecasted value of X_{n+h} is

$$\hat{X}_{n+h} = \mu + \phi_1\left(\hat{X}_{n+h-1} - \mu\right) + \phi_2\left(\hat{X}_{n+h-2} - \mu\right) + \cdots + \phi_p\left(\hat{X}_{n+h-p} - \mu\right).$$

Although we would prefer an explicit formula, the recursive formula is easy to implement for an observed time series $x_1, x_2, \ldots, x_n$. As in the case of the AR(1) and AR(2) models, long-term forecasts for a stationary AR(p) time series model tend to μ as the time horizon $h \to \infty$.

We would like to pair the point estimator $\hat{X}_{n+h}$ with an interval estimator, which is a prediction interval in this setting. The prediction interval gives us an indication of the precision of the forecast. In order to derive an exact two-sided $100(1 - \alpha)\%$ prediction interval for X_{n+h}, it is helpful to write the shifted AR(p) model as a shifted MA(∞) model. The coefficients $\theta_1, \theta_2, \ldots$ of a stationary shifted AR(p) model written as an MA(∞) model

$$X_t = \mu + Z_t + \theta_1 Z_{t-1} + \theta_2 Z_{t-2} + \cdots$$

are given in terms of $\phi_1, \phi_2, \ldots, \phi_p$ as was illustrated for $p = 4$ in Example 9.23. Consider this model at time $t = n + 1$. Since the error terms $Z_n, Z_{n-1}, Z_{n-2}, \ldots$ are unknown but fixed because they are associated with the observed time series $x_1, x_2, \ldots, x_n$, the conditional population variance of X_{n+1} is

$$V[X_{n+1}] = V[Z_{n+1}] = \sigma_Z^2$$

because the population variance of μ is zero and Z_{n+1} is the only random term in the model. The error terms at time n and prior are observed and can therefore be treated as constants. Likewise, considering the MA(∞) model at time $t = n + 2$, the conditional population variance of X_{n+2} is

$$V[X_{n+2}] = V[Z_{n+2} + \theta_1 Z_{n+1}] = \left(1 + \theta_1^2\right)\sigma_Z^2.$$

Similarly, the conditional population variance of X_{n+3} is

$$V[X_{n+3}] = V[Z_{n+3} + \theta_1 Z_{n+2} + \theta_2 Z_{n+1}] = \left(1 + \theta_1^2 + \theta_2^2\right)\sigma_Z^2.$$

Continuing in this fashion, the conditional population variance of X_{n+h} is

$$V[X_{n+h}] = \left(1 + \theta_1^2 + \theta_2^2 + \cdots + \theta_{h-1}^2\right)\sigma_Z^2.$$

If we assume that the white noise terms in the MA(∞) representation of the AR(p) time series model are Gaussian white noise terms, then X_{n+h} is also normally distributed because a linear combination of mutually independent normal random variables is also normally distributed. So an exact two-sided $100(1-\alpha)\%$ prediction interval for $\hat{X}_{n+h}$ is

$$\hat{X}_{n+h} - z_{\alpha/2}\sqrt{1 + \theta_1^2 + \theta_2^2 + \cdots + \theta_{h-1}^2}\; \sigma_Z < X_{n+h} < \hat{X}_{n+h} + z_{\alpha/2}\sqrt{1 + \theta_1^2 + \theta_2^2 + \cdots + \theta_{h-1}^2}\; \sigma_Z.$$

In most practical problems, the parameters in this prediction interval will be estimated from data, which results in the following *approximate* two-sided $100(1-\alpha)\%$ prediction interval provided next.

Theorem 9.24 For a stationary shifted AR(p) time series model, a forecasted value of X_{n+h} can be calculated by the recursive equation

$$\hat{X}_{n+h} = \hat{\mu} + \hat{\phi}_1\left(\hat{X}_{n+h-1} - \hat{\mu}\right) + \hat{\phi}_2\left(\hat{X}_{n+h-2} - \hat{\mu}\right) + \cdots + \hat{\phi}_p\left(\hat{X}_{n+h-p} - \hat{\mu}\right),$$

where $\hat{X}_{n+1} = \hat{\mu} + \hat{\phi}_1(x_n - \hat{\mu}) + \hat{\phi}_2(x_{n-1} - \hat{\mu}) + \cdots + \hat{\phi}_p(x_{n-p+1} - \hat{\mu})$. An approximate two-sided $100(1-\alpha)\%$ prediction interval for X_{n+h} is

$$\hat{X}_{n+h} - z_{\alpha/2}\sqrt{1 + \hat{\theta}_1^2 + \hat{\theta}_2^2 + \cdots + \hat{\theta}_{h-1}^2}\; \hat{\sigma}_Z < X_{n+h} < \hat{X}_{n+h} + z_{\alpha/2}\sqrt{1 + \hat{\theta}_1^2 + \hat{\theta}_2^2 + \cdots + \hat{\theta}_{h-1}^2}\; \hat{\sigma}_Z,$$

where $\hat{\theta}_1, \hat{\theta}_2, \ldots$ are the estimated coefficients in the MA(∞) model associated with the estimated AR(p) model.

Example 9.31 For the time series of annual lynx pelt sales $x_1, x_2, \ldots, x_{55}$ from Example 9.29, forecast the next five values (for years 1912–1916) in the time series and give approximate 95% prediction intervals for the forecasted values assuming that the transformed values in the time series arise from a shifted AR(4) time series model with parameters estimated by maximum likelihood.

The R code below uses the `ar` function to estimate the parameters in the shifted AR(4) time series model to the natural logarithm of the time series values via maximum likelihood estimation. The `predict` function implements Theorem 9.24 to calculate the forecasted values and associated standard errors for the fitted AR(4) model. These standard errors can be used to calculate approximate 95% prediction interval limits. The R `exp` function is used to convert the forecasted values $\hat{X}_{t+h}$, whose units are the natural logarithm of the annual number of lynx pelts sold, back to the original time series, whose units are the annual number of lynx pelts sold.

```
y = c(23362, 31642, 33757, 23226, 15178, 7272, 4448, 4926, 5437,
       16498, 35971, 76556, 68392, 37447, 45686, 7942, 5123, 7106,
       11250, 18774, 30508, 42834, 27345, 17834, 15386, 9443, 7599,
       8061, 27187, 51511, 74050, 78773, 33899, 18886, 11520, 8352,
       8660, 12902, 20331, 36853, 56407, 39437, 26761, 15185, 4473,
       5781, 9117, 19267, 36116, 58850, 61478, 36300, 9704, 3410, 3774)
x        = log(y)
z        = qnorm(0.975)
model    = ar(x, order.max = 4, aic = FALSE, method = "mle")
forecast = predict(model, n.ahead = 40)
xhat     = exp(forecast$pred)
pred.lo  = exp(forecast$pred - z * forecast$se)
pred.hi  = exp(forecast$pred + z * forecast$se)
```

The results for the first five forecasted values are summarized in Table 9.13.

Time	$t=56$	$t=57$	$t=58$	$t=59$	$t=60$
Year	1912	1913	1914	1915	1916
Forecast	5750	14,639	41,540	74,000	75,380
Lower prediction bound	3078	6642	17,962	31,907	31,031
Upper prediction bound	10,742	32,263	96,067	171,620	183,116

Table 9.13: Forecasts and 95% prediction intervals for the annual lynx pelt sales.

For the first time, we have encountered prediction interval bounds which are not symmetric about the point estimate because of the exponentiation of the forecasted values and their prediction intervals. The widths of the prediction intervals in Table 9.13 tend to increase with the predicted value. The first forecasted value has 95% prediction interval

$$3078 < Y_{56} < 10,742$$

and the fifth forecasted value has the considerably wider 95% prediction interval

$$31,031 < Y_{60} < 183,116.$$

Figure 9.34 shows (*a*) the original time series $y_1, y_2, \ldots, y_{55}$ as points (•) connected by lines, (*b*) the first 40 forecasted annual lynx pelt sales $\hat{Y}_{56}, \hat{Y}_{57}, \ldots, \hat{Y}_{95}$ as open circles (∘), and (*c*) the 95% prediction intervals as a shaded region. There are six key observations concerning Figure 9.34.

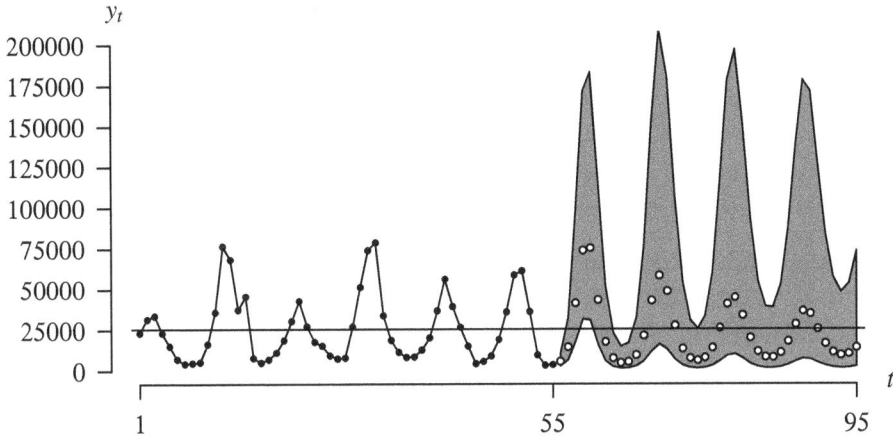

Figure 9.34: Annual lynx pelt sales forecasts and 95% prediction intervals.

- The forecasted values exhibit a similar periodicity to that of the original time series.

- The forecasted values seem to be reasonable estimates of the future values of the time series for the first cycle or two.

- The widths of the 95% prediction intervals associated with $h = 7, 8, \ldots, 12$ are considerably narrower than the width of the 95% prediction interval at $h = 5$. So unlike the previous two time series (the active beaver temperatures fit to the AR(1) model in Example 9.10 and the Lake Huron levels fit to the AR(2) model in Example 9.21), the prediction interval widths do not increase monotonically in h.

- The amplitude of the cycles of the forecasted values decreases with time. As the time horizon h increases, it becomes less certain where the time series is in a cycle, so the forecasts converge to $\hat{\mu} = \bar{y} = 25,600$ pelts sold annually. Depending on the application, this might not be a welcome aspect of the forecasted values. Using a time series model that explicitly contains a cyclic component might be more appropriate for forecasting in this setting.

- The random sampling variability which is evident in the observed time series values $y_1, y_2, \ldots, x_{55}$ is less evident in the forecasted values $\hat{Y}_{56}, \hat{Y}_{57}, \ldots, \hat{Y}_{95}$. Observed time series values tend to exhibit the typical random sampling variability; forecasted values for a stationary shifted AR(4) time series model of the transformed time series tend to be smooth.

This subsection has introduced the AR(p) time series model. The important results for an AR(p) model are listed below.

- The standard AR(p) model can be written algebraically and with the backshift operator B as

$$X_t = \phi_1 X_{t-1} + \phi_2 X_{t-2} + \cdots + \phi_p X_{t-p} + Z_t \qquad \text{and} \qquad \phi(B)X_t = Z_t,$$

where $\phi(B) = 1 - \phi_1 B - \phi_2 B^2 - \cdots - \phi_p B^p$ is the characteristic polynomial and $Z_t \sim WN\left(0, \sigma_Z^2\right)$ (Definition 9.3).

- The shifted AR(p) model can be written algebraically and with the backshift operator B as (Theorem 9.20)

$$X_t - \mu = \phi_1 (X_{t-1} - \mu) + \phi_2 (X_{t-2} - \mu) + \cdots + \phi_p (X_{t-p} - \mu) + Z_t \qquad \text{and} \qquad \phi(B)(X_t - \mu) = Z_t.$$

- The AR(p) model is always invertible; the AR(p) model is stationary when the solutions of $\phi(B) = 0$ all lie outside of the unit circle in the complex plane (Theorem 8.3).

- The AR(p) population autocorrelation function is a mixture of damped exponential functions, associated with real roots of $\phi(B)$, and damped sinusoidal functions, associated with complex roots of $\phi(B)$ (Theorem 9.18).

- The AR(p) population partial autocorrelation function cuts off after lag p (Theorem 9.19), making its shape easier to recognize than the population autocorrelation function for the statistical counterparts associated with a realization of a time series.

- The stationary shifted AR(p) model can be written as a shifted MA(∞) model (as illustrated in Example 9.23).

- The $p + 2$ parameters in the shifted AR(p) model, μ, ϕ_1, ϕ_2, ..., ϕ_p, and σ_Z^2, can be estimated from a realization of a time series $x_1, x_2, \ldots, x_n$ by the method of moments (Theorem 9.21), least squares (Theorem 9.22), and maximum likelihood. The point estimators for μ, ϕ_1, ϕ_2, ..., ϕ_p, and σ_Z^2 are denoted by $\hat{\mu}$, $\hat{\phi}_1$, $\hat{\phi}_2$, ..., $\hat{\phi}_p$, and $\hat{\sigma}_Z^2$, and are typically paired with asymptotically exact two-sided $100(1 - \alpha)\%$ confidence intervals (Theorem 9.23).

- The forecasted value $\hat{X}_{n+h}$ in a shifted AR(p) model is a function of the last p values in an observed time series $x_1, x_2, \ldots, x_n$ and can be calculated by a recursive formula. The forecast approaches $\hat{\mu} = \bar{x}$ as the time horizon $h \to \infty$. The associated prediction interval has width that increases as h increases and approaches a limit as the time horizon $h \to \infty$ (Theorem 9.24).

9.1.4 Computing

The R time series functions used in this section are summarized here. The `ARMAacf` function computes the population autocorrelation function or the population partial autocorrelation function for an ARMA(p, q) time series model. The generic version of the function is

```
ARMAacf(ar = numeric(), ma = numeric(), lag.max = r, pacf = FALSE)
```

where `ar` is a vector containing the autoregressive coefficients $\phi_1, \phi_2, \ldots, \phi_p$, `ma` is a vector containing the moving average coefficients $\theta_1, \theta_2, \ldots, \theta_q$, `lag.max` contains the number of lags required, and `pacf` is a logical object. The function returns $\rho(0), \rho(1), \ldots, \rho(\text{lag.max})$ when `pacf` is `FALSE`, or $\rho^*(1), \rho^*(2), \ldots, \rho^*(\text{lag.max})$ when `pacf` is `TRUE`. The `ARMAacf` function is illustrated in Example 9.25.

The `arima.sim` function generates a simulation of a time series. The generic version of the function is

```
arima.sim(model, n, rand.gen = rnorm, innov = rand.gen(n, ...),
          n.start = NA, start.innov = rand.gen(n.start, ...),
          ...)
```

where model is a list with components ar containing the autoregressive coefficients $\phi_1, \phi_2, \ldots, \phi_p$, and ma containing the moving average coefficients $\theta_1, \theta_2, \ldots, \theta_q$, n is the length of the simulated time series to be generated, rand.gen is a function to generate the white noise terms, n.start is the length of the warm-up period, and start.innov is a time series of white noise terms used in the warm-up period. The returned value is a vector containing the n simulated time series values $x_1, x_2, \ldots, x_n$. The arima.sim function is illustrated in Examples 9.2, 9.13, and 9.18. The warm-up period associated with the arima.sim function can be avoided by generating initial values from the appropriate multivariate distribution. For an AR(1) model with Gaussian white noise error terms, the rnorm function, whose generic syntax is

```
rnorm(n, mean = 0, sd = 1)
```

where n is the number of random variates to generate, mean is the population mean, and sd is the population standard deviation, can be used to seed the simulated time series. The rnorm function is illustrated in Example 9.1. For an AR(p) model, with $p > 1$, with Gaussian white noise error terms, the mvrnorm function from the MASS package, whose generic syntax is

```
mvrnorm(n = 1, mu, Sigma, tol = 1e-6, empirical = FALSE, EISPACK = FALSE)
```

where n is the number of random vectors to generate, mu is the population mean vector, and Sigma is the population variance–covariance matrix, can be used to seed the simulated time series. The mvrnorm function is illustrated in Examples 9.12 and 9.26.

When determining parameter estimates that cannot be expressed in closed form, the optim function provides general-purpose optimization capability that can be applied to minimizing the sum of squares to find the least squares estimates or maximizing the log likelihood function to find the maximum likelihood estimators. The generic syntax for optim is

```
optim(par, fn, gr = NULL, ...,
      method = c("Nelder-Mead", "BFGS", "CG", "L-BFGS-B", "SANN", "Brent"),
      lower = -Inf, upper = Inf, control = list(), hessian = FALSE)
```

where par is a vector containing initial parameter estimates and fn is the function to be minimized (by default). The optim function is illustrated in Examples 9.5, 9.6, and 9.16. A parameter estimation function that is exclusively for autoregressive time series models is ar. The generic format for ar is

```
ar(x, aic = TRUE, order.max = NULL,
   method = c("yule-walker", "burg", "ols", "mle", "yw"),
   na.action, series, ...)
```

where x is a vector containing the observed time series values, aic is a logical variable (TRUE means that the Akaike Information Criterion is used to choose the order of the model and FALSE means that an autoregressive model of order order.max is fitted), order.max is maximum order of the autoregressive model to fit, method is the estimation method ("yule-walker" or "yw" for Yule–Walker, "burg" for Burg's algorithm, "ols" for least squares, "mle" for maximum likelihood), and na.action indicates how to handle missing values in the time series. The ar function is illustrated in Examples 9.7, 9.8, 9.10, 9.17, 9.19, 9.21, 9.27, 9.28, and 9.31. The arima function also estimates parameters from an observed time series. The generic format for arima is

```
arima(x, order = c(0L, 0L, 0L),
      seasonal = list(order = c(0L, 0L, 0L), period = NA),
```

```
xreg = NULL, include.mean = TRUE, transform.pars = TRUE,
fixed = NULL, init = NULL, method = c("CSS-ML", "ML", "CSS"), n.cond,
SSinit = c("Gardner1980", "Rossignol2011"), optim.method = "BFGS",
optim.control = list(), kappa = 1e6)
```

where x is a vector containing the observed time series values, order is a vector containing the values of p, d, and q, include.mean is a logical variable (TRUE includes estimation of a population mean term μ and FALSE estimates just the parameters in the standard model), and method is CSS (conditional sum of squares) or ML (maximum likelihood). The arima function is illustrated in Examples 9.9, 9.18, 9.20, and 9.29.

Three functions were introduced in this section for assessing model adequacy. The Box.test function computes the Box–Pierce or Ljung–Box test statistic and associated p-value. The generic syntax is

```
Box.test(x, lag = 1, type = c("Box-Pierce", "Ljung-Box"), fitdf = 0)
```

where x is a vector containing the observed time series values, lag is the number of sample autocorrelation function values to be used in the test, type is either "Box-Pierce" or "Ljung-Box", and fitdf is the number of degrees of freedom to be subtracted in the case of x being a time series of residuals. The Box.test function is illustrated in Examples 9.8, 9.19, and 9.30, along with the hist and qqnorm functions, which are helpful in visually assessing the normality of the residuals.

Forecasting can be performed automatically using the generic predict function, which calculates predicted values of a time series from a fitted function. The predict function is illustrated in Examples 9.10, 9.21, and 9.31.

More details on the R functions used in this section can be found using the help function. Sample invocations of the functions are displayed using the example function.

This concludes the introduction to the autoregressive time series model, with subsections devoted to the AR(1), AR(2), and AR(p) models. An analogous treatment of moving average models is contained in the next section.

9.2   Moving Average Models

Moving average models for a time series will be introduced in this section. A moving average model of order q is a special case of an ARMA(p, q) model with no autoregressive terms (that is, $p = 0$) and q moving average terms, specified as

$$X_t = Z_t + \theta_1 Z_{t-1} + \theta_2 Z_{t-2} + \cdots + \theta_q Z_{t-q},$$

where $\theta_1, \theta_2, \ldots, \theta_q$ are real-valued parameters and $\{Z_t\}$ is a time series of white noise. Rather than diving right into an MA(q) model, we first have separate subsections for the MA(1) and MA(2) models because the mathematics are somewhat easier than the general case and some important geometry and intuition can be developed with these restricted models. In the subsection on the MA(1) model that follows, we will

- define the time series model for $\{X_t\}$,

- determine the values of the parameters associated with an invertible model,

- derive the population autocorrelation and partial autocorrelation functions,

- develop algorithms for simulating observations from the time series,

- inspect simulated realizations to establish patterns, and

- estimate parameters from a time series realization $\{x_t\}$.

The important steps of model assessment, model selection, and forecasting future values of the times series are left as exercises because they follow along the same lines as those steps for the autoregressive models covered in the previous section.

The purpose of deriving the population autocorrelation and partial autocorrelation functions is to build an inventory of shapes and patterns for these functions that can be used to identify tentative time series models from their sample counterparts by making a visual comparison between population and sample versions. This inventory of shapes and patterns plays an analogous role to knowing the shapes of various probability density functions (for example, the bell-shaped normal probability density function or the rectangular-shaped uniform distribution) in the analysis of univariate data in which the shape of the histogram is visually compared to the inventory of probability density function shapes.

In the MA(1) subsection that follows, a single example of a time series will be carried through the various statistical procedures given in the list above. Stationarity plays a critical role in time series analysis because we are not able to forecast future values of the time series without knowing that the probability model is stable over time. This is why the visual assessment of a plot of the time series is always a critical first step in the analysis of a time series. Fortunately, all MA(q) time series models are stationary.

9.2.1 The MA(1) Model

The moving average model with one term is the simplest of the ARMA family of time series models in terms of the ability to derive probabilistic properties.

Definition 9.4 A first-order moving average time series model, denoted by MA(1), for the time series $\{X_t\}$ is defined by

$$X_t = Z_t + \theta Z_{t-1},$$

where θ is a real-valued parameter and $\{Z_t\}$ is a time series of white noise:

$$Z_t \sim WN\left(0, \sigma_Z^2\right).$$

An observed value in the time series, X_t, is given by the current white noise term, plus the parameter θ multiplied by the white noise term from one time period ago. No subscript is necessary on the θ parameter because there is only one θ parameter in the MA(1) model. So there are two parameters that define an MA(1) model: the coefficient θ and the population variance of the white noise σ_Z^2.

Some authors prefer to parameterize the MA(1) model as

$$X_t = \theta_0 Z_t + \theta_1 Z_{t-1},$$

where θ_0 and θ_1 are real-valued parameters. We avoid this parameterization because the θ_0 parameter is redundant in the sense that the population variance of the white noise σ_Z^2 is absorbed into the θ_0 parameter. Also, some authors use a $-$ rather than a $+$ between two terms on the right-hand side of the model.

To illustrate the thinking behind the MA(1) model in a specific context, let X_t represent the monthly unemployment, as a percentage, in month t. The MA(1) model indicates that this month's unemployment, denoted by X_t, equals θ multiplied by last month's random white noise term, θZ_{t-1}, plus this month's random white noise term Z_t.

MA(1) models are used less often than autoregressive models, and this is partly due to more limited potential shapes for the population autocorrelation function, as will be seen next.

Stationarity and the Population Autocorrelation Function

One initial important question concerning the MA(1) model is whether or not the model is stationary. Rather than appealing to Theorem 8.4, we show this below using first principles. Recall from Definition 7.6 that a time series model is stationary if (a) the expected value of X_t is constant for all t, and (b) the population covariance between X_s and X_t depends only on the lag $|t - s|$. The expected value of X_t is

$$E[X_t] = E[Z_t + \theta Z_{t-1}] = E[Z_t] + \theta E[Z_{t-1}] = 0$$

for all values of the parameters θ and σ_Z^2, and all values of t. Using the defining formula for population covariance, the population autocovariance function is

$$
\begin{aligned}
\gamma(s, t) &= \text{Cov}(X_s, X_t) \\
&= E\left[(X_s - E[X_s])(X_t - E[X_t])\right] \\
&= E[X_s X_t] \\
&= E\left[(Z_s + \theta Z_{s-1})(Z_t + \theta Z_{t-1})\right] \\
&= E[Z_s Z_t] + \theta E[Z_{s-1} Z_t] + \theta E[Z_s Z_{t-1}] + \theta^2 E[Z_{s-1} Z_{t-1}] \\
&= \begin{cases} V[Z_t] + \theta^2 V[Z_{t-1}] & |t-s| = 0 \\ \theta V[Z_t] & |t-s| = 1 \\ 0 & |t-s| = 2, 3, \ldots \end{cases} \\
&= \begin{cases} (1+\theta^2)\sigma_Z^2 & |t-s| = 0 \\ \theta \sigma_Z^2 & |t-s| = 1 \\ 0 & |t-s| = 2, 3, \ldots. \end{cases}
\end{aligned}
$$

Since $E[X_t] = 0$ for all values of t and the population autocovariance function depends only on the lag $|t - s|$, we conclude that the MA(1) time series model is stationary. Furthermore, the population autocovariance function can be expressed in terms of the lag k as

$$\gamma(k) = \begin{cases} (1+\theta^2)\sigma_Z^2 & k = 0 \\ \theta \sigma_Z^2 & k = 1 \\ 0 & k = 2, 3, \ldots. \end{cases}$$

Dividing by the population autocovariance function by $\gamma(0) = V[X_t] = (1+\theta^2)\sigma_Z^2$ gives the population autocorrelation function

$$\rho(k) = \begin{cases} 1 & k = 0 \\ \theta/(1+\theta^2) & k = 1 \\ 0 & k = 2, 3, \ldots. \end{cases}$$

This derivation constitutes a proof of the following result.

Theorem 9.25 The MA(1) time series $\{X_t\}$ is stationary for all values of the parameters θ and σ_Z^2 with population autocorrelation function

$$\rho(k) = \begin{cases} 1 & k = 0 \\ \theta/\left(1+\theta^2\right) & k = 1 \\ 0 & k = 2, 3, \ldots . \end{cases}$$

So the population autocorrelation function consists of a single nonzero value at lag 1 for a nonzero parameter θ and zero values thereafter. Six important observations concerning this population autocorrelation function are given below.

- The sign of $\rho(1)$ is the same as the sign of θ.

- The population autocorrelation function cuts off after lag 1 for an MA(1) time series model. The time series model has a "memory" of just one time period. Figure 9.35 illustrates the relationship between the white noise values $\{Z_t\}$ and the MA(1) time series observations $\{X_t\}$. Observations of the time series that are two or more time periods apart, such as X_2 and X_4, have no white noise terms in common, so the lag 2 population autocorrelation, $\rho(2)$, is zero. The third observation in the time series X_3, for example, shares the white noise term Z_2 with X_2 and the white noise term Z_3 with X_4, but is not affected by any white noise terms before Z_2 or after Z_3.

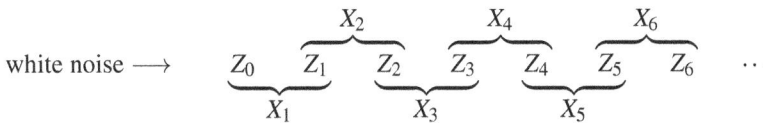

Figure 9.35: Relationship between white noise $\{Z_t\}$ and $\{X_t\}$ for an MA(1) model.

- The lag 1 population autocorrelation $\rho(1) = \theta/\left(1+\theta^2\right)$ can be written as a quadratic equation in θ as

$$\rho(1)\theta^2 - \theta + \rho(1) = 0.$$

For nonzero values of θ, the two roots of this quadratic equation are both positive or both negative. Furthermore, a little algebra reveals that the product of the two roots of this quadratic equation equals 1. Figure 9.36 shows the parabolas associated with this quadratic equation for $\rho(1) = 2/5$ (with associated roots $\theta = 1/2$ and $\theta = 2$) and $\rho(1) = -2/5$ (with associated roots $\theta = -1/2$ and $\theta = -2$).

- The value $\rho(1)$ must lie in the interval $-1/2 \leq \rho(1) \leq 1/2$. This can be seen in the plot of $\rho(1) = \theta/\left(1+\theta^2\right)$ versus θ given by the solid curve in Figure 9.37, which indicates that $\rho(1)$ is minimized at $\rho(1) = -1/2$ when $\theta = -1$ and maximized at $\rho(1) = 1/2$ when $\theta = 1$. This constraint means that the MA(1) model is more limited in application than the autoregressive models from the previous chapter. In order to fit an MA(1) model to observed time series values $x_1, x_2, \ldots, x_n$, it must be the case that (a) the length of the time series n is large enough (about $n = 50$ or $n = 60$) to use an ARMA model, (b) the sample autocorrelation function has a single statistically significant spike at lag 1, and (c) the statistically significant spike at lag 1 satisfies $-1/2 \leq r_1 \leq 1/2$ to be compatible with the constraint $-1/2 \leq \rho(1) \leq 1/2$.

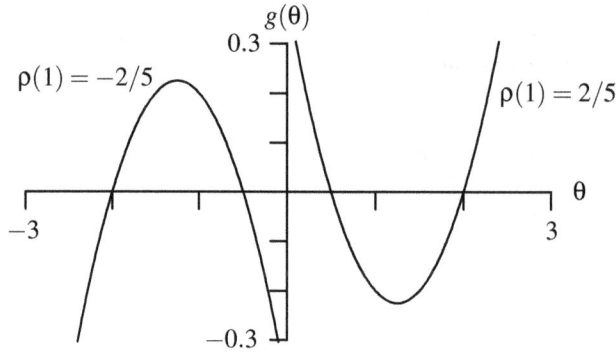

Figure 9.36: The parabola $g(\theta) = \rho(1)\theta^2 - \theta + \rho(1)$ for $\rho(1) = 2/5$ and $\rho(1) = -2/5$.

• Figure 9.37 also reveals a more subtle aspect of the population lag 1 autocorrelation. Notice that for $\theta = 1/2$, the population lag 1 autocorrelation is $\rho(1) = 2/5$. But for $\theta = 2$, the population lag 1 autocorrelation is also $\rho(1) = 2/5$. The geometry associated with these two values of θ resulting in the same value for $\rho(1)$ is indicated by the dashed lines in Figure 9.37. This problem is not just limited to $\theta = 1/2$ and $\theta = 2$; there are an infinite number of pairs of θ values that will result in the same population lag 1 autocorrelation function value. More generally, the MA(1) model

$$X_t = Z_t + \theta Z_{t-1}$$

and the MA(1) model

$$X_t = Z_t + \frac{1}{\theta} Z_{t-1}$$

have identical population autocorrelation functions. This means that there is not a one-to-one

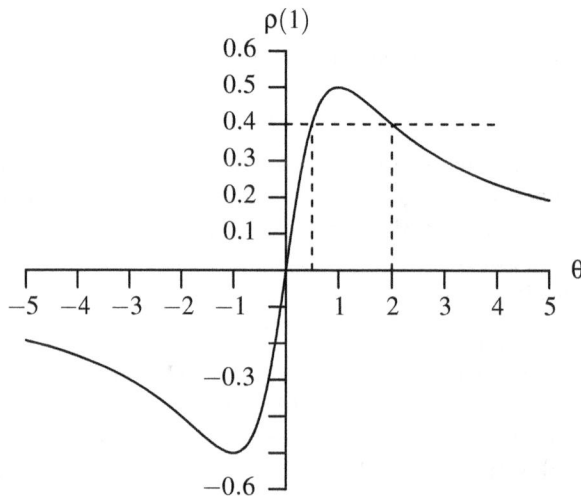

Figure 9.37: Graph of $\rho(1) = \theta / (1 + \theta^2)$ versus θ for an MA(1) time series model.

correspondence between a particular value of θ and the associated value of $\rho(1)$. This brings up the notion of *invertibility*, which was defined in Definition 8.3. An invertible time series model has a unique value of θ in the MA(1) model corresponding to a particular population autocorrelation function.

Invertibility

All MA(1) models are stationary per Theorem 8.3 because there are a finite number of moving average terms in Definition 9.4. Recall from Definition 8.3 that an ARMA(p, q) time series model for $\{X_t\}$ is *invertible* if the white noise term at time t can be expressed as

$$Z_t = \sum_{j=0}^{\infty} \pi_j X_{t-j},$$

where the coefficients π_j satisfy

$$\sum_{j=0}^{\infty} \pi_j^2 < \infty.$$

There are no restrictions on θ necessary to ensure stationarity for an MA(1) model. However, it can be advantageous to restrict the values of θ in order to achieve invertibility. Returning to Figure 9.37, we can use the definition of invertibility to determine whether we use $|\theta| < 1$ or $|\theta| > 1$ for the invertibility region for an MA(1) model.

Just as we were able to write an AR(1) time series model as an MA(∞) time series model in Section 9.1.1, we now perform the algebraic steps necessary to write an MA(1) time series model as an AR(∞) time series model. We want to write Z_t in terms of current and previous values of X_t as shown in Definition 8.3. To begin, recall that the MA(1) model given by

$$X_t = Z_t + \theta Z_{t-1}$$

can be shifted in time and is equally valid for other t values, for example,

$$X_{t-1} = Z_{t-1} + \theta Z_{t-2}$$
$$X_{t-2} = Z_{t-2} + \theta Z_{t-3}$$
$$\vdots \quad = \quad \vdots$$

These formulas can be solved for $Z_{t-1}, Z_{t-2}, \ldots$ as

$$Z_{t-1} = X_{t-1} - \theta Z_{t-2}$$
$$Z_{t-2} = X_{t-2} - \theta Z_{t-3}$$
$$\vdots \quad = \quad \vdots$$

Making successive substitutions into the MA(1) model results in

$$X_t = Z_t + \theta Z_{t-1}$$
$$= Z_t + \theta \left(X_{t-1} - \theta Z_{t-2} \right)$$
$$= Z_t + \theta X_{t-1} - \theta^2 Z_{t-2}$$
$$= Z_t + \theta X_{t-1} - \theta^2 \left(X_{t-2} - \theta Z_{t-3} \right)$$

$$= Z_t + \theta X_{t-1} - \theta^2 X_{t-2} + \theta^3 Z_{t-3}$$

$$\vdots$$

$$= Z_t + \theta X_{t-1} - \theta^2 X_{t-2} + \theta^3 X_{t-3} - \theta^4 X_{t-4} + \cdots .$$

This can be recognized as an AR(∞) time series model.

Theorem 9.26 An MA(1) model with parameter θ can be written as the AR(∞) model

$$X_t = Z_t + \theta X_{t-1} - \theta^2 X_{t-2} + \theta^3 X_{t-3} - \theta^4 X_{t-4} + \cdots .$$

Representing an MA(1) model as an AR(∞) model is known as *duality*. Solving this equation for Z_t gives

$$Z_t = X_t - \theta X_{t-1} + \theta^2 X_{t-2} - \theta^3 X_{t-3} + \theta^4 X_{t-4} - \cdots ,$$

which is the form required for Definition 8.3. So the coefficients $\pi_0, \pi_1, \pi_2, \ldots$ for the MA(1) model from Definition 8.3 are

$$\pi_0 = 1, \pi_1 = -\theta, \pi_2 = \theta^2, \pi_3 = -\theta^3, \pi_4 = \theta^4, \ldots ,$$

or in general, $\pi_j = (-\theta)^j$, for $j = 0, 1, 2, \ldots$. Definition 8.3 requires that

$$\sum_{j=0}^{\infty} |\pi_j| = \sum_{j=0}^{\infty} \left|(-\theta)^j\right| = 1 + \theta + \theta^2 + \theta^3 + \cdots < \infty$$

to achieve stationarity. This summation is a geometric series that converges when $|\theta| < 1$, so this is the invertibility region for an MA(1) model.

Theorem 9.27 The MA(1) time series model is invertible when $-1 < \theta < 1$.

The invertibility criterion $-1 < \theta < 1$ ensures that each value of θ in the interval corresponds to a unique MA(1) time series model. Stated in another fashion, invertibility implies that there is a one-to-one correspondence between the value of the θ parameter and the population autocorrelation function.

The MA(1) time series model can be written in terms of the backshift operator B as

$$X_t = (1 + \theta B) Z_t = Z_t + \theta Z_{t-1}.$$

Doubling up the use of θ as a function name, the expression

$$\theta(B) = 1 + \theta B$$

is the *characteristic polynomial* for the MA(1) model. Notice that the MA(1) model is invertible when $|\theta| < 1$, which corresponds to the root of $\theta(B) = 0$ falling *outside* of the interval $[-1, 1]$. Solving $\theta(B) = 1 + \theta B = 0$ results in $B = -1/\theta$. This notion will be generalized in the next two subsections for higher-order MA models as the roots of $\theta(B) = 0$ falling outside of the unit circle in the complex plane to establish invertibility.

Now that stationarity for all MA(1) time series models has been established, the condition for invertibility has been established, and the population autocorrelation function has been derived, we turn to determining the partial autocorrelation function.

Population Partial Autocorrelation Function

The population partial autocorrelation function can be determined by using the defining formula in Definition 7.4. The lag zero population partial autocorrelation is $\rho^*(0) = 1$. The lag one population partial autocorrelation is $\rho^*(1) = \rho(1) = \theta/\left(1 + \theta^2\right)$. After a little algebra, the lag two population partial autocorrelation is

$$\rho^*(2) = \frac{\begin{vmatrix} 1 & \rho(1) \\ \rho(1) & \rho(2) \end{vmatrix}}{\begin{vmatrix} 1 & \rho(1) \\ \rho(1) & 1 \end{vmatrix}} = \frac{\rho(2) - [\rho(1)]^2}{1 - [\rho(1)]^2} = -\frac{[\rho(1)]^2}{1 - [\rho(1)]^2} = -\frac{\theta^2 \left(1 - \theta^2\right)}{1 - \theta^6}.$$

The lag three population partial autocorrelation is

$$\rho^*(3) = \frac{\begin{vmatrix} 1 & \rho(1) & \rho(1) \\ \rho(1) & 1 & \rho(2) \\ \rho(2) & \rho(1) & \rho(3) \end{vmatrix}}{\begin{vmatrix} 1 & \rho(1) & \rho(2) \\ \rho(1) & 1 & \rho(1) \\ \rho(2) & \rho(1) & 1 \end{vmatrix}} = \frac{[\rho(1)]^3}{1 - 2[\rho(1)]^2} = \frac{\theta^3 \left(1 - \theta^2\right)}{1 - \theta^8}.$$

This pattern generalizes to the lag k population partial autocorrelation

$$\rho^*(k) = \frac{(-1)^{k+1} \theta^k \left(1 - \theta^2\right)}{1 - \theta^{2(k+1)}}$$

for $k = 1, 2, \ldots$, which can also be written as

$$\rho^*(k) = \frac{(-1)^{k+1} \theta^k}{1 + \theta^2 + \theta^4 + \cdots + \theta^{2k}}$$

for $k = 1, 2, \ldots$. This constitutes a proof of the following result.

Theorem 9.28 The invertible MA(1) time series model for $\{X_t\}$ with $-1 < \theta < 1$ has population partial autocorrelation function

$$\rho^*(k) = \frac{(-1)^{k+1} \theta^k \left(1 - \theta^2\right)}{1 - \theta^{2(k+1)}}$$

for $k = 1, 2, \ldots$.

When $\theta = 0$, both the population autocorrelation function and the partial autocorrelation function have just a single spike at $\rho(0) = \rho^*(0) = 1$; the MA(1) model reduces to just white noise in this case. When $0 < \theta < 1$, $\rho(1) > 0$ and $\rho^*(k)$ tails out and alternates in sign. When $-1 < \theta < 0$, $\rho(1) < 0$ and $\rho^*(k)$ tails out and is negative for $k = 1, 2, \ldots$.

Example 9.32 Plot the first eight lags of the population autocorrelation function $\rho(k)$ and the population partial autocorrelation function $\rho^*(k)$ for an MA(1) model with $\theta = 0.9$.

The values of the population autocorrelation function and the population partial autocorrelation function can be calculated using the formulas in Theorems 9.25 and 9.28, respectively, or by calling the built-in R ARMAacf function as shown below.

```
ARMAacf(ar = 0, ma = 9 / 10, 8)
ARMAacf(ar = 0, ma = 9 / 10, 8, pacf = TRUE)
```

The plots of these functions are displayed in Figure 9.38. As expected for an MA(1) model, the population autocorrelation function cuts off after lag 1 and the population partial autocorrelation alternates in sign and tails out.

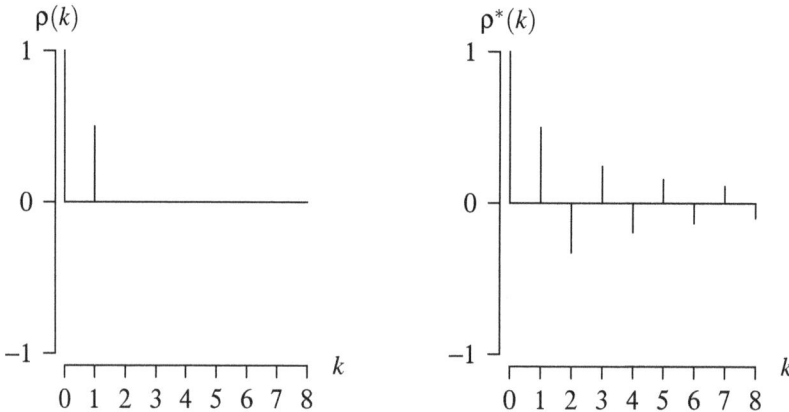

Figure 9.38: Graphs of $\rho(k)$ (left) and $\rho^*(k)$ (right) for an MA(1) model with $\theta = 9/10$.

The Shifted MA(1) Model

The population mean function for the MA(1) model is $E[X_t] = 0$, which is not of much use in practice because most real-world time series are not centered around zero. Adding a third parameter μ to overcome this shortcoming results in the enhanced MA(1) model

$$X_t = \mu + Z_t + \theta Z_{t-1},$$

which has population mean function $E[X_t] = \mu$ and population autocorrelation function and population autocorrelation function given in Theorems 9.25 and 9.28 because population variance and covariance are unaffected by a shift in the time series model. There are now three parameters for the time series model: μ, θ, and σ_Z^2.

Theorem 9.29 A *shifted first-order moving average model* for the time series $\{X_t\}$ is defined by

$$X_t = \mu + Z_t + \theta Z_{t-1},$$

where μ, θ, and $\sigma_Z^2 > 0$ are real-valued parameters and $\{Z_t\}$ is a time series of white noise. This model is invertible when $-1 < \theta < 1$. This model is stationary for all values of the parameters μ, θ, and σ_Z^2. The expected value of X_t is $E[X_t] = \mu$. The population autocorrelation function and partial autocorrelation function are given in Theorems 9.25 and 9.28.

Simulation

An MA(1) time series can be simulated by appealing to the defining formula for the MA(1) model from Definition 9.4:

$$X_t = Z_t + \theta Z_{t-1}.$$

The algorithm given below generates an initial white noise value Z_0, and then uses an additional n white noise terms $Z_1, Z_2, \ldots, Z_n$ to generate the time series values $X_1, X_2, \ldots, X_n$ using the MA(1) defining formula. Indentation denotes nesting in the algorithm.

$t \leftarrow 0$
generate $Z_t \sim WN\left(0, \sigma_Z^2\right)$
while $(t < n)$
$\quad t \leftarrow t + 1$
$\quad$ generate $Z_t \sim WN\left(0, \sigma_Z^2\right)$
$\quad X_t \leftarrow Z_t + \theta Z_{t-1}$

The three-parameter shifted MA(1) time series model that includes a population mean parameter μ can be simulated by simply adding μ to each time series observation generated by this algorithm. So to generate a realization of an MA(1) time series model in R, we must define (a) the value of θ, (b) the distribution of the white noise, (c) the value of σ_Z^2, and, if this is a shifted MA(1) model, (d) the value of the shift parameter μ.

Example 9.33 Generate a realization of $n = 100$ observations from an MA(1) time series model with $\theta = 9/10$, Gaussian white noise terms, and $\sigma_Z^2 = 1$.

The parameter $\theta = 9/10$ just barely falls in the invertibility region $-1 < \theta < 1$. This choice of θ results in a population lag 1 autocorrelation that is very close to its largest possible value. The population lag 1 autocorrelation function value associated with this model is

$$\rho(1) = \frac{\theta}{1 + \theta^2} = \frac{9/10}{1 + (9/10)^2} = \frac{90}{181} \cong 0.4972,$$

so we expect a nearby value for r_1 from the simulated time series values. The R code below generates $n = 100$ simulated time series values and places them in the vector named x.

```
set.seed(37)
n     = 100
z     = rnorm(n + 1)
x     = numeric(n)
theta = 0.9
for (t in 1:n) x[t] = z[t + 1] + theta * z[t]
```

Use the plot.ts function to plot the time series contained in x, the acf function to plot the associated correlogram, and the pacf function plot the associated sample partial autocorrelation function.

```
layout(matrix(c(1, 1, 2, 3), 2, 2, byrow = TRUE))
plot.ts(x)
acf(x)
pacf(x)
```

The time series plot of the realization, the associated correlogram, and the associated sample partial autocorrelation function for $\theta = 0.9$ and standard normal noise terms are given in Figure 9.39. A horizontal line is drawn on the time series plot at $E[X_t] = 0$.

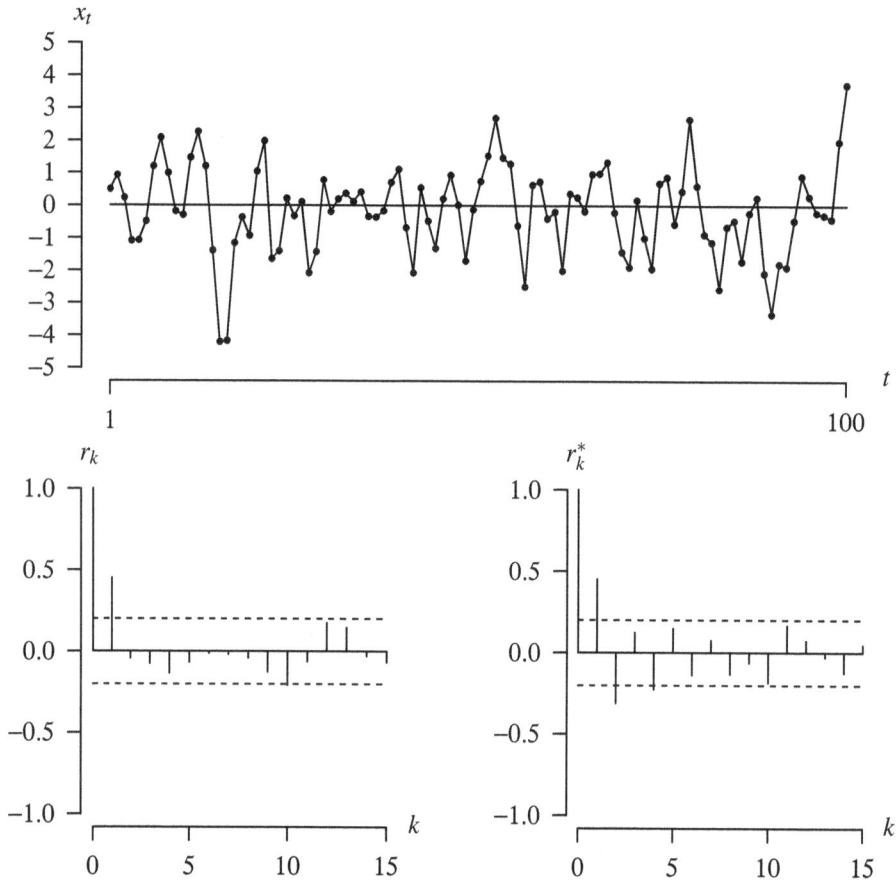

Figure 9.39: Time series plot, r_k, and r_k^* for $n = 100$ simulated values from an MA(1) model.

The time series contains *short runs* above and below the population mean, which are consistent with the statistically significant lag 1 sample autocorrelation function value $r_1 = 0.450$. This value is slightly smaller than the associated population value. The sample autocorrelation function values at lags 2 through 15 do not differ from zero by a statistically significant amount, except for the sample lag 10 autocorrelation function value $r_{10} = -0.207$. This value falls just slightly outside of the 95% confidence bounds

$$\pm \frac{z_{\alpha/2}}{\sqrt{n}} = \pm \frac{z_{0.025}}{\sqrt{100}} = \pm \frac{1.96}{10} = \pm 0.196.$$

This spike in the correlogram is not considered statistically significant because (a) there is nothing about the time series that would indicate that a lag of 10 is a special lag, (b) the spike in the correlogram at lag 10 falls just slightly outside of the confidence bounds, and (c) we expect 1 in 20 of the correlogram spikes to fall outside of the 95% confidence bounds because of random sampling variability, even if the MA(1) model were perfect (as it is in this case). The graphs of $\rho(k)$ and $\rho^*(k)$ mirror their population counterparts in Figure 9.38. You are encouraged to place these R statements in a `for` loop (of course, with the `set.seed` call outside of the loop) that generates multiple realizations of the MA(1) time series with a call to `Sys.sleep` to provide a short time delay for you to inspect the trio of plots. This will give you a feel for how this MA(1) time series, its correlogram, and its sample partial autocorrelation function vary from one realization to the next.

Another way to think about a realization of an MA(1) model is to make scatterplots of adjacent observations and observations that are two time units apart. The left-hand plot in Figure 9.40 illustrates the positive sample correlation between x_{t-1} and x_t for the realization, which is consistent with the positive lag 1 population autocorrelation. The $n - 1 = 99$ pairs of points plotted are the adjacent values in the realization of the time series, (x_{t-1}, x_t). The population autocorrelation function cuts off after lag 1 for an MA(1) time series model. This is supported by the right-hand plot in Figure 9.40, which shows the $n - 2 = 98$ pairs (x_{t-2}, x_t) for the realization, which appear to be independent. The regression lines have been added to each plot. The additional R statements below indicate that the p-values for the statistical significance of the slopes associated with the two plots are $p = 9.3 \cdot 10^{-7}$ and $p = 0.65$, respectively.

```
summary(lm(x[2:n] ~ x[1:(n - 1)]))
summary(lm(x[3:n] ~ x[1:(n - 2)]))
```

These p-values indicate that the slope of the line in the left-hand plot differs significantly from zero, while the slope of the line in the right-hand plot does not differ significantly from zero.

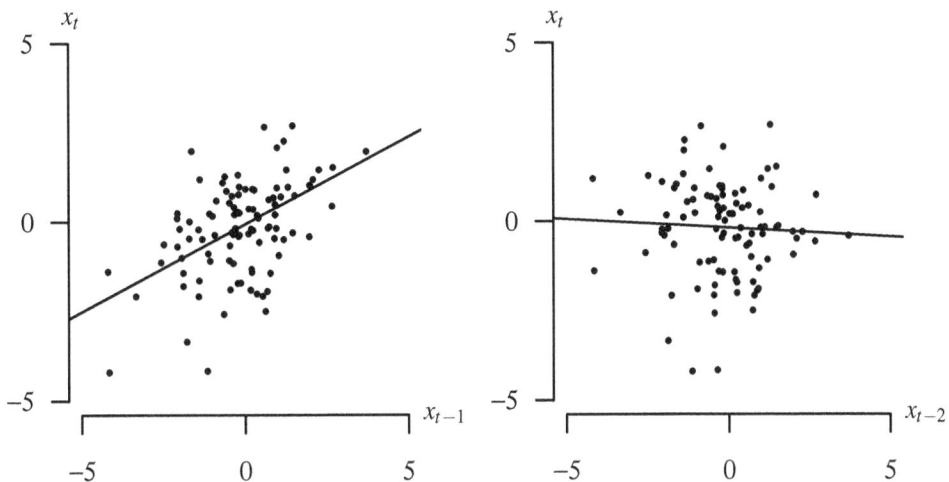

Figure 9.40: Scatterplots of pairs of simulated MA(1) observations.

Example 9.34 Generate a realization of $n = 100$ observations from a shifted MA(1) time series model with $\mu = 20$, $\theta = -9/10$, Gaussian white noise terms, and $\sigma_Z^2 = 1$.

This time series corresponds to the opposite extreme case of the MA(1) because the coefficient $\theta = -0.9$ also (barely) falls in the invertibility region $-1 < \theta < 1$. We again assume that the Gaussian white noise has population variance $\sigma_Z^2 = 1$, but now include a shift parameter $\mu = 20$. The population lag 1 autocorrelation function value associated with this model is

$$\rho(1) = \frac{\theta}{1 + \theta^2} = \frac{-9/10}{1 + (-9/10)^2} = -\frac{90}{181} \cong -0.4972.$$

So this choice of θ results in a population lag 1 autocorrelation that is very close to its smallest possible value. The R code below simulates $n = 100$ observations from this shifted MA(1) model. This code differs from the previous code in that it avoids the use of a `for` loop, which is a more efficient way to generate observations in R. The time series plot of the realization of $n = 100$ observations and the associated correlogram for $\theta = -0.9$ is given in Figure 9.41.

```
set.seed(37)
n      = 100
z      = rnorm(n + 1)
```

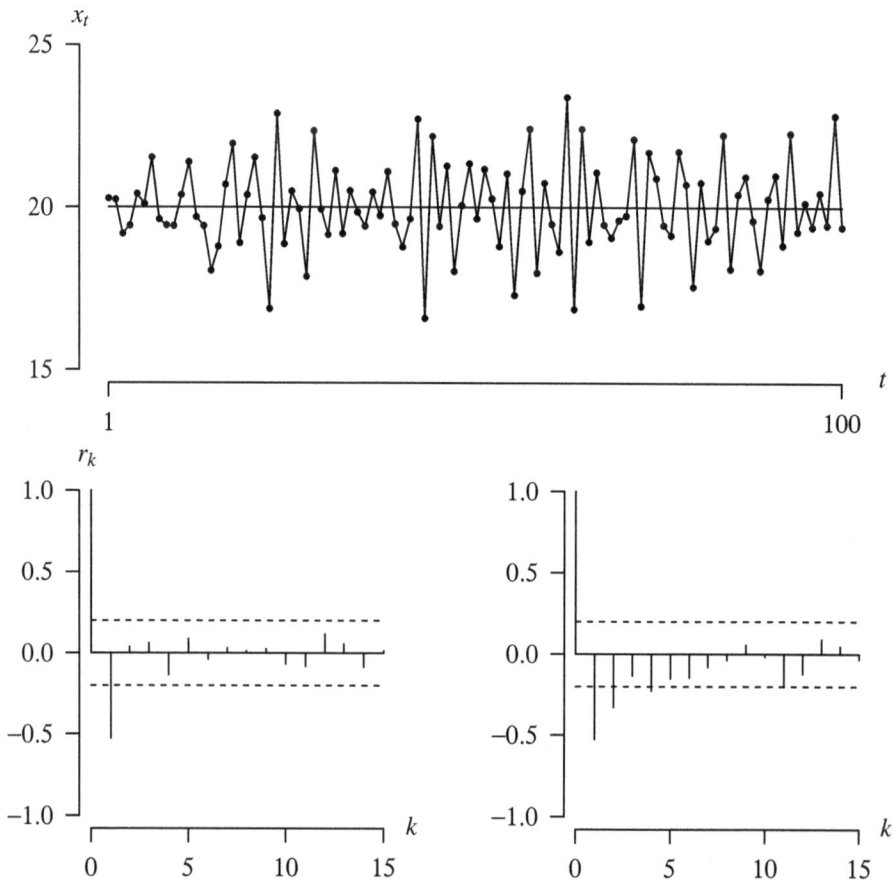

Figure 9.41: Time series plot, r_k, and r_k^* for $n = 100$ simulated values from an MA(1) model.

```
theta = -0.9
x      = z[2:(n + 1)] + theta * z[1:n]
x      = x + 20
```

The time series plot reveals that adjacent observations in the time series tend to be on opposite sides of the population mean, which is consistent with the sample lag 1 autocorrelation $r_1 = -0.524$. The sample autocorrelation function values for lags 2 through 15 do not differ from zero by a statistically significant amount. So the sample autocorrelation cuts off after lag 1 as expected. Furthermore, the sample partial autocorrelation tails out as expected. The `arima.sim` function can also be used to generate a realization of an MA(1) time series model using fewer keystrokes. The R single statement

```
x = 20 + arima.sim(list(ma = -0.9), n = 100)
```

generates 100 values from a shifted MA(1) model with $\theta = -0.9$, $\mu = 20$ and $\sigma_Z^2 = 1$. The default probability distribution for the white noise terms is Gaussian white noise.

Having established the probabilistic properties of the MA(1) model, we now turn to statistical topics, beginning with the estimation of the model parameters.

Parameter Estimation

There are several techniques for estimating the parameters in an MA(1) model; as was the case for the autoregressive models, we look at the method of moments, least squares, and maximum likelihood estimation techniques separately. Parameter estimation is more difficult for moving average models, as numerical methods are typically required to calculate the parameter estimates.

Approach 1: Method of moments. We begin with the shifted MA(1) model from Definition 9.4:

$$X_t = \mu + Z_t + \theta Z_{t-1}.$$

We want to estimate the three unknown parameters μ, θ, and σ_Z^2 from an observed time series $\{x_t\}$. In the case of the shifted MA(1) model, we match the population and sample (a) first-order moments, (b) second-order moments, and (c) lag 1 autocorrelation. These will be written with upper case values X_t although these will be replaced with numeric values x_t for a particular observed time series. Placing the population moments of the left-hand side of the equation and the associated sample moments on the right-hand side of the equation results in three equations in three unknowns:

$$E[X_t] = \frac{1}{n} \sum_{t=1}^{n} X_t$$

$$E[X_t^2] = \frac{1}{n} \sum_{t=1}^{n} X_t^2$$

$$\rho(1) = r_1$$

or

$$\mu = \bar{X}$$

$$V[X_t] + E[X_t]^2 = \gamma(0) + \mu^2 = \left(1 + \theta^2\right)\sigma_Z^2 + \mu^2 = \frac{1}{n} \sum_{t=1}^{n} X_t^2$$

$$\frac{\theta}{1 + \theta^2} = r_1.$$

The third equation is a quadratic equation in θ:

$$r_1 \theta^2 - \theta + r_1 = 0,$$

which corresponds to the parabolas in Figure 9.36, except that $\rho(1)$ is replaced by r_1. Using the quadratic formula, the product of the two roots

$$\theta = \frac{1 \pm \sqrt{1 - 4r_1^2}}{2r_1}$$

equals 1, so the root that falls within the invertibility region $-1 < \hat{\theta} < 1$ should be chosen. Some algebra shows that this can be done by always selecting the minus in the $\pm$ portion of the formula. Once the point estimator $\hat{\theta}$ has been chosen, the first two equations can be solved as

$$\hat{\mu} = \bar{X} \qquad \text{and} \qquad \hat{\sigma}_Z^2 = \frac{(1/n)\sum_{t=1}^{n} X_t^2 - \hat{\mu}^2}{1 + \hat{\theta}^2}.$$

It appears that we have closed-form solutions to the method of moments estimators, but there is a subtle wrinkle in this derivation. Because of random sampling variability there is a chance that the lag 1 sample autocorrelation r_1 might be greater than $1/2$ or less than $-1/2$, even if the population time series model truly is a shifted MA(1) model satisfying the invertibility criterion $-1 < \theta < 1$. In this case the quadratic formula yields complex roots. So the method of moments parameter estimation approach is recommended for initial parameter estimates only if the constraint $|r_1| < 1/2$ stated in the result that follows is met.

Theorem 9.30 The method of moments estimators for a shifted MA(1) model from a time series $x_1, x_2, \ldots, x_n$ with $|r_1| < 1/2$ are

$$\hat{\mu} = \bar{X}, \qquad \hat{\theta} = \frac{1 - \sqrt{1 - 4r_1^2}}{2r_1}, \qquad \hat{\sigma}_Z^2 = \frac{(1/n)\sum_{t=1}^{n} X_t^2 - \hat{\mu}^2}{1 + \hat{\theta}^2}.$$

Thus, the method of moments point estimators in Theorem 9.30 should only be used for determining *initial* estimators of μ, θ, and σ_Z^2 from $x_1, x_2, \ldots, x_n$ when the following criteria are met:

- the number of observations in the time series is greater than about $n = 60$ or $n = 70$,

- the time series appears to be stationary,

- the sample autocorrelation function has a single statistically significant spike at lag 1,

- the sample partial autocorrelation function tails out, and

- $-1/2 < r_1 < 1/2$.

The method of moments estimators are generally used to find initial point estimates for the parameters in an MA(1) model, which are subsequently used in an iterative scheme to find the least squares or maximum likelihood estimators. This will be illustrated next on a time series consisting of chemical yields.

Example 9.35 Consider the time series consisting of the production record of $n = 210$ consecutive yield values in a chemical production process in Table 9.14. The entries are read row wise. Determine an appropriate model for this time series and find the method of moments initial estimates for the parameters.

The first step is to plot the time series, sample autocorrelation function, and sample partial autocorrelation function. The following R code uses the `plot.ts`, `acf`, and `pacf` functions to produce the graphs of the time series and correlogram given in Figure 9.42. The raw time series values are stored in a file named `yields.dat`.

```
x = scan("yields.dat")
layout(matrix(c(1, 1, 2, 3), 2, 2, byrow = TRUE))
plot.ts(x)
acf(x)
pacf(x)
```

The time series appears to be stationary. The time series frequently jumps from one side of the sample mean $\bar{x} = 84.1$ to the other, indicating a negative sample correlation between adjacent values in the time series. The sample lag 1 autocorrelation $r_1 = -0.289$ falls outside of the 95% confidence bounds, so we can conclude that there is a single statistically significant spike at lag 1. There are marginally statistically significant spikes at lags 2 and 6, which we will attribute to random sampling variability. The sample partial autocorrelation function has statistically significant spikes at lags 1, 2, and 3 which are negative and decrease in magnitude. Since $-1/2 < r_1 < 1/2$, the time series plot and the shapes of r_k and r_k^* indicate that the shifted MA(1) model

$$X_t = \mu + Z_t + \theta Z_{t-1}$$

with a negative value of θ might be appropriate. The R statements that follow are used to estimate the model parameters using Theorem 9.30.

85.5	81.7	80.6	84.7	88.2	84.9	81.8	84.9	85.2	81.9	89.4	79.0	81.4	84.8
85.9	88.0	80.3	82.6	83.5	80.2	85.2	87.2	83.5	84.3	82.9	84.7	82.9	81.5
83.4	87.7	81.8	79.6	85.8	77.9	89.7	85.4	86.3	80.7	83.8	90.5	84.5	82.4
86.7	83.0	81.8	89.3	79.3	82.7	88.0	79.6	87.8	83.6	79.5	83.3	88.4	86.6
84.6	79.7	86.0	84.2	83.0	84.8	83.6	81.8	85.9	88.2	83.5	87.2	83.7	87.3
83.0	90.5	80.7	83.1	86.5	90.0	77.5	84.7	84.6	87.2	80.5	86.1	82.6	85.4
84.7	82.8	81.9	83.6	86.8	84.0	84.2	82.8	83.0	82.0	84.7	84.4	88.9	82.4
83.0	85.0	82.2	81.6	86.2	85.4	82.1	81.4	85.0	85.8	84.2	83.5	86.5	85.0
80.4	85.7	86.7	86.7	82.3	86.4	82.5	82.0	79.5	86.7	80.5	91.7	81.6	83.9
85.6	84.8	78.4	89.9	85.0	86.2	83.0	85.4	84.4	84.5	86.2	85.6	83.2	85.7
83.5	80.1	82.2	88.6	82.0	85.0	85.2	85.3	84.3	82.3	89.7	84.8	83.1	80.6
87.4	86.8	83.5	86.2	84.1	82.3	84.8	86.6	83.5	78.1	88.8	81.9	83.3	80.0
87.2	83.3	86.6	79.5	84.1	82.2	90.8	86.5	79.7	81.0	87.2	81.6	84.4	84.4
82.2	88.9	80.9	85.1	87.1	84.0	76.5	82.7	85.1	83.3	90.4	81.0	80.3	79.8
89.0	83.7	80.9	87.3	81.1	85.6	86.6	80.0	86.6	83.3	83.1	82.3	86.7	80.2

Table 9.14: Production record of $n = 210$ consecutive chemical yields.

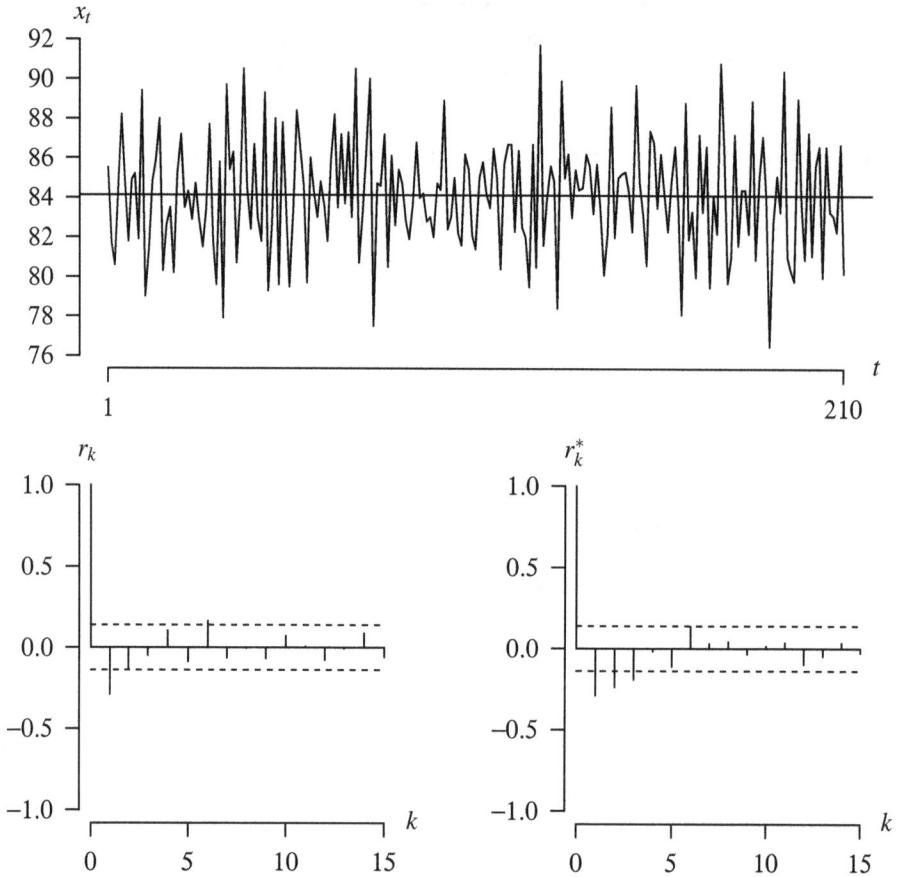

Figure 9.42: Time series plot, r_k, and r_k^* for $n = 210$ chemical yields.

```
x         = scan("yields.dat")
r1        = acf(x, plot = FALSE)$acf[2]
muhat     = mean(x)
thetahat  = (1 - sqrt(1 - 4 * r1 ^ 2)) / (2 * r1)
sigma2hat = (mean(x ^ 2) - muhat ^ 2) / (1 + thetahat ^ 2)
```

This code yields the following method of moments point estimators for the three parameters:

$$\hat{\mu} = 84.1 \qquad\qquad \hat{\theta} = -0.318 \qquad\qquad \hat{\sigma}^2 = 7.50.$$

This value of $\hat{\theta}$ falls within the invertibility region $-1 < \theta < 1$ for the shifted MA(1) time series model.

Approach 2: Least squares. Consider the shifted stationary MA(1) model

$$X_t = \mu + Z_t + \theta Z_{t-1}$$

from Theorem 9.29. For least squares estimation, we first establish the sum of squares S as a function of the parameters μ and θ. Numerical methods are required to determine the least squares estimators of μ and θ. Once these least squares estimators have been determined, the population variance of the white noise σ_Z^2 will be estimated.

Solving the shifted MA(1) model defining formula for Z_t results in

$$Z_t = X_t - \mu - \theta Z_{t-1}.$$

Seeding this recursive formula with $Z_0 = 0$ gives the residuals

$$Z_1 = X_1 - \mu$$
$$Z_2 = X_2 - \mu - \theta Z_1$$
$$Z_3 = X_3 - \mu - \theta Z_2$$
$$\vdots \quad = \quad \vdots$$
$$Z_n = X_n - \mu - \theta Z_{n-1}.$$

Thus, the sum of squared errors is

$$S = \sum_{t=1}^{n} Z_t^2 = (X_1 - \mu)^2 + \sum_{t=2}^{n} (X_t - \mu - \theta Z_{t-1})^2.$$

Numerical methods are required to find the parameter estimates. This will be illustrated next for the time series of chemical yields.

Example 9.36 Find the least squares estimators for the time series of chemical yields given in Example 9.35.

The R code below uses the `optim` function to perform a search to find the least squares parameter estimates for μ and θ. The first statement reads the time series values into the vector `x`. The second statement uses the `length` function to calculate the number of observations in the time series. The third statement defines the function `s`, which calculates the sum of squares. The last statement calls the `optim` function with the method of moments estimators as initial parameter estimates as its first argument. The `optim` function *minimizes* the function in its second argument by default.

```
x = scan("yields.dat")
n = length(x)
s = function(parameters) {
   mu    = parameters[1]
   theta = parameters[2]
   z     = numeric(n)
   z[1]  = x[1] - mu
   for (t in 2:n) z[t] = x[t] - mu - theta * z[t - 1]
   sum(z ^ 2)
}
optim(c(84.1, -0.318), s, method = "L-BFGS-B")
```

The `optim` function is being called in this case to perform a two-dimensional search without any derivative information being supplied. In order to visualize what the `optim` function is up against, picture yourself standing blindfolded on the side a mountain. The position where you are standing corresponds to the initial estimates for the parameters. In order to find the *least* squares parameter estimates, you want to take steps that lead you to the bottom of the valley, where the height corresponds to the sum of squares of the residuals. For *maximum* likelihood estimation, you want to take steps that lead you to the peak of the mountain, where the height corresponds to the likelihood function. But you are not given any gradient information from the function about the best direction to proceed for your next step. Regardless of the argument selected in the `method` argument, the internal algorithm in the `optim` function converges to roughly the same parameter estimates, as shown in Table 9.15. All five of the methods round to $\hat{\mu} = 84.13$ and $\hat{\theta} = -0.483$. The columns in Table 9.15 give the method, the number of calls to the function, the number of calls to evaluate the gradient, the least squares parameter estimates of μ, the least squares parameter estimates of θ, and the associated sums of squares S.

Method	Function	Gradient	$\hat{\mu}$	$\hat{\theta}$	S
Nelder-Mead	67	NA	84.12932	−0.48259	1484.989
BFGS	22	6	84.12942	−0.48261	1484.989
CG	180	23	84.12942	−0.48261	1484.989
L-BFGS-B	17	17	84.12942	−0.48261	1484.989
SANN	10000	NA	84.13053	−0.48269	1484.990

Table 9.15: MA(1) least squares parameter estimates for the chemical yields.

There is a large difference between the method of moments estimator of θ, which was $\hat{\theta} = -0.318$ from Example 9.35, and the least squares estimator of θ, which is $\hat{\theta} = -0.483$ from Table 9.15. This is an instance of why the method of moments estimators are only used for initial estimates for iterative schemes for finding least squares estimates or maximum likelihood estimates. The population variance of the white noise can be estimated using the method of moments formula as

$$\hat{\sigma}_Z^2 = \frac{(1/n)\sum_{t=1}^{n} X_t^2 - \hat{\mu}^2}{1 + \hat{\theta}^2} = 5.61.$$

Approach 3: Maximum likelihood estimation. We use the `arima` function to do the heavy lifting with respect to the estimation of the parameters in the MA(1) time series model via maximum likelihood. In addition to the point estimates, confidence intervals are based on the asymptotic distribution of the maximum likelihood estimator $\hat{\theta}$, which for large values of n is

$$V\left[\hat{\theta}\right] \cong \frac{1-\theta^2}{n}.$$

So when the parameter θ is estimated by maximum likelihood from a time series, an asymptotically exact $100(1-\alpha)\%$ confidence interval for θ is given in the result below. It is based on the consistency and the asymptotic normality of maximum likelihood estimators, which in this case implies that

$$\hat{\theta} \overset{a}{\sim} N\left(\theta, \frac{1-\theta^2}{n}\right).$$

Replacing θ by its maximum likelihood estimator in the variance yields the following result.

Theorem 9.31 For an invertible MA(1) time series model, an asymptotically exact two-sided $100(1-\alpha)\%$ confidence interval for θ is given by

$$\hat{\theta} - z_{\alpha/2}\sqrt{\frac{1-\hat{\theta}^2}{n}} < \theta < \hat{\theta} + z_{\alpha/2}\sqrt{\frac{1-\hat{\theta}^2}{n}},$$

where $\hat{\theta}$ is the maximum likelihood estimator and $z_{\alpha/2}$ is the $1-\alpha/2$ fractile of the standard normal distribution.

The formula for the confidence interval from Theorem 9.31 will be illustrated for the chemical yield data from the previous two examples.

Example 9.37 Find the maximum likelihood estimators for μ, θ, and σ_Z^2 for fitting a shifted MA(1) time series model to the time series of chemical yields from Example 9.35. Give a 95% confidence interval for θ and μ.

The following R code uses the `arima` function to calculate the maximum likelihood estimators for μ, θ, and σ_Z^2, estimate the variance–covariance matrix of the standard errors, and display the standard errors of the estimators.

```
x = scan("yields.dat")
fit = arima(x, order = c(0, 0, 1), include.mean = TRUE, method = "ML")
fit$coef
fit$var.coef
fit$sigma2
sqrt(fit$var.coef[1, 1])
sqrt(fit$var.coef[2, 2])
```

The resulting fitted MA(1) time series model to significant digits via maximum likelihood estimation is

$$X_t = \hat{\mu} + Z_t + \hat{\theta}Z_{t-1}$$

or

$$X_t = 84.13 + Z_t - 0.480Z_{t-1},$$
$$(0.0958) \qquad (0.0667)$$

where Z_t is white noise with estimated population variance $\hat{\sigma}_Z^2 = 7.071$. The numbers in parentheses just below the parameter estimates are the estimated standard errors of the associated parameter estimates. Using the standard errors from `arima`, the associated approximate 95% confidence intervals are

$$-0.611 < \theta < -0.349$$
$$83.94 < \mu < 84.32.$$

The 95% confidence interval for θ using Theorem 9.31 is slightly narrower than that produced by the `arima` function: $-0.599 < \theta < -0.362$.

The parameter estimates using the method of moments, least squares, and maximum likelihood estimation from the previous three examples are summarized in Table 9.16. Notice that the least

Method	$\hat{\mu}$	$\hat{\theta}$	$\hat{\sigma}_Z^2$
Method of moments	84.1	−0.318	7.50
Ordinary least squares	84.1	−0.483	5.61
Maximum likelihood estimation	84.1	−0.480	7.07

Table 9.16: Point estimators for the MA(1) parameters for the $n = 210$ chemical yields.

squares and maximum likelihood estimates of θ differ significantly from the associated method of moments estimator of θ.

In the interest of brevity, we leave the model assessment, model selection, and forecasting steps of the process for the chemical yields time series as an exercise. The derivations for these procedures follow along the same lines as those for the autoregressive models from the previous section.

This subsection has introduced the MA(1) time series model. The key results for an MA(1) model are listed below.

- The standard MA(1) model can be written algebraically and with the backshift operator B as

$$X_t = Z_t + \theta Z_{t-1} \qquad \text{and} \qquad X_t = \theta(B)Z_t,$$

where $Z_t \sim WN\left(0, \sigma_Z^2\right)$, $\sigma_Z^2 > 0$, and $\theta(B) = 1 + \theta B$ (Definition 9.4).

- The shifted MA(1) model can be written algebraically and with the backshift operator B as (Theorem 9.29)

$$X_t = \mu + Z_t + \theta Z_{t-1} \qquad \text{and} \qquad X_t = \mu + \theta(B)Z_t.$$

- The MA(1) model is stationary for all finite real-valued parameters θ and σ_Z^2 (Theorem 9.25).

- The MA(1) model is invertible when $-1 < \theta < 1$ (Theorem 9.27).

- The MA(1) model can be written as an AR(∞) model when $-1 < \theta < 1$ as (Theorem 9.26)

$$X_t = Z_t + \theta X_{t-1} - \theta^2 X_{t-2} + \theta^3 X_{t-3} - \cdots$$

- The MA(1) model lag 1 population autocorrelation is $\rho(1) = \theta / \left(1 + \theta^2\right)$, and $\rho(k) = 0$ for $k = 2, 3, \ldots$ (Theorem 9.25). The lag 1 population autocorrelation satisfies the inequality $-1/2 \leq \rho(1) \leq 1/2$ (Figure 9.37).

- The MA(1) lag k population partial autocorrelation for $-1 < \theta < 1$ is

$$\rho^*(k) = \frac{(-1)^{k+1}\theta^k \left(1 - \theta^2\right)}{1 - \theta^{2(k+1)}}$$

for $k = 1, 2, \ldots$ (Theorem 9.28).

- A time series of $n + 1$ white noise values $Z_0, Z_1, Z_2, \ldots, Z_n$ can be converted to n simulated observations $X_1, X_2, \ldots, X_n$ by using the MA(1) defining formula $X_t = Z_t + \theta Z_{t-1}$.

- The parameters in the MA(1) model can be estimated via the method of moments, least squares estimation, and maximum likelihood estimation.

9.2.2 The MA(2) Model

The additional term in the MA(2) model gives it increased flexibility over the associated MA(1) model.

> **Definition 9.5** A second-order moving average time series model, denoted by MA(2), for the time series $\{X_t\}$ is defined by
> $$X_t = Z_t + \theta_1 Z_{t-1} + \theta_2 Z_{t-2},$$
> where θ_1 and θ_2 are real-valued parameters and $\{Z_t\}$ is a time series of white noise:
> $$Z_t \sim WN\left(0, \sigma_Z^2\right).$$

An observed value in the time series, X_t, is given by the current white noise term, plus the parameter θ_1 multiplied by the white noise term from one time period ago, plus the parameter θ_2 multiplied by the white noise term from two time periods ago. So there are three parameters that define an MA(2) model: the coefficients θ_1 and θ_2, and the population variance of the white noise σ_Z^2. As was the case of the MA(1) model, some authors use a $-$ rather than a $+$ between three terms on the right-hand side of the model.

The probabilistic properties and statistical methods associated with an MA(2) model are straightforward generalizations of those properties and methods for the MA(1) model. Rather than deriving these results from first principles, we simply state several of these results without proof and then conduct a Monte Carlo simulation experiment which highlights issues that arise in model selection.

- The standard MA(2) model can be written algebraically and with the backshift operator B as

$$X_t = Z_t + \theta_1 Z_{t-1} + \theta_2 Z_{t-2} \qquad \text{and} \qquad X_t = \theta(B)Z_t,$$

where $Z_t \sim WN\left(0, \sigma_Z^2\right)$, $\sigma_Z^2 > 0$, and $\theta(B) = 1 + \theta_1 B + \theta_2 B^2$.

- The shifted MA(2) model can be written algebraically and with the backshift operator B as

$$X_t = \mu + Z_t + \theta_1 Z_{t-1} + \theta_2 Z_{t-2} \qquad \text{and} \qquad X_t = \mu + \theta(B)Z_t.$$

- MA(2) models are stationary for all finite, real-valued parameters μ, θ_1, θ_2, and σ_Z^2.

- Just as the stationarity region for the AR(2) model has a triangular shape, the invertibility region for the MA(2) model also has a triangular shape defined by the three constraints

$$\theta_1 + \theta_2 > -1, \qquad \theta_2 - \theta_1 > -1, \qquad \theta_2 < 1.$$

This region is an upside-down version of the region for an AR(2) time series model depicted in Figure 9.12. In other words, the triangles are equivalent when reflected vertically about the origin.

- The population autocovariance function is

$$\gamma(k) = \begin{cases} \left(1 + \theta_1^2 + \theta_2^2\right)\sigma_Z^2 & k = 0 \\ \left(\theta_1 + \theta_1\theta_2\right)\sigma_Z^2 & k = 1 \\ \theta_2\sigma_Z^2 & k = 2 \\ 0 & k = 3, 4, \ldots \end{cases}$$

- The population autocorrelation function is

$$\rho(k) = \begin{cases} 1 & k = 0 \\ \left(\theta_1 + \theta_1\theta_2\right)/\left(1 + \theta_1^2 + \theta_2^2\right) & k = 1 \\ \theta_2/\left(1 + \theta_1^2 + \theta_2^2\right) & k = 2 \\ 0 & k = 3, 4, \ldots \end{cases}$$

- The population partial autocorrelation function of an MA(2) model can be determined by using the defining formula from Definition 7.8.

- A simulated realization $X_1, X_2, \ldots, X_n$ of a time series from an MA(2) model is generated by the following algorithm.

$t \leftarrow -1$
generate $Z_t \sim WN\left(0, \sigma_Z^2\right)$
$t \leftarrow t + 1$
generate $Z_t \sim WN\left(0, \sigma_Z^2\right)$
while $(t < n)$
    $t \leftarrow t + 1$
    generate $Z_t \sim WN\left(0, \sigma_Z^2\right)$
    $X_t \leftarrow Z_t + \theta_1 Z_{t-1} + \theta_2 Z_{t-2}$

- The parameters of an MA(2) time series model can be estimated by the method of moments, least squares, and maximum likelihood estimation. As shown in the next example, the `arima` function can be used in R to calculate these parameter estimates.

The previous subsections have analyzed n observed values of a time series in order to determine an AR(p) or MA(q) model which adequately describes the probabilistic mechanism governing the observed time series. Instead of following this same pattern, we instead conduct a Monte Carlo simulation experiment that highlights weaknesses in the model selection process.

Example 9.38 Consider a standard (unshifted) MA(2) model with parameters $\theta_1 = 0.4$, $\theta_2 = 0.6$, and $\sigma_Z^2 = 1$. For a realization of $n = 60$ observations from this time series fitted by maximum likelihood estimation, use Monte Carlo simulation to estimate the probability that the correct model is identified if the AIC criterion is used to determine the correct model.

This Monte Carlo simulation experiment answers an important question in time series analysis. If we have just a single realization of a time series (this is often the case in practice) which is governed by an approximately ARMA model, what is the probability that we correctly identify the p and q values associated with the ARMA(p, q) time series model which generated the observations?

The MA(2) model is the population time series model in this particular simulation experiment. The choice of parameters $\theta_1 = 0.4$ and $\theta_2 = 0.6$ falls in the invertibility region, so this particular population MA(2) model is both stationary and invertible. Furthermore, the parameters $\theta_1 = 0.4$ and $\theta_2 = 0.6$ have been chosen so that the population autocorrelation function is

$$\rho(k) = \begin{cases} 1 & k = 0 \\ \left(0.4 + 0.4 \cdot 0.6\right)/\left(1 + 0.4^2 + 0.6^2\right) & k = 1 \\ 0.6/\left(1 + 0.4^2 + 0.6^2\right) & k = 2 \\ 0 & k = 3, 4, \ldots \end{cases}$$

or

$$\rho(k) = \begin{cases} 1 & k = 0 \\ 0.42 & k = 1 \\ 0.39 & k = 2 \\ 0 & k = 3, 4, \ldots. \end{cases}$$

As anticipated, the population autocorrelation cuts off after lag 2. So we expect that the first two values in the sample autocorrelation function computed from a realization of this time series model, r_1 and r_2, will be statistically significant, and the others will fall between the confidence bounds $\pm 1.96/\sqrt{60} = \pm 0.25$. The coefficients θ_1 and θ_2 have been chosen so that the first two values in the population autocorrelation function, $\rho(1) = 0.42$ and $\rho(2) = 0.39$, both fall outside of the confidence bounds $\pm 1.96/\sqrt{60} = \pm 0.25$ associated with the sample autocorrelation function. This choice of parameters has been made to give the ARMA modeling procedure a good chance of correctly identifying the underlying population MA(2) time series model.

The R code below generates realizations of 1000 time series of length $n = 60$ from an MA(2) model with $\theta_1 = 0.4$, $\theta_2 = 0.6$, and $\sigma_Z^2 = 1$. The `arima.sim` function is used to generate each realization, and the simulated values are placed in the vector `x`. The two inner nested `for` loops fit all ARMA(p, q) models to the simulated values, for $p = 0, 1, 2, \ldots, 5$ and $q = 0, 1, 2, \ldots, 5$ using the `arima` function. The AIC for each of the fitted models are stored in the a matrix. Finally, the `which.min` function is used to determine which of the $6 \cdot 6 = 36$ models has the lowest AIC value.

```
set.seed(3)
nrep = 1000
a    = matrix(0, 6, 6)
r    = matrix(0, 6, 6)
for (i in 1:nrep) {
  x = arima.sim(list(ma = c(0.4, 0.6)), n = 60)
  for (p in 0:5)
    for (q in 0:5)
      a[p + 1, q + 1] = arima(x, order = c(p, 0, q), method = "ML")$aic
  j = which.min(a)
  r[j] = r[j] + 1
  print(a)
  print(j)
  print(r)
}
100 * r / nrep
```

Table 9.17 contains the estimated probabilities of the selection of the various models expressed as percents. The good news is that the MA(2) model is the one that is chosen most often of the 36 models based on the AIC criterion. The associated bad news is that the MA(2) model is chosen less than half of the time. So in the practical case of analyzing a single time series of $n = 60$ values, there is a better than 0.5 probability that this procedure will identify the *wrong* population time series model. This illustrates the unwelcome effect of random sampling variability in model selection.

Bear in mind that the estimated probabilities given in Table 9.17 only apply to the use of the AIC criterion and the parameter settings $\theta_1 = 0.4$, $\theta_2 = 0.6$, $\sigma_Z^2 = 1$, and $n = 60$.

	$q = 0$	$q = 1$	$q = 2$	$q = 3$	$q = 4$	$q = 5$
$p = 0$	0.2%	0.0%	40.1%	4.9%	2.4%	1.9%
$p = 1$	0.7%	0.0%	3.1%	6.0%	1.4%	1.6%
$p = 2$	0.7%	0.9%	3.7%	1.3%	5.3%	1.5%
$p = 3$	3.2%	1.6%	2.6%	0.7%	1.9%	2.2%
$p = 4$	0.9%	0.9%	2.0%	1.4%	1.5%	1.1%
$p = 5$	0.6%	0.6%	0.8%	0.7%	1.0%	0.6%

Table 9.17: Estimated probabilities of selection based on AIC criterion.

Changing any one of these parameters will alter the probabilities. The purpose of this example is to highlight the pitfalls associated with fitting a time series model to a single realization of n time series values. The probability of an incorrect selection is high, and this is an argument for collecting a longer time series when possible. In addition, if a time series is collected periodically (for example, n values collected annually), then the fits to various realizations should be compared.

To summarize the models considered so far in this chapter, the AR(1), AR(2), MA(1), and MA(2) models are parsimonious in the sense that they have significant explanatory power with few parameters. By deriving the population autocorrelation function and partial autocorrelation function for these models, we now possess an inventory of possible shapes that guide us toward one particular time series model or another. Figure 9.43 gives examples of these shapes for various values of the parameters.

9.2.3   The MA(q) Model

The MA(1) and MA(2) models introduced in the previous two subsections generalize to the MA(q) model defined in this section.

Definition 9.6 A moving average time series model with q terms, denoted by MA(q), for the time series $\{X_t\}$ is defined by

$$X_t = Z_t + \theta_1 Z_{t-1} + \theta_2 Z_{t-2} + \cdots + \theta_q Z_{t-q},$$

where $\theta_1, \theta_2, \ldots, \theta_q$ are real-valued parameters and $\{Z_t\}$ is a time series of white noise:

$$Z_t \sim WN\left(0, \sigma_Z^2\right).$$

An observed value in the time series, X_t, is given by the current white noise term plus a linear combination of the q previous white noise terms. So there are $q + 1$ parameters that define an MA(q) model: the coefficients $\theta_1, \theta_2, \ldots, \theta_q$, and the population variance of the white noise σ_Z^2. As was the case of the MA(1) and MA(2) models, some authors use a $-$ rather than a $+$ between terms on the right-hand side of the model.

The probabilistic properties and statistical methods associated with an MA(q) model are determined in the usual fashion. Here are several of these results stated without proof.

- The population mean and variance of X_t are easily calculated by taking the expected value and

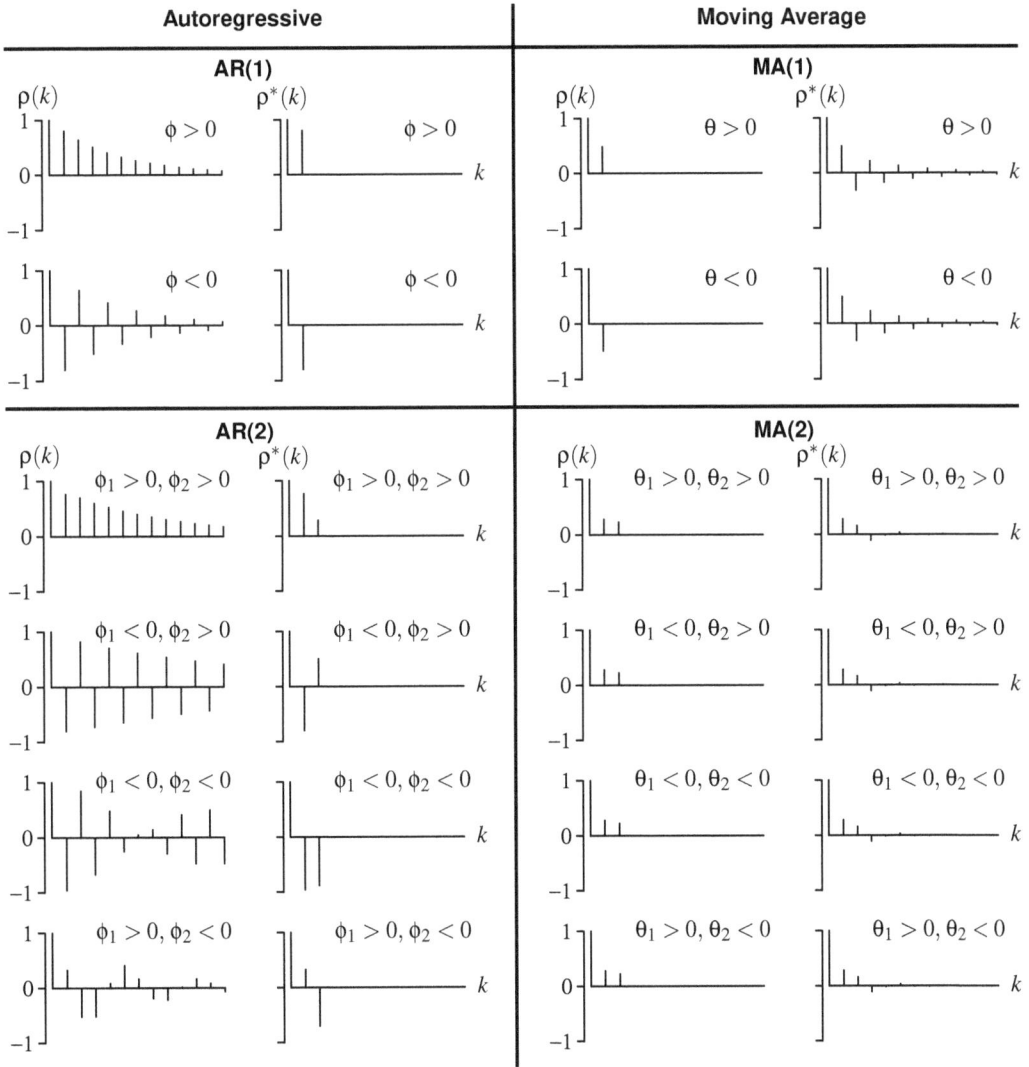

Figure 9.43: Characteristic shapes of $\rho(k)$ and $\rho^*(k)$ for AR(1), AR(2), MA(1), and MA(2) models.

the population variance of both sides of the equation given in Definition 9.6:

$$E[X_t] = E[Z_t + \theta_1 Z_{t-1} + \theta_2 Z_{t-2} + \cdots + \theta_q Z_{t-q}] = 0$$

and

$$V[X_t] = V[Z_t + \theta_1 Z_{t-1} + \theta_2 Z_{t-2} + \cdots + \theta_q Z_{t-q}] = \left(1 + \theta_1^2 + \theta_2^2 + \cdots + \theta_q^2\right)\sigma_Z^2.$$

- The standard MA(q) model can be written algebraically and with the backshift operator B as

$$X_t = Z_t + \theta_1 Z_{t-1} + \theta_2 Z_{t-2} + \cdots + \theta_q Z_{t-q} \qquad \text{and} \qquad X_t = \theta(B)Z_t,$$

where $Z_t \sim WN\left(0, \sigma_Z^2\right)$, $\sigma_Z^2 > 0$, and $\theta(B) = 1 + \theta_1 B + \theta_2 B^2 + \cdots + \theta_q B^q$.

- The shifted MA(q) model can be written algebraically and with the backshift operator B as

$$X_t = \mu + Z_t + \theta_1 Z_{t-1} + \theta_2 Z_{t-2} + \cdots + \theta_q Z_{t-q} \qquad \text{and} \qquad X_t = \mu + \theta(B)Z_t.$$

- MA(q) models are stationary for all finite, real-valued parameters μ, θ_1, θ_2, $\ldots$, θ_q, and σ_Z^2.

- MA(q) models are invertible when the q roots of the characteristic equation

$$\theta(B) = 1 + \theta_1 B + \theta_2 B^2 + \cdots + \theta_q B^q = 0$$

all lie outside of the unit circle in the complex plane.

- The population autocovariance function is

$$\gamma(k) = \begin{cases} \left(1 + \theta_1^2 + \theta_2^2 + \cdots + \theta_q^2\right)\sigma_Z^2 & k = 0 \\ (\theta_1 + \theta_1\theta_2 + \theta_2\theta_3 + \cdots + \theta_{q-1}\theta_q)\sigma_Z^2 & k = 1 \\ (\theta_2 + \theta_1\theta_3 + \theta_2\theta_4 + \cdots + \theta_{q-2}\theta_q)\sigma_Z^2 & k = 2 \\ \vdots & \vdots \\ \theta_q \sigma_Z^2 & k = q \\ 0 & k = q, q+1, \ldots. \end{cases}$$

This can be written more compactly as

$$\gamma(k) = \begin{cases} (\theta_k + \theta_1\theta_{k+1} + \theta_2\theta_{k+2} + \cdots + \theta_{q-k}\theta_q)\sigma_Z^2 & k = 0, 1, 2, \ldots, q \\ 0 & k = q, q+1, \ldots, \end{cases}$$

where $\theta_0 = 1$.

- The population autocorrelation function is

$$\rho(k) = \begin{cases} (\theta_k + \theta_1\theta_{k+1} + \theta_2\theta_{k+2} + \cdots + \theta_{q-k}\theta_q) / \left(1 + \theta_1^2 + \cdots + \theta_q^2\right) & k = 0, 1, \ldots, q \\ 0 & k = q, q+1, \ldots. \end{cases}$$

As expected, the population autocorrelation function cuts off after lag q.

- The population partial autocorrelation function of an MA(q) model can be determined by using the defining formula from Definition 7.8.

- A simulated realization $X_1, X_2, \ldots, X_n$ of a time series from an MA(q) model is generated by the following algorithm.

> generate $Z_{-(q-1)}, Z_{-(q-2)}, \ldots, Z_0 \sim WN\left(0, \sigma_Z^2\right)$
> $t \leftarrow 0$
> while $(t < n)$
> $t \leftarrow t + 1$
>   generate $Z_t \sim WN\left(0, \sigma_Z^2\right)$
> $X_t \leftarrow Z_t + \theta_1 Z_{t-1} + \theta_2 Z_{t-2} + \cdots + \theta_q Z_{t-q}$

- The parameters of an MA(q) time series model can be estimated by the method of moments, least squares, and maximum likelihood estimation. The `arima` function can be used in R to calculate these parameter estimates for particular values of a time series $x_1, x_2, \ldots, x_n$.

Table 9.18 shows some of the symmetry between autoregressive and moving average models. When one aspect of the time series model is easier to derive for one of the models, it is often more difficult to derive for the analogous time series model. The population autocorrelation function for an MA(q) model is closed form, for example, but the population autocorrelation function for an AR(p) model requires solving the Yule–Walker equations. As a second example on the statistical side, the least squares estimators for the AR(1) model are closed form, but the least squares estimators for the MA(1) model require numerical methods.

	Autoregressive: AR(p)	Moving Average: MA(q)
Model definition	$\phi(B)X_t = Z_t$	$X_t = \theta(B)Z_t$
Characteristic polynomial	$\phi(B) = 1 - \phi_1 B - \phi_2 B^2 - \cdots - \phi_p B^p$	$\theta(B) = 1 + \theta_1 B + \theta_2 B^2 + \cdots + \theta_q B^q$
Stationarity condition	$\phi(B) = 0$ roots outside of unit circle	always stationary
Invertibility condition	always invertible	$\theta(B) = 0$ roots outside of unit circle
Equivalent model	MA(∞) when stationary	AR(∞) when invertible
General linear model π weights	finite series	infinite series
General linear model ψ weights	infinite series	finite series
Shape of $\rho(k)$	tails out	cuts off after lag q
Shape of $\rho^*(k)$	cuts off after lag p	tails out
Simulating a realization	warm up period needed	no warm up period needed

Table 9.18: AR(p) versus MA(q) models.

9.3 ARMA(p, q) Models

The autoregressive and moving average models outlined in the previous two sections often prove to be inadequate time series models in a particular application. Occasions arise in which the best model for a time series involves both autoregressive and moving average terms. Recall from Definition 8.4 that an ARMA(p, q) time series model with p autoregressive terms and q moving average terms is

$$X_t = \overbrace{\phi_1 X_{t-1} + \phi_2 X_{t-2} + \cdots + \phi_p X_{t-p}}^{\text{autoregressive portion}} + \underbrace{Z_t + \theta_1 Z_{t-1} + \theta_2 Z_{t-2} + \cdots + \theta_q Z_{t-q}}_{\text{moving average portion}},$$

where $\{X_t\}$ is the time series of interest, $\{Z_t\}$ is a time series of white noise, $\phi_1, \phi_2, \ldots, \phi_p$ are real-valued parameters associated with the AR portion of the model, and $\theta_1, \theta_2, \ldots, \theta_q$ are real-valued parameters associated with the MA portion of the model. The ARMA(p, q) model can be written more compactly as

$$\phi(B)X_t = \theta(B)Z_t,$$

where $\phi(B)$ and $\theta(B)$ are the characteristic polynomials defined by

$$\phi(B) = 1 - \phi_1 B - \phi_2 B^2 - \cdots - \phi_p B^p$$

and

$$\theta(B) = 1 + \theta_1 B + \theta_2 B^2 + \cdots + \theta_q B^q.$$

This model on its own is of little practical use because most real-world time series are not centered around $E[X_t] = 0$. Using the compact notation for the ARMA(p, q) time series model, a shift parameter μ is easily added:

$$\phi(B)(X_t - \mu) = \theta(B)Z_t.$$

So there are $p + q + 2$ parameters that define a shifted ARMA(p, q) time series model: the p autoregressive coefficients $\phi_1, \phi_2, \ldots, \phi_p$, the q moving average coefficients $\theta_1, \theta_2, \ldots, \theta_q$, the shift parameter μ, and the population variance of the white noise σ_Z^2.

Recall from Table 9.7 in Example 9.20 that the ARMA(1, 1) model fitted by maximum likelihood estimation gave a slightly lower AIC than the associated AR(2) model when applied to the Lake Huron level time series. This section will consist of one long example that concerns the fitting and assessing this ARMA(1, 1) model to determine whether it is an adequate model for the Lake Huron levels. Rather than deriving all of the probabilistic properties and statistical methods for the ARMA(1, 1) model, the `arima` function in R will be used to perform the fitting, leaving the details to the reader. By default, the `arima` function (a) ignores external regressor variables, (b) ignores seasonal variation, (c) includes a shift parameter μ, (d) uses the same parameterization for the ARMA(p, q) process as that used in this text, (e) transforms the AR parameters $\phi_1, \phi_2, \ldots, \phi_p$ if necessary so that they stay in the stationarity region, and (f) uses a conditional sum of squares method as initial parameter estimates, then returns the maximum likelihood estimators.

> **Example 9.39** Fit the ARMA(1, 1) model to the $n = 98$ annual Lake Huron levels from 1875–1972 described in Example 9.14. Assess the model adequacy of the fit and predict the level of Lake Huron for the next five years (1973–1977).
>
> The first R statement below fits the ARMA(1, 1) model to the Lake Huron levels. The next four statements extract the estimated coefficients, estimated white noise variance, estimated variance–covariance matrix of the coefficients, and the residuals.

```
fit          = arima(LakeHuron, order = c(1, 0, 1))
coefficients = fit$coef
variance     = fit$sigma2
variancecov  = fit$var.coef
residuals    = fit$residuals
```

> The parameter estimates (to five-digit accuracy that is inherent in the time series values) are

$$\hat{\phi}_1 = 0.74490 \qquad \hat{\theta}_1 = 0.32059 \qquad \hat{\mu} = 579.06 \qquad \hat{\sigma}_Z^2 = 0.47494.$$

The estimated variance–covariance matrix of $\hat{\phi}_1$, $\hat{\theta}_1$, and $\hat{\mu}$ is

$$\begin{bmatrix} 0.0060296 & -0.0046761 & 0.0017655 \\ -0.0046761 & 0.0128889 & -0.0020637 \\ 0.0017655 & -0.0020637 & 0.1225691 \end{bmatrix}.$$

Using the square roots of the diagonal elements of the variance–covariance matrix as standard error estimates, the following additional R commands give approximate two-sided 95% confidence intervals for the parameters.

```
coefficients[1] + c(-1, 1) * qnorm(0.975) * sqrt(variancecov[1, 1])
coefficients[2] + c(-1, 1) * qnorm(0.975) * sqrt(variancecov[2, 2])
coefficients[3] + c(-1, 1) * qnorm(0.975) * sqrt(variancecov[3, 3])
```

The approximate 95% confidence intervals are

$$0.59271 < \phi_1 < 0.89709 \qquad 0.09808 < \theta_1 < 0.54310 \qquad 578.37 < \mu < 579.74.$$

Since none of these confidence intervals contains zero, we continue to entertain this tentative ARMA(1, 1) model and transition to an analysis of the residuals.

The following R commands plot the residuals as a time series, along with the associated sample autocorrelation function and sample partial autocorrelation function.

```
zhat = arima(LakeHuron, order = c(1, 0, 1))$residuals
layout(matrix(c(1, 1, 2, 3), 2, 2, byrow = TRUE))
plot.ts(zhat)
acf(zhat, lag.max = 40)
pacf(zhat, lag.max = 40)
```

The results are displayed in Figure 9.44. From the top graph, the residuals do not appear to have any cyclic variation, trend, or serial correlation. The sample autocorrelation function values for the residuals do not have any values that fall outside of the 95% confidence limits. Likewise for the sample partial autocorrelation function values.

Since there are no sample autocorrelation function values that fall outside of the 95% confidence limits $\pm 1.96/\sqrt{n}$ in the plot in Figure 9.44 of the first 40 sample auto-correlation function values associated with the residuals, and we expect $40 \cdot 0.05 = 2$ values to fall outside of these limits in the case of a good fit, we fail to reject H_0 in this case. The independence of the residuals is not rejected by this test. The tentative fitted ARMA(1, 1) model is not rejected by this test.

The R code below calculates the Box–Pierce test statistic and the Ljung–Box test statistic and the associated p-values using the built-in Box.test function.

```
zhat = arima(LakeHuron, order = c(1, 0, 1))$residuals
Box.test(zhat, lag = 40, type = "Box-Pierce", fitdf = 3)
Box.test(zhat, lag = 40, type = "Ljung-Box", fitdf = 3)
```

The Box–Pierce test statistic is 17.4 and the associated p-value is $p = 0.997$. The Ljung–Box test statistic is 23.0 and the associated p-value is $p = 0.966$. We fail to

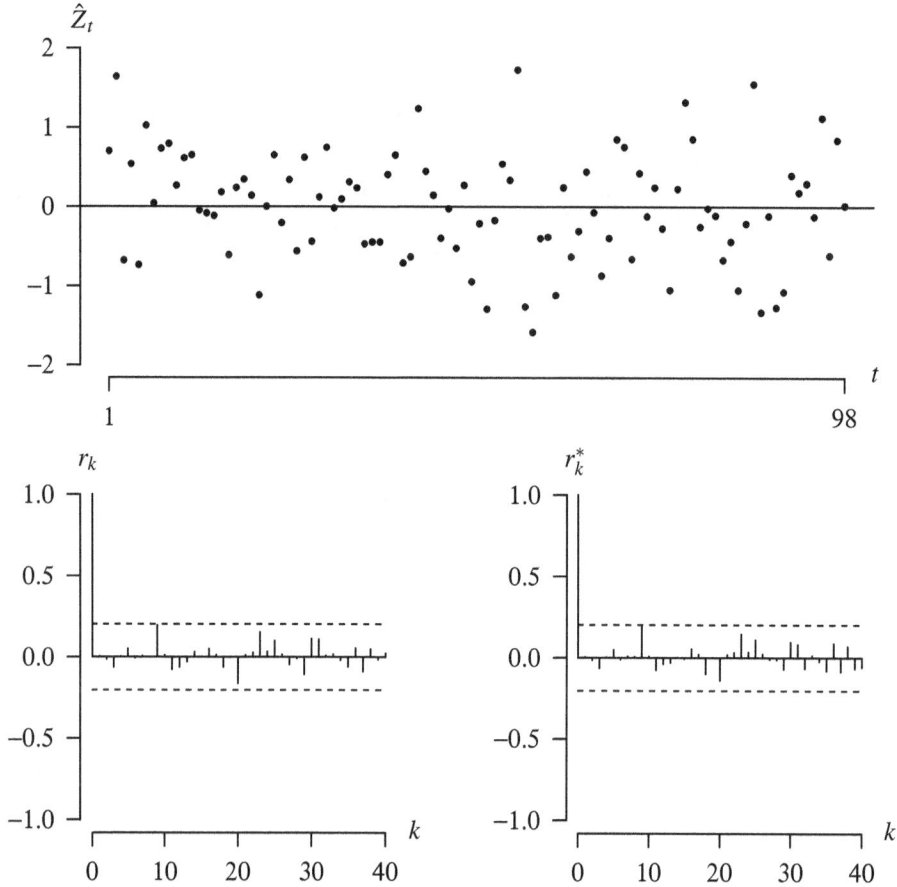

Figure 9.44: Time series plot, r_k, and r_k^* for the residuals from a fitted ARMA(1, 1) model.

reject H_0 in both tests based on the chi-square critical value with $40 - 3 = 37$ degrees of freedom. The independence of the residuals is not rejected by this test. The tentative fitted ARMA(1, 1) model is not rejected by these tests.

The following R code calculates the test statistic and the p-value for the turning point test applied to the time series consisting of the residual values for the ARMA(1, 1) fit to the Lake Huron time series.

```
zhat = arima(LakeHuron, order = c(1, 0, 1))$residuals
n = length(zhat)
m = (2 / 3) * (n - 2)
v = (16 * n - 29) / 90
T = 0
for (i in 2:(n - 1)) {
  if ((zhat[i - 1] < zhat[i] && zhat[i] > zhat[i + 1]) ||
      (zhat[i - 1] > zhat[i] && zhat[i] < zhat[i + 1])) T = T + 1
```

```
}
s = (T - m) / sqrt(v)
2 * (1 - pnorm(abs(s)))
```

The tail probability is doubled because the alternative hypothesis is two-tailed for the turning point test. The test statistic s is 1.21 and the p-value is $p = 0.23$. The turning point test found that there were $T = 69$ turning points in the time series of the residuals, and that is just slightly higher than the number that we expect to have if the residuals from the fitted ARMA(1, 1) model were mutually independent random variables. We again fail to reject the null hypothesis in this case. The independence of the residuals is not rejected by this test. The tentative fitted ARMA(1, 1) model is not rejected by this test.

The residuals are standardized by dividing by their sample standard deviation. The following R statements plot a histogram of the standardized residuals using the `hist` function and a QQ plot to assess normality using the `qqnorm` function.

```
zhat = arima(LakeHuron, order = c(1, 0, 1))$residuals
hist(zhat / sd(zhat))
qqnorm(zhat / sd(zhat))
```

The plots are shown in Figure 9.45. The histogram shows that all standardized residuals fall between -2.5 and 2.5 and exhibit a roughly bell-shaped probability distribution, with the exception of a deficit of residuals falling between -1.5 and -1.0. The horizontal axis on the histogram is the standardized residual and the vertical axis is the frequency. The QQ plot is approximately linear, indicating a reasonable approximation to normality for the standardized residuals. The horizontal axis on the QQ plot is the standardized theoretical quantile and the vertical axis is the associated normal data quantile. Although a formal statistical goodness-of-fit test (such as the Shapiro–Wilk or the Kolmogorov–Smirnov test) should be conducted, it appears that the assumption of Gaussian white noise is appropriate for the ARMA(1, 1) time series model based on these two plots.

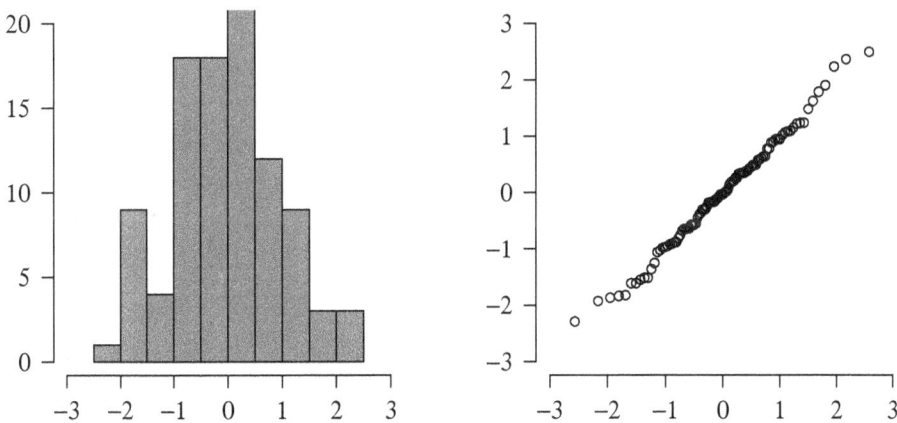

Figure 9.45: Histogram (left) and QQ plot (right) of the fitted ARMA(1, 1) standardized residuals.

We have seen a number of indicators that the ARMA(1, 1) time series model with Gaussian error terms seems to be an adequate model for the Lake Huron lake level time series, with the exception of a linear trend apparent by viewing the time series in Figure 9.19. The ARMA(1, 1) model has not been rejected by any of the model adequacy tests.

The final fitted shifted ARMA(1, 1) model with maximum likelihood estimates for the parameters is given by

$$X_t = 579.06 + 0.74490\,(X_{t-1} - 579.06) + 0.32059Z_{t-1} + Z_t,$$

where Z_t is a sequence of independent and identically distributed $N(0, 0.47494)$ error terms.

With the shifted ARMA(1, 1) model established, we now consider forecasting future values of a time series. In the case of the Lake Huron time series, this corresponds to the one-step-ahead forecast for 1973, the two-steps-ahead forecast for 1974, the three-steps-ahead forecast for 1975, etc. The code below uses the R predict function to generate the forecasted values and their standard errors.

```
fit      = arima(LakeHuron, order = c(1, 0, 1))
forecast = predict(fit, n.ahead = 5)
lower    = forecast$pred - qnorm(0.975) * forecast$se
upper    = forecast$pred + qnorm(0.975) * forecast$se
```

These standard errors can be used to calculate approximate two-sided 95% prediction interval limits on the forecasted values. The results are summarized in Table 9.19. Notice that the forecasts trend monotonically toward $\bar{x} = 579$ and the standard errors increase as the time horizon h increases. The increasing standard error is consistent with having less precision in the forecast as the time horizon h increases.

Time	$t=99$	$t=100$	$t=101$	$t=102$	$t=103$
Year	1973	1974	1975	1976	1977
Forecast	579.73	579.56	579.43	579.34	579.26
Standard error	0.689	1.007	1.146	1.216	1.254
Lower prediction bound	578.38	577.59	577.19	576.95	576.81
Upper prediction bound	581.08	581.53	581.68	581.72	581.72

Table 9.19: Forecasts and 95% prediction intervals for the Lake Huron time series.

Figure 9.46 shows (a) the original time series $x_1, x_2, \ldots, x_{98}$ as points ($\bullet$) connected by lines, (b) the first 10 forecasted lake levels $\hat{X}_{99}, \hat{X}_{100}, \ldots, \hat{X}_{108}$ as open circles ($\circ$), (c) the 95% prediction intervals as a shaded region, and (d) the next 10 actual average lake level values in July for the years 1973–1982 taken from the NOAA Great Lakes Experimental Research Laboratory website,

580.98, 581.04, 580.49, 580.52, 578.57, 578.96, 579.94, 579.77, 579.44, 578.97,

as points ($\bullet$) connected by lines. The forecasted values as well as the prediction intervals given in Figure 9.46 associated with the fitted ARMA(1, 1) model are very similar to those in Figure 9.25 from Example 9.21. The two models are clearly close competitors for modeling the Lake Huron levels.

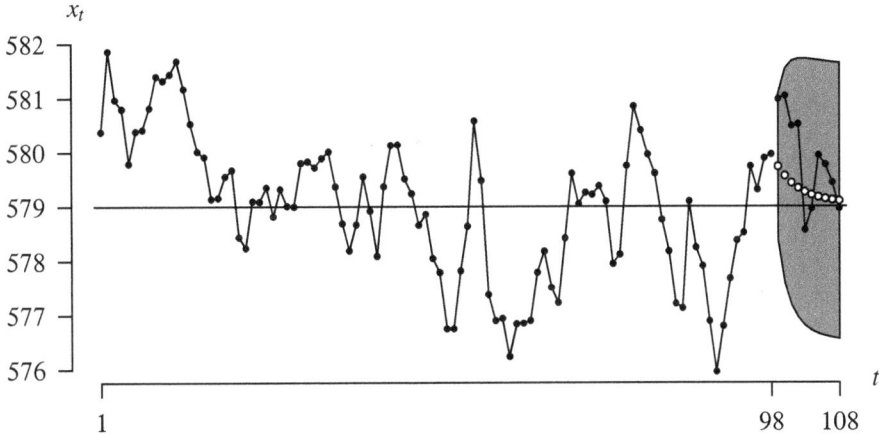

Figure 9.46: Lake Huron level forecasts and 95% prediction intervals from an ARMA(1, 1) model.

ARMA modeling can achieve population autocorrelation function and population partial autocorrelation function shapes that are not possible with just AR(p) and MA(q) models alone. For an ARMA(p, q) model with $p > 0$ and $q > 0$, both the population autocorrelation function and the population partial autocorrelation function tail off; neither of the two cut off after a certain lag.

An inherent weakness of ARMA modeling is that it requires stationarity. Many time series which occur in practice are not stationary, and the next section gives techniques that can be used to overcome this weakness.

9.4 Nonstationary Models

There are two commonly-used strategies for converting a nonstationary time series to a stationary time series in order to use ARMA modeling (or some other model which requires stationarity) on the resultant stationary time series. The first strategy is known as *detrending*. In this case, the modeler estimates the trend, and then fits a stationary time series model to the difference between the raw time series data and the estimated trend. The second strategy is known as *differencing*. In this case the modeler differences the time series one or more times, resulting in a stationary time series. Differencing carries the added benefit that no parameters are required other than the number of differences to take. The following two subsections consider these two strategies.

9.4.1 Removing Trends Via Regression

Although regression is not the only way to detrend a time series, it provides an adequate roadmap on how to proceed with the detrending process that generalizes to other mechanisms. This subsection illustrates detrending with a single example. We return for a third time to the Lake Huron levels which were fit to an AR(2) model in Section 9.1.2 and fit to an ARMA(1, 1) model in Section 9.3.

Example 9.40 We again consider the construction of a time series model from the $n = 98$ annual observations of the level of Lake Huron (in feet) between 1875 and 1972 that was first encountered in Example 9.14. The observations are stored in a time series in R named LakeHuron. The scatterplot of the lake levels depicted in Figure 9.47 includes a regression line showing the downward trend in the lake levels over time. The p-value for the statistical test for significance of the slope of this regression line is $p = 4 \cdot 10^{-8}$, providing strong evidence of a downward trend over time, even though the usual assumptions associated with simple linear regression with normal error terms are not perfectly satisfied in this setting. Although the AR(2) and ARMA(1, 1) models have been successfully fitted to this time series treating it as stationary, this tiny p-value prevents us from fully embracing either of these models. The purpose of this example is to explicitly consider this downward trend by fitting the residuals from this simple linear regression model to an ARMA time series model.

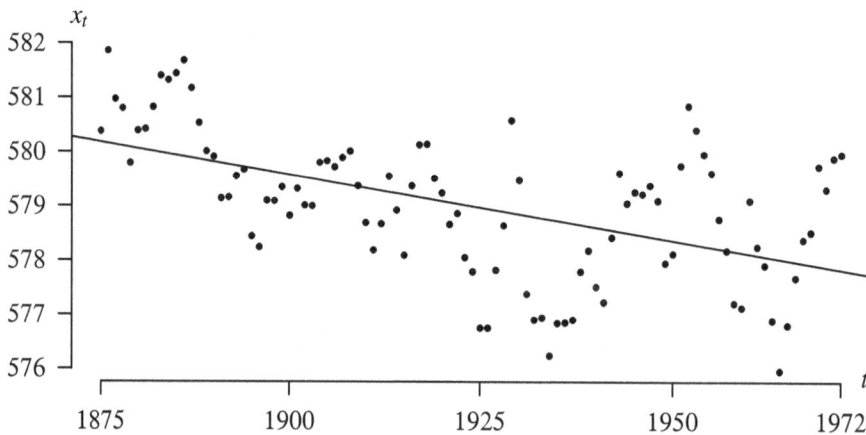

Figure 9.47: Lake Huron levels (1875–1972) with regression line.

The residuals from this simple linear regression form a new time series which will be denoted by $\{y_t\}$, where $y_t = x_t - \hat{x}_t$ and $\hat{x}_t$ is the fitted value in the simple linear regression. The R statements below generate plots of the time series, the sample autocorrelation function, and the sample partial autocorrelation function for the residuals of the simple linear regression.

```
y = lm(LakeHuron ~ seq(1:98))$resid
plot.ts(y)
abline(h = mean(y))
acf(y, lag.max = 40)
pacf(y, lag.max = 40)
```

These plots are displayed in Figure 9.48. The time series of the residuals appears to have no trend and also appears to be centered around zero. In fact, the time series is *exactly* centered around zero because the residuals of this regression must sum to zero via Theorem 1.6. This means that there is no need to include a shift parameter μ in the ARMA model that we develop for the residuals. The sample autocorrelation function of the residuals appears to be tailing out and the first two sample partial autocorrelation

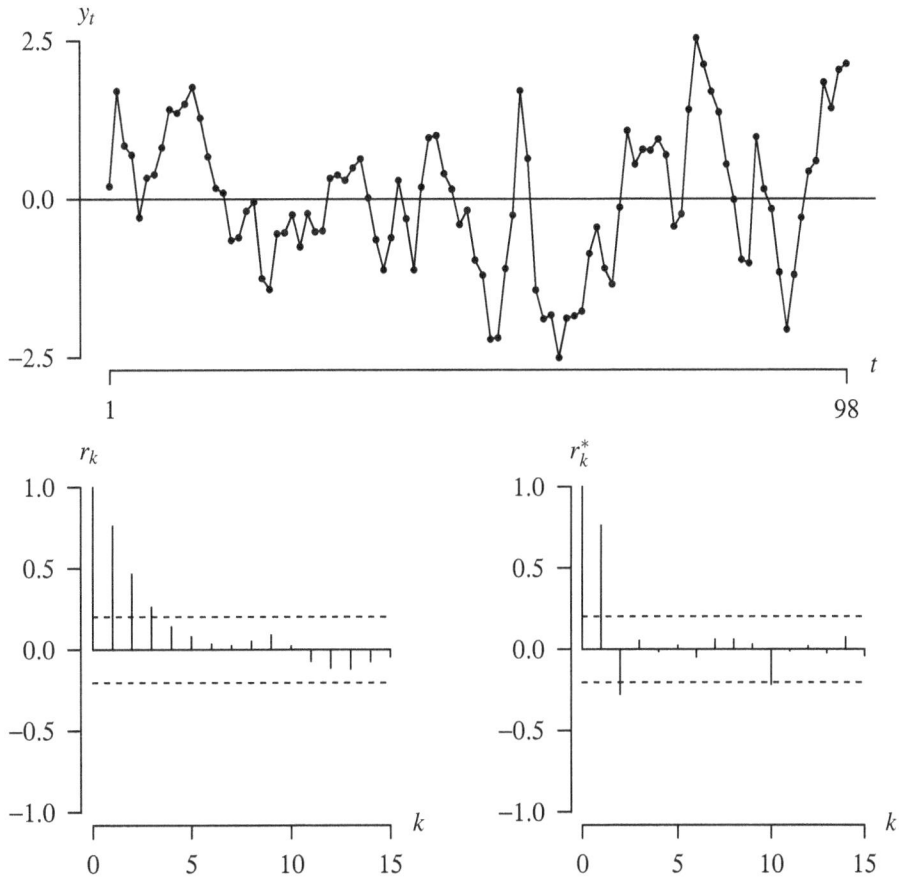

Figure 9.48: Residuals plot, r_k, and r_k^* for Lake Huron lake levels.

function values are statistically significant. This is strong evidence that an AR(2) model is an appropriate tentative model for the residuals.

The following R statements fit the AR(2) model via maximum likelihood estimation to the time series of residuals.

```
y = lm(LakeHuron ~ seq(1, 98))$resid
model = ar(y, order.max = 2, method = "mle")
```

This model should be subjected to all of the model assessment tests that have been applied to all previous time series analyzed in previous examples. The residuals of the estimated AR(2) model to the simple linear regression residuals result in large p-values for the Box–Pierce test and the Box–Ljung test, along with a bell-shaped histogram and an almost perfectly linear QQ normal plot. This evidence confirms the evidence in the plots of r_k and r_k^* which pointed to an AR(2) model for the residuals of the simple linear regression. The predict function can then be used in the usual fashion to forecast the residuals into the future. The plot of the residuals, the next 10 forecasted residuals, and the associated 95% prediction intervals is given in Figure 9.49. As expected, the forecasted values are smooth and converge to zero.

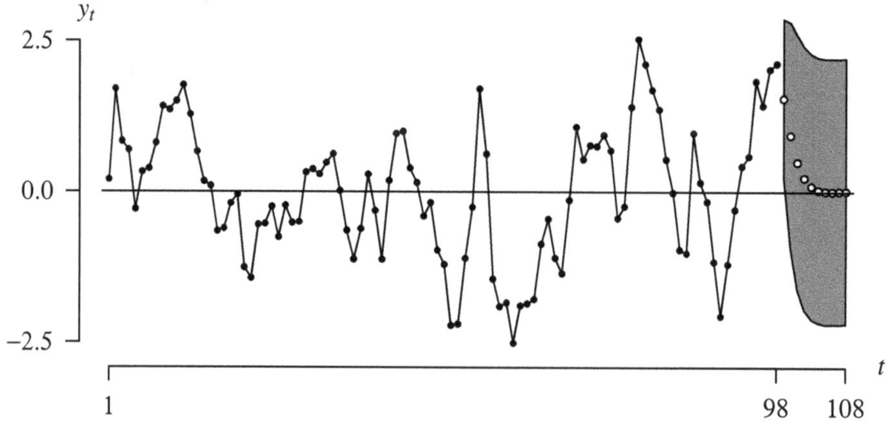

Figure 9.49: Lake Huron residuals, forecasts, and 95% prediction intervals.

Finally, the last step is to translate Figure 9.49 back to the raw time series observations. Figure 9.50 shows (a) the original time series $x_1, x_2, \ldots, x_{98}$ as points (•) connected by lines, (b) the regression line associated with the original time series, (c) the first 10 forecasted lake levels $\hat{X}_{99}, \hat{X}_{100}, \ldots, \hat{X}_{108}$ as open circles (◦), (d) the 95% prediction intervals as a shaded region, and (e) the next 10 actual average lake level values in July for the years 1973–1982 taken from the NOAA Great Lakes Experimental Research Laboratory website,

580.98, 581.04, 580.49, 580.52, 578.57, 578.96, 579.94, 579.77, 579.44, 578.97,

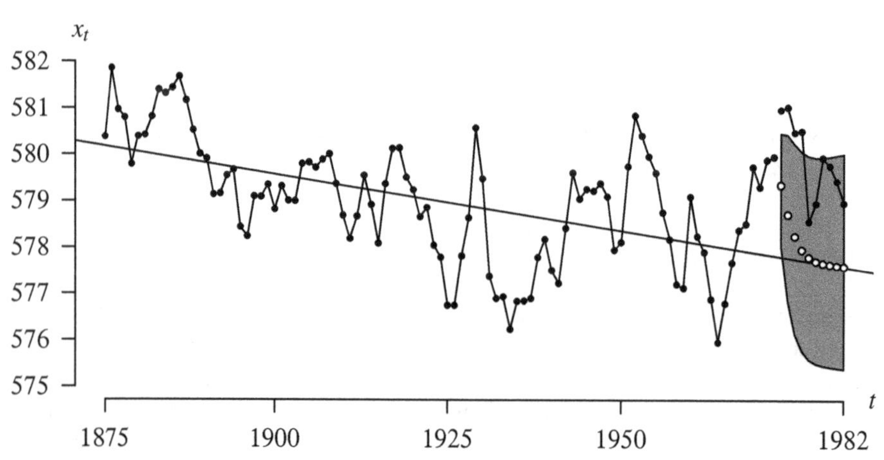

Figure 9.50: Lake Huron levels (1875–1972) with regression line and forecasts.

as points (●) connected by lines. Notice that the forecasted values converge to the regression line as anticipated. Notice also that the actual values all exceed the forecasted values and the first four forecasted values fall outside of the prediction limits. If we did not know of these actual values, we would be satisfied with this detrended AR(2) time series model. However, these actual values call into question whether the lake levels are truly decreasing over time.

So far, there have been three different approaches to constructing a time series model for the Lake Huron levels:

- the shifted AR(2) model from Examples 9.14, 9.15, 9.16, 9.17, 9.18, 9.19, 9.20, and 9.21,

- the shifted ARMA(1, 1) model from Example 9.39, and

- the AR(2) model applied to the residuals from a simple linear regression from Example 9.40.

Which approach is preferred? Although the shifted AR(2) and shifted ARMA(1, 1) models fitted to the raw time series are roughly comparable and give nearly-identical forecasts, the shifted ARMA(1, 1) model has a slight edge for the following two reasons. First, from Table 9.7, the AIC value is 215 for the shifted AR(2) model and the AIC value is 214 for the shifted ARMA(1, 1) model. A smaller value implies a better fit. Second, the sum of squared residuals for the shifted AR(2) model is 46.9 and the sum of squared residuals for the shifted ARMA(1, 1) model is 46.5. A smaller sum of squared residuals for two models with an equal number of parameters is preferred. Both models have four parameters. These two sums of squared residuals for the two models are computed with the R statements

```
sum(arima(LakeHuron, order = c(2, 0, 0))$residuals ^ 2)
sum(arima(LakeHuron, order = c(1, 0, 1))$residuals ^ 2)
```

Although the differences between the AIC values and the sums of squares is small, the shifted ARMA(1, 1) model holds a slight edge.

The detrended model from Example 9.40, on the other hand, is preferred over the two stationary models because it explicitly models the decreasing lake levels over time. However, the fact that all of the forecasted values in the detrended model are low relative to the actual values in the years 1973 to 1982 is troubling. Could it be the case that there was no downward trend after all? At this point, some serious detective work is in order to see if the early values in the raw time series were elevated by some external influence and should not be included as a part of the time series. A rigorous search should be conducted for any external cause which might elevate the early values in the time series: excess rainfall, elevated temperatures, dredging, bridge projects, flow control projects, etc. As a particular instance, if the first 20 values of the time series can be eliminated due to the identification of an assignable cause for the years 1875–1894, for example, the p-value from simple linear regression testing for the statistical significance of the slope increases from a highly significant $p = 4 \cdot 10^{-8}$ to a nonsignificant $p = 0.11$. The downward trend would now be slight and a stationary model could be fitted to the remaining values in the time series.

Detrending has proved to be an effective method for transforming a nonstationary time series to a stationary time series. The second technique involves differencing.

9.4.2 ARIMA(p, d, q) Models

George Box and Gwilym Jenkins devised a time series modeling methodology known as *ARIMA modeling*. The I between AR and MA stands for *integrated*. These models are sometimes referred to as Box–Jenkins models. An ARIMA(p, d, q) time series model is one in which the dth-differenced times series, $\nabla^d X_t$, is an ARMA(p, q) time series. So ARIMA time series modeling uses repeated differencing of the raw time series in order to achieve a time series which appears to be stationary. ARMA modeling can be then applied to the resulting stationary time series.

Definition 9.7 An ARIMA(p, d, q) time series model for $\{X_t\}$ is one in which the dth-differenced times series, $\nabla^d X_t$, is an ARMA(p, q) time series for some nonnegative integer d. An ARIMA(p, d, q) model can be written in compact form as

$$\phi(B)\nabla^d X_t = \theta(B)Z_t,$$

where $\phi(B)$ and $\theta(B)$ are the usual characteristic polynomials for an ARMA(p, q) model and $Z_t \sim WN\left(0, \sigma_Z^2\right)$.

Three key parameters in an ARIMA model are p, d, and q, which are all nonnegative integers. The parameter p is the number of coefficient parameters in the autoregressive portion of the model. The parameter d is the number of differences that are applied to the original time series in order to achieve stationarity. The parameter q is the number of coefficient parameters in the moving average portion of the model. So the general format for specifying an ARIMA model is ARIMA(p, d, q). In addition to the parameters p, d, and q, there are $p + q + 1$ parameters that define an ARIMA(p, d, q) model: the p autoregressive parameters $\phi_1, \phi_2, \ldots, \phi_p$, the q moving average parameters $\theta_1, \theta_2, \ldots, \theta_q$, and the variance of the white noise σ_Z^2. As in the case of ARMA models, a shift parameter μ can be included in the model. If one or more of these parameters is zero, they are omitted from the specification. An IMA(2, 1) model, for example, has $p = 0$ autoregressive terms, $d = 2$ differences, and $q = 1$ moving average term. If a model only involves, for example, the autoregressive portion of the model with two terms (that is, no differencing and no moving average terms), then this model is specified as an AR(2) model. An ARMA(p, q) model is a special case of an ARIMA(p, d, q) model when $d = 0$.

ARIMA modeling will be illustrated by a simulation example that will reveal what a realization of an ARIMA process looks like, along with the R code required to fit these simulated values to an ARIMA model.

Example 9.41 Simulate a realization of $n = 100$ observations from an ARI(1, 1) time series model with $\phi = 0.8$ and $\sigma_Z^2 = 4$. Fit the resulting simulated values to an ARIMA model.

This problem gives one instance of what an ARIMA model with a nonzero value for d looks like. The R code below uses the `arima.sim` function to generate a realization of an ARIMA(1, 1, 0) time series model, which is more commonly known as an ARI(1, 1) model. Even though 99 observations are requested, a total of 100 will be generated because the differencing operator is being undone within `arima.sim`. The code also plots the sample autocorrelation function and the sample partial autocorrelation function of the simulated realization.

```
set.seed(1)
x = arima.sim(list(order = c(1, 1, 0), ar = 0.8), n = 99, sd = 2)
layout(matrix(c(1, 1, 2, 3), 2, 2, byrow = TRUE))
plot.ts(x)
acf(x, lag.max = 40)
pacf(x, lag.max = 40)
```

The results are shown in Figure 9.51. The realization is clearly generated from a nonstationary time series model with an overall meandering upward trend. This conclusion is supported by the graphs of r_k and r_k^*.

The augmented Dickey–Fuller test can be used to assess the stationarity of the simulated time series. It has been implemented in R in the adf.test function in the tseries package. There is no need to run this test for this particular realization of the time series; the time series plot clearly shows that this is a nonstationary time series. Now consider fitting this time series realization to an ARIMA(p, d, q) model. Since the time series realization exhibits a meandering linear increase, it is possible that a single difference might be adequate for transforming this time series to achieve stationarity. Although it is in some sense cheating because we know that the realization was generated from an ARI(1, 1) time series model, the R code that follows takes a single difference of the time series depicted in Figure 9.51 and plots the differenced series $y_t = \nabla x_t = x_t - x_{t-1}$, the associated sample autocorrelation function, and the sample partial autocorrelation function.

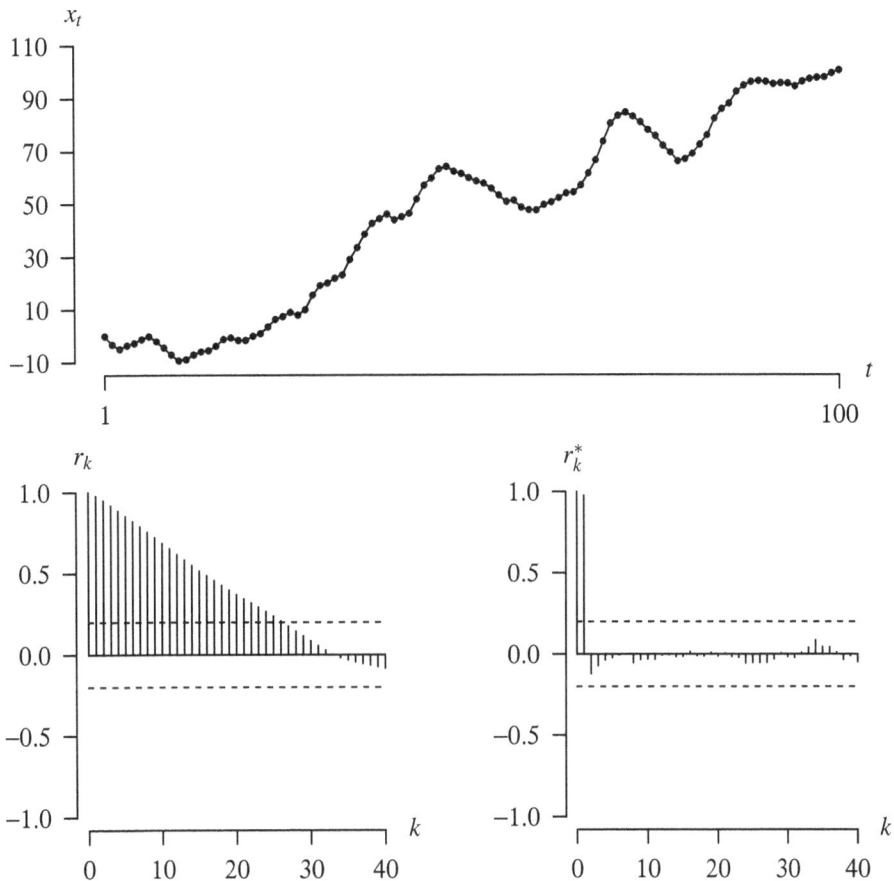

Figure 9.51: Time series plot, r_k, and r_k^* for a realization of a simulated ARI(1, 1) model.

```
set.seed(1)
x = arima.sim(list(order = c(1, 1, 0), ar = 0.8), n = 100, sd = 2)
y = diff(x)
layout(matrix(c(1, 1, 2, 3), 2, 2, byrow = TRUE))
plot.ts(y)
acf(y, lag.max = 40)
pacf(y, lag.max = 40)
```

Figure 9.52 shows a graph of the differenced time series and the associated graphs of r_k and r_k^*. The differencing has achieved its goal; the differenced values appear to be stationary. Furthermore, the sample partial autocorrelation function has a single statistically significant value at lag 1 and then cuts off. (The statistically significant value at lag 18 is attributed to random sampling variability because we expect that 2 of the 40 r_k^* values will lie outside of the 95% bounds by chance.) The sample autocorrelation appears to be gradually tailing out. This is evidence that supports an AR(1) model for

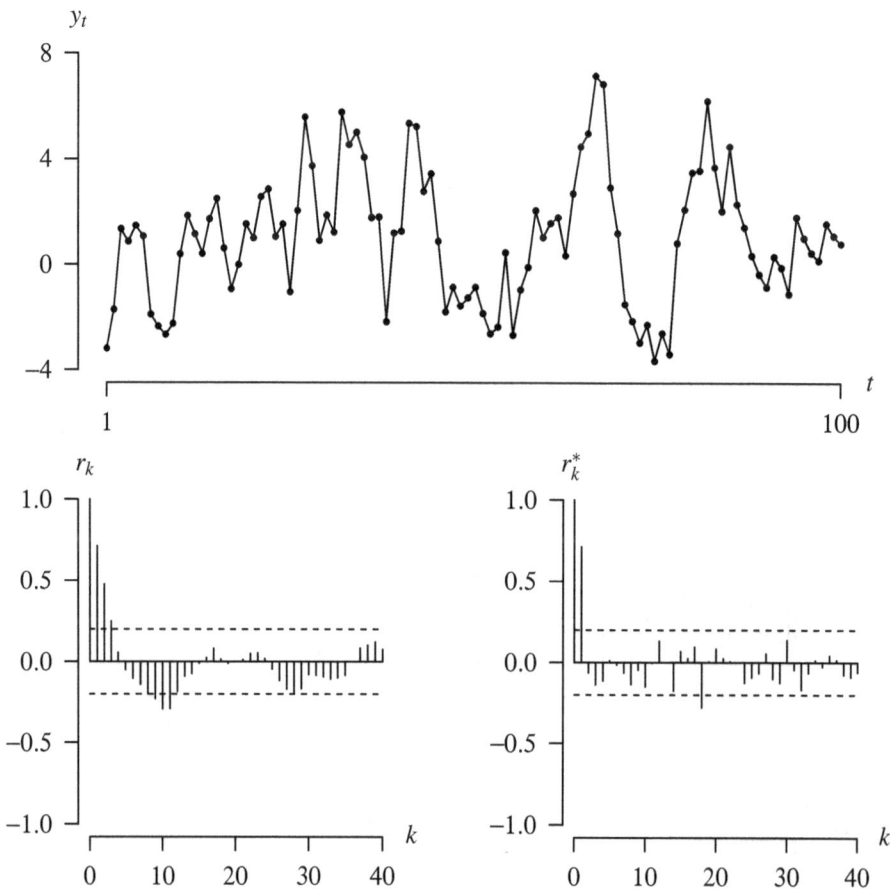

Figure 9.52: Time series plot, r_k, and r_k^* for differences of a realization of the simulated values.

the differenced values from the ARI(1, 1) realization, just as we suspected would be the case.

The R statements

```
set.seed(1)
x = arima.sim(list(order = c(1, 1, 0), ar = 0.8), n = 100, sd = 2)
y = diff(x)
fit = arima(y, order = c(1, 0, 0), method = "ML")
fit$coef[1]
fit$sigma2
```

return an estimated coefficient $\hat{\phi} = 0.73$ (which is near the population value $\phi = 0.8$) and estimated white noise variance $\hat{\sigma}_Z^2 = 2.9$ (which is near the population value $\sigma_Z^2 = 4$). A Monte Carlo simulation could be conducted to see how far these estimated values stray from their population counterparts. Increasing the length of the time series will make these estimates closer to their associated population values on average.

The ARIMA modeling process is adequate for nonstationary models but is not well-suited to handling cyclic variation. The SARIMA (seasonal autoregressive integrated moving average) model has been formulated to overcome this weakness.

Definition 9.8 A seasonal ARIMA time series model for $\{X_t\}$, denoted by a SARIMA model of order $(p, d, q) \times (P, D, Q)_s$ with seasonal order s, is given in compact form by

$$\phi(B)\Phi(B^s)\nabla^d\nabla_s^D X_t = \theta(B)\Theta(B^s)Z_t,$$

where p, d, q, P, D, Q, and s are nonnegative integers,

- $\phi(B) = 1 - \phi_1 B - \phi_2 B^2 - \cdots - \phi_p B^p$,

- $\Phi(B) = 1 - \Phi_1 B - \Phi_2 B^2 - \cdots - \Phi_P B^P$,

- $\theta(B) = 1 + \theta_1 B + \theta_2 B^2 + \cdots + \theta_q B^q$,

- $\Theta(B) = 1 + \Theta_1 B + \Theta_2 B^2 + \cdots + \Theta_Q B^Q$,

and $\{Z_t\} \sim WN\left(0, \sigma_Z^2\right)$.

An ARIMA model is a special case of a SARIMA model when $P = D = Q = 0$. The ∇^d term in the SARIMA model is associated with an ordinary difference; the ∇_s^D term is associated with a seasonal difference. Consider the inside portion of the SARIMA defining formula, $\nabla^d\nabla_s^D X_t$, in a modeling setting in which monthly data is being collected and the modeler believes that there is cyclic annual variation, so $s = 12$. In the case of $d = 1$ ordinary difference and $D = 1$ seasonal difference, this portion of the SARIMA defining formula becomes

$$\begin{aligned}\nabla\nabla_{12}X_t &= \nabla\left(\nabla_{12}X_t\right) \\ &= \nabla\left(X_t - X_{t-12}\right) \\ &= \left(X_t - X_{t-12}\right) - \left(X_{t-1} - X_{t-13}\right).\end{aligned}$$

The ∇ operator is being used to eliminate a linear trend and the ∇_{12} operator is being used to eliminate seasonality. The seasonal AR term $\Phi(B^s)$ and the seasonal MA term $\Theta(B^s)$ in Definition 9.8 provide autoregressive and moving average terms for observations that are s units distant in time.

Example 9.42 Forecast the next three years of international air travel based on the `AirPassengers` time series from Example 7.2.

The plot of the time series is given in Figure 9.53. As indicated in Example 7.32, the annual cycle associated with international air travel over this period does not appear to be sinusoidal in nature. The peak months for international travel are in July and August when school is not in session and the low month for international travel is November, as seen in Figure 7.30. This time series provides a challenging modeling exercise because it exhibits a nonconstant variance, a trend, and periodicity. These three modeling challenges will be addressed in that order, one by one.

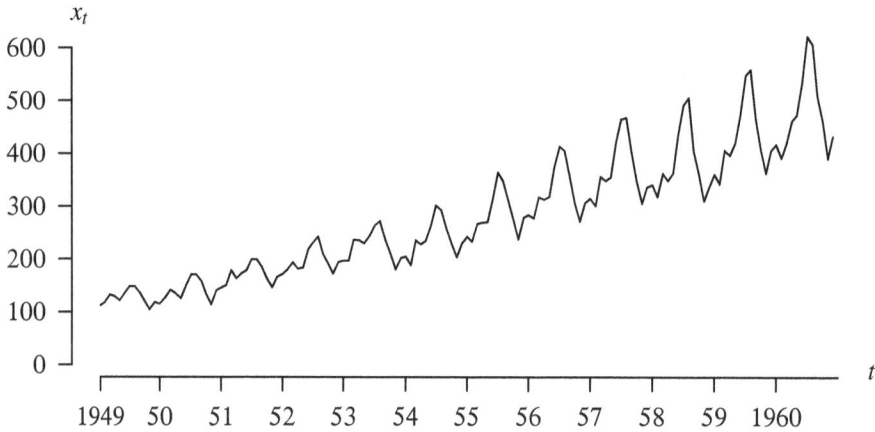

Figure 9.53: International airline passengers (in thousands) 1949–1960.

We begin by addressing the nonconstant variance. Since the variance appears to be increasing over time, a logarithmic transformation is reasonable transformation to apply to the time series. Let $\{x_t\}$ denote the original time series and let $y_t = \ln x_t$. The R statement

```
ts.plot(log(AirPassengers))
```

plots the natural logarithm of the raw time series. The plot of the transformed time series is given in Figure 9.54.

The transformation appears to be effective. The variance of the logarithms of the raw time series observations is now close to constant over time. The next step is to address the trend. Since the trend of the transformed time series depicted in Figure 9.54 is approximately linear, a single difference ($d = 1$) is taken. The differenced time series is

$$w_t = \nabla y_t = \nabla \ln x_t = \ln x_t - \ln x_{t-1}.$$

(The resulting time series is not named z_t to avoid any conflict with the white noise terms.) This transformed and differenced time series, along with the associated sample autocorrelation function and sample partial autocorrelation function are graphed with the R statements

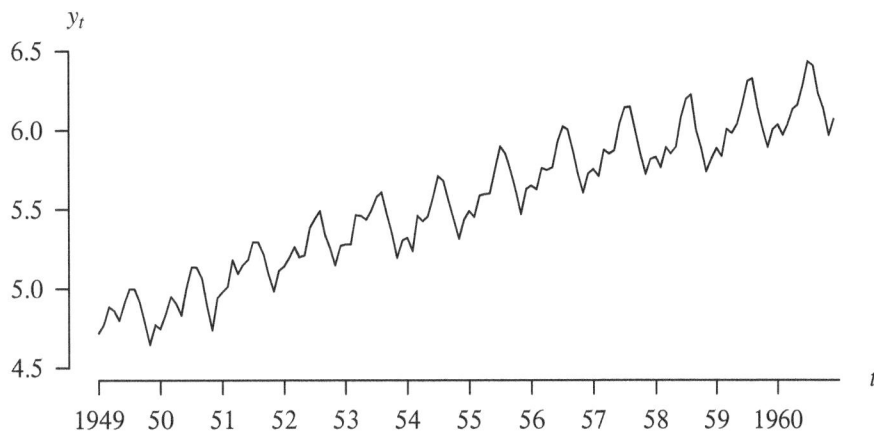

Figure 9.54: Logarithm of international airline passengers (in thousands) 1949–1960.

```
w = diff(log(AirPassengers))
layout(matrix(c(1, 1, 2, 3), 2, 2, byrow = TRUE))
plot.ts(w)
acf(w, lag.max = 40)
pacf(w, lag.max = 40)
```

The associated graphs are displayed in Figure 9.55. Instead of the usual $12 \cdot 12 = 144$ observations from the previous two figures, the differencing operation leaves only 143 observations, which is reflected in the labels on the horizontal axis of the plot of the differences of the logarithms of the original time series values. The differencing has proved to be successful. The time series plot of w_t appears to be stationary. The strongly statistically significant sample autocorrelation function values at lags 12, 24, and 36 are a reminder that even though the nonconstant variance and trend have been addressed, the cyclic variation has not been addressed. ARMA modeling is *not* appropriate at this point because r_k is neither tailing out nor cutting off. There is still an annual cyclic component present in $\{w_t\}$. A reasonable way to proceed is to employ a seasonal ARIMA model to account for the cyclic variation. Backing up one level, we would like to fit a SARIMA $(p, d, q) \times (P, D, Q)_{12}$ model to the natural logarithms of the raw passenger counts. The choice $s = 12$ for the seasonal order is to account for the monthly collection of the passenger counts which exhibit a clear annual cycle; the choice of $d = 1$ is appropriate based on the fact that the time series $\{w_t\}$ in Figure 9.55 appears to be stationary. But what about the other parameters (p, q, P, D, and Q)? An exhaustive search using the arima function in R to locate the smallest value of AIC results in the following settings:

$$d = P = Q = 1$$

and

$$p = q = D = 0.$$

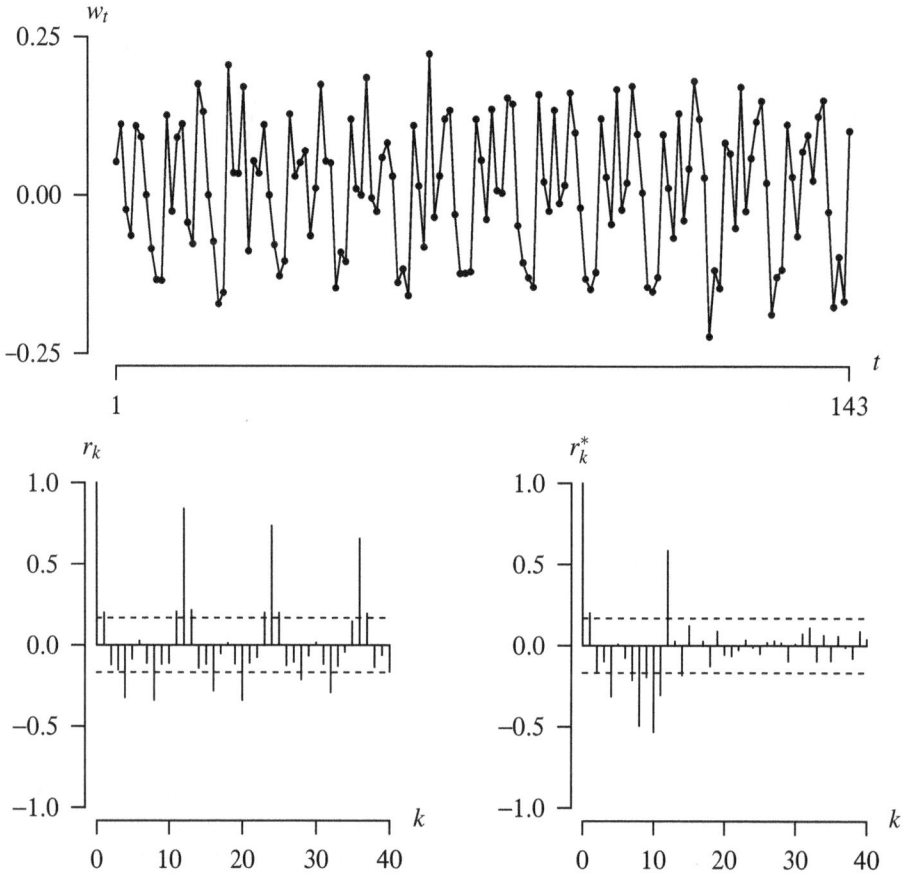

Figure 9.55: Logarithm of international airline passengers (in thousands) 1949–1960.

The single R statement below fits the SARIMA $(0, 1, 0) \times (1, 0, 1)_{12}$ model to the logarithms of the passenger counts in the `AirPassengers` time series.

```
fit = arima(log(AirPassengers),
            order = c(0, 1, 0),
            seasonal = list(order = c(1, 0, 1), period = 12))
```

The maximum likelihood estimates of the parameters are $\Phi = 0.9877$, $\Theta = -0.5935$ and $\hat{\sigma}_Z^2 = 0.001526$. So the fitted model is

$$\left(1 - 0.9877B^{12}\right) \nabla Y_t = \left(1 - 0.5935B^{12}\right) Z_t,$$

where $Y_t = \ln X_t$, and $Z_t \sim WN(0, 0.001526)$ This fitted SARIMA model achieves an AIC value of -486.9953. There are several other competing SARIMA models with nearby AIC values.

The final step is to use the fitted SARIMA model to forecast international airline travel for the subsequent three years (36 months). The R code below fits the SARIMA model with the `arima` function, uses the `predict` function to calculate the forecasted values and their standard errors, and then plots the original time series, the forecasted values, and the 95% prediction intervals.

```
fit      = arima(log(AirPassengers),
                 order = c(0, 1, 0),
                 seasonal = list(order = c(1, 0, 1), period = 12))
n        = length(AirPassengers)
h        = 36
forecast = predict(fit, n.ahead = h)
alpha    = 0.05
crit     = qnorm(1 - alpha / 2)
lo       = forecast$pred - crit * forecast$se
hi       = forecast$pred + crit * forecast$se
beginx   = time(AirPassengers)[n]
deltax   = deltat(AirPassengers)
xval1    = seq(beginx + deltax, beginx + h * deltax, length.out = h)
xval2    = c(xval1, rev(xval1))
yvals    = exp(c(lo, rev(hi)))
ts.plot(AirPassengers, exp(forecast$pred), ylim = c(0, max(yvals)))
polygon(xval2, yvals, col = "gray50")
points(xval1, exp(forecast$pred), pch = 16, cex = 0.7, col = "white")
points(xval1, exp(forecast$pred), pch = 1, cex = 0.7)
```

This graph of the original time series and the 36 forecasted values is given in Figure 9.56. The forecasts from the SARIMA model show that the nonconstant variance, trend, and cyclic variation have been adequately captured by the model. Since there are $144 + 36 = 180$ points squeezed so tightly together in the plot, a second graph of the forecasted values for just the last three cycles of the observed time series and the forecasted values is given in Figure 9.57.

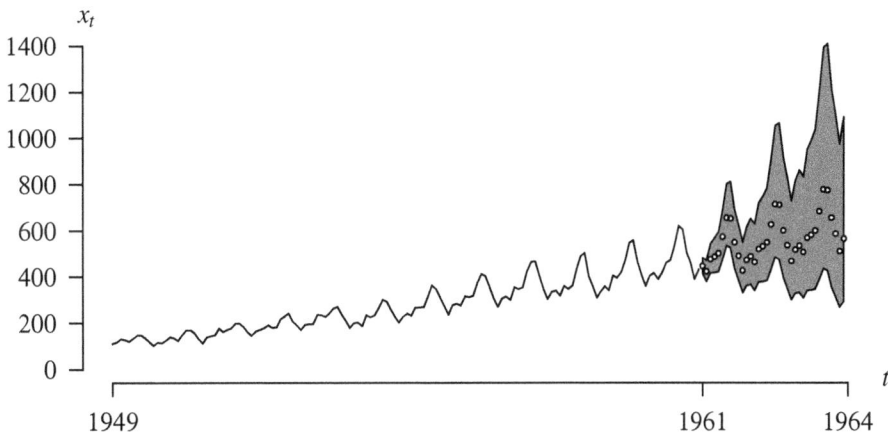

Figure 9.56: Forecasted international travel and 95% prediction intervals.

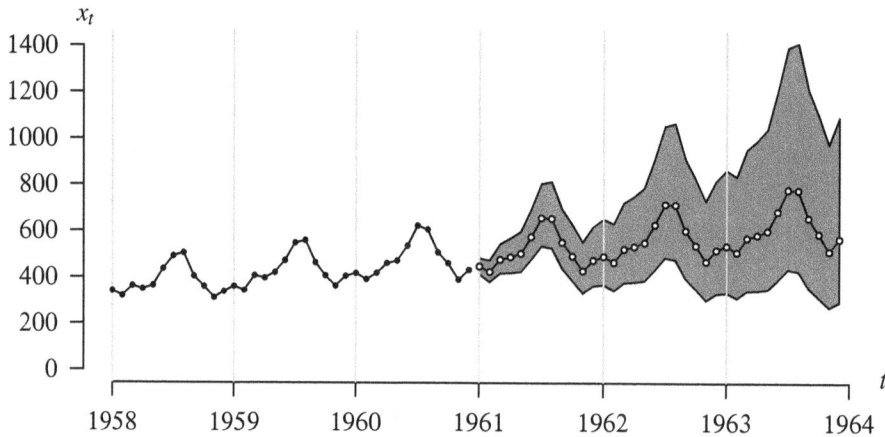

Figure 9.57: Forecasted international travel and 95% prediction intervals.

These forecasts are crucial for airports, airline manufacturers, and associated businesses as they can predict the impact of growth on supply chains, personnel requirements, and logistics necessary to support air travel.

To summarize this section, the modeling of a nonstationary time series involves the following steps.

- Plot the time series, noting any trends, seasonality, and nonconstant variance.

- Make the variance stable by applying appropriate transformations if necessary.

- Use detrending (possibly regression) or repeated differencing (to use an ARIMA model) to create a stationary time series.

- Plot the stationary time series along with its sample autocorrelation function and sample partial autocorrelation function.

- Hypothesize a tentative ARMA model for the stationary time series model. If there is a seasonal component, consider a SARIMA model on the transformed time series.

- Fit the tentative ARMA or SARIMA model. Perform the model assessment tests on the tentative time series model. If the fitted tentative ARMA or SARIMA model fails these tests, then hypothesize a new tentative model.

- Perform overfitting in the final model selection process to ensure that the best model has been selected.

- Apply the final time series model in the fashion dictated by the problem setting (this is often forecasting future values of the time series).

As illustrated in Example 9.42, time series modeling can be thought of as a step-by-step process of identifying a removing causes of variation in the time series (for example, trend, cycles, autocorrelation) until all that remains is white noise.

9.5   Spectral Analysis

In practice, many time series exhibit cyclic variation. The first two examples in Chapter 7 concerning monthly residential power consumption and monthly international airline travel both contain a cyclic component. The time series models derived from the general linear model do not explicitly consider cyclic variation; these models exist in what is known as the *time domain*. Spectral analysis considers modeling in the *frequency domain*. Spectral analysis decomposes a stationary time series into sinusoidal components (that is, sine and cosine functions) in order to identify frequencies associated with periodic components. Just as autoregressive models use regression on previous values of a time series in the time domain, spectral analysis uses regression on sine and cosine terms in the frequency domain.

Table 9.20 presents some new terminology that arises in spectral analysis and presents some analogies with known data analysis techniques. The column headings indicate that the three subsequent rows contain three application areas, three probability constructs, and their three statistical counterparts.

- The first row concerns the analysis of univariate data. In probability theory, several commonly-used probability distributions (for example, the exponential, normal, and binomial distribution) are investigated in order to build an inventory of potential probability distributions that might adequately describe a univariate data set. When an analyst encounters a univariate data set, one of the early steps in the analysis is to plot a histogram and compare its shape to the inventory of probability density functions associated with known probability distributions.

- The second row concerns time series analysis in the time domain. Shapes of the population autocorrelation function are derived for several commonly-used time series models (for example, the AR(2), MA(1), and ARMA(1, 1) models) in order to build an inventory of shapes such as those given in Figure 9.43 that might adequately describe the time series. When a time series analyst encounters time series observations, one of the early steps in the analysis is to plot the correlogram (a.k.a., the sample autocorrelation function) and compare its shape to the inventory of known population autocorrelation functions.

- The third row concerns time series analysis in the frequency domain. Shapes of the *spectral density function* are derived for several commonly-used time series models in order to build an inventory of shapes that might adequately describe the periodic nature of a time series. When a time series analyst encounters time series observations, one of the early steps is to plot the *periodogram* and compare its shape to the inventory of known spectral density functions.

The next two subsections will focus on the spectral density function and its statistical counterpart, the periodogram.

Application area	Probability construct	Statistical counterpart
univariate data analysis	probability density function	histogram
time series analysis: time domain	population autocorrelation function	correlogram
time series analysis: frequency domain	spectral density function	periodogram

Table 9.20: Population versus sample representations.

9.5.1 The Spectral Density Function

The emphasis in the spectral analysis of a time series is the identification of the frequencies associated with cycles. The frequencies will be denoted here by ω. Just as the population autocorrelation function is the natural tool for identifying and quantifying autocorrelation in the time domain, the spectral density function is the natural tool for identifying and quantifying the frequencies associated with cyclic variation in the frequency domain. As seen in the following definition, the spectral density function can be written in terms of the population autocovariance function.

Definition 9.9 Let $\{X_t\}$ be a stationary time series with population autocovariance function $\gamma(k)$. The *spectral density function* $f(\omega)$ is

$$f(\omega) = \frac{1}{\pi}\left[\gamma(0) + 2\sum_{k=1}^{\infty}\gamma(k)\cos(\omega k)\right] \qquad 0 < \omega < \pi.$$

The interpretation of the spectral density function is that $f(\omega)\Delta\omega$ reflects the contribution of frequencies in the interval $(\omega, \omega + \Delta\omega)$ to the variance of X_t for small values of $\Delta\omega$. When $f(\omega)$ is high, then frequencies near ω have a large impact on X_t. When $f(\omega)$ is low, then frequencies near ω have a small impact on X_t. The upper limit of the support of the spectral density function, π, is known as the *Nyquist frequency*. Frequencies that exceed π are not captured by the spectral density function. This is not a universal choice for the definition of the spectral density function or the upper limit of its support. There are many valid alternative choices. A common alternative choice for the upper limit of the support is $1/2$.

The first example illustrates the calculation of a spectral density function for one of the most basic time series models.

Example 9.43 Find the spectral density function for an ARMA(0, 0) time series model.

An ARMA(0, 0) model is simply white noise, so the population autocovariance function is

$$\gamma(k) = \begin{cases} \sigma_Z^2 & k = 0 \\ 0 & k = 1, 2, \ldots \end{cases}.$$

Using Definition 9.9, the spectral density function is

$$f(\omega) = \frac{1}{\pi}\left[\gamma(0) + 2\sum_{k=1}^{\infty}\gamma(k)\cos(\omega k)\right] = \frac{\sigma_Z^2}{\pi} \qquad 0 < \omega < \pi.$$

Figure 9.58 shows the spectral density function for the ARMA(0, 0) process. Since there is no cyclic variation whatsoever in the ARMA(0, 0) time series model, no frequency stands out over another, so the spectral density function is uniformly distributed of the frequencies between 0 and π. Each frequency on the interval $(0, \pi)$ contributes equally to the variance of X_t. Neither high frequencies nor low frequencies play a dominant role in the in terms of cyclic variation of this process. Notice that the area under $f(\omega)$ between 0 and π is $\sigma_X^2 = \sigma_Z^2$.

The next example calculates the spectral density function of an MA(1) model. This particular model was chosen because it has an autocovariance function that cuts off after lag 1, which means that the summation given in Definition 9.9 consists of just a single term.

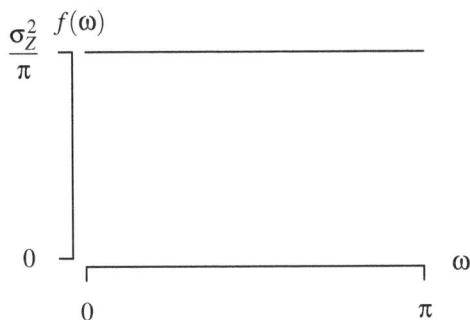

Figure 9.58: Spectral density function for an ARMA$(0, 0)$ model.

Example 9.44 Find the spectral density function for an MA(1) time series model.

As derived in Section 9.2.1, the population autocovariance function of an MA(1) time series model is

$$\gamma(k) = \begin{cases} \left(1 + \theta^2\right)\sigma_Z^2 & k = 0 \\ \theta\sigma_Z^2 & k = 1 \\ 0 & k = 2, 3, \dots . \end{cases}$$

Using Definition 9.9, the spectral density function is

$$f(\omega) = \frac{1}{\pi}\left[\gamma(0) + 2\sum_{k=1}^{\infty}\gamma(k)\cos(\omega k)\right] = \frac{\sigma_Z^2}{\pi}\left[1 + \theta^2 + 2\theta\cos\omega\right] \qquad 0 < \omega < \pi.$$

In order to develop some intuition about the spectral density function, consider two special cases of the MA(1) model: $\theta = 9/10$ and $\theta = -9/10$. These two values of θ correspond to stationary and invertible MA(1) time series models.

When $\theta = 9/10$, the spectral density function reduces to

$$f(\omega) = \frac{\sigma_Z^2}{\pi}\left[\frac{181}{100} + \frac{9}{5}\cos\omega\right] \qquad 0 < \omega < \pi.$$

Figure 9.59 shows the spectral density function for an MA(1) model with $\theta = 9/10$. Since $\theta > 0$, the lag 1 population autocorrelation is positive, which means that a realization of this time series will linger above the mean for a few observations and then linger below the mean for a few observations. But the number of observations that the sequence lingers above or below the mean is random. This is the behavior that we saw in the simulated values in Example 9.33. In the simulated realization, sometimes the time series only lingers above or below the mean for just 2 or 3 simulated observations. In other cases, the time series lingers above or below the mean for 6 or 7 simulated observations. In other words, there is low-frequency variation in this time series, but it does not have a single consistent frequency. This pattern of lingering on one side of the mean corresponds to *low frequency* cycles, and those low frequency cycles correspond to smaller values of ω. This is reflected in the spectral density function in Figure 9.59, where the lower frequencies have larger values of $f(\omega)$ than the higher frequencies. Lower frequency variation dominates.

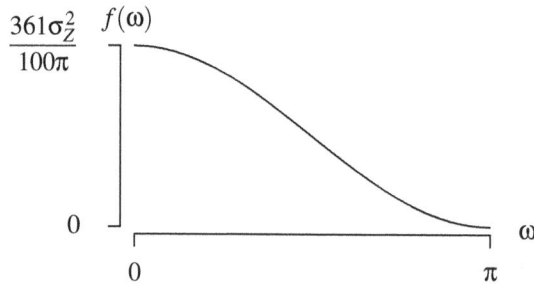

Figure 9.59: Spectral density function for an MA(1) model with $\theta = 9/10$.

When $\theta = -9/10$, the spectral density function reduces to

$$f(\omega) = \frac{\sigma_Z^2}{\pi} \left[\frac{181}{100} - \frac{9}{5} \cos \omega \right] \qquad 0 < \omega < \pi.$$

Figure 9.60 shows the spectral density function for an MA(1) model with $\theta = -9/10$. Since $\theta < 0$, the lag 1 population autocorrelation is negative, which means that the observations in a realization of this time series will often jump from one side of the mean value to the other. This is the behavior that we saw with the simulated values in Example 9.34. In most cases, when one observation was on one side of the mean, the next observation was on the other side of the mean. Occasionally, however, the time series lingered for 2 or 3 observations on one side of the mean. Once again, this behavior is random and does not correspond to a single consistent frequency. This pattern of adjacent observations jumping from one side of the mean to the other corresponds to *high frequency* cycles, and those high frequency cycles correspond to larger values of ω. This is reflected in the spectral density function in Figure 9.60, where the higher frequencies have larger values of $f(\omega)$. Higher frequency variation dominates.

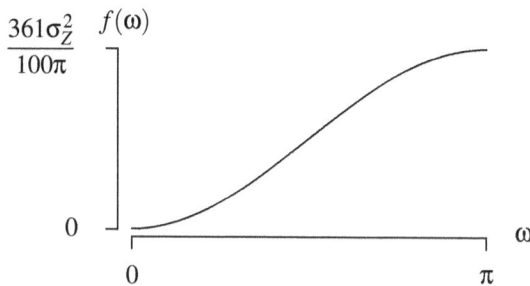

Figure 9.60: Spectral density function for an MA(1) model with $\theta = -9/10$.

One common element from the spectral density functions given in the two previous examples is that they both integrate to σ_X^2. This is true in general. Some time series analysts prefer to divide the spectral density function by σ_X^2 so that it will integrate to 1, making it a true probability density function. The *normalized spectral density function* is given by

$$f^*(\omega) = \frac{f(\omega)}{\sigma_X^2} \qquad 0 < \omega < \pi.$$

Dividing both sides of the equation in Definition 9.9 by $\gamma(0) = \sigma_X^2$ gives

$$f^*(\omega) = \frac{1}{\pi} \left[1 + 2 \sum_{k=1}^{\infty} \rho(k) \cos(\omega k) \right] \qquad 0 < \omega < \pi.$$

The associated *normalized spectral cumulative distribution function* is defined on the support of ω in the usual fashion as

$$F^*(\omega) = \int_0^{\omega} f^*(w) \, dw \qquad 0 < \omega < \pi.$$

One advantage of normalizing these two functions is that there is now a clean interpretation of $F^*(\omega)$. For frequencies ω_1 and ω_2 satisfying $0 < \omega_1 < \omega_2 < \pi$, the expression $F^*(\omega_2) - F^*(\omega_1)$ denotes the *proportion* of the variance in $\{X_t\}$ accounted for by frequencies on the interval (ω_1, ω_2).

9.5.2 The Periodogram

The periodogram is the statistical counterpart to the spectral density function. The periodogram estimates the spectral density function for all frequencies between 0 and π. The shape of the periodogram reflects the frequencies that correspond to significant cyclic variation in a time series. Peaks in the periodogram reveal the dominant frequencies associated with cyclical components in an observed time series.

One topic that is crucial in time series analysis in the frequency domain is how often a time series should be sampled. Consider sampling the outdoor air temperature, for example, in Washington, DC. There are two significant cyclic components that should become apparent in such a time series. First, there is a daily temperature cycle. Temperatures are warmer during the day and cooler at night. This corresponds to high frequency variation. Second, there is an annual temperature cycle. Temperatures are warmer during the summer and cooler during the winter. This corresponds to low frequency variation. There is a factor of 365 (well, actually 365.24219) that separates the frequencies of these two cycles which should be accounted for in how often the time series is sampled. The following illustrations provide instances of sampling this time series too often, sampling this time series not often enough, and sampling this time series at about the right intervals to capture these two frequencies in a periodogram.

- Let's say you sample 1000 outdoor air temperatures at Reagan National Airport in Washington DC *every second* beginning at noon on July 20, 1969. This data collection will be over very soon because 1000 seconds is only about 17 minutes. But you have not covered a daily cycle or an annual cycle, so the frequencies for these two cycles cannot be detected from this sample. The sampling is too frequent.

- Let's say you sample 100 outdoor air temperatures at Reagan National Airport in Washington DC *annually* beginning at noon on July 20, 1969. This experiment will take you a long time to collect because the last value collected will be at noon on July 20, 2068. Even though you have collected the observations through 100 annual temperature cycles and tens of thousands of daily temperature cycles, neither the daily nor the annual cycle can be detected. All observations were made during the summer and during the day. The sampling was too infrequent.

- If you desire to detect both the daily and the annual outdoor air temperature cycles at Reagan National Airport, then a sampling interval between the two extremes (every second and every year from the previous two illustrations) must be used. So if you begin sampling *hourly* data at noon on July 20, 1969 and collect this data for three years, you will have collected outdoor temperature observations over three annual cycles and about a thousand daily cycles. This requires $3 \cdot 365 \cdot 24 = 26,280$ outdoor air temperatures to be collected. This time series allows an analyst to detect both daily and annual cycles. The periodogram, which estimates the spectral density function will have a peak associated with the low frequency (annual) cycles and a second peak associated with the high frequency (daily) cycles.

The details associated with computing the periodogram are left for a full-semester class in time series analysis. Some of the fundamental ideas will be presented here in order to give a sense of the development of the estimator. As has been the case in regression and survival analysis, we begin with a model for a time series having cyclic behavior. One such model is

$$X_t = c \cdot \cos(\omega t + \phi),$$

where c is the amplitude of the cyclic variation, ω is the frequency of the cyclic variation, ϕ is a phase shift parameter, and the angle is measured in radians. (The ϕ used here has nothing to do with ϕ from the autoregressive time series models in the time domain.) Unfortunately, this model does not contain any random terms, and such a time series only occurs rarely in practice. So adding a time series of white noise $\{Z_t\}$ results in the much more practical model

$$X_t = c \cdot \cos(\omega t + \phi) + Z_t.$$

Since the phase shift parameter can be tedious in parameter estimation, it is common practice in spectral analysis to apply the trigonometric identity $\cos(x+y) = \cos x \cos y - \sin x \sin y$ to this model, which results in

$$X_t = a\cos(\omega t) + b\sin(\omega t) + Z_t,$$

where $a = c \cdot \cos(\phi)$ and $b = -c \cdot \sin(\phi)$. This result is symmetric in the two primary trigonometric functions sine and cosine. The derivation thus far has only involved a single frequency ω. As in the previous outdoor air temperature example, it is often the case that there are multiple frequencies of interest. The current time series model can be generalized by summing over the k frequencies $\omega_1, \omega_2, \ldots, \omega_k$:

$$X_t = \sum_{j=1}^{k} \left(a_j \cos(\omega_j t) + b_j \sin(\omega_j t) \right) + Z_t,$$

where the amplitudes a_j and b_j reflect the contribution of frequency ω_j to the variability of X_t. For example, if $a_j = b_j = 0$ for one particular index j, then the associated frequency ω_j makes no contribution to the variability of X_t. The three remaining loose ends are (a) the number of frequencies k to consider, (b) which frequencies $\omega_1, \omega_2, \ldots, \omega_k$ to consider, and (c) how to estimate the amplitudes $a_1, a_2, \ldots, a_k$ and $b_1, b_2, \ldots, b_k$. These loose ends are easier to navigate if the number of elements in the time series n happens to be even, which is assumed for now. If so, then the usual practice is to let $k = n/2$ and space the ω_j values uniformly between 0 and π as

$$\omega_m = 2\pi m/n \qquad m = 1, 2, \ldots, n/2.$$

The lowest frequency that can be detected by the periodogram is $\omega_1 = 2\pi/n$ and the highest frequency that can be detected by the periodogram is $\omega_{n/2} = \pi$, the Nyquist frequency. The periodogram can be calculated in R with the `spectrum` function, which is available in the base language. Periodograms often contain significant sampling variability and do not provide a consistent estimator of the spectral density function, so time series analysts often use various techniques to smooth the raw periodogram values.

Example 9.45 Conduct a Monte Carlo simulation experiment with 1000 replications that averages the periodograms associated with an MA(1) model with $\theta = -0.9$ and $\sigma_Z^2 = 1$ from a time series of $n = 100$ observed values.

This example uses the average of 1000 periodograms to estimate the spectral density function. The R code below uses the `arima.sim` function to generate 1000 time series of length $n = 100$ from an MA(1) time series model with $\theta = -0.9$ and $\sigma_Z^2 = 1$. The realizations of the time series are stored in the vector named x. The periodogram values are computed by the `spectrum` function and their cumulative values are stored in the vector named s. Setting the `plot` argument to `FALSE` suppresses the plots of the individual periodograms in the `for` loop. Setting the `spans` argument to a vector of odd integers smooths the periodogram values. The `spec` component of the list returned by the `spectrum` function returns the smoothed periodogram values. Finally, the `plot` function is used to plot the periodogram values. Since the `spectrum` function returns a support of $(0, 1/2)$, this is stretched to yield a support of $(0, \pi)$ in the final plot. The curve that is plotted is $k = n/2 = 100/2 = 50$ segments connecting the spectral density function estimates for the frequencies $\omega_1 = 2\pi/100$, $\omega_2 = 4\pi/100$, ..., $\omega_{50} = \pi$, which are contained in the `freq` component of the list returned by the `spectrum` function.

```
set.seed(3)
s = numeric(50)
nrep = 1000
for (i in 1:nrep) {
  x = arima.sim(list(order = c(0, 0, 1), ma = -0.9), n = 100)
  s = s + spectrum(x, plot = FALSE, spans = c(3, 5))$spec
}
f = spectrum(x, plot = FALSE, spans = c(3, 5))$freq
plot(2 * pi * f, s / nrep, type = "l")
```

The average of the 1000 periodograms is plotted in Figure 9.61. As anticipated, the shape of the average of the periodograms is about the same as the shape of the spectral density function in Figure 9.60. The smoothness of the periodogram displayed in Figure 9.60 is deceiving. It is smooth because it is an average of 1000 periodograms. The individual periodograms generated within the `for` loop are very noisy, particularly when the smoothing parameters in the call to `spectrum` are eliminated. In a time series application, you will seldom work with the average of 1000 periodograms.

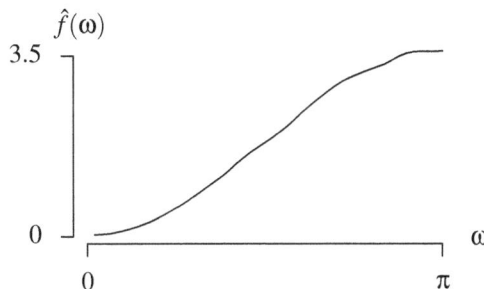

Figure 9.61: Periodogram averages for an MA(1) model with $\theta = -9/10$.

The previous example showed that the periodogram for the target MA(1) process, on average, appears to converge to the associated spectral density function. As illustrated in the final example, you will typically be working with just a single periodogram, which is typically quite noisy.

Example 9.46 Plot the periodogram associated with the annual lynx pelt sales at Hudson's Bay Company in Canada from 1857 to 1911. This data set was first encountered in Example 9.29.

The time series plot is given in Figure 9.62. There is a clear periodic component to the time series with a spike in sales every 9 or 10 years, which should be captured by the periodogram. The quickest way to generate a periodogram is with the R statements given below.

```
pelt = c(23362, 31642, 33757, 23226, 15178, 7272, 4448, 4926,
         5437, 16498, 35971, 76556, 68392, 37447, 45686, 7942,
         5123, 7106, 11250, 18774, 30508, 42834, 27345, 17834,
         15386, 9443, 7599, 8061, 27187, 51511, 74050, 78773,
         33899, 18886, 11520, 8352, 8660, 12902, 20331, 36853,
         56407, 39437, 26761, 15185, 4473, 5781, 9117, 19267,
         36116, 58850, 61478, 36300, 9704, 3410, 3774)
spectrum(pelt)
```

This code can be embellished a little to (*a*) avoid the special treatment of the Nyquist frequency, (*b*) extend the horizontal axis to π, (*c*) avoid the use of a logarithmic vertical axis, and (*d*) include some smoothing of the periodogram with the following additional R statements.

```
spec = spectrum(pelt, spans = c(3, 5), plot = FALSE)$spec
freq = spectrum(pelt, spans = c(3, 5), plot = FALSE)$freq
plot(2 * pi * freq, spec, type = "l")
```

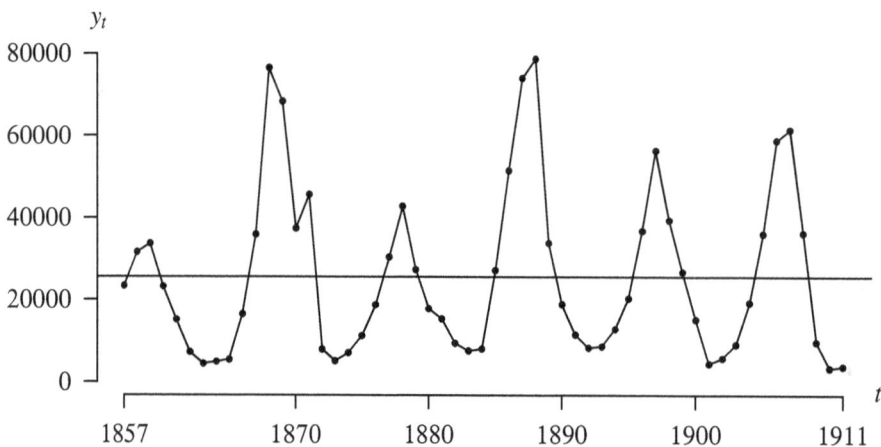

Figure 9.62: Time series plot for $n = 55$ annual lynx pelt sales (1857–1911).

The smoothed periodogram is given in Figure 9.63 There is a pronounced spike corresponding to a frequency of $\omega = 0.1 \cdot 2 \cdot \pi = 0.6283$. This corresponds to a period of

$$\frac{2 \cdot \pi}{0.1 \cdot 2 \cdot \pi} = 10,$$

which is consistent with the time series plot from Figure 9.62, which clearly displays a cycle of length 10.

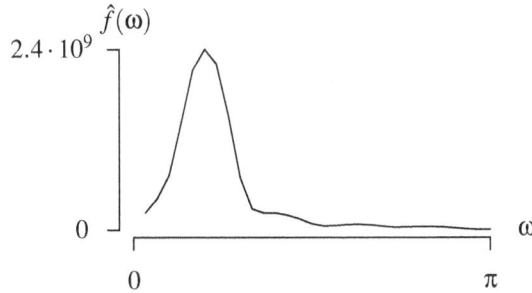

Figure 9.63: Periodogram for the lynx pelt sales.

9.6 Exercises

9.1 For a stationary AR(1) model, find $V[\bar{X}]$. Give an approximation for $V[\bar{X}]$ for large values of n.

9.2 Implement a Monte Carlo simulation that evaluates the method of moments, least squares, and maximum likelihood estimation techniques for an AR(1) model with $n = 100$ observed values and population parameters $\phi = -3/4$ and $\sigma_Z^2 = 1$ and identify the technique that has the smallest mean square error for estimating ϕ.

9.3 Consider a shifted AR(1) time series model with known parameter values μ, ϕ, and σ_Z^2. One realization of the time series $x_1, x_2, \ldots, x_{100}$ has been observed. Perform Monte Carlo simulation experiments that provide convincing numerical evidence that the exact two-sided 95% prediction intervals for X_{101} and X_{102} are indeed exact prediction intervals for parameter settings of your choice.

9.4 Consider a stationary shifted AR(1) model defined by

$$X_t = \mu + \phi X_{t-1} + Z_t,$$

where μ, $-1 < \phi < 1$, and $\sigma_Z^2 > 0$ are fixed known parameters and Z_t is Gaussian white noise. Find expressions for

(a) $\lim_{h \to \infty} E[X_{n+h} | X_1 = x_1, X_2 = x_2, \ldots, X_n = x_n]$

(b) $\lim_{h \to \infty} V[X_{n+h} | X_1 = x_1, X_2 = x_2, \ldots, X_n = x_n]$.

9.5 Find the limiting half-width of a exact two-sided $100(1-\alpha)\%$ prediction interval for $E[\hat{X}_{n+h}]$ as the time horizon $h \to \infty$ for an AR(1) time series model with all parameters known.

9.6 For a stationary shifted ARMA(p, q) time series model with population autocorrelation function $\rho(k)$, the population variance of the sample mean is

$$V\left[\bar{X}\right] = \frac{\sigma_X^2}{n}\left[1 + 2\sum_{k=1}^{n-1}\left(1 - \frac{k}{n}\right)\rho(k)\right].$$

This result was proved in Section 8.2.1. Use this result to find an approximate 95% confidence interval for μ for the beaver data from Example 9.3 for a fitted shifted AR(1) time series model with Gaussian white noise error terms.

9.7 The built-in R time series lh consists of $n = 48$ observations of the luteinizing hormone in blood samples from a woman taken at 10 minute intervals.

(a) Plot the time series, the sample autocorrelation function and the sample partial autocorrelation function.

(b) Suggest an ARMA(p, q) model based on your plots.

(c) Make a scatter plot of the data pairs (x_{t-1}, x_t).

(d) Compute the method of moments estimates of the parameters in the model suggested in part (b).

(e) Compute the maximum likelihood estimates of the parameters in the model suggested in part (b).

(f) Compute an approximate 95% confidence interval for ϕ.

(g) Forecast the next three values in the time series and report 95% prediction intervals for the three forecasts.

(h) Perform some research on the luteinizing hormone and indicate some scientific evidence that the time series model you suggested in part (b) is plausible.

9.8 Report the test statistic and p-value for the turning point test applied to the time series of beaver temperatures in their active state from Example 9.3. Comment on the sign of the test statistic and the magnitude of the p-value.

9.9 Consider the time series of $n = 70$ consecutive yields from a batch chemical process (from Box, G.E.P., and Jenkins, G.M. (1976), Time Series Analysis: Forecasting and Control, Revised Edition, Holden–Day, page 32) given in Example 7.20.

(a) Plot the time series, the sample autocorrelation function and the sample partial autocorrelation function.

(b) Suggest an ARMA(p, q) model based on your plots.

(c) Make a scatter plot of the data pairs (x_{t-1}, x_t).

(d) Compute the method of moments estimates of the parameters in the model suggested in part (b).

(e) Compute the maximum likelihood estimates of the parameters in the model suggested in part (b).

(f) Compute an approximate 95% confidence interval for ϕ.

(g) Forecast the next three values in the time series and report 95% prediction intervals for the three forecasts.

9.10 Consider an AR(1) model with population parameters $\phi = 0.8$ and $\sigma_Z^2 = 1$, and Gaussian white noise. Let r_1, r_2, r_3 denote the sample autocorrelation function values of the *residuals* of the fitted time series associated with $n = 100$ observations. Use Monte Carlo simulation to estimate the population mean vector and population variance–covariance matrix, to one-digit accuracy, of r_1, r_2, r_3 when maximum likelihood estimation is used to estimate the parameters.

9.11 Let B_1 and B_2 be the roots of the characteristic equation $\phi(B) = 1 - \phi_1 B - \phi_2 B^2 = 0$ for an AR(2) time series model

$$X_t = \phi_1 X_{t-1} + \phi_2 X_{t-2} + Z_t.$$

Let $G_1 = B_1^{-1}$ and $G_2 = B_2^{-1}$. A general solution for the lag k autocorrelation is (see Box, G.E.P., and Jenkins, G.M. (1976), Time Series Analysis: Forecasting and Control, Revised Edition, Holden–Day, page 59)

$$\rho(k) = \frac{\left(1 - G_2^2\right) G_1^{k+1} - \left(1 - G_1^2\right) G_2^{k+1}}{(G_1 - G_2)(1 + G_1 G_2)}$$

for $G_1 \neq G_2$. Show that calculating the population autocorrelation in this fashion is the same as using the recursive equation for the first five lags for an AR(2) process with parameters

(a) $\phi_1 = 1/2, \phi_2 = 1/3$,

(b) $\phi_1 = 1, \phi_2 = -1/2$.

9.12 Create a plot like the one in Figure 9.13 for an AR(2) model stationary region with $\rho(1) = -0.9, -0.8, \ldots, 0.9$ and $\rho(2) = -0.9, -0.8, \ldots, 0.9$. No labels are necessary on your plot.

9.13 A stationary AR(2) time series model can be written as an MA(∞) time series model. The coefficients $\theta_1, \theta_2, \ldots$ in the MA(∞) model can be calculated in four fashions. First, they can be calculated using the recursive formulas in Theorem 9.12. Second, they can be written explicitly as (Cryer, J.D. and Chan, K–S, *Time Series Analysis: With Applications in R*, 2008, Springer, page 75):

$$\theta_i = \begin{cases} (i+1)G_1^i & \phi_1^2 + 4\phi_2 = 0 \\ \left(G_1^{i+1} - G_2^{i+1}\right)/(G_1 - G_2) & \phi_1^2 + 4\phi_2 > 0 \\ R^i \sin\left[(i+1)\Theta\right]/\sin\Theta & \phi_1^2 + 4\phi_2 < 0 \end{cases}$$

for $i = 1, 2, \ldots$, where B_1 and B_2 are the roots of $\phi(B) = 1 - \phi_1 B - \phi_2 B^2$, $G_1 = B_1^{-1}$, $G_2 = B_2^{-1}$, $R = \sqrt{-\phi_2}$, and $\cos \Theta = \phi_1/(2R)$. Third, the coefficients can be calculated by using the factored form of the characteristic polynomial, and writing the model in terms of X_t and equating coefficients. Fourth, the coefficients can be calculated by using the ARMAtoMA function in R. Calculate the first eight coefficients of the MA(∞) model, $\theta_1, \theta_2, \ldots, \theta_8$, using these four methods for the following sets of AR(2) parameters:

(a) $\phi_1 = 1, \phi_2 = -1/4$,

(b) $\phi_1 = 1/2, \phi_2 = 1/9$,

(c) $\phi_1 = 1, \phi_2 = -1/2$.

These three parameter combinations correspond to one real root with multiplicity two, two distinct real roots, and two complex roots of the characteristic equation $\phi(B) = 0$.

9.14 For an AR(2) time series model, the asymptotic variance–covariance matrix of the maximum likelihood estimates $\hat{\phi}_1$ and $\hat{\phi}_2$ is

$$\frac{1}{n}\begin{bmatrix} 1-\phi_2^2 & -\phi_1(1+\phi_2) \\ -\phi_1(1+\phi_2) & 1-\phi_2^2 \end{bmatrix}.$$

What is the asymptotic population correlation between $\hat{\phi}_1$ and $\hat{\phi}_2$?

9.15 Consider an AR(2) time series model with $\phi_1 = 1$, $\phi_2 = -1/2$, and $\sigma_Z^2 = 1$. For a realization of $n = 100$ observations $X_1, X_2, \ldots, X_{100}$ from this AR(2) model, give convincing numerical evidence that the forecasted value for X_{103} is unbiased and that the 95% prediction interval for X_{103} is exact.

9.16 Implement Theorem 9.17 on the R built-in `LakeHuron` time series to calculate the first five forecasted values and associated prediction intervals. Do not just use the `predict` function.

9.17 Consider a standard AR(2) model for an observed time series of $n = 100$ values. The last two values in the time series are $x_{99} = 3$ and $x_{100} = 4$. The estimated coefficients in the AR(2) model are $\hat{\phi}_1 = 1$ and $\hat{\phi}_2 = -0.5$. Compute the next ten forecasted values $\hat{X}_{101}, \hat{X}_{102}, \ldots, \hat{X}_{110}$ and comment on the shape of the forecasted values.

9.18 Consider a realization $x_1, x_2, \ldots, x_n$ of a stationary shifted AR(2) time series model with fixed known parameters μ, ϕ_1, ϕ_2, and σ_Z^2. Write a formula for $\hat{X}_{n+3}$ in terms of x_{n-1} and x_n.

9.19 Consider the annual Lake Huron water level heights from 1875 to 1972 given in the R built-in data set `LakeHuron`, appended by the next ten observations,

$$580.98, 581.04, 580.49, 580.52, 578.57, 578.96, 579.94, 579.77, 579.44, 578.97,$$

for the years 1973 to 1982. Give the p-value associated with a test of the statistical significance of the slope of the simple linear regression line for the augmented time series.

9.20 Consider the AR(3) model with coefficients

$$\phi_1 = 3/2 \qquad \phi_2 = -1 \qquad \phi_3 = 1/4.$$

    (a) Is this model invertible?

    (b) Is this model stationary?

    (c) Calculate the first six coefficients of the associated MA(∞) model.

9.21 Two necessary, but not sufficient, conditions for stationarity of an AR(p) time series model are (Cryer, J.D. and Chan, K–S, *Time Series Analysis: With Applications in R*, 2008, Springer, page 76):

$$\phi_1 + \phi_2 + \cdots + \phi_p < 1 \qquad \text{and} \qquad |\phi_p| < 1.$$

    (a) Show that these conditions hold for the stationary AR(4) time series model with

$$\phi_1 = \frac{21}{20}, \phi_2 = \frac{1}{20}, \phi_3 = -\frac{23}{40}, \phi_4 = \frac{3}{10}.$$

    (b) Graphically or algebraically, show that these conditions are necessary but not sufficient for falling in the triangular-shaped stationary region from Theorem 9.9 for an AR(2) time series model.

9.22 Consider the AR(4) time series model with characteristic polynomial

$$\phi(B) = 1 - \frac{21}{20}B - \frac{1}{20}B^2 + \frac{23}{40}B^3 - \frac{3}{10}B^4$$

and Gaussian white noise with population variance $\sigma_Z^2 = 1$. Conduct a Monte Carlo simulation experiment that provides convincing numerical evidence that $\gamma(0) = 3520/819$.

9.23 The R vector `phi` contains the parameters $\phi_1, \phi_2, \ldots, \phi_p$ in an AR(p) model. Write an R function named `is.stationary` with a single parameter `phi` that returns `TRUE` if the AR(p) model is stationary and `FALSE` otherwise.

9.24 The R code below takes initial p autocovariances $\gamma(0), \gamma(1), \ldots, \gamma(p-1)$ for an AR(p) model, which are stored in the vector `gam`, and places them in a variance–covariance matrix `GAMMA` (denoted by Γ in the text).

```
GAMMA  = matrix(0, p, p)
for (i in 1:p) {
  for (j in 1:p) {
    GAMMA[i, j] = gam[abs(i - j) + 1]
  }
}
```

The code makes this conversion by using two nested `for` loops. Heather can do this calculation without using `for` loops. How does she do it?

9.25 Consider a time series that is governed by an AR(4) model with characteristic polynomial

$$\phi(B) = 1 - \frac{21}{20}B - \frac{1}{20}B^2 + \frac{23}{40}B^3 - \frac{3}{10}B^4$$

and Gaussian white noise with population variance $\sigma_Z^2 = 1$. Conduct a Monte Carlo simulation experiment that provides convincing numerical evidence that the 95% confidence interval for ϕ_3 based on the maximum likelihood estimators for an AR(4) time series model is asymptotically exact.

9.26 For logarithms of the $n = 55$ annual lynx pelt sales time series from Example 9.29, find the values of p and q associated with the ARMA(p, q) model that minimizes the AIC statistic. Assume that the models are fitted by maximum likelihood.

9.27 Fit the AR(4) model to the logarithms of the $n = 55$ annual lynx pelt sales time series from Example 9.29 by maximum likelihood. Simulate the fitted model to generate $n = 55$ random annual lynx pelt sales from the fitted model. View a dozen or so such realizations and comment on your faith in the fitted AR(4) time series model. Repeat the experiment for a fitted ARMA(2, 3) time series model and comment.

9.28 Show that the MA(1) model

$$X_t = Z_t + \theta Z_{t-1},$$

and the MA(1) model

$$X_t = Z_t + \frac{1}{\theta}Z_{t-1}$$

have the same population autocorrelation function.

9.29   Show that $-1/2 \le \rho(1) \le 1/2$ for an MA(1) model.

9.30   Derive the population autocorrelation function for the MA(1) model with arbitrary mean value μ given by

$$X_t = \mu + Z_t + \theta Z_{t-1},$$

in a similar fashion to the derivation for the standard MA(1) model.

9.31   Conduct the following Monte Carlo simulation experiment. Generate $n = 100$ observations from an MA(1) time series model with $\theta = 0.9$ and standard normal white noise terms. Estimate the expected value and standard deviation of r_1 and r_2. Run enough replications to that you can report your estimates to two significant digits.

9.32   Consider an MA(1) model with $\theta = -0.9$ and Gaussian white noise with $\sigma_Z^2 = 1$. Generate a dozen realizations of this time series for $n = 100$ observations each. Plot the time series and the associated correlogram, using a call to `Sys.sleep` between each realization to view the graphs. Write a paragraph that describes what you observe in the dozen realizations.

9.33   Consider an MA(1) time series model

$$X_t = Z_t + \theta Z_{t-1},$$

where $\{Z_t\}$ denotes Gaussian white noise. Let $\hat{\theta}_{\text{MOM}}$ be the method of moments estimator of θ and let $\hat{\theta}_{\text{MLE}}$ be the maximum likelihood estimator of θ. One way to compare these two estimators is the asymptotic relative efficiency, defined as

$$\lim_{n \to \infty} \frac{V\left[\hat{\theta}_{\text{MOM}}\right]}{V\left[\hat{\theta}_{\text{MLE}}\right]}.$$

Brockwell and Davis (2016, page 129) give the population variance of $\hat{\theta}_{\text{MOM}}$ and $\hat{\theta}_{\text{MLE}}$ as approximately

$$V\left[\hat{\theta}_{\text{MOM}}\right] \cong \frac{1 + \theta^2 + 4\theta^4 + \theta^6 + \theta^8}{n\left(1 + \theta^2\right)^2} \qquad \text{and} \qquad V\left[\hat{\theta}_{\text{MLE}}\right] \cong \frac{1 - \theta^2}{n}.$$

Write a Monte Carlo simulation that confirms these two formulas for $n = 400$, $\theta = 1/2$, and $\sigma_Z^2 = 1$.

9.34   The $n = 45$ daily average number of defects per truck at the final inspection at a manufacturing facility (from Burr, 1976, *Statistical Quality Control Methods*, Marcel Dekker, New York), read row-wise, are given below.

1.20	1.50	1.54	2.70	1.95	2.40	3.44	2.83	1.76
2.00	2.09	1.89	1.80	1.25	1.58	2.25	2.50	2.05
1.46	1.54	1.42	1.57	1.40	1.51	1.08	1.27	1.18
1.39	1.42	2.08	1.85	1.82	2.07	2.32	1.23	2.91
1.77	1.61	1.25	1.15	1.37	1.79	1.68	1.78	1.84

Fit these data values to a shifted MA(1) time series model by the method of moments, least squares, and maximum likelihood estimation.

9.35 The formula for the population variance of the sample mean for a stationary time series model (which was proved in Section 8.2.1) is

$$V[\bar{X}] = \frac{\sigma_X^2}{n}\left[1 + 2\sum_{k=1}^{n-1}\left(1 - \frac{k}{n}\right)\rho(k)\right]$$

Show that this is approximately

$$V[\bar{X}] \cong \frac{\sigma_X^2}{n}\left[1 + 2\sum_{k=1}^{\infty}\rho(k)\right]$$

or, equivalently,

$$V[\bar{X}] \cong \frac{\sigma_X^2}{n}\left[\sum_{k=-\infty}^{\infty}\rho(k)\right]$$

for large values of n whenever the autocorrelation function values decay rapidly enough with increasing k such that

$$\sum_{k=1}^{\infty}|\rho(k)| < \infty.$$

Index

www.ingramcontent.com/pod-product-compliance
Lightning Source LLC
Chambersburg PA
CBHW081210220326
41598CB00037B/6736